The Discovery of Dynamics

Long-exposure photograph of star trails over the Kitt Peak National Observatory, Arizona. The apparent circular motions of the stars were a potent and initially very helpful factor in the formation of early concepts of celestial motions. They also supplied an extraordinarily accurate and convenient clock, without which the laws of planetary motion, and through them the laws of dynamics, could never have been found. The recognition that the diurnal motion of the stars could be explained by rotation of the earth was also a major factor in the debates about the relativity of motion, the central topic of this book. (*Photo by Paul Shambroom/Science Photo Library.*)

The Discovery of Dynamics

A study from a Machian point of view of the discovery and the structure of dynamical theories

JULIAN B. BARBOUR

OXFORD
UNIVERSITY PRESS

2001

OXFORD
UNIVERSITY PRESS

Oxford New York
Athens Auckland Bangkok Bogotá Buenos Aires Cape Town
Chennai Dar es Salaam Delhi Florence Hong Kong Istanbul Karachi
Kolkata Kuala Lumpur Madrid Melbourne Mexico City Mumbai Nairobi
Paris São Paolo Shanghai Singapore Taipei Tokyo Toronto Warsaw

and associated companies in
Berlin Ibadan

Published by Oxford University Press, Inc.
198 Madison Avenue, New York, New York 10016

Originally published as *Absolute or Relative Motion?* Volume 1,
The Discovery of Dynamics, Cambridge University Press 1989

Library of Congress Cataloging-in-Publication Data
Barbour, Julian B.
The discovery of dynamics / Julian B. Barbour.
p. cm.
Includes bibliographical references and index.
ISBN 0-19-513202-5 (pbk.)
1. Motion. 2. Dynamics. 3. Astronomy. I. Title.
QC133.B36 2000
531′.11—dc21 00-062340

1 3 5 7 9 8 6 4 2

Printed in the United States of America
on acid-free paper

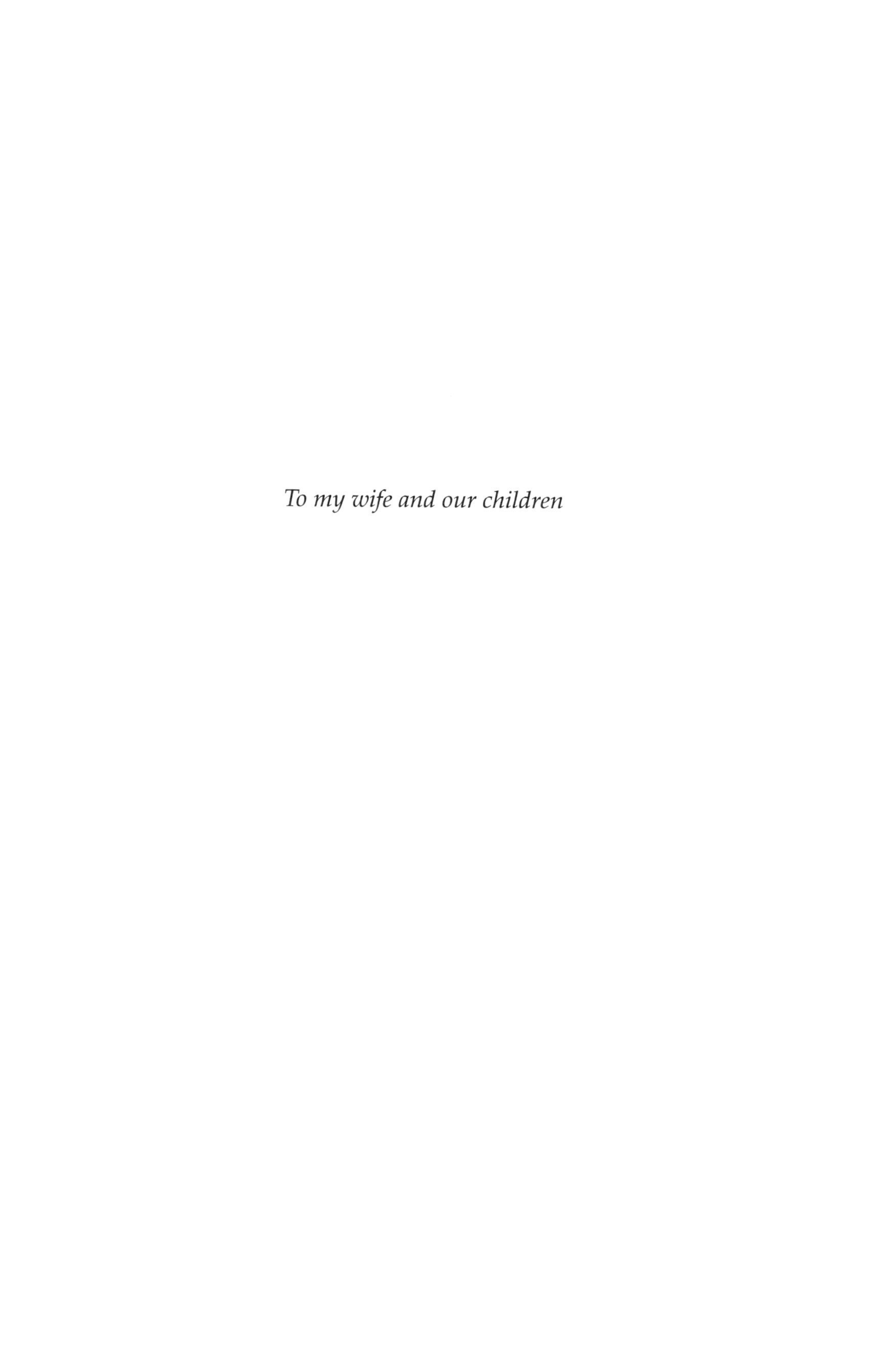

To my wife and our children

PREFACE TO THE PAPERBACK EDITION

1. About this Book

This book was originally published in 1989 as *Absolute or Relative Motion?* Volume 1, *The Discovery of Dynamics*. It was my hope that it would be followed within a few years by a second volume with the provisional subtitle *The Frame of the World.* The two volumes would together have given a comprehensive account of what I call the absolute/relative debate. This is closely related to Ernst Mach's daring idea (Mach's principle) that inertia arises from the combined effect of the universe and not from a straitjacket imposed by Newton's absolute space. What is at stake is the foundation of dynamics and our view of the cosmos: Does the universe exist within an invisible absolute framework, as Newton argued, or is it some holistic and self-contained relational system, as proposed by Leibniz and others? Closely related are the questions of the nature of time and motion, and how these issues relate to Einstein's theories of special and general relativity.

In the event, I have so far failed to complete the second volume—though for personally encouraging reasons, as I explain in Section 3 of this Preface. Following discussions with my present and previous publishers, it has now been decided to retitle the already published book *The Discovery of Dynamics* and issue it as a paperback. This is appropriate, since the book is mainly the story of how dynamics came to be discovered. The book has been widely read as such and not just as a monograph on the absolute/relative debate.

Except for this Preface, the book is being reissued exactly as originally printed. I find no reason to undertake any significant revisions, and this of course has helped to keep the cost down. It does have the one disadvantage that the reader will encounter various references to the now never-to-be-published Volume 2. However, a substitute of sorts is already to hand, as I explain later, so I hope the reader will not find this irksome.

The remainder of this Preface is divided into two parts. In Section 2, I take the opportunity to respond to some points raised by reviewers when the book was originally published, and to mention some new studies by other authors that relate to and complement the subject of *The Discovery of Dynamics*. In Section 3, I explain why its companion volume has not yet appeared, and draw attention to already published material that covers much of the planned content of that volume. The fact that a good proportion of the material is already available is a further justification for the independent publication of the present work in paperback.

2. Response to Reviewers and Additional Comments

By and large, the reception of this book on first publication was most encouraging for a first-time author. However, several reviewers felt its value was diminished by being written from an exclusively modern perspective. Historians distrust "Whig history," that is, history conceived under the assumption that humanity through the centuries has necessarily been progressing toward the great ideals of liberal democracy and enlightenment. My most severe critic was the late Eric Aiton, who wrote [in Band 672 (1989) of *Zentralblatt für Mathematik und ihre Grenzgebiete*]) that "the author's arguments are flawed, as a result of a deep-rooted anachronism. Everything is judged in relation to what the author regards as the correct theory: that is, the explanation accepted today. . . . The idea of a linear progression towards the modern world view . . . has long been abandoned by historians of science." Now I am sure there is some truth in what Aiton says, and I am happy to leave the reader to judge whether or not the "Everything" is fair. However, I do feel Aiton failed to grasp the fact that the book had several aspects and aims, only one of which could be called history of science. And I am not at all sure Aiton's own position is so secure. All too many modern historians of science seem to believe there are no criteria of good science.

Indeed, in his own book on Cartesian vortex theory (*The Vortex Theory of Planetary Motions*), Aiton claimed that "Any evaluation of the vortex theory . . . requires consideration of the currents of scientific thought of the time, and especially of such aspects as the prevailing ideas on the nature and purposes of scientific theories. It is only in this context that questions concerning the originality of the theory, its degree of success in relation to its aims, the force of the criticisms against it and the extent of its acceptance become meaningful." Well, that is fine if you want to study such things (and they are quite interesting), but the simple fact is that the vortex theory failed because its practitioners did not yet appreciate the essence of good science, which needs both accurate observations and theories and models that truly describe them. The vortex theory had neither. It took into account only the coarsest aspects of astronomical observations and sought to account for them by pictures, not mathematical equations (or the geometrical demonstrations that Newton employed).

There are criteria of good science. And few working scientists are interested in the dead end of vortex theory. Mostly they want books about real science and why it works. You get that in particular from the study of Newton. Descartes's great importance for Newton was his embryonic dynamical scheme and almost perfect formulation of the law of inertia, as I argue in chapters 8–10 of this book. That is where the real interest for science lies, not in the minutiae of vortex theory.

I have set out my own position on these matters in the essay "Reflections on the aims, methods and criteria of the history of science" in *Contemporary Physics*, 38, 161–166 (1997), which includes a review of S. Chandrasekhar's *Newton's Principia for the Common Reader* (which was published by Oxford University Press in 1995 and about which I shall say something below). I concluded that essay review with the words: "I remain convinced that the history of science represents a rich organic development in which success has reinforced success and that the primary task of the historians should be to tell that great story as it is. We need more work on a broad canvas." This is my apologia for the present work.

Since the original publication of *The Discovery of Dynamics* in 1989 I have done very little work on the period it covers and so am not in a position to add much of significance to it now. However, I do regret that in the original I did not cite the remarkable essay by Clifford Truesdell in his founding issue of the *Archive for History of Exact Sciences*, in which he outlined "A Program toward Rediscovering the Rational Mechanics of the Age of Reason." The main point he made was that a whole age of the development of much of rational mechanics as we now know it had been lost from sight through the towering prominence given to Newton's *Principia*, published in 1687, and Lagrange's *Méchanique Analitique*, published a century later in 1788. In the course of this essay he makes some very interesting and provocative comments on Newton's *Principia.*

Some I feel are off the mark, especially the claim: "It is small oversimplification to say that . . . the *Principia* is a retrospective work which selects, marshals, and formalizes the achievements of the century before it." I really cannot see how Truesdell can say this of a work that introduces the clear notion of force and with total clarity foresees the next three hundred years of physics and expresses it in 92 words: "I wish we could derive the rest of the phenomena of Nature by the same kind of reasoning from mechanical principles, for I am induced by many reasons to suspect that they may all depend upon certain forces by which the particles of bodies, by some causes hitherto unknown, . . . cohere in regular figures. . . . These forces being unknown, philosophers have hitherto attempted the search of Nature in vain; but I hope the principles here laid down will afford some light either to this or some truer method of philosophy." That is hardly a retrospective sentiment. But Truesdell's essay is a fine read and a worthy tribute to his hero Euler.

Another quite different matter concerns the moon, which might have figured more prominently in this book than it actually does. Among the motions of the celestial objects that move in the sky relative to the background of the fixed stars, those of the moon are by far the most obvious yet at the same time the most baffling. In fact, some argue that the attempts to understand these motions created most of dynamics and many of the most powerful methods of modern mathematical physics. There is much truth in this view, but in my estimation it was nevertheless observations and interpretation of the simpler motions of the sun and the planets that unleashed the Copernican Revolution, led Kepler and Galileo to some of their greatest discoveries, and then provided the most telling clues for Newton in his creation of dynamics. Happily there are, for more advanced readers, two excellent accounts of lunar studies that I can recommend wholeheartedly, both by Martin Gutzwiller. The first is the part of his *Chaos in Classical and Quantum Mechanics* (Springer-Verlag, 1990) devoted to the moon. The second is his review article on the lunar three-body problem in *Reviews of Modern Physics* (70, 589, 1998).

I should also like to draw the reader's attention to the useful book *The Key to Newton's Dynamics: The Kepler Probe* by J. B. Brackenridge (University of California Press, 1995), and to two very interesting articles by Michael Nauenberg: "Hooke, Orbital Motion, and Newton's Principia" (*American Journal of Physics*, 62, No. 4, 1994) and "Newton's Early Computational Method for Dynamics" (*Archive for History of Exact Sciences*, 46, No. 3, 1994). Nauenberg, like the great Russian mathematician V. I. Arnol'd [whose little book *Huygens and Barrow, Newton and Hooke* (Birkhäuser, Boston, 1990) is strongly recommended], shares my belief, expressed in this book, that Hooke's contribution to the elaboration of the law of universal gravitation (and much else) has not been adequately recognized. Nauenberg's second paper seems to me to be a genuine and most interesting contribution to our understanding of the development of Newton's dynamical methods. The paper is recommended to those who like a bit of good detective work coupled with a fine feeling for dynamics.

I should also like to quote from an otherwise positive review of *The Discovery of Dynamics* by Bruce Brackenridge (*Isis* 82, No. 3, 1991): [I]n the discussion of the *Principia* and of its immediate development, there is not one reference to D. T. Whiteside's monumental eight-volume . . . *The Mathematical Papers of Isaac Newton* (CUP, 1967–1981). . . . It is so extensive, in fact, that it is often a work more referenced than read. Instead, Barbour seems to rely almost exclusively upon the works of John Herivel and R. S. Westfall, and his references to Whiteside are limited to a few published articles." Brackenridge is right; I looked at only small parts of Whiteside's great work and did not use them in this book. Partly this was due to lack of time, but it was mainly because I was only attempting to capture the main developments, and I felt that here it was more appropriate to consider Whiteside's papers (though I actually disagree with one of his conclusions, as the reader will see). However, the lack of prominent mention of the *Mathematical Papers* was inexcusable. My apologies to Whiteside, who has done Newton scholarship a magnificent service.

Now a brief word about Chandrasekhar's book on the *Principia,* which was published shortly before his death. This has become something of a bestseller and has certainly made many people much more aware of Newton's great achievements. There are many fine things in the book, but even Homer nods occasionally—there are some embarrassing misunderstandings and misrepresentations. More seriously, I feel Chandrasekhar has completely misread the historical development. I had some interaction with him when he lectured in Oxford while writing the book. One of the main problems is Chandrasekhar's unshakeable conviction that Newton had the notion of universal gravitation with complete clarity by 1666. I asked him (but to no avail) to read Curtis Wilson's fascinating paper on the subject (See ref. 57 in chap. 10 of this book), which to my mind demonstrates beyond all reasonable doubt that this cannot be true. Chandra was curiously blind to the absorbing story of Newton's discoveries, and I think this distorts his account of the *Principia* and even dehumanizes Newton. I have written about this at some length in my essay review mentioned above.

3. Why Volume 2 Did Not Appear, and the Substitute for It

There are three main reasons why Volume 2 has never made it to the press. The first is tucked away in the footnote on page 5, in which I comment that the 'time aspect of the problem of the relativity of motion has been curiously neglected'. I predicted that time would play a large role in the second volume. It has certainly played a large role in my life since I made that prediction. There are two fundamental issues related to time in classical physics: What is 'duration'—in other words, what is the theoretical justification for saying that a second today is the same as a second yesterday (or immediately after the Big Bang?) And how is one to define simultaneity at spatially separated points?

Einstein's brilliant answer to the second question led to the creation of his two relativity theories, but for some reason he ignored the first question. Surprisingly, there is, however, a beautiful theory of duration hidden away within the mathematics of his general theory of relativity. In fact, time and duration arise within that theory from an arena in which there is no time at all. For a variety of historical reasons Einstein was unaware of this fact, and the same is still true of most people who work in relativity. The timeless basis of general relativity has profound implications, especially for attempts to make it compatible with quantum mechanics and create a quantum theory of the universe. It is distinctly possible that quantum cosmology will be *static*—that time will cease to play any role in the foundations of physics.

This conviction grew on me very strongly in the years immediately following completion of *The Discovery of Dynamics.* It resulted in several papers, which I have listed with brief commentaries on my website (www.platonia.com), and in the recent publication of my book *The End of Time: The Next Revolution in Physics* (Weidenfeld & Nicolson, London, 1999; Oxford University Press, New York, 2000). A large part of this book, which substantiates the statements made in the previous paragraph, is also concerned with the absolute/relative debate, and much of the material in it, including historical matters, would otherwise have appeared in the planned second volume.

The second thing that deflected me from my purpose was an invitation in 1991 from Herbert Pfister, of the University of Tübingen, to organize with him a conference specifically devoted to Mach's principle. This took place in 1993 in Tübingen, and was attended by many of the physicists and historians and philosophers of science who have a serious interest in Mach's ideas on the foundations of dynamics. The conference proceedings, edited by Herbert and myself and with full transcripts of the lively discussions, were published by Birkhäuser in 1995 as Volume 6 of their Einstein Studies series with the title *Mach's Principle: From Newton's Bucket to Quantum Gravity*. This book has sold gratifyingly well and is now regarded as the standard source for studies of Mach's principle. It too is a substitute for my second volume, with the great added benefit of the many contributions by authors other than myself.

The third and final reason for the absence of the second volume is related to ongoing work that I find most exciting and encouraging. In the 'Introduction to Volumes 1 and 2' with which *The Discovery of Dynamics* begins, I mention the work by Bruno Bertotti and myself in which we showed how Mach's principle could be implemented in the mechanics of point particles by a technique that I now call 'best matching'. It enables one to construct a dynamical theory in which only the separations between the bodies in the universe play a role in its dynamics. In contrast to Newtonian theory, it is impossible for any imagined overall rotation of the universe to play a role. This is why a theory based on best matching implements Mach's principle. Bertotti and I also showed that best matching is one of the two key properties of the deep mathematical structure of general relativity when it is treated as a dynamical theory of the evolution of three-dimensional geometry (as opposed to a theory of four-dimensional spacetime treated as a 'block'). The other key property of general relativity is its timeless basis, as I have already mentioned.

All this is very satisfactory from the Machian point of view, but it would be highly desirable to go a stage further and create a dynamical theory in which 'the overall size of the universe' has no meaning. It has long been the dream of theoretical physicists to create a *scale-invariant dynamics*. In the period from 1996 to the middle of 1998, I gave much thought to this problem and succeeded in taking a first step towards such a goal. This involved a generalization of the notion of best matching. However, I then had to put aside the work in order to complete *The End of Time*, and only returned to the problem at the beginning of 1999, when I succeeded in formulating a scale-invariant generalization of the Machian particle dynamics that Bertotti and I had created 20 years earlier. Because of the welter of new developments to which this led, I have still not yet found time to publish this theory. The most important thing about the new theory is that it showed how a similar generalization of general relativity could be attempted

At this point I was extremely lucky to join forces with Niall Ó Murchadha of University College, Cork. Niall has great expertise in precisely the kind of mathematics (three-dimensional conformal geometry) that was needed to push forward the idea of best matching in a form appropriate for a modern theory of gravitation. He added a crucially important ingredient to my proposal, and very soon we had the outlines of *conformal gravity*. This is a putative new theory of gravitation that is nevertheless extremely similar to general relativity, and therefore may give the same predictions as Einstein's theory in the domains where it has been well tested. Deviations from Einstein's theory are, however, to be expected in domains where it remains to be tested, as may well happen in the coming years.

This work has led to one surprise after another, and is still undergoing rapid development. To me it now seems possible that it will transform our understanding of both relativity and gauge theory, which describes the interaction of the various different forms of matter that exist in the universe. At the very least this work should demonstrate how far Machian ideas can be developed and applied usefully to the description of the universe. As yet, the new work has been published only electronically (cited on my website). It would be premature to go into further details at this stage, since the new ideas have yet to be exposed to thorough peer review and are still developing so rapidly. It would be equally premature to attempt any sort of summary of the absolute/relative debate until full clarity on this latest development has been achieved. So this is the third reason why the second volume has not appeared. I have hopes that what does finally appear will be a considerably more valuable and interesting work than anything I contemplated as possible back in 1989. Meanwhile, regularly updated detail about these developments and the earlier work can be found on my website.

Julian Barbour
South Newington, March 2001

PREFACE

At the deepest level, the present remarkable understanding that has been gained by natural scientists into the working of the physical world is based to a very large degree on the general theory of *dynamical systems* and the numerous particular examples of such systems that are found to be realized in nature. These range from the hydrogen atom, through more complicated atomic and nuclear systems to things of immense practical importance for everyday life, on to huge systems such as star clusters and galaxies, and even, it seems, the universe itself. Many of the most characteristic features of dynamical systems first came to light when Newton published his *Mathematical Principles of Natural Philosophy* in 1687. In a very real sense this event can be said to mark the *discovery of dynamics*.

The present book is an attempt to explain to any reader interested in these absorbing matters how Newton was able to make his monumental discovery. Three quarters of the book deal with the preparatory work in astronomy and the mathematical study of terrestrial motions that made Newton's work possible. The final quarter describes and analyses Newton's own discoveries, his synthesis of a viable scheme of dynamics, and his introduction of the concept of universal gravitation. The book is however much more than just a history of the discovery of dynamics. For it attempts to put this discovery in the perspective of as yet unresolved questions relating to the basic concepts of space, time, and motion. It is about the continuing and already quite ancient search for *the foundation, or frame, of the world*. This aspect of the work, which is to be continued with a second volume (*The Frame of the World*) covering the period from Newton to Einstein, is explained at some length in the Introduction that opens this book, and so I will say no more about it here.

In the remainder of this Preface I should like to address some words to the potential reader who has had the curiosity to read this far. My original intention was to write a specialist work of interest primarily to professional scientists in the narrow field of relativity and the history and philosophy of science. However, I soon became aware of the interest of the material I was treating for a very much wider readership. The scope of the book was extended accordingly. As a rule, working scientists know rather little about the historical origins of their discipline. Nevertheless, they do generally appreciate anything they can learn on the subject, particularly as they get older and a sense of the mystery and fascination of the world grows upon them. As Eliot says in his *East Coker*: 'The world becomes stranger, the pattern more complicated.' Thus, in addition to the select audience for which the study was originally intended, I hope scientists working in many disciplines (and not only physics and astronomy) will find much of interest in the book.

In fact, in view of the explosion of interest during the last decade in books dealing with fundamental questions of modern science I have attempted to make the work accessible to the interested layman by opening chapter 1 with a review of Newtonian dynamics at an elementary level. The mathematics required for understanding this book is minimal—little more than the most basic facts of trigonometry, a bit of vector analysis and algebra, and a few results on the geometry of triangles, circles and ellipses (which are explained where necessary). Indeed, one of the especial attractions of the early history of astronomy and dynamics is the number of interesting and highly nontrivial results that can be understood with such simple mathematics. In fact, although some passages in the book are rather technical (and just a few, especially in the Introduction, refer to advanced modern developments) the book could be read by motivated 17- or 18-year old pupils at school who are thinking of specializing at university in physics, mathematics or astronomy, especially if given a little encouragement by their teachers. This brings me to the readers I am most keen to attract—students (both undergraduate and graduate) of these disciplines at universities. For what I should above all like to do is awaken their interest in both the 'knowledge of the celestial things' (to use Kepler's words) and the ways by which they were discovered and give them a stimulus, as they pass through their studies (and later in life), to question received wisdom and think for themselves. It is only such people who, combining historical awareness with radical innovation, will advance our understanding of these fascinating matters.

South Newington,
February 1988

ACKNOWLEDGMENTS

There are many people whom I have to thank for assistance in one form or another in connection with this book. Its overall structure was strongly influenced before any writing commenced by discussions extending over many years with several friends and colleagues, in particular Bruno Bertotti, Karel Kuchař, Michael Purser, Derek Raine, and lee Smolin. Early encouragement to work in this field was given by Prof. Peter Mittelstaedt, for whom I did a PhD thesis in Cologne. The lectures and writings of Dennis Sciama and Sir Hermann Bondi were also an important stimulus, as was the encouragement of Reinhard Breuer and Charles Misner. A first draft of this book was read complete or in part by the five people first listed and also by Peter Hodgson, John Hyman, Michael Lockwood, Caroline Miles, Sydney Railton, and my brother Peter. Their comments and criticisms were invaluable and much appreciated. A considerably enlarged draft was then written, and I approached a number of professional historians of science with the request to read and comment on individual chapters or groups of chapters. They were (in alphabetical order) S. Drake, A. Gabbey, J. W. Herivel, S. Schaffer, B. Stephenson, G. J. Toomer, and R. S. Westfall. I also received technical advice on Chap. 12 from Mr C. A. Murray of the Royal Greenwich Observatory. I am quite certain that the comments, criticisms and corrections of errors made by these people have very greatly improved the book. I am particularly grateful to them for the time they were all prepared to take over such a request from someone quite unknown to them. My debt to other historians of science whose works I consulted will be evident from the text. It goes without saying that any residual errors (of which I would greatly welcome being informed—I fear there must be at least some in a book of this size) are entirely my responsibility. I am also grateful to my family, in the first place my wife, and non-scientist friends for encouragement over the years and during the writing of this book and would like to mention here especially my brother-in-law Peter Blaker and Charles Stainsby, who sadly died some time after I had started work on the book. I should also like to thank

C. Isham for introducing me to my editor, Dr Simon Capelin, whose patience and sympathetic encouragement has not failed despite more than one change of plan from the original proposal. A first-time author could not have hoped for a more helpful editor. Finally, I should like to thank my sub-editor, Mrs Susan Bowring, who has had to cope with substantial late revisions and also Mrs Keri Holman, Mrs Kate Draper and Mrs Joyce Aydon for their typing of the text and Mrs Kathy Smith for drawing some of the diagrams.

Last, but by no means least, a word of thanks to someone I have never met. When she had finished her thesis on Kepler some years ago, Dr A. E. L. Davis deposited on loan with the Radcliffe Science Library at Oxford her copies of the German translations by Max Caspar of Kepler's *Astronomia Nova* and *Harmonice Mundi*. It was only through her copy of the *Astronomia Nova* that I came to study Kepler at all and learnt about his marvellous work in astronomy. If there is anything of value in the astronomical part of this book, much of it derives from this opportunity I had to read Kepler at first hand. At this time of increasing pressure on library resources, I hope this mention of Dr Davis's generosity will encourage others to follow her example.

Throughout the work the reader will find extensive quotations from numerous primary and some secondary sources. I am most grateful to the following individuals and publishers for permission to quote from copyright material:

Mrs I. E. Drabkin, from: *G. Galilei, On Motion,* translated by I. E. Drabkin (published with *On Mechanics* by University of Wisconsin Press);

S. Drake, from: *Discoveries and Opinions of Galileo* by S. Drake, Doubleday Anchor Books; *Galileo* by S. Drake, O. U. P.; *Galileo at Work,* University of Chicago Press;

J. W. Herivel, from: *The Background to Newton's 'Principia'*, by J. W. Herivel, Clarendon Press, Oxford;

G. J. Toomer, from: *Ptolemy's Almagest,* translated and annotated by G. J. Toomer, Duckworth;

The University of Wisconsin Press, from: *The Science of Mechanics in the Middle Ages* by M. Clagett;

Springer–Verlag, from: *A History of Ancient Mathematical Astronomy,* 3 Vols., by O. Neugebauer;

Edizioni di Storia e Letteratura, from: *Die Vorläufer Galilei's im 14. Jahrhundert* by Anneliese Maier;

Rowman and Littlefield, from: *Dictionary of Philosophy,* ed by D. D. Runes;

Columbia University Press, from: *N. Copernicus, Three Copernican Treatises,* translated by E. Rosen;

Methuen, from: *The Astronomical Revolution: Copernicus–Kepler–Borelli,* by A. Koyré;

Blackie & Son, from: *Kepler,* by M. Caspar, translated by C. Doris Hellman, Abelard–Schuman;

Harvard university Press, from: *Concepts of Force* by M. Jammer; *A Source Book for Greek Science* by M. R. Cohen and I. E. Drabkin;

Open Court, from: *The Science of Mechanics* by E. Mach, translated by T. J. McCormack; *History and Root of the Principle of the Conservation of Energy* by E. Mach;

HarperCollins, from: *Essays in Honour of Gilbert Murray,* Allen and Unwin;

Macmillan, from: *N. Copernicus, On the Revolutions,* translated by E. Rosen;

The University of Chicago Press, from: *Medieval Cosmology* by P. Duhem, translated by R. Ariew;

Kluwer Academic Publishers, from: *René Descartes, Principles of Philosophy,* translated by F. R. Miller and R. P. Miller;

Abaris Books, Inc. from: *René Descartes, The World Le Monde,* translated by M. S. Mahoney;

Cambridge University Press, from: *The Correspondence of Isaac Newton,* Vol. 2, ed by H. W. Turnbull; *Unpublished Scientific Papers of Isaac Newton,* ed and translated by A. R. and M. B. Hall;

The Regents of the University of California and the University of California Press, from: *Newton's Principia,* in 2 Vols., Motte's translation revised by F. Cajori; *Galileo Galilei, Dialogue Concerning the Two Chief World Systems,* translated by S. Drake;

William Heinemann and Harvard University Press, from: *Loeb Classical Library* [*Aristotle IV* (*Physics Books I–IV*), *Aristotle VI* (*On the Heavens*), *Plato IX* (*Timaeus*)];

Random House, Inc. from: *J. Kepler, Epitome of Copernican Astronomy,* in: *Great Books of the Western World,* Vol. 16, ed B. M. Hutchins;

The Johns Hopkins University Press, from: *From the Closed World to the Infinite Universe* by A. Koyré;

CONTENTS

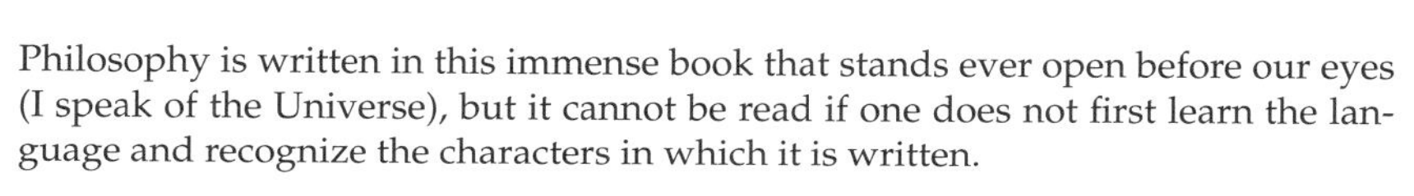
Philosophy is written in this immense book that stands ever open before our eyes (I speak of the Universe), but it cannot be read if one does not first learn the language and recognize the characters in which it is written.

Galileo

And it seems to me that the ways by which men arrive at knowledge of the celestial things are hardly less wonderful than the nature of these things themselves.

Kepler

Yet the thing is not altogether desperate.

Newton

The historical investigation of the development of a science is most needful, lest the principles treasured up in it become a system of half-understood prescripts, or worse, a system of *prejudices*. Historical investigation not only promotes the understanding of that which now is, but also brings new possibilities before us, by showing that which exists to be in great measure *conventional* and *accidental*. From the higher point of a view at which different paths of thought converge we may look about us with freer vision and discover routes before unknown.

Mach

The Discovery of Dynamics

INTRODUCTION TO VOLUMES 1 AND 2

'How differently the clouds move across that lofty, limitless sky!' Tolstoy's words, said by the wounded Prince Andrei Bolkonsky as he drifts into consciousness on the battlefield of Austerlitz, express the spirit of this book. It is about how concepts of motion have changed in the past and could change in the future. It is a subject with a never-ending fascination; for every change in our conception of motion amounts to a change in our deepest conceptions of things. Each change in our concept of motion opens the door into a new world.

If a stone is thrown at the stars with sufficient force it will travel through the universe forever – or at least until the end of time. We know that, relative to the observable matter in the universe, such a stone follows a definite path with great accuracy, *but we do not know what determines the path.* Is it space, or is it matter, or some combination of the two, or what? We see the undoubted effect but cannot put our finger on the cause. This puzzle is the central subject of this study.

By its very ubiquity, motion ceases to strike us as particularly marvellous or mysterious. But the seemingly simple is complex and subtle. The discovery of the law of inertia in the seventeenth century showed that all the motions we observe around us are merely fleeting sections of immense journeys through the universe. Everything is caught up and participates in some huge flux, of which what we observe as motions are but small details. Our little local motions have a deceptive appearance of simplicity because they are seen on the background of the relatively stable earth and even more stable starry heavens. However, we have now learnt enough to realise that motion is subject to the dominion of something far greater than the earth or the few thousand stars that man could see in the heavens before the discovery of the telescope. We can sense the throb of the pulse; but we cannot locate the heart.

Confronted with a restless, shifting universe that stretched seemingly to infinity but also with the undoubted existence of inertial motion, Newton identified absolute space and time as the ultimate framework of all motion. In this framework, he asserted, undisturbed bodies move along straight lines with uniform speed. Newton called space the *sensorium* of God and saw in it the explanation to all the mysteries of motion. In the decades immediately following the publication of the *Principia* in 1687, Newton's concepts of absolute space and time were severely criticized, above all by Huygens, Leibniz, and Berkeley. For, space being invisible, how can one say how a body moves relative to space? And, space being nothing physical, how can it influence actual motions of physical bodies?

But Huygens died soon and neither Leibniz nor Berkeley could produce any sort of theory to rival Newton, and their objections were gradually forgotten, until they were rediscovered in the second half of the nineteenth century by Ernst Mach.* Mach, although himself a most reluctant theorizer, proposed one of the most radical ideas in the history of science. He suggested that inertial motion here on the earth and in the solar system is *causally determined* in accordance with some quite definite but as yet unknown law by the totality of the matter in the universe.

It is worth stating precisely the new element in Mach's proposal. Newton, in common with all thinkers both before and after his time, accepted that motion could only be *observed* relative to other bodies. However, Newton nevertheless held that motion actually takes place in absolute space and time, which he assumed to exist irrespective of the presence of bodies in the universe. Thus, a solitary body would still have a motion even if there were no other bodies in the universe. In contrast, Mach asserted that motion does not exist except as a change of position relative to other bodies and that the law which governs the changes in relative position must be expressed directly in these same relative terms. He anticipated that such a law would lead, under certain conditions at least, to objective and observable differences from Newton's laws.

Mach expressed his ideas in a book on the history of mechanics,†[1] which was very widely read and had a considerable influence on twen-

* The *Mach number* is, of course, named after Mach (1838–1916) on account of his important work on shock waves, in which he developed a brilliant method of flash photography. In psychology *Mach bands* are also named after him. Mach's extraordinarily wide range of interests and his unusual personality are well covered in J. T. Blackmore's biography: *Ernst Mach, His Work, Life, and Influence* (University of California Press, Berkeley, 1972). He was famous, if not notorious, for his opposition to the idea of atoms. Einstein saw Mach's greatest strength as his 'incorruptible scepticism and independence' (see Ref. 5). Mach acquired such a following for his philosophical ideas about the nature of science that he even influenced political developments. Lenin's most important philosophical work is his *Materialism and Empirio-Criticism* (1909), a violent attack on Mach's antimaterialistic philosophy, which was supported by several Russian socialists whom Lenin opposed.

† References to the literature sources are indicated by superscripts and are given all together, chapter by chapter, at the end of the book. Most readers will have no cause to consult them. All footnote material, which has been kept to a minimum, is given directly at the foot of the corresponding page.

tieth-century physics. In the period from 1907 to 1918 Einstein worked with feverish enthusiasm to discover the mysterious law of nature that Mach had postulated. He coined the expression *Mach's Principle*[2] for the conjecture that the inertial properties of local matter are determined by the overall matter distribution in the universe and was convinced that his general theory of relativity, which took its definitive form in 1915,[3] would give full expression to the principle.

The outcome was a decided paradox. Einstein was forced to conclude that although matter in the universe did clearly influence inertia within the framework of his theory, his theory was nevertheless unable to demonstrate that inertia is completely determined by matter. After one attempt[4] to save the situation (which itself had a most ironic consequence – it laid the foundations of cosmology as a modern science), Einstein reluctantly concluded[5] that his attempt had failed. Despite this failure, general relativity was immediately recognized as one of the supreme achievements of the human intellect. More recently there have also been some most impressive experimental confirmations of predictions of the theory,[6] and today few working physicists doubt its essential correctness, though many believe that general relativity is itself only one aspect of a more comprehensive theory that embraces all the forces of nature.

The present status of Mach's Principle can only be described as confused. It has been said that there are as many Mach's Principles as there are people who have worked on the subject. Twenty or thirty years ago it was the subject of very lively discussion, and Machian theories were advanced by among others Sciama,[7] Hoyle and Narlikar,[8] Brans and Dicke,[9] and Treder.[10] Much of this work, to which he has himself contributed, has been reviewed by Raine,[11] who gives a useful bibliography. These theories aroused considerable interest, though it is fair to say that none has achieved widespread acceptance. Another development about the same time was a reinterpretation of general relativity by Wheeler,[12] who argued that Einstein did, in fact, give expression to Mach's ideas. In his view Machian ideas only make sense in the case of a closed, i.e., finite, universe, and it is Wheeler's contention that in a closed universe Einstein's theory is in fact perfectly Machian, though perhaps not quite in the way Einstein originally envisaged and certainly not in the way Mach did. Wheeler too has been only partially successful in persuading his fellow relativists of the correctness of his interpretation. In the last ten years or so the question has passed somewhat out of vogue, partly because it seemed so difficult to make progress or even achieve agreement on the nature of the problem, even more probably because of a dramatic surge of interest in some exciting new theories and ideas that offer tantalizing prospects for the unification of all the forces of nature (gauge theories, supersymmetry, grand unification theories, and, more recently, superstrings).

It might seem that this therefore is not the best time to come forward with a relatively lengthy monograph on the subject. I would argue, on the contrary, that the opposite is the case.

The first point is that the problem has not been solved. It is only in abeyance and remains the central enigma of motion; it will surely come to the fore again. For a start, if and when *the* grand unified theory is discovered – i.e., the ultimate law of nature, which some would have us believe is only just over the horizon – one will surely want to see how it stands *vis-à-vis* this great problem of the foundations of dynamics, indeed our very concept of the universe, space, and time.

Second, it is widely agreed that the relationship between the quantum theory and general relativity is probably the most baffling problem currently on the agenda of theoretical physics. This problem goes under the name *quantization of general relativity*, which means roughly that gravitational effects, like all other physical phenomena, should be made subject to the general laws of quantum theory. Although much here is wrapped in obscurity, it is at least clear that it is, in fact, the most fundamental and characteristic property of Einstein's theory, its *general covariance* (as the property is known), that presents the biggest obstacle to its quantization.[13] But general covariance was precisely the property that Einstein invoked to implement Mach's Principle.[3] There is therefore a most intimate connection between Mach's Principle, at least as perceived by Einstein, and the obstacles to quantization. Our attitude to the quantization problem must at the very least be influenced by Machian considerations.[14]

A third reason for this being a good time to review the history of Mach's Principle is precisely the fact that not too much is happening in the field at the present time. Even the great surge of work on the quantization of general relativity has notably flagged in the last few years, as could be noted in the much more reflective nature of the third of the Oxford conferences on quantum gravity (held in March 1984).[15] The calm between storms is a good time to take stock. Moreover, the lack of progress in both fields may well have a common origin, as has just been argued. Let us look back on what has been achieved, see if we can identify the weak spots, but above all let us attempt to identify the core problem and at least see if we can agree on its precise nature. This is the main task which I have set myself in this study, about the origins of which a few words may be appropriate, because they supply the fourth reason for the writing of this book at this time.

My interest in the problem dates back to 1963 and was stimulated by reading Bondi's book *Cosmology*;[16] this led to a reading of Mach's *Science of Mechanics*,[17] which made on me, as on so many others, a profound and lasting impression. Mach's arguments for the relativity of not only space but also time* appeared to be quite undeniable. If neither time nor space exist as true entities, it would seem that the entire theory of motion needed to be recreated *ab initio*.

However, careful reading of Mach's book and Einstein's papers written while he was working on the creation of general relativity led me to a surprising conclusion: that although Einstein professed great admiration for Mach and claimed to be intent on solving the problem of inertia as laid bare by Mach, he did in fact have a significantly different understanding – one could go so far as to call it a misunderstanding – of the problem. Above all, Einstein appeared to confuse two quite distinct uses of the word *inertia*. When Mach spoke of the problem of inertia, he was referring exclusively to Newton's First Law of Motion, the statement that bodies subject to no forces move through absolute space with a uniform rectilinear motion. Mach insisted that the law as formulated was an epistemological nonsense – since it made statements incapable of objective verification (for the reasons to be explained more fully on p. 8 ff), and such statements could not provide true grounds of knowledge for an understanding of nature (epistemology is the study of the grounds of knowledge) – and a physical implausibility. He called for a proper operational definition of the phenomenon and a physical and causal explanation of it in terms of something real, i.e., observable (at least in principle). Although he criticized the other use of the word inertia, to describe the *inertial mass* which appears in the Second Law of Motion, Mach had no criticism of the concept itself. He felt it was entirely unproblematic, especially after his own operational definition had been substituted for Newton's circular definition of mass. (Mach's clarification of Newton's concept of inertial mass is discussed in the final chapter of this book.)

In contrast, examination of Einstein's papers reveals that he used the word *inertia* indiscriminately to describe both phenomena and seemed to regard the existence of both as requiring physical explanation. A typical example of Einstein's interpretation (or rather misinterpretation) of Mach is the following statement made in 1913:[19] '. . . one must require that the appearance of inertial resistance be due to the relative acceleration of a body (with respect to other bodies). We must require that the inert resistance of a body increase solely because in its neighbourhood there are unaccelerated inert masses . . . .' (It is worth mentioning here that the allegedly Machian variation in the inertial mass which Einstein believed

* The time aspect of the problem of the relativity of motion has been curiously neglected. In fact, it will be argued in Vol. 2 that there are *two* Mach's Principles: the First, relating to the problem of space and the relativity of motion; and the Second, relating to the relativity of time.[18] What is normally called Mach's Principle is to be identified with the first of these. The Second Mach's Principle will play a very large role in Vol. 2.

he had incorporated in the general theory of relativity has since been shown to have no genuine physical reality. It is what is known as a coordinate effect.) Much of the confusion surrounding Mach's Principle stems from this curious interpretation; it is surprising that this fundamental shift by Einstein from Mach's original position has never been clearly pointed out. Moreover, it will be argued in this study that, on this point at least – namely, that the concept of inertial mass is unproblematic – Mach was right and Einstein wrong. This is a first major point that must be made.

A second such point is that, in developing general relativity, Einstein did not make a frontal attack on the Machian problem of finding a dynamical explanation for the law of inertia. Einstein himself commented[20] that the simplest way of realizing the aim of the theory of relativity would appear to be to formulate the laws of motion directly and *ab initio* in terms of relative distances and relative velocities – nothing else should appear in the theory. He gave as the reason for *not* choosing this route its impracticability. In his view, the history of science had demonstrated the practical impossibility of dispensing with coordinate systems. He therefore adopted an indirect approach and was guided, it seems, more by gut intuition than a clear formulation of principles that would of necessity lead to the realization of his aims. This was in striking contrast to the means he adopted to achieve the other main aims he had in developing general relativity – the creation of a field theory of gravitation consistent with the fundamental principles of the special theory of relativity and the inclusion of geometry as an integral part of dynamics (made possible by Riemann's demonstration in 1854 that Euclidean geometry was just one amongst many possible geometries that the world could possess). Here his approach was crystal clear, entirely logical, and could not fail to achieve its aim if pushed with sufficient vigour – as it was.

When I read Einstein's papers more than twenty years ago (many of which have not yet been published in English), I was struck by the fact that in attacking the triple problem of the origin of inertia, the relativization of Newtonian gravity, and the inclusion of geometry in a dynamical framework, Einstein was working on foundations that had been very unequally developed. The Machian concept of inertia as the outcome of interaction with distant matter in the universe was but a vague hunch, whereas special relativity, created by Einstein in 1905,[21] was already a highly developed theory. Above all, it had been cast into a beautifully perspicuous form by Minkowski[22] in his creation in 1908 of the space–time concept, a concept that was tailor-made for the introduction of a variable Riemannian geometry, which was by then a highly developed discipline and perfectly suited to the task Einstein had in hand. In addition Maxwell's electrodynamics, translated into relativistic form, gave

Einstein several clear hints of the way he should proceed. Einstein's carriage was being drawn by a Machian donkey and a Minkowskian and Maxwellian stallion. It's hardly surprising the stallion won out!

Reflection on these matters led to the conclusion that one ought to go right back to first principles in the Machian problem and attempt the route which Einstein had said was impracticable. In particular, the problem might not appear so insuperable if, as a first approximation, it was attacked in a nonrelativistic approach.* After all, Mach had identified and formulated the problem of inertia in the prerelativistic world. Might it not be possible to solve the *nonrelativistic* Machian problem? If this could once be cracked, one would at least have some definite theoretical models on the basis of which the full relativistic problem could be attacked. The development of a few simple Machian models would also be of great value in demonstrating clearly, first, the actual possibility of a Machian origin of inertial motion and, second, the kind of effects and structures one should expect in a more realistic and fully relativistic theory. It was important to establish the plausibility of the idea and get a feeling for its consequences.

Some years pondering these matters culminated in the formulation of what appeared to be the appropriate framework for constructing theories that are Machian of necessity, i.e., by virtue of the basic principles on which they are constructed. Publication of a short paper in *Nature*[23] in 1974 outlining the basic principles and illustrating them by a particularly simple model led to a collaboration with Bruno Bertotti, which lasted for about six years and in which the broad aims were first to develop to the full the nonrelativistic theory and then to incorporate the basic facts of special relativity in a Machian manner.

The outcome of this work[24] was not what we had expected. It began as an attempt to find an alternative to general relativity, which was, as explained, felt to be not truly Machian, but our final conclusion was that Einstein's theory was actually much more Machian than we had believed; it was in fact Machian precisely in the sense required by our general principles! No detailed attempt will be made here to justify this conclusion, to which we were led in considerable part by the intervention of Karel Kuchař; the necessary explanations will be given in Vol. 2. The point of mentioning this work now is the explanation that it provides (a) for the writing of this book at this time and (b) for the overall view adopted

* It should be pointed out that the word *relativity* in the special theory of relativity does not at all have the same meaning as in Mach's assertion of the relativity of motion. This point will be clarified on p. 9 and in Sec. 1.2. The significance of the special theory of relativity for the Machian problem is that it introduced a quite new problem to do with the nature of time, since it showed that the naive concept of absolute simultaneity – that there is a unique 'now' defined throughout the entire universe – must be radically revised. By a nonrelativistic formulation of the Machian problem, I mean a formulation in which the concept of simultaneity is still allowed.

in this study, which, broadly in line with Wheeler's position,[12] is that general relativity is a true implementation of many of Mach's most important ideas but that this has been obscured by accidental and historical reasons and by Einstein's attempt to find an allegedly Machian explanation of inertial mass (instead of concentrating on the real problem, the origin of Newton's First Law). In the very broadest terms, the grounds for asserting the 'Machianity' of general relativity are as follows.

Mach's ideas were put forward at a time in which the field concepts developed by Faraday and Maxwell had not yet supplanted in the minds of physicists the Newtonian concepts of material bodies which act on each other through long-range forces. But by the time that Einstein came to attack Mach's problem he was working in a climate of opinion dominated by field-theoretic concepts. Many people, including Einstein himself in his later years,[25] believe that the transition from a matter-dominated concept of the world to one in which fields are the primary entities makes the original Machian idea obsolete. Let us look at this more closely, for it will take us straight to the heart of the problem. We consider first a Newtonian type situation, in which material bodies constitute the entire content of the universe.

Thus, imagine an infinite Euclidean space; for simplicity of visualization assume it to have only two dimensions instead of the three of the real world. Imagine this space populated with material bodies that we can see (although the space itself remains invisible). The relative distances between the objects, which we assume to be n in number and to be so small that they can be regarded as points, are well defined and satisfy all the geometrical relationships that follow for such distances in the framework of Euclidean geometry. Suppose we observe that the relative configuration of the bodies changes, i.e., the mutual separations of the bodies change. From what we observe, is it possible to deduce that any particular body which we might choose to consider has a definite motion in space? This immediately brings us up against what may be called the *fundamental problem of motion*.

Suppose we take a 'snapshot' of the instantaneous configuration of the bodies at some moment. It will show a pattern of dots; their positions could be as indicated by the crosses in Fig. I.1. A little later we take a second 'snapshot'; the new relative positions could now be as in Fig. I.2, in which the positions are indicated by the symbols ⊙. The relative positions are not quite the same. On the basis of this information – and we have nothing else at our disposal – what can we say about motions of the individual bodies?

Here we confront the invisibility of space. Suppose we take the second snapshot and place it on top of the first, obtaining the situation shown in Fig. I.3. We could then say that the motions of the individual bodies are

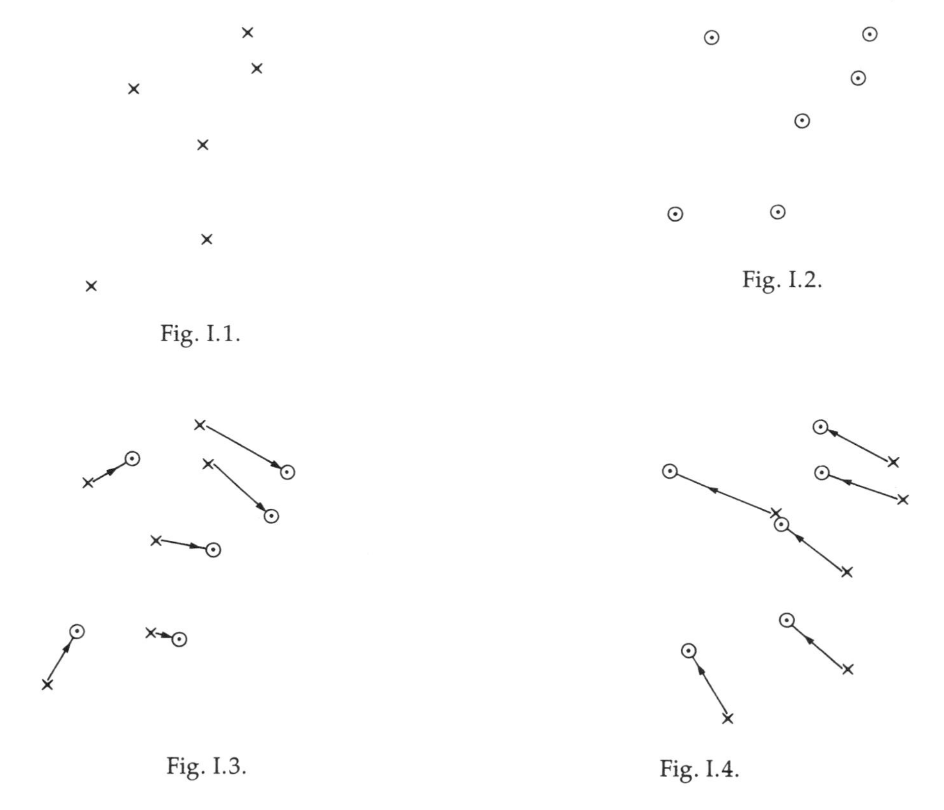

Fig. I.1.

Fig. I.2.

Fig. I.3.

Fig. I.4.

represented by the little arrows. But this is clearly arbitrary; we could just as well have placed the second snapshot as in Fig. I.4 and then the arrows showing the displacements would be quite different, as we see. It would appear from this simple illustration that the notion of a body as having a quite definite motion is untenable. In particular, it would seem to be absurd to say of a body that it moves through space along a straight line.

Although Mach never expressed himself precisely in these terms, we have here the essence of the epistemological problem with which he was so concerned. Motion of one body can only be observed relative to others, and if all bodies are simultaneously in motion we obviously face an acute problem if we wish to attribute a definite motion to any particular body. Because everything is in motion, all connection between one 'time slice' (one snapshot) and the next is broken. We clearly cannot say where any of the bodies 'have gone'. If motion is to be defined as something directly observable, then it is transparently relative.

It is here appropriate to make a brief but vital digression (announced in the footnote on p. 7) into the uses that are made of the word *relativity*. We shall see that time and again confusion of meanings plays a large part in the seeming obscurity of much of the absolute/relative debate. The two meanings of inertia have already been mentioned. The relativity of motion as made manifest by the two-snapshot illustration is normally called *kinematic* (or *optical*) *relativity*. It must be very clearly distinguished from *Galilean relativity*, which will be discussed briefly in Sec. 1.2 and extensively in the second half of Vol. 1, and also from the use of the word *relativity* (and also the related adjective *relativistic*) in Einstein's special theory of relativity (which is very closely related to Galilean relativity). The connection between kinematic relativity and the use of the word relativity in Einstein's *general theory of relativity* will occupy us through most of Vol. 2. Throughout this study, the expression *relativity of motion* will be used in the kinematic, i.e., Machian, sense.

After this digression let us consider the situation in field theory; we shall see that in essence it is no different. In field theory the dynamical problem can be posed typically as follows. Imagine a pattern of intensities in two dimensions (suppressing again the third dimension of space for better visualization) and once again suppose a snapshot taken of the intensities. A little later the pattern of intensities has changed everywhere by a certain amount. We take a second snapshot. Now the aim of a dynamical field theory, expressed in these terms, is to formulate laws which say how the intensity at each point of space changes with the passage of time. But again we confront the invisibility of space. Given our two snapshots, the only way that we can determine how much the intensity has changed is by comparing the one pattern of intensities with the other. *But how is the one snapshot to be placed with respect to the other*? We lack all objective criteria for making any definite placing of one snapshot relative to the next but for every particular placing we choose we obtain in principle different changes in intensity. No less than in the case of material particles, the universal change that takes place between the two snapshots simultaneously severs all connection between the two time slices.

Thus, the *Machian problem* exists just as acutely in the one scenario as in the other. This conclusion is not altered by the fact that Einstein considered an even more difficult situation in which allowance has to be made for the difficulties of defining simultaneity and the geometry of space is allowed to change as well. Note also that the Machian problem outlined above clearly has no relationship at all to the masses of bodies; the mass concept, i.e., inertial mass, does not even exist in the field scenario. It is for this reason I disagree with Einstein's summary of the problem given late in his life:[25]

Mach conjectures that in a truly rational theory inertia would have to depend upon the interaction of the masses, precisely as was true for Newton's other forces, a conception which for a long time I considered as in principle the correct one. It presupposes implicitly, however, that the basic theory should be of the general type of Newton's mechanics: masses and their interaction as the original concepts. The attempt at such a solution does not fit into a consistent field theory, as will be immediately recognized.

Einstein's final comment 'as will be immediately recognized' is to be read in the light of the fact that for a long time he achieved no such recognition. There does therefore seem to be some grounds for believing Einstein was confused on this particular subject. Moreover, we shall see in Chap. 2, in which the immediate continuation of the above passage is quoted, that as late as 1949 Einstein was still very loath to abandon Mach.

Let me now outline *what might have happened* to the Machian problem. Had Mach been a less reluctant theorizer, he might well have solved the problem in the 1870s in his original Newtonian scenario. That is, he might well have found means for formulating dynamical laws that overcome the problem posed by the severance of the connections between the two successive time slices. It would of course involve the masses of the universe, since it would have solved the Machian problem in the context of the then prevailing concepts of the nature of matter. It would also have used the concept of absolute simultaneity. But such a theory might then have served as a paradigm for overcoming the almost exactly analogous problem in field theory and the even more ambitious theory involving variable geometry and an absence of absolute simultaneity that Einstein actually created in general relativity. Of course, in such later developments, material bodies could not play the decisive role that they would have done in the original Machian theory, for the very concept of material bodies is to a large degree superseded in these later developments. But the commonality of the problem outlined above, equally acute in the two quite different scenarios, suggests that its solution would not in essence depend upon any particular theory of the *contents* of the universe. The problem arises, not because of the specific contents of the universe, but because of the fact that they are constantly changing.

Mach, of course, did not develop any such theory and Einstein's theory was created without any of the preparatory work I have just outlined. Many people are convinced that it bears little or no resemblance to the sort of theory that Mach advocated. In one sense, they are perfectly correct. But only, I contend, because the concept of the world's contents has changed out of all recognition. The part played by the triumphant success of the field concepts developed by Faraday and Maxwell in transforming the *context* of the Machian problem is highly significant even though the underlying Machian problem, the core problem, is still basically the same. This I think is where the work that Bertotti and I did during the 1970s has its value. For I believe that we did succeed in finding the solution to the Machian problem in its original Machian context (subject to the crucial condition that the universe is assumed to consist of a *finite number of bodies* moving in either an infinite or finite space). It was then in the process of extending the basic idea which solved the problem in the original Machian context to first the case of field theory and then on to the more difficult case of variable geometry that we realized that the *basic principles* we were using were in essence the principles that underlie general relativity.* (A word of explanation to relativists: the principles here referred to are those discovered by Arnowitt, Deser, and Misner[28] and discussed in connection with Mach's Principle by Wheeler.[29] The 'Machian principles', developed by Bertotti and myself quite independently of this earlier work in general relativity, were shown by Kuchař to be in essence the same principles as used in general relativity, though at a lower level of sophistication. It may also be worth pointing out here that general relativity is extremely special in the way in which it fulfils the Machian criteria just mentioned. The general scheme could therefore be used to look for other Machian alternatives to general relativity.)

It is in this sense that I assert general relativity to be basically Machian: its dynamical laws are based on a principle (to be explained in Vol. 2) that enables one to overcome the lack of connection between successive time slices of a universe all of whose contents – whatever they may be – are constantly changing.

These therefore are my grounds for believing that the time is ripe for a comprehensive discussion of the entire history of the absolute/relative debate from its earliest beginnings through to the present time. And what more appropriate time to write such a study than at the tercentenary of the publication of Newton's *Principia* with the famous Scholium on absolute and relative motion which initiated the whole subsequent debate. The aim of this study is to trace systematically the evolution of the major strands in the history of dynamics which terminate in general relativity.

* If general relativity is Machian in this very basic sense, it must be admitted that it is at best partially Machian in another sense of the word. Mach had an extremely radical philosophy of science; he asserted that the sole task of science was to establish correlations between *directly observed phenomena*. In the period of his youth, an age still dominated by the Newtonian concept of material bodies, it was much easier to assume everything of dynamical significance (i.e., all the masses of the universe) to be presented directly to the human senses. However, the revolutionary developments in physics that occurred in the second half of Mach's life led, through the creation of field theory, to a mathematical description of physical phenomena in which the basic theoretical concepts are very much further removed from direct sense perception. It cannot be denied that these developments, which will be discussed in Vol. 2, present us with severe interpretational problems and make general relativity rather un-Machian if judged as a theory which correlates sense perceptions *directly*. Short of a very drastic revision of ideas, in which mind and perceptions are taken to be primary and matter secondary, such as has been advocated, for example, by Schrödinger in the epilogue of his famous *What is Life*? and, implicitly, in his *Mind and Matter*,[26] and by Wigner,[27] and for which, perhaps, hints are to be found here and there in the history of the absolute/relative debate, I do not see how this situation can be readily changed.

The original intention was to treat the span from Newton to Einstein, following broadly the historical development of the debate about the absolute or relative nature of motion but with insertion of the work just described in order to supply an organic link between the nonrelativistic Newtonian world-view in which Mach advanced his ideas and the field-theoretic relativistic framework that Einstein used to create general relativity.

However, it did seem appropriate to look into the grounds that led Newton to formulate the concepts of absolute space and time in the first place, and this led me to an examination of Galileo's writings. But Newton is truly a watershed: forward – there is no stopping until you get to Einstein (and no doubt we shall go further; indeed there are unmistakable signs that the caravan is already on the move); backward – well willy-nilly there is no logical stopping place before Aristotle or even a little earlier.

This historical research I felt compelled to make has added very greatly to my appreciation of the problem and convinces me that the historical perspective is the only correct one in which to present the subject. There are few subjects with a longer or more absorbing history than the theory of motion. Few problems have given rise to such profound reflections and speculations. And few have brought such a rich harvest of successes. No one can survey the whole field from the pre-Socratics through to the first tentative grand unified field theories of the present decade without being struck by the depth of sophistication that has been achieved or by the extraordinarily roundabout ways in which some of the key ideas have gradually become established. For this reason alone the historical approach is to be recommended.

There are other good reasons too. One is that it demonstrates the extent to which Newton put himself out on a limb over the question of absolute space and time. His position was exceptional both with respect to what went before and what came after. There is a good case for arguing that absolute space and time were to a large degree the accidental products of the historical development, which is a very intriguing story in its own right, especially as regards the part played by the show-down between Galileo and the Inquisition over the question of the earth's motion and the effect it had on Descartes. Newton's Scholium on absolute space and time was in fact a thinly veiled polemic against Descartes.

Equally relevant is the fact that 'Machian' ideas were very prevalent long before Mach and even Newton and actually played a highly important role in the discovery of dynamics. There was a strongly Machian strand to the thinking of both Aristotle (whom Copernicus followed in this respect) and, above all, Kepler. Indeed, Kepler had a most interesting pre-Newtonian form of Mach's Principle, and it played a significant part in his work. Just as the Machian stimulus can take half the credit for the psychic motivation which led to general relativity, essentially the same idea can take credit for the discovery by Kepler of the laws of planetary motion.

Moreover, realization of the relativity of motion played a crucial role in leading to the acceptance of the Copernican proposal of a two-fold motion of the earth (daily around the polar axis and annually around the sun). Several important features of Newtonian dynamics actually stem from Galileo's reflections on the relativity of motion and the consequences of Copernicus's proposal. Finally, a deep conviction of the relativity of motion was a major factor that enabled Huygens to find the correct laws of elastic collisions between bodies, which was another most important development in the creation of dynamics. There is a very real sense in which all these important advances can be seen as due to 'Machian' principles (admittedly of differing degrees of purity, as we shall see).

These 'Machian' successes are well worth emphasizing, since many people, exasperated by seemingly fruitless discussions about Mach's Principle in the framework of general relativity, are inclined to dismiss it as idle philosophy. In fact, the Machian ideal is ultimately indistinguishable from one of the highest aspirations of the natural scientist – to show that the world we observe has an essentially rational structure in which *there are observable causes of observable effects and everything fits together into a coherent whole.* As Einstein put it:[30] the chain of cause and effect is closed. This ideal has several times been the driving force in the discovery of highly important laws of nature, and there is no reason to doubt its ability to be so again. Its force is not spent.

A further strong argument for the historical presentation of the subject has already been indicated – through the whole development of the problem there runs a common thread, the core problem, as I have called it. The root difficulty is that of defining unambiguously and uniquely what one means by motion or more generally change in a context in which everything else which might be used for reference purposes to define and quantify that change is itself changing. We are thus faced with the problem of quantifying and mathematizing something that seems incapable of unique definition. And, as we have seen, this problem reappears in different guises and must at any period be attacked within the framework of the prevailing conceptions and level of knowledge. It is quite clear that an historical approach is needed to do justice to such a constantly changing problem.

A final argument for the historical approach is that it permits an organic presentation of some of the ideas thrown up in the course of the debate about the nature of motion but not as yet successfully incorporated in any theory. A genuine seeker after truth must surely always be open to the possibility that our view of the world may need significant adjustment. Recent experience with the history of Yang–Mills (gauge) theory shows how a powerful concept can remain more or less dormant for quite a long period before some further element is added, which only then makes progress possible. It was just the same with the Copernican revolution. Copernicus's original proposal had a quite modest effect on astronomy until Kepler added a quite new (and strongly Machian) dimension. For this reason I have felt that this book should not only report the ideas that have been successfully incorporated into physical theories but also some others that, to me at least, seem to have as yet unrealized possibilities; in particular, they might have a bearing on the difficulties mentioned in the footnote on p. 12. For this reason a certain amount of space is devoted to fundamental philosophical questions, in particular the very basic divide that developed during the seventeenth and early eighteenth centuries between the philosophy of materialism, on the one hand, and idealism (in which mind and perceptions are primary) on the other [broadly speaking, the materialists (or realists) were identified with the absolute (Newtonian) approach to motion, the idealists with the relational (Machian) approach, though there were notable exceptions, above all Einstein himself]. Closely related to this divide, and directly relevant to our subject, is the antithesis between the *concrete* and the *abstract*.

Provided these matters are competently handled, and here I must leave judgement to the reader, no apology needs to be made for this inclusion, on a limited scale, of as yet unresolved philosophical disputes. General relativity itself is one of the most dramatic examples of the transmutation by genius and empirical input of what were initially very abstract philosophical ideas into a concrete physical theory of both motion and geometry capable of experimental testing to a degree that is nothing less than amazing. The clouds are no doubt the same as they were in the seventeenth century, that heroic age of science, but indeed how very differently they are now perceived to move. I rest my case for the historical and philosophical presentation of the subject and express the hope that the airing of these questions in the historical perspective will not only help to clarify the problem of the origin of inertia but also encourage the reader to think afresh about these matters and ask questions as fundamental as those posed by Copernicus, Kepler, Newton, Mach, and Einstein.

The inclusion of the pre-Newtonian history and the fact that Newton marks such a clear dividing line make it convenient to divide the complete study into two volumes, the first of which, *The Discovery of Dynamics*, presented here, treats the pre-Newtonian origin of dynamics, its formulation by Newton, and the clarification in the second half of the nineteenth century of certain key conceptual aspects of dynamics that Newton left in a somewhat obscure and confused form. It is in the first place a study of how the prevailing concepts of space, time, and motion influenced the final structure of Newtonian dynamics at each of the intermediate stages that led to the ultimate Newtonian synthesis. But this makes it simultaneously a history of the discovery of dynamics itself. In order to ensure that Vol. 1 can be read in its own right as such a history, the original scope of the book has been significantly extended and contains much material not strictly essential to the absolute/relative debate, particularly in the astronomical part. However, almost all aspects of dynamics and its discovery have some bearing on the question. Naturally, the main stress in this book is on conceptual developments. For this reason, as explained in the preface, I hope it will have much to say to working scientists, in casting light on the fascinating origins of their discipline, and that it will also appeal particularly to anyone interested in philosophical and historical questions. It is also my hope that *The Discovery of Dynamics* will fill a certain gap in the historical literature. The ground it covers has of course been extremely well trodden, but it seems to me that there remains a gap between the numerous books that treat the general development of the history of ideas, such as Kuhn's *The Copernican Revolution*[31] and Koyré's *From the Closed World to the Infinite Universe*,[32] to mention only two, and the equally numerous books by highly professional historians that treat either specialized subjects (for example Jammer's three books on the concepts of force, mass, and space,[33] and Westfall's *The Concept of Force in Newton's Physics*[34]) or concentrate on particular periods or particular scientists. I do not think there is any book that quite covers the same field, certainly not in the way in which the discovery of dynamics is examined simultaneously with the evolution of the ideas of absolute and relative motion. By restricting the ambit to the discovery of dynamics as opposed to the larger scientific revolution in our view of the world, I have been able to concentrate on the development of certain key concepts, which undoubtedly has a unique and absorbing fascination. Indeed, it seems to me that in the discovery of dynamics there were about a dozen key events, almost all of which were associated with the breakthrough to precise mathematical formulations of empirical facts about particular observed motions. The task of a history of the discovery of dynamics, as I have conceived it here, is to identify clearly these pivotal events, show how the insights were achieved, and put them in the proper perspective within the overall picture.

In fact, making a precise count, I identified thirteen, divided neatly into six associated with the astronomical study of celestial motions, six associated with the study of terrestrial motions, and the odd thirteenth, making up a baker's dozen, being the mathematical insight behind the Copernican revolution, which linked together for the first time the theory of celestial and terrestrial motions and was therefore a key event in the eventual emergence of dynamics. These thirteen insights, which will be noted as we proceed through the book, do not include the discovery of the geometry of the three-dimensional world nor the final synthesis of all the various elements by Newton that created dynamics, which may, in a sense, be regarded as the geometry of the four-dimensional world.

It is in this respect that the present study will be found to differ most strongly from 'history of ideas' approaches, which often tend to avoid going in any depth into technical details. However, as this book will show, it is precisely through the technical details, which become most absorbing when seen in the light of their full significance, that the true electric current of discovery runs. If we do not follow the current closely, we shall not really understand what happens and why. To use another image: the technical details are like the hinges of a door opened on a new vista. The change in view they make possible seems out of all proportion to what, objectively speaking, they represent. But their immense strength comes from the fact that, like hinges, they are mounted on a secure support: successful mathematical description of observed phenomena. And we cannot understand why the door swings unless we see how the hinges work.

This, in particular, is the reason why a comparatively large amount of space is devoted to ancient astronomy, for it provided the foundation of much of what followed. In fact, Chaps. 3, 5, 6 of the present book represent a more or less self-contained history of astronomy from antiquity to the discovery by Kepler of his laws of planetary motion, though the material is, of course, integrated into the overall structure of the book and has been written very much with an eye to the significance of the astronomy for the subsequent dynamical interpretation of Kepler's discoveries. The purely astronomical material of these chapters could, for example, serve as the basis of a semester-length course on the history of astronomy of the kind that is so popular in American universities.

If the first volume of *Absolute or Relative Motion*? should be easy enough going for the reader with a good grounding in the rudiments of Newtonian dynamics, the second volume, which treats the reaction to Newton and then the extraordinary culmination of the debate in the creation of general relativity, will of necessity get progressively tougher. I fear that my readers will fall away chapter by chapter as we get nearer the Holy Grail. There is, alas, very little that can be done about this. By no stretch of the imagination is the general theory of relativity anything but very sophisticated, a fact that in no way detracts from, but rather enhances, its conceptual appeal and beauty. By concentrating throughout on the bare essentials of the conceptual thread, I hope to keep the reader with me as far as is humanly possible. There will be no gratuitous mathematics, but nor will it be shirked when unavoidable for the undistorted presentation of essential arguments. All I can say to the reader is, keep going as long as you can. It's worth it to get even an inkling of the *dénouement* which that unsurpassed genius, Einstein, has bequeathed us as our best understanding to date of the central problem of motion – a problem that already vexed the greatest minds in the cradle of our civilization about two and a half millennia ago.

So, then, let us start. But a word of warning. Before the reader embarks on the journey from Aristotle to Einstein, he or she may want to know the conclusion that can be expected at the end of this trek through the millennia. If Frodo is to undertake the journey to Mordor,[35] will there at least be a positive outcome? In a word: *Is motion absolute or relative*? No definite answer to this question can be given. In one sense motion is unquestionably relative, since no experimentalist can possibly measure a motion that is not relative. It will be shown that the real distinction is in the structure and predictive power of the theory that is found to describe motion. When applied to a finite universe, Newton's theory is found to have less predictive power than a Machian theory. Newtonian theory can,in fact, be recast in purely relative terms, but when this is done it is found to have not only an ungainly and somewhat arbitrary structure but also to be less predictive than theories with a more obviously Machian structure.[36] The positive conclusion of this study, reached at the end of Vol. 2, is that general relativity suffers from no such defect. At least as regards its *basic structure*, general relativity can be said to be designed in such a way as to make it almost as predictive as one could imagine: it is a *ne plus ultra* (though there are some worrying technical details about the mathematics).* So much at least can be promised to the traveller who will persist to the end. But there looms an even larger difficulty: the possible infinity of the universe. If the world is truly infinite, the chain of cause and effect can never be closed. New influences can always 'swim in' from across our most distant horizon and it would not appear to be possible under such circumstances to close the circle. Even Frodo, his mission at Mordor accomplished, set out again once more into the unknown. Whether that is our fate remains to be seen.

* For the benefit of relativists, this is a reference to the fact that Wheeler's original 'thin-sandwich' conjecture does not appear to be an appropriate way of approaching the initial-value problem in general relativity.[37] There are also of course major uncertainties related to the part played by quantum theory in the whole absolute/relative question, to say nothing of the various theories that are putative successors to general relativity. Finally there remain the difficulties mentioned in the footnote on p. 12.

1

Preliminaries

1.1 Newton's laws and their conceptual framework

We begin at the end – by recalling Newton's laws and the concepts used to express them. For the modern reader, the most illuminating approach to the history of the discovery of dynamics is probably to start with a clear understanding of what we now know and then trace the gradual clarification and emergence of the key concepts and results from their earliest beginnings. Even the reader who is extremely familiar with Newtonian dynamics may find this survey of value, since it will stress aspects of particular relevance both for the discovery of dynamics as well as for the absolute/relative question. In this summary, detailed references will not be given; the quotations are from Ref. 1.

Of all the Newtonian concepts, those of *absolute space and time* are the most important, for they provide the framework of everything else. Newton imagined his absolute space as rather like a block of perfectly translucent glass stretching from infinity to infinity. Of course, it is only a conceptual block; objects can move through it perfectly freely. The essential purpose of absolute space is to provide a definite frame of reference: each and every body, however it may move, is always at some quite definite point. All the relations of Euclidean geometry hold in the block; above all, any two points are joined by a unique straight line. Conceptually, at least, it is therefore meaningful to say of a body that it follows a definite path in absolute space.

Newton conceived absolute time as, 'from its own nature', flowing 'equably without relation to anything external'. To have a definite picture, one can imagine God holding in his hand a watch that keeps perfect time and observing the various bodies in the universe as they travel through absolute space. To God at least it is therefore meaningful to say that a particular body is moving at a particular time at a certain speed in a certain direction.

Newton specifically introduced the concepts of absolute space and time in order to overcome the problem of kinematic relativity sketched in the Introduction and to have a precise and unambiguous concept of the motion of any particular body. After space and time, motion is the most fundamental concept in his scheme.

It is implicit in this scheme that space, time, and motion are conceptually prior to the actual laws of motion – God might have chosen different laws of motion but he had no alternative but to place and move bodies in space and time.

Throughout this discussion of Newtonian dynamics, we shall ignore the problem of kinematic relativity and assume that, in some manner, absolute space and time have been made directly accessible to human senses, so that all the Newtonian concepts have a well-defined meaning.

A highly significant feature of Newtonian dynamics is that the motion of actual bodies has very far reaching *directional* aspects. The mere *speed* of a body in absolute space has only a restricted dynamical significance; a more significant concept is *velocity*, i.e., the speed in a definite direction. This could not possibly be deduced from the mere concepts of space, time, and motion and was a realization that came very late in the discovery of dynamics.

The most natural language for expressing the directional aspects of physically realized motions is by means of vectors, which are characterized by both magnitude and direction; for example, the speed of a body in a certain direction gives its velocity vector. Even though the theory of vectors was not developed formally until the middle of the nineteenth century, long after the discovery of dynamics, several parts of this book will be much clearer if expressed by means of vectors.

In the overall scheme of Newtonian dynamics, the most important characteristic of the motion of actual bodies is that it has a dual nature. In Newton's scheme, any particular body is at any instant responding simultaneously to two influences of quite distinct natures. The natures of these influences, which are exerted on the one hand by absolute space and time and, on the other, by the remaining bodies in the universe, are characterized by their effect on the instantaneous velocity with which the body under consideration is moving through absolute space at the time considered. We must look at this in some detail.

According to Newton's First Law (*Lex Prima*), the consequence of the first influence, that exerted by space and time, can be expressed as follows: if the second influence were not present, i.e., if external bodies were to cause no disturbance, the instantaneous velocity that a body has at any particular moment of time would persist forever; thus, a body set in motion initially in a given direction with a given speed would continue forever in that given direction at that same given speed were it not for the influence of other bodies. The First Law states specifically: 'Every body continues in its state of rest, or of uniform motion in a right line, unless it is compelled to change that state by forces impressed upon it.'

To define the nature of the second influence, we must first introduce some further fundamental concepts, above all that of the mass of a body: in Newtonian dynamics each body is characterized by a positive number, the *mass*, which Newton says is the *quantity of matter* that the body possesses. At this stage we shall not attempt to look more closely at this concept, which, for all its seeming transparent simplicity, requires a rather sophisticated definition. In Newtonian dynamics, mass is neither created nor destroyed. The mass of a body can only change if new mass is added to it from some other body or alternatively some of its mass is removed.

By means of the mass concept we form what is perhaps the most important of the specifically dynamical concepts in the Newtonian scheme, that of the *momentum* of a body, or its *quantity of motion*, to use Newton's expression. This is defined as the product of the mass of the considered body and its velocity. Because velocity is a vector, while mass is simply a number without any directional attributes (a scalar), momentum is also a vector. If m is the mass of a given body (which for simplicity we assume to be so small that it can be regarded as a mass point) and $\mathbf{v}$ is its instantaneous velocity, then the momentum $\mathbf{M}$ of the body is defined as

$$\mathbf{M} = m\mathbf{v}. \tag{1.1}$$

By Newton's First Law, $\mathbf{v}$, and therefore $\mathbf{M}$ too, remains constant in time unless the body under consideration is acted upon by some other body. Newton's Second Law (*Lex Secunda*) tells us how the momentum of the body we consider is changed by other bodies. According to Newton, this can happen in one of two ways: through direct contact in a collision or through a definite influence exerted through space by a distant body. In both cases, Newton says that the momentum $\mathbf{M}$ of the considered body is changed by the application of a force. In the first case there is an abrupt change in $\mathbf{M}$ by a finite amount: in the second case the change in $\mathbf{M}$ is continuous. It is implicit in the scheme of Newtonian dynamics that there are definite rules of nature which determine the forces that act on any given body for any definite configuration and state of the remaining bodies in the universe. For the moment, we shall consider the simplest applications of Newton's Second Law.

Force is an essentially vectorial quantity. Thus, a force acts in a definite direction and has a definite magnitude or strength. Let us consider the force produced by a collision. In modern terminology this produces what is known as an *impulse*; we shall denote such an impulse by $\mathbf{F}$. As a result of collision with some other body, the body we consider is subjected to the instantaneous effect of the impulse **F**. Newton's Second Law tells us what the consequence is.

The consequence can be thought of in two different ways: (1) the body retains the momentum **M** it had prior to the collision and, as a consequence, has the same tendency as it did before the collision to continue in the original direction of the velocity **v** with unchanged speed v; however, there is superimposed on the body a second tendency. The body has been subjected to the given impulse **F**. This impulse will produce a tendency for the body to move in the direction of **F** at speed equal to F/m. The resultant motion is obtained by the law of vector addition. We shall express the result first in terms of velocities. The first tendency – to continue with unchanged momentum **M** – causes the body to have a velocity component along **M** of magnitude $\mathbf{M}/m = \mathbf{v}$. As a result of the collision, there is added to this a second tendency: to move with speed F/m in the direction of **F**. This additional velocity component $\mathbf{F}/m$ must be added vectorially by the parallelogram rule of vector addition to $\mathbf{M}/m$ in order to find the resultant velocity. Let $\mathbf{v}'$ be this new, post-collision velocity and let $\mathbf{F}/m = \delta\mathbf{v}$. Then

$$\mathbf{v}' = \mathbf{v} + \delta\mathbf{v}. \tag{1.2}$$

However, Newton does not express his Second Law in terms of velocities but in terms of the momentum and the force. The reason for this is that in any given situation the force has a quite definite direction and magnitude but the change in the velocity which it produces depends on the mass of the body to which it is applied: the larger the mass, the smaller the change in the velocity. In order to obtain a universal form of expression for the Second Law, i.e., a form that does not depend on the particular mass of the body considered, Newton expressed the law in terms of forces and momenta. As a result of the collision the original momentum **M** is changed by $\delta\mathbf{M}$, $\mathbf{M} \rightarrow \mathbf{M} + \delta\mathbf{M}$, and

$$\delta\mathbf{M} = \mathbf{F}. \tag{1.3}$$

In Newton's own words: 'The change of motion [in Newton's terminology *motion* means momentum] is proportional to the motive force impressed [*motive force* means here **F**, i.e., the magnitude of **F**]; and is made in the direction of the right [i.e., straight] line in which the force is impressed.'

The law (1.2) is shown in Fig. 1.1. It might seem that (1.2) is a result of pure mathematics or at least kinematics, i.e., a consequence of the mere concept of motion as a vectorial quantity. However, this is by no means the case. At least two points must be emphasized: (1) as a result of the collision, the tendency for the body to move in the original direction is not lost. This is a most important point; the 'slate is not wiped clean' by the

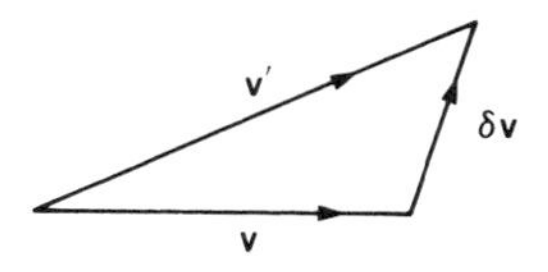

Fig. 1.1.

collision. Indeed, the very fact of persistence with or without collision is highly significant; (2) as we have seen, in any given situation the force (or impulse in the case of a collision) is given, but the effect it produces in terms of an observable change in the velocity depends on the mass of the body to which **F** is applied. Both of these results are pre-eminently physical; in no way does either of them follow from the bare concepts of space, time, and motion.

Newton initially discovered his Second Law for the case of collisions. The much more familiar expression of the law in terms of accelerations is used in many places in the *Principia* but is not stated there as a primary law. That development was due to Euler.[2] To this form of the law we now turn. Its discovery by Newton for the case of gravitational forces was in fact the final insight that completed the discovery of dynamics. According to Newton's law of universal gravitation, bodies attract each other with a definite force. Since this force produces continuous changes in the momentum (and hence velocity) of the body to which it is applied, it is truly a force in the modern usage of the word; we shall denote such a force by **f**. Then for such a force Newton's Second Law is expressed in the form ($\mathrm{d}/\mathrm{d}t$ denotes the derivative with respect to the time t)

$$\frac{\mathrm{d}\mathbf{M}}{\mathrm{d}t} = \mathbf{f}. \tag{1.4}$$

Since $\mathbf{M} = m\mathbf{v}$, this can also be written as

$$\frac{\mathrm{d}}{\mathrm{d}t}(m\mathbf{v}) = \mathbf{f}.$$

Finally, since the mass remains constant, $m = \text{const}$ (unless mass is added to or removed from the body), and $\mathrm{d}\mathbf{v}/\mathrm{d}t$ is the acceleration, denoted by **a**, we can also write

$$m\mathbf{a} = \mathbf{f}, \tag{1.5}$$

which is the form of the law with which most readers will be familiar.

That this form is equivalent to (1.3) can be seen as follows. Let the force **f** act for only an infinitesimal amount of time δt. From (1.5) the corresponding acceleration will be $\mathbf{a} = \mathbf{f}/m$; the resulting change in the velocity will be $\delta\mathbf{v} = \mathbf{a}\delta t$. Thus, the result of the action of the force is to

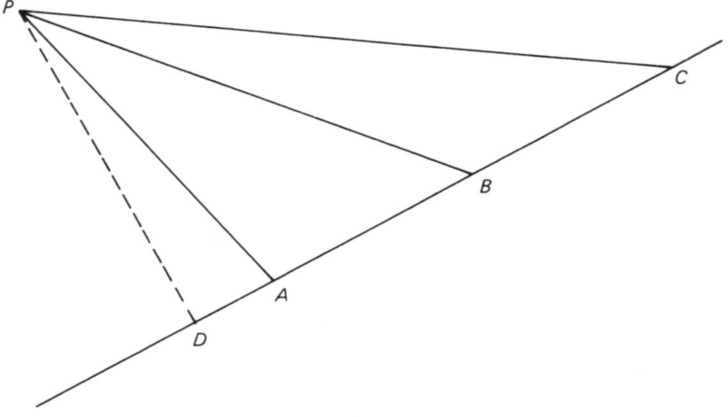

Fig. 1.2.

produce a velocity increment $\delta\mathbf{v}$ and change the initial velocity $\mathbf{v}$ to $\mathbf{v} + \delta\mathbf{v}$. The effect of the force $\mathbf{f}$ acting for time δt is, in the limit when $\delta t \rightarrow 0$, equivalent to an instantaneous impulse $\mathbf{F} = \mathbf{f}\delta t$.

As a simple application of Newton's first two laws, we prove the so-called *area law*. This will not only demonstrate the power of the laws in a nontrivial example but also enable us to understand the significance of many of the astronomical discoveries made in the pretelescopic period.

Consider first a single particle moving inertially along a straight line ABC, in which AB = BC. Thus, it takes equal times to traverse AB and BC. Consider also any point P that does not lie on ABC (Fig. 1.2). Drop the perpendicular PD from P onto ABC. Then since triangles PAB and PBC have the same height PD and equal bases AB and BC their areas are equal. Thus, as the particle moves along ABC the radius vector from P to the instantaneous position of the particle sweeps out *equal areas* in the plane PABC in equal intervals of time. For purely inertial motion, this is true for any point P not on ABC. Let the unit of time correspond to the time taken to traverse AB (or BC). Then AB represents the magnitude of the velocity of the particle.

Now suppose that when the particle reaches B it is subjected to an impulse exactly along the direction to P. Let the impulse be of such strength that, in the absence of the inertial component of the motion along ABC, the velocity the particle acquires as a result of the impulse would take it to X (Fig. 1.3) in the unit of time. Let us now find the resultant of the motion due to the original inertial component and the impulse. To do this, we describe CX′ parallel to BX and of equal length. Then, since BXX′C is a parallelogram, it follows from the parallelogram law of vector addition that the actual path taken by the particle after the impulse will be

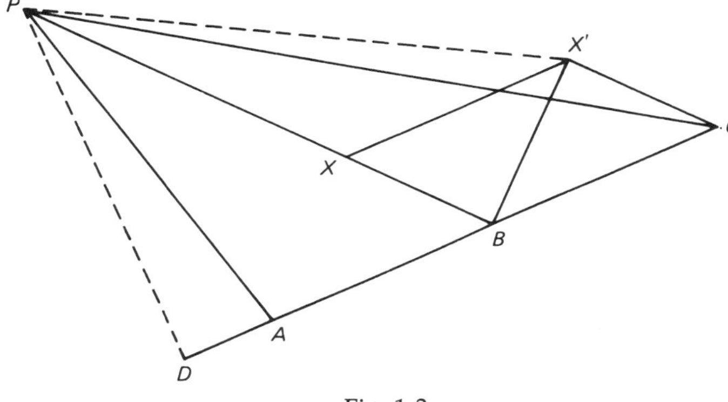

Fig. 1.3.

along BX′, and it will arrive at X′ after the second unit of time. Now triangles PBX′ and PBC have the same base PB while their vertices X′ and C are on a line parallel to this common base. Therefore the areas of these two triangles are equal. But PAB and PBC have equal areas, so that PAB and PBX′ also have equal areas.

This shows that the equal area result is still true with respect to P for an impulse directed exactly along the direction of P (it may, of course, be either towards or away from P). Thus a particle which moves inertially except for a series of instantaneous impulses directed exactly along the line to a given fixed point P at the corresponding instants of time will always sweep out equal areas with respect to P. Equally important is the fact that the motion remains forever in the plane defined by P and the direction of the initial velocity.

To pass from the case of discrete impulses to a force generating an acceleration continuously, Newton supposed that the impulses occur much more frequently and that their magnitude is simultaneously decreased, so that the resulting polygonal figure described in space approaches ever closer to a continuous curve. For the curve obtained in the limit, Newton argued that the equal area result will still hold, i.e., that the radius vector from P to the moving particle will describe equal areas in equal times in the plane defined by the radius vector and the initial velocity. Moreover, this will always be true provided the instantaneous force always acts exactly along the line towards the centre P, which must, of course, be fixed. As the aim here is merely to demonstrate Newton's use of his physical concepts, we shall not attempt to discuss the rigour of his proof – the result is certainly correct.

As we shall see in Chap. 10, the discovery of this result by Newton around 1680 can, more than any other event, be said to mark the point at which the full structure of Newtonian dynamics had been found. It is a special case of what is now known as the *conservation of angular momentum*. The proof makes it clear that the area law arises almost entirely from the First Law. Although the instantaneous velocity suffers innumerable changes, the perpetual uniformity of unadulterated inertial motion is reflected in the equality of the area swept out, while the rectilinearity of pure inertial motion is reflected in the fact that the motion remains perpetually in one plane. (Of course, for this, it is essential that the acceleration be always in the line towards P – in Newton's words, it must be produced by a *centripetal force*.)

Of equal importance for astronomy and the discovery of dynamics is the consequence of the conservation of angular momentum for the earth. As this would take us into the unnecessary technicalities of rigid-body theory, which was only developed after Newton's death, let us merely state the result. Except for very small perturbations, of no consequence at all for pretelescopic astronomy, the earth rotates about its axis with constant angular velocity and the axis remains pointing along a fixed direction in space (ignoring for the moment the precession of the earth's axis). This result more or less exactly parallels the area law, the constancy of the angular velocity being a special case of the equality of the area swept out (it is obvious that for motion in a circle equality of the area swept out is converted into equality of the angular velocity), while the fixed direction of the polar axis corresponds to the fixity of the plane of the motion in the area law.

We must now complete the statement of Newton's laws of motion. His Third (and final) Law (*Lex Tertia*) states that: 'To every action there is always opposed an equal reaction: or, the mutual actions of two bodies upon each other are always equal, and directed to contrary parts.' This law only acquires a precise meaning when we know what Newton means by the words *action* and *reaction*. In fact, by action he means *change of momentum*. Thus, if there is an interaction between two bodies of masses m_1 and m_2 and no other bodies are involved, the meaning of the Third Law is as follows. Let their initial velocities be $\mathbf{v}_1$ and $\mathbf{v}_2$, so that their momenta are $\mathbf{M}_1 = m_1\mathbf{v}_1$ and $\mathbf{M}_2 = m_2\mathbf{v}_2$. Then suppose body 2, by exerting some force on body 1, causes an infinitesimal change $\delta\mathbf{M}_1$ in the momentum of body 1, so that $\mathbf{M}_1 \to \mathbf{M}_1 + \delta\mathbf{M}_1$. Then it follows, first, that there will necessarily be a change $\delta\mathbf{M}_2$ in the momentum of body 2; second, this accompanying change in the momentum of body 2 will always be exactly equal in magnitude to $\delta\mathbf{M}_1$ but in the opposite direction:

$$\delta\mathbf{M}_2 = -\delta\mathbf{M}_1. \tag{1.6}$$

Since the changes in momentum are proportional to the forces that produce them, Newton's Third Law can also be stated in the form that the forces with which bodies act on each other are equal and opposite.

The generalization of the Third Law to interactions in which more than three bodies are involved states that the vector sum of all the individual changes $\delta\mathbf{M}_i$ induced in the momentum of each body i, $i = 1, 2, \ldots, n$, is exactly zero:

$$\sum_i \delta\mathbf{M}_i = 0.$$

Newton's three laws do not say anything at all about the actual forces that occur in nature. These depend on the configurations and nature of the bodies involved. It was one of Newton's supreme achievements to separate out the three laws of motion as valid whatever the nature of the forces acting. The extreme fruitfulness of the scheme was demonstrated by his discovery of the law of universal gravitation, which provided an explicit rule for calculating in the case of gravity the force in any given situation.

For the simplest case when only two bodies are involved, Newton's law of gravity states that if there are two bodies of mass m_1 and m_2 separated by a distance r then each exerts a force on the other which is proportional to the product of the two masses, is inversely proportional to the square of the distance between them, and acts along the line joining them. The force acting on body 1 is

$$\mathbf{f}_1 = \mathbf{e}Gm_1m_2/r^2, \tag{1.7}$$

where G, a constant, is the constant of universal gravitation, and $\mathbf{e}$ is the unit vector that points from body 1 to body 2.

We should mention right away that gravitation is very exceptional among the forces of nature in that the masses m_1 and m_2 occur in (1.7) in the expression for the force. This has a remarkable consequence. By *Lex Secunda*, the acceleration of body 1 due to the force (1.7) is

$$\mathbf{a}_1 = \mathbf{f}_1/m_1$$

But since $\mathbf{f}_1 = \mathbf{e}Gm_1m_2/r^2$, it follows that

$$\mathbf{a}_1 = \mathbf{e}Gm_2/r^2.$$

Thus, the acceleration of body 1 is independent of its mass: all bodies fall in a given gravitational field with the same acceleration. This remarkable result is solely due to the double appearance of the mass: in *Lex Secunda* and also in the rule for determining the gravitational force.

Having reviewed Newton's laws and the framework in which they are formulated, let us now consider some important basic features of these laws.

Perhaps the first point which is worth making is the universality and generality of the concept of the world that underlies Newton's laws.

Throughout the entire universe matter is assumed to have completely uniform properties. Above all any body is characterized by having mass; at least as regards dynamics, the stuff of the universe is all the same. In its essential properties mass is the same wherever it is encountered. The laws it obeys are the same throughout the universe. Thus, a uniform view of the world is a most important part of dynamics.

Closely related to this is the important part played by what may be called *physical quantities*, above all the *masses* of bodies. Strictly speaking, because the masses also occur in the law of gravitational force, Eq. (1.7), they are the only physical quantities we have so far met. However, in the case of electrostatic attraction the masses m_1 and m_2 in (1.7) are replaced by the electric *charges* e_1 and e_2 of the two bodies and the electrostatic force is proportional to e_1e_2/r^2 (Coulomb's Law). The electric charge is a quantity quite independent of the mass and is a second example of a physical quantity.

To highlight the exceptional nature of the gravitational law (1.7), many authors distinguish between the two roles played by mass. The mass m that appears in the universally valid *Lex Secunda* in the form $m\mathbf{a} = \mathbf{f}$ is called the *inertial mass*, while the mass that occurs in the law of force (1.7) for gravitational forces is called the *gravitational charge*, or *active gravitational mass*.

In connection with the fundamental part played by physical quantities in Newtonian dynamics, it is significant for the subject of this book that they play no role in the formulation of the First Law, which in this respect is sharply distinguished from the other two laws.

This distinguished role of the First Law highlights a point that has already been made but bears repetition: the *dual nature* of the factors that govern any particular body's motion. Through *Lex Prima*, absolute space and time keep a body moving in a straight line at a uniform speed and would do so for ever but for the intervention of other bodies in the universe. Through *Lex Secunda* the bodies mutually deflect each other from one inertial motion to another. We note also that the law of inertia itself has two quite distinct parts: the rectilinearity of the motion and the uniformity of the motion. These correspond, respectively, to absolute space and absolute time.

The next general feature to which attention must be drawn is the fact that the laws of motion do not determine motions completely. The typical problem in dynamics is the following. At a given time a body is set in motion from a given point with a given speed in a certain direction: to find its subsequent motion. If the body is what is known as a test body, which means that its mass and various charges are so small that the body in question exerts a negligible influence on the other bodies in the universe, and if the initial positions and subsequent motions of the other bodies are known, then Newton's laws, in conjunction with the laws that determine the forces, suffice to determine the subsequent motion of the test body uniquely. This means that the motion of the considered test body is determined only up to the specification of six quantities: its three coordinates at the initial time and the three components of its initial velocity along whatever coordinate axes may happen to have been chosen in absolute space.

The need to specify both initial positions and initial velocities is a most characteristic feature of Newtonian dynamics. It reflects the fact that the fundamental law of motion, the Second Law in the form (1.5), is a law of second order in the time, i.e., it contains the *second* derivatives of the position coordinates with respect to the time. This is the reason why not only the initial coordinates but also the initial first derivatives of the coordinates with respect to the time (i.e., the velocity) must be specified in the initial condition.

The next general feature of Newtonian dynamics is of such overriding importance that it warrants a separate section.

1.2. Invariance properties of Newtonian dynamics

Newton's three laws of motion contain no reference to place or time. They are assumed to be *universally true*, that is, the laws of motion are exactly the same at all parts of absolute space and at all instants of absolute time. It is also implicit that the rules which govern the strengths of the forces that act in any given situation have the same universality. Because of this assumed universality of the laws of motion and because of their vectorial nature, Newton's laws possess some important properties, called *invariances*.

For example, suppose we choose a definite fixed point O in absolute space as the origin of a system of orthogonal Cartesian axes, which we align along definite directions (we recall that we assume we can somehow 'see' Newton's absolute space, so that O is a point at rest in it and the directions of the axes remain fixed). If we now consider a system of noninteracting bodies, which therefore move in straight lines in absolute space, they will obviously follow straight lines in the chosen Cartesian system. It is immediately obvious that if we either shift the entire coordinate system so that its origin is moved to a new point O′ without any change in the direction of the coordinate axes or we keep the origin fixed and change (once and for all) the orientation of the axes (keeping them mutually perpendicular), the bodies will still follow straight lines in the new coordinate systems even though, relative to the coordinate axes, the straight lines will no longer be the same. This is what one means by saying that the *form* of the laws of nature remain invariant. The specific description is obviously changed but the key property of rectilinearity of undisturbed motion is not affected. As is pointed out in all textbooks of dynamics, the same invariance also holds for Newton's Second and Third Laws (this is essentially because of their vectorial nature and the fact that the rules of vector addition are invariant with respect to the two considered transformations). Newton's laws are therefore invariant with respect to *translations* (shifting of the origin) and *rotations* (a once and for all rotation of the coordinate axes into new orientations).

It is very important to note that if we displace the origin or rotate the axes in some arbitrary manner which depends on the time, the motion of inertially moving bodies will clearly be neither rectilinear nor uniform in the new system, so that Newton's laws are not invariant under such transformations. There is, however, one further transformation that can be made which leaves the First Law invariant. This is the famous *Galileo transformation* (as we shall see in Chap 7, what is called the Galileo transformation is not quite the same as the transformation Galileo considered) and is as follows.

We consider first a frame of reference fixed in absolute space, which we call the *absolute frame*. In addition, we consider a second frame whose axes are always parallel to those of the absolute frame but which moves relative to the first with *uniform* speed in a *fixed* direction in absolute space. We call this the *moving frame*. A moment's reflection will convince the reader that, observed in the moving frame, bodies that move inertially in the absolute frame still appear to move uniformly and along straight lines. Thus, by examining the motion of bodies that move purely inertially we come to a rather surprising conclusion. As far as the law of their motion is concerned, we are quite unable to say whether we are observing that motion in a frame of reference that is at rest relative to absolute space or is moving uniformly in a straight line through it.

Let us now consider the situation with regard to the Second and Third Laws. Here we have to distinguish two different things; one is a matter of pure mathematics, the other is a matter of physics. The Second and Third Laws make statements about changes in motions, i.e., accelerations. Suppose we follow the motion of a body in what we have called the absolute frame and that it initially moves uniformly with velocity $\mathbf{v}$, is then subject to a change in velocity $\delta\mathbf{v}$, after which it again moves uniformly. Thus, in the absolute frame the velocity changes from $\mathbf{v}$ to $\mathbf{v} + \delta\mathbf{v}$. Now in the moving frame the velocity before the change is $\mathbf{v} + \mathbf{V}$ (where $-\mathbf{V}$ is the velocity of the moving frame in absolute space) while after the change it is $\mathbf{v} + \delta\mathbf{v} + \mathbf{V}$. Thus, in the moving frame the change is $\delta\mathbf{v}$, just the same as it is in the fixed frame. This is a purely mathematical result and shows that all changes in inertial motions, and thus accelerations, are exactly the same in the absolute and in the moving frame. Thus, if we observe a body of known mass and wish to deduce from the change in its momentum $m\mathbf{v}$ the force which acts upon it, we shall obtain identical answers in the two frames.

Let us now consider the physical side of this question and begin with Newton's law of gravity. Examination of the expression (1.7) for the force shows that it has a remarkable property; namely, the force is completely determined by the *relative configuration* of the two interacting bodies: the force acts along the line that joins the two bodies and its strength is inversely proportional to the distance between them. This law determining the gravitational force is therefore completely independent of the position and the velocities of the bodies in absolute space. Thus, the force depends only on the *relative* configuration of the bodies; it is therefore the same in both the absolute and the moving frame. But we have already seen that, by pure mathematics, the accelerations are the same in the two frames. Thus, the two sides of Newton's Second Law, $m\mathbf{a} = \mathbf{f}$, are the same in both cases, so that the Second Law is also invariant. It is easy to see that the Third Law is too.

It is important to note what are called the *passive* and *active* aspects of this transformation. As throughout this entire discussion, we hold fast to the notion of the reality of absolute space. Consider a collection of gravitationally interacting bodies whose centre of mass is at rest in absolute space and whose motion we follow from the initial time $t = 0$. We call this system A. Then transition to the moving frame means that we merely look at the interactions in system A from a frame moving with velocity $-\mathbf{V}$ relative to the absolute frame. This is the passive aspect. But now imagine in the absolute frame an identical set of bodies with *the same initial relative positions and relative velocities* but with an overall velocity $\mathbf{V}$ added to all velocities. In the absolute frame this is, physically, a quite different system, and we shall therefore call it system B. Yet, because the law of gravitational interaction depends only on the relative configuration, the accelerations of the bodies in system B will be identical to those in system A. But their initial velocities are the same as those in system A when viewed from the moving frame. Thus, the motion of system B when observed in the absolute frame will be identical to the motion of system A when observed from the moving frame.

Thus, as far as gravitational forces are concerned, we are quite unable, if we cannot actually see absolute space, to say which of the systems A or B is moving in absolute space. Such a system of bodies, which we assume is free of all disturbance from other bodies, is called a *closed dynamical system*.

Now one of the most interesting of the implicit assumptions that Newton made in the *Principia* is that the property which we have seen holds for gravitational forces, i.e., that they do not depend on an overall motion through absolute space, is universally true of all the forces of nature. He assumed[3] (without proof) that the forces which act between bodies depend either solely on the relative configuration of the bodies, or else on the relative configuration and the relative velocities of the bodies but in no case on their position or overall velocity in absolute space. In particular, he assumed that a uniform rectilinear motion common to two bodies would have no effect at all on the force that acts between them. This is *not* a consequence of his laws of motion. For example, the strength of the force that acts between two bodies could depend on the velocity of the centre of mass of the two bodies through absolute space without violating any of the three laws of motion.

If this property of forces, which is in fact confirmed by experiment, is assumed, then the result which we showed to be true for purely inertial motion is true for any motion governed by Newton's laws and forces which have the above property. This result was stated by Newton in his famous Corollary V to the laws of motion: 'The motions of bodies included in a given space are the same among themselves, whether that space is at rest, or moves uniformly forwards in a right line without any circular motion.' This property is what is now called the *Galilean invariance* (or *Galilean relativity*) of Newtonian dynamics. It clearly puts a question mark over the whole concept of absolute space. If our motion through it cannot be revealed by any experiment, of what use is the concept?

We can answer this question by pointing out that Newton did have very good grounds for the assumptions he made and they do make a great deal of sense. For example, the astronomers had by his time established very accurately how the planets move around the sun *relative to the distant stars*. In conjunction with Newton's law of gravitation, this motion demonstrated unambiguously the possibility of a description in the terms that Newton proposed provided it is assumed that the distant stars are nonrotating relative to absolute space. That is, it demonstrates that in all cases the motion of the planets can be conceived as due to the supervention of the gravitational effect of nearby bodies on a rectilinear motion relative to the stars. There is thus a pervasive and powerful constituent to actual motions that is manifestly nothing to do with local bodies. It shows up particularly in two effects: (1) a body that passes by the sun with a very high velocity tends to follow a path that approximates more and more to a straight line (ignoring relativistic effects); (2) if the earth's motion were solely governed by its interaction with the sun and nothing else (neither by absolute space nor other bodies; we ignore the manifestly small effect of the other planets), it is difficult to see why the earth should not in all cases fall straight into the sun, since it is undoubtedly attracted to it by the sun's gravity. Consider in particular an initial condition in which the distance between the earth and the sun is not changing. In the frame provided by the distant stars, such is the case when either the earth is at rest in that frame or is moving with any velocity at right angles to the line joining the earth and the sun. But as far as the earth–sun system is concerned, such a motion is nonexistent (we treat the earth and sun as mass points since their extension evidently has no bearing on the matter). Thus, on a purely relational theory of motion, when we look at things from the point of view of kinematic relativity, it is completely inexplicable why the earth falls to the sun in only the single case when the transverse velocity is zero; in all other cases a completely different motion results. Thus, there is evidently a great deal more to motion than just the earth–sun relative separation. Newton said that the 'something else' is to be attributed to absolute space, and the worth of this concept is only very slightly diminished by the problem with Galilean relativity.

This can be put quite graphically in terms of frames of reference. Take a given frame of reference and a second that moves with respect to it in an absolutely arbitrary manner. Space is certainly invisible and if it had no influence at all on the motion of bodies it is hard to see why one frame of reference should be distinguished with respect to any other. Yet we know: (a) such frames exist (the whole of astronomy proves their existence), (b) considered in terms of the freedom that the nonexistence of a dynamical role of space would allow to the frames of reference their choice is amazingly circumscribed, just the family of nonrotating frames in uniform rectilinear motion being allowed. Moreover, all bodies in the universe respect the same set of frames of reference that permit the remarkable decomposition into inertial motion and locally determined force-induced changes in inertial motion. (This last statement will require modification when we come to consider general relativity but this does not affect the basic point being made here.) Newton was wrong to imply that there is a uniquely determined frame of reference. There is not one such frame of reference but rather a family of them. But, looked at from the point of view of kinematic relativity, the family is inexplicably restricted.

We began this discussion by asking what is the use of the concept of absolute space if our motion through it cannot be detected by any experiment. The answer is that the motion must be highly special – a uniform motion remains undetectable but any acceleration can be detected. Why this is so we do not know. It is this puzzle that Mach hoped to solve with his suggestion that in some way the distant stars are what actually determine the allowed frames of reference.

As already pointed out, it is necessary in all cases to distinguish carefully between *Galilean* and *kinematic* relativity. They are clearly not the same thing. One should also be on guard against the assumption, often made, that although the two concepts are not identical Galilean relativity is at least a necessary consequence of kinematic relativity. For, it is argued, if motion is entirely relative and space plays no part in dynamics, any motion through 'space,' including uniform motion, should obviously be undetectable. This is clearly the case if the system one is considering is the entire universe, but it does not follow for the case of a relatively small system of bodies (such as the solar system) within a much larger collection of bodies taken to represent the rest of the universe. In this case one could readily imagine that a motion relative to the masses of the universe as a whole would be observable dynamically. However, this is not the case.

Finally, it is convenient to introduce some more terminology. Sticking for the moment to the useful fiction that we can 'see' absolute space, we shall define an *inertial frame of reference* (also called *inertial system*) as one that is either at rest in absolute space or moves uniformly through it in a fixed direction. One can alternatively ignore all references to absolute space and simply say that an inertial frame of reference is one in which Newton's laws hold in the form he stated them.

We can now briefly state what Vol. 1 contains as regards the history of the discovery of dynamics: It aims to cover all the key events needed for the eventual Newtonian synthesis, the synthesis itself, and then, jumping forward a couple of centuries, the final clarification of the true empirical content of Newton's laws by Neumann, Lange, and Mach. Of especial interest is the operational definition that Lange eventually found for an inertial frame of reference. We shall see that it puts the relationship between space and motion in a quite new and most intriguing perspective.

1.3. Why it took so long to find the laws of motion

It took a remarkably long time for man to discover dynamics. This section will attempt to identify some of the reasons. We begin with the purely physical aspect: What chances did nature give man? In fact, there seems to have been almost a conspiracy on the part of nature to hide the laws of dynamics from man. Einstein is on record as saying that the Lord is subtle but not malicious.[4] At times one could almost wonder. If we look at the most important features of Newton's three laws we shall see that, even without the self-imposed obstacles of preconceptions, the odds were against the discovery of any of them.

On the face of it the prospects were best for the discovery of the First Law. The behaviour of arrows and sling casts did, after all, vouchsafe man a glimpse or two of the law. But there were still several formidable obstacles to its recognition. The first and most obvious is air resistance. The forward speed of the thrown object is manifestly diminished. This presented an insuperable obstacle to mathematization; the ability to describe nonuniform motions mathematically developed only very slowly indeed. Until man obtained a very clear awareness of the possibility of free continuation of motion *forever* once commenced, mathematization of this aspect of motion was ruled out. And from what activities could man have gained such an awareness? Skating perhaps, but that was hardly a sport practised by the ancients in the Mediterranean. (One wonders if it is entirely fortuitous that the first reasonably clear statement of the law of inertia in its modern form was made by a man – Descartes – who spent most of his adult working life in *Holland*, at a time moreover in which skating had become a great vogue.) In this connection, it should not be forgotten that the development of modern means of rapid transportation have given us vastly more opportunities for observing the immediate consequences of inertial motion. Particularly important in this respect is the relatively modern insight (not clearly expressed in the literature on motion before about 1630) that motion of a transported body can continue of its own accord when the transporting body ceases to move. When the car we are travelling in comes to an abrupt halt, our bodies automatically continue their motion with consequences of which we are only too painfully aware. This affords us an insight into inertial motion quite different from that gained from watching a javelin thrower. For in that case the motion has clearly been initiated by a forceful act and this led, in medieval times, to the idea that the projectile is kept in motion by the transference of a 'force' or 'impetus' or even 'spirit' from the thrower to the projectile.

Thus, although this was a step, and quite an important one, towards the law of inertia, it remained very restricted: a particular cause was invoked for a restricted class of motions – those manifestly initiated by a thrower. There was no real awareness that the factor at work keeping the javelin moving is universal and just as much at work in all the other myriad motions of which we take cognizance.

Equally difficult to acquire was the notion that a given motion of this kind must be decomposed into two parts – the horizontal inertial component and the vertical motion decelerated and then accelerated by gravity. It was a decisive and bold step on Galileo's part to divide up projectile motions cleanly into two constituents. Crucial to this was his perception that without air resistance the horizontal motion remains perfectly uniform – that was the property that, so to speak, identified the horizontal component as something autonomous and simultaneously rendered it amenable to precise mathematization.

If we turn now to the Second Law and start by considering what was needed for the discovery of its most important feature, the central role played by acceleration, we find that here the difficulties were even greater. For let us consider what is involved in studying motion.

First, we need to see objects and follow the various paths they take. Here we are well endowed by our senses. Objects are easy to see, and both in astronomy and terrestrial physics nature provided us with an almost perfect backcloth on which to trace their motions – the fixed stars in the sky and the solid ground on the earth. Thanks to the stability of the ground, the existence of solid objects, and the excellence of our vision, we acquire a good qualitative grasp of geometrical relationships within a year or two of our arrival in the world. Arithmetic belongs to prehistory, but geometry too was developed early to a highly precise and formalized science. Measurement with rulers is one of the easiest and most reliable means of making quantitative observations. There can be little doubt that the early discovery of geometry was due to these favourable circumstances. Particularly important was the fact that the means of measurement were sufficiently accurate to point forcibly to the existence of *absolutely exact relationships*. The recognition of such relationships and their formal statement as axioms appears to be a key step in the development of science. It is only absolutely exact relationships that admit mathematization and can provide a formal theoretical framework in which the need for the measurement of particular quantities is perceived.

Thus, if we ask what it is that stimulates the development of an exact science, the answer appears to be that there must be a means of observation sufficiently accurate to alert man to the possible existence of simple exact numerical or geometrical relationships behind the empirical phenomena. In the case of geometry and pitch, excellent means of measurement are available – rulers for geometry, as we have seen, and the human ear for pitch. Even the unmusical can hear acoustic frequencies with remarkable accuracy, so the conditions for Pythagoras's discoveries in harmony were very favourable.

In addition to the means of observation there must of course be suitable phenomena to observe. They must exist and be reproducible and sufficiently striking to attract attention. Now we have seen that in the case of the First Law suitable phenomena were simply not presented to the senses. However, in the case of the Second Law it would seem that nature did provide a reproducible and most striking motion that should have been appropriate: the free fall under gravity of stones and other heavy objects. As Aristotle himself commented:[5] 'His must surely be a careless mind who does not wonder how it is that a small particle of the earth, if raised to a height and then set free, should refuse to remain where it was but begin to travel.' Why was it so long before this remarkable phenomenon gave rise to a quantitative science?

The main answer must have been: *there was no suitable clock*. The typical speeds of motions observed on the surface of the earth are measured in metres a second but the accelerations are typically measured in metres-per-second a second. For this reason, the motions themselves (unless subject to human or animal will) only last a few seconds. Whereas man found about him an abundance of highly accurate means for measuring distance (rulers etc.), so that a metre can easily be measured to an accuracy of one part in a thousand, nature signally failed to provide us with an adequate clock. It is well known that Galileo made the first discoveries in dynamics using the pulse-beat – an erratic clock with an accuracy of not better than a second. Thus, the speeds of the motions from which the rudiments of dynamics might have been learnt could be measured to an accuracy of, say, only 50%. This accuracy was of course quite inadequate for the development of quantitative study of motions, and accelerations, which ultimately unlocked the mystery of motion, were beyond measurement. This sheer inability to 'see' accelerations must have been the main factor why free fall and other terrestrial motions failed to attract quantitative study.

A graphic illustration of the extreme difficulty of arriving at accurate knowledge about terrestrial motions caused by the absence of a convenient clock is provided by one of the few passages that have survived from antiquity with a valuable insight into *nonuniform* motion. It derives from Strato, who in 287 BC became the head of Aristotle's Lyceum. He comments[6] that a falling body as it accelerates 'completes the last stage of its trajectory in the shortest time. . . . For if one observes water pouring down from a roof and falling from a considerable height, the flow at the top is seen to be continuous, but the water at the bottom falls to the ground in discontinuous parts. This would never happen unless the water traversed each successive space more swiftly.'

Note that Strato, who was obviously not the first person to note the phenomenon, is completely reliant upon a secondary effect, from which nothing more than a qualitative judgement can be made. He has succeeded in grasping only the grossest features of the motion.

It is worth mentioning specifically two further errors about terrestrial motion that were widespread before Galileo's time. The most famous is, of course, the mistake made by Aristotle and other Greeks that heavier bodies fall faster than lighter ones. The effect of air resistance does, of course, provide some sort of excuse, but this was really an extraordinarily gross error, which neither air resistance nor the absence of a suitable clock can possibly explain. As late as Galileo's time many people believed that a two pound weight would fall twice as fast as a one pound weight. We shall return to a discussion of this point shortly.

The other error is very peculiar. In Aristotle's *On the Heavens* there is a curious passage[7] which states that 'for things whose motion is that of a missile' the greatest speed occurs in midflight. In fact, it seems most probable that there is here some error of transcription or a misunderstanding. Nevertheless, in the Middle Ages it was firmly and quite widely believed[8] on the basis of this statement that a thrown projectile does not acquire its maximum speed immediately it leaves the thrower's hand but only in the middle of its flight. According to this notion, which was taken as an undoubted empirical fact, there is therefore an initial build-up of speed and only then does deceleration commence. As we shall see in Chap. 4, this was the belief of one of the greatest of all medieval natural philosophers, Nicole Oresme.

Under such circumstances, presented with a plethora of diverse motions and a focus inadequately sharp for the salient details to obtrude on the consciousness, it is perhaps not surprising that ancient and medieval physics was content with a purely qualitative description. How and when did the shift to quantitative study occur?

Working through the history of dynamics, one of the things that strikes one most forcibly is the occurrence of key events when often quite unwittingly – or at least only with a very dim awareness of the long-term consequences – someone decides to look at long known phenomena from a different point of view. The initial shift of emphasis is often apparently quite small, seemingly innocent; the end result appears out of all proportion to the initial step. In fact, subsequent events show that it is, quite literally, a step into a new world.

Such was Galileo's conscious decision to take an interest in accelerations. Having once got the idea, he was forced to use a clock to make measurements sufficiently accurate for his purposes. Galileo's originality did not lie in the particular clock he invented but rather in deciding to use it on *terrestrial motions*. It says much about the totally different ways in which astronomy and terrestrial physics developed that, according to Ptolemy,[9] water clocks were used before his time (*circa* AD 150) in an attempt – unsuccessful in fact – to measure the apparent diameter of the sun (by measuring the time taken for the sun to rise) and thereby establish in a direct manner whether the distance of the sun varies significantly. This shows clearly that Ptolemy's predecessors were already convinced that *there was something worth finding* in the heavenly motions. Conscious efforts were made to wrest the secrets of the celestial bodies from the heavens. But no one before Galileo seems to have had the idea of using accurate water clocks to study terrestrial motions. This therefore provides a further explanation for that extraordinarily gross error about the times of falling of objects of different weights. There was simply a total lack of awareness that any significant or rewarding insight was to be gleaned from attentive and *quantitative* observation of terrestrial motions. What thinking that did go into the subject concentrated instead on the seemingly much more important and interesting question of *why* the apple falls, not how it falls.

This is an appropriate place to give Galileo's own description of the clock he used in the experiment that at last set the study of terrestrial motion on the right road:[10]

For the measurement of time, we employed a large vessel of water placed in an elevated position; to the bottom of this vessel was soldered a pipe of small diameter giving a thin jet of water, which we collected in a small glass during the time of each descent, whether for the whole length of the channel or for a part of its length; the water thus collected was weighed, after each descent, on a very accurate balance; the differences and ratios of these weights gave us the differences and ratios of the times, and this with such accuracy that although the operation was repeated many, many times, there was no appreciable discrepancy in the results.

It has been remarked that a major scientific discovery can be expected whenever the accuracy of measurement of any fundamental quantity is increased tenfold. This is beautifully confirmed in Galileo's case. Galileo reports that he measured the time 'with an accuracy such that the deviation between two observations never exceeded one-tenth of a pulse-beat.'

More than enough facts were constantly under our nose but were for that very reason just out of focus.

I still have the vivid memory of watching on television the American astronauts working on the moon. By chance, one of them happened to kick up gently some moon dust, which could be seen to travel slowly about five or six metres in a perfect parabola. The motion was sufficiently slow for the mind to be forcibly struck by the uniform persistence of the horizontal (i.e., inertial) motion and the gradual deceleration and acceleration of the vertical motion. As the moon's surface gravity is about a sixth of the earth's, the transfer to the moon achieved much the same effect as Galileo's clock described above. In addition, the disturbing influence of air resistance was completely eliminated. To inhabitants of the moon, the laws of motion would have been as transparent as geometry was to the Greeks. The irony is that the sustenance of life requires an environment rich in change and motion, but this very richness and mutability prevented man seeing through to the inner essence of motion.

Are there other clues, literally before our eyes, that, even now, in this high tide of the technological era, could transform our conceptions as profoundly as Galileo's clock? The history of dynamics suggests there could be.*

We still have not completed the survey of nature's perversity in hiding the clues to the discovery of dynamics. We recall that acceleration is only part of the story of the Second and Third Laws. Just as important are the physical quantities which appear too, above all the *mass* of the accelerated body and the *charge* which causes the acceleration. What chance was there of discovering the physical essence of these two laws, that is, that bodies *accelerate each other* through *forces*. To answer this question, we have to look at the forces separately. Nuclear forces were, of course, quite out of

* As a striking example of what can escape notice (though with no suggestion that it has any bearing on fundamental questions of dynamics) the following may be mentioned. In a department store in the centre of Brussels the undersides of the escalators have been covered with mirrors at exactly 45° to the vertical. If one stands and looks at the mirrors, all the public is seen walking upside down, and the curious swaying and rising and sinking manner in which people walk is suddenly made apparent. The effect is so surprising as to be almost hypnotic.

reach. There were, however, abundant manifestations of both electrical and gravitational forces – but none in a form that could have led readily to the formulation of the Second Law. Consider electrostatic forces; on the face of it, these should have been ideal for recognizing the inverse square law. Quite the contrary; for a start, due to the existence of both positive and negative charges, almost all the electrical charges in the world are locked up in neutral complexes. The Coulomb forces are thus almost completely withdrawn from the macroscopic world and all that remains are contact forces. This greatly strengthened the natural view that motion is all a matter of pulling and pushing and put in the foreground characteristics of material bodies such as their geometrical shape or their hardness. No one looked for or suspected the existence of things such as charge. And even in the cases, well known to the Greeks, in which an electrostatic charge can be accumulated, great care and controlled laboratory conditions are needed if repeatability is to be achieved and genuine effects observed.

In the case of gravity the situation was slightly more favourable. Here at least there is no mutual cancellation, but the extreme weakness of gravity has the consequence that the only obvious manifestation of gravity is free fall to the surface of the earth. But this is just a single phenomenon; there is not the remotest possibility of recognizing directly that the strength of gravity depends upon distance or on the gravitational mass of the accelerating body. Nor is there any chance of perceiving that gravity is a force of mutual attraction: The mass and pull of the earth totally outweigh the effect of the attracted stone or apple. There is no manifestation of the law of action and reaction. Thus, although Galileo brilliantly overcame the problem of timing the descent of bodies and achieved a more or less complete understanding of terrestrial motions under the joint influence of gravity and inertia he made no progress at all towards the statement of Newton's law of universal gravitation. His insight, although crucial, was still very partial.

Thus, many factors taken together contributed to the extreme difficulty of perceiving the *physical* aspect of motion.

There was just one chink left in nature's physical armour – magnetism. In permanent magnets, the miraculous lodestones, enquiring man was granted one reliable force that came in a manageable amount and could be used in controlled experiments. The conditions were certainly nowhere near favourable enough for the discovery of the actual law of force, but they were good enough to suggest a qualitative concept of physical force capable of acting over distances. Gilbert's book on magnetism[11] published in 1600 marked a significant development and suggested the concept of force in a form capable of mathematical development. But that did not occur in magnetism. Kepler took the idea of physical force from Gilbert and applied it *in the heavens*. Here at last were motions of a kind suitable for mathematical analysis. They eventually yielded their secret.

The next section considers the cause of the curious paradox that the key to terrestrial dynamics was found in the heavens, but before that we should point out how all the reasons so far examined for the nondiscovery of dynamics were amalgamated together into a fundamental proposition that contained sufficient truth to have appeared plausible for two millennia and yet was disastrously wrong. Under the influence of Aristotle, whose ideas we shall consider in Chap. 2, there became established the doctrine that no body could move unless it was moved by some other body.[12] In the Middle Ages, this was expressed by the following words: *Omne quod movetur ab alio movetur*. However, the motion of the pushed body was believed to come to a stop as soon as the pushing stopped. This accorded with the even more fundamental doctrine that when a cause ceases to work the effect it produces disappears: *Cessante causa, cessat effectus*.

In the light of the above discussion, we can easily see how such an idea arose. Most bodies on the earth are in a state of rest. In accordance with Newton's Third Law, they can only be set in motion by the action of some other body. But air resistance, friction, and the absence of an accurate clock hide the fact that the pushing body produces an *acceleration* rather than the motion itself. Also, because resistance brings most motions quickly to a halt, it is very easy to see how the idea that motion itself must be sustained by a pusher rather than merely initiated by the pusher took such a hold. Finally, the effective absence of electrostatic forces means that in almost all cases direct contact is needed to produce motion. All these factors taken together therefore conspired to produce Aristotle's doctrine, which was all the more misleading because it was not totally wrong. As so often, partial truth is more dangerous than straight error.

Finally, there is a further very deep reason why dynamics took so long to discover. In accordance with Newton's laws, the earth is unquestionably in a state of rotation. Nevertheless, a frame of reference fixed to the surface of the earth approximates an inertial frame of reference to a remarkably good degree, and since according to Newton's own laws a state of uniform motion is quite indistinguishable from one of perfect rest the Galilean invariance of dynamics produces an extremely powerful impression of the complete immobility of the earth. In the following chapters we shall see how the belief in terrestrial immobility was a powerful factor that held back the discovery of dynamics. This only serves to emphasize the crucial role of astronomy. For it was astronomical facts, patiently accumulated in painstaking observations stretching over centuries, that finally produced the most convincing evidence that the earth rotates and moves (both with respect to the stars and the family of inertial frames of reference). It was this that broke the log-jam and more than anything else provided the hints that enabled Galileo and Newton to bring about the synthesis of the disparate parts and thereby create the science of dynamics.

1.4. Why the first breakthrough occurred in astronomy

Man was endowed by nature with two clocks. With the one, his heart-beat, he lived for millennia in a cornucopia of motions *but had not an inkling* of their secret.

The other clock was so fantastically accurate and convenient that astronomy became a highly accurate and sophisticated science well over one and a half millennia before Galileo obtained the first useful results in terrestrial dynamics. I refer, of course, to the rotation of the earth and the apparent motion of the sun and the moon. In Chaps. 2 and 3, we shall consider the nature of time from the philosophical and empirical points of view and establish what it means to say *time passes uniformly*. But for the purposes of the present discussion, let us take the concept of uniform flow of time as granted. Then this uniform flow is measured by the rotation of the earth with an accuracy of about one second in a year, i.e., one part in thirty million. Almost as important as the accuracy is the convenience. The alternation of day and night marks the 'second,' the very ticks of the clock; the sun and moon are its big and little hands, the star-studded welkin the clockface, on which the twelve signs of the zodiac mark the twelve 'hours' in the astronomical system of time keeping. No wonder Plato called the celestial bodies the 'eternal image' of time.[13]

And it was not only the existence of a suitable clock that created such favourable conditions in astronomy. The motions themselves that were available to be studied were of such a different kind. Terrestrial motions exhibit a bewildering variety – from the purposeful locomotion of animals to the falling of rain and snow, the blowing of leaves in the wind, ordered wave motion in calm weather, and the raging of the sea in a storm. Moreover, all terrestrial motions are readily distorted by extraneous – from the point of view of the subsequent theory of motion – factors, above all air resistance and friction, these themselves being highly variable, unpredictable, and differing in their action on different bodies (stream-lined or not, dense or light, rough or smooth).

How different and so simple are the celestial motions in comparison and with what sedateness and clarity are they described! Seven lights (sun, moon, and five planets) creeping more or less regularly across a black backcloth richly laid with perfect markers, the stars, which never move or fail. It is almost as if a system of polar coordinate axes were painted on the sky: azimuth measured round the zodiac, declination from the pole of the ecliptic. Thus, for those prepared to make the effort – and many were – nature in her bounty made it as easy to pluck precise data from the heavens as nuts from the autumn hedges.

These facts give some indication as to why astronomy was to prove the high road to the discovery of dynamics and how it was that ancient astronomers were able to make the significant discoveries they did.

There is another most important factor that favoured the quantitative development of astronomy and held back terrestrial dynamics. We have already mentioned the bewildering variety of terrestrial motions. But this is only a part – and perhaps the least important part of the story. In a study concentrated solely on dynamics, it is all too easy to forget that motion is but one of innumerable phenomena that take place on the earth. The variety of motions is as nothing compared with the rich profusion of earth-bound events that involve *qualitative* changes and in which the motional aspects often appear quite minor and insignificant. This is an aspect of terrestrial physics that we have not yet mentioned. Yes, the falling apple is striking but not half so marvellous as the ripening apple, the golden corn, and the fermentation of grape juice into wine. Or the mystery of fire, the freezing of water, and the deep blue of the sky giving way in the evening to the red glow of sunset. Is it surprising that under such circumstances Aristotelian and medieval physics was essentially a theory of *all qualitative changes*? Was it not the purest speculation to suppose that the quantitative aspects of motion alone, the least exciting perhaps of all these almost unbelievable transformations, should hold the key to their understanding?

In contrast, the celestial spectacle was pure ballet; attention was perforce concentrated on motion, for almost nothing else could be observed.

It is here worth quoting from the opening passages of Ptolemy's *Almagest*, written about five hundred years after Aristotle flourished and another one thousand five hundred years before Galileo. It reveals a truly remarkable awareness that the study of motions, especially celestial motions, represented pretty well the only place in which man could hope to make a genuine breakthrough to a secure understanding of things. Here are Ptolemy's words; note in particular the parts I have italicized:[14]

For Aristotle divides theoretical philosophy too, very fittingly, into three primary categories, physics, mathematics and theology. For everything that exists is composed of matter, form and motion; none of these [three] can be observed in its substratum by itself, without the others: they can only be imagined. Now the first cause of the first motion of the universe, if one considers it simply, can be thought of as an invisible and motionless deity; the division [of theoretical philosophy] concerned with investigating this [can be called] 'theology', since this kind of activity, somewhere up in the highest reaches of the universe, can only be imagined, and is completely separated from perceptible reality. *The division [of theoretical philosophy] which investigates material and ever-moving nature, and which concerns itself with 'white', 'hot', 'sweet', 'soft' and suchlike qualities one may call 'physics'; such an order of being is situated (for the most part) amongst corruptible bodies and below the lunar sphere.* That division [of theoretical philosophy] which determines the nature involved in forms and motion from place to place, and which serves to investigate shape, number, size, and place, time and suchlike, one may define as 'mathematics'. Its subject-matter falls as it were in the middle between the other two, since, firstly, it can be conceived of both with and without the aid of the senses, and, secondly, it is an attribute of all existing things without exception, both mortal and immortal: for those things which are perpetually changing in their inseparable form, it changes with them, while for eternal things which have an aethereal nature, it keeps their unchanging form unchanged.

From all this we concluded: that *the first two divisions of theoretical philosophy should rather be called guesswork than knowledge, theology because of its completely invisible and ungraspable nature, physics because of the unstable and unclear nature of matter*; hence there is no hope that philosophers will ever be agreed about them; and that only mathematics can provide sure and unshakeable knowledge to its devotees, provided one approaches it rigorously. For its kind of proof proceeds by indisputable methods, namely arithmetic and geometry. Hence we were drawn to the investigation of that part of theoretical philosophy, as far as we were able to the whole of it, *but especially to the theory concerning divine and heavenly things. For that alone is devoted to the investigation of the eternally unchanging*. For that reason it too can be eternal and unchanging (which is a proper attribute of knowledge) in its own domain, which is neither unclear nor disorderly.

Note first how Ptolemy more or less identifies mathematics with the study of motion and sees absolutely no hope of progress in either 'physics' or theology and that alone in the study of 'divine and heavenly things', i.e., the sun, moon, planets, and stars, is there any hope of genuine progress – and that because of the 'eternally unchanging' nature of what happens in the heavens.

From the point of view of the discovery of dynamics, astronomy offered other advantages, one a fortunate fluke: the existence of the moon, the eclipses of the sun it causes, and the eclipses it suffers. These, thanks to the perfect clock, which enabled them to be recorded with such accuracy and convenience, revealed a most mysterious feature of celestial motions, which is already hinted at in the above extract from Ptolemy. Although the motions themselves were not perfectly regular – even the simplest motion, that of the sun, is not perfectly uniform – the predictability of eclipses was nevertheless evidence of some deep interconnection of things. Moreover, the eclipses were such awesome events, man's curiosity could not but be drawn to ponder their origin. It was the experience of an eclipse at the age of 14 which turned Tycho Brahe into the greatest astronomer of perhaps any age. It struck him as 'something divine that men could know the motions of the stars so accurately that they were able a long time beforehand to predict their places and relative positions'.[15]

The cyclic regularity of the celestial motions was probably the single most important factor in the discovery of dynamics if not the entire scientific revolution.

Quite generally, astronomy appears to have attracted attention because of its 'divine' attributes, which awoke in man a sense of awe before the starry heavens, the perfect uniformity of the diurnal rotation, and the seeming eternity of the celestial constellations. In the case of terrestrial motions, it was an example of familiarity breeding contempt; but the celestial motions, few as they were, exerted a constant pull upon the imagination and simultaneously imposed a severe discipline on the investigator. He had no choice but to investigate their motion; there was nothing else to study. Thus, the great bulk of Ptolemy's *Almagest*, the bible of ancient astronomy, is in a terse, sober, and very factual style. Most of it could be printed in the *Physical Review* without appearing out of place. But every now and then there are lyrical passages in which Ptolemy breaks out in enthusiasm for his subject and reveals the inner drive that sustained him through all those hours of observation and computation. We have already seen how he identifies the heavenly bodies with the divine. A little later, commenting on the celestial affairs and their proximity to perfect divinity, he writes: 'from the constancy, order, symmetry and calm which are associated with the divine, it makes its followers lovers of this divine beauty'.

What was the physical cause of this inspiration in Ptolemy? What made him want to study what he was convinced was divine – and thereby lay the foundations of a science that ultimately killed the divinity he so fervently worshipped?

We shall go into this in more detail in Chap. 3, but basically the inherent advantage of celestial motions (as the key to find some of the most important secrets of dynamics) derives from the fact that the moon and planets move in the nearly perfect vacuum of interplanetary space and in the central force field of the sun, so that the effect of the law of inertia is manifested in their behaviour in an almost undiluted form, as we saw in Sec. 1.1. in the discussion of the area law. All earth-bound motions come to an end, but the celestial motions go on for ever. This is what made them appear an eternal image of time and divinity. The search for immortality was the first stimulus to the discovery of dynamics.

But even in astronomy, though the conditions were much more propitious, Newton's laws could still not be 'read off' directly from the heavenly motions. Because of the overwhelming mass of the sun in the solar system and of the earth in the earth–moon system there is no real manifestation of the Third Law. Moreover, and more seriously, the observed motions arise from the combination of the First and Second Laws, neither of which are manifested in pure form. Without hints that were finally supplied by terrestrial motions, the clear separation of the effects of these two laws in the heavenly motions could never have been made. Thus it was that Kepler achieved marvels in the actual quantitative description of the motions and had several ideas which the subsequent work of Newton showed were basically sound, even inspired, yet he completely failed to overcome the supreme impediment to the discovery of dynamics: Aristotle's dictum *Omne quod movetur ab alio movetur*.

1.5. General comments on the absolute/relative debate

The debate about the absolute or relative nature of motion revolves around two basic questions. In broad terms, the first is: how is motion (or more generally any kind of change), irrespective of its causes and particular properties, to be described? And the second is: what is the nature of motion, what laws does it obey? We shall see that the two questions are much more closely interrelated than might appear at the first blush. For what is at stake is not so much a description in any terms provided only that they are unambiguous but rather an *appropriate* description that not only describes but simultaneously puts that which is described into its most comprehensible form. For it turns out that one and the same objective state of affairs can be described in more than one way. This is the origin of the seemingly endless debate about whether the earth actually moves or not. It is here essential to distinguish between features of motion and change that exist objectively and independently of the means of description from those that vary with them.

The clarification of these questions is thus an ongoing enterprise, and the truth is that the conclusions reached at any one stage in the development of dynamics in particular and physics in general are always liable to radical revision when new objective relationships are discovered in the world. This is reflected in the fact that since the original proposal of the earth's mobility by Copernicus the question of whether the earth moves or not – and, if so, in what sense it does move – has passed through at least three different stages, each of which has put previous conclusions in a different light. Indeed, there seems to be clear evidence that the conceptual difficulties get more complex the more progress is made in understanding how the world works. The level of sophistication needed to treat these matters unquestionably increases.

Let us briefly anticipate the main stages (though this will take us beyond the period covered by the present volume). Up to and including the discovery of dynamics by Newton and its conceptual clarification in the nineteenth century there did not appear to be any severe restriction on the means of description. The main question appeared to be merely that of finding the means that were most appropriate. Nevertheless, the end result of the process was quite startling in its implications for the relationship between space and motion, as we shall see in Chap. 12. The next really significant development came with the growing awareness, which crystallized in Einstein's work, that our means of observation, on which our description must ultimately rest, are themselves governed by the laws of nature. This has the consequence that the possibility of describing motion and change meaningfully is not independent of the actual laws that govern the very motion and change one is wishing to describe. Thus, it is not possible to lay down *a priori* a framework in which motion can be adequately described come what may, i.e., whatever the actual nature of motion may turn out to be. The instinctive approach is to say that motion is to be measured by rods and clocks; but once it is realized that the behaviour of rods and clocks is itself governed by laws of nature that are not known in advance it becomes clear that the choice of the means of description goes hand in hand with the establishment of concrete empirical facts about motion as it is actually observed.

A further significant stage in our growing appreciation of the complexity of the problem of describing nature and the world is represented by quantum mechanics. Here the very existence of a unique well-defined externally existing world is questioned in the most acute form. The problems presented by quantum mechanics go well beyond the scope of the present study and will be considered at most peripherally. They are mentioned here only to indicate the further level of sophistication that is needed to think about the world.

The realization of the provisional status of all our conclusions about something so apparently simple as motion should not, however, be the occasion for pessimism. The successes already achieved belong to some of the finest accomplishments of the human intellect; they are associated above all with the names of Newton and Einstein.

After these general comments about the interconnection of the two basic questions mentioned at the beginning of this section, let us now nevertheless look at them separately to the extent which that is possible. For the topics of concern in Vol. 1, their mutual interdependence does not become acute. We begin with the first question.

If scientific discourse is to serve any useful purpose, a moment's reflection will convince one that any statement about the motion of a body will be meaningless unless it is simultaneously stated with respect to what that motion takes place. The thing with respect to which the moved thing moves must be just as unambiguously defined as the thing which moves. This is clearly an epistemological imperative, but it is nevertheless remarkable how often this simple fact has been ignored or overlooked in the history of science. What is the cause of the neglect?

There can be little doubt that it is due to a pure accident, namely, the fortunate fact that we live in a relatively very stable environment. From earliest times nature presented man with two seemingly completely stable and rigid structures – the ground beneath his feet and the unchanging appearance of the heavens. The rigid structure of the earth provided a perfect framework within which motion could be observed and studied. Because nothing significant in the framework changed, it acquired the status of the frame, or better arena, of motion. It is certain that this apprehension of the existence of an all pervading frame of reference, which for practical purposes is indistinguishable from Newton's absolute space, occurred instinctively and unconsciously. It takes a conscious effort of will to question the existence of this space, which may be called *intuitive space*. As will be seen in the next chapter, the first major figure to do this was Aristotle. He initiated the strand of philosophy that seeks to demonstrate the relative nature of motion and space. However, it will be worth pointing out here that the instinctive belief in intuitive space is very deep indeed. In this connection Anneliese Maier (writing on the subject of medieval concepts of place and space) makes the following very relevant comments:[16]

> In general, the philosophers of late scholasticism behaved in the face of concrete physical questions just like the natural scientists of all times: when the all too abstract philosophical concepts became uncomfortable, they tacitly replaced these concepts by the naive empirical concepts of prescientific thinking and in practice worked with them. This explains why all kinematic problems were treated, not on the basis of the . . . Aristotelian space and time definitions, but rather on the basis of a purely descriptive determination in which *motus localis* [i.e., motion*] is treated simply as a successive change of position, this change of position moreover being by no means relative to an *ultimum continentis* [ultimate container] in the Aristotelian sense but relative to the empirical space of practical experience; this space is the same as the one Galileo meant and which then was finally introduced officially as 'absolute space' into physics by Newton. And the change of position takes place in a time of which the same can be said.

The truth of what Maier says here about empirical space (clearly the same as what I have called intuitive space) is borne out time and again – not only in the medieval period but before that in the writings of Aristotle and then later in the work of nearly all the major natural philosophers of the seventeenth century. Only on such a basis is it possible to explain what appear to be some gross inconsistencies on the part of several of the major figures who contributed to the creation of dynamics. The point is that the

* The curious expression *motus localis* (= local motion) has the following origin. Aristotle's physics, which we shall study in the next chapter, did not give to motion the predominant status that it has acquired in the modern age since the discovery of dynamics in the seventeenth century. It therefore treated motion within an all-embracing framework that encompassed all forms of change (e.g. hot to cold, blue to red, etc.) and not just change of place. Although the Greek word *kinesis* (from which both *kinetic* and *kinematic* derive) has the primary meaning of movement, it can be used of all transitions, in particular from one qualitative state to another. This was Aristotle's practice. When he required an exclusive word for motion in the modern sense, he, like Plato, used *phora*[17] [*phoronomy* is a rarely used synonym of *kinematics*, coined in fact before the latter (in 1716 by Hermann as the German word *Phoronomie*[18])]. In Latin the word corresponding to *kinesis* is *motus*, and in this case the modern meaning *motion* was distinguished by adding the adjective *localis*. This construction persists, of course, in our word *locomotion*. Modern physicists should therefore be aware that 'local' in *local motion* (which is often found without any explanation in translations of early texts on motion) is nearly always redundant in modern English and has a meaning quite different from the same word in, say, *local* field theory.

As we shall see in later chapters, medieval physics was full of expressions like *motus ad formam, motus ad calorem, motus ad quantitatem* applied to processes which involve changes in shape, heat, and quantity, respectively.

concept of space is shaped by the problem with which any given thinker or natural scientist is actively confronted and is adapted accordingly. We must be prepared to encounter the concepts of space and motion of both the philosopher and the natural scientist and realize that they need not be the same – even when one and the same person is involved. I would only add to Maier's broad division of spatial concepts into those that are practical and those that are philosophical a third strand which runs between them and is sufficiently distinct from both to warrant special consideration, namely, the concepts used by working astronomers. By the very nature of their work, which requires the determination of the position of one object relative to others, they are constantly reminded of that epistemological imperative which might otherwise occur to only the reflective philosopher. On the other hand their work is pre-eminently practical and, at least in the tradition of theoretical astronomy initiated by the Greeks, is at the same time a fully-fledged science of motion.

This is one of the reasons why a relatively large amount of space will be given to the work of the astronomers. It is, I feel, a part of the absolute/relative debate that has been rather seriously neglected. For example, Čapek has published a quite excellent selection[19] of extracts from numerous authors in his book *The Concepts of Space and Time* yet Ptolemy, Copernicus, and Kepler are represented only in an essay on Copernicus by Koyré, and that is on the finiteness of the Copernican cosmology rather than the absolute/relative question.

Just as the stability of the earth and the heavens gave rise to intuitive space, another accident of our particular circumstances gave rise to an equally pervasive concept: that of the uniform flow of time. The origin of this notion is clearly to be sought in the rotation of the earth. It is interesting to note that an inveterate belief in the idea that there is such a thing as a *uniform* flow of time was clearly established long before secure scientific evidence was forthcoming to demonstrate that there was a sound basis in fact for the conviction. Once again we shall find in Chap. 2 that Aristotle was the first who seriously and clearly posed the fundamental question: what do we mean by the flow of time? And once again we shall find that just at the point at which philosophical enquiry becomes both interesting and difficult Aristotle lapses into intuitive practicality.

The clearest evidence that the Greeks had an instinctive awareness not only of space (reflected in the development of Euclidean geometry) but also of what may be called the uniform flow of time is to be found in the development of *kinematic geometry*, for which there is evidence that it began to emerge as early as the middle of the fifth century BC.[20] This branch of mathematics was of the utmost importance for the development of Greek astronomy and much later played a crucial role in Galileo's work. The distinctive feature of kinematic geometry is that it is used to generate curves of a higher order of sophistication than the circle, doing this by means of points and lines which are assumed to move uniformly. A classic and very important example is provided by Archimedes' *Spiral Lines*, which contains the famous definition:[21] 'If a straight line one of whose extremities is fixed turns with uniform speed in a plane, reassuming the position from which it started, and at the same time a point of that rotating line is moved uniformly fast on that line, starting from its fixed extremity, the point will describe a spiral in the plane.'

We can now briefly state what were the prevailing notions about space and time at the various stages in the discovery of dynamics. Throughout the entire period, the concept of time seems to have given very little trouble. Many pondered its nature but no one seriously doubted that it flowed uniformly. This uniformity of flow was the only thing the dynamicists needed to transform kinematic geometry into physical dynamics. In Chap. 3 we shall see that the most important practical problem of time-keeping had been solved by Ptolemy's time, and this probably explains why the nature of time, as opposed to its measurement, played such a small role in the history of dynamics right up to the beginning of this century. The early intuitive concept passed quite unscathed through the two major events that did necessitate drastic reconsideration of the concepts of space and motion: the Copernican revolution and the revival of atomism.

As regards the concepts of space and motion, we find in Aristotle, in the prescientific (or, at best, quasiscientific) period, a very interesting anticipation of the debate that raged throughout the seventeenth century and has still not subsided. It was largely a reaction to atomism. Perhaps even more than the Copernican revolution it is atomism that makes the relativity of motion into an acute issue. We have already seen that our environment provided two alternative frames of reference for describing motion: the earth and the fixed stars. As long as neither was seriously believed to change intrinsically, the problem of describing motion never really became acute. Motion was indeed recognized as relative but only relative to the frame provided by the solid earth or the firmament of stars. In the presence of an essentially unchanging framework, the question of whether motion is relative to space or matter is at most academic. It is only when the awareness dawns that the frame may not be rigid at all, that the universe may contain nothing but a seething mass of atoms, that the real dilemma of motion becomes acute.

This was what Aristotle sensed with a fair degree of clarity – and he did not like it, as we shall see. The alternative cosmology that he developed had the effect of refreezing the firmament of the stars and placing an immobile earth firmly at the centre of a spherical rotating cosmos. Although the Copernican revolution reversed the motions of the firmament and earth, it did not immediately unfreeze what Aristotle had refrozen in reaction to the atomists. It was in this immediately post-Copernican period that Galileo established many of the most important results of dynamics, using as an effective framework what he still regarded as truly fixed stars. Virtually all the major discoveries that went to make up the final Newtonian synthesis were made in the period in which kinematic geometry appeared to have a natural validity. The great early discoveries in astronomy were made in the post-Aristotelian period in which the earth provided the fiducial body at rest in an effective space in which kinematic geometry was clearly the appropriate tool to use to describe motion. The Copernican revolution merely transformed the fixed stars into the fiducial body, which was the explicit frame of reference for Kepler's great discoveries in astronomy. Galileo too had no cause to worry about the frame of reference.

The fact that Galilean dynamics seemed to work perfectly well in the intuitive space of kinematic geometry, in which Galileo instinctively believed the stars to be at rest, made it that much easier for Newton to believe in *real* space, even though he was very much aware of the potential unfreezing of the cosmos. In fact, Newton's formulation of his concepts of absolute space and time was in the nature of a blessing after the event, not the setting up of a general framework in which the laws of motion had yet to be found.

This will suffice for a general outline of the origin of the intuitive concepts of space and time and how it came about that they were used by Newton in almost exactly the same form as they must instinctively have been conceived about two thousand years earlier by Greek geometers and, deep down, even by philosophers such as Aristotle.

It will also be helpful to distinguish at this stage between some typical forms through which the laws of motion passed, as this also has a bearing on the absolute/relative debate. It will in fact be helpful to coin new words to underline the conceptual differences between the different types of laws of motion.

Let us first mention the *mechanical* (or *contact-mechanical*) theories. Of this kind, the crudest are those that posit an invisible machine or mechanical contrivance as an explanation of observed motions; for example, the solid spheres invoked by Aristotle to carry the planets, sun, and moon around the earth. Although he is not explicit, it seems that Copernicus adhered to some such conception.

Less crudely machinelike are the mechanical theories developed in antiquity and revived in the seventeenth century. They come in two broad classes: *plenum* theories, the crucial feature of which is that the world is assumed to be completely full of a fluid-like matter (which may also contain some solid matter), so that all motion results from the direct pushing of one part of the fluid on another, and *atomistic* theories, in which particles of fixed shape are assumed to move through the void, suffering intermittent collisions. The characteristic feature of early mechanical theories is that they seek to explain observed phenomena almost exclusively in *three*-dimensional geometrical terms. In histories of dynamics, such geometromechanical explanations of motion are often described as physical, but this latter word will be reserved in this book for genuinely dynamic concepts such as mass and charge, as defined in Sec. 1.1.

Next in order of development were approaches that I shall call *motionic*. We have here a class of theories that marked an important step towards fully-fledged dynamics but that is neverthless sufficiently different in outlook to warrant a special name. The hallmark of such approaches is that they concentrate on motion as such and aim to describe the observed motions, either qualitatively or quantitatively; they seek neither a hidden mechanical explanation for motion nor a visible source of forces that cause motion, as, for example, in Newton's theory of gravitation. Aristotle's approach, for instance, was by no means exclusively mechanical; it also contained much that can be dubbed motionic. The main exponents of the motionic approach were the Hellenistic astronomers and Galileo, who resolutely refused to look for physical or mechanical causes of motion. Within the general motionic approach, it is also worth introducing a special word to denote a particular type of law of motion that appeared in antiquity and has special significance for the debate about the nature of motion.

Historically, the first laws of motion clearly formulated as such were purely geometrical in nature; they were manifestly inspired by the spirit of kinematic geometry. Thus, the fundamental law of ancient Greek astronomy stated that all celestial bodies move in perfect circles at a uniform (perfectly constant) speed. In accordance with this law, the motion as such is entirely independent of all the other bodies and matter in the universe.

Of course, since this law was formulated by astronomers, they had a very clear idea that the motion was actually taking place at some quite definite position relative to their fundamental body of reference (the earth), and they attempted as best as they could to determine that position: *independent* in the previous paragraph means that the position of the centre of the circular motion and its speed were not in any way determined by the remaining bodies in the universe. Provided there is an unambiguous specification of the frame of reference and the means for measuring time, such a law is epistemologically unexceptionable. Since its formulation involves the bare minimum of concepts and is *aphysical* in that physical concepts such as mass or charge do not appear in it, we may call it a *geometrokinetic* law. Not a very elegant word, but none better comes to mind. Such a law can be contrasted with dynamical, or physical laws of motion, of which Newton's law of universal gravitation is the classic example. Such laws cannot be formulated without the introduction of essentially physical concepts such as mass and charge. These are concepts that go beyond the purely kinematic concepts of space and time and can in no way be derived from them. They belong to dynamics proper, the study of motion in relation to the forces that produce the motion.

It may be wondered why it is necessary to introduce the new concept geometrokinetic when we already have the perfectly good distinction, introduced by Ampère, between *kinematics* and dynamics. The point is that Ampère defined kinematics as the science of pure motion, considered *without reference to the matter or objects moved* or to the force producing or changing the motion (*OED*; my italics), the purely geometrical science of motion in the abstract. But this means that the adjective *kinematic* cannot properly be used to describe either the ancient Greek law of celestial motion nor, more importantly, Newton's First Law, the law of inertia. For these are laws of nature and they apply to real bodies. Moreover, they hold, if at all, as empirical laws. They cannot be obtained as geometrical theorems, in contrast to genuine kinematic relationships (such as, for example, the purely mathematical relationships for the transition from one coordinate system to another moving relative to it at some given velocity). Nevertheless, since the Greek law of astronomy and the law of inertia do not contain any reference to a force 'producing or changing the motion', one does quite commonly find both laws referred to as 'kinematic'. This is normally not serious and will probably be understood to mean that forces do not figure in the laws. However, when Greek astronomy as a whole is characterized as 'entirely kinematic' I suspect that many readers would take this to mean that astronomers like Hipparchus and Ptolemy were concerned solely to fix the positions of the celestial bodies in the heavens and dispensed entirely with theory. This would be a decided misconception; they had laws of motion, or, in essence, just one law of motion, but it was a *geometrokinetic law* – it was not physical, and we lack a word to describe it.*

* It is interesting to note in this connection the absence of an adequate word to describe Galileo's work on motion, which can, depending on the author, be found described variously as kinematic, dynamical, or even occasionally mechanical (the least appropriate). I shall mostly refer to it as *motionic*. It is worth noting that the general term *mechanics* (still widely used today) became established in the seventeenth century at the time, before the creation of dynamics by Newton, in which *mechanical* explanations of motion were assumed more or less as an article of faith. The word *dynamics*, which derives from the Greek word for power and force (cf. dynamite), was originally coined by Leibniz to describe his own theory of motion but by the mid-eighteenth century was used to mean 'the science of the motion of bodies acting on each other in any way whatever.'[22] Ampère's distinction between dynamics and kinematics came later. The distinctive characteristic of dynamics is the *interactive* nature of the motion which it describes; the word *dynamics* will be used almost always throughout this book with this connotation.

The distinction is important since, through a series of intriguing metamorphoses, which we shall trace in the following chapters, the ancient Greek geometrokinetic law of astronomy split in two and became by stages Newton's law of inertia (which remained geometrokinetic, i.e., aphysical; we recall from Sec. 1.1 how the First Law is set apart in this respect from the Second and the Third) and Newton's law of universal gravitation (which became a fully fledged dynamical, i.e., physical law). Now the question at the heart of the debate about absolute and relative motion is this: Is the law of inertia truly a geometrokinetic law or is it, like the other half of the ancient Greek law, physical too and only apparently geometrokinetic? Historical examination of the discovery of dynamics makes the question particularly relevant, since, as we shall see, Kepler used very much the same epistemological and physical arguments in performing the Herculean task that helped ultimately to prise the Newtonian dynamical law of gravitation out of its geometrokinetic antecedent in Greek astronomy as Mach did when he argued that inertia has a physical origin. It was in this sense that Kepler's discovery of the laws of planetary motion was a pre-Machian (and pre-Newtonian) triumph of Machian ideas and suggests that in Newton's formulation the phenomenon of inertia appears in geometrokinetic garb only because its physical origin was rather more thoroughly hidden than was the case for Newton's Second and Third Laws. As pointed out in the Introduction, such considerations are among the main justifications for the historical approach to these problems.

1.6. Was dynamics discovered or invented?

One of the *leitmotifs* of this book is the declaration by Mach of his belief in an ultimately indissoluble unity of the universe:[23] 'Nature does not begin with elements, as we are obliged to begin with them. It is certainly fortunate for us, that we can, from time to time, turn aside our eyes from the overpowering unity of the All, and allow them to rest on individual details. But we should not omit, ultimately, to complete and correct our views by a thorough consideration of the things which for the time being we left out of account.'

Science as an attempt to find a rational description of the world is subject to an occupational hazard to which this section is intended to draw attention – and which the reader is asked to bear in mind throughout the book.

The natural scientist is constantly seeking 'basic' elements from which the world may be built up conceptually in a coherent rational scheme. This analytic method, still followed in its essentials 2500 years after its formulation in ancient Greece, was expressed in the following way by Aristotle:[24] '. . . knowledge is always to be sought through what is primary, and the primary constituents of bodies are their elements. . . . Let us then define the element in bodies as that into which other bodies may be analysed . . . and which cannot itself be analysed into constituents differing in kind. Some such definition of an element is what all thinkers are aiming at throughout.' The word *analyse* derives from *unravel*.

The danger with this approach – to which there does not appear to be any alternative – is that any new insight into the workings of the world, in particular the recognition of 'elements' from which it appears possible to understand a large body of phenomena, is liable to crystallize and gradually assume the nature of an unquestioning belief that the world is cast in a particular mould. It then slowly becomes inconceivable that it could be any other way. While the extent to which this happens will depend on the particular historical circumstances, the danger is inherent in the nature of the enterprise.

A first and very obvious example is the early belief in a flat earth. The world around us appears flat, *ergo* the whole world is flat. The proofs of the earth's sphericity (given by Aristotle and Ptolemy) belong to some of the first solid achievements of science and demonstrated how observations of comparatively small effects combined with rational arguments using geometrical theorems could change drastically a view of the world formed by premature conclusion from its local apparent flatness. Nevertheless, as we shall see in the next chapter, Aristotle's recognition of the sphericity of the earth did not prevent him making several mistakes about the nature of the world every bit as misleading as the flat earth idea.

The flat earth mistake and these mistakes of Aristotle to be considered in the next chapter spring either from an unjustified extrapolation (local flatness extended to infinity) or an inability to see distant things clearly enough. As we shall see, this was ultimately the reason why both Aristotle and Ptolemy thought the heavenly bodies were divine. But there is also a potentially much more subtle danger and one that is also far less easy to eliminate by experimental advance. It is this: when some striking facet of the world is recognized and postulated as a basic element from which the whole is built up, the key assumption which is made is that the facet retains, when detached from its context, the essential property or properties for the sake of which it was selected. The danger here is that the facet may only have the desirable property for which it is chosen *because it is in the context of the whole.* In such a case, an attempt to regard the facet as capable of existing in its own right with all these properties and to imagine the whole built up from facets with such properties will clearly lead to a quite incorrect conception of the interrelation of the whole and the parts.

To give a very simple illustration. Suppose a cathedral were built of bricks that under pressure change their colour, the colour depending on the pressure. Because of the different pressures exerted on the bricks, the cathedral, when built, will obviously glow with many different colours. Then the naive 'atomistic' interpretation of the phenomenon is that bricks come in different colours and different proportions of bricks were used by the builders to achieve their particular end. Short of dismantling the cathedral, it might be difficult to recognize the fallacy of the theory (though not in principle impossible, for example, by observing carefully the relationship between colour and position in the overall structure). This is the problem we face with inertial motion: the actual phenomenon is not in doubt (even though it took remarkably long to discover); the problem is in the nature of its relation to the whole. It is particularly acute since the 'cathedral' in this case is nothing less than the entire universe, so there is no way in which it can be 'dismantled' to see if inertia still persists in its absence.

It is appropriate here to anticipate the criticism that Mach made of Newton's conclusion from the undoubted existence of inertia *in the presence* of the stars that inertia exists *independently* of the stars. The following quotation is from Mach's *Mechanics*, the italics are mine:[25]

> When we say that a body K alters its direction and velocity solely through the influence of another body K', we have asserted a conception that it is impossible to come at unless other bodies A, B, C . . . are present with reference to which the motion of the body K has been estimated. In reality, therefore, we are simply cognizant of a relation of the body K to A, B, C . . . If now we suddenly neglect A, B, C . . . and attempt to speak of the deportment of the body K in absolute space, we implicate ourselves in a twofold error. In the first place, *we cannot know how K would act in the absence of A, B, C . . .*; and in the second place, every means would be wanting of forming a judgment of the behavior of K and of putting to the test what we had predicated – which latter therefore would be bereft of all scientific significance.

Mach is here pointing out that underlying Newton's apparent proof of the existence of absolute space there is a major hypothesis. It may be called the *detachment hypothesis*: that a phenomenon observed within a given environment is in essence independent of the environment. Moreover, the logical possibility that the phenomenon *could* be independent of the environment is not a proof that it actually is independent.

It is worth pointing out here that the word *absolute* carries the detachment hypothesis implicitly in its etymology. The Latin *absolutum* meant originally loosened, free, separate. Moreover, the concept of absolute space is, as we shall see in Chap. 2, closely related to the atomic concept of matter, according to which the world, which is presented to us through the senses as an indissoluble whole, is dissolved conceptually into atoms and space. Atomic concepts are so familiar that we tend to forget what a huge assumption underlies them – it is arguable that many working physicists, as opposed to philosophers of science, are quite unaware that any assumption is made at all. But it should not be forgotten that we never actually see space; all we ever see is matter against the background of other matter. If we were to stick to the bare facts, all our talk would be relative – how certain matter stands with respect to other matter. With the assistance of the concept of space, the atomic hypothesis breaks the bond between matter and matter, detaches individual parts of matter from the overall material concatenation, and sets them loose in conceptual space. This discounts the possibility that the material concatenation is the cause of the very existence of the parts. Mach's Principle is a suggestion of this kind, not so much about the existence of material bodies as such but rather the motion they execute.

In the light of these comments, which clearly imply that Newton may well have given us a seriously distorted picture of the world, it may reasonably be asked why this volume is nevertheless entitled *The Discovery of Dynamics*. Does not the very title imply that dynamics is something quite definite, sitting out there in the world waiting to be discovered like America? Is it correct to equate Newton with Columbus? Is it not much truer to say that Newton put together elements in a particular way, but that there is nothing particularly sacrosanct about the actual structure he finally chose? Should not one speak of the *invention of dynamics*? And did not Einstein himself give powerful support to such a view with his assertion[26] that the basic concepts and laws of physics are 'free inventions of the human mind'?

By way of answer to these questions, put to me by Michael Purser after reading a first draft of this volume, let us look a little more closely at what Einstein actually said in his Herbert Spencer Lecture given in Oxford in 1933, from which the above quotation is taken. Einstein was countering the view that the basic laws of nature can be deduced with ineluctable logic from facts of experience. There can be no doubt that Einstein was perfectly correct to point out that this is not the case. The simple fact that his own general theory of relativity describes the phenomena of gravitation every bit as well as Newton's theory (better indeed, though in 1933 the proven differences were very few) puts his case beyond any question. Nevertheless, the expression 'free inventions,' which has subsequently been quoted many times, often misleadingly and out of context, is rather unfortunate. It seems to imply that experience puts little or no constraint on theorizing.

That this is very far from true follows from a passage a little later in Einstein's lecture, in which he says: 'Experience can of course guide us in our choice of *serviceable* mathematical concepts' (my italics). And then comes the explicit statement: 'Experience of course remains *the sole criterion of the serviceability* of a mathematical construction for physics' (again my italics). But the fact that a physical or mathematical concept, which is indeed a free construction, is actually serviceable for the description of natural phenomena is itself unquestionably a *discovery*. The construct itself is not discovered but created; but the *congruence between the construct and the phenomena* (the fact that they match to within the experimental errors) *is truly discovered*.

Seen in this light, all the various individual parts that were used by Newton in his synthesis of dynamics and the synthesis itself represent true discoveries – not of the parts and the synthesis but of the fact that they are congruent to contingent facts. The congruences are real, every bit as real as America, and it is their existence that is significant. Mathematical constructs are two a penny but such congruences, which are naturally never more exact than the accuracy to which they have been tested, are very rare indeed, as the history of science shows. It is in this sense of the discovery of congruences that it seems to me far more appropriate to speak of the discovery of dynamics, rather than its invention or creation.

Such considerations help to answer the extreme sceptics who maintain that science does not rest on objective facts, and that the impression it gives of making steady forward progress is an illusion. It is asserted that there is no such thing as a bare observation, that all observation contains an element of theory, which is in turn a 'free invention'. According to this view, we are brainwashed into seeing a 'reality' in the world that does not exist at all. Now since the whole purpose of this book is to raise questions about some of our most basic concepts – concepts that undoubtedly shape the picture we form of the world – this last view is a thesis with which I would not wish to disagree. There is undoubtedly an element of brainwashing. On the other hand, I am convinced that we are making progress in understanding the nature of our existence. How are these two viewpoints to be reconciled?

The reconciliation is to be found, I suggest, in the essentially open-ended but at the same time at least partially rational nature of our existence. Hitherto mankind has not come up against any limit to either the flux of new experiences to which it is exposed or the ability of great minds to find rational constructs congruent to these experiences. And the development of dynamics is itself clear proof that the finding of objective congruences between the rational constructs and experience is genuinely progressive. Once a real congruence has been established, it is never discarded; it is merely fitted into a new whole. Of course, in the process the view that we have of the significance of the congruence is changed out of recognition. In this sense all discovery and progress in science is a journey *into the totally unknown*, in which shock follows shock. But we must ask how it is that we then do not simply sink in the sea of our new experiences. The answer is that we are still carried by the discoveries that lie behind us, even if we now see them in a quite different light and perspective. They are still there. As long as man and man can agree that the moon is the moon and the sun the sun (unlike Kate and Petruchio[27] in

The Taming of the Shrew) there will still be a great deal of truth and utility in the Ptolemaic system, which, hardly less than the Newtonian, was also a true science.

The past discoveries of true congruences are the raft that carries us into the unknown future.

2

Aristotle: first airing of the absolute/relative problem

2.1 Brief review of the period up to Aristotle

The scientific revolution was a unique event in human history. It was created out of the intellectual activity of two great periods, of which the second is our own and began in the high Middle Ages. The first lasted from around 600 BC until the first centuries after Christ. It was almost entirely associated with Greek thinkers and Greek culture, which became very widely diffused by the activities of Alexander the Great (356–323 BC). It is convenient to divide Greek history into two basic periods – before and after Alexander. The second period is generally referred to as the Hellenistic age and will concern us in the next chapter.

The first two centuries of the pre-Alexandrian period (600–400 BC) were the age of the so-called pre-Socratic philosophers. They are noted for the boldness of their philosophical speculations, but unfortunately there is not much first-hand information available about them. The following notes are based mainly on the *Dictionary of Philosophy*;[1] fuller accounts can be found in Barnes[2] and Kirk, Raven, and Schofield.[3]

The best known figure of the entire early period was Pythagoras (*circa* 572–497 BC). He founded a school of philosophy and mathematics that was simultaneously a religious brotherhood and flourished until the end of the fourth century BC. Although the famous theorem is associated with his name on dubious grounds, he and his school did make important discoveries in harmony and mathematics (discovery of irrational numbers). Pythagoras is held by some[4] to have been the most important formative influence responsible for the scientific revolution because of the great emphasis that the Pythagoreans put on geometry and numbers. They held that all matter is literally composed of numbers and that numerical ratios underlie all sensuous phenomena. For them mathematical harmony was the sole reality behind the visible universe. They had a great influence on Plato and the early giants of the modern scientific age: Copernicus, Kepler, and Galileo.

Quite independent of the Pythagoreans was the Milesian or Ionian school of philosophy. This was developed in the sixth century BC by Thales, Anaximander, and Anaximenes. Thales is supposed to have predicted an eclipse of the sun in 585 BC. The distinctive idea of the Milesians was that a *single* elementary cosmic matter underlies all the transformations of nature. Thales declared this primordial matter to be water; Anaximenes said it was air; while Anaximander identified it with 'the unbounded'.

The next major figure, a generation later, was Heraclitus (*circa* 536–470 BC) from Ephesus. In opposition to the Milesians, he held that there is nothing in the world that persists. There is merely a constant and ceaseless flux. Change is the only reality; the appearance that things persist is an illusion created by the ordered manner in which change takes place. In this respect he foreshadowed the modern realization that the laws of nature, describing the way things change, are more important than what is actually changing.

The next step was the reaction to Heraclitus, which was taken by Parmenides, who founded the Eleatic (of Elea) school and whose life spanned the sixth and fifth centuries BC. Whereas Heraclitus put all the emphasis on change and becoming, Parmenides insisted on the supremacy of being. However, his was not simply a return to the Milesians, for he took his notion to its logical and paradoxical extreme. He held that there is only the one being without inner differentiation and that change and the apparent diversity of things is an illusion. He was followed by Melissus (of Samos; active around 450 BC), who believed in a One that is eternal, motionless and without change, and by Zeno of Elea (*circa* 490–430 BC), who devised his famous paradoxes in order to prove that motion is an illusion.

The final phase of the pre-Socratic period is marked by attempts to reconcile the two extreme and apparently irreconcilable points of view of Heraclitus and the Eleatics. Empedocles of Agrigentum (*circa* 490–430 BC) developed the idea that all individual things are produced by the mixing of the four elements: earth, air, fire and water. The elements remain the same, things arise from their mixing in different proportions. This seems quite modern but one can hardly say the same of his idea that love and hate are the cause of motion and therefore bring about the mixing of the elements.

An alternative reconciliation of Heraclitus and the Eleatics was attempted by the atomists from Abdera: Leucippus (active around 450 BC) and Democritus (*circa* 460–360 BC!). Leucippus retained Parmenides' idea that there is just one being, of itself completely homogeneous, but he supposed that it was broken up into infinitely many pieces, or atoms, of different shapes and sizes, and that their combination gave rise to all the variety of individual bodies. This atomic hypothesis was made famous by

Democritus and in antiquity was further developed by Epicurus (341–270 BC). The main source of knowledge about ancient atomism, especially in the Epicurian form, is the famous poem of the early Roman poet Lucretius (*circa* 95–55 BC) *De Rerum Natura* (*On the Nature of Things*).

The atomic hypothesis is important for this book in marking an early step towards the concept of absolute space. The point is that Leucippus not only broke up the homogeneous block of Parmenides's being into atoms but also assumed that these pieces of being are separated by that which is non-being or void, i.e., empty space. This introduction of empty space, as that in which the atoms can move (mobility is evidently crucial for the atomic hypothesis), was the outcome of a sophisticated philosophical debate, which is well described by Bailey.[5] The void is not nearly such an obvious concept as it might appear to the modern mind. Leucippus defined the void, or emptiness, by contrast with what is real and tangible. If we stretch out our hand and feel something tangible, then that is matter; but the *lack* of this positive sensation is also in its way real. Thus, nothing (≡no-thing) has a kind of reality too. It should, however, be said that the atomists seem to have developed only the vaguest of spatial concepts, getting little further than the idea of emptiness between the atoms. The reader is especially recommended the early selections in Čapek's *The Concepts of Space and Time*.[6] As we shall see shortly, the vagueness of the atomists on this question was an important stimulus to Aristotle.

The last of the pre-Socratics to be mentioned is Anaxagoras of Klazomene, who settled in middle age in Athens and was active around 430 BC. He taught that there was an infinity of simple substances, divisible into parts, as in the atomic hypotheses. All becoming is due to their combination and separation. To explain the motion of the parts he assumed the existence of a 'soul-substance' that is itself in motion and can set the normal substances in motion. In contrast to the atomists, Anaxagoras accorded purpose an important role in nature. However, in practice he did not develop this idea very much; Aristotle, for whom teleology was the supreme principle in nature, took it much further.

In philosophy, the century before Alexander was dominated by Athens and its three great philosophers Socrates (*circa* 470–399 BC), Plato (*circa* 427–347 BC), and Aristotle (384–322 BC), who was actually tutor to the young Alexander. For the subsequent development of science, Plato and Aristotle were the most important, since they shaped the general climate of thought. The two men had very different attitudes to the material world. For Plato it was largely an illusion; he distrusted the senses and held that true reality resided in eternal immaterial forms. On the other hand, he greatly emphasized the importance of mathematics, especially geometry. In 387 BC he founded a school of mathematics and philosophy in Athens, which became known as the Academy (after the name, *Akadēmos*, of Plato's garden, where the school was held). His works have been very well preserved. Largely through the writings of St Augustine (AD 354–430), the part of Plato's philosophy distrustful of the material world and the senses influenced the early Christian world and may have been a factor in the decline of science during that period. In contrast the mathematical and geometrical aspect of his teaching was very influential in the sixteenth and early seventeenth century.

Aristotle was much more 'realistic' than Plato. He took the observed world at its face value and was therefore far more empirical. The second great period of intellectual activity that produced the modern scientific revolution was initiated by the rediscovery of his works (which survive in a much less satisfactory form than Plato's – basically as lecture notes with many apparent interpolations that may be by later hands) in Christianized Western Europe in the thirteenth century. It was his cosmology and concepts of motion that provided the framework of scientific thought up to the time of Galileo and this is the reason why a complete chapter needs to be devoted to his ideas, which have a very close bearing on the absolute/relative debate – indeed they amount to the first full-scale airing of this question and there are many parallels between his writings and the discussions about the problem in the seventeenth century.

Like Plato, Aristotle founded a school of philosophy in Athens, called the Lyceum, at which he taught from 335 BC for twelve years. The school was also known as the Peripatetic, apparently because Aristotle liked to lecture in a covered walk, or peripatos. The Peripatetics, as they were called, dominated philosophy in European universities from 1300 to about 1650. Aristotle is also known as the Stagirite, because he was born at Stagira. His works constitute the oldest extant treatises that are consistently informed by an essentially scientific spirit.

The period before Alexander left few if any solid results in the study of the natural sciences. It was, in general, far too speculative and without sound empirical basis, except in the study of the geometry of the natural world. Almost all of Aristotle's output was purely qualitative and much was, with hindsight, badly wrong. The period's most permanent achievements were in mathematics, especially geometry, which undoubtedly laid the foundations of the later great advances in astronomy beginning in the Hellenistic period, in statics, also in the Hellenistic period, mainly through the work of Archimedes (*circa* 287–212 BC), and then, much later, in dynamics in the early seventeenth century with Galileo (1564–1642).

Thus, the main importance of the first period of Greek history was in awakening the spirit of rational enquiry about the world and in creating the science of mathematics, which has proved to be the *sine qua non* of all advance in the study of motion.

2.2. Aristotle: the man and his vision

In most accounts of the history of dynamics, Aristotle is regarded as having had a perverse effect and to have retarded the development of the subject. He is generally compared unfavourably with Plato and the pre-Socratics. This judgement is too facile and was certainly encouraged by the fact that the dramatic rise of modern science in the seventeenth century coincided with the overthrow of the last remnants of Aristotelianism. However, one could well argue that the largely qualitative Aristotelian phase had to precede the quantitative stage in the development of dynamics.

There is a good rough and ready way of estimating Aristotle's contribution. Newton, as we know, postulated three fundamental laws of motion. Examination of pre-Aristotelian authors reveals only traces of anything resembling these laws. On the other hand, antecedents of all three laws occur in Aristotle's main works relating to the problems of space, time, and motion. Further, it is not just that they are mentioned among much else; on the contrary, they occupy a prominent position in his work. It is true that they are subordinated to his supreme teleological principle (that[7] 'God and nature create nothing that does not fulfil a purpose') and this was completely abandoned by Newton (or it was at least as an explicit dynamical principle). However, this lofty teleology – and theology – of Aristotle's overall scheme was not in itself a hindrance to scientific progress, being far removed from the rough and tumble of the actual world. God could see the overall pattern and determine the final causes of things, but the nitty-gritty details were, so to speak, delegated to efficient causes, and it was with these that Aristotle's principles, just like Newton's laws, were concerned.

Aristotle was the great systematizer. He arrived on the scene at the end of a period of intense scientific and philosophical speculation. Making a broad survey of what had been achieved, Aristotle perceived that, with the partial exception of Plato, all his predecessors had hardly come to terms with motion. They had not even considered carefully the meaning of crucial words such as place and space; thus, their concepts of motion itself were almost nonexistent. Moreover, in the enthusiasm generated by the discovery of geometry, the atomists and Plato sought to explain virtually everything by purely three-dimensional geometry. For a variety of reasons, by no means always sound, Aristotle completely rejected this approach, according to which the majority of the phenomena associated with motion, change, and transformation quite generally have an explanation in terms of *shape*. Instead, he asserted that motion, far from being explained, must be accepted as a primary phenomenon, not reducible to three-dimensional geometry. Just how radical and consequential he was in this respect we shall see shortly; in this respect

Aristotle was far more progressive and modern than his predecessors and in many ways deserves to be regarded as the pioneer of dynamics.

With the benefit of hindsight, we can see that his biggest mistake was the failure to appreciate the crucial significance of the quantitative aspect of motion. Plato and the atomists had an exceptionally sharp three-dimensional vision; Aristotle was revolutionary in having a *four-dimensional* vision, but it was fuzzy and qualitative. He could see that the mountain range extended in a further dimension and he correctly discerned the main valleys and ridges that should be explored, but there was no sharp focus in depth, as we saw in the previous chapter. Indeed, there seems to have been no awareness at all that such focus might prove crucial and put things in a quite different light. In this respect, Aristotle was indeed a step backward. The deepest and most remarkable secret of dynamics, the connection between force and *acceleration* (rather than velocity), was entirely hidden from his hazy vision; it could only come to light when Galileo insisted that the world is as sharp and precise in four dimensions as Plato had seen it in three. Thus, there were two essential aspects that fused in the discovery of dynamics – quantitative mathematical exactitude and four-dimensionality. Recognition of the former was Plato's contribution, but it was Aristotle who pointed the way into the enigmatic fourth dimension and made Chronos the world's arbiter.

Besides this fundamental turn to a truly dynamic cast of mind, Aristotle is remarkable for his anticipation of several Machian principles of motion. His cosmology and physics were in fact founded largely on Machian, or perhaps one should say epistemological, principles. However, here too there was a fundamental flaw in his thinking; once again, it derives from Aristotle's incomplete awareness of the strong quantitative aspect of existence. Writing in the period before geometry was fully formalized, and above all before the systematic use of trigonometry, Aristotle developed a concept of space and position that was almost exclusively topological; only through the neglect of the metrical and above all trigonometrical properties of space was he able to construct an apparently consistent and complete cosmology.

From inadequate conceptions he nevertheless constructed a vision of great beauty, harmony, and reassurance. His predecessors described a world of clinical exactitude populated with aggressive sharp-edged atoms. The gentle doctor from Stagira saw things with a human eye. His was not the world of sharp edges and mathematical triangles; position was not for him a point in Euclidean space, a mathematical abstraction, but rather place, the place of wine within the bottle, the baby within the womb. His principle was: each thing to its proper place. This is what gives his philosophy its wombic reassurance and also the sense that fulfillment is possible. It also no doubt largely explains how his overall conception could survive for so long despite the fact that it was increasingly seen to be internally deficient and contradictory and was also steadily undermined (without collapsing) from within by the application of trigonometry to astronomical problems and from without by the astronomical discoveries made possible by this use of trigonometry.

In the following sections we shall look at the various reasons that led Aristotle to his physical and cosmological conceptions. But before that a brief outline of the whole.

The Aristotelian cosmology is finite and rather like an onion. At its centre is the spherical earth (Aristotle was well aware of the earth's sphericity and in fact gave the first numerical estimates of its radius), which Aristotle held to be at rest. Above the earth extends a region of air and fire, which reaches to a sphere that carries the moon. Beyond this sphere are further spheres which carry the sun and the five naked-eye planets known to the ancients: Mercury, Venus, Mars, Jupiter, and Saturn. Beyond that comes the sphere that carries the stars, which Aristotle held to be fixed and unchanging. Beyond this outermost sphere, the *ouranos*, there is, according to Aristotle, neither space nor motion nor even time. There is, in short, absolutely nothing.

All the various spheres rotate. The outermost sphere has the fundamental rate of rotation corresponding to what, since Copernicus, has been recognized as the daily rotation of the earth. Because its rotation seemed to be so much more fundamental and regular than that of the other spheres, the outermost sphere acquired particular significance. It rotates about fixed poles, corresponding to the two poles of the celestial sphere. Within the *ouranos* are numerous other spheres (over fifty), only seven of which actually carry celestial bodies. They are all concentric with the centre of the earth and the *ouranos* but rotate about different poles and at different rates. The purpose of all these spheres, which Aristotle took over from Eudoxus, was to explain the irregular motion of the planets, sun, and moon relative to the stars.

The Aristotelian cosmos was divided into two very distinct regions: the sublunary region below the sphere of the moon and the superlunary region of all the celestial spheres above it. The 'laws of physics and motion' (to use a modern anachronism) that governed these two regions were completely heterogeneous. The superlunary spheres were held to be made of material quite different from that below the moon and to move in a quite different way – always circularly and with perfect uniformity. Aristotle in fact named this material *aither* (ether), believing it to derive from a Greek expression meaning to 'run for ever'. This substance was believed to be completely unchangeable, incorruptible, and later became known as quintessence, literally the fifth (quint) essence or element, to distinguish it from Empedocles' familiar four elements earth, air, fire, and water, of which the sublunary region was composed.

A very characteristic feature of the Aristotelian scheme is the doctrine of *proper places* and *natural motions*. For example, according to this scheme the proper place of the element earth is the centre of the universe. The falling of a stone is explained by the striving of earth, the predominant element in a stone, to reach the centre of the universe, which coincides with the centre of the earth. Similarly, the proper place of fire is on the circular periphery of the world, and this explains its striving to fly upwards.

In contrast to the incorruptible heavens, the sublunary region is the domain of corruption and generation. Aristotle's physics is concerned to describe and explain all the great variety of changes that take place in this sublunary region. As already emphasized in the previous chapter, what we call motion was only one among very many other changes that take place. Aristotelian physics, like the studies to which it gave rise in the Middle Ages, really bears very little resemblance to modern physics. Its overarching principle is that everything happens for a specific (and basically good) purpose. Nature never does anything in vain. Thus, as already explained, the stone falls for a purpose, to reach its proper place. Only when there is the stone fully real, or actual. This goes some way to explaining the Aristotelian notion, so difficult to the modern mind, that all change, and not just motion, is a process of passing from *potentiality* to *actuality*.

In the whole of Aristotelian physics there is very little that can be called quantitative or mathematical. It is much more concerned with establishing the *essence* of things and their *causes*. The analysis is predominantly verbal and logical. Much attention is paid to the various *categories* of existing things and Aristotle is forever posing questions like: what is matter? what is motion? what is space? what is time? We are of course still asking these questions but seldom from the same point of view as Aristotle. The same is true of our notion of causality. The flavour of Aristotelian enquiry in this field is well caught by the following summary of Clagett:[8] 'The four causes can be explained by analogy with something artificially produced. A bed is a bed because it is made of wood (the material cause), in a given shape (its formal cause), by a carpenter (its efficient cause), for the purpose of providing slumber (final cause).'

It was probably in this logico-verbal aspect more than any other that Aristotelianism hindered the development of science in general and dynamics in particular. For all that, as we shall see in Chap. 4, the medieval study of motion that developed in the form of commentaries on Aristotle all but made the breakthrough to quantitative science. One really needs to put oneself in the position of Aristotle, an eminently practical and down-to-earth person (quite unlike the almost ethereal Plato), to realize just how implausible and unexpected it was that the quantitative study of pure motion should ultimately cast such an extraordinary amount of light on the multifarious qualitative changes with which the world confronted him. And even in this century there have been very great physicists, above all Schrödinger,[9] who insist that modern physics is still grasping and comprehending at best only a portion of reality.

The work in which Aristotle came closest to the spirit of modern dynamics is his *On the Heavens* (usually known as *De Caelo*). This, together with his *Physics*, will be the main subject of the present chapter. In it he treats motion as an autonomous spontaneous phenomenon. Thus, the heavens spin forever without the intervention of any other agency, stones fall to the centre of the universe as if drawn there by a magnetic force, and fire similarly seeks the periphery of the cosmos drawn there by a similar attraction. Not surprisingly, most of the quantitative studies that led to modern dynamics developed out of the conceptions outlined in *De Caelo*. In fact, Galileo still retained Aristotle's notion of natural motions; he merely treated them mathematically and quantitatively. This shows that it is the mathematical treatment of motion rather than the underlying concept one has of its nature that is essential for scientific advance.

In other works, notably the *Metaphysics*, which, at least in part, appears to represent a later development of his thinking,[10] Aristotle developed a scheme in which all motion is produced by an active agent. As regards the natural motion of the four sublunary elements there is little change, since they are moved by internal striving to reach their proper places. Other motions below the moon are *violent* or *enforced* (because the corresponding motion is not one to the proper place) and are caused either by an animal agent, or ultimately, in the case of most terrestrial motions, by friction between the nonrotating sublunary region and the rotating heavens, which thereby transmits its motion to the corruptible part of the universe and keeps it in a perpetual state of unrest.

But the heavens themselves are moved (or rather inspired to movement) by the divine and ultimate Unmoved Mover, which in Christian theology came to be identified with God. According to this idea, the state of perfect rest is the highest ideal, but this is achieved only by the divine. The heavens represent the next best approximation to this ideal. For the *ouranos*, by moving circularly and with perfect uniformity, is at least always perpetually in the best place in which it can be, nearest to the divine. Thus, by executing perfectly uniform circular motion the heavens emulate and are drawn to the divine, as a lover to the loved one. To achieve this aim, Aristotle felt that the heavenly spheres needed guidance and assistance by ethereal spirits. These spirits, which are quite absent in *De Caelo*, figured prominently in medieval philosophy and helped to sustain the notion that the cosmos is an animated organic structure. Kepler (1571–1630) still took Aristotle's spirits as a perfectly serious possibility, though in his mature work he inclined to mechanical or dynamical explanations.

Thus, in the final Aristotelian synthesis all motion in the world derived ultimately from the external and unchanging divine. It is interesting to note that the Aristotelian ideas about the heavens, which took such a firm grip on the human mind, sprang from a deficiency of the human circumstances exactly analogous to the absence, noted in the previous chapter, of a clock suitable for timing terrestrial motions. In this case it was the acuity of human vision and the relative brevity of human life when measured in astronomical time scales. Human vision was marvellous for the earth but just (and only just) failed to reveal to human beings that the heavens were not really all that different from the corruptible earth. With absolutely minimal artificial assistance, the human eye crosses the resolution threshold and can see, as Galileo did, mountain ranges on the moon and continually changing spots on the sun. So much for the eternal and unchanging perfection of quintessence and all things heavenly. All that was needed was again a factor of about 5 or 10, this time in the power of vision, and the world was changed out of all recognition. Similarly, had Aristotle's life span been lengthened by a similar factor – to the biblical ages of Noah and Methuselah – and had he watched the stars intently, he would have just been able to establish relative movements of the stars among themselves, which would again have destroyed his scheme.

But for Aristotle the firmament of stars swept above his head, never appearing to change intrinsically within itself though always seemingly in a state of perfectly uniform motion around the centre of the earth. Transferring the sense of awe and wonder that man feels before the sight of the heavens to the stars themselves, Aristotle made them divine and quintessential and said of his vision[11] '. . . it is the only way in which we can give a consistent account and one which fits in with our premonitions of divinity'. This vision, constructed by the naked human eye and speculating mind, held in sway even the greatest minds for nearly two millennia until it was destroyed, literally overnight, by about an ounce of carefully ground glass and a beady Tuscan eye. The vision was as fragile as the glass and eye which destroyed it during those fateful nights in Padua in December 1609 when Galileo first turned his telescope to the heavens.* Seen in the historical perspective, the world has still only just recovered from the shock of finding it is not nested within the quintessential spheres that were held to set the limit to not only our mortal existence but even space and time.

There is obviously a lesson in this for us too; for our present conceptions are, just as for Aristotle, determined by our current level of ability to 'see' and, equally and just as strongly, by our 'premonitions' of how the world

* Galileo went blind in his old age.

must be. Let us now look in a little more detail at how Aristotle's ideas were formed.

2.3. Pre-Aristotelian geometrism

The revolutionary nature of Aristotle's approach to motion can be best illustrated by looking briefly at the ideas of Pythagoras, Leucippus and Democritus, and Plato.

As noted in Sec. 2.1, the Pythagoreans appear to have believed that matter is actually *composed of numbers*.[12] They seem to have arrived at this idea from the way in which they represented squared and cubed numbers. Thus, a square number such as 9 was represented by a pattern of three rows of three dots or pebbles placed next to each other and forming thus a square. The cube number 27 was then formed by placing three such squares on top of each other. Our very words *square* and *cube* for such numbers derive from this Pythagorean notion, according to which all bodies could be built up in this way.

It is interesting to note that another early step towards what was later to become absolute space can be found in this Pythagorean construction – it is the space, or void, between the dots representing the numbers. The evidence for this remarkable notion is to be found in Aristotle:[13] 'The Pythagoreans too asserted the existence of the void and declared that it enters into the heavens out of the limitless breath – regarding the heavens as breathing the very vacancy – which vacancy "distinguishes" natural objects, as constituting a kind of separation and division between things next to each other, its prime seat being in numbers, since it is this void that delimits their nature.'

Let us now turn to the basic principles of atomic theory developed by Leucippus and Democritus. According to a standard summary:[14]

> Nothing exists but atoms and empty space; everything else is opinion. Only in opinion does sweetness exist, only in opinion bitterness, in opinion hot, cold, colour; in truth there exists nothing but atoms and empty space. The atoms are infinite in number and of infinite variety in shape. . . . The differences of all things derive from the differences of their atoms in number, shape, and arrangement; there is no qualitative difference between the atoms; the atoms have no 'inner state'·

If we compare this with Newton's theory of the microscopic nature of matter, the similarity seems at first striking. Here is what Newton says in his *Opticks*:[15]

> It seems probable to me, that God in the Beginning formed Matter in solid, massy, hard, impenetrable, moveable Particles, of such Sizes and Figures, and with such other Properties, and in such Proportion to Space, as most conduced to the End for which he form'd them. . . . And therefore, that Nature may be lasting, the

Changes of corporeal Things are to be placed only in the various Separations and new Associations and Motions of these permanent Particles.

So far the standpoints are remarkably similar. It is, however, in the ideas about motion that the real gulf between the atomists and Newton and the linking role played by Aristotle become apparent.

The fact is that the ancient atomists never really developed any very precise ideas about how their atoms actually move. There seems to have been only one clear idea – that they *fall* (even this, as we shall see in Sec. 2.5, may have been an addition of Epicurus):[14] 'In eternal falling through infinite space, the larger atoms, which fall faster, collide with the smaller; the sideward movements and vortices that then arise are the beginning of the building of worlds. Innumerable worlds arise and pass away, next to each other and one after another.'

Note (in addition to the standard Greek error according to which larger bodies fall faster than small ones) how vague the picture almost immediately becomes ('sidewards movements and vortices'). In Lucretius's great poem in praise of atomism, written nearly four centuries after the time of Leucippus and Democritus, the descriptions of the motions of the atoms are still remarkably vague, although the poem is otherwise often marvellously precise and clearly thought out. Thus, the geometrical aspect of atomism was clear, the motional aspect most vague.

By contrast, the quotation from Newton's *Opticks* continues as follows (my italics at the end): 'It seems to me farther, that these Particles have not only a *Vis inertiae*, accompanied with such passive Laws of Motion as naturally result from that Force, but also that they are moved by certain *active Principles*, such as is that of gravity.'

The first steps towards the discovery of the *Vis inertiae* (force of inertia) and the *active Principles* were taken by Aristotle, as we shall see in the next section. In the meanwhile, we continue with the pronounced geometrical bias of Plato in his attempt to understand the world.

Whereas Leucippus and Democritus sought to explain all the transformations of matter by means of atoms of all shapes and sizes, Plato had a far more definite vision. In his scheme, outlined in the *Timaeus*,[16] material body is built up from the simplest plane figures, triangles, which he used to construct what are now known as the Platonic, or perfect, solids as the 'molecular' bodies that constitute the four elements: the tetrahedron for fire, octahedron for air, icosahedron for water, and the cube for earth.* Then the greater density of earth is readily explained by the possibility of

* Unfortunately, there were only four elements but five perfect solids. Plato remarks airily:[17] 'And seeing that there still remained one other compound figure, the fifth [the dodecahedron], God used it up for the Universe in his decoration thereof.' As the translator of the *Timaeus* remarks: 'How God "used it up" is obscure: the reference may be to the 12 signs of the Zodiac.'

dense packing of the cubes while the penetrating power of fire is attributed to the sharpness and cutting capacity of the points and edges of the little pyramidal tetrahedrons of which it is composed:[18]

Firstly, then, let us consider how it is that we call fire 'hot' by noticing the way it acts upon our bodies by dividing and cutting. That its property is one of sharpness we all, I suppose, perceive; but as regards the thinness of its sides and the acuteness of its angles and the smallness of its particles and the rapidity of its motion – owing to all which properties fire is intense and keen and sharply cuts whatever it encounters – these properties we must explain by recalling the origin of its form, how that it above all others is the one substance which so divides our bodies and minces them up as to produce naturally both that affectation which we call 'heat' and its very name.*

There is even a mechanical explanation of old age:[20]

Now when the structure of the whole creature is new, inasmuch as the triangles which form its elements are still fresh, and as it were straight from the stocks, it keeps them firmly interlocked one with another. . . . But when the root of the triangles grows slack owing to their having fought many fights during long periods, . . . in this condition every animal is overpowered and decays; and this process is named 'old age'.

Although some of these passages in Plato have an occasional appearance of modernity and the *spirit* of his approach (mathematical and quantitative) is far closer to that of modern physics than Aristotle's (organic, qualitative, and teleological), the fact remains that most fundamental properties of matter – mass, weight, momentum, and energy – are quite inexplicable in terms of three-dimensional geometry. They are instead properties of the four-dimensional world in which time is included as an essential dimension. It is precisely here that Artistotle makes such a significant departure from Plato and points the true way forward. Before discussing this, it is worth reflecting on the contribution that Platonic thought made to the historical development of physics and dynamics.

It is well known that Plato is supposed to have had inscribed above the entrance to his Academy in Athens the words: 'Let no man ignorant of geometry enter here.' Whether he did or not, there is no doubt that Plato put great emphasis on geometry. It could well be argued that this was his supreme contribution to the development of science. One of the most striking features of the history of fundamental physics is the way in which mathematical theories were developed more or less completely without any idea of direct application to the world and were then found to be essential for the correct application of fundamental physical processes. These applications often occurred decades, and in extreme cases millennia (in the case of conic sections, i.e., ellipses, parabolas, and

* Plato derived the Greek word for 'heat' from the word for 'mince up'.[19]

hyperbolas), after the mathematics had been fully worked out. Eugene Wigner has written a famous essay on this subject:[21] 'On the unreasonable effectiveness of mathematics.'

Thus, the importance of Plato and his emphasis on mathematics is manifest.* Nevertheless, examination of the actual successes that were achieved shows that they were not in the field of mechanico-geometrical explanation of the transmutations of the elements and their dynamical behaviour as outlined in the *Timaeus*; such models proved to be sterile. The emphasis on geometry and geometrical relationships bore its richest fruits in other applications. Of these, probably the most important (because it represented the first real breakthrough to what can be called genuine laws of motion) occurred about 150 years after Plato's death when Hellenistic astronomers started to use geometrokinetic models to explain and describe planetary motions – and achieved spectacular success. Much later, in the sixteenth and seventeenth centuries, Plato's motto was again taken up with the greatest enthusiasm, and geometry formed the indispensable tool for first Copernicus and then Kepler and Galileo.

On the basis of Aristotle's topological concept of position (to be discussed in Sec. 2.5), the projection of man's understanding far beyond the surface of the earth would have been quite impossible. Moreover, by his emphasis on exactness,† Plato encouraged both Kepler and Galileo to expect and seek precision not only in spatial relationships but also in time – in motion. It was only then that the dynamics of terrestrial motions progressed beyond the primitive embryonic and qualitative form in which Aristotle had left it and could be unified with the much more advanced art of celestial dynamics.

Even so, science advanced in curious and ironic ways. Kepler was initially fired by a dream as purely geometrical as anything thought up by Plato: to show that the relative distances of the planets from the sun can be understood in terms of successive nesting of the five Platonic solids within one another. To the end of his life he regarded his (worthless) Platonic theory of the sizes of the planetary orbits more highly than his correct determination of the motions of the planets, a discovery that finally blew apart the world to which he too clung.

* Mention should here be made of Eudoxus, who was born circa 400 BC in Cnidus and died circa 347 BC. He was a scholar and scientist of great eminence who contributed to the development of astronomy (through his theory of the motion of the planets based on the system of concentric spheres that Aristotle adopted and modified), mathematics, geography, and philosophy, as well as providing his native city with laws. His thinking lies behind much of Euclid's *Elements*, especially Books V, VI, and XII. He worked for a while in association with Plato. Unfortunately, not a single text by his hand has survived.[22]

† 'as regards the numerical proportions which govern their [i.e., the molecules of the elements] masses and motions and their other qualities, we must conceive that God realized these everywhere with exactness.[23]

2.4. Aristotle's natural motions

> 'If you can look into the seeds of time
> And say which grain will grow and which will not'
> *Macbeth*, Act I, Scene 3

It is a little difficult to know where to step into Aristotle's closed and self-contained world. Each part depends on the rest. However, as already indicated, one of his most interesting (and eventually fruitful) innovations was in the doctrine of *natural motions*, so let us begin there.

Quite why Aristotle put such emphasis on motion is not clear to me from his work. It could be that, having examined the work of his predecessors, he came to the conviction that purely geometrical considerations were quite incapable of providing explanations for some of the simplest and most striking phenomena relating to motion. An earlier quotation demonstrated the impression that free fall made upon him. A great part of Book III of his *De Caelo* is concerned with the discussion of whether weight and lightness can be explained in geometrical or other terms. Aristotle concludes that they cannot. One source of his doubt must surely have been the multiplicity of explanations, often contradictory, that had been advanced by his predecessors for the phenomenon of weight. In addition, many explanations were revealed on examination to be mere hand waving with unbridgeable lacunae in the logical chain. For example:[24] 'There are some, e.g. certain Pythagoreans, who construct nature out of numbers. But to construct the world of numbers leads them to the same difficulty, for natural bodies manifestly possess weight and lightness, whereas their monads in combination cannot either produce bodies or possess weight.'

Again, he is critical of the indiscriminate and vague use of motion made by the atomists:[25]

> When therefore Leucippus and Democritus speak of the primary bodies as always moving in the infinite void, they ought to say with what motion they move and what is their natural motion. Each of the atoms may be forcibly moved by another, but each one must have some natural motion also, from which the enforced motion diverges. Moreover the original movement cannot act by force, but only naturally. We shall go on to infinity if there is to be no first thing which imparts motion naturally, but always a prior one which moves because itself set in motion by force.

One could see here a qualitative anticipation of Newtonian dynamics, with the natural motions that Aristotle seeks to define corresponding to inertial motion and the enforced motion that diverges from it corresponding to the force-induced deviations from rectilinear inertial motions. However, instead of just the one Newtonian inertial motion, Aristotle has several natural motions, none of them corresponding exactly to the Newtonian. Nevertheless, the tendency is in the right direction and there is for the first time an overall, potentially complete concept of motion. It was precisely this concept, with one significant alteration that Copernicus forced upon it, that Galileo developed with such dramatic effects.

Besides this partial anticipation of the overall structure of the Newtonian scheme, we should note Aristotle's conviction that concepts which belong to dynamics can only be defined in dynamical terms. This is particularly interesting in connection with the tortuous history of the concept of mass, which, as explained in the Introduction, plays an important part in the discussion of Mach's Principle. Aristotle's discussion of weight and lightness anticipates the methodology of Mach's eventual clarification of the mass concept.

It will be recalled that the atomists, like others of Aristotle's predecessors, sought to reduce weight to size. Bodies are heavier simply because they are larger; the dynamical property weight is thus to be explained geometrically in terms of extension. For example, bodies containing invisible pores, i.e., voids, will be lighter than ones of the same overall size containing no such pores. There is not much point in going into the details of Aristotle's discussion of this question (in Book III of *De Caelo*) because often both his arguments and those of the people he is criticizing are flawed by straight errors of fact. What comes out clearly is Aristotle's conviction that all such attempts are misguided and that a quite different approach is needed.

Thus, whereas, for example, in the *Timaeus* of Plato the essence of the four elements is seen (or rather conjectured) to lie in the three-dimensional structure of the small bodies of which these elements are composed, Aristotle completely dismisses such an approach and, instead of making the motions of the elements the secondary consequences of their primary geometrical properties, promotes the characteristic motions (falling of earth, ascending of fire) to the essential *defining* properties of the elements. His standpoint is essentially that earth, by definition, is that which falls:[26] 'Let "the heavy" then be that whose nature it is to move towards the centre, "the light" that whose nature it is to move away from the centre, "heaviest" that which sinks below all other bodies whose motion is downwards, and "lightest" that which rises to the top of the bodies whose motion is upwards.'

This definition thus establishes the ordering: earth, water, air, fire. It is almost an operational definition of how to obtain these elements and anticipates, in a way, the mass spectrometer, in which a fundamental dynamical property is used to separate different substances. There is no doubt that Aristotle regarded his definition as an important insight and a significant advance. He does not look for any deeper explanation of weight and falling unless it be in metaphysics. The task of the scientist is to note the existence of the characteristic phenomena, not explain them.

In this attitude, as in several other respects, he was a precursor of Mach.

The extent to which Aristotle regarded motion as primary is evident in his statement:[27] 'Of bodies some are simple, and some are compounds of the simple. By "simple" I mean all bodies which contain a principle of natural motion.'

Thus, the very essence of a simple body is its principle of natural motion. The modernity of this approach is confirmed by the fact that it is precisely the same principle which is used today to define and distinguish the various elementary particles.

This is an appropriate place to put in a good word about those much abused terms *potentiality* and *actuality*, which figure so prominently in Aristotelian physics. Originally, when Aristotle called 'weight' the unrealized potential of a body to fall, he was merely putting his finger on an important physical property and moreover one that he correctly perceived was primary. Modern science does no different when it says that a body is electrically charged when it is capable of being accelerated in an electric field. Newton himself was being characteristically Aristotelian when, in his definition of inertia, he described the faculty by which any body remains in its state of rest or uniform motion in a straight line as a *potentia*.

It is also worth noting that the modern word *potential*, as in *potential energy*, was introduced deliberately[28] in almost exactly the Aristotelian sense (existing in a latent or undeveloped state). In modern science, the expression *vis potentialis* was first used by the Bernoulli brothers and Euler around 1750. The concept of *potential function*, with multifarious uses in mathematics and physics, was introduced by Green in 1828. Finally, the expression *potential energy* was introduced in 1853 by Rankine in explicit contradistinction to *actual energy*, which last expression was then renamed *kinetic energy* by Thomson and Tait, which somewhat obscured Rankine's explicitly Aristotelian contrasting of *potential* and *actual*.

How very right Aristotle was in his insistence on the study of motion in its own right is highlighted by a further remarkable passage from Ptolemy's *Almagest*, which comes immediately after the passages quoted in the previous chapter (pp. 43–4) and reveals once more the clear influence of Aristotle. Having pointed out that the mathematical study of celestial motions offers the only true prospect for a genuine scientific understanding of things, as opposed to mere 'guesswork', Ptolemy points out that this also applies to motions on the earth:[29] 'As for physics, mathematics can make a significant contribution. For almost every peculiar attribute of material nature becomes apparent from the peculiarities of its motion from place to place. [Thus one can distinguish] the corruptible from the incorruptible by [whether it undergoes] motion in a straight line or in a circle, and heavy from light, and passive from active, by [whether it moves] towards the centre or away from the centre.'

This is so close to the spirit and inspiration of Galileo one wonders why the scientific revolution did not take off then and there. Was the explanation perhaps that the pursuit of pure understanding was the passion of such a small band of people?

From the historical point of view, the problem with Aristotle's philosophy was that it could be developed in two directions: 'upwards' in the direction of metaphysics, theology, and teleology and 'downwards' in the direction of quantitative investigation of the precise way in which potentials were in fact actualized. Perhaps not surprisingly, the medieval philosophers mostly found the 'upward' direction much more congenial. Aristotle himself gave precious little encouragement to the quantitative development of his physics. Moreover, by his espousal of *plenism* (i.e., the view that the world is a plenum rather than a combination of matter and void) as opposed to *atomism*, he put a premium, in his physics of terrestrial enforced motions, on elucidation of the *mechanisms* by which motion is communicated from one body to another rather than on the quantitative study of motion in its own right. This was, in fact, a relapse into geometrism. Thus, from embryonic science Aristotle's motionics soon degenerated into mere mumbo-jumbo (of which Aristotle himself provided a goodly portion) until Galileo injected enough Platonism into the fourth dimension to reveal its remarkable properties and spell the end of the Aristotelian organic view of the world.

And perhaps even committed scientists would allow that the detour through Aquinas and Dante (see Chap. 4) was worth the candle.

2.5. The corruptible and the quintessential

We have already seen certain affinities between the approach of Mach and Aristotle. Basically they stem from the same instinctive reaction to the perceived world – to regard it as a reliable source of knowledge, which is to be extracted by careful observation. In this respect, Mach was, however, far more single-minded and exclusively empirical than Aristotle.

Perhaps the closest direct parallel between Aristotle and Machian ideas in the narrow sense of the problem of giving meaningful expression to Newton's First Law is to be found in Aristotle's discussion of what he saw as a manifest defect in the atomists' ideas of motion in the void. His recognition of it seems to have been one of the major factors that helped to determine his own rival scheme. His point was that in an undifferentiated isotropic void, the same everywhere and in all directions, it is impossible to conceive of any definite motions at all. For he notes that terms such as up and down, left or right, are, in a geometrical space, purely relative:[30] 'The comparison of mathematical figures illustrates the point. For such figures occupy no real positions of their own, but nevertheless acquire a right and left with reference to us, thus showing that their positions are merely such as we mentally assign to them and are not intrinsically distinguished by anything in Nature.' The final words, 'not intrinsically distinguished by anything in Nature', are extremely typical of the relationists and similar passages are found repeatedly in Berkeley, Leibniz, and Mach and, before them and Newton, in Kepler too.

Thus, in a vacancy, i.e., void, it is quite impossible to conceive that motion can take place in a definite direction (as happens in the case of falling under gravity)[31] 'since no preference can be given to one line of motion more than to another, inasmuch as the void, as such, is incapable of differentiation.' And even more explicitly:[32] 'But how can there be any natural movement in the undifferentiated limitless void? For *qua* limitless it can have no top or bottom or middle, and *qua* vacancy it can have no differentiated directions of up and down (since the non-existent can no more be differentiated than "nothing" can, and the void is conceived as not being a thing, but as mere shortage).'

The parallel with Mach and indeed specifically Mach's Principle becomes almost complete when Aristotle argues from the undoubted existence of the falling of bodies that they must be falling to a *definite place*. Aristotle's position is as follows. We observe in the world several extremely striking examples of natural motions that quite clearly distinguish certain directions: these are the falling of heavy bodies, the ascent of fire, and the circular motion of the heavens. But for him it is quite inconceivable that these directions should somehow crystallize spontaneously out of the atomists' void. They must be motions tending to some quite definite goal.

Now these directions are not like the purely relative ones of mathematical space just discussed; for:[33]

These terms – such as up and down and right and left, I mean – when thus applied to the trends of the elements are not merely relative to ourselves. For in this relative sense the terms have no constancy, but change their meaning according to our own position, as we turn this way or that; so that the same thing may be now to the right and now to the left, now above and now below, now in front and now behind; *whereas in Nature each of these directions is distinct and stable independently of us*. 'Up' or 'above' always indicates the 'whither' to which things buoyant tend; and so too 'down' or 'below' always indicates the 'whither' to which weighty and earthy matters tend, and does not change with circumstance; and this shows that 'above' and 'below' not only indicate definite and distinct localities, directions and positions, *but also produce distinct effects*.

The italics in this passage are mine. They emphasize the double parallel with Mach – the epistemological (in a featureless space there are no markers to which motion can be referred) and the physical (the idea that the places not only *act as markers* but also *exert a causal effect*, serving as it were as *centres of attraction*). Of course, the parallel is not complete: the motions are not the same and the role of the markers is also subtly different. In Aristotle, there are two distinguished places: the centre of his universe, occupied by the earth, and the periphery of his 'onion', where the shells of quintessence spin. The earth and the quintessence mark these distinguished positions, and thereby define definite goals for motion, but the attractive power resides in the places themselves, not in the matter that occupies the places. In this respect Aristotle is unphysical. Mach identified the masses of the universe as the actual 'attractors' (one should really say 'governors of motion' – governor being used in its original Latin meaning of the steersman, i.e., the masses of the universe literally steer inertial motion according to Mach's idea). Further, he almost certainly assumed that their 'attracting' or 'steering' power is proportional to their masses, a physical concept of which only the first hints can be discerned in Aristotle.

Before we complete the description of the solution that Aristotle found to the problem of defining motions in a cosmological context, it is worth making a digression on the subject of the perpetual falling of the atomists, since this provides one of the most beautiful examples of a *detachment hypothesis* of the kind mentioned in Sec. 1.6. It illustrates how deeply and unconsciously the stability of the human environment influenced thought about space throughout all ages – and, no doubt, still does.

There has been considerable controversy in the literature on classical philosophy about who precisely did introduce the notion of perpetual falling (see, for example, Barnes's comments on atomism and the motion of the atoms[2]). It was at first believed that the idea originated with Leucippus or Democritus, but majority opinion now inclines to the view that it was introduced by Epicurus, possibly reacting to Aristotle's criticism that the atoms were not given definite motions.[34] Whatever the truth, the idea resulted in a most amusing paradox, one, moreover, which should not have persisted (let alone arisen if it did) after Aristotle's clear pointing out of the absence of distinguished directions in a featureless void.

For the atomists, wishing to explain the existence of our stable earth, made the daring conjecture that originally no world existed at all. Instead, there was nothing but an immense void through which the myriad atoms fell, some faster than others, their collisions giving rise to innumerable worlds, both 'next to each other and one after another'. One such world was supposed to be ours. The detachment hypothesis here is as follows. The actual world in which the atomists lived had an Up and a Down. There was a dynamical definition of up and down, through falling, and a purely geometrical definition, through the surface of the earth. Although the two agreed to a high degree of accuracy, the atomists who postulated the falling did not perceive there could be a *causal* connection between the two. Instead, transferring conceptually the effective frame of reference defined practically by the earth from the earth to the void, they imagined the earth to be no more but left in the void the disembodied Up and Down that the 'earth-to-be-explained' had defined for them. And they explained the existence of the earth by means of a phenomenon that we now know is not the cause of its (the earth's) coming into existence but is in fact caused by the very thing whose existence the atomists sought to explain – the earth. We can, of course, see the fallacy so clearly because, in effect, the atomists' 'world' has long since been 'dismantled'.

The relevance of this discussion to the Machian problem is highlighted by the following quotation from Einstein's late *Autobiographical Notes*, which follows immediately the passage quoted in the Introduction (one wonders if Einstein knew that what he described did in fact happen almost exactly as he envisaged):[35]

> How sound, however, Mach's critique is in essence can be seen particularly clearly from the following analogy. Let us imagine people construct a mechanics, who know only a very small part of the earth's surface and who also can not see any stars. They will be inclined to ascribe special physical attributes to the vertical dimension of space (direction of the acceleration of falling bodies) and, on the ground of such a conceptual basis, will offer reasons that the earth is in most places horizontal. They might not permit themselves to be influenced by the argument that as concerns the geometrical properties space is isotrope and that it is therefore supposed to be unsatisfactory to postulate basic physical laws, according to which there is supposed to be a preferential direction; they will probably be inclined (analogously to Newton) to assert the absoluteness of the vertical, as proved by experience as something with which one simply would have to come to terms. The preference given to the vertical over all other spatial directions is precisely analogous to the preference given to inertial systems over other rigid co-ordination systems.

But now to continue with Aristotle's scheme.

The spherical universe was the solution to all his problems, since, as we have seen, in such a universe there are unambiguously defined distinguished places and motions, above all the centre of the sphere and its surface as the distinguished places, and circular motion around the centre and linear motion towards the centre (downwards) and away from the centre (upwards) as the distinguished motions. Aristotle puts great stress on the fact that these are the only simple – and hence natural – motions and proceeds to construct not only his cosmology but also his 'chemistry' upon them. As far as the sublunary world is concerned, it is all built up from what goes 'up' (fire) and what goes 'down' (earth):[36] 'Every element has its proper motion, and the motion of a simple body is simple. But there is not an infinite number of simple motions, because the directions of movement are limited to two [up and down].'

For Aristotle, this is the basis of all terrestrial chemistry.

It must be said that the intermediate elements, water and air, have to be introduced rather unnaturally into this scheme:[37]

> the centre is the contrary of the extremity, and the constantly falling body of the rising. That there are two kinds of body, the heavy and the light, is thus conformable to reason, for the places are two, centre and extremity. But there is also the space between the two, which bears the opposite name in relation to each, for that which lies between the two is in a sense both extremity and centre. Owing to this there is something else heavy and light, namely, water and air.

In this respect, at least, Plato is more convincing with his perfect solids. (Many Greek thinkers appear to have placed great store in contraries, and Aristotle often introduces them on rather shaky grounds.)

This then gives Aristotle his four terrestrial elements. There is no point at all in going into his theory of transformation – generation and corruption – but it is remarkable how deeply he was impressed by motion and how he regarded motion as more significant than substance or shape. This left an indelible mark on medieval and Renaissance physics and eventually bore fruit.

Almost more valuable (by an ironic quirk of history) for the evolution of dynamics was his theory of quintessence (*aither*), the fifth element that he postulated as the substance of the celestial bodies and the spherical shells that carry them.* This was just an extension of his theory of simple (natural) motions. Aristotle was always on the lookout for *intelligible concepts*. The infinity of endless space cannot be comprehended in his view; the mind cannot grasp it. In contrast, the circle, which closes on itself, is the paradigm of intelligibility. Circular motion about the centre had two irresistible attractions for Aristotle – it was self-contained (and therefore intelligible) and, if the periphery is regarded as a proper place, it has the wonderful property of maintaining the substance that circles around such a periphery eternally in its proper place. These were grounds enough for Aristotle to postulate the existence of a fifth element, *aither*, and endow it with all the divine attributes of which he could think. The

* It is interesting that Aristotle did not postulate the crystal spheres to carry the stars and planets because he found it inherently impossible to believe that they could move of their own accord. The significant thing for him was the *manner* in which they moved.[38] The argument was as follows: first, the sun and the moon were evidently spherical. From this he concludes all celestial bodies must be so. Now in the case of the moon it could actually be seen that it neither rotated on its axis nor rolled – but these were the only two effective (natural) motions of which terrestrial spheres were observed to be capable. Extrapolating from the moon to all the celestial bodies, Aristotle concluded that they must all move by virtue of being fixed on spherical shells, since otherwise they would have to be endowed with wings or flippers to swim through the ether. One sees here how *plenism* dominated his thinking. One of the more entertaining passages in Kepler's *Astronomia Nova* is his quite serious discussion of whether the planets have flippers.[39]

following is one of the passages that came to dominate centuries of physical, philosophical, and theological thought:[40]

From what has been said it is clear why, if our hypotheses are to be trusted, the primary body of all is eternal, suffers neither growth nor diminution, but is ageless, unalterable and impassive. I think too that the argument bears out experience and is borne out by it. All men have a conception of gods, and all assign the highest place to the divine, both barbarians and Hellenes, as many as believe in gods, supposing, obviously, that immortal is closely linked with immortal. It could not, they think, be otherwise. If then – and it is true – there is something divine, what we have said about the primary bodily substance is well said. The truth of it is also clear from the evidence of the senses, enough at least to warrant the assent of human faith; for throughout all past time, according to the records handed down from generation to generation, we find no trace of change either in the whole of the outermost heaven or in any one of its proper parts. It seems too that the name of this first body has been passed down to the present time by the ancients, who thought of it in the same way as we do, for we cannot help believing that the same ideas recur to men not once nor twice but over and over again. Thus they, believing that the primary body was something different from earth and fire and air and water, gave the name *aither* to the uppermost region, choosing its title from the fact that it 'runs always' (ἀεὶ ϑεῖν) and eternally. (Anaxagoras badly misapplies the word when he uses *aither* for fire.)

The grandeur of such passages cannot be denied (though it is interesting to find the *OED* seems to support Anaxagoras with regard to the etymology of *aither*!) Part of Aristotle's grip on men's minds must have come from the way, already noted, in which he transfers the awe man feels before the welkin to the substance of the firmament itself. Thus, we attribute to the putative quintessence all that we associate with our own moments of heightened awareness and ecstasy. (In a notable passage in his *Dialogo*,[41] Galileo points out that quintessence and the never changing celestial world would in reality be excruciatingly dull.)

But the history of dynamics is nothing if not ironic. On the one hand Aristotle put an almost unbridgeable gulf between the heavens and the earth, but at the same time he attributed to the motion of the heavens pretty well all the correct properties of inertial motion, as is clear from the following passage:[42]

the circular motion in question, being complete, embraces the incomplete and finite motions. Itself without beginning or end, continuing without ceasing for infinite time, it causes the beginning of some motions, and receives the cessation of others . . . it suffers from none of the ills of a mortal body, and moreover . . . its motion involves no effort, for the reason that it needs no external force of compulsion, constraining it and preventing it from following a different motion which is natural to it. Any motion of that sort would involve effort, all the more in proportion as it is long-lasting.

It is by no means fortuitous that Aristotle does ascribe the essential properties of inertial motion to the heavens, since the actually observed diurnal motion of the heavens (which inspired him to the above passage) is a consequence of the rotational, i.e., basically inertial, motion of the earth, while the proper motion of the planets, sun, and moon are to a large degree the result of the inertial component of their motion. Indeed, although not normally interpreted that way, the Ptolemaic planetary system is an early expression of the law of inertia – incomplete but accurate in essentials. There are three aspects of unadulterated inertial motion that still evoke wonder, in ascending order: the rectilinearity, the uniformity, and the eternity. What made Galileo's introduction of inertia so breathtaking – and simultaneously amenable to mathematization – was precisely the eternal uniformity. The revolutions of the heavens did appear to persist for ever, but on the earth, in Aristotle's words (echoed by the poets of all ages) 'time itself is destructive . . . things perish without anything being stirred, and it is a kind of perishing without apparent provocation that we especially attribute to time.'[43]

Lost from view on the surface of the earth because of the host of disturbing factors and the absence of a decent clock, the most fundamental law of nature (admittedly not yet formulated in the Newtonian manner) finally forced its way into Galileo's consciousness (or, at least, into the way he presented the matter in his famous *Dialogo*) after Copernicus had boldly proclaimed the earth to be a planet. Locked in deadly combat with the Catholic Church and more or less forced to it by Copernicus's proposal of the earth's mobility, Galileo pulled down eternity from the heavens in a desperate attempt to prove that the earth too could move. He realized the implication of the earth's mobility – that through terrestrial perishing and decay there must after all run a thread which persists for ever. He made us partakers of eternity. (It's ironic that Hamlet called man the *quintessence of dust* just when Galileo was taking our 'quintessential nature' seriously.)

As we shall see, this was one of the decisive steps in the eventual recognition of the law of inertia and opened the door to mathematization of motion (eternal uniformity was a concept that could be formulated precisely before the development of the calculus). It also had the consequence that the law of inertia was formulated in geometrokinetic – indeed almost transcendental – terms. As already intimated, Mach's Principle is the attempt to recast the law in physical terms – to shake off the residual geometrokineticism which it inherited from Aristotelian metaphysics and ancient astronomy.

One is left to wonder if things had to take such a curious roundabout route. In the paragraph immediately following the one from which the quotation of Ref. 42 is taken, Aristotle dismissed the Atlas myth because he said it arose from the mistaken belief that the material of the heavens has weight and must therefore somehow be prevented from falling (according to his fundamental motion-based physics and chemistry, only the substances that fall have weight). He then went on: 'We must not then think in this way, nor in the second place must we say with Empedocles that it [the sky] has been kept up all this time by the cosmic whirl, i.e., by having imparted to it a motion swifter than that to which its own weight inclines it.'

Poor Empedocles. How near the truth he was. Two thousand years later, Hooke, newly appointed Secretary to the Royal Society, would write to Newton, posing the problem of the planets in strikingly similar terms.[44]

This completes what needs to be said about the five Aristotelian elements and their natural motions. These, on earth at least, were augmented by the concept of violent motions, enforced by some agent (either an animated being or some body which itself is kept in motion in some manner). It is again interesting and ironic that the quantitative description of such motion (the development of which eventually led to Newton's Second Law) constituted the terrestrial physics that Kepler learnt at his *Alma Mater*, Tübingen, and was imported by him into the heavens to account for planetary motions just when Galileo was importing celestial inertia into terrestrial physics in order to understand earthly motions.

Ironic yes, but there is a logic to it too. Copernicus's mobile earth bridged the Aristotelian gulf between the terrestrial and the celestial. Rather over half a century later, two intrepid robbers, each with a bag of booty on his back, could be seen venturing through the void across that flimsy arch. Kepler from the earth to the heavens, Galileo from the heavens to the earth. They exchanged a cordial word or two as they passed, but each went resolutely on his own way.

2.6. The concept of place and the self-contained universe

The key to understanding Aristotle's cosmology is his concept of place (*topos*). Aristotle presents the problems associated with the concepts of place and space very clearly in Book IV of his *Physics*, from which it is worth quoting at some length, particularly in view of the fact that the problem of the nature of space figures prominently in the discussion about absolute and relative motion. Aristotle opens the discussion as follows:[45]

> The Natural Philosopher has to ask the same questions about 'place' as about the 'unlimited'; namely, whether such a thing exists at all, and (if so) after what fashion it exists, and how we are to define it. . . .
>
> But we encounter many difficulties when we attempt to say what exactly the 'place' of a thing is. For according to the data from which we start we seem to reach different and inconsistent conclusions. Nor have my precursors laid anything down, or even formulated any problems, on this subject.

He sees the strongest argument for the reality of space in the phenomenon of 'replacement':[46]

To begin with, then, the phenomenon of 'replacement' seems at once to prove the independent existence of the 'place' from which – as if from a vessel – water, for instance, has gone out, and into which air has come, and which some other body yet may occupy in its turn; for the place itself is thus revealed as something different from each and all of its changing contents. For 'that wherein' air *is*, is identical with 'that wherein' water *was*; so that the 'place' or 'room' into which each substance came, or out of which it went, must all the time have been distinct from both of the substances alike.

Although as a plenist he does not himself believe in it, Aristotle also mentions the argument from the void:[47]

Further, the thinkers who assert the existence of the 'void' agree with all others in recognizing the reality of 'place', for the 'void' is supposed to be 'place without anything in it.'

One might well conclude from all this that there must be such a thing as 'place' independent of all bodies, and that all bodies cognizable by the senses occupy their several distinct places. And this would justify Hesiod in giving primacy to *Chaos* [=the 'Gape'] where he says: 'First of all things was Chaos, and next broad-bosomed Earth'; since before there could be anything else 'room' must be provided for it to occupy. For he accepted the general opinion that everything must be somewhere and must have a place.

But Aristotle is not happy with the idea that void has reality:[48]

And if such a thing should really exist well might we contemplate it with wonder – capable as it must be of existing without anything else, whereas nothing else could exist without it, since 'place' is not destroyed when its contents vanish.

But then, if we grant that such a thing exists, the question as to *how* it exists and what it really is must give us pause. Is it some kind of corporeal bulk? Or has it some other mode of existence?

His initial discussion of this question, which is not particularly illuminating in the context of this book, leads him to the conclusion that the 'mode of existence' of space is extremely difficult to pin down, and he concludes the chapter by remarking:[49] 'after all, we are forced by these perplexities not only to ask what a "place" is, but also to reopen the question that appeared to be closed and ask whether there is such a thing as "place" at all'.

In coming to grips with this problem, Aristotle exhibits a strongly positivistic frame of mind.* He insists instinctively, though without, I think, stating it as an explicit principle, that there must be something

* The doctrine of *positivism*, first associated with Auguste Comte (1798–1857), is the philosophical basis of Mach's approach to physics and asserts that the highest form of knowledge is obtained from direct description of sensory phenomena in mathematical terms.[50]

observable that defines place. This leads him to his fundamental concept: place, whatever may be its actual mode of existence, is defined by a material container. His favourite example is a vessel that may contain successively water, wine, or air. Very characteristic is his assertion that the 'immediate place' of a thing is the 'inner surface of the envelope [i.e., container]'.[51] This is the rock on which everything in his scheme stands or falls: unless there is some defining container, all talk of place or space is meaningless. This is why he needs a plenum.

However, this still leaves open a number of possibilities for what place actually is. We know that a body occupies a place when within a container. Then in answer to the question 'what is place?' Aristotle asserts:[52] 'it must be either (i) the form or (ii) the matter of the body itself, or (iii) some kind of dimensional extension lying between the points of the containing surface, or (iv) – if there be no such "intervenient", apart from the bulk of the included body – the containing surface itself.'

He (at least) can readily eliminate the first two possibilities. It is clearly not the matter of the body itself (ii), since that can be replaced (as we have seen). Nor should it be regarded as the form of the body. For though 'that which embraces' may suggest the moulding 'form' (since 'the limiting surfaces of the embracing and the embraced coincide') and 'both the "place" and the "form" are limits', they are not limits 'of the same thing, for the form determines the thing itself, but the place the body-continent'.

He finds possibility (iii), which he says Plato favoured by advocating the identity of 'room' and 'matter', harder to eliminate and admits:[53]

we see that what makes 'place' appear so mysterious and hard to grasp is its illusive suggestion now of matter and now of form, and the fact that while the continent is at rest the transferable content may change, for this suggests that there may be a dimensional something that stays there other than the entering and vacating *quanta* – air too contributing to this last illusion since it looks as if it were incorporeal – so that the 'place', instead of being recognized as constituted solely by the adjacent surface of the vessel, is held to be the dimensional interval within the surface, conceived as 'vacancy'.

and he also says:[54] 'it is no wonder that, when thus regarded – either as matter or as form, I mean – "place" should seem hard to grasp, especially as matter and form themselves stand at the very apex of speculative thought, and cannot well, either of them, be cognized as existing apart from the other.'

In the face of these baffling questions, he falls back on the doctrine of a plenum and completely eliminates space as a conceptual entity:[55]

But because the encircled content may be taken out and changed again and again, while the encircling continent remains unchanged – as when water passes out of a vessel – the imagination pictures a kind of dimensional entity left there, distinct from the body that has shifted away.

But this is not so; for what really happens is that (instead of anything being *left*) some other body – it matters not what, so long as it is mobile and tangible – *succeeds* the vacating body without break and continuously.

There is no point in getting involved in the arguments for or against the idea that the world is a plenum. It was the subject of great controversy both in the ancient world – in *De Rerum Natura* Lucretius referred to the supporters of the plenum as *stolidi* – and again in the seventeenth century. However, this 'great debate' in the end proved to be little more than an exercise in shadow-boxing, the reason for which will be indicated shortly. For the purposes of the present discussion, let it simply be accepted that for Aristotle the plenum was more or less an article of faith.

Having thus disposed of all the other alternatives, Aristotle concludes that place is not only defined by but actually *is* the inner surface of the container: the envelope. The inner wall of the bottle is the place of the wine, the water that touches its outside is the place of the boat. However, this is only the *immediate place* of the body. In another clear anticipation of Mach, Aristotle puts the immediate place in a larger context: he is careful not to fall victim to a detachment fallacy:[56]

And so, too, a 'place' may be assigned to an object either primarily because it is its special and exclusive place, or mediately because it is 'common' to it and other things, or is the universal place that includes the proper places of *all* things.

I mean, for instance, that you, at this moment, are in the universe because you are in the air, which air is in the universe; and in the air because on the earth; and in like manner on the earth because on the special place which 'contains and circumscribes you, and no other body'.

The hierarchical nature of place comes out explicitly in the following passage, which more or less completes Aristotle's account of how we are to understand place:[57]

And, from this point of view, if one thing is moving about inside another, which other is also in motion, as when a boat moves through the flowing water of a river, the water is related to the boat as a vessel-continent rather than as a place-continent; and if we look for stability in 'place', then the river as a permanent and stable whole, rather than the flowing water in it at the moment, will be the boat's site. Thus whatever fixed environing surface we take our reckoning from will be the place.

So the centre of the universe and the inner surface of the revolving heavens constitute the supreme 'below' and the supreme 'above'; the former being absolutely stable, and the latter constant in its position as a whole.

He has found and defined the universal and ultimate frame of reference.

This might seem facile; for it appears that the outermost heaven, the *ouranos*, should be regarded as 'placed' by the void beyond it. In fact, the problem presented by the placing of the outermost sphere played a significant role in the late medieval and Renaissance periods in undermining the Aristotelian cosmos on the basis of purely philosophical considerations. Already in Plato's time Archytas, a Pythagorean, had posed questions that we find recurring time and again from the thirteenth to the beginning of the seventeenth century (and, indeed, still are put by the layman to any relativist who attempts to explain the concept of a closed finite world):[58]

If I am at the extremity of the heaven of the fixed stars, can I stretch outwards my hand or staff? It is absurd to suppose that I could not; and if I can, what is outside must be either body or space. We may then in the same way get to the outside of that again, and so on; and if there is always a new place to which the staff may be held out, this clearly involves extension without limit.

Equally famous is Lucretius's taunt:[58]

If for the moment all existing space be held to be bounded, supposing a man runs forward to its outside borders and stands on the utmost verge and then throws a winged javelin, do you choose that when hurled with vigorous force it shall fly to a distance, or do you decide that something can get in its way and stop it? for you must admit and adopt one of the two suppositions; either of which shuts you out from all escape and compels you to grant that the universe stretches without end.

Such criticisms of Aristotle's remarkable idea that there is literally nothing beyond the outermost shell of his cosmos strengthened and anticipated the even more destructive evidence that was to be brought forward by the astronomers and were an important factor in preparing the human mind for the transition from a closed world to an infinite universe.

With the benefit of hindsight, I do not suppose that Aristotle's answer to the problem of the placing of the outermost sphere will be found by the reader to be completely convincing – for this, Aristotle would have had to anticipate non-Euclidean geometry. However, an answer of sorts can be found in the way Aristotle couples the problem of space so strongly to that of motion. In fact what partially saves Aristotle's universe from the objection that his *ouranos* must itself be ultimately contained in the limitless void is that his scheme is *dynamically self-contained*. In fact, quintessence comes to the rescue. It is not so much that Aristotle abolishes the problem of the 'place' of the 'supreme above', as that within his scheme the problem never becomes acute since all substance at the 'supreme above' moves circularly. The threat to the logical coherence of his scheme from Archytas's question is simply ignored, but with a semblance of logic and coherence since, in accordance with Aristotle's own laws of motion, the outermost sphere spins within itself and the laws of motion for the other elements say explicitly that they have no cause to pass beyond the *ouranos*; for all the places to which they strive are contained within it.

Such thoughts are implicit in the following passages:[59]

It follows that if a body is encompassed by another body, external to it, it is 'in a place'; but if not, not. As to such an 'unplaced' body (be it water or anything else) its parts may be in motion, for they embrace each other, but as a whole it can be said to 'move' only in a special sense. For as a whole it cannot change its collective place. But it may have a motion of rotation, which motion is what constitutes the 'kind of place' with respect to which the rotating *parts* move.

Now in the universe there are some parts which do not move up and down but do rotate; and others (such as are susceptible to condensation and rarefication) which can either rotate or move up and down . . . but the heavenly mass as has been said cannot change its place as a whole. Nor indeed has it a place to change, seeing that there is no body-continent embracing it. But after the fashion of its own motion it constitutes places for its own parts, since one part embraces another. . . .

Heaven therefore 'rotates', but the universe has not a 'where', for to have a 'where' a thing must not only exist itself but must be embraced by something other than itself; and there *is* nothing other than the universe-and-the-sum-of-things, outside that sum, and therefore nothing to embrace *it*. . . .

So earth is naturally surrounded and embraced by water, water by air, air by aether, aether by heaven, and heaven itself not at all.

It must be admitted that Aristotle's solution to the problem of space and its profoundly mysterious, elusive nature has many attractions. By defining place as the envelope of the contained body and insisting the world is a plenum, he was effectively able to say that space does not exist at all and was able to reduce everything to matter. Not only is matter everywhere, so that the embarrassment of the void is eliminated, but matter also, in being everywhere, is always available to define not only the immediate place but also a universal frame of reference. Moreover, from the epistemological point of view, the very diversity of matter and its differentiation was welcome in lending a certain precision to the concept of space – it is seldom difficult to say where one thing ends and another begins.

Where is the flaw in this all but self-contained whole? Undoubtedly in Aristotle's neglect of the metrical, as opposed to topological, properties of space. His concept of place is at once fundamental but primitive. Fundamental because it derives directly from the most basic of spatial concepts – contiguity, coincidence, and inclusion. Primitive because it stops there and takes almost no account of the metrical properties of the world, which although secondary nevertheless have a very real practical existence and lead to the instinctive posing of questions like the one that Archytas put. The defect with Aristotle's concept of place is that it ignores a second – and in many ways much more convenient and effective – way of determining position. This is the method in which position is determined by using the metrical properties of Euclidean space, e.g. by defining the position of a body by means of its distances from certain selected points of reference. In this method, the distances can be measured either directly, by rulers, or indirectly, by means of trigonometry.

In this connection it is interesting to note that Aristotle, despite his high-brow philosophy, also had a concept of a metric space in the manner of Euclid buried deep in his psyche. For metrical geometry does enter Aristotle's scheme in numerous places, above all in the use he makes of straight lines and circles to define natural motions. He also frequently talks about distance without making the least attempt to show how such a precise quantitative concept derives from his more primitive topological concept of place. But perhaps the most graphic evidence of his instinctive but unconscious belief in space is to be found in his firm belief that the earth is at rest while the heavens spin.[60] For all that one can observe is relative: there is absolutely no objective criterion provided by anything in Aristotle's concept of place that would enable one to say that it is the heavens that spin rather than the earth. Indeed, in Aristotelian rigour the notion that either is at rest (nonrotating) is meaningless.

Aristotle is thus a classic example of the point made by Maier and quoted in Chap. 1. Like other philosophers, he was capable of developing a sophisticated concept of place that dispenses with space, but this broke down as soon as it became necessary to treat problems involving motion. Here instinctive man, with both feet firmly on the solid ground, immediately comes to the fore and we are confronted with what seems to us to be an almost schizophrenic failure to link together the two aspects of geometry: the Aristotelian topological (and positivistic) concept of place and the Euclidean concept of space with its metrical notions adopted as true *a priori* – because self-evidently true!

Kuchař has pointed out to me that the coexistence for around two millennia of these two concepts, represented respectively by the finite Aristotelian cosmos and infinite Euclidean space, which were both developed with great precision and detail long before the end of Greek antiquity and then lived on cheek by jowl until almost the middle of the seventeenth century, when the Euclidean concept finally triumphed, is a most remarkable and almost inexplicable phenomenon. How were people able to live for so long with such contradictory notions?

I think the answer must be that it takes the human mind an extraordinary effort to put together the disparate parts of experience into a coherent and unified whole. Who knows, millennia from now our descendants may smile at the way most modern scientists live quite happily with a scientific picture of the world devoid of all qualitative sensations, with which, however, they are confronted in the most intense manner in every instant of their conscious existence. I refer again to Schrödinger's *Mind and Matter*.[9]

One thing at least is suggested by the history of the period from Aristotle to Descartes – that clarity of philosophical argument and mathematical intuition alone are not sufficient to bring about a fundamental change of concepts. Opposition to Aristotle was frequently voiced, often with striking clarity, as for example by John Philoponus (in the sixth century AD):[61] 'Place is not the adjacent part of the surrounding body. . . . It is a given interval, measurable in three dimensions; it is distinct from the bodies in it, and is, by its very nature, incorporeal. In other words, it is the dimensions alone, devoid of any body. Indeed, insofar as their matter is concerned, place and the void are essentially the same thing.' This concept of space is already extremely close to that of Newton. Moreover, it reappeared at numerous times in the Middle Ages but never with enough strength to overthrow Aristotle. (For a discussion of concepts of space in the Middle Ages the reader is referred to Grant's *Much Ado About Nothing*.[62])

What was lacking, I suggest, was really hard and ineluctable evidence, empirically based and expressed in mathematical form, suggesting that there was something *fundamentally wrong* with the Aristotelian scheme. This evidence was supplied by the astronomers.

It is for this reason that a particularly important part of this book is concerned with tracing how the alternative metrical, above all trigonometrical, concept of position determination steadily undermined the Aristotelian universe. Here again the difference between the heavens and the earth is most interesting and significant. The point is that whereas on the earth *both* methods, metrical and topological, can be used – and in practice are used – to determine position, in the heavens the topological method is, with a very few exceptions, entirely useless. It can be used to determine the apparent position of a planet on the heavens but fails completely to say how far away the planet is. It is interesting to note in this connection that in *De Caelo* Aristotle gives only the most meagre information about the distances of the planets and the size of his universe. In fact, he is content to give an estimate of the size of the spherical earth and to note the fact – established by a topological observation, an occultation of Mars by the moon – that the planets are further away than the moon. He does not even attempt to estimate the distance to the moon.

We shall see shortly how quantitative study of the heavens, first by the ancients and then by Copernicus and Kepler, simply bypassed Aristotle's philosophy and step by step reinstated geometry as an essentially quantitative science of the real world. This will also give us a key to understanding why one of the greatest of all philosophical disputes, that between the adherents of the void and plenum, generated a great deal of smoke without materially affecting the actual development of dynamics once it got into its stride. Because *de facto* planets have no visible 'envelopes', Aristotle's concept of position fails completely for them. Astronomers were forced to use trigonometry and geometrical models to determine planetary positions. But this concentrated attention directly on trigonometric position and change of position, both being treated quantitatively of necessity, since the raw data were gathered in quantitative form and could be evaluated in no other way. Thus, far more effort was expended on establishing the precise motions of the planets than on speculating about their origin and whether the planets are moved by the agency of a medium, by an active soul, or by an occult faculty – gravitation. This is true above all of Kepler, despite the fact that for him physical speculation was a consuming passion and he dabbled in all three explanations.

Thus it was that Aristotle's physics lived on, a fleshy beast with no clear contours, while astronomy, an austere science far removed from the fleshpots of the world, was forced willy-nilly to come to terms with the bony quantitative reality of celestial phenomena. This may also explain why Kepler, the astronomer by choice, made a more radical break with Aristotle than Galileo, the astronomer by accident.

We cannot conclude this section without commenting on the remarkable similarity of the closed universes constructed by Aristotle and, more than two thousand years later, by Einstein. Both were spatially spherical and infinite in both temporal directions. (Einstein's *metrically closed* universe was, of course, intellectually far more satisfying than the topologically closed universe of Aristotle, based, as we have seen, on inadequate concepts.) Even more striking is the fact that both were constructed for essentially Machian reasons – a desire to find a closed circle of observable physical causes.

This point is of the greatest importance for the understanding of Aristotle and the realization of what ultimately defeated Einstein. In common with many Greek thinkers, Aristotle abhorred the idea of the 'actual infinite.' As we have seen, he was always on the look out for 'intelligible concepts', ones that the mind could completely grasp. He literally recoiled before the idea of an infinite universe, and much of his *De Caelo* is devoted to proving that the universe cannot be infinite. The only form in which Aristotle could contemplate the infinite was as a *potentiality*, for example, as the potential possibility of infinite division of a finite piece of matter into ever more and more pieces. Looking forward to the end of Vol. 2, it is worth noting that conscious rejection of the infinite was probably one of the major factors that led Aristotle to his closed finite cosmology in the first place, whereas for Einstein realization of the danger posed by the infinite came very late, in fact only after he had developed his dynamical theory in its entirety. As the story of the absolute/relative debate unfolds, we shall see that the intervening discovery and systematic elaboration of dynamics goes a long way towards explaining the different perspectives of these two great men who dominate the history of dynamics at its respective ends.

Another significant similarity between the cosmologies of Aristotle and Einstein is that both were blown apart by that unbiddable beast appropriately called dynamics. The Aristotelian vision of the quintessential divine heavens was indeed destroyed overnight, but the Aristotelian universe lived on for another half century or so, a rusting old hull whose quintessence had lost all its sheen and inner vitality. But then came Newton and his laws of motion that burst asunder the *cordon sanitaire* with which Aristotle had fended off the nagging void. It was Newton's laws (already anticipated in some of their essential features by Descartes) that let in infinite space, the logical extension of the trigonometric space of the solar system in which Kepler had uncovered ellipses, significant steps from the finite Aristotelian circles to infinite conic sections stretching out forever into the void.

If Aristotle's spherical universe survived two millennia, Einstein's had barely two months of existence before de Sitter gave it the *coup de grâce*. Hamlet would have had as much pleasure from the end of Einstein's universe as he had from the abrupt demise of Guildenstern and Rosencrantz. Truly a case of the 'engineer hoist by his own petard'. Aristotle's universe was blown apart by Newton's laws; Einstein's by Einstein's. We shall come to all this anon.

2.7. Time in Aristotelian physics

In his discussion of time, Aristotle again exhibits characteristic reflexes that can be called 'Machian' but, as in the case of space, the explicit positivistic philosophy exists alongside abstract mathematical intuition. This section is to be seen as an introduction to some very remarkable empirical and quantitative results about the nature of time discovered by the Hellenistic astronomers. These will be discussed in the next chapter.

Innumerable phenomena give rise to a sense of awareness of the passage of time – the alternation of day and night, the seasons, the lunar cycle, birth, growth, and death. Ariotti[63] makes the interesting point that, apart from the Greeks, the people living around the Mediterranean in antiquity 'did not separate time from its contents . . . Time was not a neutral and abstract frame of reference. Time was its own contents. Events were not *in* time, they *were* time. Ancient Hebrew did not have a word for time, but for season, point in time, or eventful duration.' There were as a result multiple times. Ariotti sees a significant conceptual development in the fact that the Greek theogonists and poets 'hypostatized time into a single entity, cosmological principle or god: Chronos.' In the great flowering of Greek literature, time plays a dominant role. For Sophocles time was all-mastering. For the emergence of the science of dynamics, the transition from all the particular measures of time to the concept of a universal and uniformly flowing time was crucial.

In this process, Plato and Aristotle mark two important stages. Plato said that,[64] 'with a view to the generation of Time,' God brought the celestial bodies into existence 'for the determining and preserving of the numbers of Time'. He seems to have come quite close to identifying the passage of time with the motions of the heavenly bodies. On the other hand, Plato regarded the world as a mere play of appearances and in his hands time does not acquire a genuine concrete reality. There is also the obvious problem that the various celestial bodies move at different apparent speeds at different times. Which among them would define a unique time, if, indeed, such a thing exists? It was left to Aristotle, who, as we have seen, took a much more commonsense attitude to the world and regarded it as the only reality, to crystallize, very nearly, the concept that would endure until the end of the Middle Ages. We recall that for Aristotle motion was a primary phenomenon. However, it was not the case that Aristotle conceived motion as taking place *in* time. Rather, for him, all changes, including the special case of motion, are what is primary. Aristotle, like the medieval philosophers, and unlike scientists of the post-Newtonian age, was still much closer to the primitive concept in which time is not distinguished from its contents.

As is his usual practice, Aristotle attempts to establish what is the nature of time, a thing evidently as elusive as space. Of one thing he is certain:[65] 'Time cannot be disconnected from change; for when we experience no changes of consciousness . . . no time seems to have passed' and again: 'Since, then, we are not aware of time when we do not distinguish any change . . . it is clear that time cannot be disconnected from motion and change.' Aristotle, as always, strives to be concrete, as is evident in his remark[66] 'when any particular thing changes or moves, the movement or change is in the moving or changing thing itself or occurs only where that thing is.'

Numerous passages in which Aristotle discusses the measurement of time and motion, which we shall shortly consider, indicate that he would have preferred to think merely in terms of concrete change and a succession of states, or 'nows', dispensing entirely with an abstract notion of time. However, he is prevented from taking such a course by two things which are in reality but one – the overwhelming sense of a passage of time that goes on independently of any particular process of change one may happen to be observing and the fact that innumerable processes of change are going on simultaneously. For example, he immediately qualifies the comment just made by saying[66] 'Whereas "the passage of time" is current everywhere alike and is in relation with everything. And further, all changes may be faster or slower, but not so time; for fast and slow are defined by time, "faster" being more change in less time, and "slower" less in more.' The philosopher of the concrete here reverts to the abstract with the same instinct that told him the heavens truly rotate.

His conclusion is:[67] 'Plainly, then, time is neither identical with movement nor capable of being separated from it.'

However, his tendency to the concrete reasserts itself when he considers the question of measurement, which in his mind is closely related to the true nature of time. Suppose, as a clarification of Aristotle's ideas, we imagine ourselves taking a series of snapshots of an object moving against a fixed and unchanged background, on which much detail can be seen. We obtain, say, a hundred such pictures. Because of the detail supplied by the background, we could, even if the snapshots were handed to us in a completely mixed-up order, readily put them in the correct order. There is a clear ordering with respect to before and after. Aristotle comments that[68] 'the primary significance of before-and-afterness is the local one of "in front of" and "behind" . . . But there is also a before-and-after in time, in virtue of the dependence of time upon motion.' He seeks to make this before and after ordering, derived from spatial relationships through motion, correspond in the closest possible way to time itself: 'Motion, then, is the objective seat of before-and-afterness both in movement and in time.' He can now almost lay his hands on time:[69]

> Now, when we determine a movement by defining its first and last limit, we also recognize a lapse of time; for it is when we are aware of the measuring of motion by a prior and posterior limit that we may say time has passed. And our determination consists in distinguishing between the initial limit and the final one, and seeing that what lies between them is distinct from both; for when we distinguish between the extremes and what is between them, and the mind pronounces the 'nows' to be two – an initial and a final one – it is then that we say that a certain time has passed; for that which is determined either way by a 'now' seems to be what we mean by time. And let this be accepted and laid down.

Although this seems definite enough, it is still not quite as precise as one would wish. There is unfortunately a gap between the principle and concrete practicality. Let us first look at a few more explications of the general principle. Aristotle says:[70] 'When we perceive a distinct before and after, then we speak of time; for this is just what time is, *the calculable measure or dimension of motion with respect to before-and-afterness*' (my italics). The ability to count different stages in a process is here important:[71] 'It is in virtue of the countableness of its before-and-afters that the "now" exists,' so that:[72] 'time, then, is the dimension of movement in its before-and-afterness, and is continuous (because movement is so).' Even more explicitly:[73] 'For "before" and "after" are objectively involved in motion, and these, *qua* capable of numeration, constitute time.' And finally:[73] 'Time is the numeration of continuous movement.'

Taken literally, this would seem to suggest that time is to be identified with the counting of the snapshots we considered above, or perhaps with the distance that the object has travelled ('the dimension of movement'). There is a danger here that we then finish up with a separate time for each motion; however, this is a possibility that Aristotle explicitly rejects in a passage that simultaneously makes his overall concept of time much clearer:[74]

> But if we take one kind of change and say 'now' with respect to it, other kinds of change, each of which has a specifically different unit to be counted in, will be at a certain stage of their change at this same 'now'. Can each of them have a different time, and must there be more than one time running concurrently? No; for it is the same lapse of time that is counted by two 'nows', everywhere at once, whatever the units of movement or change; whereas the one-and-sameness of the units is determined by their kind and not by their 'at-once-ness'; just as if there were dogs and horses, seven of each, the number would be the same, but the units numbered different. So, too, of all movement-changes determined simultaneously the time is the same; one may be quick and another slow, and one a change of place and the other of quality; the time, however, is the same, if the counting has reached the same number and been made simultaneously, whether of the qualitive modification or of the change of place. So the movements or changes are different and stand apart, but the time is the same everywhere, because the numeration, if made simultaneously and up to the same figure, is one and the same.

Because Aristotle's universe is finite, the ideas expressed here can be illustrated particularly well by the snapshot device. Let us suppose snapshots taken of successive 'nows' of the *entire universe*. Within the universe, innumerable processes of both general change as well as motion are taking place. We could concentrate our attention on any one of these, establish the succession of 'befores' and 'afters' and count them. According to Aristotle, this numeration *is* time. We can do the same with any other particular process of change. But because all the processes take place simultaneously, there is a one-to-one correspondence between the 'nows' in the two parallel trains of successive 'nows'.

Although he does not say so explicitly, such a notion of a universal succession of all-embracing simultaneities in which all the different processes of the universe 'run together' is clearly what underlies the above passage.

At this point we must distinguish between *mere succession* and the notion of *uniform motion*. It is here that commonsense prejudice gets the better of the philosopher of the concrete. As we saw in the previous chapter, by Aristotle's time the Greek mathematicians had clearly developed the concept of a uniform motion and used it as the basis of their constructions in kinematic geometry. It is quite obvious from the passages which follow that Aristotle shared their belief in the existence of uniform motion. But if time is revealed by nothing else but motion and change, where do we find a criterion that tells us a motion is uniform? All that is agreed so far is that change and motion provide the evidence for the lapse of time. That is what Aristotle calls time. But uniform motion can only mean that the motion results in equal displacements in equal times. But the times are derived from the motion, so we get ourselves into a vicious circle.

In the case of a single motion, as in the first snapshot example we considered, it is manifest nonsense to adhere to the Aristotelian concept of time as constituted solely by the changes that occur and attempt in addition to say that such a motion is uniform. There is just a mere succession, nothing more and nothing less.

Aristotle in fact gets very close to solving the problem of saying what one means by uniform motion within the framework of a philosophy that admits nothing but the concrete when he considers what it means to say that one motion is quicker than another:[75] 'What I mean by one change being quicker than another is that, of two homogeneous change-movements (either both on a periphery, for instance, or both on a straight line, if it be a local movement, and *mutatis mutandis* in other kinds of change), that one is the quicker which reaches a certain determined stage or point in its course "before" the other reaches the point at the same distance from the starting-point in its course.'

From this it is just one short step to an *operational definition* of uniformity of motion, namely, one motion may be said to be uniform *relative to another motion* if the distance traversed in one motion always stands in a given fixed ratio to the distance traversed in the other.

But Aristotle does not take this final step, which would have made time both *concrete* (entirely reduced to motion and change and not conceived as an abstract or transcendent, i.e., not realizable in experience, entity) and *relative*. [It would then, of course, have been an empirical matter to establish to what extent such *mutually uniform* motions are actually realized in the world.] He introduces the concept of *uniform motion* without defining it in concrete terms:[76]

And not only do we measure the length of uniform movement by time, but also the length of time by uniform movement, since they mutually determine each other; for the time taken determines the length moved over (the time units corresponding to the space units), and the length moved over determines the time taken. And when we call time 'much' or 'little' we are estimating it in units of uniform motion, as we measure the 'number' of anything we count by the units we count it in – the number of horses, for example, by taking one horse as our unit. For when we are told the number of horses, we know how many there are in the troop; and by counting how many there are, horse by horse, we know their number. And so too with time and uniform motion . . .

But the problem with this is that a horse may be recognized without reference to anything else, whereas, as we have seen, a *uniform* motion is incapable of definition in isolation.

Aristotle's discussion reveals no evidence that he was aware of the vicious circle or that he could have avoided it by a relative definition. This, then, is another classic confirmation of Maier's thesis. Abstract instinct takes over. The remainder of his discussion is concerned solely with practicalities. Given that uniform motions do exist, by which may time be most conveniently measured? Circular motion is chosen as an obvious candidate:[77]

And now, keeping locomotion and especially rotation in mind, note that everything is counted by some unit of like nature to itself – monads monad by monad, for instance, and horses horse by horse – and so likewise time by some finite unit of time. But as we have said, motion and time mutually determine each other quantitively; and that because the standard of time established by the motion we select is the quantitive measure both of that motion and of time. If, then, the standard once fixed measures all dimensionality of its own order, a uniform rotation will be the best standard, since it is easiest to count.

This is then followed by a somewhat curious passage:[78] 'Neither qualitive modification nor growth nor genesis has the kind of uniformity that rotation has; and so time is regarded as the rotation of the sphere, inasmuch as all other orders of motion are measured by it, and time itself is standardized by reference to it.'

The translator provides no explanation of what Aristotle means here by 'the sphere'. It does not seem that Aristotle means the sphere of the fixed stars, though this would fit in very well with a passage in *De Caelo* in which he says[79] 'the revolution of the heaven is the measure of all motions, because it alone is continuous and unvarying and eternal'. In fact, intepretation of 'the sphere' as the generic sphere would not be inconsistent with the idea, found at various points in Aristotle's work (cf. the footnote on p. 81), that uniform rotation is natural to all spheres. This is certainly true of the celestial spheres, for a primary property of the Eudoxan system of concentric spheres, which Aristotle adopted and modified, is that all the spheres rotate uniformly, though not about the same poles or at the same rate. As we shall see in the next chapter, belief in the perfect uniformity and circularity of celestial motions appears to have been an article of faith from the period of Plato and Aristotle through to Ptolemy about five hundred years later.

It seems reasonable to conclude therefore that both practically and as a metaphysical principle Aristotle believed the rotation of the heavens defined and measured the uniform flow of time. In adopting this position he was not true to his philosophical instincts. It seems that the intuition of Plato and the geometers held too strong a grip on his mind. Even so, it is worth emphasizing that in attempting to define time concretely Aristotle, who was followed by the medievals,[80] developed a concept of time that, in many ways, is more modern than the concept of Newtonian absolute time, as we shall see in Chaps. 3 and 12.

3

Hellenistic Astronomy: the foundations are laid

3.1 Historical: the Hellenistic period

In the second half of the period dominated by Greek culture the most important centre of learning was Alexandria, which was founded by Alexander the Great in 332 BC near the Nile estuary. Through the encouragement of the rulers, the Ptolemies, the city acquired superb library facilities, and most scientists of any note went there at some time. The two books most influential for the subsequent development of science in Western Europe were written in Alexandria: The *Elements* of Euclid (fl. around 300 BC) and the *Almagest* by Ptolemy (*circa* AD 100–*circa* AD 170) (the name is common and does not imply a royal connection).

Alexandria was, however, not the only centre of learning. Archimedes (born 287 BC) studied there before returning to his native city of Syracuse in Sicily, where he is alleged to have been killed after its capture in 212 BC by a Roman soldier while intent on drawing a mathematical figure in the sand. For the general development of science, Archimedes' writings were almost as important as those of Euclid and Ptolemy. Archimedes is not only regarded as the greatest mathematician of antiquity; he also created the science of statics on a rigorous mathematical basis. This example was undoubtedly of the greatest value to Galileo, who was inspired to set up a similar mathematical science of terrestrial motion. However, in contrast to many historical studies that treat statics and dynamics within the general framework of mechanics and see statics as an important source of inspiration for the concepts of dynamics, this approach does not seem particularly illuminating in the framework of the absolute/relative debate nor in a specialized study of dynamics. The law of the lever and related mechanical machines (pulleys, screws, etc.) undoubtedly helped to suggest the fundamental importance of momentum, but it is questionable whether this was a real help, since it led to the identification of force with momentum rather than change of momentum, which was the decisive step. On the other hand, the concept of the *centre of gravity* of a system of bodies, which was clarified by Archimedes and used in brilliant applications, did play a significant part in the work of Huygens and Newton and could almost be reckoned among the baker's dozen of key insights (mentioned in the Introduction) that were essential for the final synthesis of dynamics. It should also be said that ideas from statics played an important role in the post-Newtonian development of analytical mechanics by continental mathematicians.

Of much more direct relevance for the present study was the work of the Hellenistic astronomers. Unfortunately, very little indeed is known about the early history of Greek astronomy as an exact and quantitative science as opposed to philosophical speculation of the kind found in Aristotle. This is not the impression that one might gain from many books written on the subject,[1] in which several brilliant ideas are attributed to early figures such as Heraclides (born *circa* 388 BC, died after 339 BC) and especially Aristarchus (*circa* 310–230 BC), who is often credited with having developed a heliocentric system almost as detailed as Copernicus's. As we shall see later, this is an unjustified speculation and tends to give credit to the early Greek astronomy at the expense of the later work. For a modern evaluation of the achievements of Greek astronomy the reader is referred to Neugebauer's monumental *A History of Ancient Mathematical Astronomy*[2] and Toomer's articles in the *Dictionary of Scientific Biography*[3] on Heraclides, Apollonius (*circa* 255–170 BC), Hipparchus, who was the first astronomer with unquestioned 'world rank' status (born in the first quarter of the second century BC and died after 127 BC), and Ptolemy.

From the meagre information that is available about the early figures, Neugebauer concludes that[4] 'there is much in the astronomy of Eudoxus, Aristarchus, and Archimedes (i.e., in the period just preceding Apollonius) that shows a lack of interest in empirical numerical data in contrast to the emphasis on the purely mathematical structure'. Eudoxus's scheme of concentric spheres has already been briefly mentioned. It was discussed and modified by Aristotle in his *Metaphysics*. According to Aristotle's commentator Simplicius (sixth century AD), the Eudoxan system was very soon abandoned because of its complete inability to explain the extreme variations in the brightness of Mars. Dreyer gives a quite full description of the system[5] as does Neugebauer.[6] Viewed in the historical perspective, its chief virtue was in being the first attempt to find a rational and rigorously mathematical description of planetary motions by the superposition of circular motions which were each separately and strictly uniform; after modification, this became a very fruitful paradigm. However, because all the Eudoxan motions were concentric, each of the planets must remain at a constant distance from the earth, which was why it could not cope with the variations of Mars's brightness. It also failed badly in the description of Mars's motion. It is for this sort of reason that Neugebauer classes Eudoxus's work as mathematical speculation.

The only two extant works from the early period are Aristarchus's *On the Sizes and Distances of the Sun and Moon,* which will be discussed in Sec. 3.2, and Archimedes' *Sand-Reckoner,* a curious work in which Archimedes calculates how many grains of sand would fill the universe. He does this to demonstrate the use of a system he had devised to represent very large numbers. It is in connection with an estimation of the size of the universe that Archimedes mentions Aristarchus's lost proposal of a heliocentric arrangement of the earth–sun system. Both of these works show a pronounced lack of concern with observational facts (Aristarchus actually gives the moon's diameter as 2°, which is four times too large) and are manifestly mathematical demonstrations rather than descriptions of methods used by working astronomers. Both works are redolent of the pure mathematician's insistence on rigour. Neugebauer comments:[7] 'As soon as pure geometry is involved both Aristarchus and Archimedes proceed without mercy and completely ignore the practical significance of the problem.'

One of the most interesting questions in the history of Greek astronomy concerns the discovery of the epicycle–deferent scheme for the description of the motions of the heavenly bodies, especially the planets. As Neugebauer points out,[8] this scheme, which will be described in Sec. 3.9, opened the way for an astronomy that was at once rational and empirical. Until recently, histories of astronomy put this discovery quite early, most attributing it to Heraclides. For this there seems to be no sound evidence at all.[9] All that is known for certain is that Ptolemy[10] attributes to Apollonius of Perga, who wrote a famous treatise on conic sections, a very important theorem in the epicycle–deferent theory, so it was obviously known by then, i.e., by about 200 BC. Neugebauer[11] in fact concludes that these models for planetary motions were most probably invented by Apollonius. However, it is not at all clear to what extent the models were actually tested by him against empirical observations.

Almost as interesting as the discovery of the epicycle–deferent model is the question of when and why the Greek astronomers began the systematic testing of this and other models against observations. It was the combination of the two strands – the empirical and the theoretical – that more than anything else gave rise to dynamics. This proved to be the very essence of the science of dynamics. The first person definitely known to have undertaken such work was Hipparchus. Very interesting here is the possible influence of 'Babylonian' astronomy, i.e., the highly-developed astronomy that appeared in Mesopotamia in the fifth and fourth centuries BC. For an account of this remarkable science and the story of the deciphering of the cuneiform tablets the reader is referred to Neugebauer.[12] The point that needs to be made here is that the Babylonian astronomers, using intricate numerical difference sequences, compiled tables giving the positions of the celestial bodies with high accuracy though without, it seems, any of the underlying geometrical theory that the Greek astronomers developed somewhat later. Hipparchus had access to Babylonian data, and may well have been stimulated by their influence to import empirical exactitude into the theory he had inherited from Apollonius. Neugebauer writes:[13] 'For us the influence of Babylonian data, accompanied of course by the sexagesimal number system, is first clearly visible with Hipparchus (around 150 BC). Now astronomy becomes a real science in which observable numerical data are made the decisive criterium for the correctness of whatever theory is suggested for the description of astronomical phenomena.'

The band of theoretical astronomers, who made such a major contribution to the discovery of dynamics, was extremely small. There are, in fact, only three well-attested major figures associated with this first development of a successful rational theory of empirically observed motions: Apollonius, Hipparchus, and Ptolemy. It should not be forgotten that theoretical astronomy was very much a side-stream of the main astronomical activity and much more work was devoted to calendric and astrological aims. (Astrology spread to the Greek cultural world as a pseudoscience from Mesopotomia in the second century BC and took root, after which it spread to the entire world.)[14] In the early history of the subjects, astrology and astronomy were not distinguished. Astrology was still a major (and respectable) activity in Kepler's time; he seems to have had some remarkable successes with horoscopes.[15] A comment by Neugebauer about the tabulation of mathematical functions and the absence in Ptolemy's work of the trigonometric tan function brings home rather graphically the minority status of the work that would eventually do so much to transform man's concept of the world:[16] 'Obviously it is the smallness of the number of people who were interested and able to undertake productive work in theoretical astronomy that is responsible for the slow progress in the mechanization of procedures. Whenever a large number of practitioners is involved as, e.g. in calendric or in astrological computations, we notice the tabulation of a variety of sometimes very complicated functions taking place.'

After a section dealing with purely geometrical topics and their relation to the philosophical questions of the previous chapter, the greater part of this chapter will be concerned with an explanation of the epicycle–deferent scheme, the demonstration of why it worked so well, and a discussion of its significance for the discovery of dynamics. This will demonstrate simultaneously the importance of the empirical input. The intimate interplay of observation and theory, so characteristic of all modern science, will be seen at work for the first time in history.

Before we start, it is worth emphasizing the striking difference between the Hellenistic and more or less contemporaneous Babylonian astronomy, the one permeated by geometrical models, the other seemingly not. This makes one wonder if science had to develop in the way it did. Neugebauer comments:[17] 'It is a historical insight of great significance that the earliest existing mathematical astronomy was governed by numerical techniques, not by geometrical considerations, and, on the other hand, that the development of geometrical explanations is by no means such a "natural" step as it might seem to us who grew up in the tradition founded by the Greek astronomers of the Hellenistic and Roman period.'

3.2. Purely geometrical achievements and the development of trigonometry

The first solid achievements of Greek astronomy were those that established the spherical shape of the earth, led to an estimate of its diameter, and proved that the stars and planets were at an immensely great distance compared with the diameter of the earth. Some of the arguments were quite sophisticated and give evidence of systematic compilation of data. For example, Ptolemy[18] noted that an eclipse of the moon commences when the shadow of the earth falls on the moon and is therefore observed at the same instant by all observers but that those further to the east invariably record the eclipse as having occurred a longer time after their local noon than the observers further to the west. From the fact that the differences in the local time are always found to be proportional to the difference in longitude, Ptolemy concludes that the earth's shape must be spherical. The diameter of the earth was, of course, deduced from observation of the altitude of the pole star at different latitudes with known north–south separation. The first such estimate had already been given by Aristotle.[19]

Particularly interesting are the arguments advanced by Ptolemy[20] to show that the earth must be at the centre of the celestial sphere which he believed carried the stars. These arguments were, of course, invalidated by the one simple fact that the stars are at a distance whose immensity was beyond the capacity of the ancient mind to conceive. However, *had* the ancients been correct in their assumption of comparative proximity of the stars, Ptolemy's arguments (some of them very elegant) for the earth's occupying the central position would have been entirely correct.

The ancients' mistake about the disposition and distances of the stars has a moral for discussions about the origin of inertia: a successful, intuitively appealing theoretical interpretation of phenomena that is confirmed by careful observations at the limiting accuracy available in a particular epoch is liable to become dogma with the passage of time, an interpretation of the world ingrained in our conceptions. Great powers of

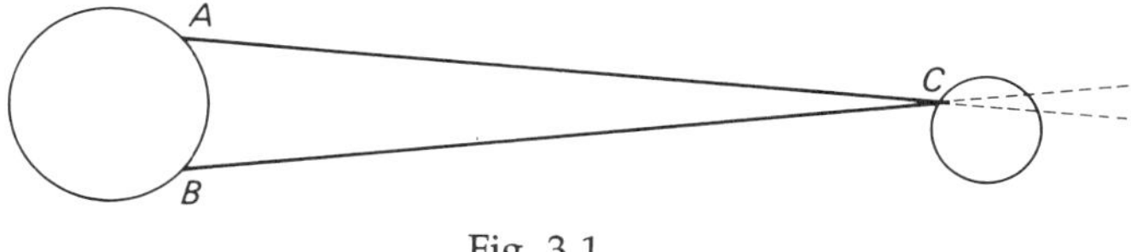

Fig. 3.1.

imagination are then required to establish an alternative conception, to separate the world from the dogma.

Apart from the great distance of the stars, the major problem that the ancient astronomers faced in forming a conception of the world was that they could recognize only very few possibilities to exploit the main tool used to build up a three-dimensional picture from a set of two-dimensional views of it: the phenomenon of *parallax*.

We can best appreciate the significance of parallax by considering how the Greeks did succeed in obtaining a relatively good estimate of the distance of the moon. Parallax is the change in the apparent position of a relatively nearby object against the background of very distant objects produced by a change in the position of the observer. In Fig. 3.1, A and B are two points on the surface of the earth and C is a point on the moon (the figure is not to scale!). The angle ACB is equal to the angle on the sky between the positions at which C is seen against the background of the very distant stars from A and B. If the distance between A and B is known (which in principle it is), the distance of the moon from the earth follows by elementary trigonometry. The parallax effect for the moon is quite appreciable, since the diameter of the earth is about 8000 miles while the mean distance to the moon is 240 000 miles. Thus, for observers on opposite sides of the earth, who obtain the maximal parallax effect for terrestrial observers, the angle ACB is about 2°. This is, of course, the apparent diameter of the earth as seen from the moon. Since the moon's apparent diameter is $\frac{1}{2}°$, it is readily understood that the moon's position against the background of the stars is very sensibly different for two observers on opposite sides of the earth.

This effect, which was used by Hipparchus in order to obtain an estimate of the distance of the moon, was probably recognized very early by the ancient astronomers, though they found it quite difficult to exploit or take into account in observations of the motion of the moon. Ptolemy's *Almagest* contains numerous references to the difficulties of calculating the parallax.[21] These were mainly due to the absence of means of instantaneous communication between observers at different points on the earth. In addition, the radius of the earth was not accurately known and all observations were effectively made from a relatively small area of the world (around the Mediterranean). It is worth mentioning here the importance of eclipses of the moon for determining lunar positions free of parallax. The astronomers knew that the sun was much further away from the earth than the moon and therefore had a practically negligible parallax. Also, as we shall shortly see, they always knew to a reasonably good accuracy the position of the sun on the sky at any given time. Now at a total eclipse the earth comes between the sun and moon and at mid-eclipse, the time of which can be readily determined, the moon must be exactly opposite the sun. From this it was therefore possible to determine the position at which the moon would be seen from an observer at the centre of the earth and thus free of parallax.

In principle there are numerous possible ways in which parallax can be used to determine the distance of the moon. In this connection it is worth noting that the determination of the moon's distance by parallax demonstrates simultaneously the strength and weakness of Aristotle's definition of the position of a body by means of the containing envelope. As we saw, Aristotle was most reluctant to define position as relative to space – for the simple reason that space can never be observed. Hence, he relied on immediate contiguous matter to define position. However, the Greek determination of the earth–moon distance shows that one must distinguish between actual (physical) contiguity and *perceived contiguity*. The point is that in all parallax observations there is an 'envelope' which plays an indispensable part. It is the background against which the object whose distance is to be determined is observed. Without the backcloth of the stars, the parallax of the moon would simply be invisible. Each observation of the moon against a particular background can be called a *primal observation*; it is an actuality without which nothing at all can be said. As in Aristotle's concept of space, this is a topological relation of contiguity: the observed background of the stars envelopes the observed moon. The great difference between the Aristotelian philosophical definition of position and that used in Hellenistic (and all later) astronomy is that the former requires actual contiguity and is based directly on primal observations, whereas the latter derives from primal observations but determines position only after the intervention of a theory of vision. (The ancient Greeks assumed in effect that if the images of two celestial bodies were seen at the same point on the sky then at the instant of observation they were situated on the same line of sight, or 'ray of vision', emanating from the observer; translated into modern terms, this resulted in an interpretation of observations that assumes an infinite speed of propagation of light along straight lines.)

As this question bears directly on the epistemological foundation of the absolute/relative debate, some general comments are here in place. All knowledge of the contingent world rests ultimately on primal observations. A primal observation is something 'that is the case'. No other datum is recognized by experimental science. That science is possible at all is based on the fact that the primal observations which we each make are not random or irregular impressions but exhibit a high degree of correlation. It is the nature of this correlation which suggests to us a representation of our successive two-dimensional views of the world by a single three-dimensional representation, which is, when all is said and done, a theoretical construct. Thus, science constructs a picture of the world by ratiocination over primal observations and the correlations which they exhibit. The method of ratiocination and the direction in which we develop it are to some extent at our disposal; in particular, we are guided by the correlations which appear to us to be significant. Unlike the primal observations, which are given and cannot be changed, the selection of the correlations that are deemed to be significant is subjective and can therefore be changed. Aristotle's remark that place *is* the inner surface of the envelope applies only to the primal observations. It is a constraint on the very possibility of saying anything meaningful. But it does not constrain the conceptual picture we form from valid observations; Aristotle's arguments about space have to do with the validity of observations, not the nature of the conclusions that can be drawn from them. In Vol. 2 of this study we shall see that similar arguments reappeared in a different guise in the present century and played a very important conceptual role in both the special and the general theory of relativity. Meanwhile, we note that the philosophical basis of Aristotle's cosmology – the concept of place as the inner surface of the container – was undermined, or rather circumvented, almost before it was formulated.

Whereas the effect of parallax for the moon is appreciable (2°), for the sun it is about 360 times smaller, the sun being 360 times further away. Such an effect was quite out of range of the ancient astronomers. Nevertheless, they did succeed in obtaining a very rough estimate of the sun's distance by two independent methods.

We shall look at one of these methods, the one in the extant work of Aristarchus already mentioned, because, though of little practical significance, it too highlights an aspect of position determination very relevant to the absolute/relative debate, as we shall see at the end of this section.

The idea is as follows.[22] In Fig. 3.2 (not to scale), E is the earth, M is the moon, and S the sun. Aristarchus made the assumption that when the moon is seen from the earth in the half-moon phase, the angle EMS is an exact right angle. Since the earth–moon distance EM can be measured by one of the parallax methods and angle MES can in principle be measured, the earth–sun distance can be estimated. Except for the purposes of obtaining a very rough estimate – that the earth–sun distance is very much greater than the earth–moon distance – Aristarchus's proposal is impracticable because it is extremely difficult to determine the exact

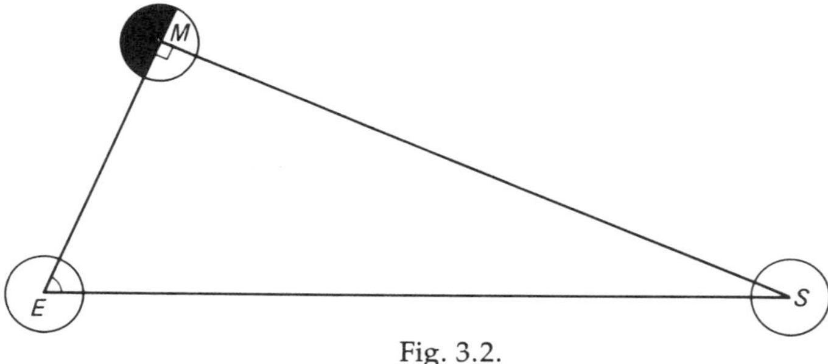

Fig. 3.2.

instant at which the moon is at one of its halves, and it is also very difficult to measure the angle MES with sufficient accuracy. Not surprisingly, Aristarchus seriously underestimated the distance and size of the sun.

Hipparchus later developed a more realistic method that made use of both lunar and solar eclipses. An important part was played in this case by the apparent diameters of the sun and moon and also of the shadow of the earth where it is intersected by the moon in a lunar eclipse. The method is discussed in the works already quoted,[23] and I shall merely quote the results that Ptolemy obtained after analysis and recalculation of his predecessor's determination of the distances of the moon and sun. He concluded that the best estimates were (in units of the earth's radius):[24] mean earth–moon distance 59, mean earth–sun distance 1210 (i.e., about 5 million miles). These values were still accepted as basically correct as late as Brahe's time. Thus, although the ancients were wrong by a factor of nearly 20 in their estimate of the distance to the sun, they did at least have some comprehension of the distances involved and certainly knew that the sun was vastly larger than the earth (170 times the earth's volume according to Ptolemy's estimate). It has been suggested[22a] that Aristarchus may have been encouraged to develop his anticipation of Copernicus's heliocentric astronomy through his awareness of this fact.

It is here worth making a comment about the astronomers' development of systematic trigonometry, which means, of course, measurement of triangles; these may be either plane or spherical. Neugebauer comments[25] that Pythagoras's theorem had been used to solve (plane) right triangles and to break up general triangles into two right triangles since Old Babylonian times. But that is not yet the essence of trigonometry. As Neugebauer says: 'What constitutes real progress, however, is the decision not to solve individually every problem as it arises but to tabulate the solutions of right triangles once and for all as functions of one of its angles.' He thinks that this decisive step may well have been taken by Hipparchus, possibly through his familiarity with Babylonian numerical triangles.

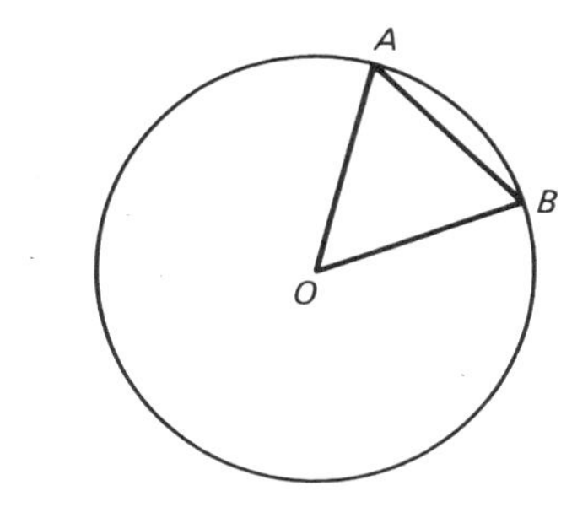

Fig. 3.3.

What Ptolemy did is known because it is all explained fully in the *Almagest*.[26] He had a single trigonometric function, which did not correspond to any of those currently used. It was the *chord function*. That is, Ptolemy calculated the lengths of chords of a circle that subtend given angles at the centre of the circle. In Fig. 3.3, AB is the chord and O the centre of the circle. As he used the sexagesimal system (analogous to the decimal system with 60 used as base instead of 10), he took the radius OB to have the length 60 and calculated the length of chord AB for angles AOB from 0 to 180° at intervals of half a degree.

The development of trigonometry not only provided a means for man to probe the universe far from the regions to which he had direct access. It also opened up an entirely new way of thinking about motion (though it was not one that was developed until the modern age).

As already noted in Chap. 1, it is clear that the earliest conceptions of space, position, and motion were strongly influenced by the (fortunate) accident of our happening to live on the earth. Geometrical relations were discovered on its remarkably stable surface. Thus, we walk around on a part of the lawn of a garden and rapidly form a three-dimensional picture of the disposition of the bushes, trees, and borders distributed over the lawn. We can then choose a point that we have never previously visited, let ourselves be guided there blindfolded, and *predict* the two-dimensional picture we shall see when the scarf is taken from our eyes. We think of the trees and bushes as having a definite place *on the lawn*. We ourselves move across the lawn. Not surprisingly, man hypostatized space, took it to be a real thing with all the geometrical properties of the lawn, removing only the tactility and visibility. The environment was divided into two – the lawn and the objects on it. The concept of *intuitive space*, the container of the material objects in the world, developed very naturally as a consequence of the very special nature of our environment.

All the major contributors to the development of dynamics, above all Galileo and Newton, worked with an underlying conception of motion as taking place in space. Even Einstein, as we shall see in Vol. 2, could not entirely shake off this way of thinking. But the real lesson of trigonometry is that the establishment of spatial relationships is in no way dependent on the lawn; the trees and bushes, and, very important, the light by which we see them, are sufficient in themselves. Figure 3.2 contains just three bodies – the earth, sun, and moon; the relationships of relative size and mutual distance which Aristarchus deduced for them owe nothing to 'space'. Even the backcloth of the stars is not required (the observation of the angle MES must of necessity be made in daylight, when the stars are not visible). Why then must we conceive of them as moving in a space that plays no part at all in the operational determination of these relationships? Should we not do better to concentrate directly on what is objectively given?

It is the ascent into the heavens that forces upon man a revision of his concepts of space. The further we progress, the more the old familiar framework recedes. The surface of the earth is far below us, but somehow we seem to feel the need to take it, or rather its surrogate, space, with us – a safety net for the first hesitant attempts in celestial acrobatics. But the acrobat keeps his eye on the trapeze and on his partner; he cannot even see the net. Aristarchus pointed the way. Kepler learnt the trick. But not until Poincaré perfected the art of celestial dynamics at the end of the nineteenth century was the 'lawn' revealed for what it was: a convenient but dispensable aid to conceptualization. We shall come to this too in Vol. 2.

3.3. Astronomical frames of reference

To begin the discussion of the discovery of the laws of planetary motion, let us distinguish three frames of reference. (1) We begin with the geocentric and geostatic frame; for brevity, this may be called the *earth frame*. The earth is assumed to be at rest and at the centre of the universe. A point anywhere in the universe is defined by its distance along the ray from the centre of the earth to the point in question and by the point of intersection of this ray with the surface of the earth. (2) As emphasized in Chap. 2, the stars exhibited no motions relative to each that could be detected by the ancient astronomers. On the other hand, the planets, sun, and moon do move perceptibly relative to the distant stars. In many ways it is therefore convenient to 'subtract out' the diurnal rotation and refer the motion of the seven 'wanderers' directly to the stellar background. This leads us to the *geoastral frame*: the origin is still the centre of the earth, from which the distances are therefore measured, but now it is not the points of the earth's surface but rather the relatively fixed stars on the celestial sphere that are the markers of the angular position; in this frame the stars are at rest. Although the two greatest astronomers of antiquity,

Hipparchus and Ptolemy, were firm believers in the geocentric and geostatic cosmology, much of their work was in effect done in the geoastral frame, since this was so much more convenient. (3) Finally, we have the *helioastral frame*. The origin is now the sun, while the coordinates are the distance from the sun and the angular position determined by the position of the stars on the celestial sphere.

Throughout all the discussion of prerelativistic physics, we shall ignore the effects due to the finite speed of propagation of light. Thus, we assume that light propagates instantaneously along straight lines. This means that we ignore aberration (which causes a change in the apparent positions of the stars and will be discussed in Vol. 2) and also the time taken for light from the planets to reach the earth. All the associated effects are below the accuracy attainable with naked-eye astronomy.

An important property of all the three frames introduced above is that they are concretely realized by observable objects. Their conceptual status is therefore free of all the ambiguities associated with Newton's absolute space. Although the determination of the distance is problematic, the angular positions can in principle be determined directly in the earth frame and the geoastral frame. As regards the determination of position in the helioastral frame, that is very largely the story of pretelescopic astronomy. As we shall see, the greater part of this task was performed quite unwittingly by the Hellenistic astronomers as they developed a theory of motion of the planets, which although formulated nominally in the earth frame became in effect a theory of motion in geoastral space.

Indeed, the stars of the geoastral frame, in providing points of reference far removed from our immediate terrestrial environment, made possible that first loosening of man's imagination from the chains of preconception imposed by the immediate vicinity. By far the most striking observed phenomenon was the diurnal rotation, observed of necessity in the earth frame. But careful observation revealed small deviations from the dominant pattern, the slow creeping of the planets, sun, and moon across the sky relative to the stars. Both practically and conceptually it was much easier to refer this motion directly to the geoastral frame, rather than the earth frame.

Thus, the first step away from the primitive earth-based viewpoint occurred unconsciously and unwittingly. As the ancient astronomers immersed themselves more and more in the planetary phenomena, the diurnal motion was gradually lost from sight: they paved the way for the ultimate transition to the helioastral system by showing that what was initially regarded as a minor blemish (an alien disorder) on the divine perfection of the diurnal rotation was in fact a *completely autonomous phenomenon*, a spectacle played out on the heavens and not in any way tied to the 24-hour diurnal rotation.

3.4. Manifestations of the law of inertia in the heavens

In Sec. 1.4 we pointed out the great advantages but also the disadvantages of celestial motions from the point of view of the discovery of dynamics. We may summarize the discussion by saying that the celestial motions were excellent from the point of view of suggesting the existence of laws that govern them; they were less good as guides to finding the precise form of those laws.

Let us now look at individual phenomena and see how they shaped conceptions and dictated (to a large degree) the manner in which the problem of the celestial motions would be attacked.

Astronomically speaking, the most pronounced effect was produced by the least significant cause: the rotation of the earth (which to an observer outside the solar system would appear to be rather a minor matter compared with all the other motions). Coupled with the complete absence of any apparent dynamical effects of the rotation for an observer on the earth rotating with it, this created the powerful impression of a stationary earth about which the stars and other celestial bodies appeared to move in perfect circles.

The frontispiece of this book shows just how dramatic the effect is. It is a photograph of the stars taken by opening for about two hours the shutter of a camera placed on the ground at night and pointed towards the pole star. As the earth rotates, the apparent positions of the stars change, and their images leave circular tracks on the photographic film. As one gets further and further away from the pole, the lengths of the tracks become correspondingly longer. Although the ancient astronomers had no cameras, they were extremely conscious of this effect. It is not difficult to realize how the notion of perfectly uniform and circular motion made such an impact on the ancient imagination. It established itself as the norm against which everything else was compared.

Apart from their more or less fortuitous alignments in constellations, the stars do not distinguish any particular positions on the celestial sphere. This situation is completely changed by the diurnal rotation of the earth, which distinguishes the two poles of the rotation and the great circle between them that cuts the celestial sphere into two equal hemispheres, the northern and the southern. The stars on this great circle, which is called the (celestial) *equator*, have an important property. The circle on which they move is the longest of the circles described by the stars as the celestial sphere turns in its apparent rotation. Since the horizon defines a great circle, it cuts the equator at any instant into two equal semicircles. This means that a star situated on the equator takes exactly twelve hours to pass across the sky from the point at which it rises to the point at which it sets. (This ignores refraction, which causes celestial objects near the horizon to appear higher than they would in the absence of the earth's atmosphere.) It is because of this effect that the equator gets its name. When the sun crosses the equator, day and night each last twelve hours, and the equator is therefore the circle that equalizes day and night (*circulus aequator diei et noctis*), and the time at which this happens is therefore called an *equinox* (when the night, *nox*, is equal to the day). This is a phenomenon that is quite independent of an observer's position on the earth; equinox occurs on the same day throughout the world.

Because the two poles of the diurnal rotation and the equator are uniquely defined, they suggest the use of a very convenient orthogonal spherical coordinate system. The poles of this system are, of course, the points at which the earth's axis punctures the celestial sphere. Now imagine all the great circles that pass through these two poles; they are completely analogous to the lines of longitude on the earth. They cut the equator and any other circle parallel to it at right angles. Such circles are analogous to the lines of latitude on the earth. Thus the position of a star on the celestial sphere can be uniquely specified by coordinates exactly analogous to latitude and longitude. Note that whereas latitude measures the angular distance north or south of the terrestrial equator the corresponding distance on the celestial sphere from the celestial equator is called the *declination*, while the celestial analogue of longitude is called *right ascension*. (The fixing of the great circle of zero right ascension – the analogue of the Greenwich meridian – will be considered shortly.)

It seems[27] that in Hipparchus's time such a coordinate system had not yet been established and that he employed a more archaic system based on position relative to chosen constellations. However, in the *Almagest*, Ptolemy does use such a system. Since Ptolemy, in his work as a geographer and cartographer, was the person who first introduced latitude and longitude for the specification of position on the surface of the earth, Neugebauer thinks he may well have been the first to do it for the celestial sphere as well.

To summarize this part of the discussion. One simple phenomenon, the rotation of the earth, suggested to the ancient astronomers the notion of perfectly circular and uniform motion as the paradigm of celestial motion, provided an effective clock, and strongly suggested a method of position location on the sky.

The reason for this pervasive influence of the diurnal rotation is not far to seek. It is essentially conservation of angular momentum that keeps the earth spinning uniformly about a direction fixed in geoastral space. As with the area law, this, as we saw in Chap. 1, is an almost direct consequence of Newton's First Law. From the start this law was therefore shaping conceptions and imposing patterns of thought, though not ones that led directly to its discovery.

The next effect we consider is the consequence of the earth's motion around the sun. This, of course, appears as an apparent motion of the sun against the background of the stars. We immediately encounter an equally dramatic and striking consequence of inertia. The earth's motion around the sun meets the conditions under which the area law holds, since, very small perturbations apart, the motion is governed by inertia and the central attraction of the sun. The motion therefore takes place in a plane that, according to Newtonian theory, is fixed in absolute space. What we actually observe is that the plane is fixed in geoastral space, i.e., relative to the stars (again ignoring very small perturbations).

Now the only thing we can observe from the earth directly is the direction towards the sun; we cannot readily determine its distance. The sun is therefore observed to move around the heavens on a great circle (the plane in which the earth moves of necessity contains the earth and, since it is a plane, it must cut the celestial sphere in a great circle), and this circle keeps a fixed position in geoastral space. Since eclipses of the sun or moon can only occur when the moon, which normally is not situated in the plane defined by the motion of the earth around the sun, is actually in this plane, the great circle of the sun's apparent motion is called the *ecliptic*. At an eclipse of the moon, the sun is on one side of the ecliptic while the moon is at the diametrically opposite point, as we noted in the discussion of parallax. Eclipses of the moon therefore 'pick out' the ecliptic. Being such awesome events, they helped underline the striking and singular nature of the ecliptic. After the diurnal rotation of the earth, the existence of the ecliptic is the 'cleanest' and most striking reflection of Newton's First Law presented to observant man. It is therefore not surprising that, like the diurnal motion, this phenomenon soon began to impose patterns of thought and to dictate technical developments.

Because the axis of the earth's rotation is not perpendicular to the plane of the earth's orbit, the great circles of the equator and the ecliptic do not coincide but meet at a certain angle, which is called the *obliquity of ecliptic*. Its present value is a bit less than $23\frac{1}{2}°$ and varies slowly over a period of more than a thousand years, deviating by at most $1\frac{1}{2}°$.

It so happens (for reasons that obviously have to do with the origin of the solar system) that the planes of the orbits of the other planets all lie quite close to the plane of the ecliptic, and for this reason they do not, when observed from the earth, deviate very much from the ecliptic. The moon too is never too far from it. The ecliptic is, however, sharply distinguished from the paths of the planets and moon as observed from the earth since none of these other bodies follows a *fixed* great circle on the celestial sphere (the moon is observed on a great circle that moves slowly). Because the sun moves more or less steadily around the ecliptic (just how steadily we shall shortly see) and the moon and planets never deviate far from it, and, moreover eclipses always occur on it, the Greek astronomers quite rapidly found it convenient to use it too, like the celestial equator, for reference purposes. This led to the introduction of a second orthogonal spherical coordinate system, in which the analogue of the celestial equator was played by the ecliptic. The corresponding poles of this coordinate system are therefore displaced from the poles of the celestial equator by $23\frac{1}{2}°$, i.e., by an angle equal to the obliquity of the ecliptic. In the ecliptic coordinate system, the coordinates analogous to terrestrial latitude and longitude are also called latitude and longitude (*ecliptic latitude* and *longitude*), and are thus distinguished from the corresponding declination and right ascension of the equatorial celestial system. It seems likely that it was Ptolemy too who introduced systematic use of the ecliptic coordinate system.[28] Indeed, one of the most impressive features of Ptolemy's *Almagest* is his use of theorems of spherical trigonometry, which were actually discovered by Menelaus (first century AD), to show how to make a transformation of coordinates from the equatorial to the ecliptic coordinate system and vice versa. Figure 3.4 is a diagram, due to Kepler, showing the mutual disposition of the two systems on the celestial sphere. The signs of the zodiac mark the position of the ecliptic (LQ). The poles of the ecliptic are O and P and the poles of the celestial equator are B and C.

It is worth mentioning here the basic principle of an instrument which enables one to determine the instantaneous position of the ecliptic on the sky. Suppose that on the surface of the earth a shaft, which can rotate about its axis, is set up in a fixed position parallel to the line of the poles of the earth's rotation. Let a second shaft, which I shall call the ecliptic shaft, be rigidly attached to the rotating shaft, inclined to it at fixed angle equal to the obliquity of the ecliptic. As the first shaft is rotated around the line of the earth's poles, the ends of the inclined shaft will, if imagined continued to the celestial sphere, describe on it the circular tracks of the two poles of the ecliptic in their diurnal motion around the poles of the earth's motion. The actual position of the poles of the ecliptic at any instant can be determined as follows. Suppose a rigid circular hoop fixed at right angles to the ecliptic shaft (like the ecliptic ring in Fig. 3.4) as a

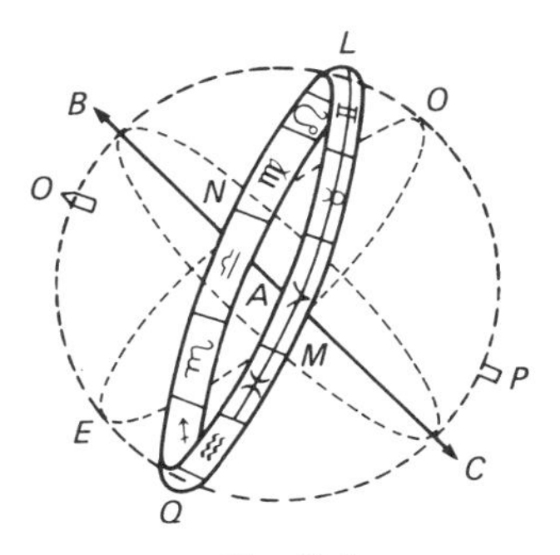

Fig. 3.4.

wheel on an axle. Suppose also the sun is shining. Let the rotating shaft (and with it the ecliptic shaft and hoop) be rotated. When it reaches the position in which the ecliptic shaft is parallel to the true line of the ecliptic poles, the ecliptic shaft must be at right angles to the direction of the sun (which defines the ecliptic) and in this (and only this) position the ecliptic ring will shade itself, i.e., the shadow of the half facing the sun will fall exactly on the opposite half. The ecliptic hoop is then parallel to the actual ecliptic and the ecliptic shaft points to the instantaneous positions of the two poles of the ecliptic.

This arrangement illustrates the theory of the *armillary sphere,* which has a skeleton of metal or wooden moving and fixed concentric rings set up within a frame aligned north–south that permits the whole system of rings to rotate about an axis parallel to the earth's rotation axis. By means of two of the rings it is possible to adjust the system to the correct instantaneous position of the ecliptic (by means of the sun or, at night, by a reference star) and then by means of a further moving ring, which is at right angles to the ecliptic ring, it is possible to determine directly the ecliptic latitude and longitude of any object in the sky. The armillary sphere, which is described by Ptolemy in the *Almagest,* was a most important instrument in astronomy up to the time of Tycho Brahe and reflects the dominant and guiding role played in early astronomy by the ecliptic.

To summarize this section: the two cleanest manifestations of the law of inertia presented to the astronomers gave rise to two different coordinate systems and helped to generate and impose two different sets of techniques for studying astronomical problems. There was not the remotest chance of 'reading off' the law of inertia directly from these phenomena. Nevertheless, they still contained so much of its essential features that they inevitably governed the way in which astronomy and dynamics would develop. In all the plethora of motions presented to man, genuine stability could be discerned in only two; that of the stars around the poles of the celestial equator and that of the sun around the ecliptic (it is, incidentally, in the opposite direction to that of the diurnal rotation of the stars, i.e., from west to east). These provided an 'eternal image' onto which the astronomers could latch themselves. These two motions, the one as perfectly steady as anyone could wish, the other nearly so (as we shall see), slowly drew the astronomers on step by step into the development of ever more sophisticated theoretical and experimental techniques for describing the motions of the sun, moon, and planets. In the process they went a very long way towards creating the science of dynamics.

3.5. The 'flaw' from which dynamics developed

Pearls grow from an imperfection, a bit of grit, within the oyster. The first step towards theoretical dynamics as a quantitative discipline controlled by empirical observation can be seen in the way Hipparchus attempted to account for a 'flaw' in the ecliptic motion of the sun. Before we consider what he did, we must look at the nature and origin of the 'flaw'.

There are four points on the ecliptic that stand out by virtue of their singular properties. At the two points at which the ecliptic crosses the equator the lengths of day and night are, as we have seen, equal. These two equinoctial points are crossed by the sun in the spring (vernal equinox, around 21 March) and in the autumn (autumnal equinox, around 21 September). The other two distinguished points occur at midsummer (21 June) and midwinter (21 December); because the sun at these times reaches its greatest distance from the equator, so that its motion in declination comes to a stop before recommencing in the opposite direction, these two points are called the summer and winter solstices, from *solstitium*, the standing still of *sol*, the sun. Because of their significance for calendric and agricultural purposes, these points entered deep into the awareness of ancient man, much more so than in the present age of convenient clocks and the inevitable present of a calendar at Christmas. They are separated from each other by exactly 90° around the ecliptic and therefore divide it into four equal quadrants. The vernal equinox in particular played (and still plays) an especially important role in astronomy since it was chosen by Ptolemy as the zero point for the measurement of both right ascension and ecliptic longitude, with positive sense of increase eastward from the vernal equinox. This convention, which has been retained ever since, completes the definition of the two coordinate systems described in the previous section. (It should be noted that because of the phenomenon known as the *precession of the equinoxes* the points at which the ecliptic cuts the equator are displaced at a very slow rate. This effect, which was discovered by Hipparchus sometime after 135 BC, will be discussed in Chap. 5 but ignored in this chapter.)

One of the earliest of the 'shocking' discoveries that the Greeks made was that the sun does not take the same number of days to pass through each of these four quadrants. Its apparent motion is *nonuniform*. (In the context of Greek astronomy, uniform motion always has the precise meaning of motion at an exactly *constant speed*. The speed may be either angular speed, as in the case of the apparent motion across the sky, in which case it is measured in degrees per unit time, or alternatively actual speed through space.) It is not known when the nonuniformity of the solar motion was discovered (it was certainly not later than the time of Aristotle), but the first attested attempt at a *theoretical explanation of reasonably accurate observations* is due to Hipparchus. Ptolemy reports that:[29] 'these problems have been solved by Hipparchus with great care'.

Hipparchus's work is to be seen as a most significant step forward in the Greek programme of finding geometrokinetic explanations for why the observed motions of the sun, moon, and planets did not fit the divine paradigm of perfect uniform circular motion. The divinity of the celestial bodies had to be saved by a geometrical explanation. This was the programme, begun by Eudoxus in the time of Plato, that the Greeks came to call *saving the phenomena* (or *appearances*). In Hipparchus's hands the programme progressed from the qualitative to the quantitative stage. Before we look at this first step on the long road to theoretical dynamics, it will be helpful to discuss Kepler's laws of planetary motion and their explanation in Newtonian terms.

3.6. Kepler's laws of planetary motions

As pointed out earlier, the overwhelming dominant force that acts on each of the planets, including the earth, is the gravitational attraction towards the sun. If, as is perfectly justified for naked-eye astronomy unless conducted over periods spanning several centuries, all the other gravitational forces acting within the solar system are ignored, the motion of any individual planet satisfies the conditions under which the area law holds. The first consequence of this is that the motion of any given planet is entirely restricted to a plane that contains the sun, through the centre of which the plane passes. This, as we have seen, is why the apparent motion of the sun always takes place exactly along the great circle of the ecliptic. It is, as we saw in Chap. 1, one of the most primitive consequences of Newton's three laws, since it holds whatever the nature of the force acting between the sun and the planet provided only the force is central, i.e., directed exactly towards (or exactly away from) the sun. The second consequence is that as the planet moves around the sun the radius vector from the sun to the planet sweeps out equal areas in equal times. This result is again quite independent of the actual force acting provided only it is central. These two properties are the closest we get in astronomical phenomena to direct manifestations of the two parts of the law of inertia; the uniform rectilinear motion of the primal law is, so to speak, bent into a curve, which, however, still possesses very remarkable properties.

Both results, which, as we have seen, take us very close to the heart of dynamics, were established in an absolutely clean form by Kepler. The first effect, that the motion of each planet takes place in a fixed plane that contains the sun, was discovered for the earth's motion in antiquity in the guise of the ecliptic. Its extension to all of the planets by Kepler is not reflected in any special designation, but it would be highly appropriate to call it *Kepler's Zeroth Law* (as it is sometimes and will be so called in this book). The area law itself is what is now known as *Kepler's Second Law*, and was announced (in a somewhat confused manner) by him in 1609, at the same time as the Zeroth Law and the First Law, to which we now turn.

Unlike the Zeroth Law and the Second Law, *Kepler's First Law* is a specific consequence of the fact that gravity is an attractive force with a strength that varies inversely as the square of the distance from the attracting centre. As Newton showed in the *Principia*, this law has the consequence that any body subject to the attractive force of the sun and no other force will always describe a curve which is fixed in space and which corresponds to one of the so-called *conic sections*. These curves can be generated as follows. Imagine two straight lines that intersect at a point A (Fig. 3.5(*a*)) and the line that bisects them (the dotted line in Fig. 3.5(*b*)). Hold this line fixed and rotate the two lines around it to obtain a circular double cone of revolution (Fig. 3.5(*b*)). Then imagine a plane, not passing through A, that cuts the surface of the cone (Fig. 3.5(*c*)). An *ellipse* is obtained when the plane makes an angle with the cone axis greater than the angle made with the axis by the conical surface. An *hyperbola* (with two branches) is obtained when the angle of the plane is less than the angle of the cone. A *circle* (a special case of an ellipse) is obtained when the plane is perpendicular to the cone axis; and finally a *parabola* is obtained when the plane makes the same angle with the cone axis as the conical surface.

Mathematically, the most convenient definition of a conic section is that it is the curve (locus) described by a point that moves in such a way that the ratio of its distance from a fixed point to its distance from a fixed line is constant. The ratio is called the *eccentricity*, denoted e; the fixed point is called a *focus*; and the fixed line is called the *directrix*. When $e = 0$, the curve is a circle, when $e < 1$ an ellipse, when $e = 1$ a parabola, and when $e > 1$ an hyperbola. There are two foci; in the case of a circle they coincide and are situated at its centre; in the case of a parabola, one of them is at

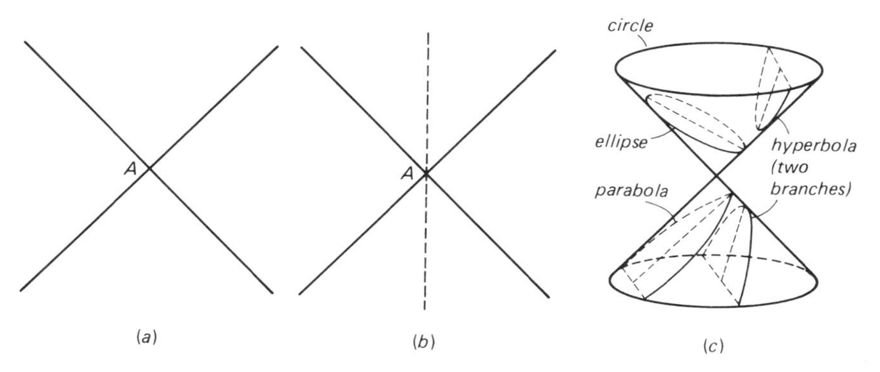

Fig. 3.5. The generation of conic sections. (Figure (*c*) is derived from *Mathematics Dictionary*, by James and James, 4th edn, Van Nostrand, New York (1976), p. 71.)

infinity. The nonmathematical reader will probably find this definition rather opaque. For the case of an ellipse, which is much the most important for our present purposes, an alternative method of generation is much more transparent.

Take a smooth board and into it push two pins. Then take a loop of string of strictly fixed length, lay it around the pins on the board, take a sharp pencil, hold it in such a way that the loop is kept taut, and then move the tip of the pencil around the pins, making sure to keep the string taut while describing a curve on the board. The resulting curve (Fig. 3.6) is an ellipse and the two pins (S and S′) are at its two foci. The point O midway between the two pins is the centre of the ellipse. The longest axis AOB is called the *major axis*; the shortest axis DOC, the *minor axis*. The segment AO (=OB) is the semimajor axis; OC (=OD) is the semiminor axis. The eccentricity as defined earlier turns out to be equal to the ratio OS/OA. Thus, if OA is taken to have unit length, the distance OS (=OS′) is the eccentricity *e*. The eccentricity is then the distance of either focus from the centre of the ellipse. We shall see shortly that this explains the origin of the word *eccentricity*.

In the *Principia*, Newton established the conditions under which the various different types of conic section are realized by bodies moving in the gravitational field of the sun. In rather simplified terms, one can say that a body which does not have sufficient speed to escape completely from the sun moves in an ellipse. If the body has just sufficient speed to escape, then it moves in a parabola. If its speed is still higher, it will move along one branch of an hyperbola. In all cases the *sun is at a focus of the corresponding conic section*. Thus, in Fig. 3.6, the sun will be at S, say, while the second focus S′ will be a void point. (In the case of parabolic motion, the sun is obviously at the focus which is not at infinity. This case is obtained from the situation in Fig. 3.6 by taking the second focus S′ further and further away from S. Another special case is obtained by bringing S and S′ closer and closer together and making them coincide at O; the ellipse then becomes a circle.)

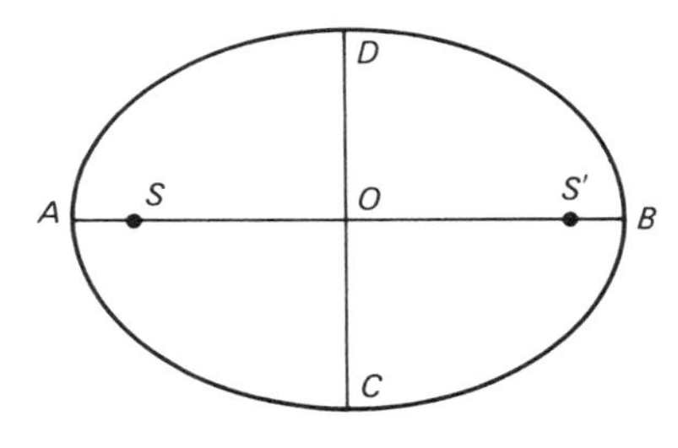

Fig. 3.6.

For the planets the conditions for elliptical motion are satisfied. This is what Kepler found empirically about 75 years before Newton discovered his theoretical explanation. Kepler's *First Law* states that the planets move in ellipses that have the sun at one of the foci.

It is worth making two comments about this law. First, the orbit is closed and fixed in helioastral space (except for the perturbations introduced by the other planets, which are very small). Thus, the planet keeps on going round exactly the same curve and returns to exactly the same point in helioastral space whenever a period of its motion has been completed. As already emphasized, without this strict periodicity it is difficult to see how the laws of motion would ever have been found. The fact that the curves which the planets follow are closed is actually very special. For a general force law, the curve does not close after every circuit around the force centre, though it fortunately does so for the inverse square law of gravitational attraction.

The second comment is that the planetary problem decomposes into the problem of the geometrical shape of the orbit and of the speed with which the motion is executed. For any given orbit, the speed within it is determined by Kepler's Second Law, the area law. The average speed is determined by the distance from the sun. Although this last aspect of planetary motion, which is governed by Kepler's *Third Law*, did not play a particularly significant role in the early history of astronomy, we shall describe it here for the sake of completeness. Like the First Law, the Third Law, which Kepler announced in 1619, is a specific consequence of the inverse square law and states that the squares of the periods of the planets are proportional to the cubes of their mean distances from the sun. If P is the period and R the mean distance of a planet, then

$$P^2 \propto R^3,$$

or

$$P = aR^{3/2},$$

where a is a constant of proportionality.

A point to note is that the actual speed of the planet is slower for the planets at greater distances from the sun. If all planets were to move at the same speed, one would simply have a direct proportionality between the period and the radius, $P \propto R$, since the distance travelled increases linearly with R. Kepler's Third Law shows that the outer planets travel slower. Thus, the mean orbital speeds of the planets in helioastral space are (to the nearest km/s) 48 for Mercury, 35 for Venus, 30 for the earth, 24 for Mars, 13 for Jupiter, and 10 for Saturn.

3.7. The zero-eccentricity and small-eccentricity forms of Kepler's laws

In many problems in physics and mathematics it is helpful to consider what are known as *small-parameter expansions*. It is very often the case that the particular manifestation of a given phenomenon depends in a characteristic way on the numerical value of some determining parameter and that significant simplifications arise when the numerical value of the parameter is small. In such cases it is very helpful to expand the functions that describe the phenomenon with respect to this small parameter, ε say. This leads to a series expansion of the form $f = f_0 + \varepsilon f_1 + \varepsilon^2 f_2 + \ldots$. Here, f_0 describes the behaviour when $\varepsilon = 0$. If ε is very small, then only the first two terms of the expansion, f_0 and εf_1, need to be taken into account. Then the term εf_1 appears as a correction to the term f_0, which is called the term of *zeroth order*. When ε gets rather larger, the term $\varepsilon^2 f_2$ must be taken into account and is to be regarded as a correction to the first-order approximation represented by $f_0 + \varepsilon f_1$.

In the planetary problem such a parameter is the eccentricity. For the early history of astronomy it is extremely illuminating to consider the approximations to Kepler's laws given by the zeroth and first approximations in the eccentricity.

In the zeroth approximation, the elliptical orbits all then have zero eccentricity, which means that they become circles with the sun at the exact centre. The orbits are than a series of concentric circles around the sun. This is what happens to Kepler's First Law. The Second Law now takes the form that the planets each move at an exactly constant speed in the orbit. The Third Law remains essentially unaltered.

Before we discuss the corresponding forms of the laws in the first-order approximation, which may be called the *small-eccentricity approximation* (in contrast to the *zero-eccentricity approximation* just discussed) we need to know one more important fact about ellipses. In Fig. 3.7, we have

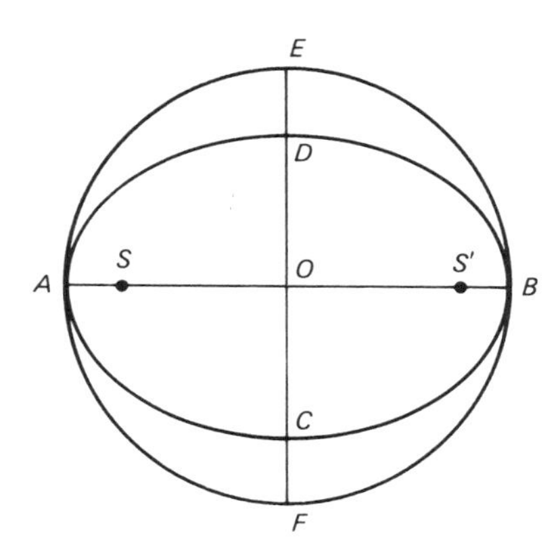

Fig. 3.7.

redrawn the ellipse from Fig. 3.6 but we have added to it the circle with centre O and radius equal to the semimajor axis OA. This circle, AFBE, circumscribes the ellipse, touching it at A and B.

As we see from Fig. 3.7 the areas between the ellipse and this circle take the form of crescents: BDAEB and BCAFB. Kepler called each of them a *lunula* (literally *little moon*). The thickness of the *lunulae* at their greatest widths, ED and CF, is a measure of the extent to which the ellipse is *elliptical*, i.e., falls short of being a perfect circle (*elleipo* means *come short* in Greek, hence *ellipsis* for words omitted from a quotation). We have already noted that if AO = 1, then OS = e. For the history of ancient astronomy a most important property of the ellipse is the fact that in the limit of small eccentricity its *ellipticity*, defined as the ratio ED/OE (or, therefore, ED/OA), is equal to a very good accuracy to $\frac{1}{2}e^2$, i.e., half of the *square* of the eccentricity. Because of the factor $\frac{1}{2}$ but even more because of the squaring, the ellipticity remains numerically very small until the eccentricity reaches quite appreciable values. This means that until the eccentricity reaches a value of about half, the lunulae remain very slender indeed. If the circumscribed circle is not shown, ellipses with eccentricities less than about 0.4 can hardly be distinguished by the eye from circles. This is illustrated by the ellipses in Fig. 3.8. The first is the degenerate case of a circle, for which the two foci S and S′ (not shown) coincide with the centre O. Then come ellipses with eccentricity $\frac{1}{10}$, $\frac{1}{5}$, $\frac{1}{3}$, $\frac{1}{2}$, and $\frac{2}{3}$. Note how far the foci move apart before the ellipticity becomes at all apparent. The displacement of the foci from the centre is an effect of first order in the eccentricity, while the ellipticity is an effect of second order.

Let us now consider what happens to Kepler's First Law in the small-eccentricity approximation. We start with a somewhat surprising result: the orbit is still a circle, since the ellipticity is a correction of order e^2. However, the sun is no longer at the centre of the circle, as it was in the zero-eccentricity approximation, but is displaced to what is in fact one of the foci, though the fact that this is a focus is not at all apparent since the ellipse is still a circle to an extraordinarily good approximation. Thus, in this approximation, which is remarkably accurate for $e < \frac{1}{5}$, Kepler's First Law says the planets move in circles with the sun somewhere in an eccentric position not too far from their centre. All the planes of the circles in which the planets move intersect at the centre of the sun (by Kepler's Zeroth Law).

We now ask what happens to Kepler's Second Law. Here the area law gives rise to a very interesting effect associated with the second (void) focus. Since the sun is displaced from the centre of what is still a circular orbit, it is immediately evident that, because of the area law, the planet can no longer move at uniform speed about the centre of the circle, i.e., its actual speed in helioastral space is nonuniform. An observer placed at

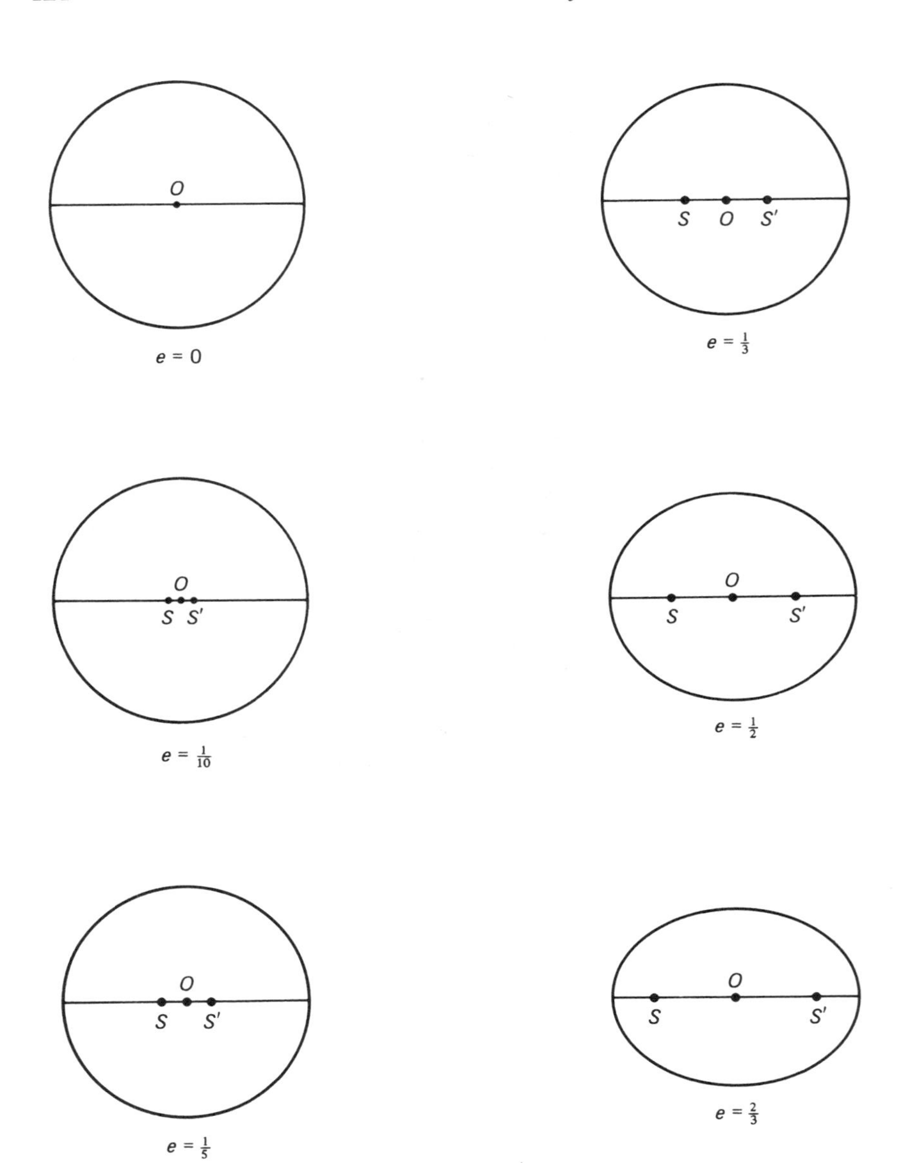

Fig. 3.8.

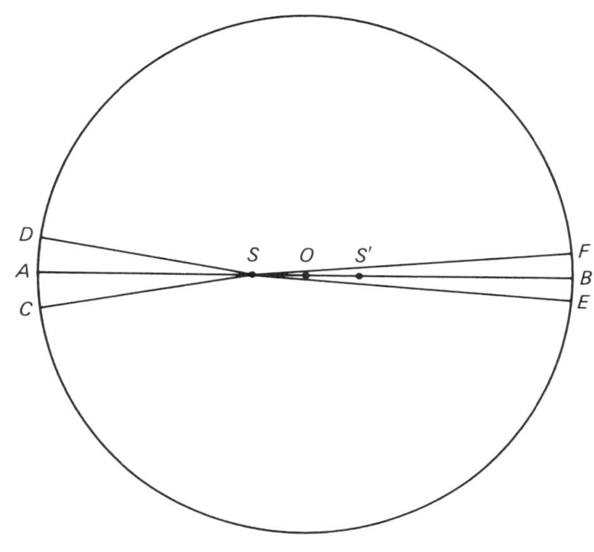

Fig. 3.9.

either the centre of the orbit or at the sun would see the planet moving relative to the stars at a nonuniform rate. However, if the observer were placed at the void point corresponding to the second focus, a striking effect would be observed. Seen from this point, the planet would appear to move with constant angular speed relative to the distant stars.

We can see intuitively how this most important effect comes about by examining Fig. 3.9, which shows the case corresponding to $e = \frac{1}{5}$. The centre of the circle is at O, the sun is at S, at distance e on one side of O (we take the radius OA $=1$), while the point S′ corresponding to the void second focus is at distance e on the opposite side of O from S. When the planet is at A it is closest to the sun. This point is called the *perihelion* (meaning nearest the sun) in helioastral space, the *perigee* (nearest the earth) in geoastral space, which corresponds to the ancient geocentric viewpoint. The distance SA is $1 - e$. When the planet is at B, it is at its furthest from the sun S. This point is called the *aphelion* in the heliocentric scheme, the *apogee* in the geocentric. The distance SB is $1 + e$. Now suppose that in unit time the planet moves from D to C. Then to a very good approximation the area that is swept out by its radius vector from the sun is $\frac{1}{2}\mathrm{DC}(1 - e)$. Now suppose that in the same length of time the planet at the other end of the orbit traverses the distance EF. Then the corresponding area is $\frac{1}{2}\mathrm{EF}(1 + e)$. By Kepler's Second Law, these two areas must be equal:

$$\tfrac{1}{2}\mathrm{DC}(1 - e) = \tfrac{1}{2}\mathrm{EF}(1 + e).$$

Therefore

$$DC/EF = (1 + e)/(1 - e).$$

Thus, an observer at O would see the arcs DC and EF as unequal, in just this ratio. But now suppose the observer moved to S′. For purely geometrical reasons, the arc DC will appear from S′ to be smaller than it appears from O by a factor $(1 + e)$. Similarly, the arc EF will appear enlarged by the same factor. Seen from S′, the two arcs will therefore appear equally large, and the planet will appear to move equally fast over these two arcs.

I leave it as an exercise to the reader, if so inclined, to prove that this result, obtained for the perihelion and aphelion, holds at all points on the orbit, that is, an observer stationed at the point corresponding to the second focus will see the planet travel around a great circle on the sky at a uniform angular speed. It must be borne in mind that the result holds *only to first order in the eccentricity* but is remarkably good in that order (just how good we shall see shortly). This then is the form that Kepler's Second Law takes to first order in the eccentricity. One can imagine a spoke with one end fixed at S′ and rotating with uniform angular speed about that point. Suppose the other end of the spoke and the planet pass through A at the same instant. Then at all subsequent times the position of the planet will almost exactly coincide with the point at which the spoke intersects the orbit.

For reasons that will become clear towards the end of the chapter, the point S′ is called the *equant point* (*punctum equans*, the equalizing point), or simply *equant*. (Strictly speaking, it is a certain circle whose centre is the equant point that is properly called the equant, but I shall follow the widespread practice of applying this word to the centre of the circle.) The points A and B are known as the *apsides* (singular *apsis*, or also *apse*, as in a church), and the line AB is called the *line of the apsides*.

Table 3.1 gives the periods, eccentricities, ellipticities, semimajor axes (in astronomical units; 1 AU is the semimajor axis of the earth's orbit), and inclinations of the orbits (relative to the orbit of the earth) of the six planets out to Saturn for the epoch 1900.[30] It should be noted that all the orbital elements of the planets change very slowly over periods of centuries. The values given in Table 3.1 are only slightly different from the values in Ptolemy's age. The most noticeable change is the precession of the lines of the apsides in helioastral space. These changed by several degrees between the time of Ptolemy and Kepler, who was well aware of these *secular* variations, as they are called.

Examination of Table 3.1 shows clearly why the approximations of zero and especially small eccentricity played such an important role in the early history of theoretical astronomy. For the earth and Venus, the eccentricities are so small that the zero-eccentricity approximation is already very good for them. For the other four planets the eccentricities,

Table 3.1. *Orbital elements of the planets*

Planet	*Semimajor axis of orbit (AU)*	*Sidereal period (tropical years)*	*Eccentricity*	*Ellipticity*	*Inclination to ecliptic*
Mercury	0.387	0.241	0.206 (~1/5)	~1/47	7°0′
Venus	0.723	0.615	0.007 (~1/147)	~1/43 478	3°24′
Earth	1.000	1.000	0.017 (~1/60)	~1/7033	—
Mars	1.524	1.881	0.093 (~1/11)	~1/227	1°51′
Jupiter	5.203	11.862	0.048 (~1/21)	~1/852	1°18′
Saturn	9.539	29.458	0.056 (~1/18)	~1/646	2°29′

especially of Mars and Mercury, are quite appreciable but the ellipticities are all very small, reaching about 1/47 in the case of Mercury alone.

Later we shall consider precisely what opportunities existed for detection of the eccentricities, ellipticities, and nonuniformities in the motions of the planets. However, one general comment can already be made. Several books on the history of astronomy (especially Koestler's[1c] but also to some extent the studies of Dreyer[1a] and Koyré[31]) give the quite erroneous impression that the early astronomers (up to and including Copernicus (1473–1543) and Tycho Brahe (1546–1601)) could not make sense of the planetary motions because they assumed through thick and thin that the planetary orbits were perfectly circular. The implication is that if only they had had a more open mind about alternative possible orbits the data would have fitted much better. However, this is quite wrong. The circularity of the orbits was, in fact, by far the best assumption that the early astronomers made, and the problem of the celestial motions would never have been solved without it. No one made more use of circles, nor more effectively, than Kepler, though he eventually discarded them with less than complimentary comments.

As we shall see, the problems the astronomers faced were of quite a different kind and had very much more to do with the specific eccentricities of the various planetary orbits. Among the books on the history of astronomy known to me, only Hoyle's little book on Copernicus[32] emphasizes the importance of the small-eccentricity limit, though the essential facts about it are given by Neugebauer.[33]

To conclude this section let us briefly mention another approximation of a quite different sort. We recall that the plane of each planetary orbit contains the sun. It therefore cuts the plane of the earth's orbit, i.e., the ecliptic, along a certain line, called the *line of the nodes*. The *nodes* are the two points at which the orbit of the planet passes through the ecliptic (the *ascending node* is the node at which the planet passes into the northern side of the ecliptic; the other node is the *descending node*). The angle between the plane of the orbit and the ecliptic is called the *inclination*. Like all the other orbital parameters, the inclinations and the positions of the lines of the nodes vary very slowly by small amounts over characteristic periods of thousands of years.

Now, as can be seen from Table 3.1, the inclinations of the planetary orbits to the ecliptic are all small or very small. This is the reason why the planets are never observed to move far in latitude from the ecliptic. As a result, the motion in longitude can to a very good approximation be treated independently of the (actually rather complicated) motion in latitude. This gives us the approximation of *zero-inclination*, which is all that we shall consider in this chapter.

3.8. Hipparchus's theory of the apparent solar motion

In considering Hipparchus's attempt to describe the apparent motion of the sun in geoastral space, let us first consider what could be observed. The eccentricity of the earth's orbit is about 1/60 (~1/57 in Hipparchus's time) and this means that the sun's apparent diameter, which is about 30 minutes of arc (30′), varies during the year by a factor of about 1/30. This was too small to be reliably detectable by naked eye methods. Thus, all Hipparchus could actually observe was the position of the sun as it moves round the ecliptic. Now the apparent motion of the sun from the earth is exactly the same as that of the earth from the sun with a phase difference of 180°. How will this motion appear? We start with purely formal considerations, the significance of which will become clearer shortly.

In Fig. 3.10 the curve with short dashes is an ellipse with eccentricity $e = \frac{1}{3}$, i.e., appreciably greater than the eccentricities of any of the planetary orbits. Its centre is at O and its two foci are at E and E′. We shall take this ellipse to represent the orbit of the sun in geoastral space. Then A is the apogee, P is the perigee, E is the position of the earth, and E′ is the position of the void focus. In later applications in the Copernican scheme we shall assume that the sun is at E and the ellipse is the orbit of a planet. For the moment we ignore the two circles in Fig. 3.10.

There is, of course, no mark on the ecliptic indicating the position of either the apogee A or perigee P of the sun's orbit around the earth. However, suppose we did know the position of A. Let us measure time in units such that the period of the sun's motion is 2π, i.e., if it passes A at time $t = 0$ it will pass P at time π and return to A at the end of its complete revolution at $t = 2\pi$. Suppose that at time t the sun has progressed to the position S indicated by the arrow in Fig. 3.10. Let γ be the angle SEA, i.e., the angle measured from the earth between the sun and the apogee. In the theory of Keplerian motion[33] this angle is a very important quantity. It cannot, in fact, be represented in closed form by means of elementary

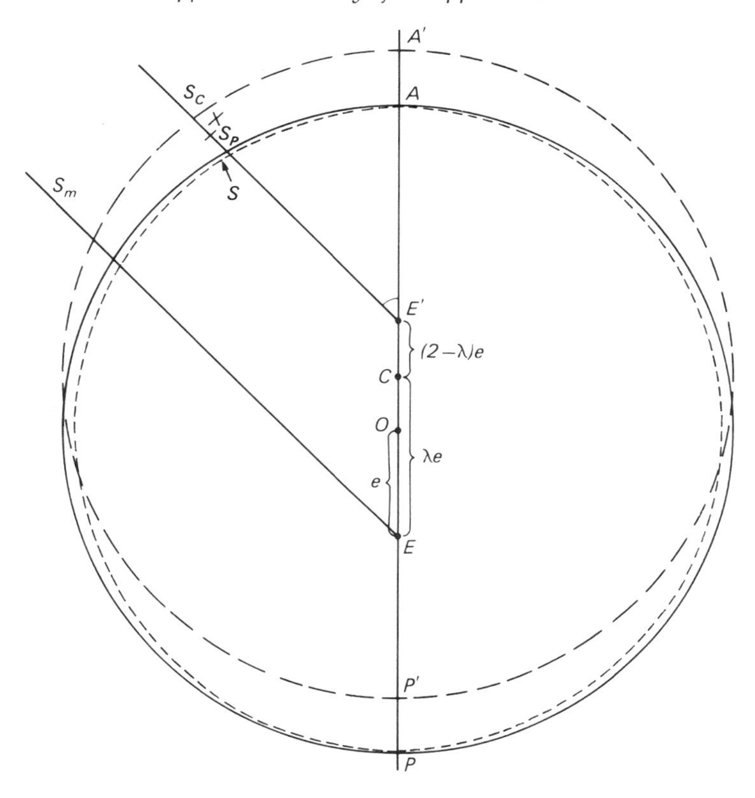

Fig. 3.10.

functions and can only be specified by means of a series expansion in powers of the eccentricity e. To the terms quadratic in e, the expansion is

$$\gamma = t - 2e \sin t + \tfrac{5}{4}e^2 \sin 2t + \ldots . \tag{3.1}$$

Let us now compare this actual motion of the sun with that of a hypothetical body called the *mean sun*, as opposed to the *true*, or *apparent, sun*. As we shall see in what follows, the mean sun was, in Ptolemy's scheme, to develop into what was perhaps the most important concept in the whole of ancient astronomy. The hypothetical body of the mean sun simply defines the position that the sun would have if it were to move around the ecliptic with an exactly uniform angular motion. Its position in Fig. 3.10 is shown by the spoke ES_m, which rotates around E. Since the angle S_mEA (measured in radians) is given directly by the time t, the angle corresponding to (3.1) for the mean sun is simply

$$\gamma_m = t. \tag{3.2}$$

From the Greek point of view the solar motion was anomalous in two respects. First, the fact that the sun moved relative to the stars at all was anomalous. This is probably the reason why the angle γ came to be called the *anomaly* (or *true anomaly*). By contrast, the angle (3.2) became known as the *mean anomaly*.* The solar motion was then doubly anomalous in that the sun not only moved relative to the stars but did so in a nonuniform manner. As a result, γ differed from γ_m. The difference was revealed by the fact that the sun took different lengths of time to pass through equal arcs of the ecliptic. This is the nonuniformity of the solar motion that Hipparchus attempted to explain theoretically. It is worth mentioning here that, in medieval European astronomy, the difference between the true anomaly and the mean anomaly came to be called the *equation*, or *equation of centre*, presumably because it equalized out the difference between these two angles. One of the main tasks of theoretical astronomy was to devise models that could reproduce the equation (and its analogue in planetary theory, to which we shall come later).

Let us now consider the orders of magnitude involved. First, since there are $365\frac{1}{4}$ days in a year and 360° in a circle, the mean sun moves through just a little less than one degree per day, i.e., it travels in a day through a distance equal to about twice its apparent diameter. As regards the magnitude and nature of the nonuniformity, these depend, as is evident from (3.1), entirely on the value of the eccentricity. As we noted, in Hipparchus's time the eccentricity of the earth's orbit was about 1/57. We can conclude immediately from this that the maximal value of the first correction term ($-2e \sin t$) in (3.1), which is reached when $t = \pi/2$ and $3\pi/2$ (the two corresponding positions of the sun are known as the *quadrants*) and $\sin t = 1$, is $\pm 2e$, i.e., just about 2° (because 1 radian is just over 57°). Since the ancient astronomers could measure to an accuracy of about 10 minutes of arc (one third of the apparent diameter of the sun or moon), the effect of this term was comfortably within reach of Hipparchus's observations. Let us now consider the term of second order in the eccentricity ($\frac{5}{4}e^2 \sin 2t$). Since this depends on $\sin 2t$, it is maximal in magnitude at the four so-called *octants*, when $t = \pi/4, 3\pi/4, 5\pi/4$, and $7\pi/4$ (or 45°, 135°, 225°, and 315°). However, since $e \sim 1/57$, the maximal value of this term is about 90 times less than the maximal value of the first term and a great deal less that anything Hipparchus could measure. In fact, it was also just too small for even Tycho Brahe to pick up. Thus, in the case of the solar motion the term of second order in the eccentricity was simply invisible.

* The expressions true anomaly and mean anomaly are still used in modern astronomy. They are, however, now measured from perigee (or perihelion) rather than the ancient practice of measuring from the apogee. The switch was made to accommodate comets, whose perihelion as they sweep past the sun is readily observable, whereas the aphelion is excessively distant and in most cases quite unobservable.

We now come to Hipparchus's observations and theory, which are reported by Ptolemy in §III.4 of the *Almagest*. Ptolemy reports that Hipparchus found the interval from spring equinox to summer solstice to be $94\frac{1}{2}$ days and the interval from summer solstice to autumnal equinox to be $92\frac{1}{2}$ days. This was clear evidence of nonuniformity of the sun's motion. These two data were sufficient for Hipparchus to find a theoretical solution to the problem that can reproduce the observations with remarkable accuracy and is also quite close to the truth. Hipparchus's theory can be represented in two forms, of which we shall choose the simpler, which was also the one adopted by Ptolemy. The alternative will be discussed briefly in connection with the epicycle–deferent theory a little later.

It will be recalled that all celestial motions were believed to be perfectly circular and exactly uniform. How could this be reconciled with the manifest nonuniformity that Hipparchus found? Hipparchus made the rather natural assumption (which belonged to the theoretical ideas already worked out before his time)[34] that the sun did move with perfect uniformity in a perfect circle but that *the centre of this circle did not coincide with the centre of the earth*, i.e., the position of the observer. He assumed that the sun moved around the earth on an *eccentric circle*. In Greek astronomy such a circle was, for this reason, called an *eccentric*, and the distance from its centre to the position of the observer became known as the *eccentricity*.*

It will be helpful at this stage to analyze a generalization of Hipparchus's model. This will give us an overall view of what his model and its generalization by Ptolemy for the planets were able to achieve in the programme of 'saving the appearances'. We attempt to reproduce the solar motion by means of a simple model that contains two essential features, the first of which is that the sun is assumed to move on an exactly circular orbit.

Namely, in Fig. 3.10 we imagine a circle of diameter equal to the semimajor axis AP of the sun's orbit in geoastral space. The centre of this circle is taken to lie at some point C situated on the line of apsides. The position of C is taken to be λe from E (the centre of the earth) and therefore $(2 - \lambda)e$ from E′, the void focus. By varying λ we shift the centre of the circle up and down on the line of the apsides. For $\lambda = 1$ the centre of the circle coincides with the centre of the ellipse and we obtain the

* This is, in fact, the origin of the word, both in normal language and for the eccentricity of conic sections.[35] It comes from the theoretical description of nonuniform apparent motions of celestial bodies by means of circular motions about centres not coincident with the observer. It was only after Kepler's discoveries that the astronomical eccentricity was identified with the key ratio in the mathematical theory of ellipses and used to denote that ratio, after which the term was extended to all conic sections to denote the ratio used in the definition of such curves in terms of the directrix.

circumscribing circle of the ellipse that, as we know from the discussion of Sec. 3.7, is such a good approximation of the actual orbit. This is the continuous circle in Fig. 3.10. The circle formed by the long dashes in the same figure has centre at the point C corresponding to $\lambda = 1.5$.

The second essential feature of the model is that the speed with which the sun (or planet, *mutatis mutandis*) is assumed to move around the orbit is regulated by a simple but special device – the equant prescription. Namely, let one end of a spoke be fixed at E′, the void focus, and swing about that point with uniform angular velocity equal to unity in the units we have adopted. This means that after time t it has swung through angle t from the direction of the apogee A if it started from that point at time $t = 0$, i.e., when the sun also passes through its apogee. Thus, the spoke remains always parallel to the direction of the mean sun from the earth and completes one revolution after time $t = 2\pi$. In this one-parameter family of models of the solar motion, the body that models the sun is assumed to move round the corresponding circular orbit in such a way that it is always exactly at the point of intersection of the equant spoke with its circular orbit. Thus, for the model with circle centered on O it is at S_P, and for the model with centre C it is at S_C.

It should be emphasized that a generalized model of this kind was quite alien to Hipparchus. Nevertheless, in the special case when $\lambda = 2$, the centre of the circle coincides with E′ and for this unique case of our one-parameter model we obtain the Hipparchan situation in which the sun is supposed to move uniformly around the centre of its circular orbit.

Since Hipparchus could observe only angles, the point of most immediate interest is the ability of such models to reproduce the observed angular positions of the sun. In a useful paper, Whiteside[36] has calculated the corresponding angles for the exact Keplerian orbit and for the generalized model to the third order in the eccentricity. However, for our present purposes we only need go to the second order (the term of third order, i.e., cubic in the eccentricity, is just about observable at the maximal accuracy achieved by Brahe in the observations of Mars, the analysis of which by Kepler was Whiteside's concern). The expression we need is

$$\gamma_\lambda = t - 2e \sin t + \lambda e^2 \sin 2t + \ldots . \tag{3.3}$$

Now, except for the appearance of λ in the term quadratic in e, this expression is identical to the Keplerian expression. A particularly important point to note is that λ does not occur in the first correction term ($-2e \sin t$), but this, as we know from the earlier discussion of orders of magnitude, is all that is effectively visible in naked-eye astronomy (for the case of the sun).

We see that, if only angles are observed, models of the solar motion of the kind considered are subject to a double and serious ambiguity. First, the overall size of the circle on which the sun is supposed to move is almost completely arbitrary (the only restriction is that it must be sufficiently large for parallax effects to be undetectable). Second, the position of the centre of the circle is equally arbitrary. The only quantity that follows unambiguously from the data at the attainable observational accuracy is $2e$, the distance of the fixed point E′of the equant spoke from the observer at E (strictly it is the ratio of this distance to the ratio of the diameter or radius of the circle that is determined).

Thus, if Hipparchus was able, and we shall see that he was, to determine the line of the apsides, it is evident from (3.3) that he would conclude, on the basis of observations and the model he adopted, that the sun moved uniformly round a circle whose centre was situated at relative distance $2e$ (where e is the eccentricity of the Keplerian orbit) from the terrestrial observer.

Before continuing with Hipparchus's model, let us take the opportunity to point out how remarkably good in quantitative terms is the small-eccentricity circle-plus-equant model of Keplerian motion of the planets described in Sec. 3.7. This is the model obtained by taking $\lambda = 1$, so that the circle which approximates the orbit is centred on O, exactly half-way between the two foci. Then

$$\gamma_{\lambda=1} = t - 2e \sin t + e^2 \sin 2t,$$

and we see that, to second order in e, the Keplerian angular positions are reproduced except for the very small error of $\frac{1}{4}e^2 \sin 2t$. Moreover, we have already noted how extraordinarily thin are the two lunulae between the circumscribing circle and actual orbit (the maximal separation between the two is $\frac{1}{2}e^2$). In fact, if we take a circle concentric with the circumscribing circle but with radius reduced by $\frac{1}{4}e^2$, the separation between this circle and the orbit is never greater than $\frac{1}{4}e^2$ (taking the radius of the circle to be unity) and the position of the model planet from the actual planet is never greater than about $\frac{1}{4}e^2$. At the octants the actual planet is exceptionally close to the approximating circle and the error is almost entirely concentrated in the angular position. At the quadrants and apsides the angles are correct but the model and actual orbits are $\frac{1}{4}e^2$ apart. There is a real beauty about the circle-plus-equant model and, as we shall see, it lends the early history of astronomy an especial fascination. For full appreciation of its subtlety it needs, as here, to be considered to second order in the eccentricity.

It is also worth emphasizing the dangers that lurk in the exclusive concentration on reproducing the observed angles at the level of the second-order term too. Suppose these angles can be measured to great accuracy. Then, in the framework of our one-parameter model, analysis of angle data alone would clearly lead one to choose the circle-plus-equant model with $\lambda = \frac{5}{4}$, for then (3.3) becomes identical to (3.1). This gets the angles exactly right to second order but shifts the centre of the circular model orbit $\frac{1}{4}e$ away from O. In Fig. 3.10 the position of the planet is then shifted from the tip of the arrow at S to X. The price that has to be paid for getting the angles correct to second order is a perceptible worsening of the spatial positions, in fact, an error of first order is introduced.

We come up here against the fundamental problem that the ancient astronomers had to overcome, namely, that theories of the motion of the sun and planets may reproduce the angular positions extremely well and yet give quite seriously incorrect results for the actual position in space of the object being studied. The overcoming of this problem and the conclusions drawn in the process are what make the astronomical problem absorbing and so significant in the context of this book. Essentially, the aim of this chapter and those on Copernicus and Kepler is to give the reader a clear grasp of all the essential steps in this process.

Let us now consider the principle of the method that Hipparchus used in the actual solution of his problem, since it became the paradigm of virtually all the techniques used in early theoretical astronomy. In Fig. 3.11, which *is* drawn to scale, the circle represents the theoretical Hipparchan orbit of the sun with centre O. According to Hipparchus's theory, the sun moves round the circle at a uniform speed. Since the year lasts $365\frac{1}{4}$ days, the sun must advance from a given point V on its orbit by $360/365\frac{1}{4} = 0.986°$ per day. Now Hipparchus found that the time taken by the sun to pass from the vernal equinox to the summer solstice was $94\frac{1}{2}$ days. Therefore let V be the point on the Hipparchan circular orbit corresponding to the vernal equinox. In $94\frac{1}{2}$ days the sun will have travelled on the orbit to the point S, the summer solstice, where the angle VOS $= 94.5 \times 0.986° \approx 93.2°$. This fixes the point S. He also found that from summer solstice to autumnal equinox (F) the sun took $92\frac{1}{2}$ days. The point F is therefore found by making the angle SOF $= 92.5 \times 0.986° = 91.2°$. Now the earth, from which Hipparchus made the observations, must clearly lie on the line VF, since, on the sky, the vernal and autumnal equinoxes are separated by 180°. Further, the earth must be at the point E obtained by dropping the perpendicular from S onto VF, since angle VES $= 90° =$ angle SEF.

Thus, E is the position of the earth, OE is the eccentricity and PEOA is the line of the apsides, and we see that three observations of the angular position of the sun separated by known intervals of time suffice to solve the problem by elementary trigonometry. Figure 3.11 gives a good idea of the smallness of the effects with which we are concerned but also demonstrates that a relatively small eccentricity can in principle be measured by such methods, though, for the reasons explained, Hipparchus's theory will always give a value for the eccentricity twice as large as it should be.

Hipparchus in fact obtained only a moderately accurate result,[37] finding

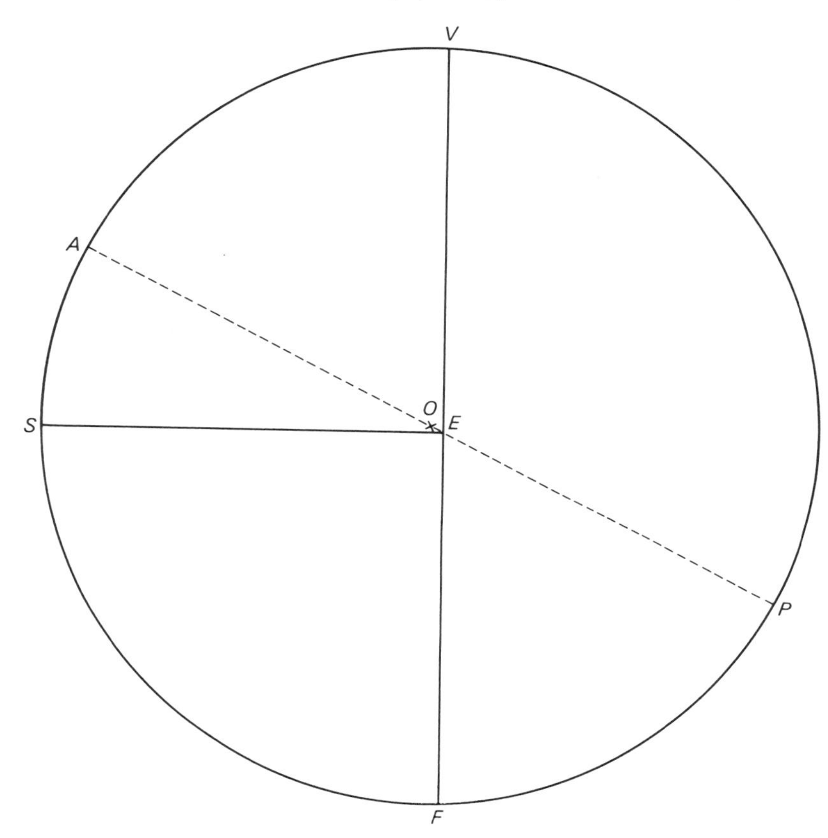

Fig. 3.11. Hipparchus's theory of the solar motion (drawn to scale). The sun is assumed to move uniformly round the circle, whose centre is at O. The arc VS corresponds to the distance travelled by a uniformly moving sun in the $94\frac{1}{2}$ days that Hipparchus found was necessary to pass through the apparent angle of 90° between the vernal equinox and the summer solstice. The arc SF corresponds to the $92\frac{1}{2}$ days between the summer solstice and the autumnal equinox. In accordance with Hipparchus's theory, the earth must be situated at the base E of the perpendicular from S onto VF. Note that the apogee A occurs near the summer solstice, so that the earth is further from the sun in the northern summer.

for the eccentricity the value $r \sim 1/24 = 2/48$ whereas he should have obtained about 2/57. However, for the line of the apsides (which, incidentally, can in principle be determined correctly from Hipparchus's theory, for the error in spatial position associated with his model merely displaced the entire orbit along the line of the apsides but does not affect the angular position on the sky of the apogee and perigee) he did get the rather good result 65°30′ for the ecliptic longitude of the solar apogee, i.e., he found it to be 65°30′ east of the vernal equinox along the ecliptic. Modern calculations show that its value at his epoch was 66°14′.[38] However, this result was something of a fluke. Indeed, it is obvious from a glance at Fig. 3.11 that the position found for the line PEOA will be rather sensitive to errors in the observations. For precise estimates, see Ref. 38.

Some three hundred years later, Ptolemy repeated Hipparchus's observations. These observations, like quite a number of others in the *Almagest*, have been the subject of much controversy. Ptolemy, in fact, is widely suspected of having faked, or at least 'doctored' his observations, quite for what reason is not clear (we shall return to this briefly in the final section). Whatever the truth, he reports in §III.4 of the *Almagest* that he obtained more or less identical results to Hipparchus, including the same ecliptic longitude of the apogee. On this basis he concluded that the solar eccentricity remains constant and that the solar apogee keeps a fixed distance from the vernal equinox. This is in fact wrong; the eccentricity does change and the solar apogee moves relative to the equinoxes (because of their precession) and also, more slowly, relative to the stars. Both effects were too small for Ptolemy to have discovered given the accuracy of his and Hipparchus's observations. We shall see in Chap. 5 that this insufficient accuracy led to numerous confusions in the Middle Ages and caused quite a headache for Copernicus.

It is here appropriate to say something about the accuracy of ancient astronomy. As we see in the case of the solar model, the accuracy was not marvellous; in fact, when $\gamma = 90°$, i.e., the sun is at the quadrants, the inaccuracy of Hipparchus's and Ptolemy's model reached about 23 minutes of arc, i.e., about $\frac{3}{4}$ of the apparent solar diameter. Now this inaccuracy permeated the whole of ancient astronomy because of the key role played by eclipses, which, as we pointed out in Sec. 3.2, were used to determine the position of the moon, which at mid-eclipse was taken to be 180° from the true sun, whose position was in turn deduced from the position of the mean sun in accordance with Hipparchan theory. Moreover, as we shall see, the theory of the moon's motion was itself developed to a very respectable level, and this made it possible to use the moon as a 'marker' to determine the positions of fixed stars from which the positions of the planets were then determined. Thus, the almost steady motion of the true sun and the perfectly steady motion of the mean sun served as the referential basis of virtually all ancient astronomy in the sequence sun–moon–stars–planets. Unfortunately, the basis was not determined through observations as accurately as it might have been.

It is worth quoting here what Neugebauer has to say on the subject of observational accuracy:[39]

> Both Babylonian and Greek astronomy are based on a set of relatively few data, like period relations, orbital inclinations, nodes and apogees, etc. The selection of these data undoubtedly required a great number of observations and much experience to know what to look for. Nevertheless, a mathematical system constructed at the earliest possible stage of the game was generally no longer systematically tested under modified conditions.
>
> This attitude can be well defended. Ancient observers were aware of the many sources of inaccuracies which made individual data very insecure. . . . On the other hand period relations, e.g. time intervals between planetary oppositions and sidereal periods, can be established within a few decades with comparatively high accuracy because the error of individual observations is distributed over the whole interval of time. If the theory was capable of guaranteeing correct periodic recurrence of the characteristic phenomena then intermediate deviations would matter little . . .

There is therefore little point in looking at the actual accuracy Hipparchus and Ptolemy achieved. It is, however, well worth considering the *potential accuracy* that might have been achieved by their models had they had the means and inclination to make regular observations at the accuracy achieved by Brahe in the late sixteenth century, which was about 2 minutes of arc, i.e., about $\frac{1}{15}$ of the apparent diameter of the sun and moon. Seen in this light the potential accuracy of the Hipparchan solar model was almost phenomenal. As we have seen, the maximal deviation from the exact law is only $\frac{3}{4}$ of a minute of arc. In fact, the Islamic astronomer Al Battani (858–929), using Hipparchus's theory, obtained a value of the eccentricity corresponding to an error of maximally just 1 minute of arc.[40] The *potential accuracy* of the ancient models is important because they went on being used essentially unchanged until Kepler's time, when of course the accuracy was much better. In fact, a large part of the story of theoretical astronomy up to and including Kepler's discovery of his laws consists of the extension and modification of the initial Hipparchan solar model.

Indeed this simple model led the astronomers almost effortlessly into what finally developed into fully-fledged dynamics. Numerous characteristic features of the theory that reached maturity with Newton's *Principia* make their first appearance with Hipparchus's seemingly modest attempt to remedy the 'blemish' of the sun's nonuniform motion. The earlier quotation from *Macbeth* in Chap. 2 about 'which grain will grow' is particularly relevant here. Indeed, all the Hipparchan seeds grew. They were: (1) The measurement of speeds by means of some reference motion taken as a measure of the flow of time (rotation of the earth, already discussed); (2) the idea that bodies (the sun in this case) follow quite definite paths in three-dimensional space and that it is the task of observation to determine the nature and position of these paths from the two-dimensional motions observed on the sky; (3) the use of the well-developed mathematical science of geometry to solve this problem; (4) the realization that the immediate deliverances of the senses must be guided by a precise theory if the primal observations are to be made intelligible; (5) and, finally, the realization that the worth of the scheme is ultimately confirmed through its ability to predict future events from a few carefully selected observations made in the past. In fact, the Hipparchan theory requires just four input data: the number of days in the year and the three observations quoted by Ptolemy. With these data Hipparchus in principle succeeded in predicting the position of the sun on the ecliptic for many years in advance with a reasonable accuracy (generally significantly better than the apparent width of the solar disk).

We see from this discussion that not only the entire body of ancient theoretical astronomy but also a great deal of later dynamics was literally 'carried' by the two favourable manifestations of the law of inertia reflected in the diurnal rotation of the heavens and the apparent motion of the sun around the ecliptic.

But there was also an adverse side to the Hipparchan theory. It was only partially correct. Out of the three features of the correct small-eccentricity theory of the earth's motion around the sun, it grasped only two: the circularity of the orbit and the eccentric position of the observer. The actual nonuniformity escaped Hipparchus because his theory automatically selected the unique model with uniform actual motion among the one-parameter family of models that all equally well reproduced the angular positions of the sun, and, as we have seen, he obtained an eccentricity about *twice* as large as he should have. We shall see that nature and mathematics conspired to play several caddish tricks on the ancient astronomers; this hidden defect of the seemingly perfect Hipparchan solar theory was perhaps the nastiest. It had no effect on the theories of either the sun or moon, but it did have serious consequences in the theory of the planets and remained undetected for over 1700 years.

There is one final point that may have escaped the reader; for Hipparchus made a significant but arbitrary assumption to do with the passage of time. Let us recall and emphasize the crucial importance of the 'ticks' provided by the earth's rotation; without them Hipparchus would have had absolutely no means of saying where the sun must be on its theoretical orbit. But we know from the philosophical discussion at the end of Chap. 2 that the speed of one motion can be measured only by means of a second motion, taken as reference. Hipparchus was concerned with just two motions – that of the diurnal rotation of the stars and that of the sun around the ecliptic. Under these circumstances it is clearly impossible to say that one or the other motion is uniform; they are either mutually uniform or mutually nonuniform. What then is the justification for saying one is uniform while the other is not? In Hipparchus's time, there was little sound scientific evidence for the choice that he made, namely, the assumption that the diurnal rotation of the stars was uniform and the apparent motion of the sun nonuniform. However, his gut intuition, guided no doubt by the overwhelming impression that the diurnal rotation does make, led him to make a choice that was in fact 'right'. We shall discuss the precise sense in which Hipparchus was 'right' after considering the Ptolemaic theory of the moon and planets.

The next topic to be discussed is the epicycle–deferent technique, which was the most characteristic part of the theoretical structure of Hellenistic astronomy. Although virtually nothing is known about the process of its discovery beyond the fact that the technique was clearly known to Apollonius, the reason for its invention is not hard to seek – it was, just like the theory of the eccentric just described, born from the attempt to explain nonuniformity of the two-dimensional motion of the celestial bodies as observed on the sky by means of *uniform circular motions* in three-dimensional space.

3.9. The epicycle–deferent theory

The basic idea of the epicycle–deferent method is very simple. Imagine a large circle (Fig. 3.12) at the centre of which an observer O is situated and on the circumference of the circle, which is called the *deferent* (this is the medieval term), imagine a second circle, smaller than the first and with its centre B exactly on the deferent's circumference. The second circle is called the *epicycle*. Now suppose that the centre of the epicycle moves at a

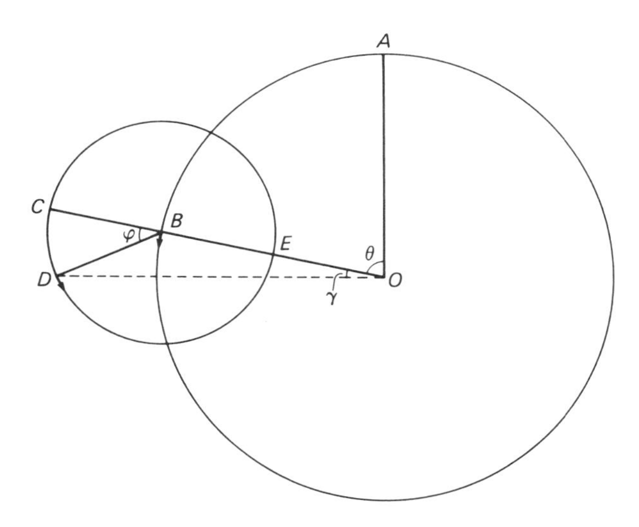

Fig. 3.12.

perfectly uniform speed around the circumference of the deferent, so that the angle θ between the fixed line OA and the moving radius OB, which we shall call the *deferent spoke,* increases at a constant rate, and that simultaneously an observable body D moves round the epicycle in such a way that the angle φ made by the radius vector BD, the *epicycle spoke,* and OBC also increases at a uniform rate. From O, the object will, of course, be observed along the direction of OD. It is obvious that the angle between OA and OD, i.e., $\theta + \gamma$, will increase at a nonuniform rate, so that a nonuniform apparent motion can be generated.

There is, in fact, a rather good physical realization of the epicycle–deferent system in the solar system – the earth–moon–sun system. Put the sun at O, the earth at B, and the moon at D. Then to a first approximation the motion of this system corresponds to the epicycle–deferent scheme. From the sun, the moon is observed to pass around the earth as the earth itself moves around the sun. However, the analogy is not complete and highlights an important feature of the epicycle–deferent scheme. In the physical example just given, the earth occupies the position B but in Hellenistic theoretical astronomy the point B, which may be called a *guide point*, is void. There is just one observable body, and that is at D.

It is quite surprising how much can be achieved by this simple model. Let us list the disposable parameters that can be adjusted in order to reproduce observed nonuniformities. Except for the moon, the absolute scale was inaccessible to the astronomers, so OA can be taken as the nominal unit of distance. Then the most important parameter of the model is the ratio of the epicycle radius BD to OA, the deferent radius. Next come the two angular velocities $\dot{\theta} = \mathrm{d}\theta/\mathrm{d}t$ and $\dot{\varphi} = \mathrm{d}\varphi/\mathrm{d}t$. (Note that either angle may increase in either clockwise or anticlockwise direction; the observed behaviour of D is clearly strongly influenced by the sense of rotation of the epicycle relative to the deferent motion.) Finally, there is the *relative phase* of θ and φ. Suppose that at time $t = 0$ the radius vector OB passes through A. At this time, the angle φ will have a certain value φ_0. This too is an adjustable parameter of the model. We shall see later how these parameters, which are all observable in principle, were determined in practice.

We can mention here already some of the simple modifications of which the basic epicycle–deferent scheme is capable and which greatly extend its ability to reproduce observed nonuniform motions. First, a further circle, or 'epiepicycle' can be added. In this case, D too becomes a void guide point, the centre of an even smaller circle that travels around the epicycle while the body itself travels around the epiepicycle. A second simple modification is to assume that the observer is not at O, the centre of the deferent circle, but at some eccentric point. This obviously combines the eccentric scheme already discussed in the Hipparchan solar theory with the epicycle technique. Finally, we have assumed that the epicycle is in the plane of the deferent. This means that the observer at O always sees the body move on a great circle on the celestial sphere. But if, as is the case with the planets, the observed body always remains close to a great circle (the ecliptic), but not exactly on it, one can attempt to reproduce this motion by tilting the plane of the epicycle at some definite angle to the deferent. Numerous other modifications are possible, and some of them were actually used.

Viewed from a certain distance, one can see that the epicycle–deferent scheme and its various modifications evolved in an *ad hoc* but very natural manner. The astronomers grasped whatever device came to hand in order to describe nonuniform apparent motions by mathematically tractable techniques, the ultimate basis of which was, in all cases, a uniform circular motion (which is particularly easy to handle), nonuniformity being generated either by superposition or by eccentric position of the observer. Once the desire to find a rational explanation for motions believed to be divine had given the initial stimulus, and the first successes had given confidence in the basic correctness of the technique, theoretical astronomy developed as a more or less autonomous science. In Kuhn's terminology,[41] it is the first example – and a very good one at that – of *normal science* at work. And the circles survived for so long for one simple and very sound reason – the orbits of the celestial bodies in helioastral space *are* circles to a very high degree of accuracy. This means that good results could be obtained with only comparatively slight modifications to the basic scheme, so that need for an alternative was not felt. Even the quite appreciable nonuniformity of the actual motions could be accommodated, as we shall see.

3.10. First application of the epicycle–deferent theory: alternative form of Hipparchus's theory

Let us now look at the application of the epicycle–deferent technique to two of the great early problems in astronomy: the solar motion and the motion of the planets. This is just to give us a first idea of its applications and power. We start with an alternative version of the Hipparchan solar theory.

In Fig. 3.13 let the large continuous circle be a deferent, the small circle an epicycle and suppose that when the guide point of the epicycle is at B, the body itself is at D and that as the guide point moves anticlockwise relative to the fixed line OB, so that it is carried to a certain point B′ at time t, the epicycle spoke moves relative to the main spoke OB′ *clockwise* at exactly the same rate, so that at time t the body will be at D′. Then obviously B′D′ will be parallel to BD and, since B′D′ = BD, it is quite clear that the path described by the body in space will be the dashed circle with

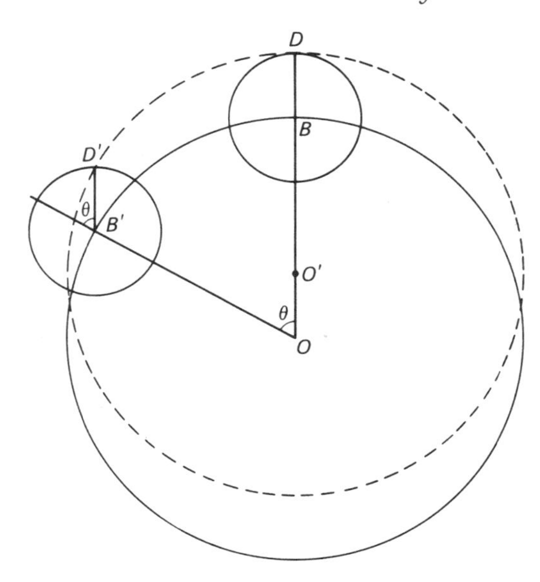

Fig. 3.13.

centre at O′. If we now look back to Fig. 3.10 and take the epicycle radius and OO′ to be 2*e*, it is quite clear that the motion of the body with respect to O (and hence its observable position on the sky) will be exactly as that of the sun in the Hipparchan solar theory described earlier. Thus, in this very special case, the eccentric scheme and the epicycle–deferent scheme are completely interchangeable.

Neugebauer points out that both the Greeks and the medieval astronomers switched frequently from the one model to the other and that this readiness to switch:[42] 'is the best indication of the fact that none of these models implied that there existed in nature a corresponding mechanical structure'. Although we shall see later that there is some doubt about the universal validity of this assertion, the techniques were certainly to a very large degree formal devices. This is often described as a weakness of Greek astronomy – as betraying a lack of interest in physical explanation. It is, however, necessary to bear in mind what was said in the previous chapter about geometro-mechanical explanations of motion. They were in fact a hindrance to progress, since they obscured the fact that, in the final analysis, motion is a *primary* phenomenon and cannot be reduced to three-dimensional geometry. The great virtue of Hellenistic astronomy, as the subsequent history showed, was that it determined the actual motions in three-dimensional geoastral space. Although the devices used to reproduce these motions were often to a high degree arbitrary, as illustrated by the frequent switching, the actual motions were the same whatever model was used. When the Copernican revolution finally came, it was merely necessary to transcribe the celestial motions from geoastral space to helioastral space in order to obtain what was already an extremely good approximation to the actual motions in helioastral space. Whether by luck or judgement, the astronomers undoubtedly chose the correct route. Before the actual motions were known to a high degree of precision, the search for a physical cause of them was wildly premature. Moreover, when the explanation did come it was a total surprise and quite unlike anything that could have been anticipated. It is thus important not to be misled by the mathematical techniques used by the astronomers to describe the celestial motions. For the discovery of dynamics, the important thing was the actual motions, not the means of their description.

One should also not be misled by a comparison, sometimes made, between the epicycle–deferent technique and Fourier's representation of arbitrary continuous functions by trigonometric series. There is unquestionably an analogy, especially in the case of the Hipparchan solar theory. In fact, examination of Eq. (3.1) shows that the effect of the epicycle is exactly reproduced by the first term of a trigonometric series, which can be seen as a small correction to the main term t. What was striking about the further development of Hellenistic theoretical astronomy was its demonstration that amazingly few 'correction' terms were needed to reproduce the observed celestial motions to a very good accuracy. Had the motions been more arbitrary, far more terms would have been needed. But in any case the comparison between Fourier analysis and the epicycle–deferent scheme is flattering rather than pejorative – and that for two reasons: (1) it correctly reflects the genuine *periodicity* that resides in the celestial motions, (2) as we have said, the most important thing in the early steps towards the rational description of motion was to describe the actual observed motions accurately and economically – in a form that could be readily surveyed. For this purpose, the epicycle–deferent technique, with its potential for easy adaptation to introduce small corrections when needed, was almost as good as Fourier's admirable technique, which still serves as a guide to innumerable discoveries.

3.11. Second application of the epicycle–deferent theory: the motion of the outer planets

Figure 3.14 shows, in the Copernican scheme, the orbits of the earth and Saturn to scale but in the zero eccentricity and zero inclination approximation, i.e., both planets are assumed to move at a perfectly uniform speed in coplanar circles that have the sun (S) at their common centre. The point P representing Saturn moves around S in about $29\frac{1}{2}$ years, while the point

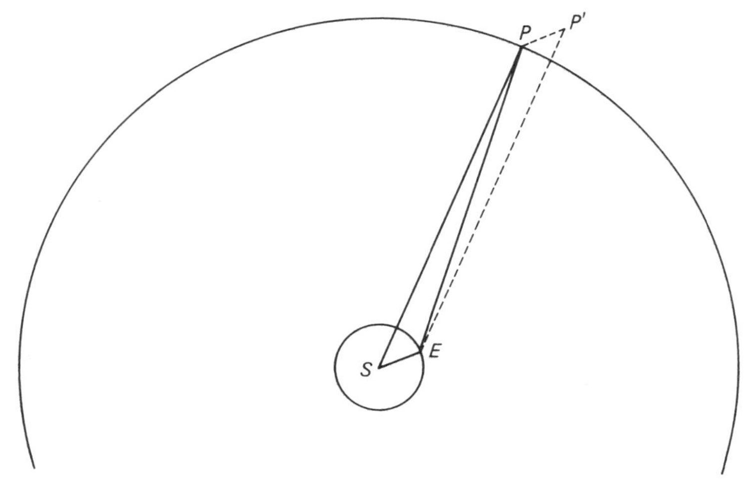

Fig. 3.14.

E representing the earth takes one year to traverse its orbit. Therefore, as seen from the sun, Saturn moves uniformly around a great circle. Our question is: how does Saturn appear to move as seen from the earth?

Since the circles are assumed coplanar, Saturn is observed from the earth to move around the same great circle as for a solar observer. However, the motion is clearly no longer uniform. It is very easy to see how it will appear. When Saturn is seen from the sun along SP, it is seen from the earth along EP. In Fig. 3.14, the point P′ makes SEP′P into a parallelogram, i.e., SE is parallel to PP′ and SP to EP′. Thus, if there *were* a real body at P′ an observer on the earth would see it at the same point on the stellar background as an observer on the sun would see the actual planet. Now as both E and P move, the motion of P about P′ is the same as that of E about S but 180° out of phase. Thus, the apparent motion of P about E can be decomposed as follows. First, the void point P′ moves around the earth with exactly the motion of Saturn about the sun. At the same time, P moves around P′ with the motion of the earth about the sun but 180° out of phase. But this last motion is the apparent motion of the sun about the earth.

Thus, the terrestrial observer will be led very naturally to picture the motion of Saturn as follows. First, a void point P′ moves round the earth with exactly the same motion as does Saturn about the sun (we recall that the stars are effectively at infinity, so that P′ is projected from E onto the same point of the stellar background as P from S). Then the actual planet P moves around P′ with exactly the apparent motion of the sun about the earth. But this is exactly what we have in the epicycle–deferent scheme with EP′ corresponding to the deferent spoke and P′P to the epicycle spoke.

Of course, what we have in effect done is to go over from the helioastral frame to the geoastral frame, i.e., we have merely shifted the origin of the coordinate system from the point S, which is at rest in the helioastral space, to the point E, which is moving in that space. The directional coordinates remain unaffected because of the great distance of the stars. We note a curious effect of this shift of the coordinate origin: the actual motion of the real planet Saturn about the sun is transformed into apparent motion of the void point P′ around the earth. If one considers just the deferent motion, then the deferent guide point seems to move around the earth in exactly the same way as the planet moves around the sun.

Figure 3.15 shows the motion (over a period of just over 2 years) in geoastral space (i.e., as obtained by the method just explained) of the three outer planets under the assumption that all eccentricities (of the orbits of the earth and Mars, Jupiter, and Saturn) are zero. The points on the curves show the positions of the planets at intervals of 30 days (the corresponding void guide points on the deferents are also shown at intervals of 30 days). Some positions of the epicycle spoke are shown for the case of Saturn. The direction of the epicycle spoke always gives the direction of the sun as seen from the earth. For example, when the planet Saturn is at O, the sun is at O′.

The figure reveals very clearly, especially for Saturn and Jupiter, the two most characteristic features of the apparent motions: their pronounced nonuniformity and the so-called *retrogression loops*. Let us, for example, follow the motion of Saturn from position m, at which it becomes visible in the morning sky, through to O when it is in what is called *opposition* (because it is then in the position of the sky opposite to the sun) and on to e, when it is still visible in the evening sky. At the start of this period the planet moves forward around the ecliptic in the same sense as the invisible guide point and also the sun. However, quite soon the apparent forward motion is slowed down, this happening for two reasons. First, because the distance the planet moves in geoastral space in equal intervals of time becomes less (the points in the figure are crowded together more closely) and, second, because the motion is no longer purely radial in geoastral space but has an ever increasing component towards the observer. Indeed, at the point S_1, the apparent forward motion comes to a halt and for quite a length of time the planet seems to stand still before starting to move backwards. This *retrograde* motion lasts until the planet reaches the point S_2, at which it again appears to stand still before once again resuming the dominant forward motion. An important point to note is that the apparent retrograde motion at O is always slower than the apparent forward motion at m and e. For the construction of an

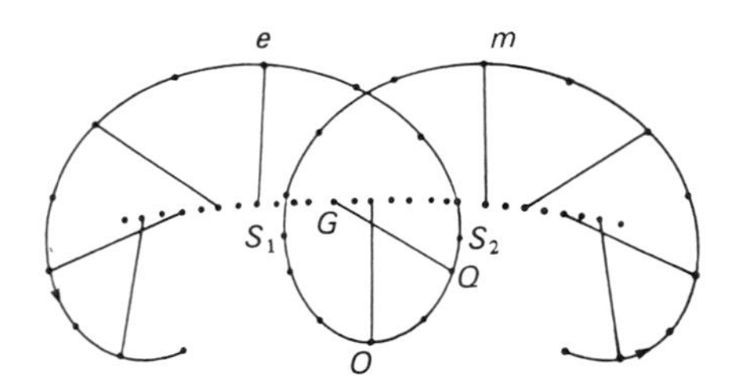

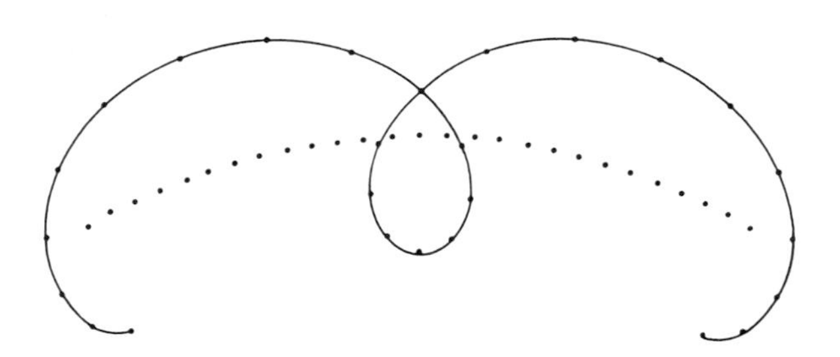

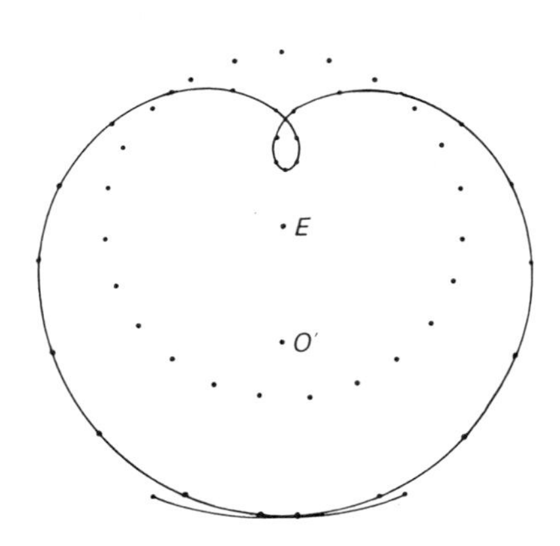

Fig. 3.15.

epicycle–deferent model, this is a clear hint that the sense of rotation of the epicycle spoke must be the same as that of the deferent spoke. Of course, the terrestrial observer sees only the projection of the motion onto the sky but a clear indication (albeit not too accurate) of the actual varying distances between the observer and planet is provided by the change in brightness, which is always greatest when the planet is in opposition. For Mars this effect is very pronounced indeed and confirms that at opposition the epicycle spoke must be pointing towards the earth.

On the basis of these very striking features one can readily appreciate how the Greeks were led to propose the epicycle–deferent model for the explanation of the planetary motions.* From the apparent motions as observed on the sky it is not difficult to deduce the basic parameters of the model. Because the centres of the retrogression loops always occur at opposition (at O for Saturn), when the sun is diametrically opposite to the planet on the ecliptic, one can deduce the crucial property that the epicycle spoke must always be parallel to the direction from the earth to the sun. It must therefore rotate once through 360° relative to a fixed direction in geoastral space in $365\frac{1}{4}$ days (the length of the year), corresponding to the mean motion of 0.9856° per day. Because the deferent spoke itself moves, the angle between the epicycle spoke and the deferent spoke increases not quite so fast: 0.9522° per day for Saturn, 0.9025 for Jupiter, and 0.4616 for Mars. It only remains to calculate the ratio of the length of the epicycle spoke to the length of the deferent spoke. This can be done, for example, as follows. Sixty days after Saturn has passed through opposition at O the void guide point will have advanced round the deferent to the point G, while the epicycle spoke will have simultaneously rotated into the position GQ. The position G can be calculated from the mean motion of Saturn and the position Q is observed. Thus, the angle QEG can be determined and from it the required ratio.

* The precise manner in which the model was found historically is not known. The account given here is merely an illustration of how the discovery could have been made; no pretence to historical accuracy is claimed.

Fig. 3.15. The motion of Saturn, Jupiter, and Mars in geoastral space, drawn to scale but under the assumption of zero eccentricity of all three planets. The earth is at E. The points on the continuous curves show the positions of the planets at intervals of 30 days. The corresponding positions of the void guide points of the deferents are also shown. Note that the ancient astronomers had no sound basis for determining the absolute dimensions. In the *Almagest*, Ptolemy set the radii of the deferents equal in all cases to the nominal value 60. In his later *Planetary Hypotheses* he assumed that the epicycle-deferent systems fitted flush next to each other, eliminating the gaps shown in the figure. Note also that the figure is drawn for the case in which all three planets are in opposition at once – a very rare occurrence.

Figure 3.15 also shows very clearly how the characteristic observed behaviour of the three outer planets changes. For Saturn about $28\frac{1}{2}$ retrogression loops are completed before the planet returns (after $29\frac{1}{2}$ years) to the same point in the ecliptic; for Jupiter the number is just under 11; while for Mars the retrogression loop is not in fact quite completed. It is worth pointing out that the form of the retrogression loops depends both on the radii of the orbits (i.e., on the ratio of the lengths of the epicycle and deferent) and on the speeds of the planets in their orbits. This is readily seen. Let R and r be the radii of the deferent and the epicycle respectively and let the angular velocity of the deferent spoke in geoastral space be $\dot{\theta}$ and the angular velocity of the epicycle spoke *relative to the deferent spoke* be $\dot{\varphi}$. Then EO $= R - r$ and the forward motion of the deferent spoke at O in a short time interval $\mathrm{d}t$ will be $(R - r)\dot{\theta}\mathrm{d}t$. But the motion of the tip of the epicycle spoke as it passes through O (when its backward motion is fastest) will carry it relative to the deferent spoke a distance $r\dot{\varphi}\mathrm{d}t$ in the opposite direction in the same time. Thus, retrograde motion will only occur if $r\dot{\varphi} > (R - r)\dot{\theta}$ or

$$\frac{\dot{\varphi}}{\dot{\theta}} > \frac{R - r}{r}. \tag{3.4}$$

It is interesting to note that the condition (3.4) is always satisfied for the planets by virtue of Kepler's Third Law. If Saturn's orbital speed were the same as the earth's, it is easy to show that $\dot{\varphi}/\dot{\theta} = (R - r)/r$ would hold, i.e., Saturn would appear to come to a halt but would not actually retrogress – after a pause, the planet would recommence its forward motion. However, it is a consequence of Kepler's Third Law (in the case of zero eccentricity) that the orbital speed is proportional to $1/\sqrt{R}$, with the consequence that retrogression is always observed, though the retrogression loops occupy a progressively smaller proportion of the complete cycle the larger the ratio r/R. This effect is clearly seen in the sequence Saturn–Jupiter–Mars (Fig. 3.15). Note, however, that, as seen from E, the apparent size of the loops decreases in the sequence Mars–Jupiter–Saturn.

For the three outer planets the retrogression loops can always be observed extremely well since the sun is in the opposite part of the sky and the planets are seen due south at midnight. These are the best observing conditions. It is therefore perhaps not surprising that the Greek astronomers with their instinctive use of geometry (which is so characteristic of their work) developed quite early a precise mathematical theory of retrogression. In fact, the theorem of Apollonius mentioned in Sec. 3.1 deals with precisely this point.[43] For given R, r, $\dot{\varphi}$, and $\dot{\theta}$ it tells one the position of the epicycle spoke relative to the deferent spoke at the moment at which retrogression commences, i.e., the forward motion stops and is then reversed.

This entire treatment so far has been based on the assumption of zero eccentricity. As we shall see in Sec. 3.14, the situation becomes significantly more difficult when the actual eccentricities have to be taken into account. Before that we need to consider the theory of the motion of the two inner planets, Venus and Mercury.

3.12. Epicycle–deferent theory for the inner planets

In Fig. 3.15 we have plotted the actual orbits in geoastral space of the three outer planets. We see that as the ratio r/R gets larger on the transition from Saturn through Jupiter to Mars the characteristic form of the orbits remains the same though the retrogression loops become less frequent. However, on the transition to the inner planets, a qualitative change takes place. The reason for this can be seen by going back to Fig. 3.14, which we interpreted as follows: the void point P′ moves around E with the same motion as P about S while P moves around P′ with the reverse of the earth's motion about the sun. There is, however, an interesting alternative way of looking at the situation, which rests on the commutativity of vector addition. Still holding on to the idea that E is at rest, one could equally well say that S, the sun, moves around E, the earth, on a small deferent circle with radius ES and that the planet P moves about S on a large epicycle with radius SP. In this case the void point P′ plays no role, but the price one has to pay for this is a 'top heavy' situation in which the epicycle has a larger radius than the deferent. Examination of Fig. 3.15 shows how unnatural it would be to adopt such an explanation for the motion of Saturn and Jupiter in particular. However, for Mars, for which the two radii are much more nearly equal, the choice between the two possibilities is not nearly so clear cut. Moreover, once we pass to the two inner planets the ratio of epicycle radius to deferent radius will have to become greater than one if we are to stick to the same scheme as for the outer planets. But at this the ancients baulked.

That the two mathematically equivalent possibilities for interpreting Fig. 3.14 do exist was undoubtedly well known to them, since Ptolemy proves Apollonius's theorem simultaneously for the two cases.[43] However, reading the *Almagest*, one gets the impression that Ptolemy was somewhat mesmerized by the matter and could not feel comfortable unless the epicycle had a radius less than the deferent. He therefore chose differently for the two cases – using the scheme explained in Sec. 3.11 for the outer planets but with the reversed situation for the two inner planets. This ensured that he had in all cases an epicycle spoke shorter than the deferent spoke but at the price of a notable heterogeneity in the system. For in the case of the inner planets it is now the deferent spoke, not the epicycle spoke, that always remains parallel to the sun.

The transcription from helioastral to geoastral space for the two inner planets is shown in Fig. 3.16(*a*). The deferent spoke ES for both planets coincides with the radius vector from the earth, E, to the sun, S, around which Mercury (M) and Venus (V) rotate on epicycles that have the sun at their centre. Note the most striking feature in the apparent behaviour of the inner planets; they are never seen at more than certain definite maximal angles from the sun, quite unlike the outer planets.

It must however be said that Fig. 3.16(*a*), which shows the correct positions in geoastral space, is seriously misleading in one respect; for the Greeks, or at least Ptolemy, did not adopt the model shown in Fig. 3.16(*a*)

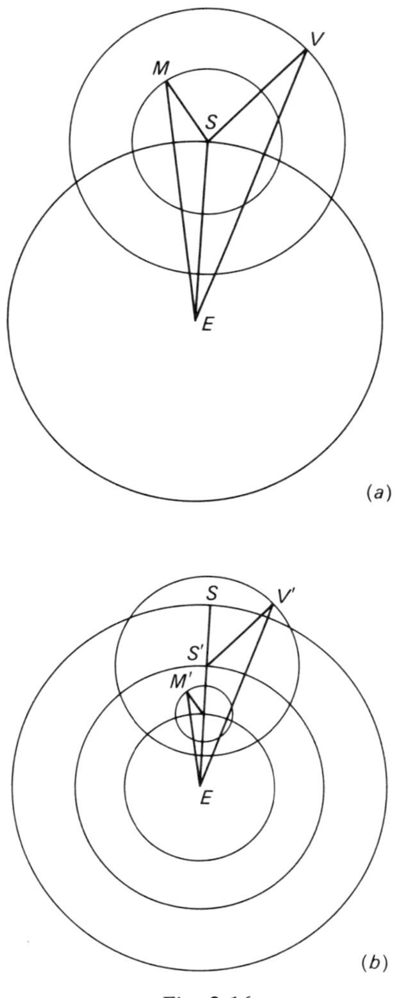

Fig. 3.16.

but one that was merely equivalent to it as regards the observations that could be made from the earth E. We recall that the astronomers were unable to measure any distances except that of the moon and, very inaccurately by means of eclipses, that of the sun. They had no means of determining the distances of the planets. This led to a *scale invariance* of the planetary problem. The epicycle–deferent theory made it possible to determine from the observations the angles of the epicycle spokes (SV and SM in Fig. 3.16) relative to ES, which were determined without any ambiguity. However, only the *ratio* of the lengths of SV and ES could be determined. This leads to a great ambiguity of possible models for the description of the same observed motions on the sky. In Fig. 3.16(*b*), S′V′ is parallel to SV in Fig. 3.16(*a*) and of length such that EV′ is parallel to EV in Fig. 3.16(*a*). It is evident that if S′ moves round on a circle of radius ES′ at the same rate as ES, and V′ swings round S′ at the same rate as V about S, then from E the point V′ will always appear on the sky at the same angular position as V. A similar alternative construction is shown for Mercury (M′). To the modern mind, shown a diagram like Fig. 3.16(*b*), it seems almost incredible that the ancient astronomers did not automatically assume that the two inner planets did actually circle the sun. For they knew that the deferent for Venus and Mercury always pointed in the direction of the sun and that neither planet was ever observed at more than a certain fixed maximal elongation from the sun, appearing now on one side of it, now on the other. What more natural explanation of this behaviour could be found than the suggestion that Venus and Mercury do actually circle the sun? This would have led to a *partially heliocentric* scheme.*

There are several reasons that can be advanced to explain why the Greeks did not adopt this idea. The first is to do with the earlier history of astronomy. We recall that Aristotle had believed the celestial bodies to be carried by quintessential spheres. But if Mercury and Venus do circle the sun on an epicycle, they must be constantly passing through the sphere that carries the sun. It could be that to avoid this possibility the ancient astronomers assumed that the entire epicycle–deferent system for each celestial body must lie either completely outside or inside the sphere of the sun. Such a notion did in fact gain support from another remarkably perverse and caddish trick that nature played on the astronomers. It seems that Ptolemy,[45] considering the great distance between the moon and sun (of which he had an estimate at least), felt that it would be

* Many histories of astronomy suggest that Heraclides actually discovered the epicycle–deferent theory for the case of Venus and that he did posit partial heliocentricity. This appears to be a classic case of historians repeating earlier historians' mistakes. According to Toomer, there is no solid foundation to the suggestion, though the arrangement was proposed in antiquity, in fact before the time of Ptolemy, who for some reason ignored the proposal.[44]

appropriate if this empty space were filled by the epicycle–deferent systems of Mercury and Venus. He therefore assumed that Mercury at its closest approach to the earth came just to the sphere of the moon. Since the moon's distance was known, this enabled him to determine the absolute scale of the epicycle–deferent scheme of Mercury. He then assumed that Venus at its *closest approach* to the earth was at the distance of Mercury when it was *furthest* from the earth. This then set the scale for Venus. When Ptolemy did his calculations, he found that Venus had a furthest distance of 1190 earth radii from the earth that was almost exactly equal to the minimum distance of the sphere of the sun as determined through the eclipse observations (1160 earth radii). This, of course, was the purest of flukes but it lent powerful support to the idea, which then remained unchallenged until the Copernican revolution.*

Taken together, the scale invariance and the epicycle–deferent inversion for the two inner planets went a long way to destroying an organic picture that Ptolemy would otherwise have obtained – a picture that would probably have suggested the heliocentric possibility more strongly than was the case. For the three outer planets there was a clearly discernible trend. In the sequence Saturn, Jupiter, Mars the periods of the deferent motion became successively shorter while the ratio of the lengths of the epicycle spoke to deferent spoke increased. This suggested strongly to the ancients what was, in fact, the correct ordering of the orbits: Mars nearest the earth, then Jupiter, finally Saturn, all three being placed beyond the sphere of the sun. But for the two inner planets, which were generally but not invariably taken to be within the sphere of the sun, the deferent period was the same and equal to that of the sun. This produced a curious and ungainly system and strengthened the impression that true universality was not to be found in the celestial motions. For the moon and sun each had a type of motion quite distinct from the planets. But even these, in their turn, seemed heterogeneous on account of the epicycle–deferent inversion. Thus, there appeared to be four broadly different types of motion: for the moon, for the sun, for Mercury and Venus, and for the three outer planets. In fact, the heterogeneity was increased still further by the details of the motion of Mercury, which, as we shall see, Ptolemy found himself forced to describe in a manner significantly different from Venus.

* This 'filling-the-empty-space' theory was not put forward in the *Almagest* but in Ptolemy's later *Planetary Hypotheses*. It is worth noting that in the *Almagest* Ptolemy merely said that Mercury and Venus must be a substantial distance beyond the moon since neither exhibit parallax effects. His later theory conflicts with this requirement, and Mercury should certainly have exhibited parallax. This is one of only two manifest errors in Ptolemy's work, i.e., examples of errors in which his theory was in clear disagreement with observation. We shall come to a more serious example shortly.

After this survey of some of the problems that still survived or were even in part created by the epicycle–deferent scheme, we should now point out that its discovery was nevertheless a triumph whose significance can hardly be overemphasized. The motions of each of the planets as observed from the earth are highly irregular with enormous deviations from uniform circular motion. Yet they are all described well, and some almost perfectly, by a clean decomposition into just two circular motions, both of them uniform. Moreover, the extremely large variations in the brightness of Mars, in particular, were simultaneously given a very good qualitative explanation – two birds hit with one stone! The epicycle–deferent theory really was a great discovery and certainly deserves to be identified as the first of the half-dozen insights from the theory of celestial motions that went into the creation of dynamics. It gave a tremendous boost to the faith in a rational explanation of the seemingly irrational motions of the seven 'wanderers',* who represented such a curious blemish when compared with the perfection of the diurnal rotation of all the stars. Once it had been discovered, it, much more than 'divine preconceptions', was clearly what sustained the astronomers in the belief that they possessed a universal law: *every celestial motion is either a uniform circular motion about a centre or else is compounded out of such motions*. We shall see shortly how well even the most irregular motion in the heavens, the moon's, seemed to satisfy the principle.

However, before we pass on to the final achievements of ancient Greek astronomy, the theory of the moon and Ptolemy's theory of the planets, it is worth emphasizing the extent to which the epicycle–deferent theory of planetary motion clearly implicated the sun as a major determining factor in the apparent motions of the planets. For, as we have seen, the epicycle–deferent decomposition of the motion of any planet always has a component along the line joining the sun and the earth. Either the epicycle spoke (in the case of Mars, Jupiter, and Saturn) or the deferent spoke (Venus and Mercury) is always aligned along a line parallel to the earth–sun direction. No matter how the decomposition of the motion is made, one leg of the decomposition – either the deferent motion or the epicycle motion – always marches in phase with the sun. It seems that the sun exerts a *partial control* on the planets in a most mysterious fashion. This was perfectly well known to the ancient astronomers and yet seems to have attracted remarkably little comment on their part. In the *Almagest*, Ptolemy states this highly important observational result in a most dry and matter-of-fact way – there is no hint of mystification or surprise. The result permeates the entire theory of the planets, but an inattentive

* Initially the Greeks referred to the sun, moon and five naked-eye planets as 'wanderers' (planets); only later were the sun and moon excluded from this designation.

modern reader might well miss the point at which Ptolemy mentions the fact explicitly.[46]

Thus, Ptolemy and the ancients were content merely to note the fact; no attempt was made to seek a cause for the commonality in the motion of the planets and the sun. It was left to Copernicus, Kepler, and Newton to do that. But, as we shall see in Chaps. 10 and 11, Newton discovered a remarkable commonality in the motion of *all* bodies in the universe – inertial motion. Moreover, there is an 'alignment' too: in an inertial frame of reference (to use modern terminology), the distant stars are observed to be at rest. The inertial motion is 'aligned' on the distant stars. Newton's explanation of this commonality by means of absolute space parallels the ancients' acceptance of alignment of one component of the planetary motion with the sun as a fact of life. He did not look for a *visible cause* of the commonality. Mach suggested that it could be found in the totality of the matter in the universe. According to his suggestion, the universe as a whole stands in a causal relation to inertial motion in much the same way as the sun to the mysterious sun-aligned component in the planetary motions.

We conclude this section by noting that Aristotle's cosmology and physics had a charmed life. The early work of the Hellenistic astronomers undermined it in two directions: first, through the development of trigonometry. This was a time bomb with a very slow burning fuse. Potentially more dangerous was the failure of Eudoxus's strictly earth-centred scheme to account for the apparent motion and brightness of the planets and its replacement by the far more successful epicycle–deferent theory. Strictly speaking, this should have demolished the entire Aristotelian structure; for it established beyond reasonable doubt that planetary motions are in no way compounded from simple uniform motions around the single centre of the universe. There was in fact no need to wait for the Copernican interpretation of the astronomical observations for the overthrow of Aristotle; the ancient system was already at variance with it.

Significantly, this was one of the points at which Copernicus chose to mount his assault on Aristotle:[47] 'nothing prevents the earth from moving. . . . For, it is not the centre of all the revolutions. This is indicated by the planets' apparent nonuniform motion and their varying distances from the earth. These phenomena cannot be explained by circles concentric with the earth. Therefore, since there are many centres, it will not be by accident that the further question arises whether the centre of the universe is identical with the centre of terrestrial gravity or with some other point.'

This poses the question of how Aristotle survived so long. The first reason no doubt was that there was still no direct evidence for any motion of the earth. A second was certainly that although the complete scheme was no longer tenable some of the most important parts of it had in fact been dramatically confirmed. This is the irony of the circles: Aristotle was only saved by the fact that the planets do indeed move in nearly perfect circles around the sun. This need not have been the case.

The circle concept arose originally from the diurnal motion. Developing trigonometric methods totally at variance with Aristotle's philosophy of place, the Hellenistic astronomers applied them to essentially new phenomena, the residual planetary motions when the diurnal motion was subtracted; miraculously, the circles appeared to work perfectly. The two-millennial survival of the Aristotelian system was based on a solid success, even if it was a fluke. Nevertheless, the first major step had been made from Aristotle's qualitative and organic geocentric world. The inexorable logic of trigonometry, coupled with accurate observation, had established beyond reasonable doubt the existence of *other* centres of revolution at immense distances from the earth. The world no longer had a single centre. The Aristotelian unity was falling apart, though the most characteristic feature of this process could already be discerned: the unity imposed by the concept of a unique centre of the universe was being replaced by a *unity of behaviour* of the parts. For, however imperfectly perceived, the behaviours of the planets considered separately still undoubtedly possessed features in common. We have here the first hints of a quite different type of unity; expressed through the laws of motion being the same everywhere and at all times rather than their being referred to a unique centre of the universe.

There was never really any danger of the whole world falling apart – it was just the first step in the process of adjusting to the awareness that unity can be expressed in more subtle and all-pervading ways. Nature is more sophisticated than the mind of man – even Aristotle's.

3.13. The theory of the moon

The techniques so far described correspond roughly to the level of sophistication to which theoretical astronomy had advanced when Hipparchus died sometime after 127 BC. It will be helpful to review these briefly in terms of orders of magnitude with respect to the eccentricities. The behaviour of the planets in zeroth order in the eccentricities was perfectly described by the epicycle–deferent theory. However, the situation with regard to the two relatively large and equally important first-order effects–the displacement of the still almost perfectly circular orbit into an eccentric position and the marked nonuniformity of the motion within the displaced orbit–was more complicated. For, deceived by a perversity of the mathematics and the specific eccentricity of the earth's orbit, Hipparchus (using either an eccentric or the epicycle–deferent scheme) had managed to reproduce at a good quantitative level the observed manifestations of both effects in the solar motion by means of a theory that took into account only the displacement to an eccentric position. The effect of the actual nonuniformity of the motion had been mimicked by a false doubling of the displacement.

Now in fact exactly the same thing happened with the moon, though here there was a curious and helpful twist. The part played by the moon, which was bound to attract attention from an early date on account of its phases, the rapid speed of its motion, and the eclipses it suffers and causes, is somewhat surprising. Its motion is complicated to say the least, since it is the only one of all the dynamical problems in the naked-eye astronomy of the solar system in which so-called *three-body* effects play a significant part. The point is that the moon's motion relative to the earth is largely determined by the earth, but the sun nevertheless exerts quite substantial perturbations. The most striking consequence is that the plane in which the moon moves around the earth itself moves. It maintains a more or less constant angle of about five degrees to the ecliptic, but the nodes, at which it cuts the ecliptic, move steadily once round the ecliptic in the opposite direction to the motion of the moon in about $18\frac{2}{3}$ years. To a first approximation, the orbit of the moon around the earth within this moving plane is nearly circular but has a relatively large eccentricity of about 1/18, significantly more than the eccentricity of the earth's orbit around the sun. The nonuniformity of the moon's apparent motion is therefore about three times more pronounced than the sun's but, rather remarkably, the eccentricity is still just small enough *not to reveal* the defects of the Hipparchan solar theory when it is applied to the moon. In addition the moon's apogee moves in the same direction as the moon in a period of 8.85 years. The model which Hipparchus adopted to describe these three effects was equivalent to having the moon move uniformly in an eccentric that itself rotated once in 8.85 years while the entire plane of the orbit moved round the ecliptic backwards in $18\frac{2}{3}$ years. As a result, without essentially changing his solar theory, Hipparchus obtained a quite good description of the moon's motion (apart from one effect to be described in a moment). Moreover, combining it with the theory of the sun, he was able to predict lunar eclipses with a very tolerable degree of accuracy. This was a notable triumph and a rather wonderful one too considering the complexity of the observed motions and the simplicity of the means employed. It marks the high point in the successful application of the 'universal law of celestial motions' (p.153). Lunar theory takes up a large portion of the *Almagest* and represents the most sophisticated body of theory that Ptolemy inherited from Hipparchus. It was nevertheless deceptive since the underlying theory took no account of the *actual nonuniformity* of the motion of either the moon or the sun (in geocentric guise). This may well have been the reason why Hipparchus never apparently succeeded in developing a satisfactory theory of the planets, whose observed motions on the sky cannot, as we shall see, be described without introducing genuine nonuniformity in one form or another into the theoretical models. In fact, although Greek astronomers were undoubtedly active after Hipparchus died, no really outstanding new discovery seems to have been made until Ptolemy began his work.

Interestingly, the first significant innovation that he made was in the theory of the moon, not the planets. This was associated with a further effect of the solar perturbations. In fact, the moon is subject to a second significant nonuniformity within the moving plane of its orbit, and its longitude θ from apogee is given to a good approximation by the formula[48]

$$\theta = \overline{\theta} - 2e \sin \overline{\theta} - c \sin (2E - \overline{\theta}), \tag{3.5}$$

where $\overline{\theta}$ is the distance of a fictitious mean (uniformly moving) moon from the apogee, e is the eccentricity of the moon's orbit, E is the angle between the sun and the moon, and c is a constant with a value of about $e/3$. Now the first correction term on the right-hand side ($-2e \sin \overline{\theta}$) is exactly analogous to the corresponding first-order correction term for the apparent motion of the sun [see Eq. (3.1)] though, as we have said, it is about three times as large. However, the other correction term is completely new and is due to the perturbation exerted by the sun. The effect it describes is known as *evection* and has rather remarkable properties – at both new moon and full moon, when $E = 0$ and π respectively, the second correction term has exactly the same effect as the first and there is a single effective correction term, $-(2e - c) \sin \overline{\theta}$. The evection term is most readily apparent at half moons ($E = \pi/2$ or $3\pi/2$) when simultaneously $\overline{\theta} = \pi/2$ or $3\pi/2$.

As we have noted, Hipparchus more or less correctly determined the $-2e \sin \overline{\theta}$ term in the lunar motion, just as he had in the solar motion, but the evection term escaped him. The reason for this is quite simply that Hipparchus based his theory solely on lunar eclipse observations. As we have seen, it is only at such eclipses that the effect of parallax can be readily eliminated, but precisely then the second correction term becomes indistinguishable from the first and can be taken into account by a change of the factor $2e$ into $2e - c$, which is what Hipparchus effectively did. It seems likely that towards the end of his life Hipparchus came to suspect the existence of the evection term, for he left behind a record of observations of the moon made at times that would reveal the effect.[49] Ptolemy[50] recounts how he himself did suspect the term and had an armillary sphere (see Sec. 3.4, pp. 115–116), which he in fact called an *astrolabe*, specially built in order to measure the angular position of the moon relative to the sun and the ecliptic. With this instrument, and also using the observations of Hipparchus, he established the existence of the evection term and measured its magnitude with good accuracy.

He then proceeded to devise a geometrical model of the motion of the moon using the basic epicycle–deferent scheme with in addition eccentric position of the observer in order to represent all three terms in Eq. (3.5). I shall not attempt to describe this model, which is ingenious but quite complicated (the reader is referred to the *Almagest*, and also Neugebauer's explanation[52]) but it is important for two reasons. First, it was in this model that Ptolemy first introduced the first new element into the armoury of theoretical tools he had inherited from his predecessors. As we have emphasized, in all of these the underlying principle is that of a circular motion that *is uniform relative to the corresponding centre of the motion*. Ptolemy recounts in the *Almagest*[53] how in his theory of the moon he was forced to abandon this principle. He found that the idea of uniform motion could still be saved, but that one had to introduce a further point, not coincident with the centre of the circular motion, with respect to which the motion was uniform. In the case of the moon's motion, this rather technical innovation on Ptolemy's part was not of any great import (except in permitting a very good description of the observations). However, as Neugebauer points out,[54] it was probably what suggested to Ptolemy the use of a similar device to describe the planetary motions. As we shall see, it had there the most profound consequences.

The second reason for the importance of Ptolemy's work on the moon was that it produced results of such potential accuracy that the very definition of time became acute. This aspect of his work will be discussed in Sec. 3.15.

It should also be said that, despite the excellence of Ptolemy's predictions for the longitude of the moon, his theory was quite wrong in the predictions it made for the earth–moon distance. According to his model, the ratio of the greatest to the least distance is almost 2 : 1 and this should be reflected in a corresponding change in the apparent diameter of the moon. In fact, the changes in the moon's apparent diameter were known to the ancients, and were within the range of estimation because the eccentricity of the moon's orbit around the earth is three times greater than that of the earth's around the sun. However, the ratio of nearly 2 to 1 was in flagrant contradiction to directly observable facts. Neugebauer comments:[55] 'Nevertheless, the longitudes are so well represented by the new theory that it was not replaced . . . before the late Islamic period and then again by Copernicus. Ptolemy himself never mentions this difficulty although he cannot have overlooked it.'

A final comment about Ptolemy and the theory and observations of the moon. Nowhere more than in Ptolemy's historical introduction to the study of the moon,[56] where he discusses attempts to find periodicities in the lunar motion made centuries before his time, does one get such a good feeling for the antiquity of astronomy as a discipline subject to a rigour worthy of modern science. The searches for periodicities involved painstaking study of a vast mass of empirical material and were in many ways remarkably like much modern work done, for example, at high-energy accelerator laboratories. There is no doubt where quantitative empirical science was born: in the study of the moon and its eclipses.

Let us now see what Ptolemy made of the problem of the planets.

3.14. Ptolemy and the small-eccentricity planetary system

It is helpful to begin this discussion by considering the accuracy Ptolemy could achieve, so that we know what effects he had a chance of discovering. The moon and sun each subtend about half a degree on the sky. Ptolemy's accuracy corresponded on average to about, say, one third of this, i.e., he made observations with an accuracy of around 10 minutes of arc. It will here be helpful if the reader refers back to Table 3.1 on p. 127, which gives the eccentricities and ellipticities of the various planetary orbits. It is these quantities that determine the observed deviations from the zero-eccentricity approximation. If this last were exactly valid, the planetary motions would all be perfectly described by the simple epicycle–deferent scheme, and Ptolemy would have had nothing to do. But, as we know, they are not – deviations arise because of the eccentricity of the orbits, the nonuniformity of the motions, and the ellipticities of the orbits. Which of these deviations did Ptolemy have a realistic hope of discovering?

A simple estimate shows that if $\delta\alpha$ is the angular accuracy of the observations in minutes of arc, then in principle an eccentricity or ellipticity of value

$$C \cdot \frac{2\pi\delta\alpha}{360 \times 60} \tag{3.6}$$

can be established (provided the correct theory is being used). Here, C is a numerical constant of order unity peculiar to each individual observation; it lies in the range 0.2 to 5 depending on the position of the earth, the sun, and the observed planet. The significance of (3.6) is that in the most favourable circumstances Ptolemy had the potential possibility of discovering eccentricities as small as 1/350. Thus, the eccentricities, even of Venus (~1/147), were readily accessible to observation.

Equally accessible were the deviations from the zero-eccentricity approximation associated with the departure from uniformity of the motion. They are in fact of exactly the same magnitude as the eccentricity deviations. However, examination of Table 3.1 shows that the ellipticities (except in the case of the very unfavourable Mercury) were quite out of reach for Ptolemy. (It might seem that Mars's ellipticity was just within Ptolemy's reach. However, in this case the geometrical factor C in (3.6) militated against him, quite apart from the fact that the errors in his scheme completely swamped the ellipticity.)

Before starting on a detailed discussion of what Ptolemy did, a general comment is in order. In the framework of the general programme to 'save the phenomena', Ptolemy's main aim was, of course, to explain the appearances observed on the two-dimensional sky by means of motions in three-dimensional space that were as simple and economic as possible. A key aspect of this programme was the recognition that the observed motions were evidently due to the superposition of two motions, one of which, as we now know, corresponds to the planet's own motion around the sun, while the other corresponds to the earth's motion. Since both effects represented deviations from the uniform diurnal motions, they were, for the reason given earlier, called *anomalies*; in the medieval and Renaissance literature they were also called *inequalities*, and Kepler calls the motion corresponding to the planet's own motion around the sun the *first inequality* and that corresponding to the earth's the *second inequality*.[57] Because of the epicycle–deferent inversion, the second inequality corresponds to the epicycle motion of the outer planets and the deferent motion of the inner planets. Because of this complication, it will be increasingly convenient, especially in Chaps. 5 and 6, to refer to the two inequalities, which have, in contrast to epicycle and deferent, a fixed significance. In these terms, Ptolemy's task was *to unravel cleanly the first inequality from the second* and to represent each by as simple a motion as possible. As we shall see, he nearly but not quite succeeded.

Let us now consider how Ptolemy probably attempted to refine the simple (zero-eccentricity) epicycle–deferent technique is inherited from his predecessors. Examination of Fig. 3.15 will tell us what the observations suggested. That figure has been drawn under the assumption that the earth and the three outer planets all have zero eccentricity. In that idealization, the retrogression loops, representing the effect of the second inequality, i.e., the earth's motion, repeat themselves with perfect regularity, the sizes and the intervals between them remaining constant. Now suppose that the earth still has zero eccentricity but that the three outer planets have their actual eccentricities. Because the epicyclic motion reflects the motion of the earth around the sun, the epicycle motion (uniform rotation about the guide point) will remain unaltered. However, the deferent motion, or first inequality, which is simply the motion of the planet around the sun, will be changed accordingly. When the planet is at its furthest from the sun, its apparent motion is slower; half of the reduction corresponds to the actual slowing down, the other half to the purely geometrical effect of the motions being seen from further away. This means that the deferent guide point will not advance so far around the ecliptic between successive retrogression loops – the successive points corresponding to the point O in Fig. 3.15 will be more crowded together. But when the planet is at the diametrically opposite point of the ecliptic, the deferent guide point will advance more rapidly, and the retrogression loops will be accordingly more widely spaced. The sizes of the loops will also be different, being smaller when the planet is further away from the sun. For Saturn and Jupiter these effects are quite readily observable if followed for a sufficient number of years.

Now because of the specific eccentricities of the orbits of the earth and the outer planets, above all the *relative smallness* of the eccentricity for the earth, the picture just described, corresponding to zero eccentricity of the earth's orbit, is very close to what is actually observed. The nonzero but very small eccentricity of the earth's orbit changes the apparent motion of the three outer planets from the pattern described in Fig. 3.15 by an amount that is only about a tenth of the change just described in the deferent motion in the case of Mars and even less in the case of Jupiter and Saturn. (The exact amounts are determined essentially by the ratios of the eccentricities and the ratio r/R, where r is the mean radius of the earth's orbit and R is the mean radius of the outer planet under consideration.)

Not surprisingly, Ptolemy concluded that the epicycle part of the theory for the three outer planets was perfectly correct and that in all three cases the retrogression loops were simply due to an epicycle spoke (of constant length) rotating at a uniform rate around a deferent guide point, just as in the simple theory. He saw the main task as being the explanation of the apparently nonuniform motion of the guide point around the deferent, i.e., around the ecliptic.

Now as regards the observed general tendency, the effect is just like what Hipparchus and Ptolemy observed in the case of the sun (and the main correction to the moon's motion), namely, the apparent non-uniformity in the deferent motion appears to be due to an *eccentric placing of the deferent circle with respect to the observer*. This rather obvious conclusion was moreover confirmed by the fact that the retrogression loops were observed to be smaller in the part of the ecliptic in which the planet is moving slower, as is to be expected if an epicycle spoke of constant length is observed from a greater distance. Thus, just as for the sun, Ptolemy must initially have concluded that the magnitude of the eccentricity and the direction of the line of the apsides could in principle be deduced from observations of the position of the deferent point at three different times. (It will be recalled that Hipparchus based his theory of the sun on observations of the times at which the sun passed through the vernal equinox, the summer solstice, and the autumnal equinox.)

However, in the case of the epicycle–deferent motion of the planets there is an important difference from the solar theory. The sun is seen directly as it travels around the ecliptic. But in the theory of the planets the deferent guide point is void. One can only see the planet, but that travels on the tip of the epicycle spoke. How did Ptolemy set about locating the exact position of the guide point? Crucial here was his assumption that the epicycle spoke rotated perfectly uniformly. Now if the eccentricity of the earth's orbit had been exactly zero, the epicycle spoke would always march exactly in step with the sun, since the epicyclic motion is simply the reflection of the earth's motion around the sun. Thus, the epicyclic spoke would always be exactly parallel to the direction from the earth to the sun. But here Ptolemy faced a problem. He had no idea of the underlying identity of the two motions. For the purposes of describing the apparent motions of the planets, perfectly uniform motion of the epicyclic spoke seemed quite adequate. But the sun did not move uniformly, as Ptolemy knew only too well. But, in that case, what determined the direction of the epicyclic spoke? Since the spoke was assumed to rotate uniformly, it could not remain parallel with the direction to the nonuniformly moving sun. Ptolemy grasped at the only straw to hand and assumed that the epicyclic spoke rotated uniformly and always remained exactly parallel to the *direction of the mean sun*. By making this assumption, Ptolemy elevated the mean sun to the most important concept in his theory, compared with which the part of the true sun was that of the second fiddle. (The main importance of the true sun in Ptolemy's overall scheme was in fixing the position of the moon at the centre points of lunar eclipses.) Kepler remarked[58] that Ptolemy did this without a sound observational basis, following instead 'the preconceived and false opinion that it is necessary to suppose the movements of the planets are regular throughout the whole circle'. This is a bit unfair, since Ptolemy was prepared to abandon the principle of uniform motion when it was absolutely necessary, and this, as we shall see, had very important consequences. However, in the case of the epicyclic motions, it just so happens that uniform motion in alignment with the mean sun describes the observations surprisingly well.

The assumption that Ptolemy made about the parallelism between the epicyclic spoke and the mean sun was all he needed to complete his model. Figure 3.17 shows what must, almost certainly, have been the form of his initial model, or *hypothesis* as he and all early astronomers called such models. (In the astronomical jargon, *hypothesis* had much the same meaning, say, as *physical model* does in modern nuclear physics.) In this figure, the earth (and observer) are at O. The centre of the deferent circle is at C, so that the line of the apsides is along OC with the perigee of the deferent guide point at B and apogee at A. Ptolemy's original assumption must have been that the centre D of the epicycle, the guide point, moves uniformly around the deferent (on the tip of the deferent spoke, shown by the broken line) while the epicycle spoke DP, at whose tip P the planet is carried, rotates uniformly relative to a direction fixed in geoastral space. Of course, Ptolemy could only observe P, not D, so *a priori*, he did not know the position of D. However, he made the assumption that **DP** is always exactly parallel to $\mathbf{O}\bar{\odot}$, the vector of the mean sun $\bar{\odot}$. He therefore looked for occasions at which the planet was in opposition to the mean sun (as in Fig. 3.17) and noted its longitude in

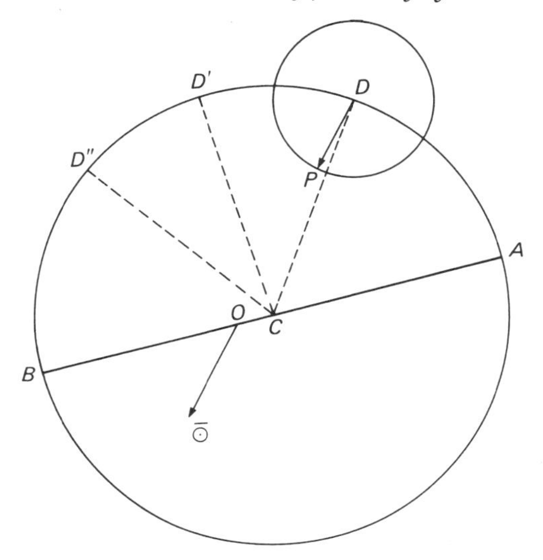

Fig. 3.17.

geoastral space (i.e., its ecliptic longitude). In accordance with his model, he knew that at the time of such an opposition the guide point would coincide with the planet's position on the sky as observed from the earth. This was the device that Ptolemy employed to make the guide point 'visible'. Such occasions occur whenever the planet is observed from the earth to be in opposition to the mean sun. Such observations were called *acronychal** observations and play a very significant role in the *Almagest* and also in the work of Copernicus, Brahe, and Kepler.

Once Ptolemy had in his possession three such acronychal observations of an outer planet corresponding to the points D, D′, and D″, in Fig. 3.17, he was in nearly but not quite exactly the same position as Hipparchus in the solar problem, for he knew the lengths of time taken by the planet to pass between these points, from which the actual lengths of the arcs DD′ and D′D″ could be calculated (since the deferent guide point is assumed to move uniformly around C and the period of this motion is known, the mean daily motion of the deferent guide point can be readily calculated). So far there is no difference from the Hipparchan theory. However, we recall that Hipparchus made his observations at the two equinoxes and the summer solstice, so that he obtained successive positions of the sun that, seen from the earth, were separated by exactly 90°. Mathematically, this gives us the *original Hipparchan problem*, which can be stated as

* The word derives from the Greek for 'at nightfall', since the superior planets rise at sunset when they are in opposition to the sun.

follows. Suppose a body moves around a circle at a given uniform speed and is observed *at three times*. The arcs of the circle traversed by the body between the first and the second observation and between the second and the third are then known. Each of these arcs is known to subtend a *right angle* from a point O situated at some unknown position not coincident with the centre of the circle. The problem, which is readily solved by elementary trigonometry, is to find this position from the given data.

We now consider the *generalized Hipparchan problem*, which corresponds to Fig. 3.17. In this case the observations again tell us the angles subtended at O by the arcs (DD′ and D′D″), but these angles are no longer right angles. This problem too can be solved uniquely by trigonometric methods though it involves considerably more work. Rather than go through the details, let me merely demonstrate its solvability by a simple heuristic argument. Suppose three coplanar lines emanating from a point O and making adjacent angles at O equal to DOD′ and D′OD″ (Fig. 3.17). Draw a circle with arcs subtended *from the centre* proportional to the times between the observations (as fractions of the orbital period). This corresponds to the assumption of uniform motion about the centre of the orbit. Now take any point on the line corresponding to OD and place it at D. Then, holding the chosen point fixed, swing the lines around until the line corresponding to OD′ passes through D′. In general, the third line will not pass through D″ but will cut the circle at some other point X. But now vary the original point chosen on the first line to coincide with D. The new point X will then be moved on the circle. Continue adjusting the point on the first line corresponding to D until the resulting point X coincides with D″. The problem is then solved, since O is at the position at which the given arcs subtend the required angles. Note that, once again, the absolute dimensions cannot be determined, so that the overall scale is arbitrary. It is also worth noting that the generalized Hipparchan problem occurred in the Hipparchan–Ptolemaic theory of the moon, in which three observations are obtained at eclipses of the moon. Such observations, like the acronychal observations of the planets, are also not separated by 90°. Finally, we may mention that Hipparchus need not have solved the solar problem with observations at the equinoxes and a solstice; he could have taken any three observations suitably spaced around the orbit. From the point of view of accuracy, this would, in fact, have been preferable, since the precise moment of a solstice, at which the sun has no motion in declination, is difficult to determine with accuracy.

Let us now return to Ptolemy and the problem of the planets. Solving the generalized Hipparchan problem with three acronychal observations, he could determine for the deferent the direction of the line of the apsides AB and the eccentricity OC (as always, Ptolemy set the length of the deferent radius CA equal to 60). Knowing then the period of the deferent motion (equal to the sidereal period of the planet being studied), he must have assumed that the deferent spoke rotated about the centre of the deferent with uniform angular speed corresponding to this period. As he also knew the time at which the guide point occupied the position D, he could then calculate its position at any other time. But he also knew the position of the epicycle spoke at the same time and how fast it too moved. The only information he lacked was the length of the epicycle spoke as a ratio of the length of the deferent spoke. But this, as we have seen, is readily determined.

It is only necessary to calculate, for example, the time required for the epicycle spoke to rotate through 90° relative to the deferent spoke. At that time the deferent guide point will have moved forward a certain calculable distance. Then from observation of the actual position of the planet at that time relative to the position calculated for the guide point the length of the epicycle spoke follows by elementary trigonometry. Thereafter, it was a matter of pure calculation to determine where the planet should be observed at any future time. In principle, the prediction should be accurate to about $\frac{1}{3}$ of the apparent diameter of the moon.

However, when Ptolemy checked the theory, he must have found the predictions were quite seriously wrong except for acronychal situations, i.e., when the planet was in opposition to the mean sun. In the light of our earlier discussion of the small-eccentricity limit of Kepler's laws the reason is evident. Ptolemy's initial model* does not reflect what actually happens. In particular, his theory takes no account of the fact that in the small-eccentricity approximation the deferent motion, which is the reflection in the epicycle–deferent model of the planet's actual motion in helioastral space, is not uniform about the centre of the deferent circle but about the point corresponding to the second (void) focus of its elliptic orbit. Thus, Ptolemy's initial deferent model suffered from exactly the same defect as the Hipparchan solar theory: he must have obtained for the deferent an eccentricity that was exactly twice what it should be. And whereas in the case of the sun this defect remained undetectable, in the case of the theory of the planets the error did show up.

The reason for this can be quite readily seen by referring back to Fig. 3.10 (p. 129). In the discussion of that figure we pointed out that provided the equant spoke always rotates uniformly about the point E′ at distance $2e$ from the observer (at E) then the position of the centre of the circle on which the observed body is assumed to move has very little influence on the position at which the model predicts it will be seen on the sky. This is because a shift in the position of the centre of the conjectured orbit merely moves the predicted position in space up or down the line $E'S_P$ (to the position S_C). But because of the conditions of observation, this motion, when seen from E, is very greatly foreshortened. It is for this

* It should be emphasized that I describe here an hypothetical reconstruction.

reason that it becomes virtually undetectable unless the eccentricity is quite large.

Now in the case of the planets the theory described by Fig. 3.10 does not give the position of observed body but only the position of the guide point around which the epicycle spoke rotates. The planet itself is at the tip of the epicycle spoke. Figure 3.18 shows the situation to scale for Jupiter, which has an eccentricity of ~1/21. The observer is at E, which is effectively the position of the sun, since the epicycle–deferent motion makes the deferent motion for an outer planet into the motion of the planet around the sun (only approximately, since Ptolemy did not separate the two motions quite cleanly). The correct position of the guide point (corresponding to the epicycle–deferent representation of planetary motion) is at D, the incorrect one deduced following the model of the Hipparchan solar theory is at D′. The points P and P′ represent the position of the planet when the epicycle spoke is at right angles to the deferent spoke in the correct and incorrect theories, respectively. Two factors help in the detection of the flaw. First, the eccentricity of Jupiter's orbit being about three times that of the earth, the angle DED′ is correspondingly larger (recall that Fig. 3.10 is shown for a greatly increased eccentricity, whereas Fig. 3.18 is to scale). Second, the position of the epicycle spoke puts PP′ in a more favourable position for observation than DD′, i.e., the epicycle spoke effectively transports DD′ into the position PP′, which is more favourable for observation from E. This effect is particularly important for Mars and Venus, since for them the epicycle spoke is nearly as long as the deferent spoke, so that the angle DEP becomes quite large (about 50°). The error shows up because the

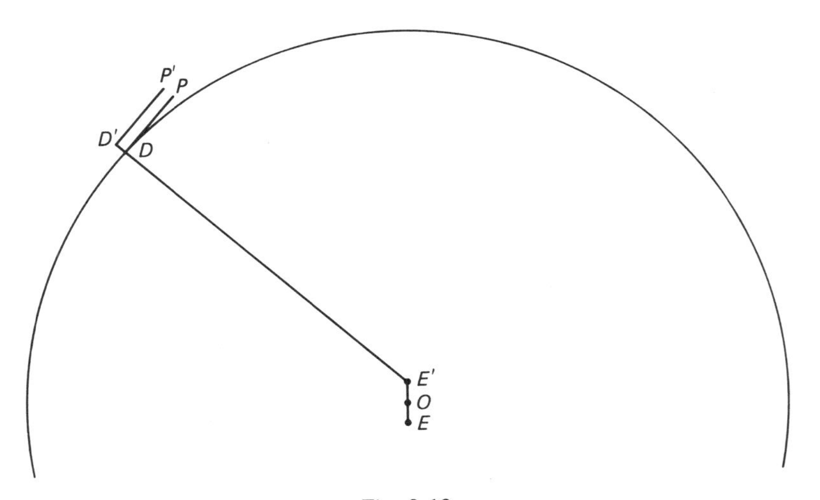

Fig. 3.18.

possibilities for using trigonometry to control not only angular positions but also actual positions in space are very greatly extended in the planetary problem.

Ptolemy was not one to admit defeat readily. It is worth quoting here what he said in the *Almagest* in connection with his realization that the theory of the moon which he had inherited from Hipparchus (and in which he had invested much effort himself) worked very well for eclipse observations but not nearly so well at the lunar quadratures:[59] 'Those who approach this science in a true spirit of enquiry and love of truth ought to use any new methods they discover, which give more accurate results, to correct not merely the ancient theories, but their own too, if they need it. They should not think it disgraceful, when the goal they profess to pursue is so great and divine, even if their theories are corrected and made more accurate by others beside themselves.'

Ptolemy does not tell us in the *Almagest* how he found the solution to his problem; he merely speaks of being compelled:[60] 'to make some basic assumptions which we arrived at not from some readily apparent principle, but from a long period of trial and application'. However, it appears that, having worked out his initial theory (presumably along the lines indicated here), he tested it by calculating how large the retrogression loops should appear when the guide point is at perigee and apogee. For purely optical reasons, the loops should appear larger at perigee, and the difference should follow from the theory. Ptolemy reports[61] however that the eccentricity deduced from the equation (i.e., from application of the Hipparchan solar theory to explain the non-uniformity of the motion of the deferent guide point) is found by observation to be about twice that derived from the size of the retrogression arcs. This presumably gave him the hint that the true geometrical centre of the deferent must be found by *halving* the eccentricity found by the Hipparchan solar theory. However, this forced him to abandon the idea that the void guide point on the deferent rotates at uniform angular speed about the *centre* of the deferent. Instead, if Neugebauer is correct,[54] he used his experience from the moon and assumed that there exists some other point about which the motion of the guide point appears to be uniform. Where could this point lie? As the reader already knows the small-eccentricity form of Kepler's laws, the result that Ptolemy obtained will come as no surprise. The point previously taken to be the centre of the orbit lost that status but was reinterpreted as the point about which the motion appears uniform. Ptolemy *found the equant*. That is, he found that all the observations, both acronychal and nonacronychal, could be explained by assuming that an observer stationed on the opposite side of the centre of the deferent from the position of the terrestrial observer and at an equally great distance from the centre would, if the guide point were visible, observe that it moves around against the backcloth of the stars with a constant angular velocity. Thus, the total distance between the observer and the equant must be precisely equal to the eccentricity obtained using the incorrect Hipparchan solar theory. Since the centre of the deferent is at the middle between these two points, the deferent must have an eccentricity only *half* as large as that deduced from Hipparchan theory. Without knowing it, Ptolemy had found the two foci of the planet's elliptical orbit together with the circle, correctly positioned with its centre at the centre of the ellipse, that gives the best approximation of the elliptical orbit. Thus, he had found, in geocentric guise, exceptionally good approximations to both of Kepler's first two laws of planetary motion; for *both* effects of first order in the eccentricity – the eccentric position of the orbit and the genuine nonuniformity of the motion – were in essence correctly described.

We must say at least a few words about the geometrical technique which Ptolemy used to determine the locations of the equant and the centre of the orbit. We may call the corresponding problem *Ptolemy's problem*. Here too he broke new ground, providing the first example in the exact natural sciences of solution of a problem by successive approximation. The most important property of the two Hipparchan problems (the original and the generalized) so far considered is the definiteness in their formulation; this derives from the fact that the body is assumed to move with exactly constant speed *about the centre of the orbit*. From the time between the observations one can then calculate directly the lengths of the corresponding arcs traversed by the body in its orbit. However, in Ptolemy's problem the centre of uniform motion is not at the centre of the orbit but at the equant, and *the position of the equant is not known in advance*. It is therefore impossible to calculate the lengths of the arcs traversed in the orbit between the times at which the observations are made from the point of observation. All the theory tells one is that it and the equant are at equal distances from the centre of the orbit, on opposite sides from the centre, and that the motion appears uniform about the equant.

The way in which Ptolemy overcame this problem appears to reflect the two successive stages that we have here conjectured his investigations to have followed. Rather than describe in detail his procedure, I shall break free somewhat from Ptolemy's exposition and merely explain the underlying ideas; in particular I omit entirely the lengthy trigonometric calculations that Ptolemy gives explicitly for each of the three superior planets. The first step in finding the orbit from three acronychal observations separated by known intervals of time is to assume that the initial theory, with equant at the exact centre of the orbit, is correct. One then has to solve a well-defined generalized Hipparchan problem. In Fig. 3.19 the continuous circle with centre C is taken as the first approximation to the orbit. The angles XCY and YCZ correspond to the arcs XY and YZ that would be traversed by a body moving around the circle at uniform speed

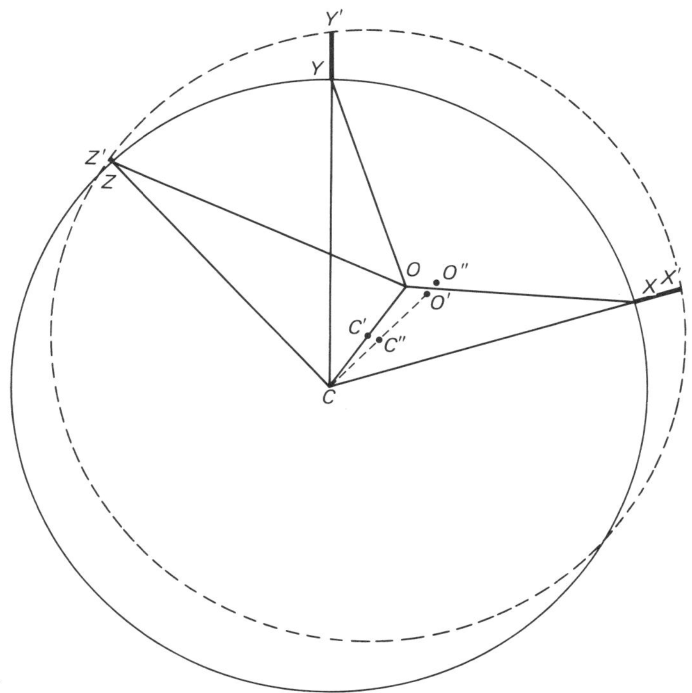

Fig. 3.19.

in the time intervals between the observations. The point O is found by solving the generalized Hipparchan problem from such data, i.e., the angles XOY and YOZ are equal to the differences of the observed longitudes at the three successive acronychal observations. This, of course, is not the solution to Ptolemy's problem, since the equant coincides with the centre of the orbit and only O, the position of observation, is eccentric. However, it can be modified to supply a reasonably good approximation to the solution. To obtain this, we join C and O and take the point C′ at the midpoint between C and O. About this point as centre we describe a second circle, of radius equal to the first. This gives us the dashed circle in Fig. 3.19. From C let CX, CY, and CZ be extended until they meet the new circle at X′, Y′, Z′. This dashed circle can then be taken as a better approximation of the orbit and C and O as the corresponding approximations of the equant and point of observation. For the centre C′ of the dashed circle is at the midpoint between C and O, as it must be in Ptolemy's model, and the angles X′CY′ and Y′CZ′ are equal to XCY and YCZ, respectively, i.e., they are the correct angles that should be observed from the equant. Only the angles at O are incorrect. For the observed angles between the three acronychal observations should be XOY and YOZ, and these are not equal to X′OY′ and Y′OZ′, respectively.

The error can be corrected in one of two ways, either of which involves renewed solution of a properly defined generalized Hipparchan problem. In the first, one can calculate the angles subtended by the arcs X′Y′ and Y′Z′ at the centre of the dashed circle and seek a new point O′ such that the arcs X′Y′ and Y′Z′ subtend from the point O′ the two angles between the observed acronychal longitudes. In this case, the Hipparchan problem is solved for the dashed circle. Alternatively – and this is the choice that Ptolemy made – one can stick to the original continuous circle (this is the *equant circle* mentioned in Sec. 3.7, p. 126) with arcs XY and YZ and solve a new Hipparchan problem, in which the two angles between the three successive acronychal longitudes are corrected by means of the solution already found. For, as Ptolemy pointed out, that solution was found under the assumption that the point X coincides with X′, Y with Y′, and Z with Z′. We recall now that the points X′, Y′, Z′, through which the planet passes in accordance with the modified model, are actually visible, whereas X, Y, Z are not. However, possessing our first approximate solution, we can say that, could we observe X, Y, Z directly, we should see them from O displaced by the angles X′OX, Y′OY, Z′OZ. These angles can be calculated by trigonometry using the first approximate solution. From the actual point of observation, which we hope to find and which can be assumed to be quite close to O, the corresponding 'correction angles' will be nearly equal to these angles. Therefore, adding them as corrections to the actually observed angles, we obtain a new Hipparchan problem with the same given arcs XY and YZ but corrected angles. It too will lead us to a revised position O′ for the position of the observer (not, in general, coincident with the O′ found by the alternative method but nevertheless very close to it).

However, we are still not at the end since the centre C′ of the dashed circle is not in general at the midpoint of O′ (found by either method) and C. Therefore, the next step is to take a new circle (not shown in Fig. 3.19) of the same radius as the first two but with centre C″ at the midpoint of O′C. The approximation procedure can then be repeated with this new circle taken as the actual orbit. Thus, the solution of Ptolemy's problem involves solution of a succession of generalized Hipparchan problems, each one leading to a better approximate solution until the observations are reproduced to realistic accuracy. Ptolemy in fact found that it was necessary to make only one or two corrections.

Having found the deferent, equant, and point of observation, Ptolemy then needed one further observation, which must be nonacronychal, to determine the relative length of the epicycle spoke. This completed his model and thus, with just four observations, he was in principle able to determine the position of the planet to a remarkably good accuracy at any other time. The simplicity, elegance and accuracy of his theory are breathtaking, and the sophistication in its practical application is impressive.

The equant was truly Ptolemy's greatest discovery and the crowning achievement of Hellenistic astronomy. Both Ptolemy's lunar theory and his discovery of the equant can be regarded as corresponding to an advanced, or second, level of theoretical sophistication. For the effects described by the Hipparchan solar theory and the simple epicycle–deferent theory of planetary motion could be read off more or less directly from sufficiently good observations. These theories are thus at the first level of sophistication. But neither evection nor the equant could become apparent before these first-level theories had been pushed to the limit of what they could achieve. Only when this had been done did these two residual effects become apparent; they were literally 'made visible' by the lower-order theories. For reasons that will later become clearer, the discovery of the equant must certainly rank as one of the most important milestones on the road to the discovery of Newtonian dynamics. It is unquestionably one of the 'baker's dozen' and the second to come from astronomy.

Great though Ptolemy's discovery was, it may have contributed to the subsequent stagnation of theoretical astronomy for nearly one and a half millennia. In this, the specific eccentricities of the individual planetary orbits undoubtedly played an important role. To see this we need to summarize what Ptolemy achieved with his discovery of the equant. At a stroke he reduced the errors of the simple epicycle–deferent theory for the three superior planets by an order of magnitude. Only for Mars does Ptolemy's final scheme lead to occasional errors significantly in excess of his observational accuracy. (From the *Almagest* it is not possible to judge to what extent Ptolemy was aware of this. He does not leave the impression of having made long and extensive series of observations to check in detail the predictions of his theory once found. There is also a passage, to be quoted in Sec. 3.16, which can be read as an admission on his part that the phenomena did not quite fit his theory in every respect.)

The unfortunate aspect of this great advance was that it obscured a particular perversity of the problem of unravelling the planetary observations by means of observations from the earth. We recall that the motion of a planet in helioastral space is a problem with a small parameter – the eccentricity of the planet's orbit. But when the planet is observed from the earth, this relatively simple theory with a single small parameter is transformed into a problem with two small parameters, for now the effect of the eccentricity of the earth's orbit is added. It is this that makes the problem in its full generality so difficult. In one sense Ptolemy was very fortunate that the eccentricity of the earth's orbit is so small, since it effectively reduced the problem for the outer planets to one with only one small parameter. This was undoubtedly the reason why Ptolemy was able to discover the equant in the first place. But he was simultaneously misled into thinking that the epicyclic part of his scheme was in essence correct – that the epicycle spoke does indeed rotate with perfect uniformity.

This led him to believe that in all cases the deferent motion must be described by an equant but not the epicycle's motion. Looked at in terms of expansions with respect to the eccentricities, Ptolemy's scheme therefore represented a hybrid – it was of first order with respect to the eccentricities of the three superior planets but of zeroth order with respect to the earth's motion.

This erroneous impression was greatly strengthened by the curious accident that the orbit of Venus, next to the earth on its inner side, has by far the smallest of the eccentricities of the earth and the five naked-eye planets. However, for this planet, like Mercury, the deferent and epicycle were interchanged. Thus, for Venus, which is in fact the only planet for which Ptolemy explicitly demonstrated[62] how the existence of the equant could be proved and its position established,* the deferent motion represented the earth's motion while the epicycle represented Venus's. Once again this made it appear that there was an equant in the deferent motion but not in the epicycle. The greatest irony here was that Ptolemy, quite unwittingly, determined the eccentricity of the earth's orbit twice – once in the theory of the sun, when, using Hipparchus's theory, he obtained a value about twice the correct one, and again in the theory of Venus, when he obtained a value much closer to the correct one, namely ~1/48. (He also found that the line of the apsides for Venus's deferent was in roughly the same direction as the line of the apsides for the solar motion. However, these coincidences did not apparently attract his attention.)

Of the five naked-eye planets, only Mercury, which has by far the largest eccentricity, failed to fit into the neat pattern Ptolemy had found for the other four (with equant in the deferent motion and uniform epicycle motion). The scheme which Ptolemy devised for Mercury was much more complicated than for the four other planets. This was not surprising, since for Mercury the eccentricities of both orbits come into play and even the ellipticity of Mercury's orbit produces a sensible effect. Thus, whereas the problem for the other four planets contained effectively only a single small parameter manifested to first order at the level of Ptolemy's accuracy, the problem for Mercury contained two with one of them, moreover, manifested in second order. To boot, Mercury, being always so close to the sun, is quite the hardest of them all to observe.

* For the inferior planets Ptolemy was able to use a different and rather more direct scheme to solve his problem than was possible in the case of the superior planets.

No wonder it is called the thankless planet. For this reason it played little positive part in the early development of theoretical astronomy.

We shall have more to say about the residual defects of the Ptolemaic system in the chapters on Copernicus and Kepler, since their elimination is closely associated with the emergence of some of the most important concepts of dynamics. In the meantime we can close this section with some general comments that are already appropriate.

First, seen in the light of the subsequent history of astronomy, the problem with the Ptolemaic system was that it was good but not quite perfect. Because it was good – in part extraordinarily good – it was able to survive for a very great length of time. Inverting Voltaire's famous aphorism, we can truly say 'the good was the enemy of the best'. The trouble was that none of the residual defects was sufficiently large or manifested with a signature sufficiently clear to suggest that anything was drastically wrong. For this the specific eccentricities of the planetary orbits coupled with the poor observability of Mercury were almost entirely responsible. Thus, for many centuries after Ptolemy the astronomers, especially the Islamic ones, busied themselves with improvements to various elements of Ptolemaic theory, making, for example, accurate determinations of the solar eccentricity. This led them to note, for example, the fact that the eccentricity for Venus's deferent was remarkably close to half the value they determined for the solar motion and this was adopted as a definite relationship,[63] without however any far reaching consequences being drawn. There was also quite a lot of what might be called aesthetic or philosophical rearrangement of the Ptolemaic models, which aimed to reproduce essentially the same motions but with different arrangement of circles etc. We shall return to this topic in Chap. 5, since it was this kind of activity that led Copernicus to his great discovery. However, because the models seemed to work so well, and with such an economy of means, no one appears to have attempted a really thorough re-examination and comprehensive testing of the entire structure of the theory presented in the *Almagest* (which, we recall, was built up in a very logical manner in the sequence sun–moon–stars–planets with each new level resting on what had gone before).

This situation actually persisted until after the Copernican revolution. It was only in the second half of the sixteenth century that Tycho Brahe (1546–1601) set about the systematic accumulation of data in sufficient quantity and with sufficient accuracy to permit a really radical re-examination of the Ptolemaic models. It was at this point that the curious failings of the models at last became obvious and it was necessary to face up to the paradox of the remarkably good accuracy of much of the system coupled with a persistent inability to get everything just right.

The Ptolemaic system in fact worked rather like a sausage machine, and indeed in two respects. First of all, it was an automatic algorithm. You fed in certain carefully selected observational data, did the calculations, and out came the information you needed: the eccentricity of the deferent, the direction of its apsides, and the ratio of the length of the epicycle spoke to the deferent spoke. Calculation of the future position of the planet at any future epoch was then more or less a matter of turning the appropriate handle. The problem was that errors associated with the mistake made in the Hipparchan solar theory, i.e., in the earth's motion in Copernican terms, and also associated with the use of the mean sun, introduced a lack of focus which had the effect of 'smearing' the predictions for the positions of the planets, especially Mars, when away from acronychal positions. Instead of producing a clean line, the algorithm produced a string of sausages: the theory matched the predictions pretty well at the acronychal points, at which the skin of the sausage pinched in, but in between the errors became larger than the observational errors.

Until Kepler joined Brahe's team, the attempts to rectify the system failed because no one grasped the direction in which the original epicycle–deferent scheme needed to be changed. It was not through the addition of extra motions (epiepicycles etc.); for (in geoastral space) the original idea was quite right: only two motions were needed. There was nothing wrong with that idea, nor even the circularity of the motion. What no one grasped was that the problem really contained not just one eccentricity and one equant (in the deferent) but two of each. Except for Mercury, *the defects of the Ptolemaic system all resided in the theory of the second inequality,* i.e., the component in the apparent motion of each of the planets with the same periodicity as the sun's apparent motion. The fruitless searching for the correct solution gave rise to considerable scepticism among Brahe's assistants about the possibility of ever 'saving the appearances'. Whatever hypothesis was chosen to sharpen the focus in one part of the system made it more blurred elsewhere. The last elusive refinements needed to perfect the system could not be run to ground.

The second comment that can be made here concerns the character of the 'laws of motion' which Ptolemy (and his predecessors) used to describe the planetary motions. Ptolemy's scheme was geometrokinetic through and through; in the *Almagest* he sought simply to describe motions – he did not invoke any sort of *causal* or *physical* explanation for them. We have already commented on the remarkably matter-of-fact way in which Ptolemy reports that the motions of the planets (and, to a lesser extent, the moon) are to a high degree controlled by the sun. But I do not think there is a single word in the *Almagest* that implies the requirement of some physical cause to explain this undoubted fact.

This comment should not be interpreted as a criticism of Ptolemy: it is certainly anachronistic to expect the Hellenistic astronomers to have had our modern notions of physical causation. Indeed, I have already emphasized the point that it was only because they did concentrate almost exclusively on the accurate mathematical representation of celestial motions that the eventual emergence of physical concepts was made possible at all. It is also worth mentioning in Ptolemy's defence that he was well aware of the apparent oddity of some of the motions that he proposed, especially in the explanation of the motion in latitude, i.e., out of the ecliptic, of the planets (this is a topic which we shall consider in Chaps. 5 and 6), but he argued, very plausibly given his basically Aristotelian outlook, that one should certainly not expect celestial bodies, which, after all, were believed to be made of an entirely different substance (quintessence) from the terrestrial bodies, to behave in the way one would expect from earth-bound experience.[64]

Finally, we may point out that closely related to the lack of a physical element in the Ptolemaic system is the prominent place accorded in it to *moving or stationary void points*. They occur in innumerable places and are an indispensable part of the theoretical scheme. Each deferent for the planets has two – the centre and the equant. To these are added the moving void points of the epicycles. Moreover, in the case of the moon the deferents themselves move about further void points. Just as with the manifest connection between the epicycle motion and the apparent solar motion, the *Almagest* does not reveal any curiosity on Ptolemy's part as to the origin of these mysterious void points. In fact, he clearly did not see them as puzzling or problematic.

Seeing a mystery in them was one of Kepler's major advances.

3.15. Time in Ptolemaic astronomy

So far we have awarded two 'buns' out of the 'baker's dozen' that went into the creation of Newtonian dynamics to Hellenistic astronomy: for the development of the epicycle–deferent scheme and for Ptolemy's discovery of the equant. In this section the case is going to be argued for the awarding of a third – for the clarification of the relationship between *time*, or rather the *measurement of time*, and the mathematical description of observed motions. And whereas the first two discoveries did not enter directly into the final synthesis of Newtonian dynamics but still needed to be transformed by the work of Copernicus and Kepler, the aspect of time that the ancient astronomers uncovered passed almost unchanged into Newton's *Principia*.

We have already seen that not only Aristotle but also the mathematicians who developed kinematic geometry had a deeply ingrained notion of the uniform passage of time. The astronomers Hipparchus and Ptolemy were clearly no exception to this instinctive standpoint. The *Almagest* contains numerous references to the passage of true time and Ptolemy clearly felt no need to elaborate philosophically in the manner of Aristotle on the nature of time. There is, however, a significant section in the

Almagest in which he discusses the practical measurement of time and identifies what he considers to be the one true source for measurement of the passage of time.

Before we start on this, it is helpful to quote what Mach said in reaction to Newton's assertion in the *Principia* that:[65] 'absolute, true, and mathematical time, of itself, and from its own nature, flows equably without relation to anything external'. In the face of this metaphysical dogmatism – which was only an explicit statement in words of what the Greeks had felt in their bones two thousand years earlier – Mach responded in his characteristically uncompromising way:[66] 'It is utterly beyond our power to *measure* the changes of things by *time*. Quite the contrary, time is an abstraction, at which we arrive by means of the changes of things; made because we are not restricted to any one *definite* measure, all being interconnected. A motion is termed uniform in which equal increments of space described correspond to equal increments described by some motion with which we form a comparison, as the rotation of the earth. A motion may, with respect to another motion, be uniform. But the question whether a motion is *in itself* uniform, is senseless.' The second half of this criticism by Mach has, of course, already been anticipated in the discussion of Aristotle's lapse into mathematical intuition following a more critical examination of what we mean by time. (It is worth noting in passing the similarity between Aristotle's 'philosophical' doctrines about time and Mach's 'time is an abstraction, at which we arrive by means of the changes of things.')

What makes the stage of theoretical astronomy reached in the *Almagest* so appropriate for the clarification of what we mean by the passage of time is that it corresponded to an age in which, unlike Mach's, there existed only one unique reliable source of time measurement (the rotation of the earth) but at the same time momentous discoveries had just been made that began to show in a precise manner (not hitherto suspected at all) how 'the changes of things' are all 'interconnected'.

Let us start by making one or two points. Until the invention of the pendulum clock by Huygens in the seventeenth century, there were no terrestrial motions or processes with a regularity remotely good enough to serve as accurate time-keepers for either short or long periods of time. They were all subject to far too many disturbances to be of any use, though water clocks provided useful measures of times accurate to a few minutes over hours or even days. This absence of a convenient clock was reflected in the fact that in antiquity the length of an 'hour' depended on the season of the year, the terrestrial latitude, and even whether it was night or day. For the time between sunrise and sunset at any particular locality was simply divided nominally into twelve equal daylight hours and the time between sunset and sunrise into twelve equally nominal night-time hours. These therefore were only of equal duration at the two equinoxes. This gave rise to the use in astronomy of *equinoctial hours* as a technical concept for the measurement of times other than at the equinoxes. By their definition, these were equal to our present hours. For most astronomical purposes, a timing accuracy of one or two equinoctial hours was perfectly adequate since the sun and the planets have on the average an apparent motion of about one degree per day; since the accuracy of measurement was about 10 minutes of arc and motion through this angle required about four equinoctial hours, an accuracy of one or two such hours was clearly adequate.

However, the moon's apparent motion is much more rapid, since it travels through a degree on the sky in about two hours, or its own apparent diameter in one hour. Thus, the unit of positional accuracy (10 minutes of arc) corresponds to about 20 equinoctial minutes, so the timing accuracy for lunar work wants to be rather better than that if, for example, the occurrence and appearance of eclipses (of great importance for astronomers) are to be accurately predicted. Now this in fact brought the question of the 'true' measure of time to a head. The reason is as follows. Assuming for the moment that the rotation of the earth does measure the 'true' passage of time (we shall return to this), the question still remains: Do we measure a day by successive return of the *sun* to the meridian or by the successive return of a particular star to the meridian? For all civil purposes, it is evidently much more convenient to use the return of the sun (*solar time*), but this leads to a difference from the time measured by the return of a given star (*sidereal time*). The difference comes about because the sun moves relative to the stars and must therefore yield a different unit of time. The most obvious difference is trivial: the mean motion of the sun around the ecliptic means that the average solar day is about four minutes longer than the average sidereal day. A more subtle difference comes from two other factors. First, the sun moves at a nonuniform rate around the ecliptic, as we know from the Hipparchan solar theory. Second, the sun moves on the ecliptic, which is inclined at an appreciable angle to the celestial equator. For this reason the solar day would not be equal to the sidereal day (less those four minutes) even if the motion of the sun around the ecliptic were perfectly uniform. The two effects are superimposed, giving a characteristic curve which measures the accumulated difference between solar and sidereal time over the course of a year (they agree again after a year, of course). This curve, which is shown in Fig. 3.20 as the heavy curve (the sum of the two components), is known as the *equation of time*. It can be seen from the figure that the maximum span of the deviation is about half an hour.

Neugebauer comments:[67] 'It is characteristic for the high level of Hellenistic astronomy that a correct determination of this correction was achieved. We do not know to whom is due this important step in the theory of time reckoning; in the sources available to us the equation of

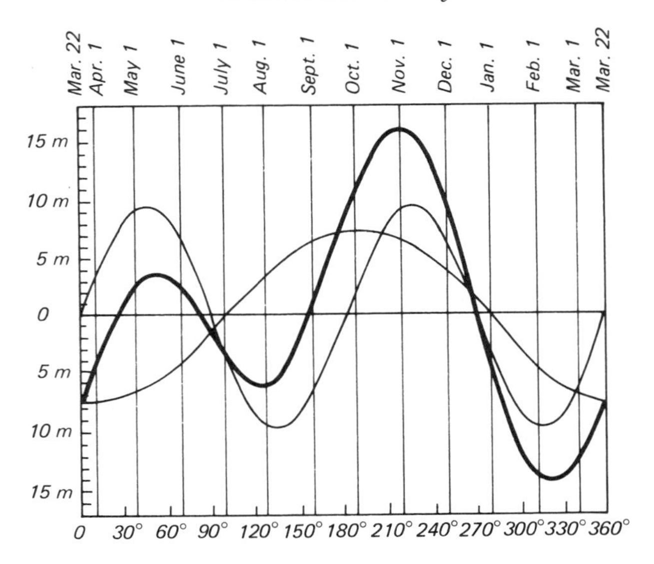

Fig. 3.20. The equation of time (thick curve) and its two components (thin curves). The component due to the obliquity of the ecliptic begins and ends at 0 on March 22. The other thin curve is the component due to the nonuniformity of the sun's motion around the ecliptic. (From *General Astronomy*, by H. Spencer Jones, Arnold & Co, London (1951), p. 47.)

time, or its equivalent, is first attested in the *Almagest*.' The particular significance of this correction is emphasized by Ptolemy himself, for having noted the magnitude of the effect, he says:[68] 'Neglect of a difference of this order would, perhaps, produce no perceptible error in the computation of the phenomena associated with the sun or the planets; but in the case of the moon, since its speed is so great, the resulting error could no longer be overlooked.'

In praising Hellenistic astronomy for discovering the equation of time, Neugebauer presumably had in mind the high level of its technical competence. It is, however, worth noting that the astronomers' work, especially Ptolemy's, had brought to light highly significant facts about the way in which the passage of time is actually manifested in the world. Aristotle may have had an intuitive notion of the uniform passage of time, but in his day none of the innumerable motions observed on the earth or in the heavens had been studied with the accuracy needed to translate this intuition into anything that could be tested against fact. By the time Ptolemy had completed his work, the comparison could be made for the celestial motions at least, and the results represented a significant modification of previous ideas. There were now sufficient data available to give a nontrivial answer to the question: What does it mean to say that a motion is uniform or nonuniform? It was possible to give precise content to the notion of a *universal and equable* flow of time.

One of the particular attractions of looking at this question on the basis of the data available to Ptolemy is the relatively small number of celestial motions that could be observed in his day. The whole field can therefore be surveyed with ease and the conceptual points brought out more clearly. Let us therefore list the available motions, bearing in mind that for time-keeping purposes a motion alone is not sufficient – the motion must be observed relative to certain marks that permit an unambiguous count to be made, as when the hands of a clock move across the marks on the dial.

There is first the apparent diurnal rotation of the stars. For time-keeping purposes, this is extraordinarily convenient. A day can be defined by the rising, setting, or passing across the meridian of any particular star as observed from a fixed point on the surface of the earth. Since the stars all appear to move together, the time defined by the apparent stellar motion is the same whatever star is chosen as time-keeper. The diurnal motions of all the stars are truly mutually uniform (though, of course, only to the extent that their proper motions are ignored). It was, no doubt, because of this fact, together with the belief that the sphere of the fixed stars represented the most important part of the universe, which was responsible for the evidently very deeply ingrained belief that the diurnal motion of the stars is uniform. As we have seen, Hipparchus appears not to have hesitated for a moment when confronted with the choice between the apparent motion of the sun around the ecliptic and the diurnal rotation. He automatically assumed the sun's motion to be nonuniform despite the fact that, strictly speaking, such a choice is meaningless if only two motions are involved. Ptolemy, for his part, made exactly the same choice and evidently felt no need to justify it. In the *Almagest* he says simply:[69] 'The revolution of the universe takes place uniformly about the poles of the equator.'

Continuing our list of observed motions, we have already mentioned the motion of the sun around the ecliptic. This in fact generates two possible measures of time because it can be considered relative to two different markers – the fixed stars along the ecliptic and its passage across some fixed local meridian. It is obvious that each of the other six 'wanderers' – the moon and the five naked-eye planets – each supply two such possibilities for time measurement. They each also provide one further possibility due to their motion in latitude, i.e., their deviation from the ecliptic. In total, therefore, the ancients had at their disposal 21 different basic measures of the passage of time ($1 + 2 + 3 \times 6$). *None of these motions are mutually uniform*. Thus, if any one is taken as the measure of time and plotted along the x coordinate of a two-dimensional system of orthogonal Cartesian coordinates (the abscissa) while the other is taken as the dependent variable y (the ordinate), then the resulting graph is not a straight line. In the light of this fact, the unquestioned choice of the diurnal apparent motion of the stars as the true measure of the uniform passage of time appears rather remarkable.

The justification comes solely from the success of the theories that Ptolemy (and, in part, Hipparchus before him) devised to account for these various motions. The distinctive feature of all these theories is a prescription that says how far the investigated body moves along a definite curve in three-dimensional geoastral space while the stars make a certain number of diurnal revolutions. Three points need to be made here: (1) the theories are all expressed in strikingly simple terms using the simplest geometrical curves (circles); (2) the prescription does not change in time over very long periods (Ptolemy was able to show that lunar eclipses recorded by the Babylonians in 720 BC led to the same theory of the moon's motion as his own observations made about 850 years later); (3) the time defined by the diurnal rotation of the stars works for the theories of the motions of *all* the observed celestial bodies but the attempt to use the time defined by any other of the motions fails; it is only when the one sidereal time is used that all motions can be simultaneously described by simple theories. This time is therefore universal.

Although some of these empirical results had been intuitively anticipated (especially the universality of the time measure) their actual confirmation was still very important. Other results had not been anticipated at all, above all the nonuniformity of the motion. The early philosophers and astronomers (Eudoxus, Aristotle, and even Hipparchus) had all assumed strict uniformity in the various elements that go into the theories. With mutual uniformity assured, the relationship between time and motion is almost trivial. Ptolemy found that nonuniformity can still be described in a constant lawful manner and thus advanced theoretical astronomy to a more sophisticated and nontrivial level.

Given the mutual nonuniformity of *all* the observed celestial motions, the fact that there is nevertheless a measure of time that is somehow distinguished in a very special way is a most remarkable fact about the world that should not escape mention. It clearly did not seem too remarkable to Ptolemy because it fitted more or less the common intuition of his time. Moreover, in the geocentric cosmology it seemed most reasonable to assume that the sphere of the fixed spheres supplied the prime definition of time. However, in the post-Copernican view of the world, the fact that the rotation of the earth (and the other planets), alone among the various motions of the planets and satellites, should give this distinguished time directly seems rather more remarkable. One of the main aims of this study must be to try and come to a deeper understanding of this fact, which will in any case be forced upon us by the later progress of dynamics, which demonstrates quite clearly that the earth's rotation cannot possibly be regarded as the ultimate measure of the passage of time.

Although Ptolemy did not bring astronomy to the level of perfection at which this need becomes acute, he did advance the art to a state in which it could cope with this problem with relative ease when it did arise; that is, he prepared astronomy for the day when it had to be recognized that there is no motion at all that realizes concretely the 'uniform flow of time'.

To see this, suppose that after Ptolemy had worked out all his theories some one presented him with a 'solar system' in which the orbital elements of the various celestial bodies had been changed somewhat but not out of all recognition and, in addition, the earth no longer rotated. The clock had stopped. In this case, all Ptolemy would be able to observe would be the positions of the seven wandering luminaries on the background of the stars. But with trial and error (and, perhaps, the assistance of a computer) Ptolemy would certainly have been able after a while to compile tables of where the various bodies would be seen at a certain 'time'. Unlike his actual tables, in which time was measured in diurnal revolutions, the 'time' in these tables would not be concretely realized by any motion but would be 'reconstructed' using the known theory and the observed motions. For example, according to his theory the longitude of the sun from apogee as it moves around the ecliptic is given by

$$\gamma = t - 2e \sin t,$$

in which t is the 'true' time that has elapsed since the passage through apogee (as earlier, we take the period to be 2π). Ptolemy would know neither the eccentricity e that appears here nor the position of the apogee, so he could not invert the formula to obtain t from the observed positions. However, by sheer brute force he could try all possible positions of the apogee and all possible values of the eccentricity and then see if any of the times then obtained made sense of the motions of the other bodies. Finally he would hit on the correct 'time', and all the motions would work out as they should according to his theory.

Nearly two millennia after he died, in the nineteenth and present centuries, astronomers did in fact construct an abstract time in this sort of fashion. This time, called *ephemeris time*, is the time according to which the tables of positions of the planets and moon are calculated. (An *ephemeris* is a table of the positions of a celestial body at future and past times as calculated in accordance with the appropriate law of motion.) Note that ephemeris time is only abstract in the sense that it is not realized by any one particular motion but is concrete in the sense that it must be determined empirically from actually observed motions.* This manner of conceiving how 'time' could be found empirically by trial and error highlights the remarkable way in which all motions do run together. It demonstrates, in Mach's words:[71] 'the profound interconnection of things'.

Because of the important part played by nonuniform motions in his scheme, Ptolemy was the first man who came to grips with this interconnection at a level that required genuine sophistication. And because he had effectively established the way in which a time parameter can be introduced (which he did by showing that there exists a common time parameter for the nontrivial laws of motion of all the celestial bodies) the transition from sidereal to ephemeris time could be made relatively painlessly when it became necessary. As a result, the concept of time never suffered any abrupt or revolutionary change analogous to the Copernican revolution between the age of Aristotle and 1905, when Einstein created special relativity and completely changed ideas about time. However, this should not obscure the important developments that occurred in the meantime, above all the clarification by Ptolemy in a nontrivial situation of the role that time plays in dynamics. As we shall see in Chap. 11, the weightiest evidence that Newton advanced for his concept of absolute time was essentially the evidence that Ptolemy produced in the *Almagest* and summarized in his discussion at the end of its third book.[72] For this reason, these two or three pages, read in the light of the theories developed in the *Almagest*, must be regarded as the most significant discussion of the nature of time as a scientific concept prior to Einstein's 1905 bombshell.

This is the reason why the third of the 'baker's dozen' should be awarded to ancient astronomy.

It is appropriate to end this section with a brief extract from the *Almagest* in which Ptolemy uses the Babylonian eclipse records mentioned a little earlier to establish the positions of the moon and sun about 850 years before his time. By means of chronological tables he knew the number of days that had elapsed from the eclipse to his own epoch. This gave him the number of 'ticks' of the clock, from which, as we shall see, he

* It is interesting to note that the practical realization of ephemeris time, which has since been replaced by the use of *atomic time*, i.e., a time measure based directly on atomic processes, actually inverted the Hipparchan choice of the motion used to measure time. For ephemeris time, which was originally known as *Newtonian time*, was based on the apparent motion of the sun around the ecliptic and was obtained by comparing the observed position of the sun with the position based on the predicted motion found by solving Newton's equations of the motion of the earth around the sun with allowance for the perturbations caused by the other planets. An important point to note is that the corresponding corrections are amenable to theoretical calculation whereas the rate of rotation of the earth is affected by frictional tidal disturbances in its interior produced by the moon, and these could not be calculated.[70]

determined the position of the sun on the ecliptic and from that the position of the moon. Note how the time of mid-eclipse is determined in order to obtain a parallax-free position of the moon. Note also the use of equinoctial hours and the conversion from Babylonian to Alexandrian local time. Here is what Ptolemy has to say about the first of the eclipses he used:[73]

> The first is recorded as occurring in the first year of Mardokempad, Thoth [I] 29/30 in the Egyptian calendar [−720 Mar. 19/20]. The eclipse began, it says, well over an hour after moonrise, and was total.
>
> Now since the sun was near the end of Pisces, and [therefore] the night was about 12 equinoctial hours long, the beginning of the eclipse occurred, clearly, $4\frac{1}{2}$ equinoctial hours before midnight, and mid-eclipse (since it was total) $2\frac{1}{2}$ hours before midnight. Now we take as the standard meridian for all time determinations the meridian through Alexandria, which is about $\frac{5}{6}$ of an equinoctial hour in advance [i.e. to the west] of the meridian through Babylon. So at Alexandria the middle of the eclipse in question was $3\frac{1}{3}$ equinoctial hours before midnight, at which time the true position of the sun, according to the [tables] calculated above, was approximately ♓ $24\frac{1}{2}°$.*

We see thus how the whole of ancient theoretical astronomy hung on the regular 'ticks' of the rotating heavens, the 'eternal image' of the law of inertia.

3.16. The achievement of Ptolemy and Hellenistic astronomy

As noted in the introduction to this chapter, the achievements of Hellenistic astronomy were for a long time underrated by comparison with the purely speculative ideas that preceded them. It seems to have suffered from the general opinion that the really great period of Greek history lasted from Homer (whenever or whoever he was!) to about 50 years after the death of Sophocles and that anything which did not come within this span was almost by definition decadent, a sad falling off from the greatness that was Athens in its prime. Hellenistic astronomy is pejoratively described as kinematic, while hints of heliocentricity in early Greek astronomy (among the Pythagoreans) are lauded for their 'intellectual vigour' and bold use of physical concepts; Aristarchus is hailed as the Copernicus of antiquity.[74] Aristarchus's suggestion will be discussed in Chap. 5, but it can already be said here that such a view does great injustice to Apollonius, Hipparchus, and Ptolemy.

* ♓ $24\frac{1}{2}°$ means the sun was at $24\frac{1}{2}°$ into the zodiacal sign of Pisces, i.e., at longitude $354\frac{1}{2}°$ (or $-5\frac{1}{2}°$) from the vernal equinox (which was why the night was approximately 12 equinoctial hours long). Note that the ecliptic was divided quite nominally into twelve equal intervals of 30° each; the working astronomers used the zodiacal names to identify these precisely defined intervals, not the corresponding constellations, which had ill-defined boundaries.

The truth is that it was the Hellenistic astronomers who patiently accumulated all the data *and* developed for the first time in history a theoretical method capable of making sense of the data. They were the ones who pioneered the true scientific method and proved it by garnering the first solid results in natural science; they discovered the first laws of nature – and they made possible the Copernican revolution. No one seized of the extent to which Newton relied on Kepler and Kepler on Ptolemy (at least as much as on Copernicus) can fail to appreciate the part played by Hellenistic astronomy in the discovery of dynamics.

For some reason the early historians of astronomy were particularly hard on Ptolemy. The first great historian of the subject, Delambre (1749–1822), dismissed Ptolemy as a hack. Neugebauer[75] discusses the way in which Delambre consistently downrated Ptolemy's work and sought to show that he got the best part of it from Hipparchus. Koestler's very widely read *The Sleepwalkers* gives a jaundiced and grossly unfair picture. He says:[76] 'There is something profoundly distasteful about Ptolemy's universe; it is the work of a pedant with much patience and little originality, doggedly piling "orb in orb".' He says that Ptolemy did not contribute 'any idea of great theoretical value' and quotes with approval the famous satirical lines from Milton's *Paradise Lost*, in which he says of the astronomers:

> . . ., when they come to model Heaven
> And calculate the stars, how they will wield
> The mighty frame, how build, unbuild, contrive
> To save appearances, how gird the sphere
> With centric and eccentric scribbled o'er,
> Cycle and epicycle, orb in orb.

Reading Koestler, and several other authors, one can easily get the impression that the Ptolemaic scheme was excessively complicated, as implied by the quotations just given and also Koestler's statement[77] that a total of no less than forty wheels (i.e., circular motions) was needed in the perfected Ptolemaic system. In fact, there are far fewer circles in the Ptolemaic scheme presented in the *Almagest* than many accounts would lead one to believe; Ptolemy was remarkably economic in his use of circular motions. For the all-important longitudinal motions, only two each were needed for Venus, Mars, Jupiter and Saturn. The sun required only one. Apart from the treatment of the motion in latitude, the only complexity was in the treatment of Mercury and the moon – and that was highly understandable for Mercury and inescapable for the moon.

It is therefore appropriate to conclude this chapter with some general comments about the overall achievement of this first great period in the discovery of dynamics, summarizing some of the points that have already been made. The first aspect to be noted is the astronomers' anticipation of the genuine scientific method, in particular the use of *theory controlled by observation*.

One should start by distinguishing two quite different kinds of theorizing: on the one hand, speculation based on plausible ideas but unsupported by quantitative comparison with proper observations and, on the other, the systematic use of theory as a means of interpreting a given body of data with, equally important, additional data being used as a control to check the correctness of the theory. Early Greek science, up to and including Aristotle, abounded in the first kind of 'theorizing', as we have seen.

The great achievement of Hellenistic astronomy was that it developed systematically and with considerable success the fruitful use of theory in which it is used to interpret observations, suggest further observations, and in turn is submitted to testing by more observations. One of the most significant passages in Ptolemy's *Almagest* is in his discussion[78] of ancient observations of the planets and his comment that they were made at times and positions unsuitable for testing and setting up theoretical schemes. By recognizing that observations had to be made at carefully chosen times in order to get the greatest value from the point of view of *understanding* the world, as opposed to just recording appearances, Ptolemy, who was himself following the tradition set by the earlier Hellenistic astronomers, above all Hipparchus, went half way towards the seventeenth-century realization that the investigator of nature must intervene actively in order to obtain the most favourable conditions of observation. And, above all, Ptolemy and his predecessors used *laws of motion* in the systematic attempt to understand the heavens. Conceptually, but with a long chronological break, the work of Copernicus, Brahe, Kepler, and even Galileo and Newton joined on continuously to what Ptolemy bequeathed to posterity in the *Almagest*.

Thus, to pass from reading, say, Aristotle's *Physics* to Ptolemy's *Almagest* is to pass from a world of philosophy rich in conceptualization but disciplined at best by the rules of logic to science based on hard fact.

With hindsight, it is all too easy to criticize Ptolemy's rigidly geometro-kinetic approach and the absence of any hint of physical causality, but this fails to do justice to his great services to science. Astronomy had to pass through that stage before it had the remotest chance of discovering, first, the deviations from uniformity of the motion and then, second, the utterly unexpected and minute deviations from perfect circularity that, taken together, at last revealed the need for physical causality and pointed the way to a true dynamic theory of motion. It would, in fact, be difficult to devise a more efficient way of discovering the secret of the solar system than the approach laid out by Hipparchus and Ptolemy.

It is also worth noting that although Hellenistic astronomy lacked our modern physical causality it was nevertheless permeated by the idea of causality. For the whole basis of ancient Greek astronomy was the idea that the irregular apparent two-dimensional motions on the sky were to be explained by regular and lawful motions in three-dimensional space. The successes of Apollonius, Hipparchus, and Ptolemy established on an extremely secure foundation one of the greatest central ideas of modern science – that behind the apparently irrational particular phenomena presented directly to our senses there is a rational but unseen basis. This idea, and the highly successful and not at all trivial demonstration of its fruitfulness, was certainly a decisive factor in the emergence of modern science.

Another point should be emphasized – Ptolemy's insistence that observation comes before theory and that any theory, however attractive, must be abandoned if it fails to describe the observations. For this attitude of mind Kepler is given great credit and deservedly, as we shall see. But the example had already been set for him by Ptolemy. We have already quoted Ptolemy's comments about the need to revise the theory of the moon's motion. It is also worth quoting what he had to say about Hipparchus's attitude to the problem of the planets:[79]

> but, [we may presume], he reckoned that one who has reached such a pitch of accuracy and love of truth throughout the mathematical sciences will not be content to stop at the above point, like the others who did not care [about the imperfections]; rather, that anyone who was to convince himself and his future audience must demonstrate the size and the period of each of the two anomalies by means of well-attested phenomena which everyone agrees on, must then combine both anomalies, and discover the position and order of the circles by which they are brought about, and the type of their motion; and finally must make practically all the phenomena fit the particular character of the arrangement of circles in his hypothesis. And this, I suspect, appeared difficult even to him.

In the light of Neugebauer's comment quoted earlier (p. 137) about the accuracy of ancient astronomy, it might seem that Ptolemy did not quite live up to the ideals that he praised so eloquently. The resolution is to be found in the failure, principally among Ptolemy's successors, to appreciate that accuracy in itself is not enough; for, as the work of Brahe and Kepler was subsequently to show, accuracy must be combined *with really systematic observational testing of the theories once found*.

Nevertheless, when we consider the cavalier attitude to observation and quantitative accuracy displayed by the Greeks, medieval philosophers, and even Descartes and other major figures in the seventeenth century in their study of terrestrial motions, we can appreciate the importance for science of this example set by Hipparchus and Ptolemy for at least the heavenly motions and their commitment to find sound theoretical explanations for accurate observations. The passage just quoted is perhaps the oldest extant statement of a systematic scientific research project. It established a paradigm that still proved remarkably fruitful when extended far beyond its original domain.

We should also mention in this connection the immense importance of Ptolemy's *Almagest*. As author of this work Ptolemy performed a service to posterity of magnitude almost as great as that of Euclid. The *Almagest*, which is more properly called the *Syntaxis*[80] (i.e., a systematic treatise on all aspects of astronomy), became a kind of *Handbuch* of ancient astronomy and was of the first importance for several reasons. First, it summarized all the major astronomical discoveries made by the ancients up to Ptolemy's time, including his own; second, it was a masterpiece of clear scientific exposition and became the standard reference work for over a millennium. By virtue of its sheer quality it rapidly became indispensable, and this ensured that the very substantial achievements of ancient astronomy survived the vicissitudes of the pre-Gutenberg world and were not lost for ever, as so much else was.

The *Almagest*, which has recently been translated into English by Toomer and is highly recommended (together with the extensive commentaries on it by Neugebauer[2] and Pedersen[81]*) to the interested reader, is written in a terse, sober, and factual style with an absolute minimum of speculation. Its approach to observation and accuracy of measurement is entirely modern (which makes some errors in the work, especially in the solar observations, and also the occasional fiddling of the data, particularly baffling). Just occasionally, there are passages of almost lyrical intensity in praise of truth and the appeal of the heavens as a subject most worthy of study. These give one an insight into Ptolemy's psyche and also demonstrate how awe of the heavens inspired ancient scientists to extraordinary labours in trying to track and interpret celestial motions while completely disdaining terrestrial motions. It was Galileo's achievement to exalt the mundane to the same status and significance as the heavenly. His application of the celestial standards set by Hipparchus and Ptolemy to the terrestrial motions was a major factor in the creation of dynamics. In a very real sense, the *Almagest* became the 'bible' of medieval and Renaissance astronomy. Just as the Sermon on the Mount set the tone

* Pedersen's *Survey* is especially recommended for its final chapter on Ptolemy's personality (he was a Stoic) and his numerous other important and influential works. There is a sympathetic discussion of Ptolemy's work on astrology, the *Tetrabiblos*, and an interesting account of the rediscovery of the part of his *Planetary Hypotheses*, written some years after the *Almagest*, in which he proposed a complete mechanical system of moving spheres to account for the motions established in the *Almagest*; this shows that despite the impression one might get from the earlier work and Neugebauer's comment quoted on p. 142, he does appear to have taken the moving spheres seriously. Pedersen also gives an account of how the *Almagest* was transmitted to posterity and includes a helpful account of the development of technical astronomical terminology in the Middle Ages. At various places he also discusses the problem, mentioned earlier, of Ptolemy's 'doctoring' of data, which is also discussed by Toomer.[82]

of Christian ethics, Ptolemy's formulation of the problem of the planets set the standards and aims of astronomy, the purest and most rigorous of the natural sciences. Ptolemy posed the great problem of astronomy. And as regards the development of dynamics and science generally it truly was a case of 'Seek ye first the Kingdom of Heaven and the rest will be given unto you.' Once the great mystery of the heavens had been cracked, the rest fell into place quite soon – though not however without vital hints supplied by the study of terrestrial motions.

Ptolemy not only posed the great problem. He also found (or used and passed on) most of the methods of solution – and many important parts of the answer. Indeed, he was as good as his word: he did succeed in making 'practically all the phenomena fit the particular character of the arrangement of circles in his hypothesis'. But the greatest prize eluded him – he did not hit on the idea of a heliocentric cosmology.

As this, more than anything else, is probably the reason why Ptolemy has so often had a 'bad press', this is perhaps a good point to review the astronomical evidence for heliocentricity as it accumulated through the Hellenistic period and, in particular, the suggestion[1b] that Aristarchus was the 'Copernicus of antiquity'. In one sense, he clearly was not: no Aristarchan revolution followed his proposal. Very little is known about the details of Aristarchus's proposal. In his extant work (discussed earlier in connection with his introduction of trigonometric techniques into astronomy), there is no trace of heliocentricity. The main source of information about Aristarchus's proposal is in Archimedes' *Sand-Reckoner*, written about a generation after Aristarchus's time. According to Archimedes:[83] 'His hypotheses are that the fixed stars and the sun are stationary, but the earth is borne in a circular orbit about the sun, which lies in the middle of its orbit, and that the sphere of the fixed stars, having the same centre as the sun, is so great in extent that the circle on which he supposes the earth to be borne has such a proportion to the distance of the fixed stars as the centre of the sphere bears to its surface.'

Archimedes has little to say about the merits of the proposal. In fact, all he really does is criticize rather pedantically the final words for being mathematical nonsense, since a point cannot bear a proportion to a surface (it seems clear that Aristarchus was only using a loose expression to mean that the stars are at an immensely great distance).

Now the part of Copernicus's *De Revolutionibus* that really established his theory was the very clear geometrical explanation he was able to provide for the retrograde motions of the planets (and indeed for the second inequality in its entirety) and the associated dramatic extension of trigonometric techniques within the solar system that it made possible (this will be discussed in Chap. 5). He spoke, as it were, directly to readers with a good mathematical intuition; such were Kepler and Galileo – and the light dawned on them. But Archimedes was just as great a mathematician. It seems hard to believe that had Aristarchus presented clearly the gist of Copernicus's arguments Archimedes would have had so little to say about them. In fact, there is no mention of the planets in the above quotation or in any of the other extant references to the proposal. This absence of any explicit reference to the planets has not prevented innumerable authors from saying that in the Aristarchan scheme *all* planets circle the sun in concentric circular orbits. Moreover, it is evident from other records that the proposal gained quite wide currency in the ancient world even though it is not mentioned by Ptolemy (who does however discuss[84] – and reject – the idea of the earth's rotation) but still it sparked no revolution. I suspect that the difference between the two men was precisely that Copernicus grasped fully the immense *significance* of the proposal (which, after all, was not originally his own), stated it clearly, and provided *solid arguments* in its support. And the reason why Copernicus could do this but Aristarchus could not is clearly to hand: all the detailed observations and analysis on which Copernicus could base his case were not available to Aristarchus; *for they were made by his successors*. They were the true creators of the heliocentric system; Copernicus only added the final touch.

For the real evidence for heliocentricity was in the demonstration by Ptolemy that in all cases the second inequality can be unravelled from the first and that when this is done it leaves behind in each case a first inequality characterized by essentially the same functional dependence and structure (circular motion with eccentric centre and equant). Without the solid basis that these factual data provided, the Copernican revolution would have been impossible.

The transition from the old cosmology of Plato and Aristotle to the new astronomy of Kepler required four major steps: liberation of astronomy from the diurnal motion of the earth, liberation from the annual motion, liberation from uniformity of motion, and liberation from circularity of the motion. The first and second of these were presented almost 'on a plate' to Copernicus by the combined efforts of the ancient astronomers. The third was entirely Ptolemy's achievement. And the fourth, in some ways the least important in the historical perspective, could never have been effected without the previous three.

Why did Ptolemy not see what he himself had made almost blindingly obvious? He had enough hints – many of his own making. He knew the sun was vastly bigger than the earth, and he knew that all the planets danced in tune in their epicyclic twirls to the steady beat of Father Sun. One important explanation was in his conception of the laws of terrestrial motion, which will be discussed in a later chapter. Another no doubt was in his reverence for his forebears, above all Hipparchus. He may too, if he did consciously ponder Aristarchus's proposal, have been simply overawed by the sheer immensity of the cosmos that it implied. But I think the truth is that he was blinded by his own success. The fate of Archimedes, killed by the Roman soldier while tracing figures in the sand, may give us the real clue. Perhaps he was simply too intent on tracing those mysterious patterns in the starry sky, too absorbed in the details of the dance.

What then was left for Copernicus to do? First, he realized how the scale invariance of the Ptolemaic algorithmic procedure could be exploited to convert the geometrical similarity of the epicyclic motion into identity, that is, he postulated that all the epicyclic radii of the superior planets should be set equal to the earth – sun distance and that the same should be done for the deferents of the inferior planets. Next, he inverted the deferent and epicycle for the two inner planets. Then he stretched out his hand, plucked the luminous planets from the tips of the epicyclic spokes and calmly placed them in the ghostly deferent guide points. Finally, he reversed the earth–sun vector. Thus he made the transition from the geoastral to the helioastral frame. The move was deft, almost cheeky; but, at a fundamental level, he did precious little else.

Ptolemy built the carousel. Long after the fair-keeper had retired to bed, Copernicus came in the night, moved the linchpin, and switched on the lights. The effect was magical. Science, already stirring, woke from its millennial torpor. Copernicus's proposal is a never-ending source of fascination, bizarre, as we shall see, in the manner he made it, incredible in the extent of its far-reaching consequences, and the parallel in science to Richard II's soliloquy on the fate of kings:[85] The merest pinprick that finds the point on which all hinges.

4

The Middle Ages: first stirrings of the scientific revolution

4.1 Introduction

In histories of dynamics written in the nineteenth century, the period between antiquity and Galileo was treated as an almost complete blank. It was believed that nothing significant had occurred during the 'Dark Ages'. This attitude was changed almost single-handed by the French physicist Pierre Duhem, who became an historian of science almost by accident.[1] In his monumental *Le Système du Monde* he argued that several basic principles of Galileo's physics were in essence already worked out in the fourteenth century, and that Galileo's work consisted more of explication and further development rather than genuine revolution.[2] The flavour of Duhem's writings is expressed by this quotation:[3]

> From the start of the fourteenth century the grandiose edifice of Peripatetic physics was doomed to destruction. Christian faith had undermined all its essential principles; observational science, or at least the only observational science which was somewhat developed – astronomy – had rejected its consequences. The ancient monument was about to disappear; modern science was about to replace it. The collapse of Peripatetic physics did not occur suddenly; the construction of modern physics was not accomplished on an empty terrain where nothing was standing. The passage from one state to the other was made by a long series of partial transformations, each one pretending merely to retouch or to enlarge some part of the edifice without changing the whole. But when all these minor modifications were accomplished, man, encompassing at one glance the result of his lengthy labor, recognized with surprise that nothing remained of the old palace, and that a new palace stood in its place.

In Duhem's view, the long process of piecemeal modification, which he set out to document in such detail, was quite erroneously seen as a revolution by those who added the final touches – and who naturally cast themselves in the roles of the revolutionaries.

Duhem's views stimulated lively discussions between historians of science, in which one point at issue was whether the role played by

Christian theology in the unique emergence of modern science was a positive or negative factor.* These discussions still continue.[5]

The considered opinion of two of the most respected historians of science in the Middle Ages, Anneliese Maier[6] and Marshall Clagett,[7] is that Duhem performed a capital service in rescuing the schoolmen from oblivion but overargued his case. In Maier's opinion Duhem was certainly correct to see the scientific ideas of the fourteenth century as a preliminary step towards and preparation of classical physics but that he had often extracted too modern a meaning from the medieval texts and had also exaggerated their importance. She points out that, considered overall, the history of the exact sciences in the Christian West was, from its beginnings in the thirteenth century through to the eighteenth century, a story of the gradual overcoming of Aristotelianism. This did not occur in a single great revolution but also not in a completely continuous process. There were two high points, the first in the fourteenth century, the second in the seventeenth.

The starting point of this process was the rediscovery in the thirteenth century of Aristotle's work. A highlight in that century was the attempt by St Thomas Aquinas (1225–1274) to synthesize Aristotelian philosophy with Christian theology. Aquinas referred to Aristotle simply as The Philosopher and his output was massive.

By helping to establish Aristotelianism at the expense of Platonic metaphysics, Aquinas must certainly have done much to create a climate favourable for the development of a scientific attitude of mind:[8] 'The Platonic man, who was scarcely more than an incarcerated spirit, became a rational animal . . . the Platonic theory of knowledge . . . was translated into a theory of abstraction in which sensible experience enters as a necessary moment into the explanation of the origin, the growth and the use of knowledge, and in which the intelligible structure of sensible being becomes the measure of the truth of knowledge and of knowing.'

This was undoubtedly a very positive and progressive development (cf. the footnote below) and helped to lay the foundations for the mini-Renaissance of science in the following century. Dante made Aquinas's synthesis the philosophical basis of his great poem *La Divina Commedia* (written

* It is worth quoting here the following passage from Whitehead's *Science and the Modern World*:[4] 'I do not think, however, that I have even yet brought out the greatest contribution of medievalism to the formation of the scientific movement. I mean the inexpugnable belief that every detailed occurrence can be correlated with its antecedents in a perfectly definite manner, exemplifying general principles. Without this belief the incredible labours of scientists would be without hope. It is this instinctive conviction, vividly poised before the imagination, which is the motive power of research: – that there is a secret, a secret which can be unveiled. How has this conviction been so vividly implanted on the European mind? . . . My explanation is that the faith in the possibility of science, generated antecedently to the development of modern scientific theory, is an unconscious derivative from medieval theology.'

1308–1321). Three centuries later, when Aristotelianism had ossified and long since ceased to be a positive development, the immense influence of these two must have done much to make the overall Aristotelian world-view seem far more unassailable than it really was.

But in the high Middle Ages, stimulated by the fresh air of Aristotelianism, many philosophers worked away on Aristotle's various books, writing commentaries, attempting to reconcile him with the works of Archimedes, which had of course very solid scientific value, and also examining the commentaries and criticisms of Aristotle that survived from antiquity. The attitude to Aristotle was by no means subservient and new points of view were steadily brought forward. Out of this busy self-confident work of the schoolmen several scientific developments important for the subject of this book were born and flourished in the fourteenth century. Perhaps the most important thing was that a branch of exact learning developed that was to a large degree independent of general philosophy and theology. It made the transition, in Maier's words, from natural philosophy to natural science. Moreover, as in the seventeenth century, the key development was associated with the study of motion. For our purposes, the most important work was done at Merton College in Oxford and at the University of Paris.

4.2. Kinematics

We start with the work done at Merton, for which I follow Clagett[9] closely. In the light of the comments made earlier about the difficulty of recognizing motion as something primary and rather special compared with general qualitative alterations (see in particular the quotation from Ptolemy, p. 43), it is interesting to note that the work at Merton, which is associated with the names of Thomas Bradwardine, William Heytesbury, Richard Swineshead and John Dumbleton and belongs to the period 1328 to 1350, grew out of the philosophical problem of how qualities change in intensity, e.g., how something becomes hotter or whiter. It seems that in the general framework of this approach what we would today call instantaneous velocity was regarded as the *intensity* of movement, while the total distance travelled in a given time was regarded as the *quantity* of the movement. The specialized study of these concepts led to important developments in kinematics. According to Clagett, the achievements of the Mertonians can be summarized as follows:

(1) Their work led to a clear-cut distinction between (in modern terms) *dynamics* and *kinematics*, i.e., between the *causes* of movement and the *effects* of movement as observed in space and time;
(2) They developed, perhaps for the first time, the idea of an 'instantaneous

velocity' having an existence conceived as distinct from the spaces traversed in given time;

(3) They defined a uniformly accelerated movement as one in which equal increments of velocity are acquired in any equal periods of time;

(4) They stated and proved a fundamental kinematic theorem, which became known as the Merton Rule.

This last is as follows. Suppose a point starts from rest and is uniformly accelerated with acceleration a for a time T. After time $T/2$, it will have speed $V_{\frac{1}{2}} = aT/2$. Then the Merton Rule states that the distance traversed by the uniformly accelerated body in the complete interval T is equal to the distance traversed in the same time T by a body moving with the speed $V_{\frac{1}{2}}$. This, of course, is just a consequence of the two fundamental results of the theory of uniformly accelerated motion:

$$v = at,$$

$$s = \tfrac{1}{2}at^2,$$

where v is the speed at time t and s is the distance traversed at time t. For a body moving with speed $\frac{1}{2}aT$ (the speed acquired at time $t = T/2$) will traverse in time T the distance $T \times \frac{1}{2}aT = \frac{1}{2}aT^2$.

These achievements of the Mertonians, especially the last two, are important because they mark the first significant steps from the Greek mathematization of *uniform* motion to the successful mathematical treatment of *nonuniform* motion; in their mathematics at least they prepared the ground for several of Galileo's most important results, especially in the clear definition of the concept of uniformly accelerated motion. It would however be a mistake to see the Mertonians as precursors of Galileo as a *physicist* and attempt to give them the credit for the discovery of the law of free fall. For, as Clagett is careful to point out, their work 'was almost entirely hypothetical and *not rooted in empirical investigations*'. Rather, we have here a classic example of the mathematics necessary for a physical discovery having been created long before it found physical application. The Mertonians no more discovered the law of free fall than Apollonius, who wrote the first great systematic treatise on conic sections, discovered the law of elliptical planetary motion. That does not, of course, detract from their mathematical achievement, nor their significance in the overall development of science.

One really needs to read the extensive extracts from the medieval documents in Clagett's work to get a feeling for the difficulty of forming the key concepts of kinematics that now come so easily to the trained scientist. Even as late as the 1630s Galileo clearly felt the need to proceed very carefully and slowly in introducing concepts such as that of uniform acceleration.[10] The medieval documents also demonstrate how precise concepts are developed in a process of gradual refinement.

The ideas of the Mertonians spread quickly and were known in France and Italy by about 1350. Clagett quotes a document written in Prague in 1360 which refers to the Mertonians as the 'Calculators'.[11] Richard Swineshead in particular was known throughout the fifteenth century as the Calculator and the technique of treating motion in the Merton manner became known as 'the Calculations'.

The spread of Merton kinematics to the continent led to an important event, namely, the application of graphing or coordinate techniques to the English concepts dealing with qualities and velocities.[12] This was in fact a partial step towards the analytic geometry of the seventeenth century initiated by Descartes, though it did not go so far as to translate algebraic expressions into geometric curves and vice versa, which is the essence of that method.

The method is normally attributed to Nicole Oresme and dated around 1350. Clagett argues that it arose slightly earlier in Italy. Interesting is his comment that the graphing concept had been applied from antiquity to cartography and astronomy and that the very terms (*latitude* and *longitude*) used by Oresme and his contemporaries suggest an intentional transfer from these older disciplines.

Also interesting in connection with the emergence of kinematics as a discipline in its own right is the fact that Oresme's text is entitled *On the Configurations of Qualities*, i.e., it treats variations of qualities in general by geometrical methods. Clagett comments that Oresme is concerned 'with a figurative presentation of hypothetical quality variations . . . totally unrelated to any empirical investigations of actual quality variations'.[13] It is therefore in the nature of an exercise to set up a general theoretical framework for the treatment of qualitative terrestrial physics, which Ptolemy, we recall, felt was such a hopeless undertaking (p. 43). However, like the Mertonians, Oresme also treated velocity changes, and so gave a geometrical method of representing them.

In view of its exceptional interest from the point of view of the modern way of thinking about motion, I quote in full the following passage from Clagett's book together with the necessary figure (Fig. 4.1):[14]

As examples of Oresme's technique let us consider the accompanying rectangle and right triangle [Fig. 4.1]. Each measures the quantity of some quality. Line AB in either case represents the extension of the quality in the subject. But in addition to extension, the intensity of the quality from point to point in the subject has to be represented. This Oresme did by erecting lines perpendicular to the base line, the length of the lines varying as the intensity varies. Thus at every point along AB there is some intensity of the quality, and the sum of all these lines is the figure representing the quality. Now the rectangle ABCD is said to represent a uniform quality, for the lines AC, EF, BD representing the intensities of the quality at points A, E, and B (E being any point at all on AB) are equal, and thus the intensity of the quality is uniform throughout. In the case of the right triangle ABC it will be

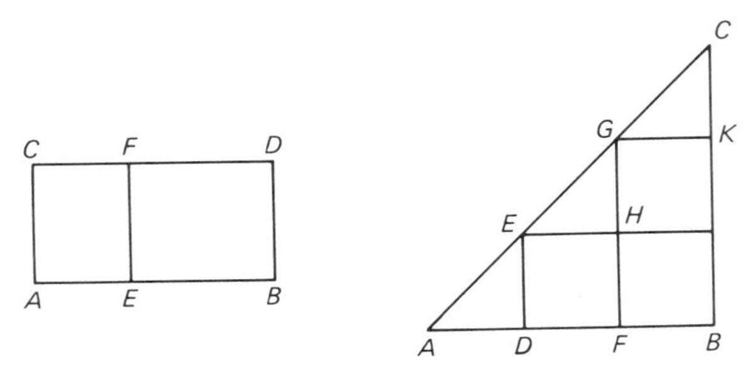

Fig. 4.1.

equally apparent that the lengths of the perpendicular lines representing intensities uniformly decrease in length from BC to zero at point A. Hence the right triangle is said to represent a uniformly difform* (nonuniform) quality. Of course the intensities could vary in an infinite number of ways and we would have a limitless variety of figures to represent other kinds of nonuniform qualities. It is worth pointing out that Oresme designated the limiting line CD (or AC in the case of the triangle) as the 'line of summit' or the 'line of intensity'. This is comparable to a 'curve' in modern analytic geometry. And thus the figures themselves in Oresme's system are comparable to 'the areas under curves'. The 'curve' or summit line of Oresme is representing a 'function' expressed verbally instead of by algebraic formula, the verbal expressions of the functions being 'a uniform quality', 'a uniformly nonuniform quality', etc. The variables in these functions of Oresme are the quantity of a quality and the extent and intensity of the quality.

It was by means of this technique, developed for qualities quite generally but applied specifically to speeds, that Oresme gave a geometrical proof of the Merton rule.

Clagett concludes his discussion of medieval kinematics with the comment:[15] 'It was, then, in the area of kinematics, and particularly in the geometrical analysis of uniform acceleration, that the medieval tradition was to be most significant for the development of modern mechanics.'

4.3. Dynamics

Of all the steps that led to the definite structure that was given to dynamics by Newton the clear recognition of the existence of inertial motion and its elevation to the first and most fundamental law of dynamics was the most important. As we shall see in the chapters on Galileo and Descartes, the process of recognition of the existence and significance of inertial motion was most complex and required the coming together of various different (and quite distinct strands). As already

* *Uniformly difform* was the medieval expression for changing at a uniform rate, i.e., *uniformly difform motion* is uniformly accelerated motion.

hinted in the chapter on Aristotle, one of these strands was provided by astronomy and came, in Aristotelian terminology, from the superlunary sphere. In the following chapters we shall see precisely how the Copernican revolution enforced the introduction of this strand into terrestrial physics.

The second strand, with which we shall be concerned here, had its origin solely in sublunary, or rather terrestrial, physics. It was this development that finally broke the tyrannical grip of the Aristotelian notion that terrestrial bodies can only move if they are constantly being pushed by something else and that if the cause of motion ceases then so too will the motion. For the heavenly motions, which manifestly persisted, this problem never arose since Aristotle made their motion directly dependent upon divine ordinances, and, as Ptolemy correctly sensed, the job of the natural scientist in this case was solely to describe the motion, not to locate some efficient cause in the shape of a material 'pusher'. But in the case of terrestrial motions, except for the natural (in the Aristotelian sense) motions of ascent and falling, the main task in their study was to locate the 'pusher'. It was for this reason that the notion of a plenum was so all important to Aristotle. His dynamical law sought (in very qualitative form) to establish how the speed of a body results from the relationship of the force of the 'pusher' to the resistance of the medium through which the pushed object is being moved.

The concept of a plenum was a severe hindrance to the recognition of inertial motion. This is most clearly seen in a rather remarkable passage in which Aristotle actually anticipated something very much like the law of inertia but used what he saw as the absurdity of such a law to supply one further argument for a plenum and against a void. Speaking of a body moving in a void, he says that, if such a thing were to happen, it would not be possible to assign a reason 'why the projectile should ever stop – for why here more than there? It must therefore either not move at all, or continue its movement without limit, unless some stronger force impedes it'.[16]

It is perhaps not surprising that it was through the consideration of the motion of projectiles that the first serious doubts about Aristotle's assertion of the invariable need for a 'pusher' for all terrestrial motions arose. Aristotle's treatment of this question, which he considered only rather peripherally, involved such manifest absurdities that they stimulated a healthy reaction which eventually developed into the second main strand leading to the clear recognition of the law of inertia.

For example, immediately before the passage just quoted, Aristotle says of projectiles that move when the body that impelled them is no longer in contact with them that this is either[17] 'due (as some suppose) to a circulating thrust [*antiperistasis*], or to the air being set by the original impact in more rapid motion than that of the natural movement of the missile towards the place proper to it'. Later, he asserts[18] 'that the prime mover conveys to the air . . . a power of conveying motion, but that this power is not exhausted when the intermediary ceases to be moved itself. Thus the intermediary will cease to be moved itself as soon as the prime mover ceases to move it, but will still be able to move something else'.

In late antiquity (sixth century AD) John Philoponus,* whose views on space we quoted in Chap. 2, criticized Aristotle's ideas, which seemed to involve two if not three theories of the phenomenon, and in his commentary on Aristotle's *Physics*, wrote:[20] 'Let us suppose that . . . according to the first method indicated above, namely, that the air pushed forward by the arrow gets to the rear of the arrow and thus pushes it from behind. On that assumption, one would be hard put to it to say what it is (since there seems to be no counter force) that causes the air, once it has been pushed forward, to move back, that is along the sides of the arrow, and, after it reaches the rear of the arrow, to turn around once more and push the arrow forward. . . . Such a view is quite incredible and borders rather on the fantastic.'

After further severe criticism, Philoponus says (italics added by translator)

> From these considerations and from many others we may see how impossible it is for forced motion to be caused in the way indicated. *Rather is it necessary to assume that some incorporeal motive force is imparted by the projector to the projectile*, and that the air set in motion contributes either nothing at all or else very little to this motion of the projectile. If, then, forced motion is produced as I have suggested, it is quite evident that if one imparts motion 'contrary to nature' or forced motion to an arrow or a stone the same degree of motion will be produced much more readily in a void than in a plenum. And there will be no need of any agency external to the projector. . . .

The italicized words contain the essence of what in the fourteenth century was to become the central idea of Buridan's *impetus* theory. Anyone remotely familiar with Newtonian dynamics will note that there has been a significant shift from Aristotle towards Newton. Also worth noting in Philoponus is the clear tendency to thinking about motion in the first place in a void rather than in a plenum. When one reads even an early work of Galileo,[21] written long before he made his most important discoveries, one of the most striking differences from Aristotle is in

* Philoponus, whose work has been the subject of a recent book,[19] is renowned for his refutation of the Greek error about the speed with which bodies fall and for the oldest extant record of an experimental test of such fall: 'But this is completely erroneous, and our view may be corroborated by actual observation more effectively than by any sort of verbal argument. *For if you let fall from the same height two weights of which one is many times as heavy as the other, you will see that the ratio of the times required for the motion does not depend on the ratio of the weights, but that the difference in time is a very small one*' (translator's italics; quoted from Ref. 20, p. 220).

Galileo's attitude to the medium through which the body under study is moving. Whereas for Aristotle the medium is the *sine qua non* of the body's motion, determining directly the most fundamental features of the motion, for Galileo the medium is merely a source of disturbance and resistance that causes a body to move otherwise than it would. In Galileo's view it is the idealized motion that the body would follow in the absence of a medium that is truly significant. This and numerous passages in other authors show that the millennia that came between Aristotle and Galileo did much to change the whole climate of opinion as to the basic attitude one should take to the study of motion. For all that it was Aristotle who drew attention to the importance of the study of motion *per se*. Virtually all the new developments grew out of commentaries on Aristotle.

We omit discussion of the extent to which Philoponus's idea and some interesting Islamic modifications of it[22] may have influenced the Paris school of dynamics founded by Buridan (this topic is very well covered by Clagett) and turn straight to Buridan's exposition of impetus theory, as it became known. Buridan, who according to Clagett[23] was probably born at Béthune around 1300 and was variously reported at the University of Paris during the period 1328–1358, was a figure rather like Bradwardine and can be regarded as the founder of a school in Paris of which the most able followers were Nicole Oresme, Albert of Saxony (who later founded the University of Vienna), and Marsilius of Inghen (who founded the University of Heidelberg).

Since the work of Buridan and the Paris school was first discovered by Duhem, it has been the subject of numerous studies, one of the fullest being Clagett's, which also gives references to many others, especially those of Anneliese Maier.

I shall quote here only some of the most striking passages from Buridan's *Questions on the Eight Books of the Physics of Aristotle*. Buridan opens with some remarks about the difficulties of Aristotle's theory that are reminiscent of Philoponus. Then follow some devastating physical arguments, chosen with a skill worthy of Galileo, that completely demolish the idea that projectiles could be pushed for a long time from behind by the air:[24]

> A lance having a conical posterior as sharp as its anterior would be moved after projection just as swiftly as it would be without a sharp conical posterior. But surely the air following could not push a sharp end in this way, because the air would be easily divided by the sharpness. . . . a ship drawn swiftly in the river even against the flow of the river, after the drawing has ceased, cannot be stopped quickly, but continues to move for a long time. And yet a sailor on deck does not feel any air from behind pushing him. He feels only the air from the front resisting.

Buridan gives numerous other examples, many of which reveal a clear awareness of the phenomena associated with inertial motion. Then comes a clear statement of his concept of impetus:[25]

> Thus we can and ought to say that in the stone or other projectile there is impressed something which is the motive force (*virtus motiva*) of that projectile. And this is evidently better than falling back on the statement that the air continues to move that projectile. For the air appears rather to resist. Therefore, it seems to me that it ought to be said that the motor in moving a moving body impresses (*imprimit*) in it a certain impetus (*impetus*) or a certain motive force (*vis motiva*) of the moving body, [which impetus acts] in the direction toward which the mover was moving the moving body, either up or down, or laterally, or circularly. *And by the amount the motor moves that moving body more swiftly, by the same amount it will impress in it a stronger impetus.* It is by that impetus that the stone is moved after the projector ceases to move. But that impetus is continually decreased (*remittur*) by the resisting air and by the gravity of the stone, which inclines it in a direction contrary to that in which the impetus was naturally predisposed to move it. Thus the movement of the stone continually becomes slower, and finally that impetus is so diminished or corrupted that the gravity of the stone wins out over it and moves the stone down to its natural place.

Clagett used italics for the sentence in the middle of the above quotation to draw attention to the way in which Buridan anticipated not only Newton's First Law but also Newton's identification of *momentum* as the most fundamental dynamical concept. We recall that momentum (*quantity of motion* was Newton's expression) is defined as the product of the mass of the body and its velocity (directed speed). It will be seen from the above sentence that Buridan made the strength of the impetus proportional to the speed of the body that is thrown. Even more remarkable is the following passage, which follows immediately after the above (again the italics are, of course, Clagett's):

> For if anyone seeks why I project a stone farther than a feather, and iron or lead fitted to my hand farther than just as much wood, I answer that the cause of this is that the reception of all forms and natural dispositions is in matter and by reason of matter. *Hence by the amount more there is of matter, by that amount can the body receive more of that impetus and more intensely. Now in a dense and heavy body, other things being equal, there is more of prime matter than in a rare and light one. Hence a dense and heavy body receives more of that impetus and more intensely, just as iron can receive more calidity than wood or water of the same quantity.* Moreover, a feather receives such an impetus so weakly that such an impetus is immediately destroyed by the resisting air. *And so also if light wood and heavy iron of the same volume and of the same shape are moved equally fast by a projector, the iron will be moved farther because there is impressed in it a more intense impetus, which is not so quickly corrupted as the lesser impetus would be corrupted. This also is the reason why it is more difficult to bring to rest a large smith's mill which is moving swiftly than a small one, evidently because in the large one, other things being equal, there is more impetus.*

As a particularly apt demonstration of his theory of impetus, Buridan says that: 'one who wishes to jump a long distance drops back a way in order to run faster, so that by running he might acquire an impetus which would carry him a longer distance in the jump.* Whence the person so running and jumping does not feel the air moving him, but [rather] feels the air in front strongly resisting him.'

Clearly, we have come a long way from Aristotle. Indeed, *on this particular aspect of dynamics*, it is not until Newton himself that we come to the formulation of fundamental dynamical insights of such clarity and cogency. No wonder Duhem raised a triumphant cry when he discovered the works of Buridan and the Paris school about 80 years ago.

The grounds for seeing Buridan as the true formulator of Newton's First Law appear to be still further strengthened by his theory of celestial motions:[26]

> Also, since the Bible does not state that appropriate intelligences move the celestial bodies, it could be said that it does not appear necessary to posit intelligences of this kind, because it would be answered that God, when He created the world, moved each of the celestial orbs as He pleased, and in moving them He impressed in them impetuses which moved them without his having to move them any more except by the method of general influence whereby he concurs as a co-agent in all things which take place; 'for thus on the seventh day He rested from all work which He had executed by committing to others the actions and passions in turn'. And these impetuses which He impressed in the celestial bodies were not decreased nor corrupted afterwards, because there was no inclination of the celestial bodies for other movements. Nor was there resistance which would be corruptive or repressive of that impetus. But this I do not say assertively, but [rather tentatively] so that I might seek from the theological masters what they might teach me in these matters as to how these things take place.

Despite all the remarkable anticipations of Newton's First Law in these quotations from Buridan, we are, *pace* Duhem, still a long way from Newton's First Law stated in the full consciousness of the fact that it is the *first law of motion* (with universal applicability to all motions of all bodies in the universe), and with full understanding of what can be achieved with it. For this we still have to wait more than three centuries (Clagett dates Buridan's *Questions* somewhere between 1340 and 1357;[27] there is some possibility that Buridan died from plague in 1358).

A fuller discussion will have to wait until later, but two points can be made already.

(1) There is no clear statement that, in the absence of resistance and gravity, impetus would carry a body on a *straight line* at a uniform speed for ever. For example, in the case of the celestial motions, Buridan must clearly have had uniform *circular* motions in mind. We note also the

* As the French indeed say: *reculer pour mieux sauter*.

reference in one of the quotations to a smith's mill, and in this case too a circular impetus may be implied, as indeed Buridan says explicitly: 'in the direction toward which the mover was moving the moving body, either up or down, or laterally, or *circularly*' (my italics). Thus, Buridan's main contribution was to overcome the Aristotelian idea that no motion is possible without a constant 'pusher'. He definitely seems to have entertained perpetual uniformity of the motion but lacked the breakthrough to *universality* of uniform *rectilinear* motion. Buridan still worked in the overall framework of Aristotelian physics and cosmology, as is exemplified by his use of the Aristotelian terms *natural* and *enforced* motion.

(2) Buridan makes only one application of his concept of impetus that goes beyond noting how impetus can sustain motion after the immediate cause of motion has ceased and how it is nevertheless gradually corrupted to nothing by resistance. This is in the following passage, which comes immediately after the passage in which he so clearly anticipates Newton's concept of momentum and before the reference to the long jumper:[28]

> From this theory also appears the cause of why the natural motion of a heavy body downward is continually accelerated. For from the beginning only the gravity was moving it. Therefore, it moved more slowly, but in moving it impressed in the heavy body an impetus. This impetus now [acting] together with its gravity moves it. Therefore, the motion becomes faster; and by the amount it is faster, so the impetus becomes more intense. Therefore, the movement evidently becomes continually faster.

Now there is one sense in which Buridan is here actually closer to Newton than Galileo, and we shall discuss it in Chap. 7. However, more telling are the two respects in which Buridan does not match Galileo. Although the kinematics needed to describe free fall had already been developed to a large degree by the Mertonians and was well known to Buridan, there is no attempt by Buridan to give a precise and quantitative mathematical treatment. Furthermore, just as in the case of the Mertonians, there still seems to be a complete lack of any idea of tying the physical concepts he has developed to experimental measurements. The time just does not seem to have been ripe for that sort of undertaking.

The above quotation is all we shall have to say about the question of free fall at this stage. Other medieval work on that particular problem will be briefly reviewed in the chapter on Galileo. There too we shall bring the history of impetus theory in the centuries between Buridan and Galileo and Descartes up to date. All that we shall say here is that impetus theory suffered a fate rather like that of impetus itself as conceived by Buridan. The only really great idea that Buridan had was that, in the absence of resistance, the impetus would last for ever with undiminished vigour. But already Buridan's brilliant successor Oresme in fact gravely weakened the clarity and clear cut nature of Buridan's concept by introducing the idea that the impetus first of all builds up and then dies away spontaneously. Oresme's reason for this curious theory was the enigmatic passage in Aristotle, suggesting that a projectile reaches its greatest speed in mid flight, to which reference was made on p. 37.[29] This greatly diminished the value of the concept; for Buridan's concept was amenable to precise expression in mathematical terms whereas Oresme's was not. Nevertheless, impetus theory gained very widespread acceptance, and by 1600 had almost completely supplanted the Aristotelian theory, albeit in a form according to which the impetus slowly died away spontaneously.

4.4. Cosmology and early ideas about relativity

As we saw in Chaps. 2 and 3, the possibility that the earth moved was discussed in antiquity several times, in most cases as a mere hypothesis to be rejected. Two motions were considered: daily rotation about the axis to account for the daily rising and setting of the sun, moon, stars, and planets, and an annual motion of the earth around the sun (Aristarchus). I hesitate to say that the latter motion was proposed to account for any particular phenomenon since what is known of Aristarchus's proposal gives no indication of why he proposed it.

Following the example of the ancient authors, above all Aristotle, many medieval authors discussed the possibility of the earth's daily rotation. The two most notable examples are Buridan and Oresme, to whose discussions we now turn. There is no evidence that any medieval philosopher proposed an annual motion of the earth around the sun.[30] As we shall see in the next chapter, this was the really decisive step that Copernicus took.

We start with some extracts from Buridan's *Questions on the Four Books on the Heavens and the World of Aristotle*. In it he says:[31]

> many people have held as probable that it is not contradictory to appearances for the earth to be moved circularly . . . and that on any given natural day it makes a complete rotation from west to east. . . . Then it is necessary to posit that the stellar sphere would be at rest, and then night and day would take place through such a motion of the earth, so that that motion of the earth would be a diurnal motion (*motus diurnus*). The following is an example of this [kind of thing]: If anyone is moved in a ship and he imagines that he is at rest, then, should he see another ship which is truly at rest, it will appear to him that the other ship is moved. This is so because his eye would be completely in the same relationship to the other ship regardless of whether his own ship is at rest and the other moved, or the contrary situation prevailed.

Here we have a clear statement of the principle of relativity, certainly not the first in the history of the natural philosophy of motion but perhaps expressed with more cogency than ever before. The problem of motion is beginning to become acute. We must ask ourselves: is the relativity to which Buridan refers *kinematic relativity* or *Galilean relativity*? There is no doubt that it is in the first place kinematic; for Buridan is clearly concerned with *the conditions under which motion of one particular body can be deduced by observation of other bodies*. He is referring to what I called the epistemological imperative (p. 47). Meaningful interpersonal communication about motion cannot be made unless reference is made to at least some comparison bodies. But this of necessity introduces a degree of reciprocity that in the case of just two bodies is complete ('his eye would be completely in the same relationship to the other ship regardless of whether his own ship is at rest and the other moved, or the contrary situation prevailed.') But at the same time an element of Galilean relativity is mixed into Buridan's statement implicitly (without Buridan's being at all aware of it) because the sailor carried along by a ship *moving uniformly* has no evidence from within his own ship to suggest that he is in motion and therefore has to look outside the ship. If ships in uniform motion were not closed dynamical systems, in the sense defined in Chap. 1 (p. 31), the sailor would not be forced to rely on external objects to supply criteria of motion. This, therefore, is the hidden dynamical reason (as yet, quite unrecognized) why ships at sea were so readily chosen to exemplify kinematic relativity. Henceforth, whenever we come across any reference to the relativity of motion we must closely examine the sense in which it is meant and try to establish the extent to which the author in each given case was explicitly aware of the two aspects of relativity.

Two more comments of Buridan are particularly interesting because they reappear with significant developments in Copernicus. They are made in the form of two further 'persuasions' for diurnal rotation of the earth rather than the highest celestial sphere:[32]

> To celestial bodies ought to be attributed the nobler conditions, and to the highest sphere, the noblest. But it is nobler and more perfect to be at rest than to be moved. Therefore, the highest sphere ought to be at rest. . . . The last persuasion is this: Just as it is better to save the appearances through fewer causes than through many, if this is possible, so it is better to save [them] by an easier way than by one more difficult. Now it is easier to move a small thing than a large one. Hence it is better to say that the earth, which is very small, is moved most swiftly and the highest sphere is at rest than to say the opposite.

Despite these arguments, Buridan says: 'But still this opinion is not to be followed'. He opens with an argument that implies the *conventionality* of science; this argument keeps on reappearing through the centuries and is still with us to this very day. Its gist is that correct saving of appearances by a particular hypothesis is no guarantee for the correctness of that hypothesis. It may simply be convenient because it provides a simple explanation. Tussling with this type of argument and attempting to describe precisely the unquestioned empirical facts in a form that is capable of withstanding sustained epistemological probing is the toughest of all tasks in a book such as the present one. It is Faust confronting stubbornly recalcitrant nature ('Wo fass ich dich unendliche Natur?'*) while Mephistopheles plays his role as the spirit that continually denies. Here is Buridan:[33]

But still this opinion is not to be followed. In the first place because it is against the authority of Aristotle and of all the astronomers (*astrologi*). But these people respond that *authority does not demonstrate*, and that it suffices astronomers that they posit a method by which appearances are saved, whether or not it is so in actuality. Appearances can be saved in either way; hence they posit the method which is more pleasing to them.

Buridan then lists various practical objections to rotation of the earth. These may all be termed *dynamical* arguments. I give all three that Clagett includes in his translation, not only for their intrinsic interest but also to show, from the manner in which Buridan writes, that rotation of the earth really must have been quite a lively topic of discussion in Paris in the mid fourteenth century:[34]

If anyone were moving very swiftly on horseback, he would feel the air resisting him. Therefore, similarly, with the very swift motion of the earth in motion, we ought to feel the air noticeably resisting us. But these [supporters of the opinion] respond that the earth, the water, and the air in the lower region are moved simultaneously with diurnal motion. Consequently there is no air resisting us.

Another appearance is this: Since local motion heats, and therefore since we and the earth are moved so swiftly, we should be made hot. But these [supporters] respond that motion does not produce heat except by the friction (*confricatio*), rubbing, or separation of bodies. These [causes] would not be applicable there, since the air, water, and earth would be moved together.

But the last appearance which Aristotle notes is more demonstrative in the question at hand. This is that an arrow projected from a bow directly upward falls again in the same spot of the earth from which it was projected. This would not be so if the earth were moved with such a velocity. Rather before the arrow falls, the part of the earth from which the arrow was projected would be a league's distance away. But still the supporters would respond that it happens so because the air, moved with the earth, carries the arrow, although the arrow appears to us to be moved simply in a straight line motion because it is being carried along with us. Therefore, we do not perceive that motion by which it is carried with the air. But this evasion is not sufficient because the violent impetus of the arrow in ascending would resist the lateral motion of the air so that it would not be moved as much as the air. This is similar to the occasion when the air is moved by a high wind. For then an arrow projected upward is not moved as much laterally as the wind is moved, although it would be moved somewhat.

* From Goethe's *Faust*: 'Where can I grasp you, infinite nature?'

I do not propose to discuss here these dynamical arguments of Buridan, which finally persuaded him that the earth does not rotate, because it will be more appropriate to consider them when we come to Galileo. I will only mention that, had Buridan had Galileo's insight, he already possessed an idea that would easily have permitted him to refute this last argument of Aristotle – Buridan's own greatest contribution to dynamics: impetus theory. This failure shows more clearly than anything else that genuine discovery in dynamics requires not only the recognition of an idea but nontrivial demonstration of how it can be used.

In Oresme's *On the Book of the Heavens and the World of Aristotle* we find several interesting advances on Buridan, admittedly often in nuances rather than explicitly. He starts with a bald assertion about the possibility of the earth's diurnal rotation, asserting that[35] 'one could not demonstrate the contrary by any experience'. It is also worth noting the clear understanding of the importance of kinematic relativity:[36] 'I make the supposition that local motion can be sensibly perceived only in so far as one may perceive one body to be differently disposed with respect to another.'

It is in supporting this assertion of *kinematic relativity* that he introduces one of his more significant nuances that takes him some way towards stating explicitly the fundamental condition that must hold in the case of *Galilean relativity* (the italics in the quotation are mine): 'If a person is in one ship called *a* which *is moved very carefully* [i.e., without pitching or rolling] – *either rapidly or slowly* – and this person sees nothing except another ship called *b*, which is moved in every respect in the same manner as *a* in which he is situated, I say that it will seem to this person that neither ship is moving.'

We also find a first tentative hint of a line of argument that Copernicus advanced, and Galileo improved out of all recognition, to meet the objection which Buridan regarded as telling most against rotation of the earth:[37]

> concerning the arrow or stone projected upward etc., one would say that the arrow is trajected upwards and [simultaneously] with this trajection it is moved eastward very swiftly with the air through which it passes and with all the mass of the lower part of the universe mentioned above, it all being moved with a diurnal movement. For this reason the arrow returns to the place on the earth from which it left. This appears possible by analogy: If a person were on a ship moving toward the east very swiftly without his being aware of the movement, and he drew his hand downward, describing a straight line against the mast of the ship, it would seem to him that his hand was moved with rectilinear movement only. According to this opinion [of the diurnal rotation of the earth], it seems to us in the same way that the arrow descends or ascends in a straight line.

Expanding on this argument, Oresme comes very close to anticipating some of Galileo's greatest insights; they may seem rather trivial but their implications, when fully grasped (which they certainly were not by Oresme), were momentous. Here are Oresme's words:[38]

Also, in order to make clear the response to the third experience, I wish to add a natural example verified by Aristotle to the artificial example already given. It posits in the upper region of the air a portion of pure fire called *a*. This latter is of such a degree of lightness that it mounts to its highest possible point *b* near the concave surface of the heavens. I say that just as with the arrow in the case posited above, there would result in this case [of the fire] that the movement of *a* is composed of rectilinear movement, and, in part, of circular movement, because the region of the air and the sphere of fire through which *a* passes are moved, according to Aristotle, with circular movement. Thus if it were not so moved, *a* would ascend rectilinearily in the path *ab*, but because *b* is meanwhile moved to point *c* by the circular daily movement, it is apparent that *a* in ascending describes the line *ac* and the movement of *a* is composed of a rectilinear and a circular movement. So also would be the movement of the arrow, as was said. Such composition or mixture of movements was spoken of in the third chapter of the first book [of the *De Caelo*]. I conclude then that one could not by any experience whatsoever demonstrate that the heavens and not the earth are moved with diurnal movement.

In anticipation of the discussion in the chapter on Galileo let it merely be noted here that the crucial idea, enforced by the Copernican revolution when it finally came, was that natural motions are *compounded*. The key phrase in this passage of Oresme's is: 'composed of a rectilinear and a circular movement'.

Like Buridan, Oresme argues that on grounds of economy one would argue for rotation of the earth rather than the heavens, emphasizing even more than Buridan the speed with which the heavens would have to rotate:[39] 'one could not imagine nor conceive of how the swiftness of the heaven is so marvellously and excessively great. It is so unthinkable and inestimable.'

Oresme's final summary of the question of the earth's rotation is particularly interesting in showing just how close he was to coming out and saying roundly that the earth rotated. It is to be noted that he finally allows himself to be persuaded by a scriptural argument despite the fact that a little earlier in his treatise he had shown with great tact towards the Scriptures how such arguments could be defused completely:[40]

It is apparent, then, how one cannot demonstrate by any experience whatever that the heavens are moved with daily movement, because, regardless of whether it has been posited that the heavens and not the earth are so moved or that the earth and not the heavens is moved, if an observer is in the heavens and he sees the earth clearly, it (the earth) would seem to be moved; and if the observer were on the earth, the heavens would seem to be moved. The sight is not deceived in this, because it senses or sees nothing except that there is movement. But if it is relative to any such body, this judgement is made by the senses from inside that body, just as he [Witelo] stated in *The Perspective*; and such senses are often deceived in such cases, just as was said before concerning the person who is in the moving ship. . . . Yet, nevertheless, everyone holds, and I believe, that they (the heavens), and not the earth, are so moved, for 'God created the orb of the earth, which will not be moved' (Ps. 92:I), notwithstanding the arguments to the contrary. . . . But having considered everything which has been said, one could by this believe that the earth and not the heavens is so moved, and there is no evidence to the contrary. Nevertheless, this seems *prima facie* as much, or more, against natural reason as are all or several articles of our faith. Thus, that which I have said by way of diversion (*esbatement*) in this manner can be valuable to refute and check those who would impugn our faith by argument.

I will leave the reader to make what he or she can of this final argument. It seems, to say the least, a curious way of bolstering faith.

It is worth noting that both Buridan and Oresme clearly have a notion of absolute space: both are convinced that *either* the earth *or* the heavens move. The idea that there is simply nothing but relative motion is quite foreign to both. This is revealed by many tell-tale expressions.

It was asserted in the Introduction that the discovery of dynamics hinged on a dozen or so insights that all had to do with the mathematical description of empirically observed motions. The clearest reason why the fourteenth century was not the true century of the scientific revolution is to be seen in its failure, for whatever reason, to produce one single such insight. The ideas were there and the mathematics had been developed. But for some reason or other they were not put together. Nowhere is the crucial role of the *application* of mathematics more evident than in the contrast between Buridan and Galileo in their discussions of the dynamical arguments for and against the earth's rotation or between both Buridan and Oresme, on the one hand, and Copernicus, on the other, in discussing the bare possibility of the earth's rotation. Oresme, in particular, is often much more incisive than the man who lived nearly two centuries later but it was Copernicus who revolutionized the world, not Oresme. He did it by an insight that no one (with the possible exception of Aristarchus) before him ever had. And because it involved *nontrivial mathematics and empirically observed motions* Copernicus spoke with urgency and the authority of one who knows. Thus, Copernicus stated that the earth does truly move whereas Oresme merely advanced the idea as a 'diversion'.

Let us now see why Copernicus could speak with such authority.

5

Copernicus: the flimsy arch

5.1 How Copernicus came to make his discovery

It would be a nice party game for historians of science to compile lists of worthy scientists born too early to receive that ultimate of accolades, the Nobel Prize. Who, one asks, should be awarded the first of these posthumous Nobel Prizes in the field of physics? Given the subject of this chapter, the reader will no doubt expect the nomination of Copernicus. In fact, the case will be argued for Ptolemy. The reasons for this will become clearer as the book proceeds, but we can already anticipate them in this imagined citation: 'For the discovery of the equant and the important stimulus which this gave to the correct solution of the planetary problem by Copernicus and Kepler and to the development of the dynamical conception of motion by Kepler and Newton.'

We recall that the rather *ad hoc* introduction of the equant was made by Ptolemy *in extremis* when all traditionally accepted means to reconcile the data had failed. Ptolemy felt he had to apologize for his radical innovation and said he was compelled 'to make some basic assumptions which we arrived at not from some readily apparent principle, but from a long period of trial and application'. Readers familiar with the early history of the quantum theory will note here a striking parallel with Planck's discovery of the quantum of action by the *ad hoc* idea of quantization of emission and absorption processes. Both men were conservative by inclination and only reluctantly took a step that observations simply forced upon them. Neither could have made their discoveries had they not deeply immersed themselves in the nitty-gritty details of the problem and, in both cases, pushed the existing theories, both of which contained large elements of truth, to the absolute limits of what the theories were capable. They both added an extraneous and incongruous device to laws of motion hallowed by centuries of successful use – an addition apparently made as a desperate last resort to save appearances that obstinately refused to fit the old mould but in reality were the harbingers of a new dispensation.

But if Ptolemy and Planck were remarkably similar in the manner of their discoveries, the development of the discoveries initially unfolded in quite different ways. Planck's was carried forward within a few years in giant leaps by two of the most daring intellects in the history of science – Einstein and Bohr. In contrast, Ptolemy's was advanced many, many centuries later in a roundabout manner by a man with a decidedly conservative, if not to say pedantic, cast of mind, who literally stumbled on the heap of gold amassed – without their realizing it – by the Hellenistic astronomers while he was in fact attempting to put back the clock by finding an alternative to the equant.

Copernicus (1473–1543) could not abide the idea of celestial bodies moving nonuniformly. He had the greatest admiration for Ptolemy, of whom he wrote:[1] 'Claudius Ptolemy of Alexandria, who far excels the rest by his wonderful skill and industry, brought this entire art [the study of celestial motions] almost to perfection', but Ptolemy's equant and nonuniform motion would not leave him in peace. In his *Commentariolus*, circulated in manuscript form many years before the publication in 1543 of his famous *De Revolutionibus Orbium Celestium* (*On the Revolutions of the Celestial Spheres*), he wrote:[2]

> these theories were not adequate unless certain equants were also conceived; it then appeared that a planet moved with uniform velocity neither on its deferent nor about the center of its epicycle. Hence a system of this sort seemed neither sufficiently absolute nor sufficiently pleasing to the mind.
>
> Having become aware of these defects, I often considered whether there could perhaps be found a more reasonable arrangement of circles, from which every apparent inequality would be derived and in which everything would move uniformly about its proper center, as the rule of absolute motion requires. After I had addressed myself to this very difficult and almost insoluble problem, the suggestion at length came to me how it could be solved with fewer and much simpler constructions than were formerly used.

He recounted in *De Revolutionibus*, in the preface addressed to Pope Paul III, how he came to make his actual discovery:[3]

> I undertook the task of rereading the works of all the philosophers which I could obtain to learn whether anyone had ever proposed other motions of the universe's spheres than those expounded by the teachers of astronomy in the schools. And in fact first I found in Cicero that Hicetas supposed the earth to move. Later I also discovered in Plutarch that certain others were of this opinion.
>
> Therefore, having obtained the opportunity from these sources, I too began to consider the mobility of the earth. And even though the idea seemed absurd, nevertheless I knew that others before me had been granted the freedom to imagine any circles whatever for the purpose of explaining the heavenly phenomena. Hence I thought that I too would be readily permitted to ascertain

whether explanations sounder than those of my predecessors could be found for the revolution of the celestial spheres on the assumption of some motion of the earth.

It will be recalled from Chap. 3 that the problem of the planets reduced to the unravelling and interpretation of two inequalities: the first, which in the helioastral frame corresponds to the proper motion of the individual planets around the sun, and the second, which corresponds to the earth's motion around the sun. Ptolemy had represented the latter by perfectly uniform epicyclic motion (for the superior planets) and the former by eccentric deferents and the *ad hoc* concept of the equant with the associated necessity of nonuniform motion in space of the guide point as it moves around the deferent.

Copernicus set out to rectify what he perceived as a defect – the equant – in the theory of the first inequality.

He found an *interpretation for the second inequality* (in terms of mobility of the earth) and realized well enough that he had struck gold. He exploited to the full the geometrical and trigonometrical potential of his discovery but went no further. Instead, the bulk of his effort was still expended on rectifying the defect in the first inequality. He appears to have believed that his discovery of an interpretation of the second inequality in terms of terrestrial mobility represented an advance in the theory of the first. In fact, it left it exactly where it was. Planetary theory itself was barely advanced an iota by Copernicus himself.

Copernicus's attitude to Ptolemy and nonuniformity of motion reminds one of the Venerable Bede and his concern about the Irish Christians – such excellent people if only they would desist from celebrating Easter on a noncanonical date.[4] And, like Bede, he cannot get the fault out of his mind. He keeps coming back to the point. Only about a fifth of *De Revolutionibus* is about the true Copernican revolution; much of the remainder is an attempt, which does not lack a certain ingenuity, to undo the equant. This is what gives his discovery its bizarre element.

Another bizarre – or perhaps one should say disconcerting – aspect of his work is that his attempt to undo the equant was anticipated in remarkable detail some two centuries before his time by Islamic astronomers of the so-called Marāgha School, which flourished under the leadership of Nṣīr al-Din al-Ṭūsī (1201–74) at the astronomical observatory at Marāgha in Iranian Azerbaijan. I say 'disconcerting' because for each of the technical devices that Copernicus employed to eliminate Ptolemy's various violations of the golden rule of ancient astronomy (p. 153) – that every celestial motion should be a uniform circular motion or alternatively be compounded of two or more uniform circular motions – a more or less exact counterpart can be found in the writings of the Marāgha School or the related work of the later astronomer Ibn al-Shāṭir (1304–75/6), who worked in Damascus and devised, among other things, a model of the moon's motion that reproduced Ptolemy's lunar longitudes but was a great improvement on Ptolemy's model in that it greatly reduced the predicted variation in the moon's apparent diameter – the one really gross flaw that can be found in the *Almagest*. Copernicus's own model for the moon is, apart from minor differences in the parameters, identical to that of Ibn al-Shāṭir. So many coincidences naturally arouse the suspicion that Copernicus knew of the earlier work, yet *De Revolutionibus* contains no hint of acknowledgement. It must also be said that no evidence has yet come to light to prove that the work of the Marāgha School was transmitted to Renaissance Western Europe. This question, to which I shall return briefly, is discussed by Roberts, Kennedy, Swerdlow, and Neugebauer.[5–9]

It will be helpful to conclude this section with some brief historical details, in which I follow the comprehensive monograph on *De Revolutionibus* by Swerdlow (the main author) and Neugebauer.[9] Although much astronomical work was done by Arabic astronomers and a certain amount of it is reflected in *De Revolutionibus* (either openly, especially in the case of Islamic determinations of elements of the solar theory, or, possibly, as plagiarism from the Marāgha School), the solid foundation of Copernicus's work is the *Almagest*. The transmission of this work to Western Europe was therefore of the very greatest importance for the Copernican revolution. Significant astronomy in Europe only commenced with the translation of astronomical texts, including the *Almagest*, in the twelfth and thirteenth centuries. The really important development came however in the middle of the fifteenth century with the work of Georg Peurbach (1423–61) and his student Johannes Müller of Königsberg (1436–76), who was called Regiomontanus. Peurbach wrote a very popular work, *Theoricae novae planetarum* (1454), with detailed models of the movements of the planets by means of spheres, and he commenced work on an exposition of the *Almagest*, the *Epitome of the Almagest*, which was completed in brilliant fashion by Regiomontanus following Peurbach's death. The *Epitome* was published in 1496 and was the book that Copernicus followed, even often it seems in preference to the *Almagest* itself.[8] Regiomontanus's early death is widely deplored, but Swerdlow and Neugebauer point out that his excellent grasp of Ptolemaic astronomy and its lucid presentation in the *Epitome* mark the effective rebirth of the *Almagest* and were crucial for the Copernican revolution. Swerdlow emphasizes[8] especially the fact that Regiomontanus draws explicit attention in the *Epitome* to the possibility of treating all planets in a uniform manner without the unfortunate epicycle–deferent inversion for the two inner planets, the use of which by Ptolemy did so much, as we noted in Chap. 3, to obscure the true unity of the planetary motions. Swerdlow believes that this was a most important preparation for Copernicus's innovation.

Many readers will no doubt be familiar with the details of Copernicus's life. He was born at Toruñ in eastern Poland and is generally regarded as a Pole, though precise nationalities are not so easy to determine for that period and region, a fact that has resulted at times in some unedifying nationalistic disputes between Poles and Germans. He studied in Cracow from 1491 for several years and was then sent by his uncle, a powerful bishop of the Catholic church, to study in the University of Bologna. In Italy he studied Greek, mathematics, law, and medicine (in later life he practised as a doctor among other things). He visited Rome in 1500, studied for almost four years at Padua, but actually obtained a degree at Ferrara. He returned to Poland in 1503 and worked for his uncle until the bishop's death in 1512. He settled permanently at Frauenberg in Ermeland on the Baltic coast and worked until his death as a canon of the Catholic church, carrying out numerous functions, some of them in difficult and even dangerous conditions.

His interest in astronomy dated from the period in Cracow, where he acquired several astronomical treatises, including the so-called *Alfonsine Tables*. It is not known when he had the idea that, at a stroke, transformed man's conception of the cosmos: it was certainly before 1515 and may have been as early as 1510. Fairly soon after this he appears to have written the *Commentariolus*, which contains a brief outline of his scheme. Realizing that a convincing presentation of his ideas, including his alternative to the equant as well as the new cosmology, would require careful revision of Ptolemy's work and redetermination of orbital elements, he left the *Commentariolus* unpublished (though it appears to have circulated in manuscript) and embarked on an ambitious programme of observations and calculations that in the event occupied him until very nearly the end of his life. Swerdlow and Neugebauer emphasize what a great labour this was on Copernicus's part. In the preface to *De Revolutionibus*, Copernicus himself commented that he had mulled over his idea 'not merely until the ninth year but by now the fourth period of nine years'. In fact, the task was really too much for him, and he never succeeded in getting his entire theory into a satisfactory shape.

Meanwhile, knowledge of his revolutionary proposal spread quite widely and attracted to Frauenberg the young, mercurial, and enthusiastic Georg Joachim Rheticus, who, together with other friends, finally persuaded the extremely hesitant Copernicus – whose main fear was that he would be 'laughed at and hissed off the stage', though the unsatisfactory state of some of his calculations must also have been a factor – to publish his work. In fact, before *De Revolutionibus* appeared in 1543 a first account of the Copernican system was published by Rheticus (with Copernicus's approval) as the *Narratio Prima* (1540). Copernicus himself died just at the time *De Revolutionibus* was published – report has it that he was handed a copy of the book on his deathbed.

It is surprisingly tricky to explain the precise details of what Copernicus actually proposed. We shall therefore attack the problem in stages.

5.2. What Copernicus did: first approximation

A rather curious aspect of the Copernican revolution was that it did not occur at the stage in the development of astronomical knowledge at which, with hindsight, it might naturally have been expected. This was the point at which the epicycle–deferent technique had been developed, probably around the time of Apollonius or a bit earlier. As we have seen, this technique gives a perfect explanation of the zero-eccentricity behaviour of the planets and it is clear from Ptolemy's account (p. 186) that Hipparchus at the latest knew it described the motions of the planets in their broad details. As the simple epicycle–deferent theory, without the modifications that Ptolemy made, is also by far the easiest point of departure for the transition to the heliocentric system, one may reasonably have expected its discovery to prompt Hipparchus to heliocentricity whereas in fact the discovery of the equant by Ptolemy was what eventually moved Copernicus. Why did Hipparchus fail to make the transition?

It may be significant that, as Ptolemy reports, Hipparchus was aware that the simple epicycle–deferent theory alone was inadequate for the exact description of the planetary motions. Ptolemy says Hipparchus knew one had to cope with problems like the nonconstant size of the retrogression loops of the planets and the variability of the motion corresponding to the first inequality. The difficulty with all these 'messy' aspects of the overall problem was that they added no further hints at all of the attractions of a heliocentric cosmology. Indeed, they made it rather less attractive; for such a cosmology loses a lot of its attraction once it is realized that the sun cannot in fact be at the exact centre of the planetary orbits, as we well know from the small-eccentricity limit of Kepler's laws. We shall never know whether Hipparchus seriously considered Aristarchus's suggestion and with it a heliocentric cosmology; if he did, he may simply have concluded it did not match the observations and for that reason had little attraction. Meanwhile the more urgent task seemed to be to account for the puzzling deviations from the simple epicycle–deferent scheme. In a way, both Hipparchus and Ptolemy were held back from a heliocentric cosmology by knowing too much. Their failure to adopt one is probably a classical example of a failure to see the wood for the trees. It is exactly the same problem that makes it difficult to say what precisely Copernicus did do.

Therefore, to reveal the *mathematical core* of the Copernican revolution, we shall go back to the state of knowledge represented by Apollonius's theorem, i.e., the pure epicycle–deferent model corresponding to zero eccentricity of all the planets (and to zero inclination of their orbits). The sun is therefore assumed to be strictly at the centre of concentric and coplanar circular planetary orbits. To be specific, we consider the Copernican explanation of the motion of one of the outer planets.

Of course, everyone knows what Copernicus proposed: that the earth rotates around an axis and simultaneously travels around the sun (he also proposed a third motion of the earth to account for the precession of the equinoxes in a rather awkward manner; this was shown by Kepler to be superfluous and will not be considered in this book). But it was not so much these motions and the rather obvious explanations that they provided for the apparent diurnal motion of the stars and annual motion of the sun that made Copernicus's suggestion into a true revolution.

At the heart of the revolution was a simple but nontrivial mathematical insight. The Greeks had represented the motion of the outer planets in a first approximation by the motion of the planet riding on the one end of a rotating epicyclic spoke whose other end moves uniformly round a deferent circle, at the centre of which the observer is placed. Copernicus's first great insight was that, if the stars are sufficiently far away, the motions observed on the sky that result from this epicycle–deferent arrangement are identical to the motions that will be observed if both the earth and the planet move in circular orbits of different radii about a common centre.

A key point here is that the ratio of the epicycle radius to the deferent radius must be equal to the ratio of the radius of the circle in which the earth moves to the radius of the circle in which the planet moves. Otherwise the mathematical equivalence of the two schemes is not present. Equally, the rotation periods must match. The transition is shown in Fig. 5.1. Shorn of all the messy aspects of the real solar system, these insights are the essence of Copernicus's revolution. He found an alternative explanation, as precisely mathematical as that of the ancients, of the nonuniform motion of the planets, above all the retrograde motions.

Before we continue, an important point should be underlined. The success of the explanation of the retrograde motions of the planets represented by Fig. 5.1 is in no way dependent on the presence of the sun at the exact centre of the two circles when the Copernican arrangement is adopted. As far as the explanation of the apparent motions of the planets is concerned, the point O in Fig. 5.1 can be void. This fact should be borne in mind by the reader. The Copernican system was not nearly so heliocentric as the reader might imagine. It is much more aptly called a *theory of the earth's mobility*. This point is anticipated by Copernicus in his preface to the pope, in which he says[10] that all the phenomena of the planetary motions can be explained 'if the motions of the other planets are correlated with the orbiting of the earth'. The somewhat curious and

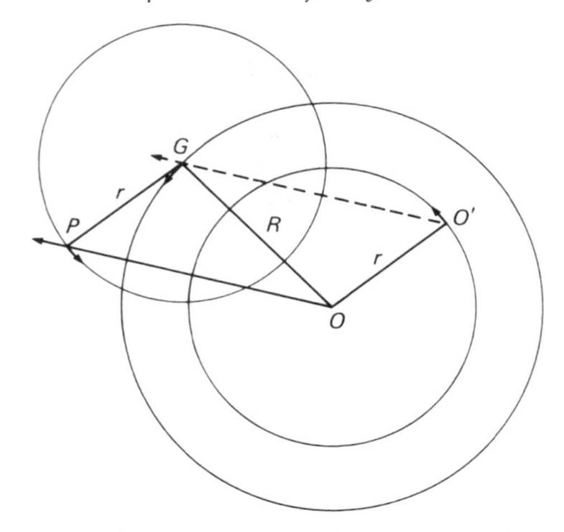

Fig. 5.1. The mathematical core of the Copernican revolution. In the Ptolemaic arrangement for an outer planet (shown for zero eccentricity of both orbits) the terrestrial observer is at O, the deferent guide point at G and the planet at P. The planet is seen along OP. In the Copernican arrangement the centre of the earth's orbit is at O, the earth is at O′ and the planet is at G. Because Copernicus takes OO′ = PG, the planet is seen from the earth along O′G, which is parallel to OP. The observed phenomena are therefore the same.

ambivalent role of the sun in the scheme will become clear as the chapter proceeds.

We can see that Copernicus's insight was nontrivial by comparing it with the two other observable phenomena that his twofold motion of the earth explained. The first is the apparent diurnal rotation of stars. It is obvious that a much smaller feat of the imagination is required to arrive at the conclusion that this could be created by rotation of the earth and not rotation of the heavens. This proposal was indeed made by Heraclides and was discussed by both Aristotle and Ptolemy. We have seen in the previous chapter that it must also have been quite widely discussed in the Middle Ages. Somewhat more advanced in its level of sophistication is the alternative explanation for the apparent motion of the sun around the ecliptic. Here it is not simply a matter of the earth rotating around its axis. One must postulate that it moves in a circle around the sun. This then is sufficient to explain the apparent motion of the sun (provided the stars are sufficiently distant for them to reveal no effect of the earth's motion).

It seems to be completely clear from Archimedes' account that Aristarchus had the clear realization that the apparent motion of the stars and the sun could be explained if the earth rotated about an axis while simultaneously moving around the sun in a circle of a rather large radius. But, to the best of my knowledge, there is no evidence beyond plausible conjecture to support the suggestion that he also had Copernicus's insight, namely, that the same motion of the earth around the sun could simultaneously explain the retrograde motions of the planets. This, it should be noted, is what first makes Copernicus's proposal into something of genuine scientific value. Putting the earth into a spin to explain the apparent motion of the stars is merely to exchange one motion for another. To put the earth into motion round the sun rather than the sun into motion around the earth is to do no better. These are merely *trivial* geometrical examples of kinematic relativity. The only advantage gained is the rather marginal one pointed out clearly by Buridan and Oresme: if the earth rotates rather than the heavens, the phenomena can be saved by a much slower motion of a much smaller object, which accords with the notion that nature never achieves a given effect by superfluous means. But to explain the apparent motion of the sun and stars and simultaneously, without having to introduce any extra motion, to explain all the retrograde motions of the planets – that was a real coup and a nontrivial advance. There is nothing like this in the writings of Buridan and Oresme; it was the first *sophisticated* exploitation of kinematic relativity.

That is why Copernicus gets the extra odd 'bun' that makes up the 'baker's dozen' and at the same time, as we shall see in Chap. 7, provides the crucial link between the heavens and the earth.

Copernicus's first great insight was the key to his second, which was hardly less important. He saw that if one of the two components in the nonuniform observed motion of each of the planets (the second inequality) is due to the motion of the earth around the sun then the radii of the circles in which the other planets move around the sun can all be deduced from the observed motions and expressed in terms of the radius of the earth's circular orbit. For, as was noted above, a necessary condition for the equivalence of the two representations is that the ratio of the lengths of the epicycle and deferent be reproduced in the ratio of the radii of the two circular orbits in the Copernican scheme. It followed immediately from this that the orbits of Mars, Jupiter, and Saturn must lie outside the earth, in that order, and that Mercury and Venus, for which it was evidently necessary to invert the epicycle and deferent, must have orbits inside the earth's, with Mercury nearer the sun.

Thus, the aesthetic guesswork on which Ptolemy had based his ordering of the celestial bodies was replaced by rigorous theory. The single assumption that the second inequality in the motion of the five naked-eye planets is actually a reflection in the sky of the earth's motion around the sun led immediately to an unambiguous ordering of the

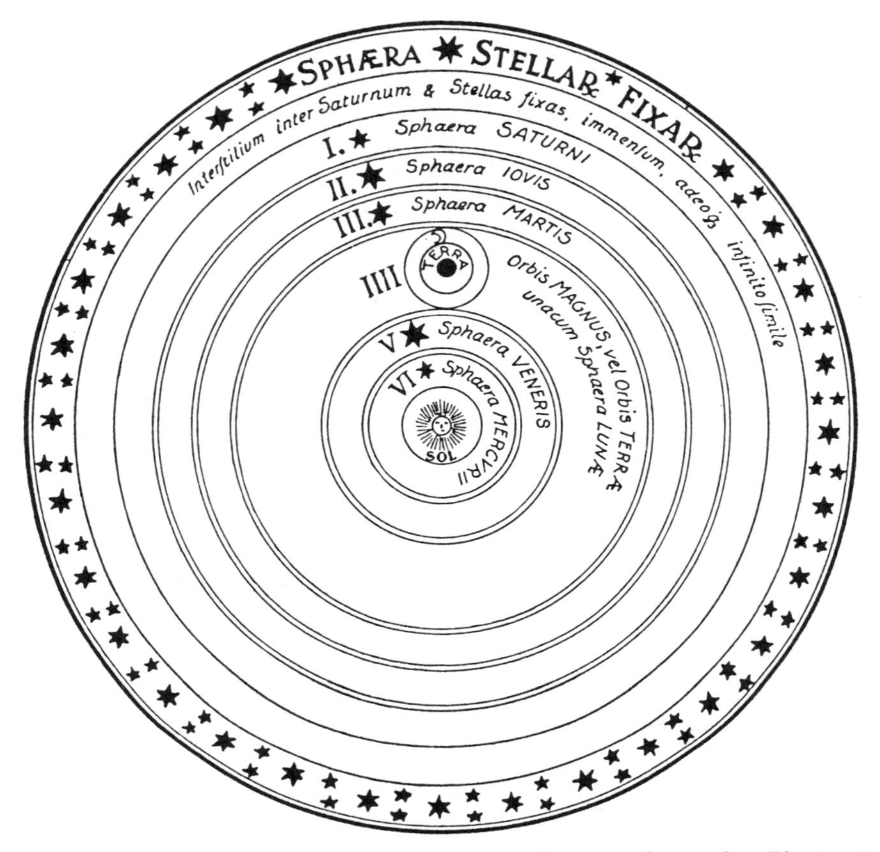

Fig. 5.2. The Copernican arrangement of the spheres as depicted in Rheticus's *Narratio Prima*. Reproduced from *Johannes Kepler Gesammelte Werke*, Vol. 1, C.H. Beck'sche Verlagsbuchhandlung, Munich (1938), p. 103.

planets and to a fixing of all distances apart from one common scale factor, the radius of the earth's orbit. The Copernican arrangement is shown in Fig. 5.2.

Thus far, Copernicus's insights were purely mathematical. However, he found a beautiful confirmation of the correctness of his conjecture when he saw the consequences of what mathematics and observation forced upon him. Whereas Ptolemy had the earth at rest and a distinctly heterogeneous system of six bodies divided into three distinct types (Mercury and Venus; the sun; Mars, Jupiter, and Saturn), Copernicus had the sun (known to be much larger than the earth) at rest and six planets moving round it with each of them having the same basic motion – simple circular. The incongruous epicycle–deferent inversion was abolished and so was the odd fact that the sun exhibited only one motion but all the planets two. But even more impressive was the correlation that Copernicus found between the positions of the planets and the speeds at which they moved in their orbits – Mercury had the fastest motion and was nearest the sun, then came Venus in the second place with the next fastest motion, the earth in the third place with the third fastest motion and so on.

The recognition of this organic structure, which was a qualitative anticipation of Kepler's Third Law, gave Copernicus immense confidence that he was on the right track and prompted him to his famous characterization of the Ptolemaic system. In his preface to *De Revolutionibus,* he said that the principal failing of the astronomers was that they could not deduce 'the structure of the universe and the true symmetry of its parts'. He continued:[11] 'On the contrary, their experience was just like some one taking from various places hands, feet, a head, and other pieces, very well depicted, it may be, but not for the representation of a single person; since these fragments would not belong to one another at all, a monster rather than a man would be put together from them.'

This is a good metaphorical description of the Ptolemaic system; the theories of the individual planets are 'very well depicted'. Ptolemy simply failed to link them together, though the potential was always there. Copernicus found the missing link and was able to knit the planetary system into a symmetric whole.

It is worth quoting here Kepler's comment on the difference between the ancient Ptolemaic school of astronomy and the new one founded by Copernicus:[12] 'The former treated each planet separately and identified the causes of its motion in its corresponding orbit. The latter compares the planets with one another and derives that which is common in their motion from a common cause . . . from motion of the earth.' Thus, Copernicus perceived the unity hidden in the Ptolemaic details, and he was able to tell the pope proudly that not only do all the phenomena follow from his system:[13] 'but also the order and size of all the planets and spheres, and heaven itself is so linked together that in no portion of it can anything be shifted without disrupting the remaining parts and the universe as a whole'. Later, in Chap. 10 of Book I he expands on this:[14]

> In this arrangement, therefore, we discover a marvelous symmetry of the universe, and an established harmonious linkage between the motion of the spheres and their size, such as can be found in no other way. For this permits a not inattentive student to perceive why the forward and backward arcs appear greater in Jupiter than in Saturn and smaller than in Mars, and on the other hand greater in Venus than in Mercury. This reversal in direction appears more frequently in Saturn than in Jupiter, and also more rarely in Mars and Venus than in Mercury. . . . All these phenomena proceed from the same cause, which is in the earth's motion.

In exploiting the assumption of terrestrial mobility to show how all the bodies of the solar system could be arranged in an harmonious order that explained the astronomical phenomena most satisfactorily, Copernicus did indeed solve at a stroke what was probably the supreme problem that concerned the astronomers of the late Renaissance: what is the form of the world? where precisely are all its parts? The quiet self-confidence with which he answers these questions in the body of his book was clearly a major factor in winning adherents to his proposal in an age characterized by great doubt and scepticism on matters astronomical.

One further point he might have made in the passage just quoted but probably did not because it is a shade technical concerns the *phase* of the epicyclic motion in the Ptolemaic system for the outer planets and of the deferent motion for the inner planets. Why did they always march exactly in phase with the motion of the sun? In principle, after all, the epicyclic spoke could rotate at any uniform rate as the epicycle moves around the deferent. Why did it always move at exactly the rate of the mean sun, always pointing moreover exactly in its direction (matching of the phase)? This Copernicus was able to explain at a stroke, as it was a necessary consequence of his alternative mathematical model for explaining the retrogression loops of the planets provided the sun is close to the centre of the planetary system. Thus, what appeared as a remarkable and inexplicable coincidence in the old representation was a simple and necessary consequence in the new. In essence, the epicyclic motions of the planets and especially their retrogression loops were revealed as simple parallax effects caused by the motion of the earth around the sun. Since the one motion around the sun produced all the epicyclic motions, it was obvious that all had to be mutually in step with one another as well as with the sun.

Let us conclude this section by considering once more the question: did Aristarchus share Copernicus's more sophisticated insight into the origin of the retrogression loops? In this connection it is interesting to compare the fate of Aristarchus's lost heliocentric treatise with the *Commentariolus*, also circulated in manuscript form. The former was read by the greatest mathematician of antiquity, himself a practising astronomer. Yet all Archimedes bothered to say about Aristarchus's treatise was a pedantic quibble; other commentators said even less. In contrast, Copernicus's *Commentariolus* created sufficient stir for a cardinal in Rome to write to Copernicus in 1536, warmly encouraging him to publish his work. The cardinal's brief summary of the heliocentric system, received at second hand, is considerably more detailed than Archimedes' first-hand account of what Aristarchus had said. Here is the cardinal:[15]

I had learned that you had not merely mastered the discoveries of the ancient astronomers uncommonly well but had also formulated a new cosmology. In it you maintain that the earth moves; that the sun occupies the lowest, and thus the central, place in the universe; that the eighth heaven remains perpetually motionless and fixed; and that, together with the elements included in its sphere, the moon, situated between the heavens of Mars and Venus, revolves around the sun in the period of a year. I have also learned that you have written an exposition of this whole system of astronomy, and have computed the planetary motions and set them down in tables, to the greatest admiration of all.

(The cardinal's letter is quoted in full by Copernicus before the preface to *De Revolutionibus*.)

The prominent references by the cardinal to the planets and their absence from the various extant accounts of Aristarchus's proposal does provide some support for the conclusion that Copernicus was the first man in history to realize (and certainly to persuade others) that a heliocentric cosmology could explain not only the apparent motion of the sun but simultaneously the principal oddity in the apparent motion of all the five planets.

5.3. Kinematic relativity in *De Revolutionibus*

Being based on a nontrivial application of kinematic relativity, and having simultaneously such dramatic and startling implications, the Copernican revolution put the question of the relativity of motion squarely in the forefront of discussion, a position that it had not hitherto occupied. Although Copernicus was not the first person to draw attention explicitly to the relativity of motion and we find in his writings several points anticipated in the Middle Ages by Buridan and Oresme, his were the first important pronouncements on the subject, since it was *De Revolutionibus* that was to become the seminal book for the development of modern dynamics. Relativity was an issue which, of course, Copernicus could not duck. His whole argument for the mobility of the earth hinged upon it. Copernicus invoked kinematic relativity, which may also be called *optical relativity*. Commenting on the apparently nonuniform motion of the planets, he first notes that their distances from the earth must be assumed to vary and he then says:[16] 'Hence I deem it above all necessary that we should carefully scrutinize the relation of the earth to the heavens lest, in our desire to examine the loftiest objects, we remain ignorant of things nearest to us, and by the same error attribute to the celestial bodies what belongs to the earth.'

He then states the principle on which he will argue that motion of the earth is at least conceivable as an explanation of celestial phenomena, however ridiculous or downright stupid it may appear:[17]

Every observed change of place is caused by a motion of either the observed object or the observer or, of course, by an unequal displacement of each. For when things move with equal speed in the same direction, the motion is not perceived, as between the observed object and the observer, I mean. It is the earth, however, from which the celestial ballet is beheld in its repeated performances before our eyes. Therefore, if any motion is ascribed to the earth, in all things outside it the same motion will appear, but in the opposite direction, as though they were moving past it.

The present study is to a large extent about the difficulty of forming adequate concepts of motion and the true nature of the world. Perhaps more than anything else, this passage should give us cause for reflection. For Copernicus taught us something about the earth by looking in exactly the opposite direction. *He persuaded men that the earth moves by getting them to examine the heavens*. The most important evidence can be found in the place we least expect it.

Later on Copernicus supports this approach with the illustration with which we are already familiar from the previous chapter and which was to become one of the most common devices in physics, repeatedly used by Einstein and innumerable textbooks on relativity. Although not the first time used, it was certainly highly significant in the context of *De Revolutionibus*:[18]'when a ship is floating calmly along, the sailors see its motion mirrored in everything outside, while on the other hand they suppose that they are stationary, together with everything on board. In the same way, the motion of the earth can unquestionably produce the impression that the entire universe is rotating.'

On this basis he argues:[18] 'Why should we not admit, with regard to the daily rotation, that the appearance is in the heavens and the reality in the earth? This situation closely resembles what Vergil's Aeneas says: "Forth from the harbor we sail, and the land and the cities slip backward" [*Aeneid*, III, 72].'

As used by Copernicus, the principle of relativity is purely kinematic. There are fewer hints of Galilean relativity than in Oresme. He merely states the fact that if two bodies share the same motion the one will seem to be at rest relative to the other. Except in one highly important passage, the elaboration of which was to occupy a significant proportion of Galileo's working life, and which we shall consider in the chapter on Galileo, Copernicus did not address himself to the question of how it can come about that two objects, for example, a sailor and a ship, *can* share a common motion. How *does* it come about that the sailors 'suppose that they are stationary, together with everything on board'? As we have seen, there are two aspects of this matter – the dynamic and the kinematic. Of the two, the dynamic is the more profound: What specific property of nature and its laws is it that makes it possible for the ship, all its contents, and the sailors to be carried along together with no appearance at all (if observation is restricted to the ship) of being in motion? Most of the early discussion of Copernicus's suggestion of the earth's mobility did not attack this question but rather took the existence of such a state of affairs as a fact and concentrated on the observable consequences it had. The principle of kinematic relativity deals with the problem of finding the evidence in external phenomena for the motion of such a ship or spaceship earth, for which the ship serves as simile, given the fact that internal evidence for motion is lacking. The emergence of the dynamic element of the principle of relativity, the first tentative beginnings of which we saw in Oresme (p. 206), is closely associated with the growing understanding of the nature of motion, a process that will occupy us much throughout the remainder of this volume and the next and has probably not yet come to an end.

5.4. Preliminary evaluation of the significance of Copernicus's discovery

Copernicus's most important insight was that if the earth is assumed to move around the sun, the diameter of the earth's orbit can be used as a trigonometric baseline. As we have seen, the peculiar retrograde motions of the superior planets then arise as a parallax effect from the motion of the earth.

All the early advances in astronomy were associated with the development of new trigonometric techniques. Copernicus's was the most important. It secured for astronomy the trigonometric baseline which it still uses. It was Copernicus who measured the heavens. He effectively introduced the astronomical scale of distance, the astronomical unit.* About a century before Galileo turned his telescope on the stars, Copernicus surmised that the lower limit for the distance to the stars must be greater than the amount suspected by Ptolemy by the ratio of the earth–sun distance to the earth's radius (8000 miles). Even with the too small value for the astronomical unit that Copernicus accepted, the distance to the stars was increased at least two or three thousandfold. Let us quote Copernicus again:[20]

> Yet none of these phenomena appears in the fixed stars. This proves their immense height, which makes even the sphere of the annual motion, or its reflection, vanish from before our eyes. For, every visible object has some measure of distance beyond which it is no longer seen, as is demonstrated in optics. From Saturn, the highest of the planets, to the sphere of the fixed stars there is an additional gap of the largest size.

Although, as we shall see, Copernicus retained an essentially Aristotelian cosmology, his dramatic magnification of its size was almost certainly one of the factors that led to its ultimate replacement, about a century after

* As an official unit of distance, the *astronomical unit* (= mean earth–sun distance = semimajor axis of earth's orbit) was actually introduced by Gauss (1777–1855).[19]

Copernicus's death, by the notion of an infinite space. Note how, in Fig. 5.2, the *interstitium* between Saturn and the fixed stars is described as 'immense and even like the infinite'. This was only one of the ways in which Copernicus forced upon thinking man a radical revision of basic concepts. In the light of the coexistence of two different spatial concepts in Aristotle's philosophy (p. 90), it is interesting to see here explicitly how the practical astronomers imposed the trigonometric viewpoint at the expense of the philosophers.

It is worth noting that in his own work Copernicus was able to draw very little additional advantage from his insight into the possibility of a further extension of trigonometry. He was not able to point to any *additional* confirmation of his claim to have correctly measured the distances to the planets and to have arranged them in the correct order, apart from the discovery of Kepler's Third Law in qualitative form, as already explained.

The dramatic confirmation of the Copernican system of triangulation came with Galileo's telescopic discoveries about 70 years after Copernicus died. The phases of the planets, visible through the telescope, especially in the case of Venus, provided strong confirmation of the distances that Copernicus had postulated and demonstrated beyond all doubt that Venus orbited the sun. Figure 5.3 is Kepler's illustration of how the phases arise for an earth-based observer. There was now a second, independent way of determining the distance (albeit only qualitative), and the two methods agreed. What is interesting is that this significance of Galileo's observations was almost completely lost sight of in the furore about whether the earth truly moves or not. Galileo was convinced that,

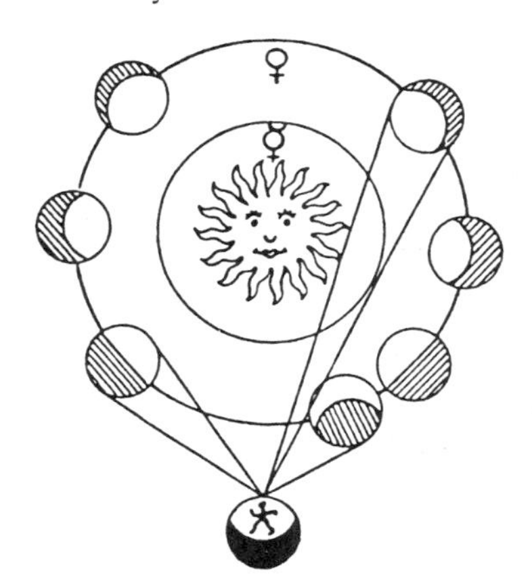

Fig. 5.3. Kepler's explanation of the phases of Venus as seen by a terrestrial observer. The orbit of Mercury is also shown. Reproduced from *Johannes Kepler Gesammelte Werke,* Vol. 7, C.H. Beck'sche Verlagsbuchhandlung, Munich (1953), p. 309.

in confirming Copernicus's prediction, these observations proved the earth's mobility. In fact, they were still compatible with what one might call the 'essential' Ptolemaic system. As emphasized in Chap. 3, the Ptolemaic theory left six free parameters that had to be fixed by guesswork. No violence was done to the essentials of the Ptolemaic theory by fixing these in such a way that the deferents of Mercury and Venus were taken equal to the earth–sun distance and the deferents of the superior planets to their actual distances from the sun. This choice has the consequence that the geometrical arrangement of the Copernican system (when treated as here in the zero-eccentricity approximation) is *exactly* reproduced, the only difference being that in one system the earth is at rest, in the other the sun. This, in fact, is the system which Tycho Brahe (1546–1601) proposed. In it, the sun goes round the earth, while the planets circle the sun as it circles the earth. As far as astronomical observations are concerned, the *Tychonic system*, which is a *special case* of the Ptolemaic one, is kinematically identical to Copernicus's except in its relationship to the distant stars. Initially, it proved to be much more popular with both astronomers and nonastronomers, since it avoided the dynamical problems of a moving earth and the much larger cosmos implied by the Copernican arrangement.

All that Galileo's observations in fact confirmed were the *geometrical* relationships that Copernicus had deduced; this, of course, was pointed out to Galileo by his opponents. Meanwhile, the unquestioned triumph of Copernicus was almost forgotten. No one doubted that his *distance relationships* were essentially correct.

It is worth pointing out just how significant an advance this was and simultaneously drawing attention to a recurrent problem in the philosophy of science closely related to the absolute/relative debate: to what extent do physical theories give us a true picture of reality? It is particularly appropriate to raise this question here in view of the notorious foreword that was appended anonymously to *De Revolutionibus* by a certain Andreas Osiander, who was entrusted with seeing the book through the final prepublication stages. Fearing that the book would cause an uproar, the officious Osiander added a foreword that was taken by many (until Kepler established and published the truth in his *Astronomia Nova*[21]) to be Copernicus's own. Echoing sentiments that we have already encountered in the previous chapter and derived from the scepticism engendered by the numerous different ways in which the astronomers attempted to 'save the phenomena', Osiander sought to soften the radical import of the book by asserting that it did not necessarily say anything about 'reality':[22] 'For it is the duty of an astronomer to compose the history of the celestial motions through careful and expert study. Then he must conceive and devise the causes of these motions or hypotheses about them. Since he cannot in any way attain to the true causes, he will adopt whatever suppositions enable the motions to be computed correctly from the principles of geometry for the future as well as for the past. The present author has performed both these duties excellently. For these hypotheses need not be true nor even probable. On the contrary, if they provide a calculus consistent with the observations, that alone is enough.' (This passage must have rather mystified the early reader of the book, since if one thing shines through its pages it is Copernicus's belief in the actual mobility of the earth.)

There has, in fact, always been a sizeable minority of philosophers and even working scientists who have had a very sceptical attitude to the more dogmatic claims of science to be able to tell us how the world actually is. Since physics, which is regarded as the most basic of the natural sciences, has passed through no less than four major revolutions in less than half a millennium (the Copernican, the Newtonian, the Einsteinian (relativity), and the quantum), each of which has changed our view of the world almost out of recognition, a certain degree of scepticism and caution does not seem out of place.

Thus, even now, three and a half centuries after Galileo's condemnation by the Inquisition, it is still remarkably difficult to say categorically whether the earth moves and, if so, in what precise sense. The basic standpoint of this book, hinted at in Chap. 1, is that correct insights into the interconnections of things are apt to suggest concepts of the world and reality that go far beyond the objective facts from which they spring and are suggested. Sooner or later they are shown to be gross distortions of the truth even though they may have done sterling service in the meantime and helped to uncover numerous further objective interconnections between observed phenomena.

Good theories contain a high truth content even though they do not tell us the final truth about the world. The measure of their truth content is their ability to make predictions. Ptolemy's theory of the planetary motions had a high truth content because, on the basis of past observations, he was able to predict, with very reasonable accuracy, how the heavens would appear at any time in the future *as seen from the surface of the earth*. The really dramatic advance that the Copernican revolution brought was that it extended the ability to predict the appearance of the heavens at any date in the future from the surface of the earth *to any point in the solar system* (in principle, in fact, to the entire universe). Thus, the astronauts knew what the universe would look like from the moon before they got there. This helps to put residual difficulties about the problem of stating the precise sense in which the earth does or does not move into their proper perspective – while also emphasizing that these very same difficulties often give hints of the direction in which new theories will develop, usually with the most profound consequences.

5.5. What Copernicus did: second approximation

Copernicus's one great original idea achieved an almost miraculous simplification of the original Ptolemaic system; at a stroke, it eliminated all the Ptolemaic epicycles, providing simultaneously a wonderfully simple explanation for their appearance. Equally important, it gave the planetary system a far greater coherence and intrinsic symmetry. The original inversion of epicycle and deferent for the two inner planets was eliminated. All five planets, to which the earth was now added, had similar orbits. Several curious and inexplicable features of the old system found very simple explanations in the new. Above all, the scales of the observed retrogressions were unambiguously linked to the rotation periods and the distances from the centre of the system. The exact alignment of all the epicycle vectors with the direction of the sun was equally well explained.

Despite all these advantages, the final Copernican system was, in fact, surprisingly messy and, to modern eyes, full of incongruities. To understand why, we need to have a clear grasp of the principles in accordance with which Copernicus chose to work and of the mathematical tools he had at his disposal. It is then quite easy to see how certain given features of the solar system, above all the specific eccentricities of the six planets' orbits, coupled with the type of observations that Copernicus inherited and made himself, led him to the system that he presented to the world in *De Revolutionibus*.

Copernicus was a purist and he had a single inviolable principle – the golden rule of pre-Ptolemaic ancient astronomy that we quoted earlier: every celestial motion should be a uniform circular motion or alternatively be compounded of two or more uniform circular motions. This therefore had the status of a rigorous law of motion, from which Copernicus was determined not to waver. It will be helpful to consider first what its practical application amounted to in ancient astronomy, in which the earth provided a unique centre of the universe. Expressed (anachronistically) in modern vector terminology, it meant that one had to suppose the motion of each celestial body to be governed by an assemblage of linked vectors that all rotate uniformly. The end of the first vector is fixed at the position of the observer, while the other end rotates in a circle with constant uniform speed. At its end there is attached another vector of fixed length which also rotates at a uniform rate. The chain can in principle be extended by as many such vectors as one pleases. At the far end of the final vector the celestial body is placed. The body is seen along the resultant line of sight from the observer and the main task of the astronomer was to devise an arrangement of vectors that matched the observed motions. Besides its philosophical and aesthetic appeal, this principle had other great advantages: it fitted very well with the idea that the celestial bodies were transported physically by revolving spheres, it provided a straightforward and tractable computational algorithm, and it was simultaneously governed by mathematical rigour. In its geocentric form, this was exactly the approach adopted by the Marāgha school.[5–7] Copernicus's approach was the same except that he placed the initial point of all the vector assemblages at a point near the sun, not at the centre of the earth. As we shall see soon, the precise point he chose had a decisive and malign influence on his system.

One of the major sources of ambiguity in ancient astronomy was a simple mathematical fact that we have already noted in Chap. 3 – the commutativity of vector addition. Provided the lengths and rotation rates (relative to the stars) of the vectors in the assemblages just described are unchanged, the motions generated by the complete linkages are completely unaffected by the order in which the vectors are taken. From Apollonius on, all the major astronomers in our story showed themselves aware of this fact to a greater or lesser extent – Copernicus's own clear recognition of it may owe a lot to Regiomontanus, as noted earlier. This freedom allowed Copernicus and his Islamic predecessors to juggle their circles with sometimes surprising results.

The ingenious but purely technical innovations made in Ptolemaic theory by the Marāgha School and Copernicus, by means of which it proved possible to eliminate the objectionable violations by Ptolemy of the uniformity principle, all rest on the special results that can be obtained by the combination of two vectors in which one rotates with a period that is an exact multiple of the other's. We can complete the account of the mathematical tools at Copernicus's disposal by considering these possibilities; this will simultaneously show what the Marāgha School achieved in this technical respect. We begin with the description of what Kennedy[7] has called the *Ṭūsī couple* after the man mentioned in Sec. 5.1 (Naṣīr al-Din al-Ṭūsī), who was apparently the first to employ it in astronomy. The Ṭūsī couple was originally obtained as a geometrical theorem in the following form. Let a circle of radius $R/2$ roll without slipping around the inside of another circle of radius R. Then any fixed point on the circumference of the smaller circle will move up and down on a diameter of the larger circle, executing simple harmonic motion on that diameter. The Ṭūsī couple is illustrated in Fig. 5.4. It is readily seen to be equivalent to a linkage of vectors OA and AB of equal length $R/2$ with AB rotating (in space) exactly as fast as OA but in the opposite direction. The successive positions of the two ends are A, A′, A″ and B, B′, B″, the latter generating the rectilinear motion. The Ṭūsī couple is thus a reciprocating device for translating uniform circular motions into rectilinear motion and thus shows how the latter can nevertheless be regarded as compounded from uniform circular motions.

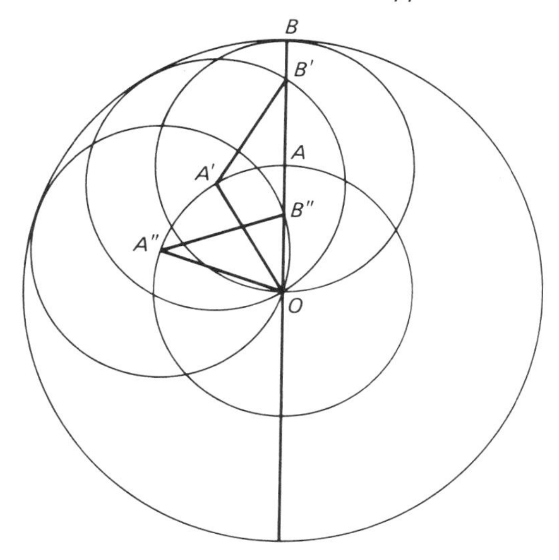

Fig. 5.4. The Ṭūsī couple.

It was used by its inventor to eliminate the equant as follows. In Fig. 5.5, D is the centre of a Ptolemaic deferent, the observer is at O, and EC is the rotating spoke with the equant point E at one end and the planet C at the other. Although the equant spoke rotates uniformly, the basic astronomical 'law of motion' is violated by the fact that the length of EC is variable, being shortest at apogee, A, and longest at perigee, P. Now for relatively small eccentricity the variation in length of the equant spoke can be represented to good accuracy by a simple harmonic motion, that is, its length l can be expressed as

$$l \approx R - e \cos \alpha,$$

where α is the mean anomaly (measured from apogee), and R is the radius of the Ptolemaic deferent. Such an effect can obviously be achieved by having an equant spoke of fixed length R but placing at its end a Ṭūsī couple, with radius of the smaller circle equal to $e/2$, as shown in Fig. 5.5. Naṣīr al-Din showed that the resultant motion is not quite circular, the point determined by the couple lying just outside the Ptolemaic deferent, though coinciding exactly with it at A and P, of course. Because the couple swings round with the equant spoke, the resultant point is always seen to move uniformly around E (it merely moves up and down on the equant spoke). Thus, this property of the Ptolemaic arrangement is exactly preserved, and for small eccentricity the observed motion is effectively identical to the Ptolemaic motion.

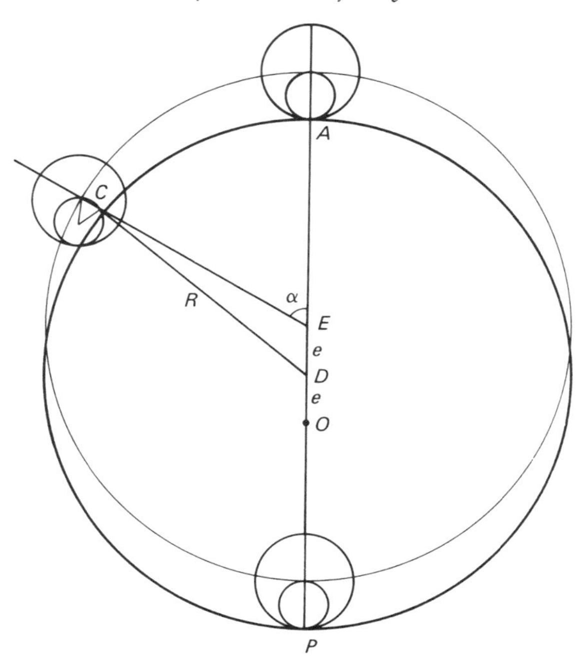

Fig. 5.5. Use of the Ṭūsī couple to reproduce Ptolemaic equant-type motion to a high accuracy. The heavy circle is the Ptolemaic deferent carrying the planet C. The alternative arrangement leads to a position for the planet that is on EC but just outside the deferent circle except at A and P.

An alternative to this device is found in the writings of Quṭb al-Dīn al-Shīrāzī (1236–1311). This is as follows.[7] In the Ptolemaic scheme a new point H is introduced (Fig. 5.6) at the *midpoint* between D and E, i.e., at $\frac{3}{2}e$ from the observer. Around this point rotates uniformly the vector HQ with radius equal to the (nominal) radius of the Ptolemaic deferent and the orbital period of the planet. At its end the vector QC, which has length $\frac{1}{2}e$, rotates in the same sense and at twice the rate of HQ. When the planet is at apogee (A), QC points towards H; two other positions of the linkage are shown. In this arrangement the Ptolemaic eccentric deferent and equant are replaced by a simple eccentric and epicycle. This device leads to exactly the same motion as the use of the Ṭūsī couple and, therefore, reproduces the Ptolemaic (and actual) motion just as well. It is important to note that both of these devices represent only formal elimination of the equant. There is no suggestion by the innovators that Ptolemy failed to describe the observed motions accurately; the only quibble is with the means of description.

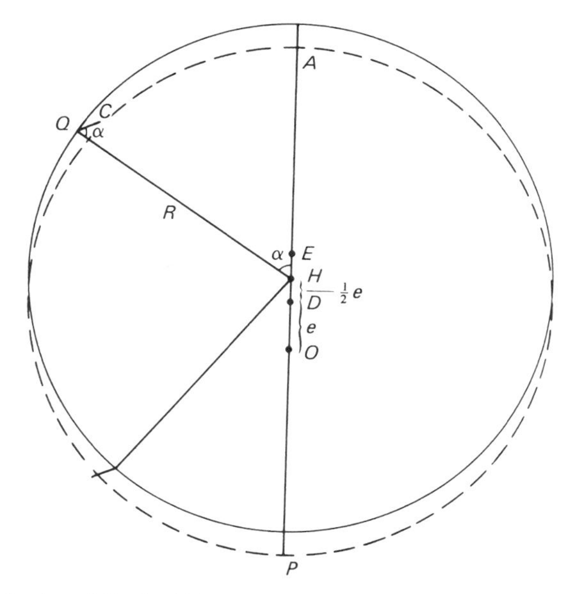

Fig. 5.6. The device of Quṭb al Dīn and Copernicus for replacement of the equant mechanism by eccentric with centre H at $3e/2$ from the observer O and epicycle with spoke QC of length $e/2$. The position of C is very close to the Ptolemaic deferent (shown by the dashed circle).

It is because Copernicus uses both of these devices and on several different occasions (with some variations between the *Commentariolus* and *De Revolutionibus*) that he comes under the suspicion of plagiarism. In *De Revolutionibus* his method of eliminating the equant is identical to Guṭb al-Dīn's, while the Ṭūsī couple is used both in his theory of precession and in his model of Mercury's motion. As has already been noted, his lunar theory is essentially that of Ibn al-Shāṭir. All that I think one can say in Copernicus's defence is that independent rediscovery (indeed, multiple rediscovery) is a commonplace in science. It is especially prevalent in periods of what Kuhn calls normal science (p. 141), i.e., at a time in the development of a science in which clear ground rules have been laid down and there are more or less well-defined problems that call for solution. This is precisely the situation created by Ptolemy's great discoveries. Indeed, it could well be argued that it would have been surprising if no attempt had been made to reconcile the facts of planetary motion brought to light by Ptolemy with the hallowed ancient law of uniform circular motions. Any reasonably competent mathematician – and that, at least, Copernicus was – was almost bound to come up with essentially the same solution. Nevertheless the grounds for doubt are clear, and the purpose of this review was not to put in a plea for Copernicus but rather to outline for the reader the main mathematical tools which he had at his disposal – and also to note that although his avowed intention was to eliminate the equant, the effect which Ptolemy had described by the equant remained an integral part of Copernicus's scheme, being described by an almost exactly equivalent mechanism. On Copernicus's part this was truly a case of having one's cake *and* eating it. No wonder he was pleased with the solution.

This brings us appropriately to the last point that we must consider before turning to the details of Copernicus's scheme, namely, his attitude to Ptolemy's work. We recall that after the death of Ptolemy no one had apparently had the idea of a really comprehensive observational testing of the Ptolemaic system in its entirety. What had been done, often quite successfully and accurately (especially in the case of the solar theory), was redetermination of particular parameters of the models. Now, in the first place, these had not been determined all that accurately by either Hipparchus or Ptolemy; secondly, many of them had in the meanwhile changed quite appreciably (mostly due to the precession of the equinoxes, so that, for example, the lines of the apsides had all shifted by about 20° relative to the vernal equinox in the time between Ptolemy and Copernicus; in addition there were smaller but still in some cases significant shifts relative to the stars). As a result of both these factors, subsequent observers, including Copernicus himself, had often found deviations, sometimes large, from the parameter values that Ptolemy and Hipparchus had determined. Copernicus, who was always too ready to believe the observations of others, especially Ptolemy, took these deviations at their face value and assumed that the parameters of the Ptolemaic models varied with the time. In this he was quite correct, but, by treating earlier observations – and his own – with too little consideration for their possible errors, he vitiated much of the potential value of his conscientious attempt to redetermine Ptolemy's parameters, finding and mixing up as a result quite spurious as well as genuine changes. What Copernicus did not do was question the basic correctness of the Ptolemaic models. Instead, he set himself a three-fold task: to convert the Ptolemaic models from geocentric to heliocentric form, to eliminate Ptolemy's violations of the uniformity rule, and to redetermine the parameters of the Ptolemaic models after they had been adapted in accordance with his first two aims. Since the first two aims were more or less purely formal in nature, it follows that Copernicus made no attempt to change the essentials of Ptolemy's work. He seems to have given absolutely no consideration to the possibility of genuine errors in Ptolemy's models, or to the possibility that the presence of the sun near the centre of the planetary system might have very far reaching implications beyond the purely kinematic consequences of which he was very well aware. He thereby missed a great opportunity, and this prompted Kepler to a famous remark:[23] 'Unaware of his own riches, Copernicus merely set himself the task of reproducing Ptolemy, not the nature of things, to which however he had come the closest of all.'

Let us now see how Copernicus set about his task. Having got hold of the idea that the earth revolves around the sun rather than the sun around the earth, Copernicus's first step was to transcribe the earth–sun motion.[24] It was here that he made his biggest mistake, for he simply inverted the Hipparchan solution. Just as Hipparchus and Ptolemy believed that the sun circled uniformly about a point at some distance from the earth, Copernicus believed the earth did the same about a point at some distance from the sun. This led him to make two mistakes. He was correct in displacing the centre of the earth's orbit from the centre of the sun but followed Hipparchus and Ptolemy in making the eccentricity (the displacement from the centre) too large by a factor of two. Second, he assumed that the earth's motion around this displaced point was perfectly uniform. He postulated that the earth moves uniformly in a circle, which he called the *orbis magnus* (the great circle), around a point at twice the distance of the actual centre of the earth's orbit.

A clear understanding of the precise location of this point is crucial – expressed in Keplerian terms, it is the second (void) focus of the earth's orbit (cf. p. 125). This is the measure of Copernicus's mistake; for, as we know, the centre of the earth's orbit in helioastral space is exactly at the *mid-point* between the two foci. Copernicus therefore misplaced the centre of the earth's orbit by its eccentricity, an error of about 1.8 solar diameters.

It is worth noting here what happens to the Ptolemaic concept of the mean sun. In Ptolemy's scheme it is merely a variable direction that points always from the earth in the direction that a uniformly moving body, coincident with the true sun in the apsides, would have. In the Copernican arrangement, these lines all pass through the position of the second focus of the earth's orbit, and one may therefore now think of the mean sun as a fictitious body occupying the void focus. The *Copernican mean sun*, as we may call it following Swerdlow and Neugebauer,[9] is at the centre of the *orbis magnus*.

There is no word in *De Revolutionibus* to indicate whether or not Copernicus ever contemplated any alternative to his chosen arrangement. One supposes he must have realized that an equant-type motion for the earth was at least a possibility. As this question is tied up with his treatment of the other planets, I defer further discussion and merely comment that his solution, which describes the observed solar motions excellently, fitted his predilection for circular motions perfectly and this must certainly have helped to still any doubts he may have had.

This error on Copernicus's part, which he probably made around 1510, had already plagued astronomy (in its geocentric guise) for a millennium and a half and continued to do so for nearly another century until Kepler finally spotted the offending bit of grit and took it out of the works.

The mistake was disastrous for the new astronomy, since, as Copernicus realized more clearly than anyone before him, correct interpretation of the spectacle unfolding on the heavens required above all a correct understanding of the part that the observer himself plays in the process. For if the observer himself is moving, the whole aspect of the dance will be changed. Copernicus in fact made precisely the mistake against which he had himself offered such good advice:[25] 'Hence I deem it above all necessary that we should carefully scrutinize the relation of the earth to the heavens lest, in our desire to examine the loftiest objects, we remain ignorant of things nearest to us, and by the same error attribute to the celestial bodies what belongs to the earth.' Thus, several of the motions that Copernicus had to invent to explain the planetary motions had exactly the same origin as the epicycles of the original Ptolemaic system – a lack of awareness of the true position of the earth. But these motions stand to the original epicycles in much the same way as the aftershocks to the original earthquake – secondary adjustments to the new equilibrium position. They are for all that still a considerable source of worry – they certainly drove Brahe and his assistants to distraction.[26]

So far, we have only considered the relationship of the earth and the sun. It was in his placing of the planets within his overall cosmology that Copernicus introduced his most incongruous ideas.[27] It was here that the specific values of the planetary eccentricities played their little trick again. Let us begin with the three outer planets. On the face of it, Ptolemy had already solved the problem; for, as we have seen, the epicycle–deferent construction had the effect of mapping the motions of these planets around the sun into the motion of the deferent guide points around the earth. It would appear that all Copernicus had to do was convert the Ptolemaic deferent motion around the earth into real planetary motion around the sun. Moreover, the deferent half of the epicycle–deferent theory was already very nearly correct. But not quite – and that 'not quite' was Copernicus's undoing. The deferent theory, possessing as it did equant and eccentricity, would have been perfectly correct (except for the minute ellipticity corrections, which, let it be said again, were the least of Copernicus's worries) had it not been for the fact that in the Ptolemaic scheme everything was aligned on the mean sun, not the true sun. Here, as in so many things, Copernicus simply followed Ptolemy. Thus, the formal and faithful transcription of the Ptolemaic scheme for the three outer planets led Copernicus to align their apsidal lines on the mean sun too. That is, he assumed that the lines through the centres of the orbits of Mars, Jupiter, and Saturn and the corresponding equants of these orbits

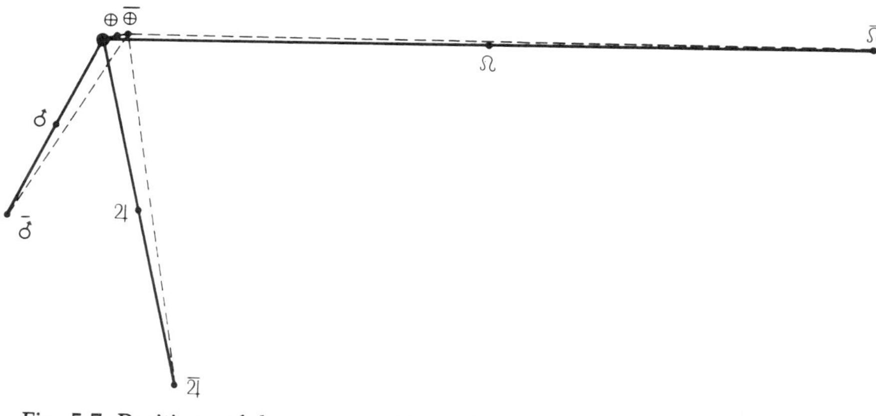

Fig. 5.7. Positions of the centres of the orbits and the equants (indicated by the barred symbols) for the earth (⊕), Mars (♂), Jupiter (♃) and Saturn (♄). The continuous lines are drawn through these points; they are the 'knitting needles' that all converge at the centre of the sun, which is represented by the small black circle and is drawn to scale. Copernicus incorrectly assumed that the centre of the earth's orbit is at $\bar{\oplus}$, not ⊕ and that the apsidal 'knitting needles' for the three outer planets also converge on $\bar{\oplus}$. This arrangement is shown by the dashed lines (it will be explained in Sec. 6.2 why the positions of the equants are unchanged). It can be seen that the resulting errors are relatively small, which is the reason they could not be readily detected.

(which Copernicus effectively retained as auxiliary concepts for computational purposes) converged like knitting needles on the centre of the *orbis magnus*. And there, in the ghostly body of the mean sun, and not in the true heart of the solar system, this crucial point was placed, an egregious insult to the sun of which Copernicus was blissfully unaware.

It is worth looking at the physical reason why Copernicus could make this remarkable mistake, which was not spotted for 60 years despite monumental efforts at the end of the period on the part of Brahe. As already hinted, it is to be sought in the specific eccentricities of the planetary orbits. Figure 5.7 shows the sun and the positions of the centres of the orbits and the equants for the earth and the three outer planets. The lines of the apsides (continuous lines) all converge on the sun. The dashed lines show the situation that Copernicus obtained by his machinelike transcription of the Ptolemaic system. The false position of convergence is the true measure of the residual errors in the Ptolemaic system. One can clearly see why they escaped detection for so long, since these errors are far smaller than the main deviations from the zero-eccentricity epicycle–deferent scheme, the elimination of which was Ptolemy's great achievement. These main deviations correspond, of course, to the displacements of the centres and equants of the planetary orbits from the centre of the sun, which are much greater than for the earth.

The same diagram explains simultaneously the other residual error of the Ptolemaic epicycle–deferent scheme for the outer planets – the perfectly uniform rotation of the epicycle spoke of constant length about the void deferent guide point. We recall that Ptolemy's scheme was a very accurate approximation of *first order* in the eccentricity of the orbits of the outer planets but only *zeroth order* in the eccentricity of the earth's orbit. To have made the scheme have first order in both eccentricities, Ptolemy would have had to have devised a very much more complicated scheme involving eccentric and nonuniform equant-type motion in the epicycle too. All these complexities simply escaped him because, first, the earth's orbit has an eccentricity much less than the orbits of the three outer planets and, second, the observability of that eccentricity in the epicyclic motions of the outer planets is additionally reduced in the ratio of the semimajor axes of their orbits to the semimajor axis of the earth's orbit. Thus, in the apparent motions of the outer planets, the eccentricities of the planetary orbits result in observable effects that are on average 8.5, 15, and 31 times larger for Mars, Jupiter, and Saturn, respectively, than the effects due to the eccentricity of the earth's orbit. This was why the Ptolemaic models could be so good and yet very misleading for the unfortunate Copernicus, who saw in the uniform epicyclic motions of the three outer planets perfect confirmation of his transcription of the Hipparchan–Ptolemaic solar motion; for they too clearly indicated that the earth must circle uniformly about the mean sun, just as he had already found from the solar model. Thus, the grounds for identifying the mean sun with the linchpin of the entire planetary system were greatly strengthened. The Copernican transcription for the outer planets is shown in Fig. 5.8.

We have not yet considered the two inner planets. Before we do so, it will be well to emphasize an important constraint under which Copernicus worked but from which Ptolemy was free. We have already hinted at it at the end of the last paragraph, namely, the Ptolemaic system treated the motion of all the celestial bodies independently; *a priori* there was no reason to suspect that there were absolutely necessary connections between the various observed motions (cf. Kepler's remark quoted earlier on p. 219). However, the observations showed that nevertheless there were apparent correlations – the epicyclic motions of the outer planets and deferent motions of the inner planets seemed to be strongly correlated to the motion of the sun. However, because no deeper reason for this correlation was looked for, the Ptolemaic system could (and did) tolerate a certain mismatch between these motions. However, Copernicus's radically new suggestion of terrestrial mobility posited that these seemingly separate motions were all merely the reflection of a single motion of the earth. This had the inescapable consequence that all must be exactly correlated.

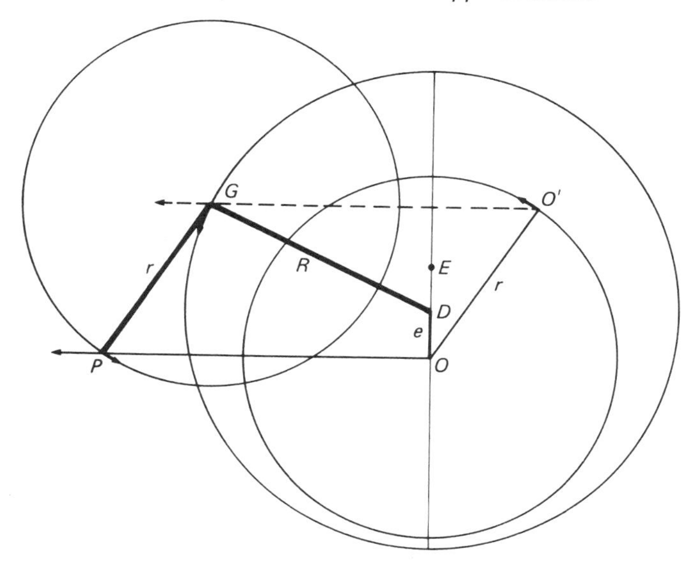

Fig. 5.8. Copernicus's formal transcription of the Ptolemaic scheme for the outer planets. In the Ptolemaic scheme the observer is at O, the centre of the deferent is at D, the equant point at E, the epicycle guide point at G, and the planet at P. Under the transcription O becomes the centre of the *orbis magnus* (the Copernican mean sun at the second (void) focus of the earth's orbit), the earth is at O′, and the planet is at G. Since OO′ is parallel to PG and of equal length *r*, O′G is parallel to OP, and the planet is seen along the same line to the distant stars. In the Copernican scheme the orbit of the planet is therefore aligned on the mean sun.

This was a point that Copernicus clearly understood very well but in interpreting the Ptolemaic system as he addressed the problem of its transcription to his own his attitude of mind to the accuracy and essential correctness of the Ptolemaic system was crucial. As we have noted, he did not question them. This led him into severe difficulties with the inner planets, to which we now turn.

Here the eccentricities led Copernicus into a very odd error. This was especially ironic since it was precisely for Venus that Ptolemy's model came nearest the truth and might have given Copernicus his most valuable hint. We recall that, quite unwittingly, Ptolemy had determined the elements of the earth's orbit twice – once incorrectly in the solar theory (getting a value for the eccentricity twice what it should have been) and once more or less correctly in the construction of his model of Venus's own motion. Here exactly the same effect occurred as between the earth and the outer planets. The eccentricity of Venus's orbit is 2.45 times less than that of the earth's and the ratio of the semimajor axes (1.39) means that in the apparent motion of Venus the eccentricity effects of the earth's orbit show up about $3\frac{1}{2}$ times more strongly on the average than those of Venus's orbit (both effects are moreover rather small compared with the outer planets). This again led to the impression that the epicyclic motion of Venus corresponded to perfectly uniform circular motion. However, in this case, the deferent motion, which in the Ptolemaic system did have eccentricity and equant, should correspond to the earth's motion. Taken at its face value the straightforward transcription of such motion would have led Copernicus to the conclusion that the earth moved on an eccentric with motion governed by an equant. Moreover, the equant must be at the position of the (Copernican) mean sun and the centre of the eccentric must be displaced from it by the corresponding eccentricity.

Such a transcription would have taken Copernicus very close to the truth. But this was where the Ptolemaic system gave him contradictory hints. Such motion could not be reconciled with the earlier transcriptions for the sun and the three outer planets, which had all told him the earth had no equant in its motion and simply circled uniformly around the mean sun as its centre. But now he found that the Ptolemaic system, transcribed mechanically, required the centre of the earth's orbit to be in two different places simultaneously! Since that could not be, Copernicus was forced to make a choice. One wonders how long he pondered his awkward choice. He must surely have felt rather like Buridan's ass, forced to choose between two equally distant heaps of hay. However, in Copernicus's case the alternatives did not appear equally balanced – the sun, Mars, Jupiter, and Saturn were all ranged in complete unanimity against the dangerous allures of the goddess of love. Further grounds for distrusting her evidence were also probably supplied in Copernicus's mind by Mercury, for which the transcription suggested yet another placing of the centre of the earth's orbit (for the details of the intricate orbit chosen for Mercury the reader is referred to Ref. 9).

But if the earth does circle the mean sun, how can this be reconciled with the apparent motion of Venus? Copernicus resorted to a remarkable dodge. He kept his previous solution for the earth's motion and simply transferred the 'excess motion' to Venus! That is, whereas strictly Ptolemy's Venus model required the earth to be described (in accordance with the scheme used by both Quṭb al-Dīn and Copernicus) by an eccentric and epicycle rotating around the guide point on the eccentric at *twice* the angular velocity of the guide point around the eccentric, Copernicus simply shifted the correction to uniform circular motion of the earth produced in this way to Venus and accordingly required the centre of the circle in which Venus moved to itself move around in a (very small) circle *twice in the time that the earth required to move around its orbit.* Copernicus's transcription for Venus and the crucial role played in it by the commutativity of vector addition are illustrated in Fig. 5.9.

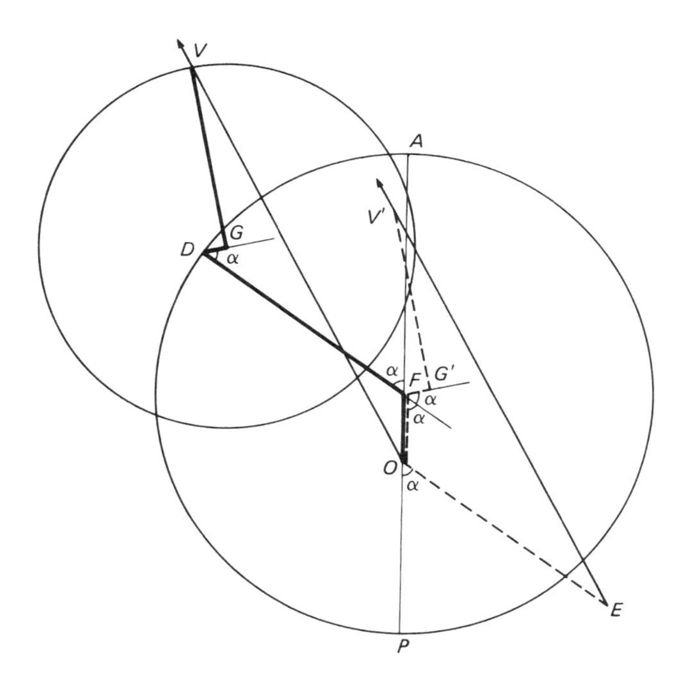

Fig. 5.9. Copernicus's formal transcription of the Ptolemaic scheme for Venus. In the Ptolemaic scheme (heavy lines) the deferent arrangement has been replaced by Copernicus's almost exactly equivalent arrangement in which the observer is at O, F (at distance $3e/2$ from O) is the centre of circle AD, D is the centre of the small epicycle of radius $e/2$, and G is the centre of the large epicycle which carries the planet V. The position of G is so close to the Ptolemaic guide point (not shown) that the two arrangements are effectively identical (if e is sufficiently small). Under the Copernican transcription O becomes the centre of the *orbis magnus*, the earth is at E and Venus, V′, circles G′, which itself circles F. The planet is seen along the same direction to the distant stars because EO is parallel to FD and of equal length and FG′ is parallel to DG and equal to it in length. If the heavy links OF, FD, DG and GV are represented by the vectors **p**, **q**, **r**, **s** respectively, then $\overrightarrow{OV} = \mathbf{p} + \mathbf{q} + \mathbf{r} + \mathbf{s}$ and $\overrightarrow{EV'} = \mathbf{q} + \mathbf{p} + \mathbf{r} + \mathbf{s}$. Strictly, the vectors **q**, **p**, **r** should all be associated with the earth's motion and only **s** with Venus. However, Copernicus lumped **p** and **r** in with **s** so that he could say, as for the outer planets, that the earth circles uniformly around O.

It should be said that Copernicus does not discuss any of these problems with his readers. He simply announces that 'this planet differs somewhat from the others in the pattern and measurement of its motions'[28] and then gives the results of his transcription without any explanation. In the case of Mercury's motion he was also forced to introduce a curious correlation with the motion of the earth on the *orbis magnus*; in fact, he required two motions, one in a small circle and one along a diameter of another small circle and controlled by a Ṭūsī couple, both to be executed twice in a terrestrial year.

We are still not yet done with the oddities of the Copernican system; for we must say at least a few words about the particularly bizarre mechanisms which Copernicus devised to account for the planetary motions in *latitude*, i.e., out of the ecliptic. This is a motion that we have hitherto almost completely ignored, but it proved to be very important in Kepler's work. The planes of the planets' orbits are, as noted in Chap. 3, inclined at small angles to the plane of the earth's orbit, i.e., the ecliptic. For this reason, the planets are sometimes to the north and sometimes to the south of the ecliptic. Moreover, because of their nonuniform motion in longitude their rate of motion from north to south is also nonuniform. In particular, the planets do not remain for equal lengths of time on the two sides of the ecliptic. In additon, as the earth moves around its orbit, its approach to and receding from the other planets create the impression of an apparent additional deviation in latitude. This effect is particularly pronounced at opposition to one of the outer planets, at which the planet usually appears to deviate more strongly from the ecliptic at the same time as performing its retrograde motion, thus describing a loop on the background of the stars. Some examples of such loops for Jupiter in different parts of its orbit are shown in Fig. 5.10.

In principle this was an effect that should have provided some of the most convincing arguments for the heliocentric theory, since the latitude motions, just like the epicyclic motions of the outer planets and the deferent motions of the inner planets, can all be explained purely kinematically as a reflection of the earth's motion by the simple assumption that the planes of the planets' orbits are inclined at certain fixed angles to the plane of the earth's orbit and pass through the centre of the sun. Although Copernicus made them pass through the mean sun, this error alone would have had comparatively little influence on the small latitude effects. He was in fact well aware of the new possibility of explaining the latitude motions in such a natural and pleasing manner but was completely led astray by the unsatisfactory state of Ptolemy's latitude theory as presented in the *Almagest*. At the time the *Almagest* was written, Ptolemy must have been working with very inaccurate data, since his latitude theory contains numerous defects, and he was forced to introduce, for example, variable tilts of the planes of the epicycles relative

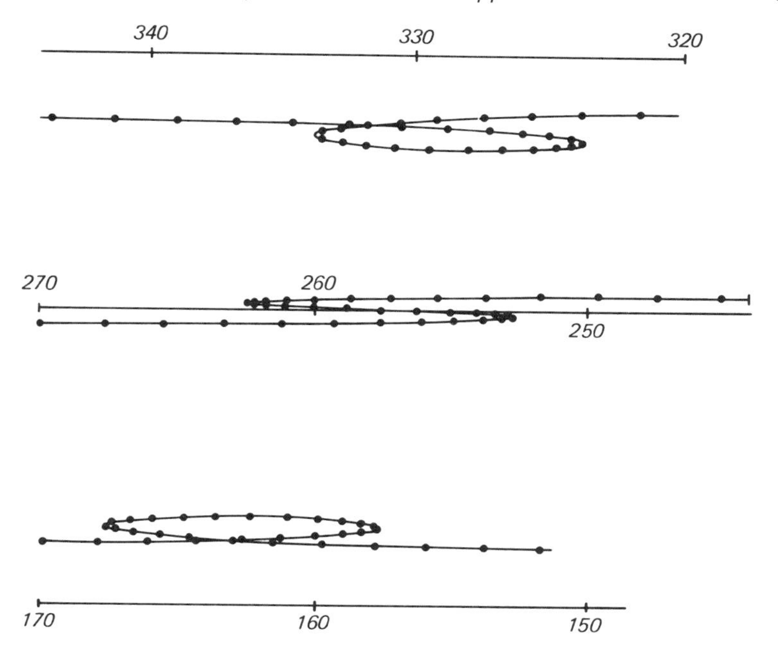

Fig. 5.10. Retrogression loops of Jupiter at different positions in the ecliptic (the numbers give the ecliptic longitude). The black points show the observed position of the planet against the background of the stars at intervals of 10 days (the planet moves from right to left apart from the retrogression). The scale of the vertical direction (latitude) has been multiplied by two to show the effect more clearly. (Derived from: O. Neugebauer, *A History of Ancient Mathematical Astronomy*, 3 vols., Springer, Berlin (1975), p. 1256.)

to the planes of the corresponding deferents. These tilts varied with the position of the sun in the ecliptic. It is evident that Ptolemy continued working conscientiously on the latitude theory after completing the *Almagest*, since his *Handy Tables* and *Planetary Hypotheses* contain a greatly improved theory – one indeed that corresponds very well to transcription from the heliocentric to the geocentric system. In particular, the variable tilts were eliminated.[29]

It appears that this later work of Ptolemy never became widely known and Copernicus reveals no awareness of its existence. He therefore merely did his best to reproduce the geocentric theory of the *Almagest* in his own system. Part of the latitude motion could be explained perfectly correctly, but for the spurious residue due to the defects inherited from the *Almagest* he invented a wobble of the planes of the planetary orbits which made them have a tilt that *varied in phase with the motion of the earth in its annual revolution*! This mistake on Copernicus's part may actually have helped to advance science, since Kepler, already recognizing it as a monstrosity,[30] decided quite soon in his investigations to test the alleged effect against Brahe's observations and very readily demonstrated its nonexistence. This work led him to establish two highly significant facts: (1) the planes of the planets' orbits have a fixed inclination in helioastral space, (2) they all pass *through the sun* and not the centre of the *orbis magnus*, as Copernicus believed. This is the result which was designated in Chap. 3 as Kepler's Zeroth Law, and it played a very important part in establishing the law of inertia and the other foundations of dynamics, as well as showing that they are valid over astronomical scales of distance.

By trusting Ptolemy so implicitly in the question of the latitude motions, Copernicus missed a very great opportunity to achieve a radical simplification in the lesser as well as the greater motions of the celestial bodies. It is when *all* the pieces come together in a coherent and accurate picture that a theoretical scheme really acquires a power to persuade acceptance. This Copernicus failed to achieve. It was, in fact, in connection with the lamentable latitude theory of *De Revolutionibus* that Kepler made his remark about Copernicus not having appreciated his own riches (*divitiarum suarum ipse ignarus*).

The numerous errors in the Copernican system – the transcription of the Hipparchan solar theory for the earth's motion, the identification of the point of convergence of the apsidal lines with the centre of the *orbis magnus*, the curious correlations between the longitudinal motions of the two inner planets and the motion of the earth, and the even more curious latitude motions (all born of overzealous fidelity to Ptolemy) – were responsible for the most remarkable feature of all in the Copernican system. For although Copernicus sensed the importance of the sun and, equally important, made the earth a planet like the others, his final scheme remained to a remarkable degree geocentric. First, the motion of the earth was quite different from that of the planets. Alone among them it moved uniformly in helioastral space. Second, the motions of the five planets were not coordinated on the sun but on the centre of the earth's orbit, the *orbis magnus*. Moreover, they were strongly correlated to the earth's orbital period. Only now can we understand the true meaning of Copernicus's comments to the pope that he will[31] 'correlate the motions of the other planets . . . with the movement of the earth'.

As we noted earlier, Copernicus pondered the heliocentric system 'not merely until the ninth year but by now the fourth period of nine years'; it seems never once to have entered his head that the earth should behave in the same way as the other planets or that the motions of the planets should be centred on the sun rather than on the centre of the earth's orbit. But before we laugh and hiss the scheme off the stage, let us recall that Brahe and other astronomers worked happily with it for a couple of generations without suspecting there was anything wrong. Nearly a century after Copernicus's death, even Galileo did not suspect there was anything amiss. More clearly than anything else, this shows us that the *conception of motion* was not yet ripe for change. This point will be taken up in Sec. 5.7 and in the chapter on Kepler.

Also we should recognize that, seen from Copernicus's perspective, his scheme was rather a success. Quite apart from the heliocentric hypothesis and the genuine improvement of Ptolemy's lunar theory, the abolition of the Ptolemaic equant, that affront to divine aesthetics which had a much more intimate relation to heliocentricity than Copernicus ever realized, must have given him much satisfaction, especially if the elegant device by which it was achieved was his own independent rediscovery. The importance of the equant as a stimulus to thorough re-examination of the Ptolemaic models should not be forgotten; indeed, we have to thank Copernicus's Quixotic mission (for he is a somewhat Quixotic figure) to purify the art of astronomy for the greatest revolution in history.

As he grappled with the problem of the celestial motions over all those years, Copernicus had to contend with several problems. Some were his own preconceptions, many were due to the almost complete lack of observations of the number and quantity needed to pin down the numerous lesser effects that he felt obliged to accommodate, and quite a number more were the result of the particular eccentricities of the planetary orbits. But, in addition to these, he faced throughout his work the problem of kinematic relativity – the apparent motion of any given body may contain a component that could be attributed to motion of the observer rather than of the observed object. This was a problem that, as we have seen, he himself highlighted and exploited brilliantly but which nevertheless got its revenge in many ways, most noticeably in the case of the longitudinal motion of Venus. It is ironic that what Copernicus failed to note in the case of Venus (and Mercury and the latitude motions) he did clearly recognize in yet another curious feature in his system. This has to do with the question of the exact position of the centre of the universe.

As we shall see in the next section, Copernicus had a rather precise notion of position and the ordering of the world. It is therefore somewhat surprising to find that he was at a loss where to place its exact centre. He was floored by the same combination of factors as in all his other work: inaccurate observations, the residual defects of the Ptolemaic system, and his own curious blend of heliocentricity and geocentricity.

His first and most important task had been to transcribe the earth–sun motion. As we have seen, this led him to the concept of the *orbis magnus* and its distinguished centre. This, however, only disposed of the gross (annual) motion of the sun. In addition to this there were the very slow secular changes that had taken place since Ptolemy's time. In the observational record at Copernicus's disposal these effects, together with

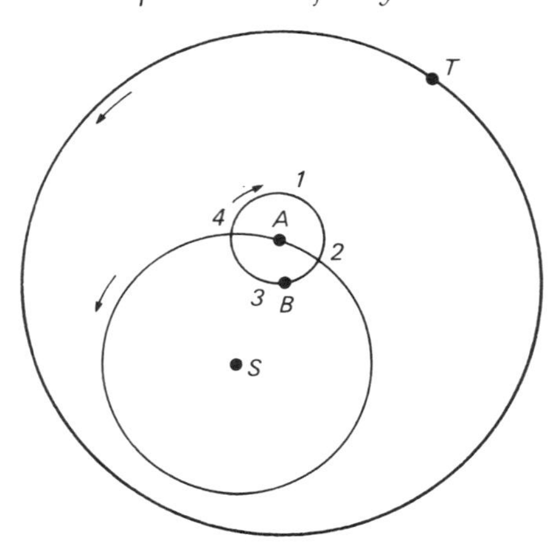

Fig. 5.11. The scheme which Copernicus devised to explain the slow variations of earth–sun motion which he believed to have taken place since antiquity. In this arrangement the sun is at S, the point A travels very slowly about a circle centred on S while the point B, the centre of the earth's orbit, travels very slowly around A. (Reproduced from: *J. L. E. Dreyer, A History of Astronomy from Thales to Kepler,* Cambridge University Press (1906) (republished by Dover Publ. Inc. (1953)).)

the effects of the precession of the equinoxes, could not possibly be cleanly separated from observational errors. Copernicus was therefore attempting to transcribe into heliocentric terms an amalgam of effects, some genuine, some the pure artefact of inaccurate observations, above all an apparent irregularity in the apsidal precession. The scheme he devised is shown in Fig. 5.11. The sun is at S, around which the point A moves in a circle from west to east in about 53 000 years (corresponding to a mean annual apsidal precession of somewhat more than 24 seconds of arc). Around A moves the point B in the small circle in the opposite direction in 3434 years. The moving point B is the centre of the orbit of the earth's annual motion, the centre of the *orbis magnus*.

The uncertainty in Copernicus's mind to which I referred just now was that he was not in his mind entirely sure – about which he was engagingly frank – whether to take the sun absolutely at rest and let the centre of the *orbis magnus* ride around as indicated in Fig. 5.11 or take the *orbis magnus* fixed and let the sun execute analogous motions, which would have the same observational consequences. Thus, at this one point he threw in the sponge and gave up the attempt to beat the impartiality of relativity. Depending on the choice, the position identified as the formal centre of the universe would be changed.

Thus, in Book I, Chap. 9, Copernicus says that 'the sun occupies the middle of the universe', but a little later in Chap. 10, he says that 'near the sun is the centre of the universe', which he here appears to identify with the centre of the *orbis magnus*. Later, in discussing the details of this theory of the earth–sun system, he says:[32]

> Nevertheless I am also not unaware that if anybody believed the center of the annual revolution to be stationary as the center of the universe, while the sun moved with two motions similar and equal to those which I explained in connection with the center of the eccentric [III, 20], all the phenomena would appear as before – the same figures and the same proof. Nothing would be changed in them, especially the phenomena pertaining to the sun, except the position. For then the motion of the earth's center around the center of the universe would be regular and simple (the two remaining motions being ascribed to the sun). For this reason there will still remain a doubt about which of these two positions is occupied by the center of the universe, as I said ambiguously at the beginning that the center of the universe is in the sun [I, 9, 10] or near it [I, 10].

We find in this frank admission of defeat on Copernicus's part an anticipation of the debate that will rage during the second half of the seventeenth century and which has still not subsided. What is motion? By what criteria are rest and motion defined? For if motion is defined by relationship to specific identifiable objects, there are as many motions as there are reference objects. Copernicus was only able to look on this problem with equanimity because he still had an ultimate fallback position: the fixed stars. In his mind, the uncertainty arose simply because the corresponding effects were much too small to show up in parallax displacements. In principle he believed he had nothing to fear. The threat posed by the realization that the 'fixed stars' might not be fixed at all was still a century ahead. We shall see in the next section how this influenced his concept of place.

This completes what we need to say about the technical aspects of the Copernican system. There is quite a lot more material in *De Revolutionibus* that is entirely omitted from this account, including his theory for the moon and Mercury.

Viewed purely in terms of improvement in the models of the celestial motions, Copernicus's contribution was very modest if the heliocentric theory is left out of account. Only his theory of the moon marked a significant advance over Ptolemy. Nevertheless, he performed a useful service to astronomy in updating Ptolemy's work, especially in recomputing orbits and determining new parameter values. His elements were employed in the compilation of new astronomical tables, the *Prutenic Tables*. Most of his immediate successors saw this as his main achievement, together with his ingenious proposal for eliminating the equant. Gingerich[33] has shown that working astronomers in the second half of the seventeenth century did make use of his work, which, despite claims to the contrary by Koestler ('the book that nobody read'[34]), was apparently quite widely read by the professionals. But, intriguingly, none of them would risk a word in print about the mobility of the earth until the 25 year old Kepler came out in the open with eloquent support more than 50 years after *De Revolutionibus* had appeared.

We now consider some more general aspects of Copernicus's work and the bearing that they have on the discovery of dynamics and the absolute/relative debate.

5.6. Copernicus's concept of place and the ultimate frame of reference

We begin this section by drawing attention to the typically Renaissance concept of cosmology that we find in a most pronounced form in Copernicus, Kepler, and Galileo. It is that of the *well-ordered cosmos*, a concept that derives from Pythagoras. Entranced by the beauty of the world, he called it *cosmos*, the Greek word meaning primarily order, but also decoration, embellishment, or dress (*cosmetic* derives from the same origin), the Latin *mundus muliebris*, the ornament of women. Thus, *cosmos* and *mundus* were conscious coinings to express the 'perfect order and arrangement' of the world, perceived as an indissoluble unity.[35]

This concept, revived strongly in the Renaissance, had a twofold significance in that period. On the one hand, it provided an extremely important psychic stimulus to the study of astronomy and the celestial motions, a stimulus that was all the more fruitful for being enriched by the reawakening of the Platonic ideals of geometrical exactitude and perfect symmetry. On the other hand, it helped to shape the concept of motion, which was very much seen as an integral expression of the overall harmony of the cosmos. Since the cosmos was believed to derive this harmony from the felicitous mutual disposition of its parts, such a concept inevitably tended to emphasize the relational aspect of position and motion. The cosmos-based concept of motion was therefore a relational and matter-based concept. Natural motions at least had to maintain the overall harmony created by the disposition of the parts. This attitude is very marked in both Copernicus and Galileo. In view of the fact that what he regarded as his major work, the *Harmonice Mundi*,[36] is a hymn in praise of Pythagorean harmony, it is rather curious that Kepler's own conception of celestial motions was much more modern than either Copernicus's or Galileo's even though his cosmology was at least as Pythagorean and Aristotelian as theirs.

If Copernicus's concept of motion can be seen as a kind of prescientific or artistic Mach's Principle, Kepler's concept of motion is a clear anticipation of a modern and fully scientific (i.e., physical) form of the principle. One of the main aims of this and the following chapter is to show how this dramatic change in the concept of motion occurred. The key element in this transition from aesthetic geometrokineticism was Ptolemy's discovery of the equant, the very thing that Copernicus was so anxious to abolish. However, before we get on to this topic, let us look more closely at Copernicus's concept of place.

In the light of the subject of this book, the Aristotelian element in Copernicus is particularly interesting. Copernicus is very concerned that his concept of motion should be epistemologically sound. It is clear that he rejected the idea that motion takes place relative to space – it must be relative to observable matter. Like Aristotle, he invokes the idea of a universal frame of reference; the only difference between them is that for Aristotle the frame of reference was conceived to spin, whereas for Copernicus it is, almost by definition, at rest.

For example, quite early in Book I, in discussing the possibility that the earth could move, he says:[37] 'since the heavens, which enclose and provide the setting for everything, constitute the space common to all things, it is not at first blush clear why motion should not be attributed rather to the enclosed than to the enclosing, to the thing located in space than to the framework of space.' By the 'heavens' Copernicus meant, of course, the immensely distant sphere of the fixed stars.

The above concept of position, and hence motion, clearly comes from Aristotle. Nevertheless, despite this strongly Aristotelian element, the centuries of practical astronomy left their mark. Copernicus needs a container, but one is sufficient; his world is not an onion in which a succession of layers are needed to give meaning to position all the way from the outermost layer right through to the centre. Copernicus combines Aristotle's ultimate *ouranos* with the Platonic concept that Aristotle rejected, namely, that space is 'some kind of dimensional extension lying between the points of the containing surface' (see (iii) on p. 86). The space within the container is no longer a nest of topological envelopes but a tautly spanned region of trigonometric relations, an invisible membrane spanned by the rim of the drum provided by the fixed stars.

This concept of space is rather well illustrated by Fig. 5.12, which actually is due to Kepler but this is not relevant since on this particular point his views were almost identical to Copernicus's. The stars ('studs') are represented by the signs of the zodiac around the rim of the 'drum'. The successive points on the rim represent successive conjunctions of Jupiter and Saturn. The lines within the rim highlight the way in which metrical geometry is used within an ultimate frame of reference provided by the stars. Copernicus had an open mind about what lay beyond them and said:[38] 'Let us therefore leave the question whether the universe is finite or infinite to be discussed by the natural philosophers.' Copernicus never forgot that the studs on the rim are what ultimately define position.

Fig. 5.12. An illustration by Kepler of successive great conjunctions of Jupiter and Saturn. Reproduced from: *Johannes Kepler Gesammelte Werke*, Vol. 7, C.H. Beck'sche Verlagsbuchhandlung, Munich (1953), p. 127.

He could not possibly afford to, since he asserted that the earth moves. And the earth frame, the solid ground on which Ptolemy stood, is the one frame that is completely useless to demonstrate mobility of the earth.

There is a curious side to Copernicus's proposal which tells us something about his concept of motion, indeed, the difficulty of anyone in forming a conception of motion. He wanted to set the earth loose from its bearings, but could not do this without simultaneously making sure it was securely fenced in. Without the fence, he cannot assert that the earth moves at all. As a result, the overall impression one gets from *De Revolutionibus* is that Copernicus inhabited a world more claustrophobic than did Ptolemy; it is almost as if he were confined in a medieval courtyard. Ptolemy stands on the earth and looks outward. Copernicus first checks the walls are there and then starts working inward. With Aristotle, he is prepared to accept that 'beyond the heavens there is . . . no body, no space, no void, absolutely nothing.'[39] He is only interested in what is inside the outermost rim and he opens the account of his cosmology with these words:[40] 'The first and highest of all is the sphere of the fixed stars, which contains itself and everything, and is therefore immovable. It is unquestionably the place of the universe, to which the motion and position of all the other heavenly bodies are compared.'

There are several other passages in *De Revolutionibus* in which he is as explicit.

It is perhaps worth making the point here that four great men working in the capacity of theoretical astronomers (i.e., actually working on the problem of describing celestial motions) made important contributions to the development of dynamics: Ptolemy, Copernicus, Kepler, and Poincaré. Without exception, when forced to ask themselves the question 'What is motion?' all instinctively referred it to observable matter. In contrast, none of the three giants of dynamics – Galileo, Newton, and Einstein – worked actively on astronomical problems in the way the other four men did.* Interestingly, all three – even including Einstein – instinctively thought of motion as taking place relative to space. To these three Maxwell, hardly less of a giant, can also be added. We shall return to this point in Vol. 2.

This is perhaps a good point at which to review the concept of the frame of reference as it developed up to Copernicus's time. As mentioned in Chap. 1, primitive man must instinctively have referred all motion to the surface of the earth. In the first rush of speculative enthusiasm, the early atomists dissolved the earth and plunged enthusiastically into the limitless void. But, in reality, their void was just the framework of the earth made invisible; they even left in Up and Down. Aristotle called a halt to this heady enthusiasm and not only insisted on epistemological decorum but actually went further, endowing properly defined place (i.e., place defined by visible matter) with dynamical power of a sort: the ability to attract the elements to their proper places.

This incipient dynamism was not developed by the Hellenistic astronomers. Instead, they concentrated on the more mundane matter of accurately charting and interpreting the motions of the celestial bodies. The early astronomers worked effectively with two equivalent frames of reference: the earth and the fixed stars, the one being related to the other by a simple and straightforward uniform rotation. However, there then came a crisis of sorts, which may in fact have helped to tip the balance in favour of a geostationary view of things. This was the discovery of the precession of the equinoxes sometime after 135 BC by Hipparchus.[41]

Let us briefly recall the modern (i.e., post-Newtonian) account of the precession of the equinoxes. As we have seen in Chap. 3, the orbit of the earth defines a plane in helioastral space which cuts the celestial sphere in the ecliptic. The plane of the ecliptic varies only very slowly and within a narrow range and can be assumed to be fixed for the purpose of this discussion. The rotation axis of the diurnal motion of the earth is inclined to the axis of the poles of the ecliptic at an angle of about $23\frac{1}{2}°$ (the obliquity of the ecliptic). Were the direction of the rotation axis fixed in space, the

* This statement requires clarification in the case of Newton, who did do such work. However, it was done in earnest only after he had put together all the elements of his dynamics and was done more with a view to confirming details of his theory of universal gravitation.

four distinguished positions in the earth's orbit separated by exactly 90° and corresponding to the vernal and autumnal equinoxes and the winter and summer solstices would correspond to fixed positions of the sun in the ecliptic, i.e., the vernal equinox would always occur when the sun is at exactly the same position in the ecliptic. However, because of the oblateness of the earth's shape and the tilt of its axis, the sun's gravity exerts a torque on the spinning earth, which causes the axis of rotation to precess uniformly, describing a circle on the celestial sphere about the pole of the ecliptic. The radius of this circle is about $23\frac{1}{2}$°, i.e., equal to the obliquity of the ecliptic, and the circle is described uniformly in about 26 000 years. This means that the equinoctial positions of the sun precess around the ecliptic, moving round the ecliptic in the opposite direction to the annual motion of the sun. As a result, the so-called *tropical year*, defined by the equinoxes and solstices, is about twenty minutes shorter than the *sidereal year*, defined by the period required for the sun to return to the same ecliptic position.

This phenomenon, discovered by Hipparchus by comparing observations over a period of many years, caused something of a crisis and made Ptolemy think very hard about what might be called the 'ultimate frame of reference'. The point is that the tropical year is defined using the earth frame; if you set up a gnomon at a fixed point on the earth, the tropical year corresponds to the interval between the successive times at which the sun's shadow at noon has its shortest length. In contrast, the sidereal year is defined in the geoastral frame by the passage of the sun along the ecliptic against the backcloth of the stars and back again to its starting point.

The problem facing Ptolemy was this. How is the length of the year defined? And, more fundamentally, is there any unambiguously defined frame of reference for describing motion? It seems that Ptolemy was swayed in the end by practical considerations. He opted for the tropical year:[42]

We must define the length of the year as the time the sun takes to travel from some fixed point on this circle back again to the same point. The only points which we can consider proper starting-points for the sun's revolution are those defined by the equinoxes and solstices on that circle. For if we consider the subject from a mathematical viewpoint, we will find no more appropriate way to define a 'revolution' than that which returns the sun to the same relative position, both in place and in time, whether one relates it to the [local] horizon, to the meridian, or to the length of the day and night; and the only starting-points on the ecliptic which we can find are those which happen to be defined by the equinoxes and solstices. And if, instead, we consider what is appropriate from a physical point of view, we will not find anything which could more reasonably be considered a 'revolution' than that which returns the sun to a similar atmospheric condition and the same season; and the only starting-point one could find [for this revolution] are those which are the principal means of marking off the seasons from one another (i.e. solsticial and equinoctial points].

Ptolemy was clearly an astronomer with his feet firmly on the ground. He accordingly measured everything relative to the earth and assumed that in addition to the diurnal rotation all the stars (and, with them, the entire orbits of the planets (though not the sun)) are subject to the additional slow motion of precession. It is interesting to note that Ptolemy anticipated the problem at the heart of the discussion of Mach's Principle that became inescapable once the idea gained ground that the stars are all in a state of relative motion. How can you define an ultimate frame of reference if all the bodies in the universe are in motion relative to one another? For he said:[43] 'One might add that it seems unnatural to define the sun's revolution by its return to [one of] the fixed stars. . . . For, this being the case, it would be equally appropriate to say that the length of the solar year is the time it takes the sun to go from one conjunction with Saturn, let us say (or any other of the planets) to the next. In this way many different 'years' could be generated.'

As there are several thousand stars visible in the sky and only the five planets, sun, and moon exhibited any observable relative motions, this objection appears something of a pedantic quibble on Ptolemy's part. For all that, Ptolemy had a valid point. Once the realization dawns that *all* the stars are in ceaseless relative motion among themselves, the heavens lose all their reassurance as the ultimate frame of reference. We come face to face with the central question: what is motion?

It is particularly interesting in this connection that both Hipparchus and Ptolemy harboured a certain distrust towards the reliability of the stars. In fact, Hipparchus at one stage believed that the stars near the ecliptic could be very slowly moving planets since he found that their longitudes increased very slowly with the passage of time; however, he later inclined,[44] though only tentatively in view of his sparse evidence, to the (correct) view that the effect was common to all stars and was simply the precession just described. Because Ptolemy had at his disposal the accurate observations of Hipparchus made about three centuries earlier, comparison with his own observations revealed clearly the true nature of the precession, and he was able to vindicate completely the revised conjecture of Hipparchus.[45] Hipparchus is justly famous for the first systematic compilation of a star catalogue (Ptolemy reports[46] that very little work of such kind was done before Hipparchus), and Ptolemy went to considerable trouble in an attempt to establish, on the basis of this catalogue, whether the so-called fixed stars exhibit *proper motions*, i.e., move relative to each other on the heavens. In particular, he compiled a list of alignments of three stars on a single line. He compared the positions in his time with those in Hipparchus's and found no visible changes had occurred in the intervening three hundred years. But[47] 'in order to provide those who come after us with a means of comparison over a longer interval' he carefully reproduced Hipparchus's list of alignments and added some more of his own. Eventually his labour paid off. In 1718, Halley[48] concluded that both Arcturus and Sirius had moved southwards since the time of Ptolemy by about 1°.

By Copernicus's time the situation with regard to the length of the year and the precession of the equinoxes had become very complicated. Inaccurate observations, including some by Ptolemy himself (who believed the precession of the equinoxes had a period of 36 000 years), had created the impression that the precession of the equinoxes was nonuniform. Copernicus firmly believed this, being too prepared, as we noted, to trust the accuracy of his predecessors' observations. To account for these phenomena, various astronomers had proposed the existence of additional spheres whose motions were intended to explain these slow variations, some of which were entirely nonexistent. Copernicus firmly rejected these proposals and attributed all these effects to motions of the earth. This was much easier to conceive than the addition of one sphere on another in the great vault of the heavens.

Writing nearly two centuries before Halley's discovery and about a century before Descartes first sowed serious doubt about the stability of the astral frame (Chap. 8), Copernicus was quite sure that the fixed stars represented the only appropriate frame of reference and said that:[49] 'we must not heed Ptolemy in this regard'.

5.7. Copernicus's concept of motion

In one very important respect the Copernican revolution is to be seen as a triumphant culmination of ancient theoretical astronomy rather than its destruction and replacement by a quite new order. It will be recalled from Chap. 3 that although Hellenistic astronomy lacked a notion of physical causality it was nevertheless permeated by a kind of causality – the idea that the two-dimensional motions observed on the sky are to be explained by laws of motion operating in three-dimensional space. So far as one can judge, it was this aspect of Greek theoretical astronomy that so significantly distinguished it from the contemporary or slightly earlier Babylonian astronomy. Seen in this light, Copernicus's insight is to be regarded as the crowning triumph of the programme to 'save the appearances' that began with Eudoxus and saw such significant advances in the hands of Apollonius, Hipparchus, and Ptolemy. For the explicatory power of motion to explain observed phenomena is the very essence of the Copernican revolution. In terms of showing how much could be achieved by the assumption of certain motions in space, its importance was at least equal to that of the discovery of the epicycle–deferent scheme, to which, of course, it bears the closest connection.

Thus, although Copernicus would probably have disagreed strongly with Osiander's argument that the idea of the earth's mobility should not be taken literally, he would have had much less objection to the basic principle of theoretical astronomy as enunciated by Osiander,[22] namely to adopt 'whatever suppositions enable the motions to be computed correctly *from the principles of geometry* for the future as well as for the past' (my italics). Let us now look in a little more detail at the manner in which Copernicus conceived motion and, above all, at the role he allotted to the sun.

We may begin by remarking that although Copernicus's thinking was extremely clear and radical when it needed to be, his concepts of motion (and its causes) looked back more to Aristotle and Plato than did Ptolemy's; he strikes one as actually more cramped in his thinking than Ptolemy. For example, geometry plays a more prominent part in its own right in *De Revolutionibus* than in Ptolemy's *Almagest*, imposing patterns of thought, rather as it does in Plato. For Ptolemy, geometry was a tool of the trade; for Copernicus, it was much more the mystical Platonic key to the essence of nature. Plato is mentioned many times in *De Revolutionibus* but not once in the *Almagest*.

Copernicus's somewhat mystical attitude to geometry is, no doubt, part of the explanation for his insistence that all observed celestial motions must be explained by a superposition of exactly circular and exactly uniform motions. Thus, Copernicus would still have disagreed with the last quotation from Osiander because of what he would have seen as the excessive licence implied in the 'whatever suppositions'. This aspect of Copernicus's thinking can also in part be explained as an historical accident – he was led to make one of the greatest discoveries in science through his insistence on the principle of perfectly circular and uniform motions in the first place. It is probably asking too much of human nature to expect Copernicus to have seen that his idea for replacing the equant had in fact nothing to do with his really great insight.

Much more relevant for the central theme of this book was an equally potent factor in the explanation for Copernicus's continued adherence to the archaic geometrokineticism of perfect uniform circular motion, namely that, as pointed out in the previous section, he did not suspect a relative motion of the stars among themselves. For Copernicus, the illusion of the seeming fixity of the stars spread, as it were, a sheet of ice over the abyss and tended to make him content with a geometrokinetic concept of motion. For on ice you can skate and trace the most exquisite patterns. Copernicus looked down from his newly-won high vantage point and had no reason to suspect that God had anything more in mind than the tracing of beautiful circles in the ice. Not such a bad idea either when you stop to think of it, especially in the man who, first among men, saw clearly for the first time the grandeur of the arena designed by God to stage the spectacle.

Thus, although Copernicus's own work was a spectacular demonstration of the power of the principle of kinematic relativity, he himself took only the first step in the direction of the thoroughgoing relativity of motion implicit in Aristarchus's introduction of trigonometry into the study of celestial motions. Only with Mach is the principle taken to its logical conclusion – that position and motion are not merely revealed and defined by the bodies in the universe (the insight for which we have to thank above all Copernicus) but are actually determined by them. Before that had to come the awareness of the acute problem of the definition let alone determination of motion in a situation in which there are no bodies truly at rest in the universe. But this insight is still a century ahead of Copernicus, who had no reason to doubt the existence of the sphere of truly fixed stars. We have seen already how it assumed an almost mystical significance in his mind. It also had a decisive effect on the way in which he thought about motion and helps to explain the curious fact that heliocentricity was an incidental rather than central feature of Copernicus's scheme.

It is significant that Copernicus begins the account of his cosmology – in the passage quoted earlier on p. 248 – with the description of the sphere of the fixed stars. This is, in fact, all he needs. Although invisible, space is perfectly well defined within the rim that the stars pick out.

In this invisible space Copernicus (or God) can describe any circle he pleases. For, by virtue of the stars around the edge, it is well defined and has a real place. Both the circle and its centre are truly somewhere. Such a circle can then become the locus of either a moving planet, which is presumed to move uniformly around the circle, or a void guide point, itself the centre of an epicycle carried around the deferent and upon which the planet rides, again moving uniformly around the epicycle. Everything is reduced to uniform motion in circles and this motion is defined with respect to the ultimate frame of reference. The existence of the fixed stars, believed to be at rest relative to each other, is what enables Copernicus to feel completely happy with a purely geometrokinetic concept of motion. He changed the frame of reference but not the concept of motion.

Thus, Copernicus believed in the reality of motion but pointed out that it will appear very differently depending on the point of view from which it is observed. In some cases it may even seem to disappear. Copernicus went only half way to a fully relative concept of motion; he still used a safety net. The motion still exists even if we do not observe it. It is very hard to accept that there is no ground at all under our feet. It is another irony that the post-Copernican defenders of Aristotelian cosmology in the late sixteenth and early seventeenth centuries in fact pushed the principle of optical relativity to its extreme; for just as Copernicus invoked the principle of relativity to show that the earth could move, even if it seemed to be at rest, they argued that the same principle implied equally well that the earth could be at rest and the remainder of the universe in motion. They took refuge in the impartiality of relativity. What they failed to appreciate was that they were espousing a doctrine with ultimately much more disconcerting implications than the scheme Copernicus actually proposed. In his cosmology, you did at least know where you were; he was much closer in spirit to Aristotle and the reassuring idea that wine belongs in a bottle, or the candle in a lantern.

Which brings us back (in stages) to the sun. For Copernicus, the problem of the motion of the planets (with the earth now included among them) was to describe it (not explain it) and demonstrate that it consisted of a superposition of uniform circular motions. All other considerations were subordinated to the last requirement; if that were met, you could rest satisfied with a job well done. As he said proudly at the end of the *Commentariolus*:[2] 'Then Mercury runs on seven circles in all; Venus on five; the earth on three, and round it the moon on four; finally Mars, Jupiter, and Saturn on five each. Altogether, therefore thirty-four circles suffice to explain the entire structure of the universe and the entire ballet of the planets.'

If he has any concept of motion that goes beyond geometrokineticism or ballet, it is *mechanical*. In his preface to the pope he says:[50] 'I began to be annoyed that the movements of the world machine, created for our sake by the best and most systematic Artisan of all, were not understood with greater certainty by the philosophers, who otherwise examined so precisely the most insignificant trifles of this world.'

Although he says remarkably little explicitly on the subject, it appears that Copernicus did believe the planets were carried on spheres of a world machine. This question, a key factor in which is the sense in which Copernicus used the Latin word *orbis* (it could, apparently, have meant several things, including both sphere and circle), has in fact been the subject of considerable controversy,[51] in which Rosen and Swerdlow took opposite sides. The debate arose from a suggestion by Swerdlow in his commentary on the *Commentariolus*[2] that Copernicus probably discovered the Tychonic system (p. 225) at the same time as his own Copernican system. Why, Swerdlow wondered, did Copernicus choose his own system in preference to the Tychonic one, which avoids all the dynamical problems of terrestrial mobility, to say nothing of the theological problems? Swerdlow concluded, tentatively in Ref. 8 (and with more conviction in Ref. 9 after the exchange with Rosen), that Copernicus was strongly swayed by purely mechanical considerations to do with his acceptance of the theory that the planets are carried by material spheres. For in the Tychonic system Mars would have to pass at some points in its motion through the sphere of the sun, and Swerdlow believed that Copernicus must have found this an insuperable difficulty, therefore opting for the intellectually much more daring heliocentric system with mobile earth.

On this particular point, I do not find Swerdlow persuasive. It is true that in Book 1 of *De Revolutionibus* (Chap. 10) Copernicus seems to be on the point of advancing the Tychonic system as an explicit possibility; however, he immediately draws attention to the advantages of his own system in terms that, to my mind at least, do not lend support to Swerdlow's suggestion. I am much more inclined to take Copernicus's writings at their face value and assume that he was primarily swayed by the harmony of his own system as expressed in his qualitative discovery of Kepler's Third Law. In the Tychonic system the neat ordering of the earth's orbital period between those of Venus and Mars is lost and some of the objectionable heterogeneity of the Ptolemaic system reintroduced. Also, in common with many commentators, I can see no reason why Copernicus should not have been quite strongly swayed by Neoplatonic sympathies to see the centre of the planetary system as an ideal location for the sun (see below). Swerdlow has little sympathy with such considerations.

Nevertheless, it does seem to me to be a very difficult question to decide precisely what either Copernicus or Ptolemy thought about the question of physical mechanisms of the planets' motions (cf. p. 142 and p. 187fn). At the least, Osiander's preface to *De Revolutionibus*, which we quoted earlier and appears to represent views quite widely held at that time, seems to suggest that specific hypotheses about the celestial motions (and, presumably, the mechanisms which were supposed to bring them about) were the subject of considerable sceptical doubt in Copernicus's age. Swerdlow[51] argues that the mechanical aspect does not loom large in either the *Almagest* or *De Revolutionibus* because these two books are primarily concerned with the mathematical description of the planetary motions but that this does not imply any lack of support for mechanical mechanisms. My own impression is that in both books the mathematical description is paramount, i.e., both authors concentrate first of all on finding what they feel is a satisfactory mathematical description ('satisfactory' for Ptolemy means simple, a point he emphasizes several times in the *Almagest*; for Copernicus it means consonant with the principle of uniform circular motions); after the mathematics has been sorted out, one may then put one's mind (if so inclined) to the elaboration of a mechanical model that reproduces the mathematics. This certainly seems to be what Ptolemy did in his *Planetary Hypotheses* (p. 187fn).

As this insistence on mathematical description by Hipparchus, Ptolemy, and Copernicus was the decisive factor from the point of view of the discovery of dynamics, the attitudes of Ptolemy and Copernicus to the mechanical spheres seem to me of much lesser significance than their commitment to careful observation coupled with rigorous mathematical description. It was this attitude, and it alone, that finally made dynamics possible. It was, moreover, an attitude found only very rarely before the late seventeenth century. Nevertheless, Swerdlow[51] is certainly quite right to emphasize that the definitive 'destruction' of the solid spheres by Brahe's observations of a comet in 1577, which, he argued, showed that it must have passed clean through successive planets' spheres, was a very great stimulus to both Brahe and above all Kepler to rethink the whole question of what made the planets move.

Whether or not Copernicus did believe firmly in the reality of the mechanical spheres, the Copernican system does in fact remind one of the threshing machines which figured so prominently in cereal farming until the combines displaced them about two generations ago. Wheels of different sizes all over the place, with the drive being transmitted from one part of the machine to another. As we have seen, the correlations are bizarre, especially in the case of the inner planets, which have circles turning at twice the rate of the earth's annual motion. And what controls it all?

The modern reader, coming to *De Revolutionibus* with so many preconceptions, automatically expects everything to revolve around the sun as linchpin. But already in the preface to the pope there are indications which, if recognized, indicate that this is not so. One such indication has already been mentioned; in the following passage he is more explicit (my italics):[52]

> In the first book I set forth the entire distribution of the spheres together with the motions which I attribute to the earth, so that this book contains, as it were, the general structure of the universe. Then in the remaining books I correlate the motions of the other planets and of all the spheres *with the movement of the earth* so that I may thereby determine to what extent the motions and appearances of the other planets and spheres can be saved if they are correlated with the earth's motions.

Most revealing of all – and quite disorientating for the modern reader – is the fact that in the diagrams in which Copernicus explains the motions of the planets *the sun is not shown at all*, and it is not at all easy to work out where it is. Instead, everything is centred on the centre of the *orbis magnus*.

You look in vain for the engine which drives the Copernican threshing machine. It is not the sun. The planets twirl at God's behest in the framework of the space that the stars pick out. Old Saturn dances slowest not so much because he is furthest from the sun but because he is the nearest to the stars, an echo of the ancient Aristotelian (and, as we saw in the previous chapter, medieval) idea that perfect rest is the state which most closely approaches the divine ideal. The planets dance for our delectation. The sun does not cause and govern their motion; it is rather a lantern hung in the middle to illuminate the dance and all the mighty temple that houses it:[53]

At rest, however, in the middle of everything is the sun. For in this most beautiful temple, who would place this lamp in another or better position than that from which it can light up the whole thing at the same time? For, the sun is not inappropriately called by some people the lantern of the universe, its mind by others, and its ruler by still others. [Hermes] the Thrice Greatest labels it a visible god, and Sophocles' Electra, the all-seeing.

In this fervour of Neoplatonic sun worship, a rare outburst in a text otherwise as sober and restrained as the *Almagest*, Copernicus allows himself one sentence which suggests he might have had in mind an even greater role for the sun: 'Thus indeed, as though seated on a royal throne, the sun governs the family of planets revolving around it.' But despite Kepler's characteristically generous comment[54] that Copernicus 'when he speculated' shares his own belief that the sun actually controlled the motions of the planets, Copernicus's book belies the suggestion. It demonstrates mobility of the earth, not dominion of the sun.

5.8. The significance of the Copernican revolution: second evaluation

Several commentators, including Kuhn and Koestler,[55] tend to emphasize the unrevolutionary and even archaic aspects of Copernicus's system – that although he eliminated one set of epicycles he introduced a new set and finished up with a system that overall was as complicated as Ptolemy's and, moreover, one that was no more accurate than his predecessor's. It seems to me that this emphasis fails to distinguish sufficiently clearly the two parts of the planetary problem – the theories of the first and second inequality. The Copernican revolution was about the second inequality; his conservatism related to his treatment of the first inequality, in which he made no positive contribution at all.

Looked at in one way, what Copernicus did was very little, a pinprick only. But he did it extremely well. On the question of the earth's motion at least, *De Revolutionibus* gets the central message across with clarity. Moreover, the preface to Pope Paul III is extremely effective, not to say artful, in building up the expectation of a momentous announcement.

On the really big claim that Copernicus made, he did deliver the goods. He complained to the pope that his predecessors could not:[56] 'elicit or deduce from the eccentrics the principal consideration, that is, the *structure of the* universe and the true symmetry of its parts' (my italics). If you had asked Ptolemy what the universe looks like from Saturn, he would have been quite unable to give you any answer based on a sound theory. He did not know how far away Saturn is; he had no way to overcome the scale invariance of his solution to the problem of planetary motions; his trigonometry failed him. By his extension of trigonometry, Copernicus put in man's hand a tool that, in his mind's eye, enabled him *to travel in space*. He could deduce what the world looks like from unvisited points at vast distances from the earth. Historically, Copernicus performed his task. He recounted his insight to his contemporaries in lucid language and, for the experts, provided the solid mathematical arguments which substantiated his vision. For the first time in history, man had a tolerably accurate picture of the solar system and at least a comprehension of the immensity of the universe. Moreover, unlike the early Greek hints of heliocentricity, Copernicus's picture was based on solid observation and theory (even if little of this was his own apart from the shifting of the linchpin).

The defects of Copernicus's system were entirely confined to secondary effects and were caused by his almost quirky refusal to take the equant at its face value and his excessive trust in the observations of others. This led him to reintroduce a remarkable profusion of epicycles, etc. There is nevertheless a very good criterion which shows what a major success Copernicus achieved. We have already mentioned the far greater coherence of the Copernican system, at least as regards the gross features. Attention should also be drawn, as it was by Kepler,[57] to the fact that the paths of the planets in the geoastral space of the Ptolemaic system were never-ending spirals which wandered all over the place. Figure 5.13 shows the orbit of Mars according to Brahe's observations from 1580 to 1596 as drawn by Kepler under the assumption that the earth is at rest, i.e., it is the orbit of Mars in geoastral space (note the signs of the zodiac around the rim).[58] Kepler describes the orbit as *in figura panis quadragesimalis* – in the figure of a lenten bun, or *Fastenbrezel* for those familiar with the German language and bakery! The intricate pattern is produced by the superposition of the Ptolemaic epicycles on top of the deferent motions. The eternal tangle of the resulting spirals is generated by the noncommensurability of the two motions (the period of the earth's revolution bears no simple relationship to those of the various planets). In contrast, the largest of the Copernican epicycles all had periods that were simple multiples of the dominant motion: this meant that *the Copernican orbits were essentially fixed in helioastral space*. The innumerable corrections that Copernicus felt obliged to introduce only had the effect of shifting the planets about their mean orbits by what were very small amounts. In themselves these deviations were not particularly remarkable. It was only the *mechanisms* producing them that seemed so daft. The actual orbits they

DE MOTIB. STELLÆ MARTIS

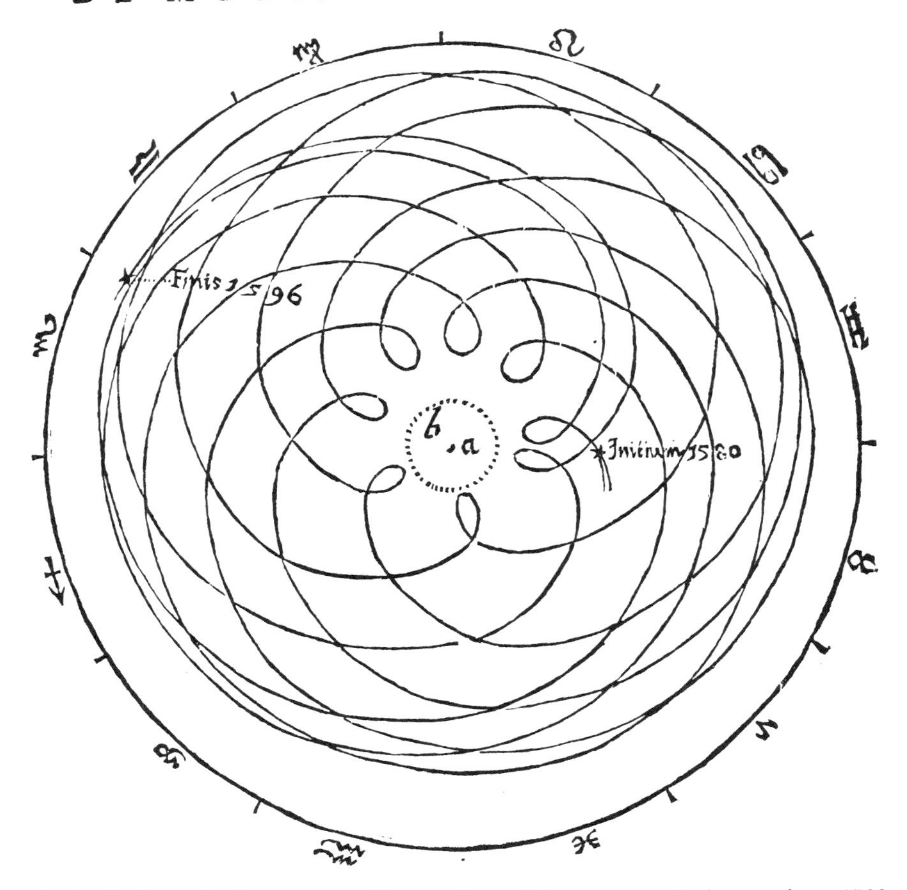

Fig. 5.13. Kepler's illustration of the motion of Mars in geoastral space from 1580 to 1596. (Reproduced from the 1609 edition of *Astronomia Nova* by courtesy of The Beinicke Rare Book and Manuscript Library, Yale University.)

produced were quite accurate and deviated only slightly from the actual fixed positions. This was an immense gain in clarity.

In a way then, it is irrelevant to find fault with the remainder of Copernicus's work and to criticize his theory of the first inequality. Copernicus happened to publish all his work in one book, and this tended to obscure the issue. It must also be said that Copernicus was partly to blame. He certainly does imply[59] that his real revolution, the magnitude of which was not in doubt in his mind (though he could not have had the remotest notion of all its consequences), had a significant bearing on his theory of the first inequality and would help to achieve the elimination of the unwanted equant. But, as we have said, it had not the slightest effect on the problem.

One of the particular fascinations of the Copernican revolution, which was indeed a cataclysmic event, is that it was based exclusively on a fully developed and already existing theory. It was not preceded by any major observational discovery on Copernicus's part. It was pure reinterpretation of something long known. This is undoubtedly what gave it much of its dramatic impact. The history of physics shows us only one other comparable event – Einstein's creation in 1905 of the special theory of relativity. Like Copernicus's proposal, it too consisted of pure reinterpretation – of Maxwell's electrodynamics in Einstein's case. And the consequence of one deft move was just as disconcerting – the overthrow of a concept of the world held for millennia as an instinctive article of faith. Appropriately enough, both revolutions were intimately related to relativity – in both its kinematic and Galilean aspects. Planck[60] was quite correct to say that Einstein would come to be seen as the Copernicus of the twentieth century on account of his 1905 paper. The comparison is most apt.

An equally fascinating aspect of the Copernican revolution, with which we shall be concerned in the following chapters, is the consequences it had for the theory of motion, both celestial and terrestrial. In the case of the celestial motions it led to what may be called the second or Keplerian stage of the Copernican revolution, which will be the subject of the next chapter. In the case of terrestrial motions it destroyed the basis of Aristotelian theory, according to which all natural terrestrial motions take place towards or away from the centre of the earth, which is assumed to coincide with the centre of the universe. It was here that the diurnal *rotation* of the earth, which in *De Revolutionibus* had the status of almost an incidental remark compared with the astronomical significance of the annual motion around the sun, revealed its full implications for dynamics. In the chapters on Galileo and Huygens we shall see how the merest hints, forced upon Copernicus by his momentous discovery and in fact anticipated by Oresme, led to follow-on revolutions in the science of motion every bit as dramatic and surprising as the original revolution.

If the work of Galileo and Kepler can be seen as second stages in the Copernican revolution, it still only partly undid the Aristotelian division between the heavens and the earth. In their lifetime, Copernicus's proposal still remained only a flimsy arch between two distinct parts of the universe. The final elimination of the gulf between the two parts represents the third stage of the Copernican revolution. At the philosophical level this readjustment of the overall conceptual picture was to a large degree the work of Descartes; at the level of a viable dynamical theory of motion it was the work of Newton.

As regards all these consequences, Copernicus clearly had much less idea of the significance of his revolution. This should not occasion any surprise, since clear and essentially correct ideas about geometry crystallized in ancient times whereas the corresponding process did not occur for motion until well over a century after Copernicus's death. The Copernican revolution was in fact the single most important event in bringing about this crystallization of concepts of motion.

As a preparation for the following chapter, it will be worth reviewing in conclusion precisely why it was that Copernicus created a theory of the mobility of the earth that was simultaneously *heliostatic* rather than *heliocentric*, the latter being used here in the sense that the sun is accorded a dominant role in actually controlling the motions of the planets rather than merely being a static illuminator of their motions.

The origin of the oddity of the Copernican revolution has already been hinted at earlier in this chapter. It has to do with the relative magnitude of two effects, both produced by the sun: the circularity of the planetary orbits and the small-eccentricity corrections to zero-eccentricity behaviour. The first is by far the larger and more striking phenomenon and was actually all that was needed for the Copernican revolution. In fact, it was a purely historical accident that Ptolemy discovered the equant and Copernicus heliocentricity. Logically, judged by the magnitude of the corresponding effects, the discoveries should have happened the other way round. In Chap. 3 it was suggested that the remarkable *successes* of the Ptolemaic system were probably the main reason why Ptolemy did not hit on heliocentricity. For the successes were very great. Ptolemy departed the scene basking in a well-earned triumph. There seemed to be very little left to do.

Another reason why Ptolemy may not have discovered heliocentricity is that the impact and interpretation of the zero-eccentricity behaviour of the planetary orbits is actually obscured by the small-eccentricity effects, which cause the centres to be displaced from the common point at the centre of the sun. Instead, they are scattered around the sun in a seemingly random way, some at considerable distances (Saturn's is over half way from the sun to the earth's orbit). If the centres alone are considered, it appears that the sun just happens to be placed in the same region as the centres of the planetary orbits. This is what Copernicus was forced to conclude. There is no apparent causal connection between the position of the sun and the various centres. This indeed is in line with the modern understanding of the situation. The centres of the planetary orbits are determined by the initial conditions, not by the law of gravity.

The true significance of the sun is, however, revealed by the less striking manifestation of the sun's dominion – the equant phenomenon. The centres of the orbits are not the only points associated with the planetary motions. The planets move nonuniformly in their orbits but, to first order, uniformly about their equants. These points are just as unambiguously determined by the observations as the centres of the orbits. (It is a highly nontrivial property that such points exist.) That the planets recognize the sun as lord is seen in the fact that, for each planet, the line joining the equant and the centre of the orbit passes through the centre of the sun. Moreover, the equant's eccentricity is exactly twice the centre's. So subtle – indeed subliminal – is the dance. Sol is worshipped with sophistication worthy of a choreographer who achieves his effects in ways not directly apparent to the eye. Not for him the mere mechanical geometrokineticism of uniform circles described on ice or in a courtyard.

Thus the irony was this. In the equant phenomenon, Ptolemy actually discovered the lesser effect of the sun's dominion but hidden in geocentric guise. Because it disturbed the old geometrokinetic harmonies, Copernicus tried to eradicate this harbinger of a quite new order of things. In the process he accidentally stumbled on the much more obvious effect of the sun's dominion – the circular orbits around the sun as approximate centre. However, he was completely misled by the accident of the specific eccentricity values. He never could abide the equant phenomenon and therefore had no particular wish to tie it into his new found heliocentricity. Ptolemy's theories, which no one had ever bettered, seemed to suggest that the system was correlated on the centre of the *orbis magnus*. Why should he demur? As long as the geometrical relationships fitted, Copernicus had no desire to better Ptolemy.

As Kepler remarked:[23] 'Copernicus merely set himself the task of reproducing Ptolemy.' Therefore, it did not occur to him that the equant phenomenon was a direct consequence of the presence of the sun and that one should accordingly look for a specifically heliocentric explanation of it. Instead he looked to cogwheels and cam shafts.

Thus the old obsession with the equant blinkered his vision. The amazed Copernicus, a student in Padua a century before Galileo came there as a professor, opened the door of night into a medieval courtyard and found the courtiers dancing, Mother Earth among them. Dazzled by the spectacle and instinctively applying the pedestrian rules of the outmoded discipline of his student days, he failed to catch the new rhythms and unwittingly demoted the lord of the dance to a mere lantern. What he made with one hand he all but unmade with the other – but it would have been superhuman to have done better. He had performed his task, prompted by the right effect but for the wrong reason. It led him to open the door into a new world; he saw enough to ensure that the door would never be closed again. Aristarchus had probably only peeped through the keyhole.

6

Kepler: the dominion of the sun

6.1 Brahe and Kepler

Flamboyant, haughty, and renowned for the gold and silver nose that he had fitted to replace the real one he lost in a duel, the Danish nobleman Tycho Brahe (1546–1601) is the most colourful personality that appears in our story.[1] He witnessed the solar eclipse whose prediction so stirred his imagination (p. 44) that he took up astronomy in Copenhagen in 1560. Equally decisive was his observation of a conjunction of Saturn and Jupiter in 1563, which occurred on a date differing from the prediction of the *Alfonsine Tables* by about a month and still by a few days from that of the much newer *Prutenic Tables*, which used Copernicus's models.

In all the centuries since Ptolemy there must surely have been astronomers who had realized that the Ptolemaic models and parameters were not quite perfect. Yet it seems that not until Brahe was an astronomer with talent stimulated to carry out a really extensive programme of observations and comprehensive testing of the models of the celestial motions. Brahe threw himself at the task with immense energy. As a young man he travelled widely and established a network of contacts with many other astronomers in Europe. Rather appropriately, his reputation was established by a new star which appeared in the constellation Cassiopeia in November 1572. This was the first of two supernovae (the second was in 1604) which occurred fortuitously at the dawn of the scientific revolution and did much to cast doubt on the ancient Aristotelian doctrine of the unchanging and incorruptible heavens.

Brahe's observations of the new star, made with great accuracy, established that it exhibited no parallax – its position with respect to the neighbouring stars in Cassiopeia changed neither diurnally nor annually. Thus it was situated at a distance much greater than the moon and, almost certainly, the sun and planets. Brahe published an account of his observations in 1573, which, together with those of others, initiated a profound rethinking of cosmological ideas. Another very important step in the same direction were Brahe's observations of a comet in 1577. The absence of diurnal parallax showed clearly that the comet too must be much further away than the moon (thereby disposing of the ancient idea that comets were sublunar 'exhalations'). Brahe concluded that it was at the distance of the planets and, most important of all, that it must have passed through the putative spheres that were meant to carry the planets.

Brahe was greatly influenced by Copernicus, whose lead he followed in many respects, but he could not accept the Copernican cosmology, which he rejected (and replaced by his own Tychonic system (p. 225)) partly on the authority of scripture, even more on his disbelief that the earth could rotate without our being forcibly made aware of the fact, but above all on the sheer immensity of the cosmos that it implied. His own observations, made with unprecedented accuracy, failed to reveal the stellar parallax that one would have expected to be associated with motion of the earth in the *orbis magnus*. The huge distances to the stars that this implied proved especially difficult to accept on account of a curious fact of which the modern reader will be quite unaware: before Galileo made his telescopic observations, astronomers firmly believed that the stars and planets exhibited perceptible circular disks like the sun and moon. The brighter stars and planets were assumed to have diameters of about two or three minutes of arc.

In the Copernican scheme this implied for the stars an incredible diameter – greater than the diameter of the *orbis magnus*. The resulting misconceptions strongly influenced cosmological ideas (including Kepler's) until Galileo demonstrated that they were based on an optical illusion.[2] As a half-way house to full-blown Copernicanism, the Tychonic system played an important role in the history of ideas, easing in acceptance of the more difficult doctrine at a time of especial confusion and great diversity of views in astronomy. (It was not the only modification of the Copernican system proposed in the second half of the sixteenth century; in addition, some philosophers were still struggling with the problem of reconciling Ptolemaic epicycles and eccentrics with the strict concentricism of Aristotelian cosmology.[3])

Brahe's best observational work was done during the years 1576 to 1597 on Hven, the island in the sound north of Copenhagen between Denmark and modern southern Sweden. In a remarkable example of royal sponsorship of science, King Frederick II of Denmark enabled Brahe to build on Hven his observatory, Uraniborg (heavenly castle), and equip it with huge instruments. Over the years Brahe trained several assistants, who stayed with him for extended periods. He developed a passion for accuracy (both in observations and their mathematical representation) and went to great lengths to eliminate errors, not always with complete success, especially in the case of refraction, which he did attempt to take into account but not sufficiently. Brahe believed he achieved accuracies of better than a minute of arc. Kepler estimated the accuracy of the planetary observations more soberly as probably not better than two minutes of arc,[4] still a phenomenal achievement of naked-eye astronomy when we recall that the moon's apparent diameter subtends only about 30′. Brahe made very substantial contributions to the theory of the moon's motion and discovered several important effects. He took great pains to determine star positions with accuracy, which he used to determine planetary and lunar positions, freeing himself thereby from the ancient overreliance on position measurements relative to the moon, whose position was in turn related to the sun's by means of the solar and lunar theories. Aiming at a complete revision of the whole of astronomy, he was aware of the need to make observations over very long periods of time and at all positions of the orbits of the planets and moon. This comprehensiveness of Brahe's programme was to prove to be almost as important as the accuracy of the observations. Brahe was a formidable calculator and prolific if somewhat conservative developer of theory. Through his very extensive correspondence he did a great deal to revive astronomy throughout Europe.

As the years passed, and especially after the death of his royal sponsor in 1588, Tycho came into progressively greater difficulties on account of his arrogant manner and arbitrary treatment of his tenants on Hven. The young King Christian IV did not share the old king's enthusiasm for astronomy, and Brahe eventually felt obliged to quit Hven in 1597. The huge observing instruments were never again used properly. After an unsettled period, Brahe found a new royal patron in the person of Rudolph II, the Holy Roman Emperor, who appointed him Imperial Astronomer and invited him to Prague, where he arrived in June 1599. Benatky Castle, 22 miles northeast of Prague, was put at his disposal. In fact, Brahe achieved little of significance in the two years of life that in the event were all that remained to him, except that is for the happy stroke which led him to invite Kepler to join him as an assistant, an event in which Kepler at least saw the hand of divine providence at work.

Johannes Kepler (1571–1630) was born at the little town of Weil der Stadt in southern Germany. His family was impoverished, having seen better times, but the obviously intelligent boy was educated at the expense of the Duke of Württemberg. In 1589 he went to study at the University of Tübingen, where he came under the influence of Michael Mästlin (1550–1631), a more than competent astronomer, who, as Kepler reported,[5] enchanted him with his account of Copernican astronomy. However, at that stage astronomy was only one of Kepler's many interests:[6] 'As soon as I had reached the age in which I could sample the sweet delights of philosophy, I grasped it as a whole with great eagerness without being especially concerned with astronomy.'

In 1594, he took up an appointment as teacher of mathematics and astronomy at Graz in Austrian Styria (Steiermark). Here he had time for thought, and he fell to pondering cosmological questions. Above all, he sought the causes of three things – the particular number of the planets (six when the earth is included), the diameters of their orbits, and the speeds of their motions. One sees here the profound influence of the Copernican revolution, for these were questions that could hardly be asked in ancient astronomy, which had signally failed to provide a reliable ordering of the planets let alone a theory connecting the distances and speeds of the different planets. Dissatisfaction with this failure had clearly been one of Copernicus's main justifications for seeking a better alternative. Copernicus's success thus enabled the young Kepler to ask questions that had never hitherto been posed with any real insistence. It is clear from numerous passages in Kepler's writings that he was also deeply affected by the Pythagorean and Platonic overtones of Book I of *De Revolutionibus*, especially Copernicus's exalted praise of the sun and the identification of the centre of the universe as a position worthy of the great luminary. Just as important was Copernicus's qualitative sensing of Kepler's Third Law and the harmony it established between the positions and speeds of the planets. Kepler was fired by the suggestive words that Copernicus had used to close the account of his cosmology:[7] 'In this arrangement, therefore, we discover a marvellous symmetry of the universe, and an established harmonious linkage between the motion of the spheres and their size, such as can be found in no other way.' To this day the Third Law is occasionally referred to as the *harmonic law*.

To most modern minds, Kepler began his career in astronomy with an approach that could barely be regarded as scientific. He was totally convinced that God had created a supremely beautiful and sparkling world with the express purpose of delighting the mind and senses of man. Nevertheless, this attitude of mind did prove to be scientifically valuable as developed over the years by Kepler, since he was equally convinced that the beauty and sparkle were the outward expression of a deep and hidden harmony that God had put into the construction of this precious ornament, the Pythagorean cosmos. For Kepler, the basis of harmony was to be sought in mathematical relationships – in numbers and in geometry. Moreover he believed equally passionately that God in his goodness had left clues in his handiwork that would enable enquiring man to see through to the deeper underlying principles on which the masterwork was based. This was the optimistic outlook that gave Kepler the confidence to ask those questions never hitherto asked – and to expect to find the answers.

Throughout his life he sought for archetypal (and architectonic) and harmonic relationships, but alongside this decidedly Platonic and

Pythagorean bent, for which the most natural modes of expression were numerical patterns and the synthetic geometry of the ancients, he developed surprisingly modern concepts. Thus, in Koestler's phrase, he was truly a watershed between the ancient and the modern.[8] Already in his first work published in 1596 and called significantly and characteristically the *Mysterium Cosmographicum* (*The Cosmic Secret*), the more purely physical ideas make their first tentative appearance.

To give the reader an idea of the fertility of Kepler's imagination and the enthusiasm with which he put it to work, it will be worth retelling how the *Mysterium* came into existence. He had sought long and unsuccessfully for numerical relationships that might hold the key to the distances between the planets and was close to despairing but then, on the summer's day of 9 July 1595 (old style), his first great 'inspiration' occurred. He was showing pupils how the great conjunctions of Jupiter and Saturn are always separated by eight signs of the zodiac and pass gradually from one triangle to another. He was drawing part of the figure shown in Fig. 6.1 (Fig. 5.12 deals with the same question) when he noted that the triangles inscribed in the large outer circle circumscribed the smaller inner circle and that the ratio of the radii of the two circles corresponded roughly to the ratio of the radii of the orbits of Jupiter and Saturn. In a brainwave that was to transform his life, he concluded that it was not numerical relationships but the successive circumscribing and inscribing of circles to geometrical figures which provided the key to the successive planetary distances. Putting a square between Jupiter and Mars seemed to lead to a reasonable fit for those two planets, and so did a pentagon between Mars and the earth. But the problem with this was that polygons with successively increasing numbers of sides form an infinite sequence but the planets were just six in number. He could find no reason 'why there were 6 rather than 20 or 100 planets'. Finally he hit upon his 'Platonic inspiration', the idea that the five perfectly regular Platonic solids (rather than plane figures) are inscribed and circumscribed between the spheres of the planets. At a stroke this solved his greatest problem – for it could be proved by mathematics that there existed just five perfect solids, so these could fit between six, and only six, planetary spheres. He had found the 'secret of the cosmos'.

In a state of the highest excitement he wrote down almost immediately the arrangement to which he proudly adhered to the end of his days (never suspecting, it seems, that it was the purest *fata morgana*):[5] 'The [sphere of the] earth is the measure for all the other spheres. Let it be circumscribed by a dodecahedron; the sphere that circumscribes this is Mars's; let the sphere of Mars be circumscribed by a tetrahedron; the sphere that circumscribes this is Jupiter's. Let the sphere of Jupiter be circumscribed by a cube; the sphere which circumscribes this is Saturn's. Now place an icosahedron within the earth's sphere; the sphere which is

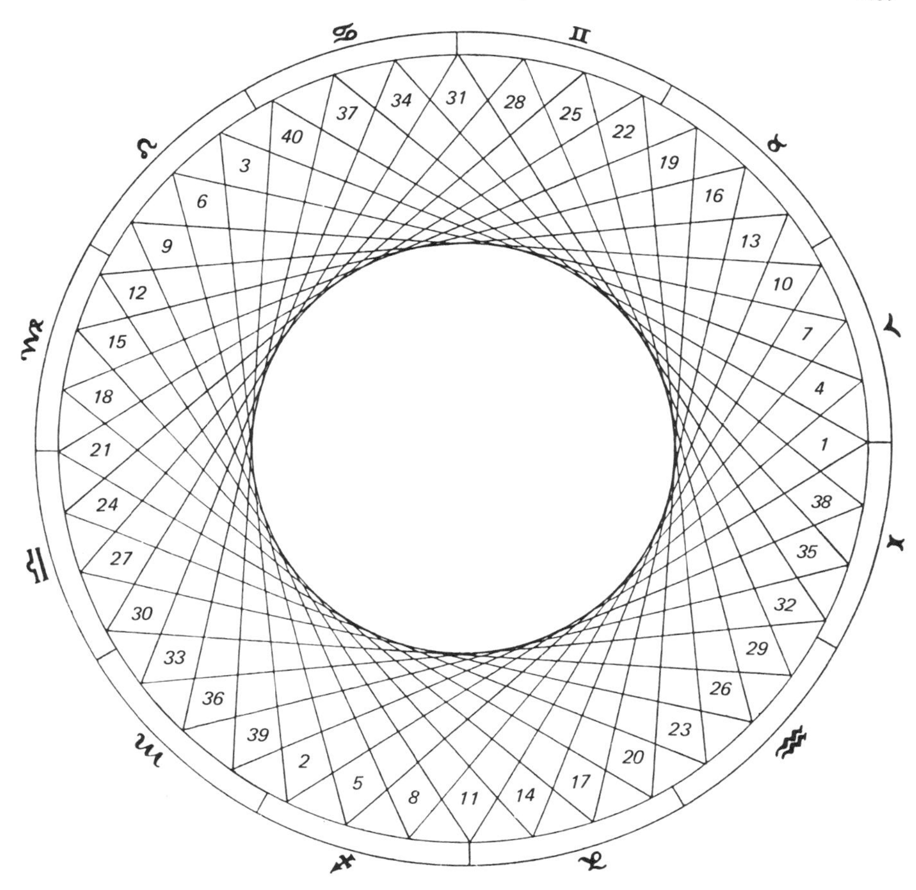

Fig. 6.1. Reproduced from *Johannes Kepler Gesammelte Werke*, C.H. Beck'sche Verlagsbuchhandlung, Munich, Vol. 1, p. 12.

inscribed to it is Venus's. In Venus's sphere place an octahedron; the sphere inscribed to it is Mercury's. You have there the reason for the number of planets.'

It is a moot point whether or not the reasonably good fit between this prescription and the actual distances is to be seen as a fortunate circumstance in Kepler's life. It sustained him through years of extraordinarily laborious astronomical calculations that bore the richest fruit but it seriously misled him in the evaluation of the importance of the various strands of his own work, causing him perhaps to underemphasize his genuine discoveries. This may be one of the reasons why his work received such tardy recognition and even now does not get its due.

Characteristic is Laplace's comment:[9] 'It is depressing for the human mind to have to see how this great man dwells with delight on his chimerical speculations and regards them as the soul and life of astronomy.' A modern physics Nobel Laureate recently wrote[10] that he found Kepler 'incomprehensible in motivation and approach' and remarked that 'in modern times, Galileo is readable, while Kepler is not'. It is perfectly true that Kepler never abandoned the mystical, speculative, and *a priori* side of his scientific personality. His comparatively late publication *Harmonice Mundi* (1619), which sets out to show that the world is constructed on harmonic principles (*harmonice* is to be translated as the 'science of harmonics' rather than simply 'harmony'), is replete with the early ideas. Though he was finally forced to admit that the observations did not quite match his Platonic solution, he attributed the discrepancies to the fact that, in laying down the overall design of the cosmos, God wished to realize musical harmonies in the eccentricities of the orbits and the rhythms of the planetary motions (as expressed in his Third Law), and these could not be completely reconciled with the Platonic proportions. This was the explanation he provided in notes that he added to his *Mysterium Cosmographicum* when he republished it in 1621.[11]

This account will help to balance a somewhat false impression which the reader might gain from the almost exclusive concentration in this chapter on Kepler's greatest work, the *Astronomia Nova* (1609), which contains the account of how, using Brahe's wonderful treasury of observations, he found his first two laws of planetary motion by studying the motion of Mars. For, whatever may be the case with his other works, the *Astronomia Nova* is in modern times extremely readable, breathtaking even and deeply absorbing. Platonic mysticism is almost completely absent from the work; instead it reveals Kepler's complete mastery of theoretical astronomy, which he enriched with the physical ideas that I have already mentioned and which we shall consider in some detail later in the chapter. It was Kepler's misfortune that his greatest work was the one that was almost the least read. Indeed, even today (at the time of writing) a translation into English of the *Astronomia Nova* is still not available, though it is at last imminent.[12] Even the very best of Kepler has not been readable in modern times except to those with good Latin or German![13]

But we anticipate. Kepler wrote up his idea in the *Mysterium Cosmographicum*, sent it to Mästlin for his opinion, was delighted with the response, and proudly published it in 1596, though not before being persuaded by the sympathetic rector of the University of Tübingen to cut out his discussion of the compatibility of scripture with the Copernican doctrine, as this might arouse such passion that it could adversely affect the book's reception.[14] (In the *Astronomia Nova*,[15] Kepler did discuss this question with great tact. He never seems to have offended churchmen in the way that Galileo did – he would probably have secured a far wider readership for the *Astronomia Nova* if he had had it banned!) However, since Copernicanism provided the entire basis of his Platonic inspiration, he could not avoid opening the book with a marshalling of the best arguments for the Copernican cosmology. Most historians of science are agreed that the most persuasive arguments for Copernicanism are to be found in Kepler's writings. Those in the first chapter of the *Mysterium* (and elsewhere) are particularly eloquent, though largely limited to what might be called kinematic and aesthetic arguments, i.e., the elimination of unnecessary epicyles and the appropriateness of placing the sun at the centre of the universe. The bulk of the book is given over to explaining his own idea and attempting to show that it was compatible with the observations, though at the end there are three highly important chapters in which Kepler considers not just the spatial arrangement of the planetary orbits but also the planetary motions.

Perhaps the most valuable consequence of the publication of the *Mysterium Cosmographicum* was that it clearly exposed Kepler's need for more accurate data and eventually brought about his collaboration with Brahe. Keen to establish a reputation, Kepler sent copies of the book to several eminent scientists, including Galileo (which led to a first brief correspondence) and Brahe, whom Kepler, perhaps not without a certain worldly cynicism, had mentioned a few times in the book in flattering terms. Brahe's copy took longer to reach its destination (he was already underway from Hven), but eventually Tycho responded quite warmly, making tantalizing reference to some of his observations. This awoke in Kepler a burning desire to have access to the observations, so that he could test the 'messy' side of his theory introduced by the nonvanishing eccentricities and inclinations of the planetary orbits, which ruled out any simple application of the Platonic solids. A more or less regular correspondence developed, and eventually Kepler, whose position at the predominantly Catholic Graz was becoming untenable because of his refusal to abandon his Protestant faith, accepted an invitation to visit Brahe at Prague.

To cut short a story already told a great number of times,[1] Kepler joined Brahe in February 1600 and worked together with him for an initial spell of about four months. A vivid picture of the difficulties of their collaboration can be obtained from Caspar's biography of Kepler[16] and also Jardine's book quoted in Ref. 3. In fact Kepler found working under the turbulent Brahe so difficult that the collaboration almost came to an end when Kepler went back to Graz to wind up his affairs. Luckily he returned in October 1600, and, building on highly important results obtained during his first stay, had made considerable progress by the time that Brahe died unexpectedly the following October 1601. Two days later Kepler was appointed Imperial Mathematician, a post previously held by

Brahe's arch opponent Ursus,[3] and was given the responsibility for analyzing Brahe's observations.

He had initially come to Prague in the hope of obtaining from Brahe the precise data on planetary distances that he believed would crown the speculations of the *Mysterium Cosmographicum*. But they did not exist because Brahe did not yet understand the *motions* of the planets and had failed in the analysis of his data. Kepler's interest was therefore deflected into the study of the planetary motions, a subject that in Chap. 22 of the *Mysterium* he had, ironically, said was of far less interest to him than the question of the distances. The new problem, to which he was able to apply most fruitfully his forward-looking physical ideas, soon almost totally displaced his original concern with the distances, where the more backward-looking architectonic principles found application. He took up the challenge of Brahe's astronomy with enthusiasm and great confidence. He made a rash bet that the answers to the problems of the Martian motions, to which he was most fortunately assigned, would be found within eight days, but in the event the bet became an intense labour spread over more than five years (at the end of which he had found his first two laws), and, indeed, most of the rest of his life, during which time he extended his initial results obtained for Mars and the earth to all the planets, compiling tables for the prediction of their positions.

One of the most remarkable aspects of Kepler's work on Brahe's observations was his refusal to accept a model of planetary motion unless it matched the observations *perfectly and comprehensively*. Although Kepler does say in Chap. 12 of his *Mysterium* that every philosophical speculation 'must have as its point of departure the experience of the senses', it is still surprising that the avid speculator submitted himself to such rigorous empirical control, which was crucial for the final success. It will be argued later that an important factor here was Kepler's religious beliefs; however, the part played by Brahe must also be considered. He set up the whole programme (thereby imposing his discipline and aims on Kepler), had an acute awareness of how much could be learnt from observations, and set the high level of accuracy to which he expected Kepler, like his other assistants, to work. Characteristic is the comment that he made to Mästlin after reading the *Mysterium*. While recognizing Kepler's obvious talent, he questioned whether the reform of astronomy should be 'accomplished *a priori* through the dimensions of these regular bodies rather than *a posteriori* from accurate observations'.[17] This judgement was obviously sound.

Brahe was one of the most important protagonists in the discovery of dynamics, but in a book on conceptual matters it is not possible to give him space commensurate to his contribution. It is therefore fitting to end this introductory section by pointing out that Brahe's share in the discovery of the laws of planetary motions was more than just the making of the observations. The *a posteriori* side was there from the beginning in Kepler but it was Brahe and the logic of the programme on which he put Kepler to work that brought it into the perfect balance (unique among great scientists) with the *a priori* that made possible the discoveries to which we now turn.

6.2. The dethronement of the usurper

In describing the way Kepler found his laws through the study of the motion of Mars, we shall follow basically his own classic account given in the *Astronomia Nova*. On the first encounter, this book, which is written for technical *cognoscenti* with few concessions to the layman, makes a rich but rather rambling impression. Kepler takes a positive delight in leading the reader through all his own false starts and mistakes. He says that he will adopt the practice of the great explorers in writing about their journeys and entertain the reader with all the dangers and adventures through which he too passed in his encounter with Mars. This may however be something of a smokescreen to disguise a very careful structuring of the book. Stephenson[18] argues that it was written and rewritten 'to persuade a very select audience of trained astronomers that all the planetary theory they knew was wrong, and that Kepler's new theory was right. The whole of the *Astronomia Nova* is one sustained argument'. Attentive reading of the work provides ample evidence to support Stephenson's contention.

As an example we mention the way in which Kepler treats many problems *in triplicate*, showing simultaneously how they appear in the Ptolemaic, Tychonic, and Copernican world systems. The reason he gives for this is a request from Brahe on his deathbed that because he (Kepler) was a follower of Copernicus he should nevertheless not omit to demonstrate all the proofs of hypotheses for the Tychonic system as well. However, in reporting this, Kepler immediately takes the opportunity to point out that, viewed in purely geometrical terms, the three forms are completely equivalent. He thereby prepares the ground for one of the great climaxes of the book – his identification of what he sees as the only true reasons for preferring the Copernican system, which are exclusively physical or dynamical and entirely new. In meeting his self-imposed obligation to treat all three systems, Kepler seldom omits to point out the severe physical difficulties that the two rivals to Copernicus face. Thus, the Tychonic system gets its deserved dues but is simultaneously rather effectively demolished.

What distinguishes the *Astronomia Nova* most clearly from the *Almagest* and *De Revolutionibus* is its intimate blending of physical reasoning with purely astronomical argument based on massive and brilliant application of trigonometry to Brahe's observations. There is a foretaste of the physics in Chap. 20 of the *Mysterium* in which Kepler, who was influenced at that time strongly by J. C. Scaliger (1484–1558) and took the existence of moving souls and spirits very seriously, had suggested that the planets nearer the sun moved faster in space than the more distant ones because they were all moved by a simple soul which resided in the sun and whose influence and ability to move the planets weakened with the increasing distance to the more distant planets. Kepler's thinking was shifted strongly from the notion of moving souls in the direction of purely physical forces by the publication in 1600 of William Gilbert's book on magnetism that was mentioned in Chap. 1 (p. 40). He often referred to force generically as *magnetic force*. The process of replacement of moving souls by physical forces takes place before our eyes in the *Astronomia Nova*, which is the most important of texts for anyone interested in the emergence of the force concept. Its full title is loaded with significance, a point that has often been made. In the original Latin (and a bit of Greek) it reads:

ASTRONOMIA NOVA
ΑΙΤΙΟΛΟΓΗΤΟΣ
seu
PHYSICA COELESTIS
tradita commentariis
DE MOTIBUS STELLAE MARTIS
Ex observationibus G.V.
Tychonis Brahe.

Thus, the title proclaims nothing less than a *new astronomy* based on *causal explanation* or *celestial physics* treating the motions of the planet Mars deduced from the observations of Brahe. Even the subtitle, *Plurium annorum pertinaci studio elaborata Pragae . . . Joanne Keplero* ('elaborated by pertinacious study over several years at Prague by Johannes Kepler'), says a great deal about the author. Indeed, Kepler attacked the problem of the planets with a highly developed intuition but also great vigour and pertinacity. He leaves the impression of someone setting about the opening of an oyster with a can opener – by brute force if necessary but preferably by application at the right point, which he did succeed in finding.

What were the factors that led Kepler to adopt such a novel approach to astronomy? Why did he part company in such dramatic fashion from Copernicus and even Brahe, who were content to accept the celestial motions as they are provided only they could represent them by combinations of uniform circular motions? There were two main reasons: the incomplete Copernican heliocentricity and Brahe's demolition of the crystal spheres. The two went hand in hand and were greatly strengthened by the input of physical ideas, from Gilbert above all. First, Kepler was quite unable to accept the curious diminution of solar pre-eminence which the need to transcribe Ptolemy had imposed on Copernicus. Kepler just could not accept that the sun was not exactly at the centre of the world. Throughout the *Mysterium* he instinctively assumed that the sun, not the mean sun, must be at the centre of his system of Platonic solids. When he turned to consider the planetary motions, he automatically applied the same approach. If the transcription of Ptolemy suggested the sun was not the centre of the planetary motions, this was a hint that the Ptolemaic models contained unsuspected defects. We shall shortly see the far reaching consequences of this attitude of mind. However, merely putting the sun at the true centre was not for Kepler a sufficient solution to the problem of the planets. This brings us to the second point; for in demolishing the crystal spheres, Brahe simultaneously removed the props by means of which the late Middle Ages and early Renaissance had supposed the planets to be moved. The number of times that Kepler mentions this difficulty is the clearest evidence of how acutely he felt it. It was necessary to rethink completely the reasons why the planets moved at all. This was where Gilbert and his physical forces came in. If the planets were not moved by crystal spheres, they must be moved and directed by something else – why not by forces? Finally, everything was knitted together in Kepler's mind by the rather natural idea (or, at least, so it now appears) that the forces which moved the planets originated in the body of the sun. In such a case the sun would naturally be the centre of the planetary motions.

These in broad outlines were the reasons for Kepler's physical approach. What makes the *Astronomia Nova* such an absorbing book is to see how Kepler applied these ideas in practice. For although the specific mechanisms that Kepler proposed in the development of his programme were all wide of the mark, they mostly contained a sound physical element and played a most important heuristic role in suggesting to him the places at which the old Ptolemaic–Copernican system was suspect. As we have seen, Ptolemy's work had succeeded in describing all the really obvious features of the planetary motions, including the key discovery of the nonuniformity of the motion. But post-Ptolemy nothing stuck out obviously in a manner that would suggest its own solution. The dance would simply not come into perfect focus, as the frustrated Brahe and his assistants were finding. To do the *Astronomia Nova* full justice one ought, as Stephenson does,[18] to follow closely the interlocking arguments from physics and astronomy by which Kepler finally arrived at his laws.* Taking advantage of the fact that this has been done, I find it more convenient in the framework of this book to give a synoptic account of Keplerian physics and dynamics later in the chapter (since they only took final shape in the early 1620s after the discovery of the Third Law) and in the meantime merely outline as much of the physics as is needed to understand the thrust of the astronomical investigations.

So let us now take up the story at the point when Kepler first came to Prague, bringing with him the strong conviction that the sun was the true centre of the planetary system and the source of an as yet unspecified virtue or force that moved the planets around in their orbits with a speed that decreased with increasing distance from the sun. As soon as he found that Brahe, just like Ptolemy and Copernicus, was still referring everything to the mean sun, Kepler suggested that the first inequality, corresponding to the planet's own motion around the sun, should be separated from the second inequality, the reflection of the earth's motion around the sun (in the Copernican system) or the sun's around the earth (in the Tychonic system), by means of acronychal observations (p. 163) made when the planet is in opposition, not to the mean sun, but to the *true sun*. For at such times the earth lies on the line joining the sun to Mars; from the earth we have the unique privilege of stepping, as it were, into the sun's shoes. We see Mars as it appears against the backcloth of the stars from the only standpoint which could have true physical significance, that of the sun. We obtain directly its angular coordinates in the helioastral frame. (This statement is not actually entirely true, since Mars's orbit is inclined to the ecliptic, so that the latitude of Mars as seen from the earth when in opposition is still not the same as that seen from the sun except at the oppositions at which Mars happens to cross the ecliptic. Thus, the first inequality is separated from the second only in longitude, not latitude. We shall see shortly the brilliant use which Kepler made of the latitude observations.)

To give substance to Kepler's general approach, let us now consider the powerful and characteristic arguments that he advanced for taking the true sun rather than the mean sun as the point of reference of the system of planets. He put himself in the position of a planet trying to find its way around the sun. Suppose first that the planet is guided by a soul. Just imagine the problem it faces. The interplanetary space through which it moves is void of all markings. Somewhere in this featureless space is supposed to be the all important point that acts as the centre of the world, the point at which the various apsidal lines of the planetary orbits

* I should like to express my especial thanks to Dr Stephenson for letting me read the text of his book in advance of publication. It has influenced my own presentation quite strongly in several places, particularly in Sec. 6.5.

converge. Somehow or other the guiding soul must be able to identify not only this void point but also the centre of its own orbit and the equant that governs the speed of its motion (the line through these two points passes through the centre of the world). But the really odd thing is that the void centre of the world is said by Copernicus to lie just four solar diameters from the centre of the true sun, which is a huge and very luminous body. It stretches the imagination, says Kepler, to suppose that a planetary soul would refer the planet's motion to such a ghostly and invisible point rather than use the true sun. Even odder was the prescription that Tycho expected the planetary soul to follow in the Tychonic system. For, according to Tycho's transcription of the Copernican system, the Copernican centre of the world, the point of convergence of the planets' apsidal lines, is transformed into a point that moves with the sun as it travels around the earth, remaining always just four solar diameters from its centre but situated sometimes next to, sometimes below, and sometimes above the sun.

If one gives up the idea of guiding souls and assumes instead that the planets are moved by a force, one must then ask: whence comes this force? Since all the planetary orbits are clearly lined up on the centre of the world, the source of the force cannot be situated anywhere other than in that centre. But now we encounter the utterly implausible oddity that a void point, marked by absolutely nothing visible, is a source from which an immensely powerful force flows while barely four solar diameters removed from it sits the idle sun, a mere spectator and a dispenser of nothing but light. A feature of the Copernican system centred on the mean sun to which Kepler especially objected was the fact that the planet in its orbit would be moving fastest when closest to the mean sun and not when closest to the true sun. This was one of his most important insights, and was announced already in Chap. 22 of his *Mysterium* and followed on naturally from his explanation, already mentioned, of why the planets further from the sun moved slower than those nearer to it. For if the sun was the cause of the planets' motion, it was not unreasonable to assume that any given planet would move slower when further from the sun and faster when closer. This opened up the possibility of a *physical* explanation of the mysterious Ptolemaic equant and the associated nonuniformity of the actual motion to which Copernicus (and, unbeknown to Kepler, the Marāgha School) had taken such exception. Moreover, the physical explanation directly predicted, in agreement with observation, that the equant must lie on the line through the sun and the centre of the eccentric, on the farther side from the sun. Moreover, if the speed of the planet was inversely proportional to its distance from the sun, a very simple argument showed that the centre of the Ptolemaic eccentric must be exactly half way from the equant to the sun, just as Ptolemy had found (in geocentric guise). However, this beautifully simple explanation of the equant, which was the single most important idea that Kepler injected into theoretical astronomy, relied on alignment of all the apsidal lines to the true sun, not the mean sun.

Such ideas are so familiar and instinctive to the modern mind that we need to be reminded by the example of Brahe, religiously following Copernicus despite being fully aware of the revolutionary nature of much of his work, and by several great scientists who lived at the same time or after Kepler (above all Galileo, Descartes, and Huygens) but refused to follow him in ascribing to the sun an ability to move the planets that these ideas were far from obvious before Kepler advanced them.

This is also an appropriate place at which to draw attention to the way in which Copernicus's proposal put an order into the planetary motions and thereby enabled Kepler's powerful intuition to develop some very clear and precise notions. As was emphasized in Chap. 5, Copernicus eliminated a great confusion of complicated and tangled motions when he made the sun static. For Ptolemy the mean sun was a moving direction, and his system was full of moving void points. For Copernicus the mean sun became a fixed point, albeit void, onto which the imagination could lock itself. Many of the other moving points were eliminated, and they all would have been had Copernicus made a slightly better job of his revolution. Kepler instinctively assumed that, apart from the very slow secular perturbations, the paths of the planets must be completely fixed and immobile in helioastral space. This gave him a relatively simple and orderly picture in his mind, and he could readily see how details of the picture could be tested by observations carefully chosen to yield the optimal amount of information. It also immediately revealed the oddity of taking the void mean sun, a point so near (astronomically speaking) the true sun, as the centre of the world. Copernicus and Brahe were probably too close to the revolution to acquire a genuine feel for its implications (which were nearly as far reaching in the Tychonic system as in the Copernican).

Kepler also had time to digest the implications of Brahe's abolishing of the solid spheres. For that taught him to appreciate the complete irrelevance, in the post-Brahian circumstances, of all the arguments about alternative but equivalent representations of the same phenomena by different geometrical devices. Thus, in the famous example of Hipparchus's two alternative representations of the solar theory, the model with epicycle and concentric differed from the eccentric only in the geometrical means used to represent one and the same path in space. In Kepler's view, the true task, belonging to the deeper science of 'contemplative astronomy', was that of finding the genuine paths; their mere representation on paper by circles or other lines belonged to 'the inferior tribunal of geometers'.[19] Although the word was not introduced until much later, Kepler was effectively using the concept of an orbit.

Thus, Copernicus and Brahe supplied Kepler with a totally new conceptual image of the world. Although for quite a time Kepler continued to use many of the mathematical methods of Ptolemy and, when he looked at the heavens, saw exactly the same spectacle, the same bare facts – Mars and the other planets tracking their way through the sky against the background of the stars – the world as he conceived it was quite transformed. Ptolemy thought he was watching a spoke rotating in the heavens, the one end fixed to an invisible circulating guide point while the other end carried the planet at its tip. But as Kepler looked at the same objective phenomena in order to make the acronychal observations at opposition he imagined himself a voyager in space and was always looking over his shoulder, watching the sun swing round behind his back.

Brahe's answer to Kepler's arguments for the adoption of the true sun in place of the mean sun was that he might well have a point, but that he, Brahe, and his assistants, using acronychal oppositions to the mean sun, had in fact achieved great success in separating the first inequality of Mars from the second. They had observed a total of 10 Martian oppositions in the years from 1580 to 1600 (the synodic period of Mars is about 780 days, so oppositions occur on the average every two years and sixty days); taking some of these and using the Copernican modification of Ptolemy's problem (p. 168), they had constructed an hypothesis for the Martian first inequality, i.e., its eccentric orbit around the sun together with equant point (which still existed in the Copernican system despite its nominal elimination). Testing this against the remaining oppositions, Brahe's assistants proudly claimed that the hypothesis predicted the positions of all oppositions correctly to within the phenomenal accuracy of 2 minutes of arc. How, they asked, could an hypothesis that predicted so many points with such accuracy be wrong?

Kepler responded to this challenge with his first great piece of purely geometrical work done on the planetary problem, and I will describe it in some detail to illustrate his mastery of purely technical questions as well as for the light which it casts on his predecessors' mistakes. He first of all checked out the difference between the original Ptolemaic theory with eccentric centred on a point e from the observer and equant $2e$ from the observer and the Copernican modification used by Brahe with eccentric centred on a point $3e/2$ from the observer and minor epicycle of radius $e/2$. He showed that even for the comparatively large eccentricity e of the Martian orbit (for Mars, $e \sim 1/10.8$) the maximal difference between the two theories must still be below the accuracy which Tycho could achieve, and could therefore be ignored. Kepler instinctively preferred the Ptolemaic scheme to Copernicus's on account of its greater simplicity and above all for the fact that it brought out very clearly the fact which Copernicus wanted to sweep under the carpet – the nonuniformity of the planetary motions. (Despite his respect for Copernicus, which verged on hero worship, Kepler had a sharp eye for the defects in *De Revolutionibus*.)

Having disposed of that technical detail, he turned his attention to the following basic question: to what extent can one tamper with the parameters of a Ptolemaic equant-type scheme without adversely affecting its ability to predict positions on the sky correctly? In Chap. 3 (pp. 132–3) we already noted that the acronychal positions are remarkably insensitive to displacement of the position of the centre of the eccentric along the line of the apsides provided the distance between the observer and the equant point is kept fixed at $2e$. Kepler's general examination of this question showed that in fact the ability to reproduce *acronychal longitudes* is determined almost exclusively by a correct positioning of the equant point and can tolerate quite large errors in not only the position of the centre of the eccentric but also the line of the apsides.

To see this, let us examine Kepler's own diagram (Fig. 6.2). Suppose the Ptolemaic description of a planet's motion is exactly correct and that the planet moves on the eccentric $\pi\tau\xi$, which has centre at β, line of apsides $\alpha\gamma\iota$, and equant at γ. The point of observation is α. This is the point to which the terrestrial observer is translated by making the acronychal observations of the planet. Kepler assumes that the correct point α is the centre of the sun. He then asks: what would be the effect of putting the point of observation at a point δ at some distance, not too great, from the sun, at the mean sun say? To answer his question, he joins γ, the equant, and δ, the new point of observation, and bisects $\delta\gamma$ with the point ϑ.

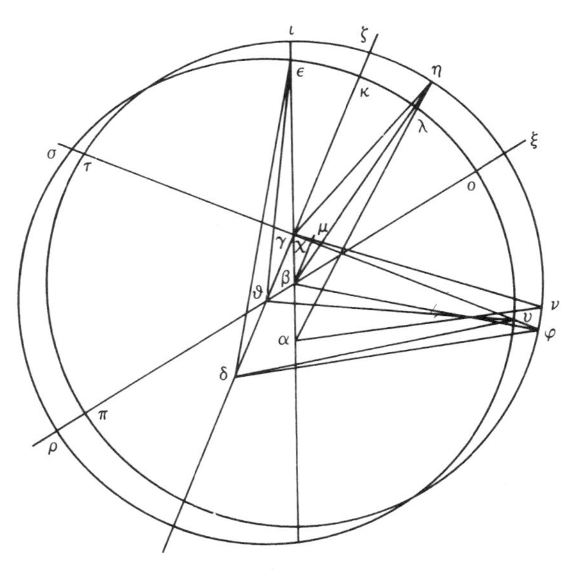

Fig. 6.2. Reproduced from *Johannes Kepler Gesammelte Werke*, C.H. Beck'sche Verlagsbuchhandlung, Munich, Vol. 3, p. 80.

About this point ϑ he describes a new eccentric $\varrho\sigma o$, keeping however the old equant γ. He has, thus, constructed a new hypothesis for which the line of the apsides, the eccentricity, and centre of the eccentric are all sensibly different from the original, correct hypothesis. Only the equant is unchanged. What will this do to the predicted positions?

Now a spoke that swings uniformly about γ, say $\gamma\tau\sigma$, will intersect the two circles at separated points τ and σ. It is immediately obvious from examination of the figure that if the spoke is either in the direction towards the apsides (either of them) or at right angles to the line joining them, roughly where the two circles coincide, the two hypotheses will yield virtually identical predictions, in the first case because the line of sight is almost directly along the spoke, so that the observer simply cannot 'see' the separation of κ and ζ, for example, and in the second case because the positions in space are almost coincident, since the quadrants are the two positions at which the circles intersect. It is obvious that an appreciable difference will only show up at positions around 45° from the line of apsides, at the four octants. By means of Euclidean geometry, Kepler showed that the greatest deviation corresponds to the spoke $\gamma\upsilon\varphi$. Substituting the actual parameters for the orbits of Mars and the earth, i.e., taking α to be the sun and δ the mean sun, so that $\alpha\delta$ has length equal to twice the eccentricity of the earth's orbit, Kepler showed that the maximal deviation would be only 5 minutes of arc, and that moreover at only one point of the orbit. It would be considerably less at almost all other points.

He then looked rather more closely at the work done by Brahe's assistants, hoping to find at least one point at which there might after all be an error greater than 2 minutes of arc and perhaps approaching 5′. In fact, he quite soon realized that their work was not nearly as good as it seemed. Simple checks against the known positions of the mean sun at the Martian oppositions, which ought to be exactly 180° opposite the acronychal positions of Mars, revealed discrepancies up to 13′. Kepler found too that the assistants had made a hash of the reduction from ecliptic longitudes of Mars to longitudes on the true Martian orbit (this tricky technical point, which concerns the treatment of the Martian latitudes, will be considered in the next section), and this must have introduced errors several times the claimed 2′. He therefore had nothing to fear from the much vaunted hypothesis of Brahe and especially his assistants. Moreover, they were forced to admit that when the hypothesis for the acronychal observations was combined with the hypothesis for the earth's motion (or, rather, the sun's motion, since they worked with the Tychonic system) it was not possible to reproduce either the acronychal latitudes or the nonacronychal longitudes. In the latter case errors of one or two *degrees* were not uncommon. Both these effects were just what Kepler had expected. Let us consider the acronychal latitudes first. In Kepler's figure (Fig. 6.2), the orbit of the earth is not in the same plane as the Martian orbit represented by one or other of the two circles according to the chosen hypothesis. Kepler had shown that precisely at opposition to Mars the errors in the *longitudes* disappear. But at opposition the earth is closest to Mars, which therefore appears to depart appreciably from the ecliptic. Since the main effect of the error introduced into the Martian hypothesis by the false placing of the position of observation must be a comparatively large misplacing of the spatial position of the Martian orbit, the resulting errors in latitude will be greatest precisely at opposition. Thus, although the acronychal longitudes are almost completely insensitive to the error, the acronychal latitudes are in principle the latitudes most sensitive to the errors (however, because the latitude effects are comparatively small at all times, the corresponding errors are numerically small). Considering now the nonacronychal longitudes, it is obvious why these can be so badly affected by an error in the hypothesis for Mars. When, taking only the acronychal observations, we shift the view point from α to δ, we still do not move any great distance from the centre of the Martian orbit. (It should be borne in mind that the eccentricities in Fig. 6.2 are appreciably magnified for the sake of clarity.) But the terrestrial observer is, in fact, a long way from the centre of the Martian orbit, so that at non-oppositional (nonacronychal) positions he can 'see' a segment such as $\kappa\zeta$ at a vastly more favourable angle – it is no longer drastically foreshortened when seen 'from the side'. Errors of orders of degrees then become apparent.

On the basis of all this preparatory work (which was already accomplished during his first visit to Brahe) Kepler knew he was closing in on the quarry. He had acquired a complete mastery of the problem and appreciated that any hypothesis must satisfy three stringent tests: it must describe simultaneously the acronychal and nonacronychal longitudes, and the acronychal latitudes, all to the nominally adopted Tychonic accuracy of 2′. Expressed in terms of the Martian orbit, not only the angles but also the *distances* must be correct. Kepler had grasped the vital truth that every hypothesis must be checked and counterchecked from every possible side, and he knew that Brahe had the observations to do it. It is true that he could not yet provide positive proof of the correctness of his alternative proposal, for other things remained to be done. But the days of the usurper, the ghostly mean sun, were numbered once its dangerous over-reliance on acronychal observations had been clearly demonstrated.

Kepler's idea of transporting the terrestrial observer to the true sun rather than the mean sun – deceptively simple for the modern mind but in reality the reflection of a profound shift – deserves to have its praises sung. For two millennia or more working astronomers had known that the observed motions of each of the five planets contain two strictly periodic components. For almost as long, certainly for nearly 2000 years, it had been known that one component was very closely related to the sun. But, floored by a mathematical perversity, not one of all these astronomers had succeeded in finding a way to separate the two components exactly. From only one point can the skin be peeled cleanly from a banana. Kepler found it.

6.3. The Zeroth Law, the vicarious hypothesis, and the demise of the old order

In Book II, Kepler recounts how he set out systematically to determine the Martian orbit in helioastral space under the assumption that the line of the apsides passes through the true sun. This book, and the following Book III, reveal the great strength of what may be called *Kepler's fundamental assumption* (Kepler did not state it as such). This was his instinctive assumption, arrived at by reflection on the residual oddities of the Copernican system and already mentioned in the previous section, that (except for very slow secular perturbations) the planets trace invariable and strictly periodic paths (I shall often call them orbits) in helioastral space in fixed planes that all pass through the true sun. This was, in fact, by far the most important assumption he made and would, applied with sufficient persistence, have sufficed to solve the problem of the planetary motions completely using nothing but trigonometry and Brahe's observations, i.e., it would have enabled Kepler to determine precisely the orbits and speeds of the planets at all points in the orbits. However, it is most unlikely that exclusive use of trigonometry and traditional astronomical techniques would have led him to discover his First and Second Laws in the precise form in which he did enunciate them and this, as we shall see in Chap. 10, was a vital factor in the discovery of dynamics.

Before he could do anything else, Kepler had to sort out the mess of the latitude theory. Brahe was well aware of the importance of treating the latitudes properly and accurately. He just did not know how to do that, and his assistants had still less idea. Kepler knew that if his fundamental assumption was correct, an observer on the sun would see the earth move round the great circle of the ecliptic while Mars would move round on another great circle inclined to the ecliptic at a fixed angle and cutting it at the two nodes (in all this discussion I ignore the very slow secular perturbations, which cause the nodes to move backwards round the ecliptic, which itself, of course, also moves relative to the stars, while the line of the apsides moves forward). At the two points 90° from the two nodes, the Martian orbit is furthest from the plane of the earth's orbit (the ecliptic); these points are called the *limits*. Neither they nor the nodes have any physical significance for the Martian orbit; they merely affect the way in which it is seen from the earth. The physically significant thing is the line of the Martian apsides, which does not coincide (except very occasionally by chance through the effect of the secular perturbations) with the line of the nodes or the line of the limits. To develop a physically meaningful theory of the Martian orbit, Kepler needed to know the longitude of Mars from the line of its apsides *in the plane of its own orbit*. This would present no problem for a solar observer but floored all the terrestrial observers before Kepler. For from the earth Mars does not move on a fixed great circle (in striking contrast to the sun) but wanders irregularly at variable distances both to the north and south of the ecliptic (cf. Fig. 5.10), tracing the characteristic retrogression loops.

The acronychal observation that was actually made at an opposition by Brahe gave the (ecliptic) latitude of Mars and the ecliptic longitude of the point on the ecliptic found by following the great circle through the poles of the ecliptic from Mars to the ecliptic. As Kepler showed, the reduction from this ecliptic longitude to a Martian longitude (i.e., in the Martian orbit as viewed from the sun) is far from trivial and must take into account the fact that a great circle through the poles of the ecliptic will not be orthogonal (viewed from the sun) to the Martian orbit except at the limits (for which it also passes through the poles of the Martian orbit). But it is not even possible to start on such a reduction until one has accurately determined the position of the line of the nodes and the inclination of the orbit. If, moreover, the plane of the orbit actually varied in the manner that Copernicus had believed, the situation was even more complicated.

Trusting in his fundamental assumption, as I have called it, Kepler first sought among Brahe's observations for occasions at which Mars crossed the ecliptic. He found Mars four times at the ascending node and noted the dates. Each occasion followed the previous occasion at an interval of 687 terrestrial days, i.e., precisely the sidereal period of the planet, a quantity well known (in geocentric guise) since antiquity. Only two observations at the descending node were on record, and Kepler found that these were separated by 3 × 687 days. He thus knew for certain that Mars passed through the ecliptic at either of the nodes at precisely regular intervals of 687 days. This was the first confirmation of his fundamental assumption. Because at each nodal observation the earth was at a different point of its orbit, Kepler did not yet know the direction from the sun to the Martian nodes. However, he already possessed (through Brahe's work) a very reasonable approximation to the solar Martian longitudes, and this established that the nodes lay diametrically opposite to each other on a line through the sun that was fixed to the accuracy he was using at the time. In particular, it did not exhibit any dependence on the position of the earth in its orbit. He had established the first part of (what I am calling) his Zeroth Law (p. 118).

Next he had to determine the angle of inclination of the Martian orbit to the ecliptic and simultaneously test for the putative Copernican wobbles.

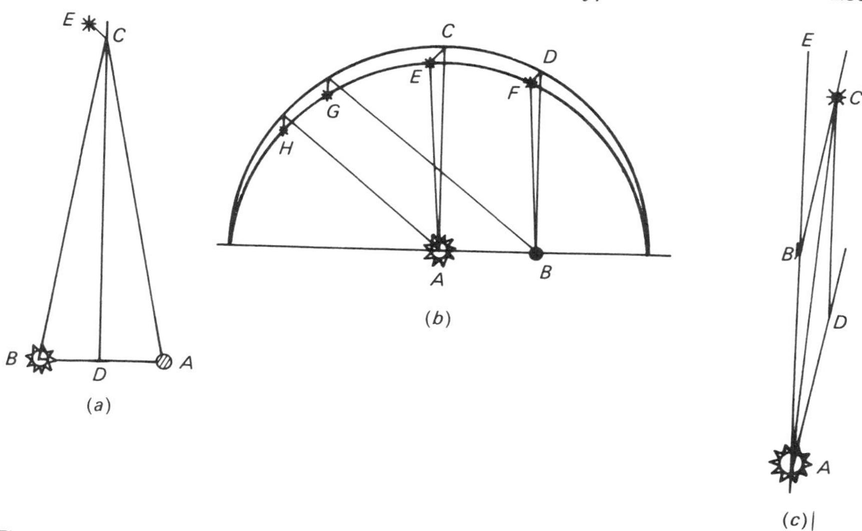

Fig. 6.3. Reproduced from *Johannes Kepler Gesammelte Werke*, C.H. Beck'sche Verlagsbuchhandlung, Munich, Vol. 3, pp. 134, 137, 140.

This was a far from easy problem, and Kepler had to choose observations extremely carefully at times at which he could be certain he was observing the true inclination directly. This brought out all his talent for judicious selection of advantageous configurations and simultaneously underlined the value of Brahe's comprehensive coverage of the orbit with observations; for Kepler needed observations at times at which earlier astronomers might never have considered making an observation. The three types of configuration he selected are shown in Figs. 6.3(*a*),(*b*),(*c*). In case (*a*), Kepler looked for occasions (again using the approximately known orbits) with Mars, E, at one of its limits and the earth, A, in a position such that the point C obtained by dropping a perpendicular from E onto the ecliptic formed an isosceles triangle with A and the sun (at B). Then EC must subtend exactly the same angle at the earth as at the sun, and the inclination would be measured directly. Another device for effectively transferring the observer to the sun! He did not find any configurations that exactly satisfied the required conditions but did have observations for cases quite near them, from which he could interpolate. In case (*b*), he considered a quite different configuration with B, the earth, on the line of the Martian nodes (passing through the sun) and Mars fortuitously placed at D with DBA a right angle; CDA is the plane of the Martian orbit, HGEF is the ecliptic and C is a limit. But in this case angle DBF (F is the perpendicular projection of D onto the ecliptic) must be equal to CAE, and that in its turn gives the inclination directly. This, in principle, is the cleanest of the methods he employed, since it required no prior knowledge of orbital dimensions but merely the position of the line of the nodes. However, due to the absence of ideal configurations he was again forced to interpolate. In the final case (*c*) he discussed, he again used reasonably well-known data and considered an opposition at which Mars, at C, was near one of its limits. From the observed angle EBC and given BA and AC he could then calculate the inclination BAC.

In all three cases, deliberately chosen in very different sun–earth–Mars configurations to test for the 'monstrous' Copernican suggestion of a Martian wobble dependent on the earth's position, Kepler found that the inclination came out very close to 1°50′. He announced his first major observational result in Chap. 14 of the *Astronomia Nova*: 'We draw from this the quite certain conclusion that the inclination of the planes of the eccentrics to the ecliptic does not change at all. (Why should we not generalize this conclusion if there is no ground for it to hold for just one planet? In fact I can also confirm it for Venus and Mercury through a proof based on observations.)' The establishment of the first of his laws, the Zeroth as we are calling it, was now complete. He had shown that Mars moved in an invariable plane in helioastral space that passed through the centre of the sun. From antiquity it was known that the earth did the same, and he had already made the obvious generalization to the other planets. It was in this chapter that Kepler made his famous remark about Copernicus being ignorant of his own riches. With his first solid triumph in the bag, he could get down to the much more arduous job of determining the orbit of Mars in the plane which he now knew passed through the sun at a fixed inclination of 1°50′ to the ecliptic.

Kepler was determined to make a success of this determination of the Martian orbit, on which he invested an immense amount of computational labour spread over a couple of years. The first thing he had to do was convert by interpolation and extrapolation Brahe's acronychal observations at opposition to the mean sun to oppositions to the true sun. Kepler described this as 'very thorny work'. To the ten oppositions used by Brahe's assistants Kepler added those of 1602 and 1604 for the account given in the *Astronomia Nova*, though the most important conclusion drawn from this work had already been reached by the spring of 1601.

Before he started he thought afresh about the Ptolemaic halving of the eccentricity, i.e., Ptolemy's placing of the centre of the eccentric exactly half way to the equant. Although, as explained earlier, he believed he had a good physical explanation for such a halving, he nevertheless thought that Ptolemy had accepted the halving on a rather inadequate observational basis. He noted too that Copernicus in his theory of Mars had not in fact held strictly to what (in his modified version) would correspond to exact halving of the eccentricity. Finally, Brahe himself had found that the acronychal observations would be better represented by assuming that the eccentricity was not exactly halved (Brahe found best agreement with his observations when the centre of the eccentric was taken closer to the equant than to the point of observation). Kepler therefore decided to leave open the question of the precise position of the centre of the eccentric and let it be determined along with the other unknowns (the distance from the sun to the Martian equant and the Martian line of the apsides) by the observations.

Compared with Ptolemy's problem (p. 168), this meant he needed to determine an extra unknown parameter. Since Ptolemy's problem required the input data of three acronychal observations it is clear that Kepler would need four. He formulated his problem using the diagram shown in Fig. 6.4, in which HI is the line of the apsides that is to be determined along with C, the equant, B, the centre of the Martian eccentric, and A, the position of the sun. The only things known in advance are the directions AF, AE, AD, and AG, which are obtained from the four acronychal observations (at opposition to the true sun *and* with latitude effects taken into account), and the angles FCE, ECD, DCG, and GCF, which must add up to 360° and be proportional to the time between the observations (modulo the Martian sidereal period); this is the equant property of C.

Kepler stated his problem thus: 'Angles FAH and FCH must be taken of such a magnitude that after their fixing we find, on the one hand, that the points F, G, D, E lie on a circle, and, on the other, the middlepoint B of this

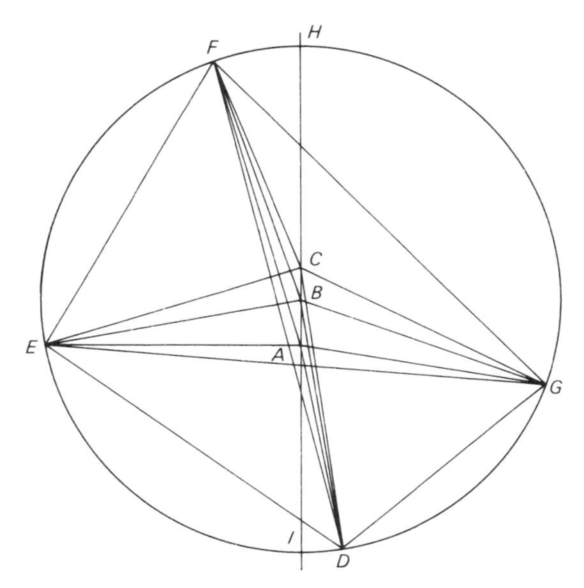

Fig. 6.4. Reproduced from *Johannes Kepler Gesammelte Werke*, C.H. Beck'sche Verlagsbuchhandlung, Munich, Vol. 3, p. 153.

circle lies between the points C and A on the line CA.' It is obvious that this problem, *Kepler's problem*, is significantly more complicated than *Ptolemy's problem*, which was already so difficult that Copernicus, shuddering no doubt at his own memory of repeating it three times (once for each of the outer planets), remarked that it involved an 'enormous mass of calculations' (*ingentem numerorum multitudinem*). Kepler remarks laconically that 'it cannot be solved geometrically to the extent that algebra is not geometrical'. Like Ptolemy he had to solve it by making an assumption known in advance to be incorrect (doubly incorrect in fact in Kepler's case) and then correct the error by successive approximation.

I shall not attempt to explain Kepler's procedure, since the discussion of Ptolemy's problem will have given the reader an idea of what is involved. Let me merely quote what Kepler said to the reader at the end of the explanation which he wrote in 1604:[20] 'If you have had enough of this wearying procedure, you will with justification feel pity for me; for I carried it out at least seventy times with great loss of time. Also you will not be surprised any longer that the fifth year already passes since I began my attack on Mars, even though the year 1603 was almost completely occupied with optical investigations.'* (Gingerich[21] has made an interesting study of Kepler's actual notebooks which throws some light on this comment.)

Kepler then proceeds to take the reader through a calculation with all its gory details. The end result that he obtained for the eccentricities, with the radius of the Martian orbit taken to be 100 000, is as follows (the first column gives Kepler's results, the second those of a recalculation by Delambre;[22] Kepler's error would have been greater but for some mutual cancellations of errors):[23]

Sun to equant (total eccentricity)	18 564	18 570	(6.1)
Sun to centre of eccentric	11 332	11 387	
Centre of eccentric to equant	7 232	7 183	

* Kepler was the most important figure in the early history of optics. His work sprang from observations of a partial eclipse of the sun which he made with a pin-hole camera in the market-place at Graz in July 1600 during his return to that city after the first visit to Prague. Soon afterwards, pondering the apparent diminution of the lunar disc in the image of the pin-hole camera, he found the principles of its operation and was led to introduce the concept of light rays. This led him on to explain how the eye functions and forms an inverted image on the retina. He was also able to provide a theory of eyeglasses, which had been in use for three centuries without an understanding of their operation (Kepler himself needed them). Later, prompted by the problems it causes in astronomy, Kepler was led to study refraction. He published his studies in 1604 in *Ad Vitellonium Paralipomena*, which had the alternative title *Astronomiae Pars Optica* (*Optical Part of Astronomy*). Following the publication in 1610 of Galileo's astronomical observations with a telescope, Kepler was able to work out the theory of the telescope within a few months and published it in his *Dioptrice* (1611).

We should quote here, for the significance it will later gain, a comment that Kepler was careful to make at this point about his monumental labour:

> Following Ptolemy I have made the following assumptions: all positions of the planet in the heavens can be arranged on the circumference of a single circle; further, the physical slowing down of the motion is greatest in the places at which the planet is furthest from the earth (Ptolemy) or the sun (according to Tycho and Copernicus), and there exists a fixed point in accordance with which this slowing down is controlled. Everything else I have proved.

After this there follows a brief digression (Chap. 17) in which Kepler looks into the question of how much the lines of the nodes and the apsides appeared to have changed since Ptolemy's time. Taking Ptolemy's values as correct and comparing them with his own determinations using Brahe's observations, he concluded that the nodes regress through about 5′17″ in 30 years and the line of the apsides moves forward by 6′29″ in the same time. Both of these are relative to the stars, not the equinoxes. The correct values are in fact about 11′ and 8′, respectively.[24] Since Brahe's observations spanned more than 20 years and were accurate to around 2′, Kepler therefore knew he had to take into account the secular perturbations even over the comparatively short period of two or three decades if he was to do Brahe's observations justice.

Finally in Chap. 18, Kepler comes to the climax of his work, the testing of all twelve acronychal positions by means of the hypothesis. The result could not have been better, and Kepler proudly announced that 'the eight other acronychal observations all fit to within 2′, an accuracy better than which one cannot expect'.

The triumph, somewhat artfully engineered in its presentation by Kepler, is short-lived. Chapter 19 opens with the words: 'Who would have thought it possible!' (*Fieri quis posse putaret?*). Kepler says bluntly that the hypothesis, calculated his way with the true sun or Brahe's way with the mean sun, is false. He established this immediately he made the additional checks of the orbit mentioned at the end of Sec. 6.2. We recall that the acronychal latitudes, in contrast to the longitudes, are sensitive to the placing of the centre of the Martian eccentric. Checking them out, Kepler found that they were incompatible with the value 11 332 (Eq. 6.1) he had found for the distance between the centre of the sun and the centre of the Martian eccentric. According to the latitudes, this distance must be in the range 8000–9943. The check against the nonacronychal longitudes suggested the same. It is true that the results were somewhat contradictory, but 11 332 for the separation between the sun and the centre of the eccentric was definitely ruled out. Moreover, Kepler noted that half of the total eccentricity, 18 654, gave the value 9282, and this lay near the middle of the range 8000–9943 indicated by the acronychal latitudes. It appeared that, after all, Ptolemy's exact halving of the total eccentricity, for which Kepler believed he had such a satisfactory physical explanation, was correct.

At this point let us recall the formulae we gave in Chap. 3 for the longitude of a planet from apogee as seen from the sun. In exact Kepler theory the longitude γ is

$$\gamma = t - 2e \sin t + \tfrac{5}{4}e^2 \sin 2t + \ldots, \tag{6.2}$$

where t is the time measured in units such that the period is 2π, i.e., t is the mean anomaly. In the Ptolemaic model with exact halving of the eccentricity

$$\gamma = t - 2e \sin t + e^2 \sin 2t + \ldots, \tag{6.3}$$

but if the centre of the eccentric deferent is taken at $5e/4$ from the point of observation and $3e/4$ from the equant, then

$$\gamma = t - 2e \sin t + \tfrac{5}{4}e^2 \sin 2t + \ldots, \tag{6.4}$$

i.e., in this modified Ptolemaic model the angles can be reproduced perfectly up to the second order in the eccentricity. Since e^2 for Mars is about 1/115, the maximal value of the term in (6.4) quadratic in e, which corresponds to $t = \pi/4, 3\pi/4, 5\pi/4$, and $7\pi/4$, i.e., at the octants, is around 37′ or 38′. Thus, whereas the term quadratic in the eccentricity was still just invisible at the level of Brahe's accuracy for the motion of the earth, i.e., for the solar theory, for the Martian motion it could be very readily observed.

Moreover, we note that if one wishes to optimize the angles and ignore the distances, which is just what Kepler was doing with his work on the acronychal longitudes, the optimal division of the total (sun–equant) eccentricity must be in the ratio $5/3 \approx 1.67$. Taking the ratio 11 332/7232 of Kepler's result from Eq. (6.1), we find that he obtained a ratio of ~1.57 (~1.59 according to Delambre's corrections), which is impressively close to the optimal value. Although Kepler did not have the benefit of Eqs. (6.1)–(6.3), he knew perfectly well what he had done – he had optimized the angles at the expense of the distances.

But he also knew something of potentially far greater importance. The whole thrust of his work in Book II, which he very deliberately emphasized in its subtitle was being done 'in imitation of the ancients', had been to reproduce the observations of Mars under the assumption that: (*a*) the Martian orbit is a perfect circle, (*b*) there exists a point somewhere on the line of the apsides from which the motion of Mars against the background of the stars appears perfectly uniform. The incredibly laborious work on the acronychal longitudes had established beyond reasonable doubt that such an orbit, if it existed at all, must have its centre at a point that divided the total sun–equant eccentricity into two appreciably unequal parts. But this, in its turn, was quite incompatible with the latitude and nonacronychal observations. If, however, one trusted the distance and latitude evidence (both of which were supported by Kepler's theoretical argument from the assumed force of the sun) and took the centre of the orbit exactly half way between the sun and the equant, then, as Kepler showed, errors of up to 8 minutes of arc appeared at the octants. To Kepler the conclusion was inescapable: one or other (or both) of the two basic assumptions, (*a*) or (*b*), must be wrong. *The kind of orbit and motion he had been seeking could not exist*. Now we appreciate why Kepler had been at such pains to state so clearly the premises on which his work on the acronychal observations was based.

This monumental discovery hinged on nothing but the difference between the term $e^2 \sin 2t$ in (6.3) and $\frac{5}{4}e^2 \sin 2t$ in (6.4) – a miserly $\frac{1}{4}e^2$ at the octants, equivalent to an angle of about $7\frac{1}{2}$ minutes of arc (8 minutes according to Kepler's calculations). (Kepler did not omit to point out that an angle of this size escaped Ptolemy because he worked to an accuracy of only 10′.)

As we shall see in the next section, Kepler had come extraordinarily close to achieving a perfect representation of the motion of all the planets by means of exact circles and Ptolemaic equants with the centre of the eccentric deferent precisely halving the total sun–equant eccentricity. That this did not happen was solely due to the fact that the eccentricity e of Mars was just big enough to reveal the minute difference. It was only by the merest whisker that Brahe and Kepler did not usher in an era in which the ancient dream of description of the celestial motions by perfect circles appeared at last to have been realized. Despite the application after 1610 of the telescope to astronomy, such an era could well have lasted for nearly a century.

But Kepler had a nose for the big discovery. Let me quote some more of his famous words:[25]

> We, whom God in his goodness has given such a careful observer in Tycho Brahe, and whose observations reveal the 8′ error of Ptolemy's calculation, should thankfully recognize this goodness of God and make use of it. That is, we should make the effort (supported by the arguments for the falsity of our assumptions) to find at last the true form of the celestial motions.

He continued:

> If I had believed that we could ignore these 8′ in longitude, I could easily have improved my hypothesis sufficiently (by halving the eccentricity). But since this error now cannot be ignored, these 8′ alone reveal the need for reformation of the whole of astronomy; they become the material for a great part of my work.

He ends Book II with the following summary: 'Thus, the house that we erected on the basis of the Tychonic observations we have now demolished with other observations of the same man.' To emphasize that the old era is drawing to a close and that astronomy faces a crisis (equivalent to the one that forced Ptolemy to introduce the equant), he says that this crisis must have come about because we made 'some plausible but actually false assumptions. This is why I have devoted such an effort to imitating the old masters.'

Kepler liked to comment on the marvellous economy of means that nature employed to bring about its effects and how it could achieve several different purposes with one and the same thing. He could with justice have applied the comment to himself as well. He used the work on the acronychal longitudes to bring down the old order. But in his later work he still retained, on a provisional basis, the orbit that he had constructed with such labour. For he knew that although it got the distances wrong, it reproduced the angles, as seen from the sun, with near perfect accuracy, and that, as we shall see, he found extremely valuable. Because he used it as a stand-in to represent the helioastral longitudes he called it the *vicarious hypothesis*.

6.4. The halving of the eccentricity of the earth's orbit

We noted in the previous section that Kepler's checking of the vicarious orbit against the acronychal latitudes and nonacronychal longitudes had certainly revealed a substantial error in the placing of the orbit but that nevertheless the data were still contradictory (quite apart from the residual 8′). Since all observation is relative, one possible explanation of apparent contradictions in the motion of Mars could be that the motion of the earth, the platform of all but the acronychal observations of Mars, did not take place exactly as had hitherto been assumed. Kepler took this as a stimulus to investigate a matter about which he had long had suspicions and to which he had already drawn attention at the very end of Chap. 22 of the *Mysterium Cosmographicum*. He had pointed out there the oddity of the Copernican system, according to which Saturn, Jupiter, and Mars had equants but the earth was supposed to move with perfect uniformity about the same eccentric point (the Copernican mean sun) at which the apsidal lines of these three planets were made to converge. Surely, he thought, the earth ought to have an equant too. He was also deeply suspicious about the Copernican models for Venus and Mercury. However, he lacked the observations to do anything about his suspicions.

His distrust of the traditional solar theory, on which Copernicus's orbit for the earth was based, was greatly strengthened by a letter he received from Brahe in 1598, in which he said:[26] 'The annual circle of Copernicus or the epicycle of Ptolemy does not appear to have the same size at all times if it is compared with the eccentric [for an outer planet]; for all three planets it gives rise to a significant variation, so that the difference angle for Mars can reach 1°45′.' Kepler's comment on this was:[27] 'Already then, when I heard that the annual circle expands and contracts, a good spirit gave me the thought that this fantastic phenomenon (*phantasma*) arises from the fact that the Copernican annual circle is not at equal distances from the point around which it is supposed to move equal angles in equal times.'

Thus, Brahe's observations suggested that the earth too was subject to the equant phenomenon, in good agreement with Kepler's intuition. This question was, in fact, one of the first that he had considered when he joined Brahe in 1600, and during those first months he had already found observational confirmation of his suspicion. It was one of the factors that fuelled his early optimism and led him to believe, for about two years it seems, that he was close to the complete solution of the problem of the planetary motions on the basis of perfectly circular orbits with lines of the apsides all passing through the sun. However, he had not investigated the matter thoroughly, and therefore had only an approximate knowledge of the earth's orbit.

Having spent so much time on the systematic study of the Martian orbit, and knowing full well that there was something exciting and quite new to be discovered there, Kepler nevertheless turned aside from the main quarry and concentrated his attention on the earth. In fact, kinematic relativity being an inescapable concomitant of observation, he had in truth no alternative to this indirect journey to the grail he actually sought. Book III of the *Astronomia Nova* is entitled: *Study of the Second Inequality*, i.e., the Motion of the Sun or the Earth, or the Key to a More Penetrating Astronomy.

It was not too difficult for Kepler, proceeding rather in the manner in which he had attacked the latitude problem, to deduce from some judiciously chosen observations that he was on the right track. He also commented that the annual variation of the sun's apparent diameter (about 30′ in summer and 31′ in winter) suggested a halving of the Hipparchan–Ptolemaic–Copernican eccentricity. But he wanted a transparent and unambiguous demonstration; even more importantly, he needed data on the position of the earth's orbit that were as accurate as he could possibly make them. How could this be done?

The solution he found is one of the master strokes of theoretical astronomy. It came straight from what I have called his fundamental assumption, and showed how astronomy, which had hitherto depended dangerously on hypotheses, could be built up on an absolute minimum of assumptions. In fact, all that Kepler used was trigonometry, Brahe's positions of the sun, and the periodicity of the Martian orbit, more precisely its sidereal period of 687 days. These were things about which, as Kepler emphasized, there could not be any doubt. (We should perhaps also mention the rotation of the earth, which provided the all-important clock.)

His idea was as simple as it was beautiful. It shows dramatically the liberation of Kepler's mind brought about by the Copernican revolution. He had learnt to wander freely in helioastral space and solved the problem by transporting himself conceptually to Mars. The space journey was done as follows. According to Kepler's fundamental assumption, Mars must (again ignoring the secular perturbations, which, however, Kepler could now handle) return to *exactly the same point of helioastral space every 687 days*. Therefore, said Kepler, let us make observations of Mars at precise intervals of 687 days. To eliminate problems with the latitude effects, let us choose the times at which Mars crosses the ecliptic. Because the period of the earth's orbit is quite different from that of Mars, we shall see Mars at such times at quite different positions on the sky. At such times the observations give us the apparent positions of both Mars and the sun as seen from the earth; from the vicarious hypothesis we also know the position of Mars as seen from the sun. Kepler's diagram is shown in Fig. 6.5. Now the sun–Mars line is fixed, since by assumption Mars returns to exactly the same point, x, of helioastral space. All the angles are known. Therefore, the positions of the earth in helioastral space can be calculated relative to the fixed positions of the sun and Mars. Given three such points (ε, η, ϑ, say) one can describe a unique circle. Given a fourth such point κ (and Kepler had one), one can check whether it lies on the same circle. Moreover, one can also determine the position of the centre of the earth's circular orbit and see where it lies relative to the sun and the already known position of the equant (the Copernican mean sun – always the easiest point to find).

It all came out just as Kepler had foreseen: the observations established beyond all doubt that the centre of the earth's orbit did not coincide with the point α about which the motion appeared uniform but lay instead on the line between α and the centre of the sun, roughly half way between these two points. The Copernican orbit (continuous circle in Fig. 6.5) was

---

Fig. 6.5. At the top: Kepler's illustration of his triangulation of the orbit of the earth (broken curve) by means of the sun and the known position, x, of Mars in helioastral space. The continuous circle is the orbit which Copernicus had assumed for the earth. In the two lower illustrations, Kepler repeats his demonstration in terms of the alternative world systems of Ptolemy and Tycho Brahe, showing the corrections to be introduced if those systems are adopted. By giving the demonstration in all three systems, Kepler highlighted their equivalence at the kinematic level and emphasized that the choice between the rival systems must be based primarily on physical and dynamical arguments. He thereby raised issues that are still very relevant to the absolute/relative debate. Part of the text of the 1609 edition of the *Astronomia Nova* is shown to illustrate the beautiful quality of the printing of this comparatively rare book. (Reproduced by courtesy of The Beinicke Rare Book and Manuscript Library, Yale University.)

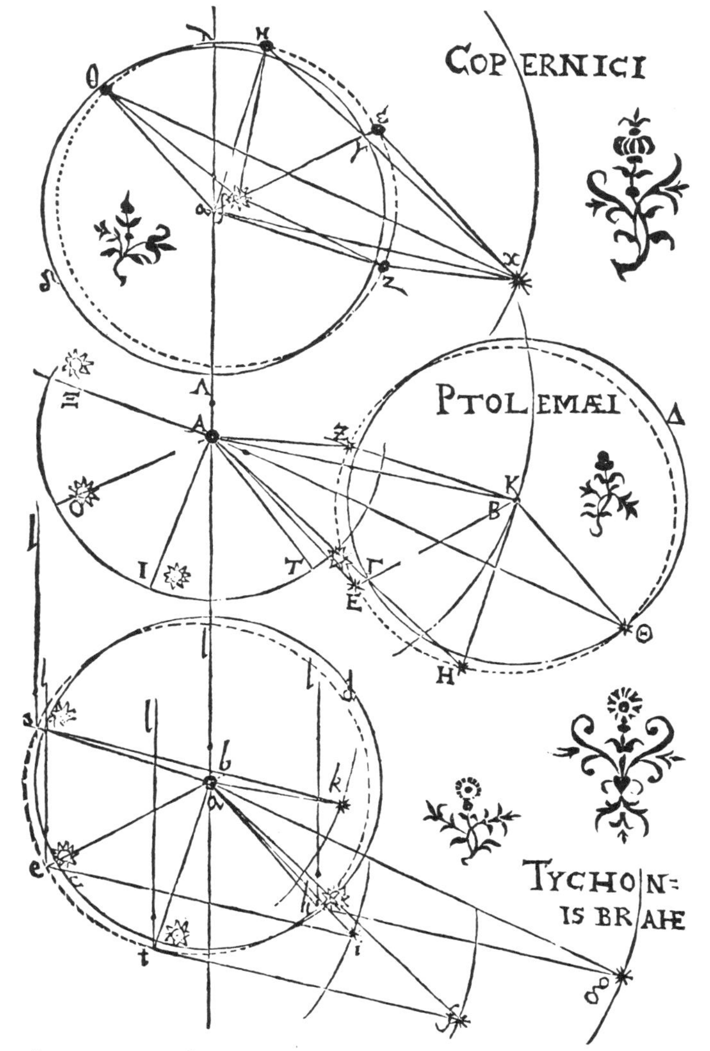

SOLIS *vergat in 5½ ♋: quamvis hunc gradum cap.* XXV *libere inquisituri sumus quasi incognitum. Et sit* TERRA *A.* MDXC *in ϑ, anno* MDXCII *in η, anno* MDXCIII *in ε, anno* MDXCV *in ζ. Et anguli ϑαη·ηαε·εαζ· æquales, quia α est punctum æqualitatis, & periodica Martis tempora præsupponuntur æqualia. Sitque Planeta his quatuor vicibus in κ, ejusque linea apsidum αλ. Est ergo angulus ϑακ secundum indicium anomaliæ commutationis coæquatæ 127°.5′.1″. Quod visum locum Martis attinet, is die* IV *antecedente hora simili fuit 24°.22′ ♈. diurnus ejus diei esset 44. Ergo ad nostrum tempus visus fuit in 25°.6 ♈. qui est situs lineæ ϑκ. Sed ακ tendit in 15°.53′.45″ ♉. Ergo ϑκα est 20°.47′.45″. Residuus igitur αϑκ ad duos rectos est 32°.7′.14″.*

Vt igitur sinus αϑκ ad ακ, quam dicemus esse partium 100000: sic ϑκα ad ϑα quæsitum. Est ergo ϑα 66774.

Quod si reliquæ ηα, εα, ζα, ejusdem prodibunt longitudinis, falsum erit quod suspicor: at si diversæ, omnino vicero.

Fig. 6.5.

therefore seriously misplaced relative to the true orbit (dashed circle). Although the observations permitted a certain latitude, Kepler unhesitatingly took the centre of the earth's orbit to be exactly half way between α and the centre of the sun. A few comments are here in order, as Wilson has emphasized.[28] The point is that, for all its conceptual beauty and simplicity, Kepler's method of triangulation, which involved the determination of three angles, was subject to observational errors. In view of the vital importance of this work, Kepler in fact employed several different methods in order to make his conclusion as secure as possible (such thoroughness was very characteristic of all his analysis of the Brahian data), but he still obtained quite a scatter of positions for the precise centre of the earth's orbit. They were all consistent with approximate halving of the eccentricity, but Kepler's decision to adopt exact halving was clearly governed by his theoretical considerations. As Wilson points out, he appears simply to have taken the eccentricity from Brahe's solar theory, which was 0.03584, halved it, and rounded off the result to 0.018. He also retained Brahe's position for the line of the apsides. Given his confidence in the theoretical proposition and its reasonably good empirical confirmation, Kepler had good grounds for doing this, since his earlier study had shown clearly how the position of the equant can be determined with much greater accuracy than the position of the centre of the orbit. Thus, halving of the empirically determined equant eccentricity was a more accurate even if indirect method of determining the orbit's eccentricity.

Kepler concluded his long and vitally important study of the earth's motion as follows: 'I believe it is now adequately confirmed that the earth–sun distances must be calculated on the basis of a halving of the eccentricity found by Tycho.' The dethronement of the mean sun, the usurper that Hipparchus had unwittingly helped to the throne all those long years ago, was complete. Kepler had the added satisfaction of confirming that this serious error had been able to remain undetected for so long solely because the eccentricity of the earth's orbit was so small. This was why Brahe's solar theory, derived ultimately from Hipparchus, was seriously defective yet nevertheless gave solar longitudes in almost perfect agreement with observation. However, the very slightest increase in the eccentricity would have led to observable effects.

In many ways, this result was Kepler's greatest triumph – there is nothing more satisfying than to announce in advance that you are going to sink the putt and then actually do so. He suspected the result and devised the stratagem to achieve it. And if his empirical evidence was still not quite as unambiguous as he might have wished, his clean halving of the eccentricity revealed the surety of the investigator who knows that a perfectly clear theoretical concept is sometimes to be preferred to the always somewhat uncertain deliverances of experiments and observations. With its bold hypotheses theoretical astronomy had always marched rather far ahead of the supporting observations. Kepler closed the gap to its absolutely irreducible minimum (at the level of the Brahian accuracy) but still trusted his theory to take him just a wee bit further than he could actually see provided the theoretical construct in his mind was at once luminous in itself and founded on a solid physical argument.

Kepler had now identified and eliminated the last of the residual defects of the Ptolemaic–Copernican models that had been able to creep into theoretical astronomy because of the particular absolute value of the eccentricity of the earth's orbit and the ratios it bore to the eccentricities of the orbits of Saturn, Jupiter, and Mars, on the one hand, and Venus, on the other. We recall that the Ptolemaic system is most accurately characterized as one that is peculiarly hybrid in the orders of the eccentricities – for the description of the apparent motions of the three outer planets it is essentially correct to *first* order in the eccentricities of their orbits but only to *zeroth* order in the eccentricity of the earth's orbit. For Venus the situation is precisely reversed. In addition, the solar theory is only correct to zeroth order in the eccentricity when distances as well as longitudes are considered. All these oddities had now been laid bare, and their origin was thoroughly understood.

We have now reached about the half way stage in Kepler's work on Brahe's observations of Mars – both conceptually and in time (the work on the earth's orbit was mostly done in the latter part of 1601) – and a certain amount of stocktaking is in order, particularly as many commentators tend to skim over the earlier work as a rather straightforward preparation for what followed. But the fact that in this part of his work Kepler had a very good idea of the direction he was going should not blind us to the originality of his work nor to the magnitude of the revolution which he had already wrought. The work of the first stage provided not only the absolutely sure astronomicotechnical foundation for what followed but also supplied Kepler with unmistakable hints of the way the sun could be expected to influence the motion of the planets. It armed him with ideas totally unavailable to his great predecessors.

Let us first have one last look at the technical aspect. Kepler proudly proclaimed that the halving of the eccentricity of the earth's orbit was to become the basis of the renewal of the entire science of astronomy. The claim is not exaggerated – for the first time since Ptolemy, struggling to make sense of the precession of the equinoxes, declared the earth to be the one secure basis of astronomy, the watchers of the sky had a firm foundation on which to work – even if it was a raft adrift in space.

There is something very appropriate about the fact that Kepler, quintessentially a child of the heady Renaissance times of new endeavour and breaking free from old bonds, was the author of an early work in the genre of scientific fiction – a dream of a journey to the moon.[29] He was indeed the first man whose spirit roamed freely in space, in this respect the mortal counterpart of Shakespeare's Ariel, a contemporary creation in the world of drama. In his *Epitome of Copernican Astronomy*, published a decade and a half later and after he had completed his work on the renewal of astronomy for all the planets, he even advanced 'space travel' as one of the reasons for believing in the Copernican world system:[30]

> For it was not fitting that man, who was going to be the dweller in this world and its contemplator, should reside in one place of it as in a closed cubicle: in that way he would never have arrived at the measurement and contemplation of the so distant stars, unless he had been furnished with more than human gifts . . . it was his office to move around in this very spacious edifice by means of the transportation of the earth his home and to get to know the different stations, according as they are measurers – i.e., to take a promenade – so that he could all the more correctly view and measure the single parts of his house.

This passage, incidentally, shows just how far the early supporters of Copernicanism were from the modern viewpoint according to which the main effect of the Copernican revolution was to demote man from the central position in the universe. They still saw things very anthropocentrically – even Newton continued that tradition.

Let us now turn to the conceptual and physical implications of what Kepler had already achieved, essentially by making the Ptolemaic schemes uniformly valid to first order in the eccentricity. The part of Book III of the *Astronomia Nova* that we have now reached almost oozes the satisfaction of an author who knows he has achieved a great synthesis (the first, in fact, in the history of the natural sciences unless geometry is counted as part of physics) and cleared up many long-standing problems that had sorely vexed the astronomers. For now nearly everything slotted into place. Kepler was virtually certain that the tidying up operation that he had carried out for Mars and above all the earth would be repeated for Venus and Mercury; as for Jupiter and Saturn they had always given the least trouble. If the minute residual problems with Mars were discounted, a perfect uniformity could be perceived: all planets moved on circular orbits on fixed planes that passed through the sun; all had equants, and all had centres of their eccentric deferents that exactly bisected the line joining their respective equants to the centre of the sun. Each planet in its orbit moved fastest when closest to the sun and slowest when furthest from it. Finally, the planets moved overall progressively slower in their orbits the further they were from the sun. At long last the sun was truly located at the exact centre of the world of the planets and most manifestly controlled their motions. The carousel which Ptolemy had built and Copernicus had so magically transformed with that one deft move but had nevertheless contrived to leave in a creaking and rickety state by making a botched job of fitting the last horse (the earth) was now in perfect working order, oiled and polished with loving care by a mechanical genius.

But Kepler was very well aware that there were many more implications of his work than simply the smooth running of a carousel. He knew that he had also transformed the debate about Copernicanism, a subject that was very dear to his heart – as many passages scattered throughout the *Astronomia Nova* (and also the whole of his *Defence of Tycho against Ursus*[3]) eloquently testify. As already pointed out, nobody understood more clearly than Kepler that, when it came to the crunch, the purely kinematic and geometric arguments that traditional astronomy could provide for Copernicanism were of no avail to settle the argument definitively. The counter put forward by Osiander in the notorious anonymous preface to *De Revolutionibus* was horribly insidious. For in its observational consequences Tycho's system, which Kepler duteously purged of its residual defects at the same time as he performed the same service for Copernicus's, was every bit as capable of *saving the appearances* as its rival.

Kepler felt so deeply about this problem that he addressed it head-on in a brief passage at the very front of his book, placed immediately after the titlepage. (It was here that he identified the hitherto anonymous Osiander.) Grasping the nettle, Kepler conceded that geometrical hypotheses alone could never settle the question; this could only be done by completing the traditional structure of theoretical astronomy by physical principles, which, if not absolutely conclusive, were at least highly plausible. Now, half way through Book III, as a complement to the purely technical chapters that had provided the astronomical arguments for the halving of the eccentricity for *all* the planets, the character of Kepler's discourse changes abruptly from a strongly innovative but basically traditional astronomical disquisition to a long stretch devoted to the arguments for Copernicanism and the exposition of his physical principles, some of which, he readily granted, were on a less secure footing than others. Not surprisingly, he started with the strongest arguments.

In the *Mysterium* he had marshalled mainly kinematic arguments. Now he came again with even more subtle kinematic arguments and followed them up with his really heavy guns, the physical arguments. First he used the refined kinematic evidence to pick off Ptolemy (Galileo had not yet made his observations of the phases of Venus, which, on the basis of direct observation, narrowed down the choice to one between Tycho and Copernicus). He had shown beyond reasonable doubt that the apparent motion of the sun – be it due to actual motion of the sun or of the earth – must be described by an eccentric and an equant with doubled eccentricity, just as was found in the first inequalities of the three outer planets and would almost certainly be found for Venus and Mercury. Moreover he had demonstrated (as yet strictly only for Mars) that this relatively complicated motion was *exactly replicated* in the second inequalities of all the five classical planets. For each and every planet it would now be necessary to replace the simple original Ptolemaic epicycle by a more sophisticated arrangement that was the perfect mirror reflection, true in every detail, of the new solar theory. It was, he asserted, simply asking too much of human credibility to deny that all these identical equantized, and eccentricated, and perfectly phased epicycles did not have a common origin – either in the motion of the earth around the sun or the sun around the earth. Either Copernicus or Tycho must be correct. Kepler confidently and correctly predicted that:[31] 'the bright sun of the truth will melt all this Ptolemaic apparatus like butter, and the followers of Ptolemy will disperse partly into the camp of Copernicus, partly into that of Brahe'. A somewhat inglorious end for the unfortunate Ptolemy, but the manner in which the equant dominates the discussion throughout the crucial Books II and III of the *Astronomia Nova* is testimony enough to the contribution he had made. The last service of all great theories is to provide the framework which isolates the very phenomena that ultimately destroy them. Every really good approximation is the basis of a better.

But now, Ptolemy dismissed, it was necessary to consider the Tychonic arrangement. Could one really regard it as plausible? He (Kepler) had proved the unquestioned existence of a unique point, the centre of the world, the point at which the apsidal lines of all the five classical planets converged (he was anticipating here a bit but was totally vindicated by the sequel). Whether the earth went round the sun or the sun went round the earth, he had proved that for these five planets this point lay right in the middle of the sun. All these five planets circled the sun in accordance with laws that implicated the sun as governor of their motions beyond all reasonable doubt: fastest precisely when closest to the sun, slowest when furthest. But what about the earth? If the sun did go round the earth, was it not simply extraordinary that it did so in accordance with exactly the same law as the five planets went round the sun, and how could the tiny earth move the huge sun and all its attendant planets? How much easier to suppose with Copernicus (duly modified) that the earth went around the sun. Then all six planets would have apsidal lines meeting at that unique point in the heart of the sun, all would follow the same law, and the earth's orbital period would fit so sweetly between the 225 days of Venus and 687 of Mars (the Copernican argument from harmony that had so wrought on Kepler's imagination). These were the physical arguments that he now had securely in hand, all of them anticipated by his powerful intuition, and all of them backed up by innumerable observations of the greatest observer in history.

This central dynamical argument for dominion and centrality of the sun (which is permanently reflected in the very language of astronomy in the characteristic words *perihelion* and *aphelion* that Kepler coined[32] for the positions in the planets' orbits when they are closest to and furthest from the sun) then served Kepler as his constant guide as he searched for principles that would enable him to understand and master those perplexing residual eight minutes of arc in the motion of Mars.

Before we start on this, the most complicated part of the story, in which Kepler, following a surprisingly tortuous route, finally arrived at the conclusion that the planetary orbits are ellipses, let us review the role that circles played in the history of astronomy prior to Kepler's momentous discovery. It has been pointed out once or twice already in this book that numerous commentators on the history of astronomy have implied that many early problems faced by the astronomers stemmed from their too rigid adherence to the tyrannical doctrine of perfect circularity of the orbits. Even such a great scholar as Koyré was capable of confusion on this point as we see from the following passage:[33]

> It so happens, that the orbit of Mars is not the most eccentric – that distinction belongs to Mercury – but it is the only one whose eccentricity [*sic*] is sufficiently large to be apparent in the observational data of astronomy before the time of Galileo, or even Tycho Brahe. This was the very reason why it was so difficult for Ptolemy, as well as for Tycho Brahe and Copernicus, to account for the orbit in terms of circular motions.

The falsity of such assertions should now be perfectly evident. Any attempt on the part of the early theoretical astronomers seriously to consider noncircular motions would have been a complete waste of time. The corresponding effects, which are only a small fraction of the readily observable effects of second order in the eccentricity, would have been completely swamped by residual defects of lower order. All these mistakes derive from the failure to distinguish between the eccentricity (which is readily observable for all the planets) and the ellipticity (which is extremely difficult to observe even for Mars and Mercury). Even if Mars had had a much smaller eccentricity, Brahe and his assistants would have had almost as much difficulty as they actually had before Kepler joined them; for their problem – and Copernicus's and Ptolemy's – was not Mars but *the earth*. The first service that Mars performed for Kepler was in holding up a mirror to the earth, in which he could perceive the correct eccentricity of its orbit. This was just as important a service as the second one, which had to follow the first, the revealing of the ellipticity.

6.5. The First and Second Laws

Now what did Kepler do about those mysterious eight minutes? He could so easily have done nothing. Who knows when or in what form dynamics would have been discovered if Kepler had thrown in the sponge at that stage in the way that the aged Copernicus more or less gave up the attempt to understand the latitudes? But Kepler was only in his mid-thirties, and, quite undaunted by the horrors of calculations, he buckled on his sword again to rejoin battle with Mars.

Since the manner of Kepler's discovery was rather surprising, any reader who does not already know how Kepler came to his first two laws might like to pause here and, before reading on, try to guess how it might have happened, bearing in mind what has already been recounted and also the fact that in his work on optics on which he was simultaneously engaged Kepler had done important work on conic sections, developing above all the theory of the foci, which had been treated only rather cursorily by Apollonius – the word *focus* was actually introduced into European literature by Kepler in this sense in his 1604 work on optics. As Wilson points out,[28] one might well expect that Kepler simply reversed the technique whereby he had used Mars to find the position of the earth, i.e., that from the now known position of the earth and the longitudes of Mars as seen from the sun (known from the vicarious hypothesis) he would have used triangulation to find out directly the positions of Mars. This would have revealed an ever so slightly elliptical orbit of Mars and, clinching the argument for a true ellipse, a focal positioning of the sun. Somehow or other he might then have been led to his area law as the rule governing the speed of motion in the orbit determined empirically in this manner.

In reality things happened very differently. The word *focus* does not even appear in the *Astronomia Nova* and the focal properties of the solar position are used only implicitly. (They are mentioned together with the word *focus* in the later *Epitome of Copernican Astronomy* (1621), where the sun is also called the focus of the universe, i.e., its hearth.) Even more remarkable is the fact that Kepler used the area law *before* he found the ellipse – indeed it played a role just as crucial, perhaps even more so, than direct triangulation in the discovery of the ellipse. The reason why Kepler did not take the route that with hindsight seems rather obvious was his passionate concern to find the true physical reasons for the motions that Mars actually made. In a rare example of physical intuition, he correctly sensed that the discrepant eight minutes of arc were not some aberrant quirk peculiar to the Martian orbit (which could be tolerated in rather the same way that Ptolemy accepted the recalcitrant Mercury as a *sui generis* oddity) but that they were a deeply significant clue to the basic physical processes by which the planetary motions were determined. Kepler undoubtedly made his discovery of the ellipse much more difficult than it might have been by taking the route through a physics that only took shape as he worked his way along (and was, in truth, often a decided hindrance). But he – and, even more so, Newton more than three quarters of a century later – got a huge bonus for all the extra labour. This was the discovery of the area law, which truly it is difficult to see how Kepler could

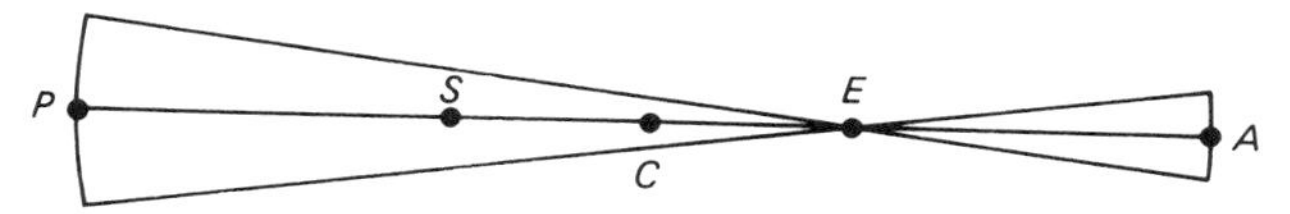

Fig. 6.6. Explanation of speed relation at apsides that follows from equant mechanism. The sun is at S, the centre of the planet's orbit is at C (e from S), and the equant is at E ($2e$ from S). In infinitesimal time dt, the 'spoke' emanating from E sweeps through angle ωdt, where ω is its angular velocity. Therefore, at A the translational speed of the planet is $\omega \cdot \mathrm{EA}$ while at P it is $\omega \cdot \mathrm{EP}$. But EA = SP and EP = SA by virtue of the particular positions of S and E relative to C. Therefore the ratio v_A/v_P of the speeds at the apsides is SP/SA, i.e., in inverse proportion to the distances from the sun.

have found by any other route than the serendipitous and very idiosyncratic one that he did take.

And yet, if you put yourself in Kepler's shoes, it was not all that idiosyncratic. One just has to appreciate the confidence that Kepler had gained from the already achieved successes which he could rightly attribute to his physical interpretation of the Ptolemaic equant. What was more natural than that he should attempt to take the success further?

Instead of turning directly to Mars, Kepler began by testing out an original idea on the motion of the earth. This is introduced in Chap. 40 of the *Astronomia Nova*. Following up his interest in the equant phenomenon, he established the important and key result that for exact equant-type motion the ratio of the orbital speeds of a planet in its orbit at perihelion and aphelion is inversely as the ratio of its corresponding distances from the sun. Thus if r_p and r_a are the distances from the sun at perihelion and apogee, respectively, and v_p and v_a are the corresponding speeds, then $v_p/v_a = r_a/r_p$. An explanation of this result in modern terms is given in Fig. 6.6. Thus, at least on the line of the apsides the speed of the planet was found to be exactly inversely proportional to its distance from the sun.

Kepler regarded this precise relationship as extremely significant and suggestive, and it completely dominated his whole approach to the problem of planetary motion; for it fitted perfectly with his physical notions, according to which the planets were driven around the sun by some force or virtue that emanated from the sun. If this were the case, then he felt that it was extremely plausible that the strength of the force should decrease in inverse proportion to the distance from the sun. Although in astronomy as hitherto practised such an approach was utterly novel, Kepler was here using an argument that any working physicist of the last three centuries would regard as perfectly sound, except for the fact that he was thereby relating force directly to speed. To fault Kepler for this mistake would be quite inappropriate; he could not possibly have anticipated the extremely subtle reinterpretation of his relationship that Newton was later to supply. Like all great theoreticians, Kepler had a very sure sense of when to generalize, and he conjectured that such a speed law held exactly not only at the apsides but throughout the entire orbit, that is, he assumed the speed v of the planet along its orbit in helioastral space to be exactly inversely proportional to its current distance from the sun. Kepler realized that if this was the case it must come into conflict with the equant law, for which the exact inverse proportionality held only at the apsides. However, he welcomed this and showed that for the earth at least, with its very small eccentricity of only just over 1/60, the two laws were effectively indistinguishable in their observational consequences at the Tychonic level of accuracy.

This permitted Kepler to conclude that the equant law was only a highly accurate approximation to the exact inverse proportionality law, which we may call the *distance law*. He then foresaw, quite correctly, that in the case of Mars, with its significantly larger eccentricity, the difference between the two laws would reach a level that could be distinguished by means of Brahe's observations. It seems clear that he hoped to explain the mysterious eight minutes of arc discrepancy by this means.

Before we go on to consider what actually happened, let us note how Kepler's new physical approach caused him to adopt what was a quite new attitude to small residual failures of the existing models of planetary motion. For Ptolemy and Copernicus they were simply mysterious curiosities, to be handled in an *ad hoc* manner as best one could. But we can now see why for Kepler they were vital clues and warranted an attention that Ptolemy or Copernicus were never led to accord them. Besides his new physical approach, a vital ingredient that made possible Kepler's heightened awareness of the significance of small deviations was the quite outstanding success he had had in clearing away all the spurious residual defects of the Ptolemaic and Copernican models. The new system was so simple and clean that the remaining deviations stood out quite clearly in a manner which they could never have done before his preparatory work. They were in fact now reduced to the eight minutes for Mars and the slow secular perturbations.

We now come to the first example of Kepler's serendipity. It is important to appreciate that Kepler's mathematical arsenal was woefully inadequate for the kind of programme on which he had embarked. He was getting into a situation in which it was necessary to sum successive small increments in order to predict the consequences of the putative law he wanted to test. For if the planet's instantaneous speed in its orbit is inversely proportional to its current distance from the sun, an integration problem must in fact be solved to determine the point in its orbit that the planet will have reached after a given time. But the calculus did not yet exist, and Kepler had to resort to a most laborious method of calculation.

Before we describe this, it is worth mentioning that although he undoubtedly had an intuitive notion of instantaneous speed, Kepler did not and could not use it directly in his work. For a start, it was considered improper in mathematics to form a ratio of heterogeneous quantities such as distance and time – one should always form ratios of like quantities. (This, incidentally, has a bearing on the question of the nature of time, to which we shall return in the final chapter of this book, p. 656.) In addition, the mathematics that Kepler would have needed simply did not exist. (Whiteside has interesting comments on this subject,[34] as does Stephenson.[18] The deficiencies of Kepler's mathematics were not always a disadvantage, as we shall soon see.)

To test his theory, Kepler needed to know how long it would take the planet, starting at aphelion, to reach a given point of the orbit. Because the speed was variable and Kepler could not solve his problem by the integration it required, he proceeded as follows. He divided the first half of the orbit, from the aphelion to perihelion, into 180 equal segments, i.e., each segment had length $\pi R/180$, where R is the radius of the orbit, which Kepler still, of course, assumed to be circular. He then calculated the distances from the initial point of each of these segments to the eccentrically located sun and assumed that the time taken by the planet to traverse this segment, the *delay* (*mora*) as he called it, was proportional to the corresponding current distance r (such an assumption being equivalent to a speed inversely proportional to the distance). The constant of proportionality was then normalized in such a way that the time required to traverse the complete orbit (found by laborious summation of the individual delays) was equal to the observationally well-known sidereal period (the year for the earth).

Kepler found this work particularly irksome, especially since to find the position at any given time it was necessary to carry out all the intermediate summations. This was intolerable, and he looked for a device to avoid such calculations. He was aware that 'there are infinitely many points on the eccentric and correspondingly infinitely many distances'. It therefore occurred to him, as he put it rather enigmatically, that 'the area of the eccentric contains all these distances'. He then recalled that 'Archimedes had once divided the circle into infinitely many triangles when he attempted to determine the ratio of the circumference to the diameter'. In this way he was led to introduce the idea that the area swept out by the line joining the planet to the sun could be proportional to the time taken by the planet to traverse the corresponding arc. It was immediately clear to Kepler that his device with the area could only be an approximation to what he took to be the exact distance law, since (expressed in modern terms) the infinitesimal area swept out by the line from the sun to the planet is exactly proportional to the delay corresponding to the distance law only when the instantaneous velocity is at right angles to the line joining the planet to the sun, i.e., only at the apsides. Nevertheless, by some rather ingenious arguments (for a discussion of which – as indeed of all this part of Kepler's work on which we are now embarked – the reader is referred to Stephenson's book[18]) Kepler was able to show that the approximation (as he thought) was remarkably accurate, leading to a significantly closer agreement with the result of the distance law than the equant approximation, though in the case of Mars he was aware that the difference between the area rule and the distance law would just about reach a level that could be detected at the level of Brahe's accuracy (maximally about four minutes of arc at certain points of the orbit).

The irony in all this is, of course, the fact that, as Kepler later discovered, the area rule is actually exact while the distance law is only approximate. Considerable confusion reigns in the literature about the extent to which and when Kepler became aware of the fact that the distance law was only approximate. There is a confused (and confusing) discussion by Kepler towards the end of *Astronomia Nova*, once he had found the correct forms of his first two laws, in which he appears to be reinterpreting the distance law to make it apply to *unequal* intervals of the orbit. In the much later *Epitome of Copernican Astronomy* he finally gave a clear and correct statement of the distance law, stating (again in modern terms) that it is only the component of the speed perpendicular to the instantaneous radius vector of the planet that is inversely proportional to the speed. As Caspar emphasizes,[35] Kepler's confusion only relates to the status of the distance law. Once he had obtained his great results, he did recognize that the *area law*, as we may now call it, represented an exact empirical result even though he still looked for a physical interpretation in terms of the (modified) distance law. For a discussion of this question the reader is referred to the studies already cited, and also two papers of Aiton.[36]

We should also mention here that the introduction of the area rule did not by any means solve all of Kepler's mathematical problems. For consider the orbit, still assumed to be circular, shown in Fig. 6.7. The centre of the orbit, with radius taken to be unity, is at O, the sun, at distance e from O, is at S; in a certain time t the planet will have moved from the aphelion A to its current position P. If we continue to measure angles in radians and take the sidereal period of the planet to be 2π, we must remember that the area of a circle of radius unity is π when we come to relate time to the area. In fact, according to the area rule, the time t will then be equal to *twice* the area ASP; but this area is the sum of the segment AOP, equal to $\frac{1}{2}\theta$, where θ is called the *eccentric anomaly* (for the circular orbit), and triangle OSP, whose base is e and height is PQ $= \sin\theta$ (PQ is the perpendicular from P onto Q), since OP $= 1$. Thus,

$$t = \theta + e \sin \theta, \tag{6.5}$$

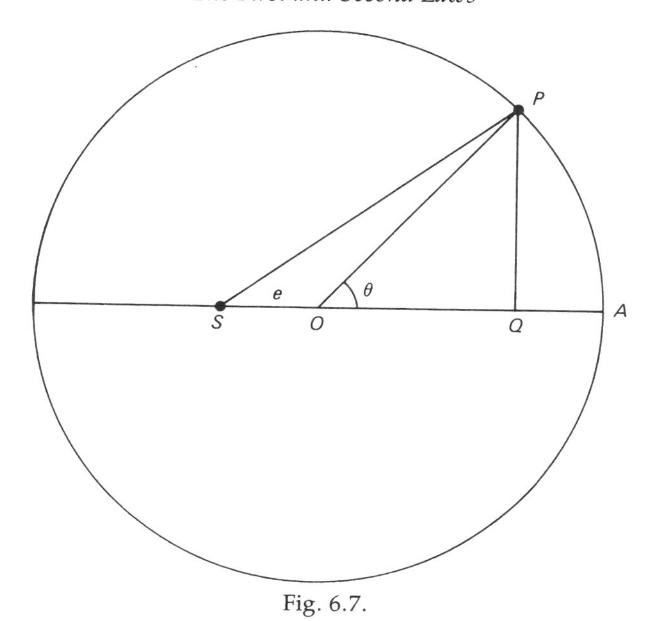

Fig. 6.7.

and to determine θ for a given t (which is the situation he always faced) Kepler had to invert Eq. (6.5), which is called *Kepler's equation*. He was (correctly) convinced that there was no simple solution to this problem, on account of the 'heterogeneous nature of the arc and the sine', as he put it, and he issued the problem as a challenge to the mathematicians of his day.[37] It became known as *Kepler's problem* (not to be confused with the same expression that I have used earlier in connection with the vicarious hypothesis) and played an important role in the history of mathematics. Rather remarkably, as we shall see, Eq. (6.5) also reappears in the theory of elliptical orbits though only with a redefinition of the eccentric anomaly.

It will deepen appreciation of the way in which the physics of the situation is expressed through the geometry and simultaneously help the reader who wishes to consult more specialized works if we introduce here one or two more of Kepler's basic concepts. Following the tradition of ancient astronomy, Kepler related all angles to the *mean anomaly*, the angle that increases uniformly with the time, has the value zero when the actual body passes through apogee (aphelion in heliocentric astronomy) and (in radian measure) reaches 2π when the actual body has completed its orbit. Since the time is given in accordance with (6.5) the mean anomaly γ will (with our choice of units) simply be equal to it:

$$\gamma = t = \theta + e \sin \theta.$$

Now were the planet to move uniformly around the centre of its orbit, one would simply have $\gamma = \theta$. Thus, $e \sin \theta$ measures the difference angle due to the actual (physical) nonuniformity. Kepler called this angle, which is represented geometrically by twice the area of triangle OSP, the *physical equation* (*equation*, we recall, is always the word used by astronomers to denote the correction between a quantity – an angle in this case – one would have expected and the actually observed quantity). However, the angle actually observed from the sun is the *true anomaly*, angle ASP; this is the angle given by Kepler's vicarious hypothesis. This differs from θ by the angle SPO, which arises for purely optical reasons (because the point of observation is at S). Kepler therefore called it the *optical equation*. Thus, to get from the mean anomaly to the observed true anomaly one had to subtract the physical equation, given by twice the area of triangle SPO, and then the optical equation, given by angle SPO. The true anomaly was therefore also called the *equated* (or *coequated*) *anomaly*. In accordance with his comprehensive programme of matching all observations, Kepler's fundamental aim was to devise theoretical schemes from physical principles that gave both the distances from the sun to Mars and the equations correctly. The distances were obviously determined by the shape of the orbit, the equations by the speed law of the planet in its orbit.

Note the beautiful way in which Kepler had managed to achieve a simple geometrical representation of all the nontrivial parts of his problem. Specify θ; then the time taken by the planet from A to P was found simply by adding twice the area of the little triangle SPO to θ. Moreover, the optical equation was simply the angle at P of the same triangle. Everything of interest was encoded in that one little triangle, and time had received a direct and perspicuous representation in purely geometrical terms. We shall see in Chap. 10 how this success of Kepler, which was only slightly modified on the transition to the true elliptical orbit, predetermined the manner in which Newton treated time in his dynamical problems.

Having tested out his ideas on the earth and shown to his satisfaction that Ptolemy's equant rule could indeed be merely an approximation to his own distance rule, Kepler at last turned his attention to Mars. We recall that he knew at least one of his basic assumptions in the derivation of the vicarious hypothesis must be wrong – either there was no equant or the orbit could not be a circle, or both. Since the equant had always been suspect and he had a ready alternative to it, it is hardly surprising that at this stage Kepler still did not seriously question the circularity of the orbit. He set to work with great vigour, applying both the area law and distance law to Mars under the assumption of a circular orbit. The result was a disappointment – he still found discrepancies at the octants that were just about as large as when he had used an equant with exactly bisected eccentricity; now, however, they were of the opposite sign. Specifically, and this is worth noting, he found when using the area law that his predicted position for the planet was about 8′ of arc ahead of the actual planet at the first octant (45° from aphelion) and the same amount behind at the second (135°). In a dejected mood he threw aside such calculations and was obviously at rather a loss.

Considering the magnitude of the step which is supposed to have been involved, Kepler says remarkably little in the *Astronomia Nova* about what the final abandonment of perfect circularity of the orbit involved for him. Perhaps his most explicit comment is the one at the beginning of Chap. 40: 'My first mistake was that I assumed the orbit of the planet to be a perfect circle. This error proved to be a particularly damaging thief of my time in having the support of the authority of all the philosophers and was in particular most agreeable to metaphysics.' One or two comments are here in order. No astronomer before Kepler had ever even contemplated that any of the seven wanderers except the sun (or earth, post Copernicus) moved in a *single* perfect circle. All that they had required was that the motion be *compounded* out of perfectly circular motions. But, as noted earlier, Kepler had long recognized the important concept to be the actual path in helioastral space, not the circular motions by means of which ingenious geometers described that path. But for noncircular paths Kepler had several precedents – the moon, Mercury and, indeed, all the planets according to the prescription of Copernicus for eliminating the equant in their first inequalities. Thus, adoption of a noncircular orbit on his part would not have been such a great step. I suspect that what held him up was much more the considerable success he had already achieved with circles. The work that Kepler did during those first few months with Brahe really were the most extraordinary vindication of the circle concept. No astronomer before Kepler had seen the prospect of adequately representing the motions of the planets with less than about 30 circles (when latitude motions are included). Kepler had reduced the number to the truly irreducible minimum of six (one for each planet) and simultaneously improved the accuracy with which the planetary motions could be predicted by about $1\frac{1}{2}$ orders of magnitude. (I am not saying Kepler had done all this in detail, but its possibility was completely clear to him.) We can therefore appreciate better the words with which Kepler opens Chap. 44 (which has the title 'That the path of the planet in the heavens is not a circle'): 'When the eccentricity and the ratio of the diameters [of the orbits of Mars and the earth] have been determined with great certainty, an astronomer might find it extraordinary that anything could remain to hinder the triumph over astronomy. *And, by God, I did triumph for two whole years!*' (my italics).

Apparently it was only in the early months of 1602, after the abortive application of the area law/distance law calculations to Mars's motion, that Kepler turned to triangulation as a means of direct determination of the orbit. What he did was determine the position in helioastral space of three points on the Martian orbit and then find the circle that passed through them by classical geometrical methods. Knowing its centre and the position of the sun, he could then determine its line of apsides. He found that this did not agree exactly with the line of the apsides of the vicarious orbit. He then repeated the procedure for further triplets of triangulated positions; each time he found a slightly different circle. The conclusion was inescapable – the orbit could not be a circle.

Kepler still proceeded in a very methodical way. Using very refined methods (including again the device of observing Mars at intervals of its sidereal period, this time at positions near the line of the apsides, and also a correction of suspect observations by using the mean motion at different locations in the orbit given by the vicarious hypothesis – this was something that could be found with much greater accuracy than individual positions), he determined as accurately as possible the precise position of the line of the apsides and the location of the sun along that line, i.e., the solar eccentricity. Both of these determinations played a crucial role in the later work, especially the eccentricity. He then imagined an auxiliary, or reference, circle that passed through the Martian aphelion and perihelion and therefore had the line of the apsides as a diameter. In his subsequent work this circle played a role almost as important as the actual orbit. He started to determine positions of Mars when away from the line of the apsides and compare them with the auxiliary circle. The first three positions, one still close to the line of the apsides, the other two around the octants, showed unmistakably that the orbit bent in from the auxiliary circle. At the middle longitudes the planet was closer to the sun than the circle. The orbit must be some kind of oval.

This step was his last that followed the marvellously limpid logic which he had employed hitherto. For at this point – apparently without making any further attempt to determine the precise orbit by direct triangulation – Kepler got completely carried away and seriously set about an attempt to determine the orbit theoretically using further physical ideas that were gradually taking shape in his mind. I think it is instructive to consider why Kepler did embark on this extraordinarily ambitious programme, which, as he ruefully admitted, led him into a fearful labyrinth.

First, we may note that the discovery of noncircularity, above all the bending in of the actual orbit within the auxiliary circle, cast an exciting and totally new light on the work with the area law that he had done on the Martian orbit. Figure 6.8 shows the auxiliary circle and the putative oval which he now expected. Consider a segment of area swept out by the radius vector from the sun, at S, to the planet when at aphelion, A. Such a segment is a smaller fraction of the area of the auxiliary circle, the orbit which Kepler had assumed for the area law work, than it is of the putative

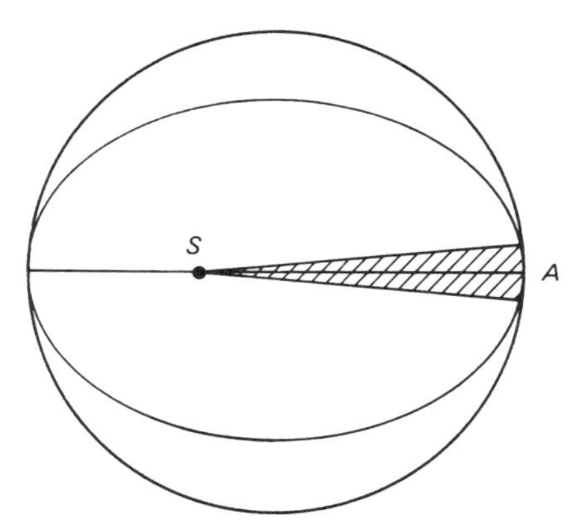

Fig. 6.8. Explanation of the effect of the area law. The hatched segment is a larger proportion of the area of the oval than of the circle. Therefore, if the oval is the correct orbit and the area law holds, the planet must spend a larger proportion of its period near aphelion, A, than in the case of a circular orbit.

oval. Thus, if the orbit actually is oval, the planet must take a *longer* proportion of its total sidereal period to traverse the apsidal regions than in the case of the circular orbit previously assumed. But this was exactly what Kepler had found – his theoretical prediction for Mars had been ahead of the real planet at the first octant, behind at the second (towards the perihelion), just as one would now expect. The distance law had also predicted quantitatively the same effect. Kepler realized that if he had only had the courage of his convictions he could have predicted not only the noncircularity of the orbit but also its bending in within the auxiliary circle.

He had at least the consolation of knowing that the area law had become a powerful tool in his hands. Any conjecture he might now make about the geometrical shape of the orbit could be tested by seeing if it, in conjunction with the area law, predicted the correct longitudes of the planet (which had to match those of the vicarious hypothesis). Expressed in traditional terms, the assumed geometrical orbit and the area law (which to Kepler's good fortune he generally used for the sake of its convenience) were required to reproduce the correct equations. Wilson[28] has emphasized the importance of the area law in the eventual finding of the elliptical orbit, noting that this has escaped many of the earlier commentators, who believed that it was found solely by triangulation. The important point here is that, generally speaking, the distance determinations were not quite so accurate nor, for Kepler, nearly so convenient as the longitudes, which the vicarious hypothesis supplied him with at any point of the Martian orbit.

A second, and perhaps more important, factor in Kepler's deciding to attempt a theoretical prediction of the orbit was a lot of prior work he had done on analyzing the reasons for the planets' *variable distances* from the sun. So far we have only considered the question of the variability of their speed in their orbits – Kepler had, as we have seen, attributed this to a weakening with distance of the solar force that drove the planets around their orbits. As we shall see in the next section, Kepler believed that this force was ultimately generated by rotation of the sun. But quite independent of this, in Kepler's view, was the question of what generated the eccentricity. How did it come about that the planet approached and receded from the sun? For him this was now the most fundamental question of all, since the speed in orbit followed as a necessary consequence once the distance had been established.

Kepler's insistence on finding a reason for the variable distances from the sun highlights once more the novelty of his approach. The purely geometrical approach of all of his predecessors had led them to look no further than a simple geometrical arrangement that could save the appearances. Such an explanation, once found, was sufficient in itself, especially if based on a circle. Inured to eccentrics by two millennia of successful saving of the appearances by means of them, astronomers did not think to ask for any further explanation.

In order to give the reader at least some idea of why Kepler got into the labyrinth he did, it will be helpful to stand back a little from his problem, and use hindsight. His first problem was a severe limitation of his mathematical equipment; this will be illustrated shortly. Much more serious was the fundamental defect from which his physics suffered – the fact that he believed force to be directly proportional to the motion which it produces. As we have already noted, Kepler could not possibly have guessed nature worked on an analogous but subtly different plan, relating forces to *accelerations*. If one compares his approach to the problem of celestial motions with that of his contemporaries and all his successors up to Newton, one has to say that Kepler had a more systematic and (with hindsight) basically correct approach to it than any of the others. What one can say is that if nature had worked in the manner Kepler believed he would surely have discovered her laws. He was very aware of the need to understand the minutest deviations from what seemed to be an almost perfect circle-based scheme. He correctly sensed that there were deeper and, so to speak, invisible or transcendent principles at work behind the beautiful geometrical structures that the traditional astronomical techniques, brilliantly consummated by his own innovations, had revealed. Most striking of all was his conviction that these deeper principles must be manifested in precise mathematical relationships. He had found one such relationship, or thought he had, in his distance law (and truly had in the area law). Now he was looking for an analogous relationship that governed the varying distances. But this was where nature's subtlety threw him – he was looking for a relationship that actually lay exactly one layer lower in terms of the differential orders of the yet to be discovered infinitesimal calculus. He was, in fact, extremely fortunate that he chanced upon the area law, for this, expressing as it does what is called a *first integral* of the motion, does involve a first derivative (the transverse speed) directly and linearly. But the other fundamental relationships of dynamics lay too deep for not only his mathematical techniques but also, perhaps more seriously, for his intuition. As a result, Kepler spent, with interruptions, about three years on a tortuous study that attempted to establish a simple and physically plausible relationship between orbital properties that the subsequent Newtonian theory showed to be relatively superficial features of the Martian motions. He did eventually find one that seemed to fit the bill, and this fortunately persuaded him that he had at last found the correct orbit. However, the discovery by Newton of the true dynamical principles underlying the Keplerian ellipses was to show that Kepler had merely stumbled across a rather remote consequence of them, devoid, in itself, of intrinsic significance. But that was the sort of luck Kepler deserved.

Now to the curious story of the deficiencies of his mathematics. Kepler, who possessed nothing like the notion of polar coordinates, which provide the natural tool for describing his orbital problem, needed a mathematical framework by means of which he could analyse the distance problem. He started on this work long before the discovery of the noncircularity and therefore naturally assumed a perfectly circular but eccentric orbit. By one of the nicest ironies of his work, he chose as analytical tool a geometrical device that was almost the most ancient that astronomy could offer: Hipparchus's epicyclic alternative for representing the solar motion. In Fig. 6.9(*a*), CDF is the eccentric circular orbit of Mars. The sun is at A, the centre of the orbit at B. We recall from Sec. 3.10 that such an orbit can be represented equivalently as follows. About A as centre, describe a circle (not shown in Kepler's diagram) with radius BC. (This circle passes through the point N.) Let an epicycle Dγ with centre at N and radius ND equal to the eccentricity AB move around the circle centred on A. If the planet D on the epicycle moves in such a way that angle γND is always exactly equal to angle CAN, then D will move around the original circle CDF.

The current distance of the planet from A is AD. Kepler transcribed these distances into the auxiliary diagram Fig. 6.9(*b*), which leaves out the large circle and merely shows the distances relative to the epicycle. That is, $\delta\alpha$ in the auxiliary diagram has length equal to DA, $\varepsilon\alpha$ length equal to EA, and so forth. Angle $\gamma\beta\delta$ is equal to angle CBD (=angle CAN), angle $\gamma\beta\varepsilon$ is equal to angle CBE, and so forth. So far, this is pure mathematics. However, Kepler now examined the two parts of the diagram to see if any physical or other reason could possibly be found to explain how such

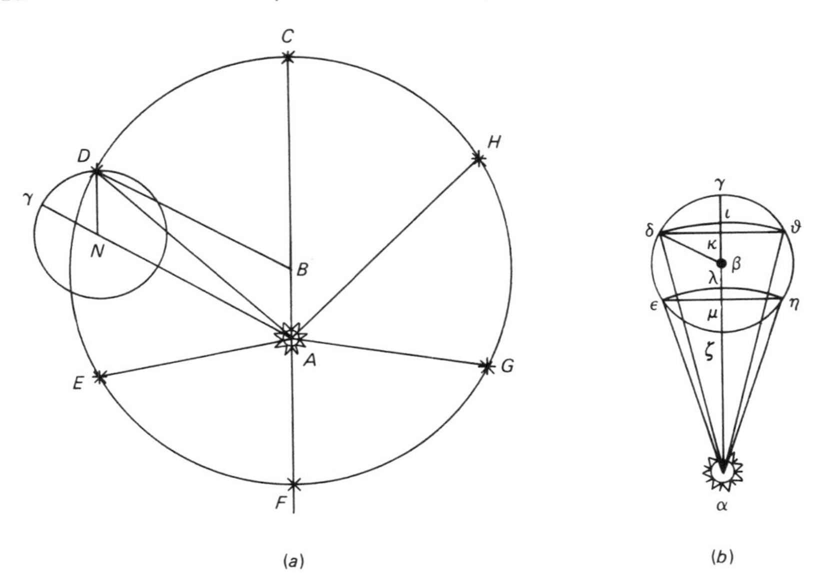

Fig. 6.9. Reproduced from *Johannes Kepler Gesammelte Werke*, C.H. Beck'sche Verlagsbuchhandlung, Munich, Vol. 3, p. 257.

varying distances could be brought about. For greater clarity, he drew circles with centre at α through δ and ϑ and through ε and η. Then as the planet moved through the points C, D, E, F, G, H, C in the main diagram (which are separated by equal 60° increments of the eccentric anomaly) the planet must have successively the same distances from the sun as the points γ, ι, λ, ζ, λ, ι, γ have from α.

No matter how Kepler looked at these two complementary diagrams, he found the evidence they gave bizarre. This was so even if he invoked purely physical mechanisms or allowed the planet's motion to be generated by some animal spirit and governed by some sort of mind. Stephenson's account[18] of this aspect of Kepler's thought is especially recommended. Most commentators seem to have preferred to pass in silence over the possible role that Kepler was apparently still prepared to accord to spirit and mind (which he carefully distinguished – animal spirit provided the force which generated the motion, mind directed it). One can clearly see a tendency on Kepler's part to rely more and more on purely physical arguments and dispense with minds. But even when Kepler does invoke minds, he seems to be using them, as Stephenson points out, mainly to test whether some particular motion is plausible or not.

Suppose for example that the epicycle has some genuine existence and the variable distances from the sun are in reality brought about by an animalistic force within the planet which drives the planet around the epicycle. The difficulty with this is that the eccentric anomaly CBD increases nonuniformly with time; since angle D$N\gamma$ must always march exactly in step in order to maintain the geometrical equivalence of the epicyclic representation, one is forced to conclude that the planet must be driven round the epicycle nonuniformly. But how could the planet's mind achieve such a feat? It would have to keep the epicyclic angle exactly equal to angle CBD (or CAN). But this involves another of these mysterious void and completely invisible points to which Kepler resolutely refused to accord any significance. The planet's mind could not possibly be expected to calculate an angle by observing the invisible point B (or alternatively N). He allowed that the mind could sense the direction ND, which is parallel to FC, the line of the apsides; for this is clearly something physical and moreover points to a point on the celestial sphere identifiable by the stars as markers. But point B is not marked by anything. As soon as any mechanism became too implausibly difficult, Kepler ruled it out.

Kepler also considered the possibility that the planet–sun distances were directly determined (rather than via an epicyclic mechanism) in some manner, either physically or by a mind. What disturbed Kepler in this case was the curious way in which the distances must then change. For in the first 60° increment of eccentric anomaly the distance is changed by the length of $\gamma\iota$, in the next by $\iota\lambda$, and in the next by $\lambda\zeta$. These three segments are all unequal. Kepler particularly distrusted the asymmetry represented by the fact that $\lambda\zeta$ was longer than $\gamma\iota$. During this early examination of the distance problem, he noted that a more plausible law of distance variation would be obtained by dropping perpendiculars from δ and ε onto $\gamma\alpha$, so that the successive changes in distance would follow the symmetric scheme $\gamma\kappa$, $\kappa\mu$, $\mu\zeta$ ($\gamma\kappa = \mu\zeta$). He called this distance law *diametral libration*, i.e., regular motion back and forth along the diameter $\gamma\zeta$ of the auxiliary epicycle.

All this initially fruitless analysis, in which, as he reports, Kepler became very well versed, had served to bring home to him a most striking fact, namely, the circular motions that seemed so satisfying to a mind accustomed to think in purely geometrical terms did not at all appear to rhyme with physical explanation. One can therefore well understand how his mind was blown by the discovery of the noncircularity. As he himself admitted, he seized on the first idea that occurred to him and charged into the labyrinth. In fact, he assumed that the planet was moved around its epicycle (which therefore remarkably reacquired a late lease of life) by a force intrinsic to the planet which always produced a *uniform* epicyclic motion (in contrast to the circular orbit, for which the epicyclic motion had to be nonuniform). However, the speed of the motion in the resultant orbit was still assumed to be governed by the distance law.

This was a problem for which the mathematics of Kepler's day was totally inadequate. No astronomer before Kepler had attempted to construct such an orbit. We cannot possibly attempt to recount here all the shifts and stratagems that Kepler employed as he struggled to come to terms with his problem. They were interrupted by the work on the optics and dragged on for about two years. In the *Astronomia Nova*, five of the most difficult chapters, in which Kepler recounts with almost masochistic delight his travails, are devoted to this phase of his work. The especially interested reader is recommended to either Kepler himself (soon available in English,[12] it is to be hoped) or Stephenson's monograph.[18] All we can do is merely note the parts of this work that finally put Kepler back on the right track.

He was able to show that his 'physical' prescription for the orbit led to one that was ever so slightly egg-shaped – wider at the top near aphelion and narrower near perihelion. However, he found calculations with it impossibly difficult. In another stroke of serendipity he was led to attempt to approximate the ovoid by an ellipse. The reason for this is an important mathematical property of ellipses which enabled Kepler to apply his area law to them directly. In Fig. 6.10, suppose the planet on the elliptical orbit adopted as approximation is at P and the sun at S. Let P′ be the intersection with the auxiliary circle of the perpendicular PQ dropped from P onto the line of the apsides SA. Kepler then exploited a result due to Archimedes: wherever P may lie on the ellipse, the ratio P′P/PQ is a constant. But, the infinitesimal areas being proportional to the respective heights, this means that the area of the elliptical segment APQ is always

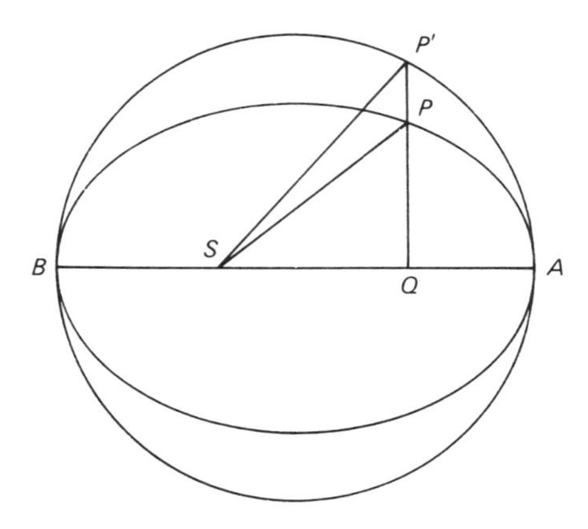

Fig. 6.10.

strictly proportional to the circular segment AQP′; the triangles SPQ and SP′Q must bear exactly the same ratio, since they have common bases and heights PQ and P′Q in the same proportion. Thus, on the assumption of an elliptical orbit and sufficient accuracy of the area law Kepler now had in his hands a geometrical prescription for finding the position of the planet at any time t after passage through aphelion. Namely, let the ratio of t to the sidereal period be p. It is then necessary to solve Kepler's equation and find angle P′SQ such that the area of AP′SA bears to the total area of the circle the ratio p. From point P′ found in this manner, drop the perpendicular P′Q onto the line of the apsides. The point P at which it cuts the ellipse is then the position of the planet at time t. Note that this construction does not require the sun to be at a focus of the ellipse.

Kepler's next piece of luck was that he was able to calculate the eccentricity of the ellipse that best approximated his theoretical ovoid (this is putting things in modern terms; he actually calculated the area and thickness of the *lunula* AP′BPA). It turned out to be an ellipse having eccentricity $\sqrt{2}e$, where e is the eccentricity of the sun from the centre of the Martian orbit which Kepler had found by his very careful triangulation. Kepler was therefore applying the area law calculation to an ellipse with $\sqrt{2}$ times the Martian eccentricity and the sun at its correct position between the Martian aphelion and perihelion but not at a focus of this ellipse. Because the maximum thickness of the *lunula* for an ellipse of eccentricity e is $\frac{1}{2}e^2$ (to good accuracy for small e), by taking an ellipse with $\sqrt{2}$ times the actual Martian eccentricity, Kepler was actually dealing with an ellipse for which the *lunula* had *twice* the thickness of the real ellipse's.

Kepler duly applied the area law and, to his disappointment, found that whereas his earlier, circle-based, calculation had given octant positions for the theoretical planet about 8′ further from the line of the apsides than the actual planet the new calculation gave octant positions that now had precisely the opposite error. Whereas before, in the circular orbit, he had moved the planet too fast at the apsides, now he was moving it too slowly. After several more abortive attempts to save the ovoid, Kepler finally had to admit that it could not save the appearances: the equations did not match those of the vicarious hypothesis. Although at this stage (in the late summer of 1604) he did note that the correct orbit could well be an ellipse half way between the auxiliary circle and the ovoid-approximating ellipse, he does not yet seem to have taken an ellipse seriously. He was, it seems, still far too intent on theoretical derivation of the law in accordance with which the planet–sun distance changed and, as yet, the ellipse offered him no illumination on that score.

Instead, he turned back to observations and at long last started systematic triangulation of positions of Mars all around the orbit and not just at the three positions initially established. The extension of these observations into the region of the quadrants, which he had hitherto omitted, confirmed the bending in of the orbit but suggested that the amount was not so much as required by the ovoid. This was in agreement with what the area law calculations had indicated. But still the key to the distance variation eluded him. Meanwhile, the emperor, Rudolph, was expecting to see results, and Kepler wrote up the outcome of his studies and had reached Chap. 51 of the *Astronomia Nova* by Christmas 1604 but still did not have the elliptical law.[38] After he had found it, he introduced some modifications into the text already written, but the bulk appears to have been left unchanged.

It was only in the early months of 1605 that he had his final major lucky break. He had calculated that, with the semimajor axis of the Martian orbit taken equal to 100 000 units, the *lunula* of his ovoid had maximum thickness 858, at eccentric anomaly $\theta = 90°$ (Fig. 6.11). Half of this number is 429; this was a number that was firmly lodged in his memory. So was another datum; this was the optical equation, angle SPO, corresponding to this position. It was 5°18′. As Kepler was idly looking through tables of trigonometric functions, he happened to note that the secant of this angle was 100 429 (the secant for his purposes was defined as the length of P′S when P′O is taken to be 100 000). A spark flashed – subtract the 429 (the width of the *lunula* that the true orbit must have) and one gets exactly the mean radius 100 000 (the radius of the auxiliary circle). But this means that at eccentric anomaly 90° the planet–sun distance PS is, to very great accuracy, 100 000. Quite unaware of the fact, Kepler had here stumbled on an important result of the geometry of ellipses, which it is worth explaining in terms of the generation of ellipses by means of a loop of string laid around two pins S and S′ (which become the foci) and held taut by the tip of a pencil (p. 120). If the semimajor axis is unity, the length of the loop of string is $2 + 2e$. At A (Fig. 6.11), the distance SA is $1 + e$.

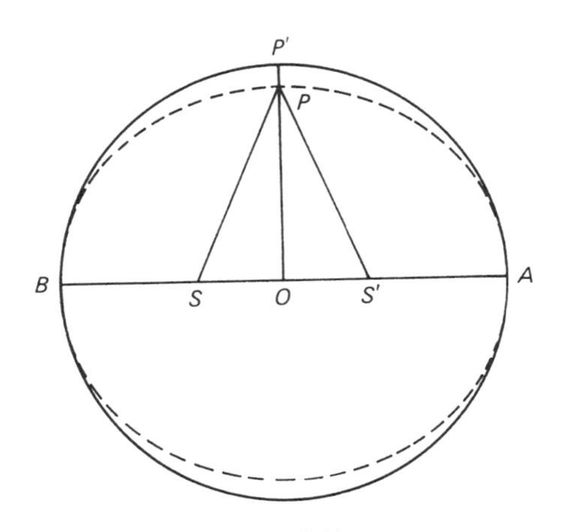

Fig. 6.11.

When the tip of the pencil is at P, $2e$ of the length of the string lies along SOS′, and the remainder is exactly divided between PS′ and PS, which must therefore each have length unity.

However, Kepler appears to have been quite unaware of this. Instead he guessed at a generalization of his result to the entire orbit. In modern terms he posited that the planet–sun distance is given by

$$r = 1 + e \cos \theta, \tag{6.6}$$

where the semimajor axis is unity and θ is the eccentric anomaly. For $\theta = 0$, this gives $r = 1 + e$ and for $\theta = 90°$ we recover the result $r = 1$ that set Kepler off on this track.

What made Kepler supremely confident of having at last found the correct law of variation of the distance was his intuition that Eq. (6.6) represented a law that could be readily derived from physical principles. He outlined in some detail an ingenious mechanism involving magnetic dipoles that he believed could give rise to such a law; we shall consider this in the next section. As this mechanism was not without at least one major difficulty, he sketched an alternative (and not too fanciful) account of how a planetary mind might be expected to bring about such a distance law. As always with Kepler, this required the mind to respond somewhat as a servomechanism (i.e., in accordance with a well-defined mathematical relationship) to a quantity that could be directly measured and involved truly observable entities. This too we will consider (in Sec. 6.7). All these matters are discussed in Kepler's long physical Chap. 57 of the *Astronomia Nova* and are fully covered by Stephenson.[18]

A final source of satisfaction to Kepler was that his law (6.6) turned out to be the libration on the diameter of the epicycle that he had long before felt was a more comprehensible law of variation of the distance than the one actually realized in a perfectly circular orbit. It must however be pointed out that the original diametral libration had been defined by means of the original eccentric anomaly corresponding to a circular orbit. But, as Stephenson points out, the eccentric anomaly, originally defined for a circular orbit, had entered a state of limbo once Kepler abandoned circularity and embarked on the study of his ovoid.

The final stage of Kepler's search for the true Martian orbit can be seen as the struggle to find an appropriate redefinition of the eccentric anomaly. Kepler's diagram (Fig. 6.12) shows the two-step process by means of which he finally arrived at the truth. It is worth explaining this diagram in detail (only the top half of which we need to consider). The sun is at A and the centre of the Martian orbit is at B (eccentricity e). The auxiliary circle with diameter formed by the line of the apsides joining the aphelion G and perihelion Q is GDPQ. The circle HKRS has centre at the sun and is the circle which carries the epicycle in the alternative Hipparchan form of the solar theory. It therefore has the same radius as

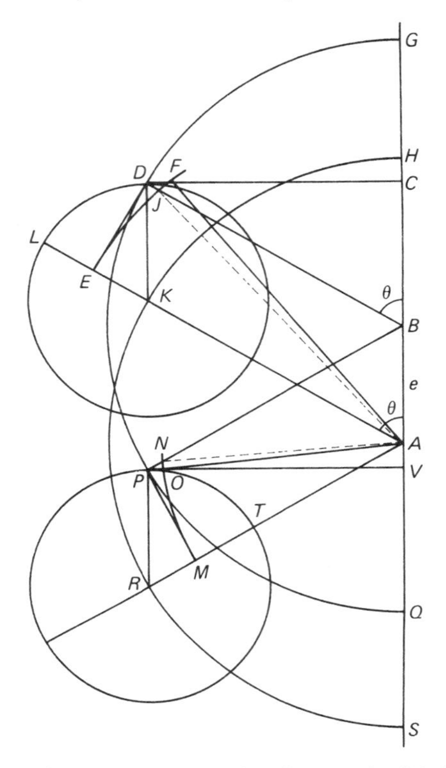

Fig. 6.12. Reproduced from *Johannes Kepler Gesammelte Werke*, C.H. Beck'sche Verlagsbuchhandlung, Munich, Vol. 3, p. 365 (the symbols θ and e have been added).

the auxiliary circle. Kepler initially constructed the orbit of the planet as follows. He specified a certain angle θ, called the eccentric anomaly, and described the straight line from B that cuts the auxiliary circle at a point D such that angle GBD $= \theta$. Parallel to BD he drew the line AK, cutting the circle with centre A. On this he imagined an Hipparchan epicycle with radius DK equal to the eccentricity AB. In the figure, angle DKL is therefore equal to θ, the eccentric anomaly. From D he dropped the perpendicular DE onto AK, obtaining the length AE $= 1 + e \cos \theta$. He then described the circle with centre at A and this length as radius. The curve EJF shows part of this circle. His first idea was that the planet should be situated at J, and that the orbit of the planet would be traced out by the locus of such points (N shows such a position in the lower quadrant). Kepler realized that this would generate a curve that, like the ovoid, would be ever so slightly thicker near aphelion than near perihelion. He called it the *via buccosa*, a 'chubby-cheeked' way. Here he had his last piece of good luck. He mistakenly believed that this orbit would lead to equations in disagreement with the longitudes known from the vicarious hypothesis. However, as Whiteside has pointed out,[34] the *via buccosa* differs so minutely (for Mars) from the true orbit that had Kepler applied the area law correctly he would not have been able to distinguish the two observationally.

After this, Kepler finally became convinced that the orbit must be a perfect ellipse; he was obviously strongly swayed by the consideration that the true orbit must exactly straddle the *lunula* between the auxiliary circle and his ovoid-approximating ellipse, since the area law had shown these to give equal and opposite errors at the octants. At long last, just after Easter 1605, he finally found the result he needed (it involved Kepler's finding a geometrical relationship which Wilson[28] believes was hitherto unknown). Instead of taking the planet to be at J, Kepler dropped the perpendicular DC from D onto the line of the apsides and posited that the planet would be at F, so that distance FA would be $1 + e \cos \theta$ with θ as previously defined. Rather remarkably, this does in fact generate an ellipse, as Kepler was able to show, and his distance law was thereby accommodated in a geometrical orbit which he already knew would give the correct equations in conjunction with the area law. The final step was to check that the predicted distances agreed with the ones that Kepler found by triangulation for 28 points distributed around the orbit. Bearing in mind the uncertainties of the distance determinations, Kepler concluded that the agreement was perfectly satisfactory.

To find the position of the planet at time t, Kepler proceeded in exactly the manner already described, solving his equation to find the eccentric anomaly (whose meaning was now clarified – it had only a mediate connection with the actual position of the planet in its orbit) corresponding to the required time. The key idea, that of dropping the perpendicular from D and then taking its intersection with the circle described from A with radius $1 + e \cos \theta$, was clearly suggested by the earlier work that had enabled Kepler to extend convenient use of the area law on a circular orbit to an elliptical orbit, initially used only for the purposes of approximation.

We see here again what a vital role was played by the area law, which Kepler now accepted as empirically exact even though he still made a confused attempt to save the law according to which the speed in orbit is inversely proportional to the distance from the sun. Wilson[28] believes that Kepler's finding of the ellipse owes rather more to theory (via the area law) than direct triangulation. This may slightly overstate the case. After all, the sun's eccentricity (i.e., its asymmetric position on the line of the apsides) was determined by very careful triangulation, and this too played its part, admittedly through a very intricate chain, by ensuring that the ovoid-approximating ellipse with eccentricity $\sqrt{2}e$ overcorrected the circular orbit by precisely a factor 2, as was then revealed by the area law and comparison of the resulting predicted equations with the vicarious hypothesis. It is, however, manifest that the area law and ellipse were discovered in an inseparable amalgam, and that Kepler's approach interwove theory and observation to such an extent that it becomes almost impossible to say whether any particular result is empirical or theoretical. The vital importance of the theory of the earth's motion and the way in which Kepler played tennis between Mars and the earth (being unable to get at either separately) should also be borne in mind. All his conclusions were empirical-cum-theoretical, and in the end the area law, the ellipse, and the theory of the earth's motion were all established together.

At the end of this chapter an attempt will be made to digest the significance of Kepler's *tour de force* in discovering his first two laws and to put them in the perspective of the complete history of naked-eye astronomy, which came to an end in the very year 1609 that Kepler published the *Astronomia Nova* (the delay of four years was due to a dispute with Brahe's heirs over the right to his observations). However, the following comparison will serve as an illustration of Kepler's achievement: If the ellipse representing the Martian orbit is drawn with semimajor axis equal to the width of the printing on this page and the circle that best approximates it is superimposed in the position which gives the closest fit, the greatest deviation between the two nowhere exceeds a quarter of a millimetre! The angular positions are equally well represented by the equant device.

Can we, in the whole history of physics, find another such example as we have in Kepler of dogged solo searching, extending over years, for a radically new conception of physics that, against all the odds and about two generations ahead of its time, is finally found? I think there is only one other example – Einstein's discovery of the general theory of relativity. As will be argued in Sec. 6.7, the stimuli to these two marvellous accomplishments had a significant element in common.

6.6. Kepler's physics and his Third Law

In the previous sections we have mentioned Kepler's physical and dynamical notions only to the extent that they were needed to highlight and explain the remarkable new directions into which he directed astronomy. Their greatest and most lasting significance was indeed that they guided Kepler to the correct laws of planetary motion. However, they also influenced the subsequent development of ideas in their own right, though to an extent which it is difficult to pin down precisely. They are also well worth considering because Kepler was the first major figure since Aristotle to attempt the construction of a genuinely new dynamical scheme. It is particularly interesting to see how very close he came to Newton's scheme in many important respects yet lacked the one vital element that was the key to Newton's success.

Aristotle's physics had been based almost completely on the idea of a unique centre of the universe. For it was around this unique centre that the celestial bodies were supposed to make their eternal circular motions, and it was to the same centre that bodies were supposed to strive in the phenomenon of gravity. Although Ptolemaic astronomy, with its multiplication of centres of motion far from the centre of the earth, put the Aristotelian scheme under great pressure, it nevertheless retained a very definite centre of the universe, so that the Aristotelian explanation of gravity was not challenged. But when Copernicus made the earth a planet and Kepler took him seriously, the old explanation became quite untenable. The earth was manifestly no longer the centre of the universe, so the phenomenon of gravity on its surface could not possibly be attributed to a striving to a centre of the universe. A new explanation of gravity would have to be found. Moreover, although both Copernicus and Kepler retained a cosmology that was strikingly Aristotelian in having a unique centre (the sun instead of the earth) and a sphere of distant fixed stars, the motions within this sun-centred cosmology bore little or no resemblance to the motions of the previous earth-centred cosmology. Kepler did not see any realistic possibility of salvaging anything of Aristotle's doctrine of natural motions – the astronomers had long since abandoned the simple circular motions of the Aristotelian quintessence, and now Copernicus had deprived the natural motions of the four ordinary elements of the unique centre which had provided their only justification.

Kepler felt it was necessary to recast the theory of motion *ab initio*. He therefore abandoned completely the notion of natural motions and assumed that *all* matter has an inherent tendency to remain at rest. In applying this assumption to all matter Kepler took an important step towards the universal concept of matter and motion which Descartes was to introduce. However, as in so many things, Kepler remained poised between the past and the future; for what he meant by 'all matter' was something rather different from what we would understand today. Kepler's cosmology (for a more detailed discussion of which see Ref. 39) consisted of three quite distinct parts: the sun, the 'movables' (by which he meant primarily the planets but also the objects on them and comets), and the sphere of the fixed stars. For a variety of reasons Kepler did not think the stars were simply other suns. For him they were quite literally situated on a huge distant sphere that formed the boundary of the universe and simultaneously, just as with Copernicus, provided a frame of reference with respect to which motion and rest were defined. The sun too played a distinguished role; for it not only provided the universe with heat and light (as it did for Copernicus) but was also the motor of all the motions within the universe.

What Kepler in fact attempted to do was set up a dynamical scheme that applied to the 'movables'. Only to that extent was it universal.

There were two basic elements of his scheme. First, he posited a universal and inherent tendency of matter to remain at rest (this idea almost certainly derived from medieval notions of the passivity and deadness of matter as opposed to spirit[40]) and, second, a notion of force, which overcomes a body's tendency to remain at rest and causes it to move. Of the two concepts, the force concept is found earlier in his writings. We have seen how it first appeared in the *Mysterium*,[41] as a spiritual and animalistic 'virtue' (*virtus*) by which the sun is capable of moving the planets around in their orbits, and then, under the strong influence of Gilbert's book on magnetism, was transformed into a purely physical 'force' (*vis*) in the *Astronomia Nova*. By the end of that work, Kepler had clarified his thoughts about three different types of force, two of which were rather remarkable anticipations of the forces that then appeared in Newtonian dynamics – except for the one crucial difference that Kepler followed medieval Aristotelian physics in relating forces directly to speeds rather than accelerations, as in Newtonian dynamics. This, as we have noted, was the fatal flaw that Kepler, radical innovator in so many ways, retained in his scheme. Let us now see the consequences of this mistake, beginning with the manner in which he supposed force to act.

Kepler lived long before the time in which laws of motion were expressed by means of equations in the modern fashion, but it is clear that he envisaged a relationship between speed v and force F of the form

$$v \propto F, \tag{6.7}$$

i.e., the speed increases in direct proportion to the force acting.

This idea is most fully elaborated for the force which Kepler believed carried the planets around the sun. It will be worth going into this in some detail particularly in order to draw attention to an aspect of the solar system that misled the seventeenth-century scientists quite grievously. This is the fact that all the bodies in it which could be readily observed in that century have the same sense of rotation about the sun (in the case of the planets) or about the individual planets (in the case of the moon and the satellites of Jupiter and Saturn). The agreement is very far reaching: the sun and planets rotate around their axes in the same sense as the planets orbit the sun and the moons the planets. This striking regularity, which modern science unhesitatingly attributes to the conditions under which the solar system was formed, caused the early investigators to believe that the common sense of rotation reflected an essential feature of the underlying dynamics, whereas modern science merely attributes it to the initial conditions. Even after Newton had found the correct laws of motion, he refused to believe that the regularity could be the result of natural processes and believed God had set up the solar system in that very special way.[42]

Considering the concordant circular motions of all the planets, Kepler was led to a rather remarkable prediction – that the planets move around the sun so regularly because the sun itself rotates about an axis that must point approximately in the direction of the poles of the ecliptic and that this rotation generates a force that, so to speak, sweeps round in the ecliptic, carrying the planets with it. One can imagine immaterial spokes sticking out from the sun in its equatorial plane and rotating with the sun like the spokes of a bicycle wheel. As they turn, the spokes push, as it were, the planets in the azimuthal direction, so that they orbit the sun. Kepler did not speak of spokes but rather of an immaterial *species* emanating from the sun; as used by Kepler, *species* is a word that is very difficult to translate into modern English. Even Kepler himself when writing in German did not attempt to translate the Latin word. Stephenson[18] suggests that *image* is the most appropriate translation. That is, there is an immaterial something transported (instantaneously) to the position of the planet which reflects the fact of the sun's rotation and causes the planet to follow the rotation.

At the time Kepler made his original proposal, there was no direct evidence at all for rotation of the sun, but very soon after he had published the *Astronomia Nova*, in which he made a confident prediction of the rotation, Galileo (and others) observed the famous spots on the surface of the sun and from their apparent motion deduced a rotation of the sun at a rate that fitted Kepler's prediction remarkably well. The rotation axis also pointed in about the right direction. Kepler naturally regarded this as a triumph, and his belief in the existence of an azimuthal force generated by the rotating sun was strengthened. (We may mention in passing that the confirmation by Galileo of Kepler's prediction seems to have had little or no impact on Galileo or any of Kepler's other contemporaries. The discovery of the sunspots did however greatly undermine confidence in the general correctness of Aristotelian philosophy.)

Both in the *Astronomia Nova* and the later *Epitome of Copernican Astronomy* Kepler indulges in quite extensive speculation as to the strength that one should expect this force to have. Analogy with light might lead one to expect its strength to decrease with the square of the distance from the sun. However, as we have seen, Kepler needed a force that decreased as the distance only, and he succeeded in persuading himself that this should be so. Stephenson[18] discusses this rather involved topic, which is interesting in its own right but is too specialized to consider here.

Of much more relevance for the development of the fundamental concepts of dynamics was Kepler's introduction of the concept of inertia as the quantitative measure of a body's tendency to remain at rest. Out of it, following a very significant transmutation by Newton, the modern concept of the *inertial mass* of a body arose. So far as I can understand Kepler, he was led to introduce it more or less as a corollary to the manner in which he conceived the rotational force generated by the sun to act. For if one were to take the idea of rotating spokes seriously, one would expect the planets to be carried round exactly as fast as the spokes themselves. But this would mean that the transverse speed of a planet would increase linearly with increasing distance from the sun and all the planets should have the same orbital periods, equal to the rotational period of the sun. But this was quite clearly not the case.

Kepler therefore posited not only that all bodies had an innate tendency to remain at rest but went beyond this qualitative notion in assuming also that they resisted motive forces applied to them by a quite definite amount, the power of resistance being an intrinsic quantitative characteristic associated with each body. The basic idea is expressed very clearly in the *Astronomia Nova*:[43]

> Thus the driving force [of the sun] is ready to impart to the planet a speed that is so great as its own. But the planet's speed is not so great because either the medium, i.e., the material of the celestial aether, or the disposition to remain at rest of the body itself which is being moved (*dispositione mobilis ipsius ad quietem*) offers a resistance. . . . The period of revolution of the planet comes about from the interaction of these factors with the impulse of the driving force.

Although Kepler was clear in his mind about the need for the introduction of such a concept, he was, perhaps not surprisingly, rather vague about how his 'resistance' should be conceived. In introducing the concept of disposition *ad quietem* in the passage above, he says: 'Others would call it the weight; however, I cannot agree to that unreservedly, even in the case of the earth.' In other places, particularly in his discussion of gravity (see below) in the *Astronomia Nova* and in the first of his two important letters to David Fabricius,[44] a fellow astronomer, he seems to regard it as a measure of the quantity of matter or just the bulk (*moles*) of the body. After the discovery of his third law of planetary motion, Kepler made his concept rather more precise. We shall come to that shortly.

It is clear from the manner in which he speaks that Kepler envisaged the actual speed of a body to which a force is applied to be determined by a balance between the resistance to motion and the force applied. Several commentators[45] suggest that Kepler had in mind a relation of the form

$$v = F/m, \tag{6.8}$$

where F is the force applied, m is the measure of the resistance of the body to motion, and v is the resulting speed. There is no doubt that one particular passage in the *Epitome*, to which we shall come, does yield such an interpretation unambiguously. However, there are several other passages which point to a different interpretation; for in several places Kepler implies quite clearly that a body with no resistance at all would not be moved with infinite speed, as Eq. (6.7) implies, but with the 'speed of the force itself'. Indeed, in the passage quoted above from the *Astronomia Nova*, Kepler says that 'the driving force is ready to impart to the planet a speed that is so great as its own'. Thus, there is definite speed associated with the force (it is clearly the rotational speed of the 'spokes' at the position of the given planet), and a body without resistance would have such a speed imparted to it. This means that Eq. (6.8) must be replaced by something like

$$v = F/(1 + m), \tag{6.9}$$

so that $v = F$ if $m = 0$ but $v < F$ if $m > 0$. In the *Epitome* there is a clear passage which needs to be interpreted in this sense:[46] 'For one mover [the sun] by one revolution of its own globe moves six globes. . . . Wherefore if the globes did not have a natural resistance of fixed proportion, there would be no reason why they should not follow exactly the whirling movement of their mover, and thus they would revolve with it in one and the same time.'

A relation like (6.9) was attractive to Kepler for a reason that is worth mentioning here, if only to highlight a crucial difference between him and Galileo on the question of the earth's rotation. As a committed Copernican, Kepler had to counter the insistent arguments of anti-Copernicans, such as Brahe, who argued that rotation of the earth would cause objects thrown into the air to be left behind by the earth's rotation. A cannon ball shot vertically into the air ought to fall far to the west when it returns to earth.

In the *Astronomia Nova* Kepler has only a few rather cryptic remarks to say on this subject, but he wrote at length at that time to David Fabricius on the subject.[44] Kepler supposed that such objects are somehow bound by forces to follow the earth round as it rotates. The vertical motion is governed by gravity, but in its horizontal motion a projectile is, as it were, grasped by the earth and carried round with it. Kepler points out to Fabricius that projectiles (whose size is a minute fraction of the earth's) can be expected to have an utterly negligible resistance compared to the force of the earth, and so will be carried at exactly the same speed as the surface of the earth, so that the effect anticipated by Brahe and others would not occur. Such an argument implies that Kepler's 'fundamental law of motion', if ever he had formulated one as such, should have the form (6.9) rather than (6.8).

In his letters to Fabricius Kepler even extended such ideas tentatively to the moon; were the moon free of all resistance and were it carried around in its orbit through a force generated by the earth's rotation (as Kepler supposed the planets to be moved by the sun's rotation), then the month should be exactly as long as the day and the moon should spin around in phase with the earth. Kepler attributed the fact that the moon moved slower than followed from such an argument to diminution of the strength of the earth's force with distance from the earth and to a nonvanishing resistance of the moon. This type of argument again implies a relationship of the type (6.9).

Kepler's explanation for nonobservation of effects of the earth's rotation on projectiles thrown into the air has an interesting implication for experiments of this kind performed on a ship. Kepler's mechanism will ensure that any body in the immediate vicinity of the earth is carried round with it with the earth's rotation velocity. But suppose a ship sails with uniform motion across the seas and a heavy weight is let fall from the top of its mast. When released, its motion in the east–west direction will, according to Kepler, be controlled by the earth. However, because the ship is capable, according to Kepler,[47] of moving objects only by direct contact, and the ship is moving relative to the surface of the earth, Kepler's proposal must mean that the weight should definitely land on deck at a point displaced from the foot of the mast. It is a remarkable fact that in Kepler's time this was firmly believed to be the case. Indeed, the erroneous belief in the reality of the phenomenon was used precisely as an argument to show that, by analogy, the same would happen for the rotating earth. Kepler's forces by which the earth 'grasped' bodies were designed to defuse this argument for the rotating earth but could not be invoked for ships. We shall come back to this point in the chapter on Galileo.

An important development in the history of dynamics was the introduction by Kepler of a special name to designate this 'resistance to motion'. In the *Astronomia Nova* he did not settle on any particular distinctive name for the property, but shortly after, in a booklet[48] written in German in answer to some queries raised by one Röslin about his earlier work on the supernova of 1604, he introduced the name which stuck – *inertia*, or rather the German *trägheit* for the Latin word, which means *inactivity, idleness, laziness* (the last especially). From then on, *inertia* was the word Kepler used regularly to describe the *proprietas* of bodies that he had recognized as being necessary to formulate the laws of motion. It is worth emphasizing the Latin word *proprietas*, which draws attention to the fact that Kepler's inertia is intrinsic to the body considered; for, as explained in the Introduction, much of the confusion that surrounds Mach's Principle stems from Einstein's attempt to make inertial mass (the concept that derives from Kepler's *trägheit*) into something with an *extrinsic* origin.

Inertia entered the vocabulary of dynamics on a permanent basis when Newton used the word to characterize the inertial mass m that now appears as a coefficient in his Second Law. Newton, who appears not to have read Kepler in the original except for his work on optics, learnt about the word from some letters of Descartes[49] and in the *Principia* described the word *inertia* as 'a most significant name'.[50]

Of course, as used by Newton, inertia means resistance to acceleration, not resistance to motion itself. It really is remarkable how close Kepler came to the correct structure of Newtonian dynamics in many respects but yet was separated from it by this one difference, which was in truth an uncrossable gulf. The whole difference can be expressed by two equations of the utmost simplicity and a striking similarity of form. The first is what we have called Kepler's 'law of motion'. Rearranging (6.8) or (6.9) and introducing vectors, which, as we shall shortly see, is justified for Kepler's mature dynamics, his law (6.8) becomes

$$m\mathbf{v} = \mathbf{F} \tag{6.10}$$

[or $(1 + m)\mathbf{v} = \mathbf{F}$ if we follow (6.9)] while Newton's Second Law can be written

$$m\mathbf{a} = \mathbf{F} \tag{6.11}$$

where $\mathbf{a}$ is the acceleration of the body, not its velocity $\mathbf{v}$ as Kepler imagined, and the 'resistance' m plays an analogous role in the two cases.

In fact, pared to the bare essentials of the discovery of dynamics, one can say that the story of the first half of this book is how Kepler came to find the laws of planetary motion and attempted to explain them by (6.10), while the second half will trace the steps that led to the rejection of (6.10) and its replacement by (6.11) by Newton. The result was a reinterpretation of planetary motions whose simplicity would have left Kepler gasping.

Let us now consider how Kepler's discovery of his Third Law, more than a decade after the discovery of the first two, led to a refinement of his ideas. He first attempted to find a relationship between the periods of the planets and their distances from the sun in Chap. 20 of his *Mysterium*. This attempt was flawed by an unfortunate mathematical slip, as Kepler ruefully admitted when he republished the *Mysterium* a quarter of a century later.[11] He noted that if all the planets were to move through space with exactly the same speed, their periods should simply increase as the radii of their orbits, since the circumference (the distance to be travelled) increases thus. He noted however that the periods of the planets increased more strongly than linearly with the radii of their orbits as deduced from Copernicus's theory of the planetary distances. From this he correctly concluded that the outer planets must move through space more slowly than the inner ones. By how much? He speculated (incorrectly) that the speed through space decreased inversely as the radius R, i.e., as $1/R$. Thus, Mercury should move faster than Venus in the proportion that the orbit of Venus is further out than the orbit of Mercury, or, in symbols $v_M/v_V = R_V/R_M$ (with obvious meaning of the symbols; Kepler's simplified treatment assumes each planet remains at a constant distance from the sun). From his assumption he should have concluded that the periods P increased as the *square* of the radii (since a longer circumference is also traversed at a slower speed):

$$(P_V/P_M) = (R_V/R_M)^2. \tag{6.12}$$

Kepler in fact considered increments. Thus, he supposed $R_V = R_M + \delta R_{MV}$ and $P_V = P_M + \delta P_{MV}$ (where the subscript MV means the increment on the transition from Mercury to Venus) and tried to work out (in words used in default of symbols) how δP_{MV} should be related to the other quantities. From his assumption about the speeds he should have deduced

$$\delta P_{MV}/P_M = 2\delta R_{MV}/R_M + (\delta R_{MV}/R_M)^2, \tag{6.13}$$

which follows from (6.12). What he actually deduced (translated into symbols) was

$$\delta P_{MV}/P_M = 2\delta R_{MV}/R_M. \tag{6.14}$$

His mistake had an ironic and misleading consequence. Because the periods actually increase as $R^{3/2}$ and not R^2, the increment (6.13) Kepler would have obtained by correct mathematics would have been too large. His mathematically incorrect result (6.14) was actually nearer the empirical truth. In fact, he obtained results that were at least good enough to persuade him that more accurate observations might match his theory. Struggling to reconcile the data with his various theories he remarked wistfully:[51] 'Would that we might live to see the day when these things are brought into harmony.' In the notes added 25 years later to the republished *Mysterium* he was able, referring to his discovery of the Third Law, to remark:[52] 'We experienced this day after 22 years.'

The *Astronomia Nova* contains virtually nothing on this subject except for a solitary remark at the beginning of Chap. 39: 'If a given planet were to execute complete orbits successively at different distances from the sun, the periods would be in proportion to the squares of the distances or the circumferences.' This suggests that Kepler had in the meanwhile noted the error in (6.14) and had corrected it to (6.12) but had not bothered to check the conclusion against the empirical facts.

He finally discovered the correct relationship in the late stages of his work on the *Harmonice Mundi*, in which the Third Law is stated in Book V as:[53] 'The ratio that exists between the periodic times of any two planets is precisely the ratio of the $\frac{3}{2}$th power of the mean distance.' Compared with his other two laws, Kepler says very little about the discovery of his Third Law; it obviously required a lot less work, but it gave a tremendous boost to his belief that God had constructed the world on architectonic and harmonic principles.

It belongs most emphatically to the mystical side of Kepler's personality and is, in fact, the only direct fruit of his search for architectonic harmonies in the world that has stood the test of time.

The discovery of the Third Law showed that the speed relationships obeyed by the planets were considerably more complex than Kepler had appreciated at the time when he discovered his first two laws. In the *Epitome*,[54] Kepler attempted to reconcile them with his physical ideas. He had by then clearly realized that if one planet is considered alone in its orbit the component v_{tr} of its instantaneous velocity at right angles to the radius vector varies in inverse proportion as the distance from the sun:

$$v_{tr} \propto 1/R. \qquad (6.15)$$

As we have seen, Kepler originally believed that (6.15) held for the total speed v and, moreover, not only for a given planet but for all planets, i.e., the more distant planets moved more slowly in accordance with the same law that Kepler originally believed to hold for the total speed in a given orbit. However, the Third Law showed that this could not be. Consider, for example, the case of zero eccentricity, i.e., circular orbit with the sun at the centre so that $v_{tr} = v$. Then from $P \propto R^{3/2}$ it follows that

$$v_{tr} = v \propto 1/R^{1/2}. \qquad (6.16)$$

To reconcile the two laws (6.15) and (6.16), Kepler held fast to the idea that the solar rotatory force decreased as $1/R$. He was able to reconcile this with (6.15) by making the modified assumption (compared with the *Astronomia Nova*) that the solar rotatory force acted only on the transverse component (at right angles to the radius vector) of the planet's velocity. To satisfy (6.16) as well, Kepler indulged in some rather obvious 'hand waving'. He supposed that three factors together determined the speed that a planet would acquire when acted upon by the solar force: the strength of the force, the amount of matter in the body of the planet (which measured its inertial resistance), and the volume of the planet. He argued that the solar force would, so to speak, grasp hold of the planet more effectively the larger its body and would therefore exert an effect in proportion to the volume of its globe. He assumed (citing some rather inaccurate observations of the apparent sizes of the planets made since the application of the telescope to astronomy) that the volume of the planetary globes increased linearly with the distance from the sun. In the calculation of the periods, this would exactly cancel the effect of the greater distance that the planets would have to travel. Now his force decreased as $1/R$ and to obtain the result of his Third Law he assumed finally that the amount [*copia*] of matter, i.e., the inertia, increased as $R^{1/2}$

on the transition from planet to planet. This is incidentally the passage in the *Epitome* mentioned earlier that implies a relationship between inertial mass and force of the simple form $m\mathbf{v} = \mathbf{F}$ rather than $(1 + m)\mathbf{v} = \mathbf{F}$.

One does not suppose that Kepler took these rather transparent manipulations too seriously. I mention them here for two reasons. The first interesting point is Kepler's clear distinction between the volume of a body and the amount of matter, resistance to motion, that it possessed. We have here a clear anticipation of inertia as a dynamical concept independent of size. As already pointed out in the discussion on Aristotle, the clarification of the mass concept proved to be very difficult indeed; in my view, the history of the discovery of dynamics ends properly with Mach's clarification in the late 1860s of the mass concept as something specifically dynamical and not definable in nondynamical terms. We shall return to this in Chap. 12.

The other point I wanted to make in this connection is an amplification of a point made in the previous section, namely, that the kind of problems with which Kepler was contending all derived from the flaw in his basic 'equation of motion' (6.10). It was the existence of the very similar law (6.11) hidden at a deeper layer of things which threw up all the striking mathematical relationships that so intoxicated Kepler's spirit and quite correctly convinced him that the sun was the source of a force that influenced the motion of the planets in a very direct manner. But the one true force, acting in conjunction with inertia under the conditions of the solar system with its hugely dominant solar mass, generated many relationships at Kepler's level of inquiry. For example, because he lacked the Newtonian concept of inertia he had to try and devise a mechanism which could explain why all the planets moved in different planes inclined to each other at small angles. Modern dynamics simply attributes them to different initial conditions and a special mechanism is unnecessary. The more Kepler worked on these problems and immersed himself in the facts of the planetary motions, the more his instinct led him to remarkably sound concepts. Yet he was doomed by the one fatal shortcoming of his scheme to trying to chase up one relationship after another like a willing but inexperienced sheep dog trying to force sheep into a fold into which they just won't go. In fact, had he made a determined effort to see the precise mathematical consequences of his assumptions (rather than merely use them, as he did, as guides in the search for simple mathematical relationships in the motions of the planets), he would actually have found that they were mutually contradictory, as Treder[55] and Hoyer[45] have shown.

Let me finish this section by briefly mentioning the two most important of the other forces that Kepler considered. Both brought him exceptionally close to the spirit of Newtonian dynamics.

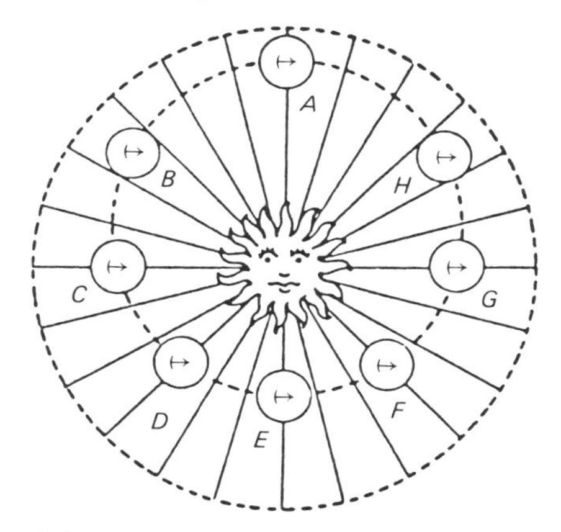

Fig. 6.13. Reproduced from *Johannes Kepler Gesammelte Werke*, C.H. Beck'sche Verlagsbuchhandlung, Munich, Vol. 7, p. 337.

Among the options that Kepler considered to explain how a planet could move towards the sun in one part of its orbit and away from it in another, his favourite by far was a mechanism that involved magnetism. In effect this required the body of the planet to be the seat of a magnetic dipole, the direction of which, lying more or less in the ecliptic, had to keep a fixed direction relative to the stars at all times. Thus, in one position in the orbit the 'north pole' of the dipole would point towards the sun, but in the opposite position the 'south pole'. At the two intermediate positions, the two poles would be equidistant. Kepler's diagram is shown in Fig. 6.13. In accordance with this proposal, the surface of the sun had to have one polarity, say 'north', and its interior the opposite. Thus when the north pole of the planet pointed towards the sun the planet would be repelled in the radial direction, but there would be attraction in the opposite part of the orbit and intermediate degrees of repulsion and attraction (which Kepler always related directly to the speed of the radial motion) at other points of the orbit. In this way Kepler supposed that the tantalizingly simple libratory radial motion which he had found could be generated. The direct influence of Gilbert's book on magnetism is nowhere more evident than here.

A point to note about this explanation, which to the modern physicist may seem rather crude and pedestrian (it also suffered from severe difficulties – as Kepler himself pointed out in Chap. 57 of the *Astronomia Nova*, one would naturally expect the dipole axis of the earth to lie along its rotation axis, but that gave an orientation which did not fit the requirements), is that it anticipates Newton's law of gravity in requiring a magnetic faculty to be present in both the sun and the planet. There is a genuine interaction – one of the most characteristic features of dynamics – between the two bodies. By comparison, the solar rotatory force was a much more archaic conception, reducing ultimately to mere pushing.

Finally, in Kepler's ideas about gravity, we have – except for the false coordination of force with speed rather than acceleration – a nearly perfect anticipation of Newton. This is what Kepler had to say about gravity in the *Astronomia Nova*:[56] 'Gravity is a mutual propensity between like bodies to unite or come together... so that the earth draws the stone to it much rather than the stone seeks the earth,' i.e., the stone does not seek the centre of the universe but is pulled to the earth, wherever the earth itself may be carried. Particularly remarkable is the following passage: 'If two stones were to be placed anywhere in the world outside the range of influence of a third similar body, then each stone, like two magnetic bodies, would come together at an intermediate point, each stone travelling towards the other a distance proportional to the bulk of the other.'

Here the interactive nature of the phenomenon is quite explicit, i.e., bodies exert a mutual influence on each other: if body A moves body B, then body B will simultaneously move body A. Moreover, when allowance for the sizes of the objects is made, the mutual influences are equal in magnitude and opposite in direction. The almost perfect anticipation of Newton's quantitative Third Law is particularly striking – so far as I know it is the earliest. Yet Kepler was able to make but little use of his concept of gravitational attraction – he used it qualitatively to account for terrestrial gravity (and thus also the existence of the earth as a compact body that nevertheless could move through space) and also the tides, which he correctly attributed to attraction by the moon (thereby earning the disapproval of Galileo, who was strongly opposed to the 'occult attractions' in which he felt Kepler was far too ready to dabble). However, lacking entirely the notion of inertial motion, Kepler had no idea that gravity could explain why the moon orbits the earth. Indeed, the mutual gravitation of the earth and moon was if anything an embarrassment and had somehow to be neutralized:[56]

> If the moon and earth were not held in their orbits by an animalistic or other force, the earth would rise up towards the moon by a 54th part of the distance between them and the moon would descend about 53 parts of the interval between them; at that point they would come together. However, it is here assumed that the substance of both is of the same density.

We see here Kepler bridging the Aristotelian gulf between the heavens and the earth by his importing of ideas taken from terrestrial physics into astronomy. But all the daring is flawed by what would later be seen to be a hopelessly inadequate conception of the manner in which forces influence the motion of bodies.

One final respect in which Kepler anticipated Newton was in his awareness that in a consistent physical scheme the very slow secular changes observed in astronomy – the advance of the apsides, the backward motion of the nodes, and the movement of the earth's axis which produces the precession of the equinoxes – all needed to be explained simultaneously by the action of the basic forces which caused the main observed motions. Needless to say, none of his explanations (which he discusses most fully in Chap. 57 of the *Astronomia Nova*) was correct but his instinct was again sound.

To summarize: what did Kepler get right in dynamics? Except for the one basic error, he had a very clear idea of the need for forces. The need was made manifest on an extremely sound basis – the extremely suggestive occurrence of the planet–sun distance (and the actual position of the sun) in all the precise laws of planetary motion that he had found. Pondering deeply the precise laws, he was led to conceive planetary motion as being determined by two basic forces, one acting radially, the other azimuthally. In the *Epitome* the motions generated by these two forces are in effect combined in accordance with the laws of vector addition. Such composition of motions was another anticipation of mature Newtonian dynamics. We have noted the early form of Newton's Third Law. Finally in trying to set up a comprehensive system of dynamics for all the 'movables' in his still rather archaic cosmology, he went quite a long way towards the universal concept of motion and matter that Descartes introduced.

What Kepler totally lacked were Newton's First Law and the correct form of his Second Law. The reader should not get the impression from this account of Kepler's physics that it was the mere misfortune of choosing one particular basic law of motion rather than another that stood between him and the insights that Newton achieved. From the astronomical phenomena in which Kepler immersed himself so deeply he could hardly even have dreamed of taking uniform rectilinear motion as the foundation of dynamics. How could the nearly perfectly circular motions of the planets ever have suggested such a thing? But even if he had had such an idea, he would still have had no stimulus to do all the extra work of elaboration needed before anything like Newtonian theory could possibly have emerged.

In astronomy Kepler fitted all the pieces together with consummate skill. In the matter of celestial dynamics he merely set the agenda.

6.7. Kepler's anticipation of Mach's Principle

It is not good for the wanderer to stray in that infinity[57]

There are obvious dangers in attempting to draw parallels between the thought of people who lived centuries apart; for in a very real sense they inhabit different worlds, and any parallels which do exist are at most partial. However, I believe that certain similarities between the approaches to the problem of describing and explaining motion that Kepler and Mach adopted are so striking that they warrant closer examination. Moreover, the issues at stake have not lost any of their relevance – we are concerned here with the perennial problem of the conceptual foundations on which physical theory is to be based.

Mach's criticisms of the Newtonian concepts of space, time, and motion had two sides: (1) a critical epistemological rejection of the idea that motion could be referred to an invisible framework of space and time conceived to exist quite independently of matter and its motion; (2) a positive suggestion about how the difficulty which led Newton to introduce his concepts of absolute space (without which he could not see any way of formulating his First Law of motion) could be overcome by ascribing to matter an as yet unsuspected capacity to influence the motion of other matter. My thesis is that, in both respects, Kepler anticipated Mach – and, moreover, that Kepler's 'Machian' instincts had very positive effects.

Let us begin with the first point. We have seen in Chap. 2 that there was a 'prerun' of the absolute/relative debate in antiquity. It was stimulated by the atomists and their introduction of the idea of atoms moving through the void. Aristotle's rejection of this idea (on the grounds that the completely empty, uniform, and featureless void made it impossible to conceive of definite motions) was a significant factor in the development of his matter-based cosmology. We have seen how well the Aristotelian cosmology (in its broad features) matched the practical needs of astronomers to define position with respect to something real, definite, and observable. In this respect Kepler was merely the last member of a distinguished quintet composed of the philosopher Aristotle and the four great practical astronomers: Hipparchus, Ptolemy, Copernicus, and himself. Kepler's ultimate notion of position is therefore virtually indistinguishable from Copernicus's. For example:[58]

The region of the fixed stars supplies the movables with a place and a base upon which the movables are, as it were, supported; and movement is understood as taking place relative to its absolute immobility.

And again we find an echo of Copernicus's rather commonsense standpoint that to conceive a motion of the system of stars as a whole is nonsense, so that the most sensible approach is to say that, *by definition*, motion is motion relative to the stars:[59]

The second argument destroys completely every movement of the sphere of the fixed stars. For it is not apparent for whose good, since nothing is outside of it, it changes its position and appearances by being moved to what place or from what place, and since it obtains by rest whatever it could acquire by any movement. For the movements of all bodies are understood from its rest, and unless it gives them a place, as it can do perfectly by being at rest, nothing can be moved.

It is particularly interesting that in the early seventeenth century Kepler was confronted with the same sort of alternative cosmologies that Aristotle had reacted against and reacted against them in much the same way, revealing a characteristically Aristotelian dislike of the vacuum and all things indefinite. For although atomism itself was not yet being revived as a serious contender, Gilbert (a hero of Kepler's, as we have seen) and Bruno had already both argued strongly for the infinity of the universe. Kepler wrote:[60] 'Gilbert's religious feeling was so strong that, according to him, the infinite power of God could be understood in no other way than by attributing to Him the creation of an infinite world. But Bruno made the world so infinite that [he posits] as many worlds as there are fixed stars.'

Kepler recoiled from the idea of a limitless space. His Aristotelian reflexes are very evident in the comment:[61] 'This very cogitation carries with it I don't know what secret, hidden horror; indeed one finds oneself wandering in this immensity, to which are denied limits and centre and therefore also *all determinate places*.'

His whole cosmology expresses the Greek abhorrence of the infinite void. Here is his description of the world:[62]

The sun is the fireplace of the world; the globes in the intermediate space warm themselves at this fireplace, and the sphere of the fixed stars keeps the heat from flowing out, like a wall of the world, or a skin or garment – to use the metaphor of the Psalm of David.

The developments that are to be described in the second half of this book show that in his cosmology Kepler was clinging to attitudes that would prove to be inimical to a general trend of thought that led to the acceptance of the law of inertia and the triumphant overcoming of the defects of Kepler's celestial physics mentioned in the previous section. In his book on this great shift of basic attitudes, *From the Closed World to the Infinite Universe,* Koyré correctly casts Kepler as a reactionary in this respect:[63] 'We have to admit that Johannes Kepler, the great and truly revolutionary thinker, was, nevertheless, bound by tradition. In his conception of being, of motion, though not of science, Kepler, in the last analysis, remains an Aristotelian.'

However, on issues of this kind, concerning basic questions such as the nature of motion, the time scale can be important. What is a harmful influence over decades may be beneficial over centuries. Kepler certainly looks back to Aristotle but he also looks forward to Mach and Einstein – and perhaps even beyond.

For on the question of the need to relate motion of a studied body to other observable matter – an aspect of the problem of motion with which

Koyré was not concerned when he made the above judgement – the astronomers were on very sure ground, none more so than Kepler. The epistemological facts of life were constantly thrust upon them by the nature of their trade. Astronomers, unlike philosophers, make measurements. Even if you have a good ruler and can put one end against the object whose position is to be measured, no measurement can be made unless there is at least one other real mark against which the other end of the ruler may be placed. In astronomy, such markers are the stars. Given this total reliance on the stars (and here we must also include the planets and the sun – just think of the role that both Mars and the sun as well as the stars played in Kepler's triangulation of the orbit of the earth!), it is hardly surprising that not only Kepler's practical work but also the theoretical ideas described in the *Astronomia Nova* are often startlingly 'Machian'. Moreover, it is easy to demonstrate that a very positive influence was here at work.

Measured simply in terms of the size of the corrections that Kepler made to the Copernican motions, his innovations do not seem all that remarkable. However, what made the true Keplerian revolution was the fact that his corrections brought all of the planetary motions into quite perfect alignment with the sun. If we ask what it was about Kepler's approach that made this possible, we find above all two related attitudes of mind that distinguish him from Copernicus and, in conjunction with the Brahian observations and his extraordinary technical competence, were responsible for his success.

The first was the remarkable degree to which Kepler, acting on his gut feeling that position is ultimately defined and determined by matter and not void space, raised his level of reliance on matter to define the position and motion of other matter. As we have seen, Copernicus relied on the stars to give him a concept of position. However the stars were enough for him. He was perfectly happy to imagine the planets wending well-defined ways (around precisely defined but void points) in the mathematical space that the distant stars defined. The fact that he sensed no difficulty in such a situation may have been in part due to an instinctive belief in spheres actually carrying the planets. But when Kepler came to the problem he found the space of the solar system swept clear by Brahe's destruction of the spheres. The ethereal air of interplanetary space is the arena of the planetary motions that is constantly presented to the reader in the *Astronomia Nova*. And the supreme problem that Kepler confronts throughout the work is this: how can it possibly come about that the planets follow definite paths through the completely featureless ether? How do they 'know' where to go? And what moves them? We have seen how such questions recur throughout the book.

The key to Kepler's success was his unhesitating belief that the planets must somehow 'use' not only the distant stars but also the more proximate bodies, above all the sun. One could say that for him the Copernican scenario, post Brahe, was like skaters on a vast lake at night who each describe an intricate but most precise and regular path about a single totally unmarked point (the ghostly Copernican mean sun), performing this feat by nothing more than an occasional glance at the distant stars. This was just too implausible for words. Kepler insisted that the sun must play a crucial role as both hub and marker about which the whole dance turned. He therefore adopted Copernicus's 'Machian' definition of position by means of the stars much more urgently and extended it to the sun. This was the first decisive shift, and it provided Kepler with a valuable heuristic principle in deciding which motions of the planets were and which were not plausible.

Time and again we find him rejecting putative motions because they require the planet to follow a path through featureless space around an entirely unmarked point. For example:[64] 'I do not deny that one can conceive a point and around it a circle. But I maintain that if the centre point exists only in thought, timeless, without outer sign, then in reality no mobile body can form a perfectly circular path around it.' The importance to Kepler of markers to define position and directions is nowhere more explicit than in his discussion, already mentioned briefly in Sec. 6.5, in Chap. 57 of the *Astronomia Nova* of how a planetary mind, if one existed, could control the sun–Mars distance in accordance with the law $r = 1 + e \cos \theta$, where θ is the eccentric anomaly, measured from the centre of the orbit. Kepler supposed that, through some sensory means, the planetary mind could gauge its distance from the sun by measuring the apparent angular diameter of the sun. Determination of the eccentric anomaly (from which the mind would 'deduce' the appropriate planet–sun distance in accordance with the above law) required on the part of the mind a capacity to sense two things. The first was the direction of the aphelion of the planet. Kepler did not find this problematic, since the line of the apsides points to a definite position on the heaven of the stars 'and the fixed stars are real bodies'. It was therefore reasonable that the planetary mind should have an awareness of this position. However, the final element in the task of steering the planet was for Kepler much more problematic; for the planetary mind also needed to have an awareness of the position of the centre of the orbit, but this was void and was therefore ruled out by the kind of argument quoted above. Kepler worried away at this problem and only declared himself content when he found that there existed a relatively simple mathematical relationship between the *true anomaly* and the apparent diameter of the sun as seen from the planet. But the true anomaly was something that the planetary mind could observe directly, since it was measured from the body of the true sun. Thus, control of the planet's motion was linked explicitly and directly to the real sun and the real stars, which between them defined the true anomaly.

The confidence which Kepler thereby gained in the correctness of his distance law was an important stimulus to his persistent and finally successful attempts to reconcile it with an elliptical orbit.

If the role of material bodies in defining position is most manifest when Kepler looks at the problem from the point of view of putative planetary minds, a more physical 'Machian' aspect comes to the fore when Kepler is thinking in terms of forces. This brings us to the second significant shift of attitude from Copernicus's standpoint. We saw in Chap. 5 that as far as planetary motions were concerned the sun played an entirely passive role in the Copernican scheme – it was a mere dispenser of light and heat, indeed so irrelevant to the problem of the planetary motions as not to warrant inclusion in the diagrams depicting them! In contrast, Kepler, as we have seen, accorded the sun an indispensable role as motor of the planetary motions. Matter is quite vitally involved in physically determining the motion of other matter. This indeed is the core of the Keplerian revolution, and is rightfully reflected by the depiction of the sun at the hub of all his diagrams in the *Astronomia Nova*. It was this conviction which led Kepler to suspect that the planetary motions were governed by simple mathematical laws in which the position of sun was the key to everything else. Without this conviction the area law, and with it the ellipse too, could never have been found.

There is a striking parallel between the way in which Copernicus accepted as merely fortuitous the presence of the sun near what he took to be the centre of the universe, i.e., near the centres of the orbits of the planets, and the way in which Newton accepted the fact that the distant stars appeared to be at rest in what he called absolute space, and we now call the family of inertial frames of reference. In forming their conceptions of motion, both discounted the possibility that the known coincidence might have profound physical significance; neither the sun in *De Revolutionibus* nor the stars in the *Principia* played more than incidental roles. This parallel between Copernicus and Newton underlines the fact that it was the same instinctive stimulus that led Kepler before 1600 to promote the sun and Mach, more than two and a half centuries later, the stars (or more generally, the matter of the universe as a whole) to the status of primary determinants of motion.

Kepler found his laws of planetary motion by a kind of first-order Mach's Principle (first-order because it considered velocities rather than accelerations). These laws later enabled Newton to peel away the law of universal gravitation from the ancient geometrokinetic law of circular motions of the Greek astronomers, transforming that part of it at least into a genuine physical law. The residuum – the law of inertia – remained geometrokinetic, circles being merely replaced by straight lines. At the end of Kepler's life serious doubts as to the stability of his ultimate markers – the stars that staked out the Copernican cosmos – were just about to be raised by Descartes. Would Kepler, faced with this threat, have followed Newton's lead and substituted space for the stars as the ultimate frame of reference?

I doubt it. He was not at all enamoured of the idea of space existing on its own as a physical entity independent of tangible and perceptible matter. To quote from the *Epitome*:[65] 'If you are speaking of void space, that is, of what is nothing, what neither is, nor is created, and cannot oppose a resistance to anything being there, you are dealing with quite another question. It is clear that [this void space], which is obviously nothing, cannot have an actual existence.' It is true that in Chap. 2 of the *Astronomia Nova* Kepler grants that a body might be able to follow a straight line through the empty ethereal air. But at the back of his mind he always had the stars to define such a motion. Thus although in Kepler's world the planets find their way by looking to markers that he believed to be fixed – the sun and stars – his conceptual framework is in fact only one short step from the solution to which Mach was led when he confronted the fact of universal motion of all matter in the universe. The natural progression from Kepler's scheme is not one in which all the bodies in the universe look to invisible space 'to see where they should go' in their motion but rather the fully Machian one in which 'all look to all' and perform coordinated motions relative to each other independently of space.

There is another affinity between Kepler and Mach that is worth mentioning, particularly in connection with an interesting difference between Kepler and Galileo in their approach to empirical facts. Although Galileo is generally given the credit for being the person who brought home to the world the need to found science on empirical fact – and there are indeed certain passages in his writings that get this message across most effectively – in their personal practice Kepler's record is much more impressive than Galileo's, and the image of Galileo as a dedicated empiricist, fostered by the story of his having dropped weights from the Leaning Tower of Pisa to disprove the most notorious error of Greek physics, is something of a distortion of the true facts. He used empirical observation to *suggest* theoretical schemes of great simplicity. This empirical input made him a great revolutionary, and in his sense for what is significant and the scope it offers for constructing a harmonious theoretical scheme he strongly resembles Einstein. But with regard to experiments and accurate observation, Galileo actually had a rather cavalier attitude – they were means to reveal the underlying simplicity and geometricity that he suspected everywhere in nature, not, in striking contrast to what Kepler made of them, precision tools for delicate probing of the secrets of nature.

The difference between the two men is highlighted by Galileo's theory of the tides, which will be discussed in the next chapter since it bears closely on the debate about the absolute and relative nature of motion. In accordance with this theory, the tides should exhibit a 24-hour periodicity, since Galileo attributed them to rotation of the earth. In fact, however, high tides recur about 50 minutes later each day, being tied to the motion of the moon. Galileo seems to have been quite capable of overlooking this embarrassing and quite large discrepancy in the most important quantity in his theory (the period of the tides). In contrast, when Kepler had brought the art of astronomy to the highest state of perfection compatible with perfectly circular orbits of the planets the residual discrepancy with the observations was minuscule yet it led Kepler to demolish the whole scheme. And what was most remarkable about this rejection of his own labour was not so much the fact that it was done for the sake of such a small quantitative failure – as we have seen, the ideal of saving the phenomena to that level of accuracy was already *de rigueur* when Kepler joined Brahe and his assistants – but Kepler's acute awareness, amply confirmed in the event even if not exactly in the manner he originally anticipated, that the minuscule discrepancy was precisely the evidence which pointed to the need for a profound reappraisal, truly from the ground up, of the entire prevailing conception of things.

Kepler could never have embarked upon his extraordinarily arduous revision of what had already been a monumental labour had he not had the deepest conviction that the deliverances of nature through the senses, however slight they might appear, were the carriers of information of the deepest significance. What led him to that conviction? It was not simply the belief that the world is organized on a rational and harmonious basis; for that he shared with all great scientists. What sets him apart is the intensity with which he saw significance in the concrete and actual manifestation of the world. It is here that we find the second link with Mach. There is a kinship in their ontology. Many of the great scientists, possibly under the influence of Plato (surely the case with Galileo), had a certain distrust of the senses and instinctively looked behind them for conceptual entities that spoke more directly to the mind. Although Kepler too provided a classic example of this with his perfect Platonic solids, he had simultaneously a religious devotion to the world as it actually is, seeing in palpable creation a direct expression of the ultimate reality, God. In a significant passage early in the *Astronomia Nova* he states as a fact that:[66] 'The divine voice which calls man to study astronomy is expressed in the visible world itself, not in words and syllables, but in the things themselves.' The same thought is expressed in his Calendar for 1604 which Caspar quotes at the beginning of his biography of Kepler:[16]

I may say with truth that whenever I consider in my thoughts the beautiful order, how one thing issues out of and is derived from another, then it is as though I had read a divine text, written into the world itself, not with letters but rather with essential objects, saying: Man, stretch thy reason hither, so that thou mayest comprehend these things.

Kepler took the writing of the divine text into the world in the form of essential objects as literal truth. It provided the underpinning of his cosmology:[67] 'The Sun represents, symbolizes, and perhaps even embodies God the Father; the stellar vault, the Son; and the space in between, the Holy Ghost.' Elsewhere he speaks of the universe as:[68] 'the image of God the Creator and the Archetype of the world . . ., there are three regions, symbols of the three persons of the Holy Trinity – the centre, a symbol of the Father; the surface, of the Son; and the intermediate space, of the Holy Ghost.'

It is on this divine ground that the planets are seen to move. Such is the attitude of mind which explains Kepler's dedication to observational facts, which he handled with a care akin to reverence and love – for him, the observations were literally as precious as jewels. No wonder he clung tenaciously to the evidence of the observations – the most faithful guides (*fidissimi duces*), as he called them.

This might all seem a far cry from post-Darwinian Mach, an anti-clerical sceptic and a ridiculer of religious interpretations of physics. Yet on closer examination there are some striking similarities. Deeply distrustful of all mechanical models of physical phenomena, Mach took the phenomena themselves, the direct deliverances of the senses, to be the only ultimate reality. Having banished models, yet deeply persuaded of the existence of order in what is perceived, Mach actually put himself in a situation remarkably like that in which Kepler found himself after Brahe had destroyed the spheres that were supposed to carry the planets. In both cases the props were all gone – it was necessary to understand the order which undoubtedly existed directly in terms of what was given. Even the mathematical relationships long recognized in the phenomena required the forging of a new framework in which they could be comprehended. It is in their instinctive reactions to the very similar problems that they faced that we see the kinship of Kepler and Mach. Moreover, Kepler's dedication to the essential objects in which he believed the divine text to be written is paralleled by Mach's dedication to the phenomena – for Mach had a true devotion to sights, sounds, and colours and wanted to show that they, and not conjectural atoms devoid of all phenomenal accidents, were the true bricks of the world.

Because Kepler and Mach share a certain childlike primitivity, we should not underestimate the vitality of their ideas. Kepler's own Platonic solids proved to be nothing but an empty shell, but with his early anticipation of Mach's Principle he found the laws of planetary motion. The more sophisticated Galileo, with his distrust of the senses, was led by his more Platonic concept of space (which contrasts with Kepler's Aristotelian notion of position – roughly the divide between the absolute and relative approaches to motion) to formulate a quite erroneous theory of the tides. And it was the Galilean line which Newton developed with his notion of inertial motion in absolute space – the very notion that led Mach to revive a principle which Kepler had used to such good effect. Let Kepler's dramatic success at least serve as ground for considering Mach's Principle seriously.

It may also be noted that in one of the great problems of today's physics – the reconciliation of quantum theory with gravitational theory in a cosmological setting – modern researchers find themselves in a situation strikingly reminiscent of the one that Kepler confronted after Brahe had removed all the spheres. For in its conventional form, quantum mechanics is formulated in a given space–time background, but in the cosmological setting such a prop really has no place or warrant. No one has yet succeeded in reformulating quantum mechanics without the old framework, though that is now as obsolete as the crystal spheres. Let me encourage the reader to ponder these questions – and take heart from Kepler's example!

The whole discussion of this section emphasizes the ebb and flow of attitudes with respect to ultimate questions. Copernicus and Kepler had an extremely clear notion of motion as being with respect to matter. But in the last decade of Kepler's life Descartes began the hatching of a new scheme which shattered the sphere of the fixed stars and led to the establishment of empty space as the ultimate referential basis of motion.

About 40 years after Kepler published the *Epitome of Copernican Astronomy*, the concept of motion had passed from being completely matter-based to completely space-based. Yet the actual observations, the objective facts from which both conceptual schemes sprang, were still exactly the same. The astronomers were still carefully tracking Mars across the backcloth of the heavens.

That was why the pendulum, having swung suddenly one way, eventually swung back again.

6.8. A last look at the astronomy and evaluation of Brahe and Kepler's achievement

Almost all of the astronomical input required by Newton to create his synthesis of dynamics was published before any of the really significant complementary work on terrestrial motions appeared (though the crucial work of both Kepler and Galileo was in fact done almost exactly contemporaneously). Therefore, before we turn to that work, it is appropriate to have one last look at the astronomy, which achieved such a wonderful state of maturity in the observations of Brahe and theory of Kepler.

Examined over its two-millennial history, the science of naked-eye astronomy can be seen to have developed very largely in an entirely predictable manner. Virtually all the significant effects were discovered in the order that one would expect on the basis of the magnitudes of their manifestations and ease of observation: first the larger effects associated with the motion of the sun and moon, then the gross effects in the motions of the planets, finally the corrections of higher order in the eccentricities of the planets' orbits. Much of the fascination of the planetary astronomy comes from the gradual uncovering of the subtle and delicate interplay of the eccentricity of the earth–sun motion with the eccentricities of the orbits of the other planets.

Superimposed on this regular predictability were certain other major events. It is not at all clear that these were bound to occur at all or, if so, when and in what order. Eclipses were such awesome events that the development of techniques to predict them does seem rather inevitable. However, the methods employed to do so do not appear nearly so inevitable in the light of the Babylonian numerical techniques. Thus, the first great unpredictable was the theoretical application of geometry by the Greeks. Equally important and uncertain was the subjecting of the geometrical models to precise observation. The brilliant idea of using geometry to represent motions observed in two dimensions by actual motions in three dimensions might well have come to nothing had not Hipparchus married it to the quantitative astronomy of the Babylonians. Toomer has written fascinatingly on the question of how Hipparchus acquired the Babylonian expertise.[69] He thinks Hipparchus may have had to travel to Babylon and get instruction at first hand. The whole history of naked-eye astronomy might therefore hang on a single decisive journey (it was certainly brought to a close by another – Kepler's to Prague). Another great unpredictable was the *time* at which the geocentric-heliocentric revolution occurred (the fascination with which can only grow the more one is exposed to its multifarious consequences). It very nearly happened before Hipparchus, yet one can see no reason why the spark that illuminated the possibility of the leap should not have occurred after the Brahian programme of ultraprecise observations to track down the last defects of the Ptolemaic models. Finally and equally mysterious, we have the appearance at particular times of the two theoretical genii: Ptolemy and Kepler. Seen in the light of these unpredictables, the history of pre-telescopic astronomy was not so much a steady and mundane accumulation of observed facts and interpretations as a fairy story that began, true to the genre, in Babylon, passed through a miraculous transformation, and ended in the city of the Winter Queen!

Almost as intriguing as these aspects is the role played by the circles. How marvellously circular the orbits are! And what a fortunate fluke was the equant and the manner in which it permitted Ptolemy to retain the mathematical tractability of circular motion in a sophisticated device that improved the accuracy of prediction so immeasurably. The conservative astronomers may have fought the equant tooth and nail but they could not do without it. Thus it remained an indispensable part of astronomy until Kepler transformed it so surprisingly into the area law (probably the supreme achievement of theoretical astronomy – and what a transformation was yet to come in Newton's hands!) Then there was the final flowering of the circles when for a brief period of about two years Kepler had put together a carousel which would have astounded Eudoxus, Calippus, Plato and Aristotle such was its simplicity (six circles) and precision. Even when Kepler pulled the whole beautiful geometrical toy apart for the sake of his seemingly mad physics, he still put the circles of Apollonius and Hipparchus to work for him. In fact, one never quite knows how he conceived the epicycle that weaves its way through the tortuous chapters of the *Astronomia Nova* in which Kepler finally found his own way to the ellipse and area law. At times he certainly accorded it physical reality, at others it and its deferent seem merely to have served as a scaffolding by means of which Kepler could clamber around on the support of the auxiliary circle to see precisely where Mars was going. Indeed, one of the nicest ironies is the way in which Kepler introduces the ellipse by means of the circles. The reader should look again at Fig. 6.12 (reproduced here as Fig. 6.14 directly from the original), which is the termination of the story. The past, represented by the epicycle and deferent, is superimposed on the future, represented by the ellipse. The whole history of theoretical astronomy is compressed into a single figure, testimony to the ingenuity of the astronomers in describing complex motions with inadequate mathematics stretched to the limit.

Finally we should mention especially the progressive development of Hipparchus's original simple but effective idea for getting from actual observations to representations of the solar inequality by a simple eccentric. We have here an organic development which is rather like the growth of polyphony out of Gregorian chant. For we have seen how the original problem (simplified by the presence of two right angles) was first generalized to arbitrary angles to treat the theory of the moon and eclipses and then drastically modified by Ptolemy into what Swerdlow and Neugebauer call[70] 'a *tour de force* of possibly the most complex and extended calculations in all of ancient mathematics': Ptolemy's problem for determination of planetary equants and apsides. They also emphasize the revival of this technique by Copernicus. Indeed, the importance of this solid piece of work on Copernicus's part is quite evident in the account which Kepler gives in the *Astronomia Nova*, where he discusses

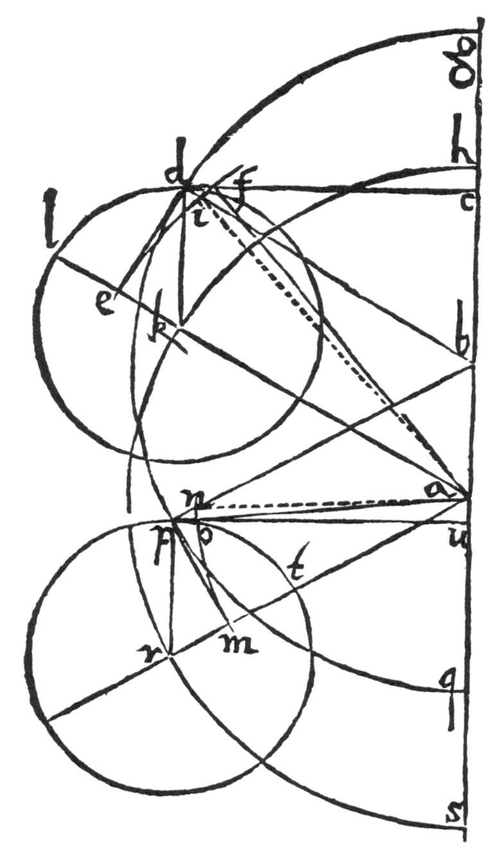

Fig. 6.14. Kepler's diagram showing the construction of the elliptical orbit of Mars as it appeared in the 1609 edition of the *Astronomia Nova*. The construction is explained on p. 319 ff. (Reproduced by courtesy of The Beinicke Rare Book and Manuscript Library, Yale University.)

first Brahe's adaptation of Copernicus's work and then his own development of the *non plus ultra* which the Hipparchan problem finally became: Kepler's problem for determining his vicarious hypothesis. This sequence of problems and their solutions represent the irreducible nontrivial core of pre-telescopic theoretical astronomy. Nearly all the great triumphs rested upon this steady evolution of theoretical techniques; in particular, the method achieved its ultimate success when it provided Kepler, through the failure of the vicarious hypothesis to match the Martian latitudes and nonacronychal longitudes, with unambiguous evidence of a breakdown of previously well-proven concepts and thereby opened the way to the discovery of ellipticity (and its own demise).

This brings us appropriately to the evaluation of the significance of Kepler's discoveries. One can often find it asserted that the greatest importance of Kepler's work was its overthrowing of the millennial tyranny that the idea of perfectly circular uniform motion had imposed on the mind of man. Kepler is held to have been the man that broke this log jam. There are several objections to such a view.

First, the notion of the tyranny of the circles carries with it the implication that their adoption was what had been holding back progress. It should by now be evident that the concept of circular motion was in fact just about the *best* possible assumption the ancient astronomers could have made. The problem of the planets could not possibly have been cracked without it. Prior to Kepler's clarification of the precise nature of the earth's motion, any use of ellipses or other curves to describe the motion of the planets would have been a *step back*. This point has already been adequately made.

A second objection to seeing Kepler's especial merit as being in the overthrow of the ancient dogma is that it tends to obscure the important fact that half of this service was performed by Ptolemy and leads to an unbalanced appreciation of the key factors at work in the history of astronomy.

In fact, if one wants a discovery that truly had an effect like the one attributed to the replacement of circles by ellipses, one should look to Ptolemy's discovery of the equant and his willingness to take it at its face value. He would undoubtedly have been able to devise schemes of the sort that Copernicus introduced and hide the fact that the sacred cow of *uniform* motion was violated in the heavens. Instead, he advertised the effect and thereby stimulated two successive revolutions: the one brought about in the attempt to undo the equant (Copernicus's) and the one that flowed from Kepler's enthusiastic welcome for the equant as a most fundamental fact of celestial motions.

So much for the role of the circles *before* Kepler made his discovery. What about the effect of the ellipses after it?

The simple truth is that for a long time they made remarkably little stir. Kepler's replacement of circles by ellipses was not followed by anything remotely comparable to the debate that followed Copernicus's proposal of terrestrial mobility. In fact, the process of assimilation and acceptance of Kepler's work was very gradual and was still far from complete when Newton found his dramatic dynamical interpretation of Kepler's laws almost exactly 50 years after the Imperial Mathematician had died. To get a clearer picture of the influence that Kepler exerted, it is necessary to distinguish the evolution of ideas among professional astronomers and among the more general community of natural philosophers. The reflection of this process in the technical astronomical literature of the seventeenth century has been charted by Russell.[71] Kepler's reputation among the astronomers was essentially established by the *Rudolfine Tables* that he published in 1627. Their accuracy was simply so much superior to any of the rivals that astronomers slowly gained confidence in Kepler's discoveries. An especially notable triumph was Kepler's rather accurate prediction of the time of a transit of Mercury across the sun in 1631, which was duly observed by Gassendi (the first such observation in history). It is interesting to note that of Kepler's three laws it was the second (the area law) that gained least acceptance, not so much because its accuracy was questioned but on account of the inconvenience of working with it. For this reason, modified equant devices remained popular right up to the time of Newton's discovery. This highlights not only the remarkable accuracy and utility of Ptolemy's discovery but also Kepler's achievement, crucial for the final synthesis of Newtonian dynamics, in breaking with the equant.

Turning now from specialized astronomy to the general evolution of ideas about cosmology and the nature of the world, I can find no evidence at all that Kepler's discoveries played any significant part in bringing about that dramatic change in outlook which is so well characterized by the title of Koyré's book *From the Closed World to the Infinite Universe*. This change was much more due to factors such as the Copernican proposal, the supernovae of 1572 and 1604, the host of telescopic discoveries that began with Galileo's observations in winter 1609/10, and the associated revival of interest in ancient rivals to Aristotelian philosophy, above all atomism. Bruno, the first great speculator who helped to bring about acceptance of an infinite universe, had been burnt alive before the ellipses were discovered. Galileo totally ignored them and anyway, as we shall shortly see, remained tied to a remarkable degree to the old cosmology. To claim that the discovery of the ellipses helped significantly to change the *climate of opinion*, one must show that it significantly shaped the philosophy of Descartes, for it was in his work more than any other that the new outlook crystallized. Here there is no evidence at all that the ellipses *per se* impressed Descartes, who tended very much to miss those sorts of minutiae. If Kepler did influence him, it was probably more through his universal concept of motion, in which he made a much more significant break with Aristotle than did Galileo. (Descartes may also have got the idea of his famous vortex theory of planetary motion – to be discussed in Chap. 8 – from a similar idea that Kepler put forward very tentatively in Chap. 57 of the *Astronomia Nova*.)

Thus, whereas in 1602 it took considerable intellectual courage on Kepler's part to reject the circles, many factors quite unrelated to his very specialized work on planetary motions had contributed within 25 years to the development of a situation in which the cosmological framework that had sustained the belief in circles was rapidly being abandoned. There were many factors that led the natural philosophers of the middle of the seventeenth century to embrace the idea of an infinite world and accept the paradigm of uniform rectilinear motions in place of the eternal circular motions of the Aristotelian universe. Kepler's detailed laws had a minimal part in the process. The circles went without the push that Kepler gave them.

There is another way of looking at this question. What would Kepler's influence have been had Brahe's accuracy been rather less than it was or, equivalently, had the eccentricity of Mars's orbit been half as great (and the ellipticity therefore only a quarter as great), so that Kepler's discovery of the ellipticity of Mars's orbit would have been impossible? His influence would undoubtedly have been almost as great. The halving of the earth's eccentricity and the replacement of the mean sun by the true sun would alone have seen to that.

In fact, consideration of these two achievements of Kepler, which, together with the Zeroth Law, date from his earliest work with Brahe and were both anticipated in the *Mysterium Cosmographicum*, give us the key to the significance of his work at the deepest level. It was, as hinted earlier, the unambiguous implication of the sun in the motion of the planets. The implication was so significant because it was established through precise quantitative laws. Approximate equality can be dismissed as mere chance; if the sun is merely *near* the centre of the world defined by the motions of the planets, no deep dynamical significance follows from that fact. When Kepler showed that the lines of the apsides of all the planets passed exactly through the centre of the sun (a result that does not depend on the precise final form of his laws) he had done enough (assuming only the continued existence of sustained human scientific endeavour) to ensure the eventual establishment of a view of nature and matter profoundly different from what had gone before. His was the first clear step towards the modern synthetic view of dynamics in accordance with which the motion of matter is *actively influenced* by the physical effect of other matter. (For those familiar with the Lagrangian formulation of dynamics, Kepler's work presaged the appearance of interaction terms in all nontrivial dynamical systems). To summarize: Brahe and Kepler's place in history did not depend on ellipticity – but it did crown an achievement both worthily deserved.

Thus, even if Kepler had ceased work when he had cleaned up and perfected the Ptolemaic scheme with perfectly circular but nonuniform equant-type motions, he would still have exerted a powerful influence at the time, about 35 years after his death, when sufficiently acute minds, armed with essential new ideas drawn from terrestrial physics, at last turned seriously and with a reasonable chance of success to the problem posed by the planetary motions. In particular, the notion of a force exerted by the sun on the planets would almost certainly have made its appearance. Without Kepler's work, it is by no means so obvious that this would have happened as might appear to the modern mind. Indeed, we shall see in the following chapters that in the years immediately following Kepler's death in 1630 majority opinion among natural philosophers swung rather decisively against the force-type notions which played such a significant and fruitful role in Kepler's work. However, in the long run it was simply impossible to ignore the solid results that Kepler had achieved. Eventually the two levels of his achievement – the clear identification of the significance of the sun and then the precise details of the planetary motions – were bound to exert a powerful influence. And because his laws of planetary motion, especially the area law, were formulated in a form that reflected the way he had found them through developing the notion of physical forces, Newton's own dynamical interpretation of these laws was of necessity strongly influenced by Keplerian ideas. This came about not so much through Newton's having read and accepted Kepler's speculative notions (it is doubtful whether Newton read any of Kepler's astronomical writings[49,71]) but rather through the fact that Kepler encapsulated the laws of planetary motion with great success and in a particular form. He provided Newton with a complete and specific formulation of the problem and thereby exerted a strong influence on the form of solution that then appeared. There is no doubt that each of Kepler's three laws belongs to the 'baker's dozen' of insights about motion expressed in mathematical form that were necessary prerequisites for Newton's synthesis of dynamics.

One last thought to end the astronomy. Kepler's work represents one of the most complete and definitive syntheses achieved in science. It is the final chapter of a long and very absorbing book. One could be excused for mistaking it for the end of the story, so much seemed to be precisely right. Yet within the outlines so clearly delineated by Kepler there lurked a fantastic and utterly unexpected secret. Imagine a boy who finds for the first time a ripe horse-chestnut with the outer shell intact. Cherishing the golden and curiously shaped object, he might take it home, quite unaware of the shiny brown and perfectly smooth conker ready to spring from the shell on application of a little directed pressure. That was Kepler's fate: he died without an inkling of what his nut really contained.

7

Galileo: the geometrization of motion

7.1 Brief scientific biography and general comments

'In the history of science there are few instances of coeval and complementary works as striking as in the case of Johann Kepler and Galileo Galilei.'[1] These words of Stillman Drake, the well-known authority on Galileo and translator of many of his works, are an apt introduction to this chapter. Kepler brought an ancient science to maturity, but completely lacked certain ideas essential for its interpretation; Galileo more or less created a new science and in the process introduced most of the seminal ideas and results which eventually did make possible the dynamical interpretation of the astronomy which Kepler so remarkably anticipated.

Galileo (1564–1642) was a few years older than Kepler and outlived him by more than a decade. His father, who was an important figure in the theory of harmony and music, wanted him to study medicine. However, Galileo became attracted to mathematics, in particular Euclid's geometry, early in his studies and, with the reluctant agreement of his father, set out on a career to make his living as a professor of mathematics. It is clear from his writings that he greatly admired aspects of Plato's philosophy, especially the emphasis on the importance of geometry. A much quoted saying of Galileo's is his comment that 'Trying to deal with physical problems without geometry is attempting the impossible'.[2] Of even greater importance than the general influence of Plato was the specific influence of Archimedes, whom Galileo never mentioned without the greatest praise. An important aspect of Galileo's work, which it will not be possible to treat here, was his development of Archimedean techniques in statics – both within statics and transferred to problems involving motion.

Galileo's credo can be summarized in three words: Nature is mathematical. The Hellenistic astronomers had mathematized celestial motions; the two aspects of their work – precise measurement interpreted by geometrical theory – were perfected to a fine art by Brahe and Kepler, respectively. Just as this astronomical work was drawing to its triumphant conclusion, Galileo applied essentially the same approach on the earth: he mathematized terrestrial motions. In a brief period of barely five or six years he achieved – single handed – results at least as important as the immense two-millennial labour encapsulated in Kepler's laws of planetary motions.

His early work on motion, done in his first post at Pisa, was Aristotelian in outlook though critical of Aristotle in many details. Characteristic of his sharp powers of observation and his ability to draw far-reaching conclusions from seemingly mundane matters was his reflection on a hailstorm that he witnessed. He noted that hailstones of very different sizes landed more or less simultaneously. Since it was reasonable to assume they had all started to fall simultaneously, the observation was in conflict with the Aristotelian teaching that larger and heavier bodies fall faster than lighter and smaller ones.[3] Despite a certain Platonic other-worldliness (which will be discussed in the penultimate section), Galileo was an extremely practical man who was very interested in engineering problems.

While at Pisa he completed his early unpublished *De Motu* (*circa* 1590), an essentially Aristotelian study of motion. This is important in proving that he had by then already read Copernicus's *De Revolutionibus* (since he uses the Ṭūsī couple) without, it seems, being persuaded at all of its truth. The cosmological conception is still entirely Ptolemaic. *De Motu* is also important in that it casts light on the way in which Galileo discovered and interpreted an early form of the law of inertia (see Sec. 7.3).

According to the reconstruction of Stillman Drake, Galileo's conversion to Copernican cosmology occurred around 1595, after he had moved to Padua as professor of mathematics. It certainly occurred before 1597, for in that year Galileo wrote to Kepler (by whom, we recall, he had been sent a copy of the *Mysterium Cosmographicum*) saying[4] 'as from that position I have discovered the causes of many physical effects which are perhaps inexplicable on the common hypothesis. I have written many reasons and refutations of contrary arguments which up to now I have preferred not to publish, intimidated by the fortune of our teacher Copernicus, who though he will be of immortal fame to some, is yet by an infinite number (for such is the multitude of fools) laughed at and rejected.'

Kepler assumed that Galileo's reference to 'many physical effects which are perhaps inexplicable on the common hypothesis' was to an explanation of the tides, and support for this assumption came to light quite recently in evidence which suggests that Galileo had developed his theory of the tides (see Sec. 7.5), for which the Copernican thesis of the earth's mobility was crucial, by about that date.[5]

The really great period of Galileo's work on motion – the discovery of the law of free fall, a restricted form of the law of inertia, and the parabolic motion of projectiles – covered the years from 1602 to 1608. These discoveries will be discussed in the following sections. Galileo was in the process of writing up these discoveries for publication when the whole course of his life was changed more or less by accident. Hearing by chance about the invention of the telescope, he set about the task of making one himself and soon succeeded in constructing an instrument greatly superior to any hitherto made. Although not the first to do so, he began to observe the heavens – the moon in December 1609, Jupiter and its moons in January 1610. Very soon after he published his booklet *Sidereus Nuncius*[6] (Galileo seems to have intended this to mean *The Message of the Stars* but it is also widely translated as *The Starry Messenger* – *nuncius* can mean either message or messenger). He exploded onto the scene and rapidly became an internationally famous figure.

The work on motion was laid aside for many years – the Latin text he had prepared before the telescopic discoveries was not published until 1638 (eight years after the death of Kepler and thirty years after it – and the *Astronomia Nova* – had been written), when it was incorporated in his *Discorsi*. Meanwhile, Galileo had returned to his native Florence and much had happened since the news of the telescope first circulated in Venice. (In Galileo's time, Padua formed part of the Venetian state, by whom Galileo was employed.) So much has been written about Galileo's telescopic discoveries and his subsequent tussle with the Inquisition, one is severely inhibited in putting pen to paper. Nevertheless, a few facts and reflections are necessary.

The telescopic discoveries were important for several reasons. For a start, they made him so famous that wide readership of his subsequent books on the Copernican system and motion was ensured. They must also have been important for establishing the mobility of the earth beyond all doubt in Galileo's mind. For the first time, he came out in the open in the *Starry Messenger* with a statement, admittedly guarded, that lent support to the Copernican system. Speaking of the moons of Jupiter which he had discovered he said:[7]

> Here we have a fine and elegant argument for quieting the doubts of those who, while accepting with tranquil mind the revolutions of the planets about the sun in the Copernican system, are mightily disturbed to have the moon alone revolve about the earth and accompany it in an annual rotation about the sun. Some have believed that this structure of the universe should be rejected as impossible. But now we have not just one planet rotating about another while both run through a great orbit around the sun; our own eyes show us four stars which wander around Jupiter as does the moon around the earth, while all together trace out a grand revolution about the sun in the space of twelve years.

The Jovian moons were, incidentally, striking confirmation of Kepler's argument that celestial bodies must circle around material bodies, not mathematical points. Kepler's intuitive shift from Copernicus's geometrokinetic concept of motion to a physical, causally-determined standpoint had found further observational support. Quite soon, Galileo discovered the phases of Venus, which unambiguously confirmed the Copernican arrangement of the planets, though it could not settle the argument between supporters of the Copernican and Tychonic systems. Very important for the destruction of the Aristotelian division of the world into the perfect heavens and the corruptible earth were Galileo's observations of mountains on the moon and spots (constantly changing) on the sun. These showed the celestial bodies to be far more like the earth than was permitted by the ossified Aristotelianism with which Galileo had increasingly to contend. Important publications that followed fast on the heels of the *Starry Messenger* were his three *Letters on Sunspots*,[8] the second of which is particularly important for us in containing his first statement in print of his form of the law of inertia (Secs. 7.3 and 7.4).

The next twenty years in Galileo's life, over which we shall skip, were dominated by the developing crisis of his dispute with the Catholic Church about the truth of the Copernican system. The crisis came to a head with his publication in 1632 of the most famous of all his books, the *Dialogo Sopra i due Massimi Sistemi del Mondo Tolemaico e Copernicano* (which I shall refer to as the *Dialogo*; it is translated as *Dialogue Concerning the Two Chief World Systems – Ptolemaic and Copernican*[9]). Written in Italian in the popular dialogue form of the period of which Galileo was such a master, it expounds his general views on motion and the world (Day 1), shows how the dynamical arguments against the diurnal rotation of the earth can be overcome by means of his theory of motion, gives astronomical arguments for the annual motion of the earth around the sun (Day 3), and ends with what Galileo himself regarded as the supreme proof of the earth's mobility – his erroneous theory of the tides (Day 4). For the subsequent history of dynamics, Day 2 is by far the most important: it formulates, albeit somewhat imperfectly and in a restricted form, the law of inertia and the principle of Galilean invariance. Together with Days 3 and 4 of the *Discorsi* (published in 1638 as *Discorsi e Dimostrazioni Matematiche Intorno a due Nuove Scienze*; translated into English as *Dialogue Concerning Two New Sciences*[10] and referred to as the *Discorsi*), it provided a secure basis for the development of a rational theory of motion. Roughly speaking, that part of Newtonian dynamics not directly related to the concept of force as the cause of motion, in particular gravity, stems from this work of Galileo, the bulk of which, as already explained, was done in the period 1602 to 1608. However, it should be emphasized that Galileo's mature work was not at all dynamic in the sense of conceiving motion as produced by forces; for unlike Kepler, Galileo made no attempt in his most important work to identify causes of motion (given the primitive state of the study of terrestrial motions, such an attempt would certainly have been premature). Instead, he simply concentrated on the mathematical description of actual motions. It is because the concept of force plays no role in his work that it is more appropriately referred to as *motionics* than as dynamics. As pointed out in Chap. 1, I use the word motionics rather than kinematics (a word often used by historians of science) because Galileo created a science of the actual motions of real bodies. (Because some of Galileo's results later become integral parts of the modern science of dynamics, it is not always possible to be entirely consistent in terminology, and I do sometimes refer to them as dynamic to emphasize that they express essential results of the modern science.)

It should also be noted that Galileo did not himself provide an overall framework for the science of dynamics. That was more the contribution of Descartes. What Galileo provided was assured quantitative results and proved techniques that Newton was able to incorporate into a quite new but only qualitative framework that Descartes supplied. In fact, we shall see that in many ways Galileo, in contrast to Descartes, stayed remarkably true to Aristotelian concepts, above all to that of natural motions. Moreover, this was actually a help rather than a hindrance to Galileo in the contribution that he made to the discovery of the law of inertia and in providing very nearly the correct explanation for the absence of dynamical effects of the earth's diurnal rotation. Kepler's more radical break with Aristotle led him to a dead end. The first genuine insights into the nature of the most ubiquitous of all phenomena – motion – were achieved by the mathematizing of Aristotle's qualitative motionics. Geometry played a key role in this process.

Galileo's faith in geometry finds expression in a memorable passage published in 1623 in his book *Il Saggiatore* (*The Assayer*).[11] The first half of this *credo in geometria* is given as one of the quotations at the start of this book. The full passage is as follows:

> Philosophy is written in this immense book that stands ever open before our eyes (I speak of the Universe), but it cannot be read if one does not first learn the language and recognize the characters in which it is written. It is written in mathematical language, and the characters are triangles, circles, and other geometrical figures, without the means of which it is humanly impossible to understand a word; without these philosophy is a confused wandering in a dark labyrinth.

Only the first half is given at the start of the book to emphasize that the 'language' in which the 'book' is written remains an open question. However, the fact that Galileo can justly be called the father of modern science is no doubt in large part due to the enthusiasm and vigour with which he applied mathematics, above all Euclidean geometry, to the problem of describing and understanding the world. The *Encyclopaedia Britannica* (15th edition) says of Galileo that 'perhaps the most far-reaching of his achievements was his re-establishment of mathematical rationalism against Aristotle's logico-verbal approach and his insistence that the "Book of Nature" is written in mathematical characters'.

Indeed, one of the most striking things about the *Dialogo* is the way that, time and again, Galileo insists that the arguments for and against Copernicanism must be examined with careful mathematical analysis. It is only *after* such an analysis that the appeal to the evidence of the senses is made. For Galileo, the senses could at times be very deceptive unless properly interpreted. What, however, he did insist upon was that observations of nature always take precedence over rival authorities – such as Aristotle or theologians. Experience is the ultimate arbiter but must be ordered by mathematics.

The reason why Galileo rather than anyone else can be called the creator of modern science is hinted at in the above quotation. Before Galileo, the theory of terrestrial motions had been dominated by the concept of cause. The important thing was to find a qualitative explanation (in terms of essential nature etc.) of why any particular body moved in the way it did. Galileo by no means threw off this way of thinking entirely. Instead, he augmented it by an approach that, at least up to the present day, has proved to be far more fruitful. He stopped looking for *causes* of motion and instead, like the early astronomers, sought merely to *describe* actually observed motions. He no longer asked: *why* does the stone fall, but *how* does the stone fall? In the *Discorsi*, he comments that innumerable books had been written explaining why bodies fall towards the ground with an accelerated motion but adds drily that[12] 'to just what extent this acceleration occurs has not yet been announced'.

Of such innocent questions are revolutions made.

Before we turn to detailed aspects of Galileo's work, it is worth pointing out that Galileo was not only a great scientist but also a great writer. He is regarded as the greatest writer of Italian prose between Machiavelli and Manzoni,[13] a span of 400 years. He is still read today in Italian schools – but more as Italian literature than for the sake of his physics. His preferred style of exposition was the dialogue form, which he adopted for his two greatest works, the *Dialogo* and the *Discorsi*. The characters are the same in both: Salviati, who is an expert and generally expresses Galileo's views, Sagredo, who is the archetypal intelligent layman, and Simplicio, who is an Aristotelian philosopher. In the preface to the *Dialogo* Galileo explains that the names Salviati and Sagredo were chosen to honour two dear but deceased friends of his Paduan period with those names and that the name Simplicio was suggested by Simplicius, the commentator on Aristotle from the sixth century AD whom we met on p. 101 and will meet once more. However, it was evidently also used by Galileo with the implication that Peripatetic philosophers were dim-wits. (For all his skill as a writer, Galileo had little sense of diplomacy. One of the conditions under which he obtained his Imprimatur for the *Dialogo* was that at the end he should include an argument against the Copernican system dear to the then pope, which was that man cannot presume to know how the world is really made because God could have brought about the same effect in ways unimagined by him (note the echo of Osiander), and he must not restrict God's omnipotence. Galileo rashly put these words in the mouth of Simplicio.) Finally, there is every now and then a reference to a mysterious, anonymous, and very brilliant Academician, who is none other than Galileo and with whom Salviati is alleged to be in regular communication.

A great advantage of the dialogue form of presentation was that it enabled Galileo to draw prominent attention to his own best ideas. The typical way in which this is done is for Salviati to report some discovery or insight of his Academician friend, whereupon Sagredo, with whom the reader (who also naturally likes to think of himself as an intelligent layman) identifies, breaks out in hymns of praise. We smile, of course, at Galileo's self-praise, expressed in this rather endearing manner, but it does simultaneously enable Galileo to overcome the problem that many great discoverers face: how can they get across the significance of what they have discovered without appearing immodest? Far more so than the authors of the three great astronomical books that preceded his dialogues – the *Almagest*, the *De Revolutionibus*, and the *Astronomia Nova* – Galileo overcame the natural inhibitions that prevent an author doing justice to his own work. The significance of the great works of astronomy was lost on all but the experts who were prepared to work their way patiently through page after page of technical details. But with marvellous flair Galileo made his insights accessible to a very wide audience and made abundantly certain that they grasped the key points. Coupled with this was a certain very intense vision which helps to give the reader a heightened awareness of where significant and exciting developments are to be expected.

This expository skill, which has never been surpassed in scientific writing, had the consequence that all of Galileo's discoveries exerted their full effect relatively quickly, even though in terms of the mathematics and observation required they were really quite modest compared with the discoveries of Ptolemy and Kepler (one could without any injustice to Galileo put them at the level of astronomy corresponding to the work of Hipparchus). The same objective results in the hands of a less skilled expositor would not have had remotely the same impact. This has had the consequence that Galileo's contribution to the development of science in general and dynamics in particular has been well and fairly reflected in histories of the subject, whereas the astronomical contribution, especially that of Hipparchus, Ptolemy, and Kepler has been, perhaps, rather unduly neglected.

7.2. Galileo's cosmology, overall concepts of motion, and the influence of Copernicus

There are dangers in writing about Galileo's overall concepts of motion and cosmology; in particular, it is necessary to reconcile some seemingly contradictory statements. Also, since Galileo's two main works were written in dialogue form, there is the problem of whether what the interlocutors say in the dialogues can be taken as his own opinion. However, the main difficulty is in the method of exposition favoured by Galileo, who was in fact always very sparing in stating the premises he needed to arrive at a particular conclusion. Not for him the grand statement of comprehensive principles adopted by Descartes and Newton. He was probably inclined to such a method of exposition by his admiration for Plato and his belief that knowledge of geometry at least is latent in the mind and only needs to be drawn out by some well-chosen Socratic questions (recollection theory of knowledge). It also accords very well with his own basic philosophy which, as Drake points out,[14] was not so much a closed system as a method – that is, mathematics and reason are to be used in an ongoing process to interpret empirical observations. Only in this way will man eventually build up a true picture of the world.

Perhaps the most interesting way to approach the problem is to see how, in the *Dialogo*, especially in Days 1 and 2 of the dialogue, Galileo, who follows Copernicus remarkably closely, amplifies and extends in one very significant respect the very bare hints that Copernicus offered as to how the dynamical arguments against the earth's diurnal rotation should be overcome. The powerful stimulus that the Copernican revolution gave to not only astronomy but also the theory of motion is then very apparent. We have already followed in the previous chapter the astronomical fork of the Copernican revolution; we now commence the descent of the fork that led to dynamics. It is appropriate to begin with Copernicus's own thoughts on the subject as expressed in *De Revolutionibus*. We shall see how the bare hints led to a veritable flood of ideas and results.

Although Copernicus was very well aware of the revolutionary nature of his proposal of the earth's mobility, he made no attempt at all to recast the entire theory of motion as it had been developed by Aristotle. This would really have been the logical course to take (and was taken by Kepler), since Aristotle's entire motionics was designed and built around the assumption of the earth's immobility at the precise centre of the universe. The overwhelming bulk of *De Revolutionibus* is purely astronomical; barely three or four pages are devoted to the problems of motion. Nevertheless, in this brief space Copernicus, like Oresme before him, succeeded in pointing the way that historically did in fact lead, through Galileo, to the foundations of the modern theory of motion. What Copernicus did was to retain the overall Aristotelian concepts of natural and violent motions but abolish the idea that a given body had a *unique* natural motion; he introduced the concept (already recognised as important by Oresme) of a *mixed* natural motion. It was in amplifying this notion, coupled with the quantitative and mathematical (as opposed to purely qualitative) treatment of free fall, that Galileo did such important preparatory work for the creation of dynamics.

In the synoptic Book 1 of *De Revolutionibus*, Copernicus outlined the main objections to the earth's mobility, especially its rotation, perceived within the framework of the pre-Copernican theory of motion. There was for a start the problem of gravity that we have already considered in Sec. 6.6. Aristotle had taught that bodies seek the centre of the universe, not specifically the centre of the earth. Were by any chance the earth to be displaced from the centre of the universe, bodies would continue to fall, according to Aristotle, to the centre of the universe. But, of course, it is not possible to put this to the test, so that speculation or enquiry on the subject did not seem to have any point. The way in which the Aristotelian–Ptolemaic cosmology had a 'freezing' effect on scientific thinking is well illustrated by a remark of Ptolemy's in the *Almagest*:[15] 'I think it is idle to seek for causes of the motion of objects towards the centre, once it has been so clearly established from the actual phenomena that the earth occupies the middle place in the universe, and that all heavy objects are carried towards the earth.' Thus, scientific enquiry is useless since we are presented with a unique situation.

Once the earth had been set in motion, the situation was entirely changed. Admittedly, Copernicus's own reaction to the new possibilities that he himself had opened up seems quaint and quite unscientific. On the question of gravity, he says the following:[16]

> the further question arises whether the center of the universe is identical with the center of terrestrial gravity or with some other point. For my part I believe that gravity is nothing but a certain natural desire, which the divine providence of the Creator of all things has implanted in parts, to gather as a unity and a whole by combining in the form of a globe. This impulse is present, we may suppose, also in the sun, the moon, and the other brilliant planets, so that through its operation they remain in that spherical shape which they display. Nevertheless, they swing round their circuits in divers ways.

This is all he has to say about the subject. Somewhat surprisingly, Galileo, writing 80 years later (and, moreover, 20 years after the publication of Kepler's much more precise and scientific ideas on the subject), appears to have been essentially content with this broad explanation; that at least is the impression given by Day 1 of the *Dialogo*. We shall go into this shortly.

The next problem Copernicus had to contend with subsequently played an important part in the clarification of centrifugal force, which was also very important in the discovery of dynamics and the debate about absolute and relative motion. Ironically, Copernicus felt obliged to answer an objection to diurnal rotation of the earth that Ptolemy had never raised – Copernicus misunderstood a poor translation. Ptolemy had in fact been countering the suggestion that the earth itself must fall. At the end of his discussion, which had nothing to do with rotation of the earth, Ptolemy said:[17] 'If the earth had a single motion in common with other heavy objects, it is obvious that it would be carried down faster than all of them because of its much greater size: living things and individual heavy objects would be left behind, riding on the air, and the earth itself would very soon have fallen completely out of the heavens. But such things are utterly ridiculous merely to think of.'

Ptolemy then followed this immediately with a discussion of the possibility, raised by 'certain people', that the earth rotates (Ptolemy did not discuss anywhere the possibility of an annual motion of the earth). His answer to this suggestion was: 'However, they do not realise that, although there is perhaps nothing in the celestial phenomena which would count against that hypothesis, at least from simpler considerations, nevertheless from what would occur here on earth and in the air, one can see that such a notion is quite ridiculous.'

Among the reasons he adduced for the suggestion being ridiculous was:

> they would have to admit that the revolving motion of the earth must be the most violent of all motions associated with it, seeing that it makes one revolution in such a short time; the result would be that all objects not actually standing on the earth would appear to have the same motion, opposite to that of the earth: neither clouds nor other flying or thrown objects would ever be seen moving towards the east, since the earth's motion towards the east would always outrun and overtake them, so that all other objects would seem to move in the direction of the west and the rear.

It was arguments like these that provided the main counter to Copernicus's proposal in the century after he made it. Few people made the effort to understand the astronomical arguments and readily agreed with Brahe that motion of the earth was unthinkable. Thomas Kuhn[18] quotes a satirical poem that circulated in Elizabethan England and put forward such arguments. Copernicus's synopsis of his mistaken understanding of Ptolemy's argument was as follows:[19]

> Therefore, remarks Ptolemy of Alexandria [*Syntaxis*, I, 7], if the earth were to move, merely in a daily rotation, the opposite of what was said above would have to occur, since a motion would have to be exceedingly violent and its speed unsurpassable to carry the entire circumference of the earth around in twenty-four hours. But things which undergo an abrupt rotation seem utterly unsuited to gather [bodies to themselves], and seem more likely, if they have been produced by combination, to fly apart unless they are held together by some bond. The earth would long ago have burst asunder, he says, and dropped out of the skies (a quite

preposterous notion); and, what is more, living creatures and any other loose weights would by no means remain unshaken. Nor would objects falling in a straight line descend perpendicularly to their appointed place, which would meantime have been withdrawn by so rapid a movement. Moreover, clouds and anything else floating in the air would be seen drifting always westward.

Copernicus immediately gave a hint of the answer he was to give to this problem. It was to suggest that the rotational motion of the earth is a *natural* motion (in the Aristotelian sense, though not hitherto included among the natural motions):[20]

if anyone believes that the earth rotates, surely he will hold that its motion is natural, not violent. But what is in accordance with nature produces effects contrary to those resulting from violence, since things to which force or violence is applied must disintegrate and cannot long endure. On the other hand, that which is brought into existence by nature is well-ordered and preserved in its best state. Ptolemy has no cause, then, to fear that the earth and everything earthly will be disrupted by a rotation created through nature's handiwork, which is quite different from what art or human intelligence can accomplish.

We then begin to understand why earlier (Book I, Chap. 3) Copernicus had been at pains to show that the earth is spherical ('it is perfectly round, as the philosophers hold') and had followed this with a chapter entitled 'The motion of the heavenly bodies is uniform, eternal, and circular or compounded of circular motions'. This opens with a very Platonic statement:[21] 'the motion of the heavenly bodies is circular, since the motion appropriate to a sphere is rotation in a circle. By this very act the sphere expresses its form as the simplest body, wherein neither beginning nor end can be found, nor can the one be distinguished from the other, while the sphere itself traverses the same points to return upon itself.'

He had thus prepared the ground for the suggestion that, the earth being round, it might well rotate itself and move in a circle around the sun. He opens Chap. 5, entitled 'Does circular motion suit the earth?' with this statement:[22] 'Now that the earth too has been shown to have the form of a sphere, we must in my opinion see whether also in this case the form entails the motion.'

On this basis he can therefore argue that a rotational motion of the earth is not only natural but actually to be expected, even if it brings with it the consequence that heavy objects on the earth have a dual natural motion:[23] 'We must in fact avow that the motion of falling and rising bodies in the framework of the universe is twofold, being in every case a compound of straight and circular.' We note how similar is this view to the one expressed by Oresme (p. 207) nearly 200 years before Copernicus. There is, however, a significant difference: Copernicus asserts that this is how things are, Oresme only how they could be.

Anticipating, let me say that if there was one single step more important than any other in the creation of modern dynamics it was Galileo's elaboration 80 years later of the concept enunciated here of dual motion of the bodies moving near the surface of the earth – a concept that was literally forced upon Copernicus once he set the earth loose. Much of dynamics was the fruit of Galileo's years of pondering the consequences of terrestrial mobility. This indirect astronomical input was perhaps of even greater importance than the direct empirical discovery of the law of free fall.

Copernicus summarizes his view of the nature of natural motions and the teleological purpose that they serve in the following passage, in which the Pythagorean concept of the well-ordered cosmos is most pronounced:[24]

> Hence the statement that the motion of a simple body is simple holds true in particular for circular motion, as long as the simple body abides in its natural place and with its whole. For when it is in place, it has none but circular motion, which remains wholly within itself like a body at rest. Rectilinear motion, however, affects things which leave their natural place or are thrust out of it or quit it in any manner whatsoever. Yet nothing is so incompatible with the orderly arrangement of the universe and the design of the totality as something out of place. Therefore rectilinear motion occurs only to things that are not in proper condition and are not in complete accord with their nature, when they are separated from their whole and forsake its unity.

Surprising as it may seem, Galileo, pioneer of the modern theory of dynamics and writing nearly a century after the publication of *De Revolutionibus*, appears to have been quite content to adopt this basic conception of motion and the purpose it served. His exposition in the *Dialogo* presents a most curious mixture of the ancient and the modern. The overall concept was almost archaic; the revolution was all in the detail. But here Galileo was on very secure ground – he had observation and powerful rational arguments to back him. He was fully aware of the far-reaching potential of the first small innovations of his quantitative motionics. They led him to the partial formulation of one of the most fundamental principles of modern science – Galilean invariance. This will be discussed in Sec. 7.4; here we complete the overall picture that Galileo describes in Day 1 of the *Dialogo*.

Galileo's picture of the world, at least as described in the *Dialogo*, was as teleological as Aristotle's, though modified. He still has the Pythagorean concept of the cosmos as the perfectly ordered work of God. Salviati says:[25] 'it [the world] is of necessity most orderly, having its parts disposed in the highest and most perfect order among themselves.' This statement is emphasized by Galileo in a marginal note, in which he states: 'The author assumes the universe to be perfectly ordered.'

The purpose of (natural) motion is to carry bodies to their proper places and then maintain them there:[26] 'We may therefore say that straight motion serves to transport materials for the construction of a work; but this, once constructed, is to rest immovable – or, if movable, is to move only circularly.'

The great virtue of circular motion is that it maintains objects in their proper place. Linear motion is quite inappropriate for this purpose:[27] 'if all the integral bodies in the world are by nature movable, it is impossible that their motions should be straight, or anything else but circular. . . . For whatever moves straight changes place, and continuing to move, goes ever farther from its starting place . . . at the beginning it was not in its proper place . . . it is impossible that anything should have by nature the principle of moving in a straight line [for ever].' And again:[28] 'only circular motion can naturally suit bodies which are integral parts of the universe as constituted in the best arrangement.' This is emphasized by another marginal note: 'Straight motion assigned to natural bodies to restore them to perfect order when they are disordered.'

Finally, we come to Galileo's views on gravity, on which, as noted in Sec. 6.6., he maintained a life-long hostility to the concept of gravitational attraction as advanced by Kepler, regarding it, especially in connection with the tides, as inadmissible invoking of 'occult qualities' that should have no place in the new science that he was helping to create. On the problem of what holds the celestial bodies together, he simply says:[29] 'just as all the parts of the earth mutually cooperate to form its whole, from which it follows that they have equal tendencies to come together in order to unite in the best possible way and adapt themselves by taking a spherical shape, why may we not believe that the sun, moon, and other world bodies are also round in shape merely by a concordant instinct and natural tendency of all their component parts?'

Thus, in all the above there is no change at all from Copernicus. Like him, Galileo uses the concept of the 'integral bodies' of the universe: parts of such bodies are brought together by linear motion and then, *integrality* having been achieved, all such motion should cease and be replaced either by perfect immobility (the most noble state of all – echoing Copernicus, who had, following Aristotle and the medievals, said:[30] 'immobility is deemed nobler and more divine than change and instability') or, at most, circular motion of the integral body as a whole.

It seems remarkable that out of this quite fallacious (in modern eyes) concept of motion Galileo should have forged the foundations of modern science. The explanation of the paradox is that what counted in the end was the use Galileo made of the details – the individual motions themselves. First, by establishing precisely the nature of these motions, he introduced the quantitative aspect that terrestrial Aristotelian physics had hitherto lacked. Equally important was the demonstration of what could be done with the individual motions. What Galileo actually created was a kind of atomic theory of motion – the atoms were not the material atoms of Leucippus and Democritus but *primordial natural motions*. Whereas the atomists (and Plato in most of the *Timaeus*) had sought to explain the phenomena of the world by geometrical shapes, Galileo sought instead to achieve the same by *geometrical motions*. This is the true significance of the passage from *The Assayer* quoted in the previous section.

Thus, although Galileo strikes one as more Aristotelian than not, especially in his retention of the concept of natural motions, he followed Copernicus's lead in extending the class of allowed natural motions, which, quite early in Day 1, he tells us are of three kinds: straight, circular, and mixed circular-straight. There is also a significant relaxation of the rigidly concentric Aristotelian world, and, equally important, an inversion of the way in which cosmology should be approached:[31]

> it appears that Aristotle implies that only one circular motion exists in the world, and consequently only one centre to which the motions of upward and downward exclusively refer. All of which seems to indicate that he was pulling cards out of his sleeve, and trying to accommodate the architecture to the building instead of modelling the building after the precepts of architecture. For if I should say that in the real universe there are thousands of circular motions, and consequently thousands of centres, there would also be thousands of motions upward and downward.

The 'precepts of architecture' – these were the primordial atomic motions that Galileo was convinced he had found: the symbols of nature's language. In the next section we recount how he found them; in the following, how he used them.

7.3. The primordial motions: circular inertia and free fall

If it had not happened, one would wonder if it were possible. The entire world – or rather the way it is perceived – turned upside down by the mere rolling of a 'perfectly hard ball' down an 'exquisitely polished plane'. That is what happened sometime around 1603. The revolution was in Galileo's mind, in the lesson he drew from the observation. He 'had experienced just once the perfect understanding of one single thing'.* He discovered and, very importantly, correctly analyzed the law of free fall. At much the same time he clarified his thoughts about what one may call a law of 'circular inertia' and how it could be combined with the law of free fall to give the law of motion of projectiles on the surface of the earth.

* From the *Dialogo*,[32] in a passage in which Galileo expresses his belief that in a few instances, in which mathematical truths are involved, the human mind can partake of divine knowledge: 'For anyone who had experienced just once the perfect understanding of one single thing, and had truly tasted how knowledge is accomplished, would recognize that of the infinity of other truths he understands nothing.'

We shall begin the account of this work with his 'law of inertia', or perhaps one should say 'laws of inertia', since he actually considered more than one such law, and, as we shall see, the whole story is rather complicated. Indeed even the expression 'law of inertia' may be misleading; for, as we have seen, Galileo spoke of *motions* (more or less in the Aristotelian sense) and inertia was a word that he did not employ at all. However, since he did clearly recognize in terrestrial motions the characteristic *persistence of motion* that is encapsulated in Newton's First Law, which today is usually called the law of inertia (though it was not by Newton), the expression is perhaps the best that is to hand, and the addition of quotation marks will signify that we have at most a partial statement of the content of Newton's First Law, and that inertia is used in the modern sense of persistence of motion (and not resistance to motion or acceleration as originally used by Kepler and Newton).

The main complication that arises in this discussion is that Galileo considered two distinct motions, both anticipations of the law of inertia, one corresponding to *natural* motion and the other to *violent* motion in the old Aristotelian sense. Moreover, the anticipation of the law of inertia corresponding to natural motion also existed in two forms: a weak one that derived from Copernicus and was necessitated by the astronomical evidence for the earth's diurnal rotation, and a stronger form that derived as an offshoot from Galileo's work on free fall and motion down inclined planes. All three forms of the 'law of inertia' are present in Galileo's *Dialogo*, introduced separately but without any discussion on Galileo's part of the relationship which he perceived between them.

To try and get an idea of how this came about, it is helpful to consider the evidence of Galileo's early unpublished tract *De Motu*,[33] which was mentioned in Sec. 7.1. This reveals that Galileo was perfectly familiar with the impetus theory considered in Chap. 4. We noted there that by Galileo's time Buridan's very clear formulation, according to which the impetus persists for ever with undiminished strength, had largely been replaced by the form in which the impetus dies away spontaneously and the body eventually comes to rest. A passage from the *Dialogus trilocuterius de possest* of Nicholas of Cusa (1401–1464), which I quote from Jammer's *Concepts of Force* shows, first, the strongly animistic component of the concept and, secondly, how far its use in the late Middle Ages was from recognition of it as a universal and truly fundamental law of motion:[34] 'The child takes the top which is dead, that is, is without motion, and wants to make it alive; . . . the child makes it move with rotational motion as the heavens move. The spirit of motion, evoked by the child, exists invisibly in the top; it stays in the top for a longer or shorter time according to the strength of the impression by which this virtue has been communicated; as soon as the spirit ceases to enliven the top, the top falls.' There was not much prospect of mathematizing that concept!

In the *Astronomia Nova*, Kepler makes a brief but graphic reference to an equally animistic form of the theory when discussing the motive power of the sun:[35] 'It might seem as if there were hidden in the body of the sun something divine, something comparable with our soul, from which that *species* springs which moves the planets, just as from the soul of the man that throws a stone there attaches itself to the stone the *species* of the motion, so that as a result the stone flies on even when the thrower has withdrawn his hand.'

Finally, it will be worth quoting a few lines from Galileo's *De Motu* to show how he too used impetus theory and to demonstrate the inferiority of that theory (at the time he used it) to his own later anticipation of the law of inertia:[36] 'In order to explain our own view, let us first ask what is that motive force which is impressed by the projector upon the projectile. Our answer, then, is that it is a taking away of heaviness when the body is hurled upward, and a taking away of lightness, when the body is hurled downward. But if a person is not surprised that fire can deprive iron of cold by introducing heat, he will not be surprised that the projector can, by hurling a heavy body upward, deprive it of heaviness and render it light. . . . The impressed force gradually diminishes in the projectile when it is no longer in contact with the projector.' Although this is not so animistic as Kepler, it is clearly of little use for a mathematical theory of motion.

Despite the manifest degeneration of Buridan's original idea, one might have thought that the law of inertia must nevertheless have developed out of such impetus-type concepts, which did indeed clearly play a prominent part in Descartes' work, as we shall see in the next chapter. However, as far as Galileo's work is concerned, it would be more accurate to speak of two initially totally distinct strands that only finally merged after Galileo's death and still persisted in independent forms in the *Dialogo*. And undoubtedly it was the other strand, which took its origin in Aristotelian cosmology and forced its way into terrestrial physics under the pressure of Copernican astronomy, that was the dominant strand in Galileo's work. To this we now turn.

The development of Galileo's thinking is fairly well documented. Besides the use of impetus theory just quoted, *De Motu* contains several important indications of how the dominant strand developed. Two chapters are relevant: Chap. 14, a 'Discussion of the ratios of the [speeds of the] motions of the same body moving over various inclined planes' and Chap. 16, 'On the question whether circular motion is natural or forced'. The first of these chapters is evidence of Galileo's early interest in the problem that would have brought him undying fame even without the telescope and Inquisition, though the emphasis is as yet on speed rather than acceleration. Galileo operates in an essentially Aristotelian world – there are innumerable references to the centre of the universe, taken to coincide with the centre of the earth – but there is a strong infusion of mathematical techniques learnt from Archimedes.

An extremely important difference from Aristotle and one which figures prominently in all Galileo's writings (and which we have already seen anticipated in Philoponus and the medieval writers) is his concept of the ideal, perfectly mathematical motion which a body would follow were it not for the effects of air resistance, roughness and friction. For Aristotle, the medium was the essential agent without which most terrestrial motions would be unthinkable – the speed of a body was the resultant of the push of the medium on the one hand and the resistance of the medium on the other. For Galileo the effect of the medium was nothing but an annoying perturbation of the perfect mathematical law that the body would follow in its absence. Here is a very characteristic passage from *De Motu*:[37]

> But this proof must be understood on the assumption that there is no accidental resistance (occasioned by roughness of the moving body or of the inclined plane, or by the shape of the body). We must assume that the plane is, so to speak, incorporeal or, at least, that it is very carefully smoothed and perfectly hard, so that, as the body exerts its pressure on the plane, it may not cause a bending of the plane and somehow come to rest on it, as in a trap. And the moving body must be [assumed to be] perfectly smooth, of a shape that does not resist motion, e.g., a perfectly spherical shape, and of the hardest material or else a fluid like water.

There is a hint here, in the 'so to speak, incorporeal', of the Platonic otherworldliness of Galileo's *weltanschauung*; we shall take up this point in Sec. 7.6.

The precise mathematical approach to the problem indicated by the above passage leads Galileo to the observation that on a plane sloping ever so gently downwards the moving body will have a slight positive tendency to motion downward while on one sloping slightly upward it will have a slight resistance to motion. Galileo interprets this entirely in Aristotelian terms – the body moves downward in a natural motion because it gets closer to its proper place while it resists upward motion, which is violent in that the body is thereby carried further from its natural place. Then comes the key passage:[38] 'any body on a [perfectly smooth] plane parallel to the horizon will be moved by the very smallest force, indeed, by a force less than any given force.'

Galileo comments that 'this seems quite hard to believe', and therefore provides a demonstration of its truth, which concludes with this statement:[39]

> A body subject to no external resistance on a plane sloping no matter how little

below the horizon will move down [the plane] in natural motion, without the application of any external force. This can be seen in the case of water. And the same body on a plane sloping upward, no matter how little, above the horizon, does not move up [the plane] except by force. And so the conclusion remains that on the horizontal plane itself the motion of the body is neither natural nor forced. But if its motion is not forced motion, then it can be made to move by the smallest of all possible forces.

Moreover, in a note added in the margin, Galileo comments that there can be no such thing as mixed (violent mixed with natural) motions with one significant exception (my italics): 'since the forced motion of heavy bodies is away from the center, and their natural motion toward the center, a motion which is partly upward and partly downward cannot be compounded from these two; unless perhaps we should say that such a mixed motion is that which takes place on the circumference of a circle around the center of the universe. But such motion will be better described as "*neutral*" than as "mixed". For "mixed" partakes of both [natural and forced], "neutral" of neither.'

In the later Chap. 16, Galileo hints that such neutral motions, in which bodies rotate without approaching or receding from the centre of the universe (i.e., their proper place), might well be perpetual:[40] 'For if its motion is not contrary to nature, it seems that it should move perpetually; but if its motion is not according to nature, it seems that it should finally come to rest.'

A little later he gives an explicit example in which he expects perpetual uniform motion:[41]

This makes clear the error of those who say that if a single star were added to the heavens, the motion of the heavens would either cease or become slower. Actually, neither of these things would happen. For since, in their view, too, the rotation of the heavens takes place about the center of the universe, the adding of a star or the further addition of any other heavy weight will neither help along nor retard the motion . . .

. . . For a star will be able to retard the motion only when it is being moved away from the place toward which it would naturally tend. But this never happens in a rotation that takes place about the center of the universe, for there never is upward and never downward motion. Therefore the motion will not be retarded by the addition of a star.

There is no suggestion at all in *De Motu* that Galileo was prompted to such thoughts by his reading of *De Revolutionibus*, which must, however, have been quite thorough for him to have picked up the detail of the Ṭūsī couple. *A fortiori* is there no suggestion that he saw any connection between his idea of a neutral motion and the problem of the earth's rotation, in which he evidently did not believe at that time. His work can be seen rather as a modification of the basic Aristotelian doctrine of natural and forced motions towards and away from the centre; the recognition of a 'neutral' motion, in which perfectly ordinary bodies might circle the centre of the universe perpetually, then obviates the need for the introduction of a special substance, quintessence, whose particular nature it is to move in a perfect circle. Thus, the concept of neutral motions is a step towards a more universal concept of motion, which, however, Galileo never took very far.

The next significant development, about which only conjecture is possible, was the conversion to Copernicus's theory of the earth's mobility and Galileo's development of a theory of the tides, the details of which we defer until Sec. 7.6. As already mentioned, Drake dates this highly important event around 1595. It is evident that from this point on Galileo's thoughts must have turned repeatedly to the dynamical problems posed by rotation of the earth.

The breakthrough came in the early years of the seventeenth century. It seems highly symbolic that the first really solid advances in the scientific revolution were made contemporaneously at the dawn of the century and that those of both Kepler and Galileo owed a great deal to the questions posed by Copernicus's proposal. The discovery of dynamics can be represented graphically in its essentials by a diagram reminiscent of split-beam experiments in quantum mechanics. The beam was split by Copernicus, each fork representing a problem: what drives the planets and how can the Aristotelian theory of motion be reconciled with motion of the earth? These problems were solved separately and in isolation by Kepler and Galileo, respectively. The reuniting of the two forks by Newton created dynamics:

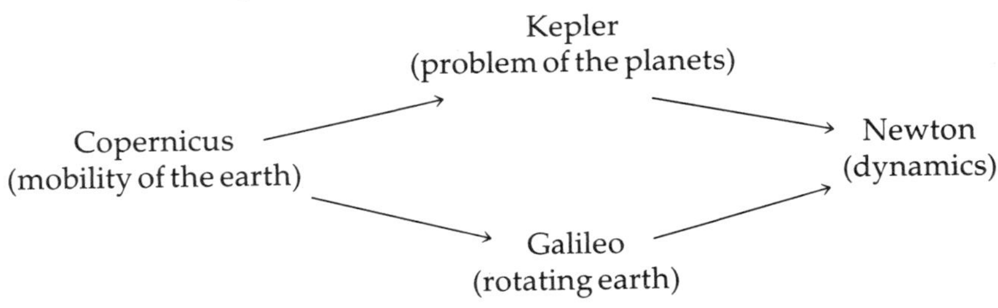

The crucial step by Galileo that made possible the eventual Newtonian synthesis occurred when he attempted to determine precisely the speed at which bodies fall freely, or rather, as that was a process too rapid for him to investigate, moved down smooth inclined planes. This drew his attention to the gradual build-up of speed. Decisive here was his interest in accelerations acquired continuously, as opposed to speeds acquired over definite finite descents. Drake believes that Galileo's attention was drawn to such accelerations by a long pendulum with which he performed experiments. For such a pendulum, the gradual build-up of speed becomes almost tangible. Drake makes the following comment:[42] 'It was probably the long and heavy pendulum that Galileo used in 1602 which called his attention to the importance of acceleration in downward motion, and to the continuation of motion once acquired – things that soon led him to an entirely new basis for his science of motion which replaced his earlier causal reasoning.' It seems to me possible that Galileo may also have been directly stimulated by a remark of Copernicus in his Chap. 8 of Book I. We have already seen the extent to which this short discussion of the problems of motion in *De Revolutionibus* appears (on the evidence of the *Dialogo*) to have shaped Galileo's overall conceptions. It is striking that the final paragraph from Copernicus quoted on p. 363 (the one ending with '. . . and forsake its unity') is immediately followed by this comment:[43] 'Furthermore, bodies that are carried upward and downward, even when deprived of circular motion, do not execute a simple, constant, and uniform motion. . . . Whatever falls moves slowly at first, but increases its speed as it drops.' Could this have provided the stimulus to the posing of that most significant of all questions in the discovery of dynamics?

Whatever the stimulus, the initial breakthrough was unquestionably empirical. By reducing the acceleration sufficiently and improving the accuracy of his timekeeping by the barest of margins, Galileo put himself in a position in which he could actually 'see' a law of motion. This is how Drake describes the crucial experiment:[44] 'In 1604 he devised a way to measure actual speeds in acceleration. For this purpose he let a ball roll from rest down a very gently sloping plane (less than 2°) and marked its positions after a series of equal times, judging by musical beats of about a half-second. These distances were then measured in units of about one millimeter.'

What Galileo discovered was a law of unsurpassed simplicity and beauty. Namely, that if the distance traversed in the first unit of time is taken as unity, then the distance traversed in the second unit of time is equal to three, that in next to five, and so on. Thus, if $s(t)$ is the distance traversed in time t, then

$$\begin{aligned} s(1) - s(0) &= 1 \\ s(2) - s(1) &= 3 \\ s(3) - s(2) &= 5 \\ &\dots \\ s(n) - s(n-1) &= 2n - 1 \end{aligned} \qquad (7.1)$$

If the poet Blake was able to see 'a world in a grain of sand', Galileo saw one too in the *odd-numbers law* (7.1), the first law discovered in the science of terrestrial rational motionics. For Galileo, this was a law of miraculous Pythagorean harmony. What clearer proof could there be of the inner mathematical harmony of nature's workings? He had seen through the deceptive outer appearances to the eternal Platonic forms – he had discovered a *four-dimensional geometry of motion*. At a stroke, the whole direction of thinking about motion was changed: Geometry is concerned with *laws* not *causes* – ergo, so too is motion.

As Drake comments,[45] once Galileo was in possession of this empirical law, he was able to make progress using mathematics alone. On the subject of Pythagoras and the connection between empiricism and mathematics Galileo himself made the following comment many years after his own discovery of the odd-numbers rule:[46]

I think it certain that he first obtained it by means of the senses, experiments, and observations, to assure himself as much as possible of his conclusions. Afterward he sought means to make them demonstrable. That is what is done for the most part in the demonstrative sciences. . . . And you may be sure that Pythagoras, long before he discovered the proof for which he sacrificed a hecatomb, was sure that the square on the side opposite the right angle in a right triangle was equal to the squares on the other two sides. The certainty of a conclusion assists not a little in the discovery of its proof . . .

The first important result that Galileo deduced from Eq. (7.1) was what is now called his *law of free fall*. He apparently arrived at it in a somewhat roundabout way and with some difficulty,[47] but for us the step involves little difficulty. For it is a matter of simple addition to deduce from Eq. (7.1) that the distance s of descent increases as the square of the time

$$\begin{aligned} s(0) &= 0 \\ s(1) &= 1 \\ s(2) &= 4 \\ &\dots \\ s(n) &= n^2. \end{aligned} \tag{7.2}$$

Thus, putting the result in its modern form, we have

$$s = \tfrac{1}{2}at^2, \tag{7.3}$$

where a, the acceleration, has the value $a = 2$ in the units adopted for the purposes of this discussion. Thus, Galileo originally discovered his law of free fall in the integrated form which states that the distance fallen is proportional to the square of the time of descent.*

It seems that it took two or three years for Galileo to deduce from the integrated form of his law the correct law that governs the increase of speed with time.[49] He initially made the rather natural assumption that the speed v is proportional to the distance of descent, i.e., $v \propto s$, and only

* One of the main reasons why Galileo had more difficulty than we do in arriving at the law (7.3) is that, in common with the practice of the time, he worked with ratios of like quantities and squaring a time and finding the distance to which it corresponded was not at all the sort of operation that would occur to him naturally.[48]

later deduced the correct result that the speed is proportional to the time t, i.e.,

$$v = at. \tag{7.4}$$

I do not propose to attempt to reconstruct here the steps by which Galileo arrived at Eq. (7.4). Some idea of the difficulty he must have had can be gauged from the amount of space and trouble he devotes in the *Discorsi* (published, as we recall, over 30 years after he had made his discoveries) to the basic concepts needed to give adequate expression to the laws (7.1)–(7.4). The treatment given in the *Discorsi* (Day 3 of that dialogue) more or less reverses the steps by which Galileo arrived at the law (7.4). For, having first of all considered the definition of uniform motion, Galileo first proceeds to define what he means by a uniformly accelerated motion, which is a motion in which equal increments of speed are added in equal increments of time (and not, importantly, distance). After that he proves (by geometrical arguments) his basic result:[50]

The spaces described by a body falling from rest with a uniformly accelerated motion are to each other as the squares of the time intervals employed in traversing these distances.

It is only then, as a corollary to this result, that he deduces the odd-numbers rule (7.1) that was his empirical point of departure.

The law (7.4) is of course the law of uniform acceleration considered by the Mertonians (Chap. 4), and much of Day 3 of the *Discorsi* is taken up with proving results that had already been obtained in the late Middle Ages, at the very least as abstract mathematical possibilities. It is therefore perhaps reasonable to ask to what extent Galileo should be given the credit for discovering the law of free fall. This question has been looked at in considerable detail by Clagett,[51] whose conclusions I shall briefly summarize. Between Aristotle and Galileo there were numerous discussions of the subject. With a few exceptions, of which the most notable is Philoponus in the passage quoted in the footnote on p. 198, there seems throughout the period to have been a remarkable lack of interest in actually attempting to determine by measurement how bodies fall. The extreme difference between astronomical practice and the attitude to terrestrial physics is nowhere more apparent than in this field. Interestingly, one of the earliest discussions of free fall to attract widespread comment was due to Hipparchus, whose ideas influenced Galileo's early thinking. In order to give a flavour of what much of the subsequent discussion was like, I quote here part of Simplicius's account of Hipparchus's lost work on the subject:[52]

Hipparchus, on the other hand, in his work entitled *On Bodies Carried Down by Their Weight* declares that in the case of earth thrown upward it is the projecting force that is the cause of the upward motion, so long as the projecting force overpowers the downward tendency of the projectile, and that to the extent that this projecting force predominates, the object moves more swiftly upwards; then, as this force is diminished (1) the upward motion proceeds but no longer at the same rate, (2) the body moves downward under the influence of its own internal impulse, even though the original projecting force lingers in some measure, and (3) as this force continues to diminish the object moves downward more swiftly, and most swiftly when this force is entirely lost.

It will be noted that in Hipparchus too there appear to be intimations of impetus theory; a similarity with the earlier quotation from Galileo's *De Motu* is evident.

In the Middle Ages, Buridan, in the passage already quoted in Chap. 4, came closest to providing an explanation of the mechanism of free fall in line with Newtonian conceptions. However, his discussion remained qualitative. As regards the quantitative description of free fall, Clagett points out that there was much confusion. Writers who followed the basic idea of Buridan's impetus theory did in some cases, not surprisingly, arrive at the conclusion that the speed of fall increased in proportion to the time of falling. But the very same people also often thought that the speed of fall increased in proportion to the distance fallen! The most notable of these were Albert of Saxony, Leonardo da Vinci, and even Galileo (as late as 1604), as already mentioned. What one lacks throughout the entire discussion is a sense of urgency. Galileo makes you feel the whole conception of nature hangs on the outcome of his discussion, and he speaks, like Copernicus on the matter of the earth's mobility and, unlike the medievals, with the authority of someone who knows. This may be contrasted with what Clagett[53] believes is the 'first statement of free fall with the infinitesimal implications of the Merton discussions explicitly applied'. It is found in a work published in 1555 by the Spaniard, Domingo de Soto. After a statement of the Merton Rule, de Soto comments, casually and in the manner of a thought experiment: 'This species of movement belongs properly to things which are moved naturally and to projectiles. For when a body falls through a uniform medium, it is moved more quickly in the end than in the beginning.'

What is perhaps most important of all in Galileo's discussion is his emphasis on the need to find the precise mathematical law which describes what happens when actual bodies fall. The attentive reader cannot fail to note the excitement and pride of someone who has made an unexpected and beautiful empirical discovery and found the mathematics which describes it adequately. Let me conclude this digression with Clagett's emphatic summary:[54] 'Regardless of how well he performed his experiments and what data came out of those experiments, Galileo's treatment was certainly the starting point of modern investigations of the problem of the acceleration of falling bodies.'

We now return to the discussion of the law (7.4), the importance of which for the development of dynamics can hardly be overestimated, as we shall see in Chaps. 9–11. It was certainly one of the most important of the baker's dozen. However, it would be wrong to conclude that Galileo immediately drew all the modern conclusions that we do from the law. In the long run his discovery opened the way to the recognition that the secrets of dynamics are revealed by examining *accelerations* and not the speed of motion or the path along which bodies move (the world that Kepler still inhabited).

Galileo was not capable of such a leap, which could not come before the development of a suitable general framework and the discovery of several more important mathematical relationships. As already hinted, what Galileo actually did was develop a kind of analytic motionics – *analytic* because he decomposed in his mind a given motion into atomic constituent motions, *motionic* because these atomic motions were taken as such without any search for a dynamic (i.e., causal) origin of them. Such motions did not arise for physical reasons; they did not have efficient causes but only the final teleological cause that Galileo seems to have been content to inherit from Aristotle through Copernicus. They were not fitted into a comprehensive scheme. But, of course, the real revolution was in the mathematization of motion and the identification and description by simple mathematical laws of motions *that do actually occur in nature.* The mathematics was like the application of a drop of easing oil to a great machine, built but never set in motion. It all started to move.

If Galileo was conservative in retaining an essentially Aristotelian and teleological overall concept of motion, he was strikingly revolutionary in postulating mathematical laws of motion. It was in this mathematization of empiricism that Galileo laid such secure foundations for dynamics.

To summarize this part of the discussion, we can say that the main significance for Galileo of the law (7.4) of free fall was that through it he felt convinced he had discovered nature's language. Secure in the knowledge of his possession of one at least of the 'symbols' in which the book of the universe is written, he could set about the task of analysing the bewildering variety of motions observed on the earth.

In fact, before he could do that he had to find his second 'symbol'. This was his 'law of circular inertia', the discovery of which was intimately linked to his work on the descent of bodies down gently sloping planes. However, before we come to that, it is worth underlining the intimate connection between the mathematization of motion and the very possibility of quantitative study of nature. It is clear from the quotations about the effects of friction and air resistance already given from *De Motu* that Galileo was strongly predisposed to the discovery of simple mathematical laws of motion years before he actually discovered them. However, it was only with the discovery of such laws that quantitative treatment of air resistance etc. became possible. The progression (7.1) is of such beauty and simplicity that – at least to some one of Galileo's bent – it just has to express the inner reality of things even though any empirical realization of descent down an actual inclined plane will not satisfy the law (7.1) exactly. The actually observed motions only approximate the progression (7.1) and other such 'exact' laws of motion, but the important thing is that the approximation is good enough for Galileo the mathematician to pick out the pure undefiled law and peel away the dross of the imperfect world. He can identify the laws bodies *would have* in the absence of the disturbing elements, and his spirit delights in their recognition. Where Kepler treasured the actual observations of planets as gems, Galileo treasured the ideal motions.

Of course, Galileo was not the first to recognize the fact that friction affects the motion of a body. But without the clear concept of a law of motion such as (7.1), there is no standard against which the disturbing effect can be measured and quantified. This remark demonstrates the truth of the statement that meaningful measurement is hardly possible without an underlying theoretical conception of what it is significant to measure. One of the most striking things about Galileo is the confidence and surety with which he writes and formulates principles of scientific method that are as valid now as in the seventeenth century. This can only stem from the confidence given him by the discovery of laws such as (7.1).

The possibility of using a precise mathematical law to 'separate' the pure state from disturbances introduced by the contingent world acquired even greater significance (admittedly after Galileo's death) in the case of the law of inertia. Galileo was thinking of disturbances introduced by imperfections of the experimental apparatus. Newton used the law of inertia to effect a clean division between primordial inertial motion and the completely uneliminable disturbances in motion introduced by forces such as gravity and magnetism. This procedure had far more radical consequences for our overall conception of the world – as drastic as Alexander taking his sword to the Gordian knot – and will lead us to the heart of the debate about the absolute or relative nature of motion.

I have emphasized this aspect of Galileo's mathematization of motion not only for its intrinsic importance in the development of the methods of modern science but also because it is characteristic of the move away from the contingent world into the perfect world of mathematics. This movement gained great strength from the success of Galileo's work and was taken still further by Descartes. These two men can be regarded as the founders of modern rationalism. Although the empirical content was much more pronounced in Galileo's thinking than in Descartes', they were united in seeing the *clarity* of the concept as all important. Such a concept was space. We shall see in the next chapters how space gradually acquired all the attributes of a perfectly clear concept. For Copernicus and

Kepler, space was truly nothing; after Galileo and Descartes it became almost palpable. The overthrow of Aristotle was complete, and space became such an ingrained part of our thought that it remains, I believe, the biggest single obstacle to a Machian, i.e., relational, conception of motion.

But we must now return to Galileo's discoveries. It is clear from his writings that Galileo was deeply struck by something which might be called 'persistence' of motion, that there is an 'amount' of motion (which is not yet Newton's momentum) and which has an existence just as real as material objects. In both the *Dialogo* and the *Discorsi* he lays great stress on the fact that the *speed* acquired by a body in descending from a given height depends on that height alone and is quite independent of the steepness of the slope down which it has descended (always assuming complete absence of friction) and, equally important, that such speed, however acquired, will suffice to carry the body back up to the same height from which it has previously descended, even if it reascends along a slope of different inclination, or even a curved slope. Mach[55] believes that Galileo discovered the essence of the law of inertia in a flash of inspiration by considering the descent of some body and its subsequent reascent along a plane inclined at a very small angle indeed to the horizontal. In such a case it would travel a very great distance in the horizontal direction before finally coming to a stop when it had regained the height from which it was originally released. From this it is a small step – which Galileo undoubtedly took – to making the assumption that on a horizontal plane the motion would persist forever. This feeling for persistence of a definite quantity of motion is strongly expressed in his writings and was clearly anticipated in *De Motu*. However, whereas in *De Motu* Galileo pointed out that a body could be put into horizontal motion by the application of the slightest imaginable force (friction of course being absent), the emphasis in his mature work is on the persistence for ever of the motion once commenced.

There are several things that must be said about this persistence of motion. The first is that in Galileo's mind a horizontal surface meant the *spherical surface of the earth*. This is repeatedly emphasized in the *Dialogo* and is implicit in the *Discorsi*. This means that for Galileo the persistence of motion was persistence in circular motion – he had a concept of 'circular inertia'. We shall return to this later. The second is that Galileo had lighted (more or less simultaneously) on a second atomic, or primordial, motion: here was another motion capable of precise mathematical formulation, another of the 'precepts of architecture'. A third point is that Galileo must certainly have been encouraged in accepting the idea of perpetual uniform motion, especially circular, from the precedent set by Aristotelian cosmology and above all Copernican astronomy and Copernicus's argument that, if the earth rotates, there just has to be some natural motion which carries the earth and all its parts around in perpetual circular motion. The precise connection with Copernicus's proposal will be discussed in the next section. The final point is that although Galileo certainly saw his 'law of inertia' as extremely important (quite contrary to Mach's puzzling assertion that he discovered it 'quite incidentally' and that it 'appears never to have played a prominent part in Galileo's thought'[56]), it did not for him become the most fundamental of all laws. Unlike Descartes and Newton, he did not elevate it to the status of *Lex Prima*, the First Law of Motion. For Galileo it was just one of the 'architectural precepts' of mathematical motionics, more or less on an equal footing with the law of free fall.

Imperfectly recognized as it was, Galileo's 'law of inertia' takes the second of the baker's dozen for terrestrial motions, though, because he failed to raise it to the status of the first law of nature, he shares the honour for its recognition with Descartes.

We shall have more to say about the manner in which Galileo discovered the 'law of inertia' in Sec. 7.5, in which we discuss the connection it had in his mind with the problem of the rotation of the earth. Before then, Sec. 7.4 will give us an idea of just how important the discovery was.

7.4. Compound motions. Parabolic motion of projectiles

We now come to the next great service that Galileo performed in creating the foundations of dynamics – the development of the technique of composition of primordial atomic motions. More than anything else, this revealed the rich potential of the law of inertia. One can say that although Galileo's concept of inertial motion was not quite correct, the use he made of it, especially in the *Discorsi*, was. He was fortunate in applying it in a situation in which the curvature of the earth's surface could be ignored, so that the difference between his law of 'circular inertia' and Newton's law had no effect. (The reason why many physicists, including Mach, formed a rather false picture of Galilean motionics is that they based their conclusions almost exclusively on the *Discorsi*, in which the subject matter is much more limited than the *Dialogo*. In particular, the fact that Galileo believed in 'circular inertia' can very easily escape a modern reader of the *Discorsi*.)

This part of Galileo's work is characterized by the masterly use that he made of geometry and his adaptation to problems in motionics of the ancient Greek technique of kinematic geometry for generating curves by the composition of different motions. The simplest example of this technique, well known to the Greeks, from whom Gaileo certainly learnt it, is as follows: if a body moves with uniform velocity along a line AB and the line is simultaneously displaced with uniform velocity in a direction not lying within the line, so that the ends A and B are displaced to C and D, respectively, while the body moves from A to B, then the resultant motion of the body will be that of uniform rectilinear motion along the diagonal AD of the parallelogram ABDC. This is the parallelogram of velocities law used by pilots when flying in a wind. Even more suggestive for Galileo was the construction by Archimedes of his well-known spiral, which we discussed in Sec. 2.7, in connection with the Greek concept of the passage of time. Galileo explicitly mentions this construction in the *Dialogo*.[57] He may also have been influenced by the astronomers' practice of superimposing motions (superposition of deferent and epicycle motion) to give the observed motion.

The more one studies Galileo's work, the more significant does his concept of physically compound motions appear. In his *Galileo at Work*,[58] Drake discusses a manuscript record of an experiment Galileo performed in Padua which served simultaneously to verify (indirectly) his 'law of persistence of horizontal motion' and the law of parabolic motion of projectiles. In this experiment, Galileo rolled balls down an inclined plane from different heights onto a horizontal table top and then let them roll across the table with their acquired speed before falling off the edge. Galileo was testing the relationship between the speed acquired in the descent and the horizontal distance the ball would travel after falling off the table before striking the floor.

This experiment is particularly interesting as an example of the rational scientific method in which Galileo delighted. We have already seen how the law of uniform acceleration in free fall was not discovered directly but through its consequence – the odd-numbers law (7.1). In the table top experiment, Galileo was able to make a similar indirect test of his 'law of inertia'. If horizontal motion acquired naturally by a previous descent remains constant in magnitude, then the horizontal distance traversed by the balls after they leave the table and before they strike the ground must simply be proportional to their horizontal speed, which Galileo could calculate, since he knew the height through which they had fallen (in the *Discorsi*, Galileo called this height the *sublimity* – literally, the amount lifted up[59]).

This experiment led to Galileo's third great discovery in motionics – the parabolic trajectory of projectiles. According to Drake,[60] this discovery was a by-product of the experiment to test the 'law of inertia' – the parabolic nature of the path can almost be seen directly. Exceptionally important for the development of dynamics was the fundamental assumption on which the whole experiment was based, namely that once the ball has passed over the edge of the table its motion consists of two *simultaneous* and *independent* primordial motions – the uniform horizontal motion and the uniformly accelerated motion of free fall. It is here that Galileo's concept of analytical motionics came into its own. Day 4 of the

Discorsi shows how the parabolic motion of projectiles arises from the superposition of these two motions.

Although, no doubt, many readers will be familiar with Galileo's derivation of the parabolic motions of projectiles, let us nevertheless look at the key passages. After all, this result must count as the third of the baker's dozen awarded to Galileo. Moreover, there are few great discoveries in science presented with the lucidity that we find put here in the mouth of Salviati, quoting directly from a text of the anonymous Academician:[61]

Imagine any particle projected along a horizontal plane without friction; then we know, from what has been more fully explained in the preceding pages, that this particle will move along this same plane with a motion which is uniform and perpetual, provided the plane has no limits. But if the plane is limited and elevated, then the moving particle, which we imagine to be a heavy one, will on passing over the edge of the plane acquire, in addition to its previous uniform and perpetual motion, a downward propensity due to its own weight; so that the resulting motion which I call projection [*projectio*], is compounded of one which is uniform and horizontal and of another which is vertical and naturally accelerated. We now proceed to demonstrate some of its properties, the first of which is as follows: A projectile which is carried by a uniform horizontal motion compounded with a naturally accelerated vertical motion describes a path which is a semi-parabola.

At this point, our intelligent layman Sagredo interrupts with a request for some mathematical details about parabolas, in which he is seconded by the good Simplicio 'for although our philosophers have treated the motion of projectiles, I do not recall their having described the path of the projectile except to state in a general way that it is always a curved line, unless the projection be vertically upwards' (which will give the reader an idea of Galileo's waspish way of making the Peripatetics look faintly ridiculous). Salviati, of course, obliges with a short discourse in which Apollonius's work naturally figures prominently. He then proceeds to the proof of the theorem, which I give together with Galileo's figure (Fig. 7.1):[62]

We can now resume the text and see how he demonstrates his first proposition in which he shows that a body falling with a motion compounded of a uniform horizontal and a naturally accelerated [*naturale descendente*] one describes a semi-parabola.

Let us imagine an elevated horizontal line or plane *ab* along which a body moves with uniform speed from *a* to *b*. Suppose this plane to end abruptly at *b*; then at this point the body will, on account of its weight, acquire also a natural motion downwards along the perpendicular *bn*. Draw the line *be* along the plane *ba* to represent the flow, or measure, of time; divide this line into a number of segments, *bc*, *cd*, *de*, representing equal intervals of time; from the points *b*, *c*, *d*, *e*, let fall lines which are parallel to the perpendicular *bn*. On the first of these lay off

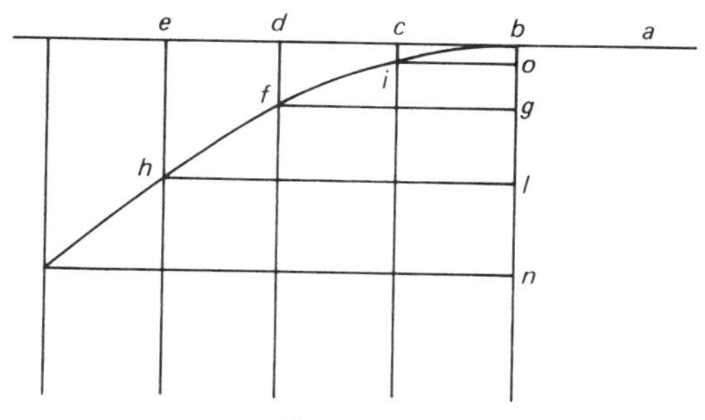

Fig. 7.1.

any distance *ci*, on the second a distance four times as long, *df*; on the third, one nine times as long, *eh*; and so on, in proportion to the squares of *cb*, *db*, *eb*, or, we may say, in the squared ratio of these same lines. Accordingly we see that while the body moves from *b* to *c* with uniform speed, it also falls perpendicularly through the distance *ci*, and at the end of the time-interval *bc* finds itself at the point *i*. In like manner at the end of the time-interval *bd*, which is the double of *bc*, the vertical fall will be four times the first distance *ci*; for it has been shown in a previous discussion that the distance traversed by a freely falling body varies as the square of the time; in like manner the space *eh* traversed during the time *be* will be nine times *ci*; thus it is evident that the distances *eh*, *df*, *ci* will be to one another as the squares of the lines *be*, *bd*, *bc*. Now from the points *i*, *f*, *h* draw the straight lines *io*, *fg*, *hl* parallel to *be*; these lines *hl*, *fg*, *io* are equal to *eb*, *db* and *cb*, respectively; so also are the lines *bo*, *bg*, *bl* respectively equal to *ci*, *df*, and *eh*. The square of *hl* is to that of *fg* as the line *lb* is to *bg*; and the square of *fg* is to that of *io* as *gb* is to *bo*; therefore the points *i*, *f*, *h*, lie on one and the same parabola.

It is quite clear from the amount of space that Galileo devotes to the question of the composition of motions in both the *Dialogo* and the *Discorsi* that he regarded this as one of the most important of his insights and as one that his readers would find particularly hard to grasp and accept. For example, Sagredo, Galileo's favoured mouthpiece for emphasizing the important insights, comments on the above proof:[63] 'One cannot deny that the argument is new, subtle and conclusive, resting as it does upon this hypothesis, namely, that the horizontal motion remains uniform, that the vertical motion continues to be accelerated downwards in proportion to the square of the time, and that such motions and velocities as these combine without altering, disturbing, or hindering each other.' Before Galileo, there was no clear conception of such superposition of motions, each conceived to have an independent physical existence.*

* The astronomers' superposition of epicyclic and deferent motion, mentioned earlier as a possible influence on Galileo, is not a true anticipation of Galileo's principle, since it was purely formal. It is only in Kepler's compounding of the two force-induced components in the motions of the planets (as he conceived them) that we find a genuine parallel to Galileo's physical principle. However, Kepler made only modest use of the principle, whereas Galileo gave it great prominence and truly demonstrated its potential.

Instead, people had thought that one of the motions would win out and completely dominate the resultant motion. According to this view, a cannon ball shot at an angle to the horizontal would at first travel in a straight path (along the line of the barrel) but near the end of its flight, when its initial impetus had been greatly weakened, gravity would take over (having originally been completely 'switched off'); it was believed that in the final section of the flight the cannon ball would drop vertically to the ground, gravity having become dominant. The earlier quotation from *De Motu* (p. 367) shows that Galileo himself initially shared similar ideas.

The advance achieved by Galileo in scientific method is very apparent when we compare his treatment of the projectile problem with the quotations from Buridan in Chap. 4. Buridan is close to the truth but remains completely qualitative – he is a logico-verbal Aristotelian. From the final two sentences of the quotation from him on p. 200 it would not be possible to predict the quantitative outcome of a projectile experiment – even one in a vacuum. The new elements introduced by Galileo are the assumption of two idealized motions, both of them described quantitatively, and the further assumption, forced upon Galileo by his belief in the rotation of the earth (see Sec. 7.5) and confirmed by the tabletop experiment, that two such motions can exist simultaneously without interfering with each other.

For the projectile problem, Galileo's assumption of an incorrect law of inertia was of no importance, and he was able to anticipate an important result of fully fledged Newtonian dynamics. As in the case of pure free fall, the result was a beautifully simple mathematical law – the path of the projectile is a parabola. More evidence of an ideal world glimpsed through the confusing senses.

It would again be wrong to conclude that Galileo had arrived in the case of the projectile problem at a Newtonian conception of the problem. It is true that, like Newton, Galileo has a generic division of motions. But Newton's is the division into free inertial motion in any direction and force-induced accelerations superimposed on the inertial motion. In Galileo's motionics, there are no forces in the Newtonian sense, and his division of motions is along different lines – into uniform and perpetual circular motion, on the one hand, and linear vertically accelerated motion on the other.

Wohlwill[64] points out a significant difference between Galileo's treatment of the vertical motion of free fall and the post-Newtonian treatment. In Newtonian dynamics, inertia is operative in both the vertical and the horizontal directions in the problem of projectile motion. In the horizontal direction there is no force acting, so the horizontal motion remains constant. In the vertical direction a constant force acts, and this keeps on adding increments to the downward velocity *which are all retained* by virtue of exactly the same principle of inertia as acts in the horizontal direction. In Newtonian dynamics, the vertical motion is itself a composite motion. Such an approach is alien to Galileo; the vertical and horizontal motions are as different as chalk and cheese. The vertical motion is a distinct primordial motion just like the circular motion. In this respect, Buridan's qualitative discussion (p. 202) was actually closer to Newton than Galileo's.

We can see here too how helpful it was for Galileo (and the development of dynamics) that he did adhere to the old Aristotelian concepts – for these suggested to him that *primordial motions do exist*. Moreover, the fact that there were for him, as for Aristotle, only two such motions, both of great simplicity, gave him the confidence to develop a primitive form of dynamics based on analysis of compound motions into their primordial constituents.

This remark also shows clearly how the Copernican revolution gave birth to dynamics. For Aristotle possessed the same two primordial motions but *they belonged to different worlds*: the circular to the heavens, the rectilinear to the earth. The two had no intercourse with each other. No compounding, no real dynamics. But when Copernicus raised the earth into the heavens and made it a planet like the others, the motions were of necessity conjoined. Dynamics was the offspring of this shotgun marriage. Copernicus caught them in the act; Galileo married them.

The work on the parabolic motion of projectiles brought Galileo to the very threshold of dynamics. In fact, on the basis of the law of free fall and the parabola law, Newton actually credited Galileo in the *Principia* with knowledge of his first two laws of motion, claiming that Galileo obtained his results by using them. This was certainly an anachronism on Newton's part but it does underline Galileo's achievement. Clear evidence of how close Galileo was to the definitive form of dynamics can be seen in the appearance in the projectile problem of one of the most characteristic features of dynamics – the determination of the actual motion of a body by the simultaneous action of the universal law governing its motion and the specific initial conditions. For, as we have already noted, Galileo considered the possibility of an arbitrary preassigned value of the horizontal velocity of the body (acquired by letting it roll down from different sublimities, i.e., heights), and this leads to a complete family of different motions, all of them, however, governed by the same principle. We see simultaneously, as pointed out by Wohlwill,[65] that there is still a severe limitation in Galileo's outlook – the initial velocity may vary in magnitude but it is always *horizontal*. Thus, the limitation of Galilean motionics is reflected in the fact that in his theorem he obtains a *semi-parabola*, the motion always commencing at the apex. There is still a vestigial Up and Down in Galileo's work, a reflection of the fact that he did not break entirely with Aristotle. In Chap. 10 we shall see the final step taken by Newton – the extension of the framework of dynamics to encompass *arbitrary initial conditions*.

To conclude this section, we again anticipate the absolute/relative debate. It was Galileo's work that made possible the concept of absolute motion. We have already seen that mathematization of motion is an essential step in the identification of the ideal, the separation of the perfect from 'accidental and special circumstances'. The decomposition of motions that Galileo demonstrated in the projectile problem was far more drastic. There is no suggestion here that either of the two constituent motions is a mere disturbance (like friction) of the other. When Galileo decomposes the composite motion into two, he is not scraping away dross; he is taking the surgeon's knife in true dissection. He showed how motions could be separated. Newton went far further. For in Galileo's mind, both motions were intimately related to the earth – circular motion around it, maintaining 'proper position', and free fall towards the earth, to 'return to the proper position – or at least get closer to it'. Both motions were still understood ultimately in teleological terms *in their relation to the contingent world*. In contrast, Newton used decomposition of motion to divide motions as observed into a part due to forces and an inertial part. For the part due to forces he provided – along Keplerian lines – an explicit dynamical explanation in terms of identifiable features of the contingent world. But the inertial part was completely detached – from *all* contingent circumstances.

The Machian thesis is that inertial motion is not at all pure motion in space on which the contingent world superimposes force-induced disturbances. It is rather that inertial motion is itself the outcome of a huge cooperative effort by the universe, with which any body in motion interacts. Mach challenges Galileo's atomic conception of motion as modified by Newton; he argues that there is no way of separating out motions pure and undefiled. Indeed, that would be to cut oneself off from the source, from the mother of all. In the Machian view, Newton used Galileo's knife to cut motion's umbilical cord to the *mater omnium*.

7.5. Rotation of the earth, different forms of the law of inertia and Galilean invariance

Galileo's aim in writing the *Dialogo*, his most famous book – though perhaps not his most influential (for the history of dynamics, that honour must be accorded to the *Discorsi*) – was *to prove that the earth truly moves*. Writing under the constraint of not being allowed openly to advocate Copernican teaching, he adopted the device of an allegedly neutral dialogue in which Salviati represents Copernicus, Simplicio defends Aristotelianism, and Sagredo represents the interested and intelligent layman wishing to discover the truth. In fact, no reader of the *Dialogo* could have been left in any doubt as to Galileo's own opinion.

There are three main planks in Galileo's argument by which he sought to meet the challenge Cardinal Bellarmine had thrown down – bring me proof positive that the earth moves.[66] Of lasting importance for the history of dynamics was the discussion in Day 2, in which Galileo countered the dynamical arguments against rotation of the earth. This will be the topic of this section. In Day 3, Galileo advanced the astronomical evidence for annual motion of the earth around the sun. This part of the dialogue is remarkable for Galileo's apparent ignorance of basic theoretical astronomy and the complete absence of any reference to Kepler's work on the orbit of Mars (beyond, in Day 4, a most cursory hint[67] that the problem of the true planetary orbits was most complicated and still far from solution). Kepler's *Astronomia Nova* had been published more than twenty years earlier; his work had been amplified in the later *Epitome of Copernican Astronomy*. Kepler had marshalled the astronomical evidence for motion of the earth around the sun with great clarity and equally great tact towards ecclesiastical sensitivity. One would like to believe that if Kepler had sat down to a quiet talk with Bellarmine, the cardinal would have come away a convinced supporter of Copernicus, or rather the Imperial Mathematician, Protestant though he was. As we have seen, Kepler immeasurably strengthened Copernicus's arguments for mobility of the earth. Above all, he had completely eliminated all the plethora of mutually contradictory hypotheses, each yielding more or less equally inadequate 'saving of the appearances', that had plagued astronomy for millennia and had done so much to foster the cynical conventionalism which Osiander expressed in the anonymous preface to *De Revolutionibus* (and in which Bellarmine at times seems to have taken refuge). He completely eliminated all uncertainty about the *relative* dispositions of the bodies in the solar system and showed that the only remaining ambiguity corresponded to the irreducible kinematic freedom associated (to use modern terms) with the choice of the coordinate system. His dynamical arguments must surely have persuaded Bellarmine (who had only asked for reasonable certainty) that the heliocentric option was the only sensible one, especially when coupled with Day 2 of Galileo's *Dialogo*, which, although it provided no positive proof of diurnal rotation, at least took all the sting from the dynamical arguments against the earth's rotation.

Yet, remarkably, Galileo chose merely to reiterate Copernicus's original argument from the retrograde motion of the planets. He said not a word of the far more telling fact that the apsidal knitting needles defined by the equants and eccentric centres of each of the planets all converge precisely on the centre of the sun and – most eloquent of all evidence – that the earth's orbit too has an equant. Galileo was, of course, able to augment the original argument of Copernicus by his own telescopic discoveries, above all the phases of Venus, which unambiguously confirmed the relative order of the celestial bodies. However, as already pointed out, the Tychonic system was just as effective from that point of view. If anything, Day 3 actually strengthened, rather than weakened, the position of the Dane, towards whom Galileo seems to have nurtured a certain antipathy ('nor have I ever set much store by Tycho's verbosity' says Salviati[68]).

Thus, positive proof, in the form of Galileo's theory of the tides, was reserved for Day 4. We shall discuss this in the next section. The theory of the tides being incorrect, the only lasting achievement of the *Dialogo* (in purely scientific terms) was to show how rotation of the earth could be reconciled with the theory of motion. This was an immense achievement and one of the most important chapters in the discovery of dynamics. There is a poem by Wordsworth that in eight brief lines succeeds in capturing, whether intentionally or not, the mystery of this whole matter of the earth's rotation – the fact that we spin with the earth and yet remain quite unaware of the fact:

> A slumber did my spirit seal;
> I had no human fears:
> She seem'd a thing that could not feel
> The touch of earthly years.
>
> No motion has she now, no force;
> She neither hears nor sees;
> Roll'd round in earth's diurnal course
> With rocks, and stones, and trees.

What Galileo did was to find a mathematical account of the matter, admittedly not quite without flaw, that did justice to the great mystery. He sensed correctly that thread of motion which persists forever and escapes 'the touch of earthly years'. In the process he very nearly succeeded in giving the correct formulation of what is, perhaps, the deepest of all the principles of physics – the principle that is now called the principle of Galilean relativity. The residual flaws in his treatment are reflected in the fact that in the process he used no less than three different forms of the 'law of inertia'.

We begin the discussion of this feat, which will simultaneously cast more light on the emergence of the concept of inertial motion (the vitally important step without which dynamics could not have appeared), with Galileo's earliest published statement of his 'law of circular inertia'. It comes in the second of his *Letters on Sunspots* (1613):[69]

> I seem to have observed that physical bodies have physical inclination to some motion (as heavy bodies downward), which motion is exercised by them through an intrinsic property and without need of a particular external mover, whenever

they are not impeded by some obstacle. And to some other motion they have a repugnance (as the same heavy bodies to motion upward), and therefore they never move in that manner unless thrown violently by an external mover. Finally, to some movements they are indifferent, as are these same heavy bodies to horizontal motion, to which they have neither inclination (since it is not toward the center of the earth) nor repugnance (since it does not carry them away from that center). And therefore, all external impediments removed, a heavy body on a spherical surface concentric with the earth will be indifferent to rest and to movements toward any part of the horizon. And it will maintain itself in that state in which it has once been placed; that is, if placed in a state of rest, it will conserve that; and if placed in movement toward the west (for example), it will maintain itself in that movement. Thus a ship, for instance, having once received some impetus through the tranquil sea, would move continually around our globe without ever stopping; and placed at rest it would perpetually remain at rest, if in the first case all extrinsic impediments could be removed, and in the second case no external cause of motion were added.

It will be noted that Galileo here follows his early *De Motu* extremely closely. Interestingly, his justification of the principle is not empirical but by more or less strictly Aristotelian arguments. Very important is the notion of rest and motion as representing intrinsic properties of bodies: the state of uniform circular motion is something intrinsic to the body, not something impressed from outside, as in impetus theory. We shall see that this is a significant conceptual advance. It clearly stems from Aristotle's concept of a natural motion.

In the *Dialogo*, published nearly twenty years later, the same principle is introduced in a much more empirical way, though even there it is through a process of Socratic questioning in which Salviati elicits from the hesitant Simplicio the facts that Galileo had observed empirically in his Paduan workshop. Salviati invites Simplicio to consider[70] 'a perfectly round ball and a highly polished surface, in order to remove all external and accidental impediments. Similarly I want you to take away any impediment of the air caused by its resistance to separation.'

By using the same sort of leading questions that Socrates employs in Plato's dialogue to extract a geometrical proof from a slave boy, Salviati leads the unfortunate Simplicio through a course of basic motionics which culminates in a heuristic derivation of Galileo's restricted law of circular inertia on the surface of the earth. In the course of this grilling, Simplicio agrees that on a horizontal plane 'there would be an indifference between the propensity [to move] and the resistance to motion' and 'I cannot see any cause for acceleration or deceleration, there being no slope upward or downward', so that 'if such a space were unbounded, the motion on it would likewise be boundless. That is perpetual.'

The concluding question of this discussion is:[71] 'Then a ship, when it moves over a calm sea, is one of these movables which courses over a surface that is tilted neither up nor down, and if all external and accidental obstacles were removed, it would thus be disposed to move incessantly and uniformly from an impulse once received?'

To which, of course, Simplicio must give his assent. Before we discuss the use that Galileo makes of this principle, which quite clearly he formulated on the basis of his Paduan experiments and his Aristotelian concept of a 'neutral' motion, let us consider the other two forms of 'inertia' that he employs.

The second form of 'inertia' is a weaker form of the principle which we have just discussed (and which is most clearly stated in the second of the *Letters on Sunspots*). The second form manifestly derives from *De Revolutionibus*. It will be recalled that Copernicus had argued that the integral bodies of the universe, being spherical in shape, would by virtue of their spherical shape have an intrinsic propensity to circular motion – either about an axis (rotation) or in revolution about the sun. Copernicus argued that parts of the earth would, simply because they are parts of the earth, have the same propensity.* This idea features prominently in the *Dialogo*, more so perhaps than the specifically Galilean form of the principle already enunciated, from which moreover it is not clearly distinguished. For example, in discussing the problem of a stone that falls from the top of a high tower, Galileo has Salviati say[74] 'the diurnal motion is being taken as the terrestrial globe's own and natural motion, and hence that of all its parts, as a thing indelibly impressed upon them by nature. Therefore the rock at the top of the tower has as its primary tendency a revolution about the center of the whole in twenty-four hours, and it eternally exercises this natural propensity no matter where it is placed.'

Note that according to this principle the only inherent property which the stone possesses is 'its primary tendency' to 'revolution about the center of the whole in twenty-four hours', i.e., strictly along a line of latitude and at a given, latitude-dependent, speed. In contrast, in Galileo's stronger form the ship can course over a calm sea in any direction

* It is interesting to note that Kepler, in the *Astronomia Nova*, explicitly mentions this theory of Copernicus in connection with his own (inadequate) discussion of rotation of the earth:[72] 'Copernicus, it is true, prefers to assume that the earth and all things terrestrial, even if detached from the earth, are directed by one and the same moving soul that, at the same time as it turns the earth, turns with it simultaneously parts detached from its body.' (The moving soul alleged by Kepler is actually completely absent in Copernicus.) It is moreover clear from the manner in which Kepler discusses this theory and his reference, noted earlier, to impetus theory that he regarded them as applying to two totally distinct phenomena. Thus both strands that led to the law of inertia passed unnoticed and distinct through Kepler's hands. This can also help to explain how Galileo could simultaneously use two different forms of inertia. It is also worth noting that at the end of his book on magnets, which Galileo read and praised (describing the book as 'great to a degree that is enviable'[73]) Gilbert argued at length for the Copernican explanation of the earth's rotation.

and at any speed. As we shall see, this strengthening was vitally important.

Finally, Galileo also introduces, though only episodically, a concept of rectilinear inertia. This is done alongside his concept of circular inertia. Baffling for the modern reader is the fact that Galileo does this without seeming to note the slightest discrepancy. There is absolutely no discussion of the connection between the two principles. The only possible explanation is that to Galileo (and, hence, by implication to his readers too) they related to phenomena that were self-evidently different. This is confirmed by the examples which are given and which strongly suggest that Galileo still adhered to the Aristotelian distinction between violent and natural motions.* The examples he discusses are examples of violent motion: stones cast by slings and bullets shot from guns. In the *Dialogo*,[76] Galileo appears to accept without question – and as something generally known – that a stone leaves a sling in a straight line along the tangent and that a cannon ball leaves the cannon along the straight line defined by the barrel. In both cases it is clear he assumes that in the absence of air resistance and weight both the stone and the cannon ball would continue forever with uniform speed along the initial line. This is, of course, the correct law of inertia. Its appearance without any clear precedent (and without particular emphasis) is puzzling. It may derive from medieval impetus theory in the pristine form given it by Buridan (rather than the form that Galileo used in his early *De Motu*). However, elsewhere in the *Dialogo* Galileo appears to dissociate himself from impetus theory, as we shall see.

It is interesting that Galileo makes no claim or implication which would indicate that he regarded himself as the discoverer of this third principle (the one actually nearest the truth). Nor, as we have said, does he supply any explanation for the apparent conflict between this law and his own law of 'circular inertia'. One can only surmise that what to a modern reader is a glaring discrepancy did not at all strike him as odd. It shows how far Galileo was from a modern understanding of dynamics; the point is that the sling-projected stone and the cannon ball are *doing* something quite different from a stone that acquires a speed by falling *naturally* before being deflected into perpetual circular motion. The cannon ball has been forced into a *violent* motion by the will of the cannoneer, while the stone, in falling, is simply striving to return to its proper place and, having reached it, can happily go into a state of uniform circular motion, which is natural because the stone remains in its proper place. In such a teleological

* It seems that this division between violent and natural motions (which even Newton[75] considered in the mid-1660s) was the last feature of Aristotelianism to die – in fact, one could even argue that the division lived on in the Newtonian distinction between inertial motion and acceleration-inducing forces.

view of motion, there is nothing odd about two objects commencing a motion at the same point, and with the same initial velocity, but then going in deviating directions although all the other external circumstances are identical. It is only from the high ground achieved by Newton, according to whom the subsequent motion – for given forces – is *uniquely determined* by the initial position and velocity, that Galileo's acceptance of two different laws appears so odd.*

In the light of these comments, Galileo demonstrates the truth of Mach's comments in the quotation at the beginning of the book about the need for an examination of science in a historical perspective. The example we are considering shows how a general philosophical framework held Galileo back from breaking through to a genuinely new concept of motion. Later in the chapter we shall see that the new mathematical and physical ideas that Galileo developed were also capable of leading him to incorrect results that are particularly illuminating in the context of the discussion in this book of the status of absolute and relative motion.

Now that we have discussed the three forms of the 'law of inertia' that appear in the *Dialogo*, we can examine some of the uses to which Galileo put them in arguing for rotation of the earth. He realized that his biggest task was to overcome ingrained prejudice. (Simplicio comments[79] that 'The crucial thing is being able to move the earth without causing a thousand inconveniences.') In the person of Salviati, Galileo comments:[80] 'Aristotle's error, and Ptolemy's, and Tycho's, and yours, and that of all the rest, is rooted in a fixed and inveterate impression that the earth stands still; this you cannot or do not know how to cast off, even when you wish to philosophize about what would follow from assuming that the earth moved.' He makes great play of the fact that many so-called proofs of the earth's immobility are logically fallacious since they assume what they are intended to prove (this leads Salviati to comment[81] that although Aristotle was the undoubted discoverer of the rules of logic that did not mean that he was himself a good logician!).

To help overcome the prejudice against terrestrial mobility, Galileo puts great emphasis on the relativity of perceived motion. Salviati's opening statement on the subject is:[82]

whatever motion comes to be attributed to the earth must necessarily remain imperceptible to us and as if nonexistent, so long as we look only at terrestrial objects; for as inhabitants of the earth, we consequently participate in the same

* This explanation of the occurrence of two different 'laws of inertia', which I formed on the basis of a first reading of the *Dialogo*,[77] is, as I found later, essentially the conclusion that Drake[78] reached too after much consideration of the matter. With one significant exception, to be discussed in Chap. 9, Galileo's examples are all consistent with such a division into violent and natural motions.

motion. But on the other hand it is indeed just as necessary that it display itself very generally in all other visible bodies and objects which, being separated from the earth, do not take part in this movement. So the true method of investigating whether any motion can be attributed to the earth, and if so what it may be, is to observe and consider whether bodies separated from the earth exhibit some appearance of motion which belongs equally to all.

This is, of course, Copernicus's point that the positive evidence for the earth's mobility must be sought in the heavens, not on the ground. For the purposes of Day 2, Galileo concentrates his attention on the disappearance of all effects of motion if we merely look at nearby objects *that move with us*, and to demonstrate that an overall motion of the earth and its parts can remain quite imperceptible on the surface of the earth, Galileo illustrates with examples the purely kinematic principle that[83] 'such motion as is common to us and to the moving bodies is as if it did not exist'. One of his most imaginative examples is of an artist sitting on the deck of a ship as it sails from Venice to Aleppo and drawing a picture.[84] Galileo asks one to imagine the *true* path of the nib of the artist's pen. It is, of course, a long line stretching from Venice to Aleppo. The movements made by the artist to produce his drawing appear as the minutest deviations around the 'mean' line. Yet the final picture reveals not the slightest trace of the long journey – which is 'subtracted' out by virtue of the fact that the artist travels with the ship. Galileo drives home his point with the comment: 'you are not the first to feel a great repugnance toward recognizing this inoperative quality of motion among the things which share it in common.'

Thus:[85] 'Motion, in so far as it is and acts as motion, to that extent exists relatively to things that lack it; and among things which all share equally in any motion, it does not act, and is as if it did not exist.' It is here worth noting the final words: *'as if it did not exist'*. We shall return to them in Sec. 7.6.

So far Galileo has merely stated a kinematic truth, admittedly with much greater urgency than the medievals or Copernicus. However, he extracts from it profound dynamical consequences through his physical assumption that all the bodies on the earth have an inherent tendency to uniform circular motion. To answer the difficulty of why a cannon ball dropped from the top of a high tower appears to fall vertically, grazing the side of the tower, and is not left far behind by the rotation of the earth, Galileo invokes his 'law of inertia' in its Copernican form:[86] 'Keeping up with the earth is the primordial and eternal motion ineradicably and inseparably participated in by this ball as a terrestrial object, which it has by its nature and will possess for ever.'

If this is granted, together with the further important principle that the motion of falling of the ball towards the centre of the earth coexists with the circular motion without in any way interfering with it, then it follows that although the actual path of the cannon ball in space will be curved, to an observer on the earth it will appear to move vertically downward. Galileo takes this opportunity to comment how deceiving the senses can be if not properly guided by reason:[87] 'With respect to the earth, the tower, and ourselves, all of which all keep moving with the diurnal motion along with the stone, the diurnal movement is as if it did not exist; it remains insensible, imperceptible, and without any effect whatever.'

In the *Dialogo*, the problem of finding the actual path of the ball as it falls to the earth is not solved exactly, but Galileo achieves a very adequate qualitative solution and states clearly and with confidence the principle of superposition of motions. As we have seen in Sec. 7.4, this principle bore its richest fruit in the solution of the projectile problem. In Chap. 10 we shall see the crucial role that this problem and Galileo's discussion of it (and others like it) played in Newton's definitive clarification of the principles of dynamics.

It is in further examples in which Galileo combines his kinematic principle of the relativity of apparent motion with his stronger form of the 'law of inertia' (for natural motions) that Galileo comes closest to the modern formulation of Galilean invariance and the related concept of inertial frames. As already mentioned, Galileo does not draw attention to the difference between his principle and the weaker Copernican principle, with the consequence that the former sometimes seems to be merely an unconscious extension of the latter. Perhaps it was, but this is hard to reconcile with the very clear statement of the stronger form in the *Letters on Sunspots* and also with the fact that, as he himself emphasizes, results far stronger than the minimal one that the Copernicans need and claim can be deduced. This comes out clearly in his discussion of a cannon ball dropped from the mast of a ship onto the deck, in one case while the ship is at rest relative to the surface of the earth, in another when the ship is under sail. In the early seventeenth century it was firmly believed that in the latter case the cannon ball would not fall at the base of the mast.

As we saw in the previous chapter, this alleged but nonexistent effect was used by the anti-Copernicans against mobility of the earth. For, they argued, if the effect occurs on a ship, it must similarly occur on a rotating earth. Therefore, since the cannon ball is seen to fall vertically when dropped from a tower, the earth must be immobile.

In the *Dialogo*, Galileo, who, if the suggestion made there can be trusted, never even bothered actually to verify the effect in an experiment ('Without experiment, I am sure that the effect will happen as I tell you'[88])*, roundly declares this to be false: 'anyone who does [the ship

* Incredible as it may seem, it appears that this crucial experiment was not performed by anyone before 1640, when Gassendi carried out the experiment on a ship in Marseille harbour.[89]

experiment] will find that the experiment shows exactly the opposite of what is written; that is, it will show that the stone always falls in the same place on the ship, whether the ship is standing still or moving with any speed you please. Therefore, the same cause holding good on the earth as on the ship, nothing can be inferred about the earth's motion or rest from the stone falling always perpendicularly to the foot of the tower.'

In fact, it is in the proof of this claim that Salviati takes Simplicio through the Socratic grilling described earlier in this section.

In the discussion of this and similar effects there may be noted a decided superiority of the Galilean conception of persistence of motion (derived at least in part, it may be said, from Aristotle) over the impetus theory of impressed force. According to the latter, force is, so to speak, transferred from the thrower to the object thrown, to which it 'attaches' itself. In the passage which now follows, in which Galileo is discussing balls thrown into the air by horsemen, he more or less explicitly rejects the idea of transfer of impetus and comes extremely close to the Newtonian (or rather, one should say, Cartesian) conception of inertial motion as a natural state rather than an enforced motion (note the 'mere opening of your hand'):[90]

When you throw it with your arm, what is it that stays with the ball when it has left your hand, except the motion received from your arm which is conserved in it and continues to urge it on? And what difference is there whether that impetus is conferred upon the ball by your hand or by the horse? While you are on horseback, doesn't your hand, and consequently the ball which is in it, move as fast as the horse itself? Of course it does. Hence upon the mere opening of your hand, the ball leaves it with just that much motion already received; not from your own motion of your arm, but from motion dependent upon the horse, communicated first to you, then to your arm, thence to your hand, and finally to the ball.

These examples of Galileo, in which riders of horses or passengers in carriages throw heavy objects vertically up into the air while in motion and are then able to catch them again because the objects keep up with the motion of the thrower, caused a considerable stir[91] in the seventeenth century; even today it is probably the case that more than half the world's population would be surprised at the demonstration of such experiments.

The great importance of these examples – and the Paduan experiments on which they ultimately rested – was that they went far beyond the bare necessities imposed by Copernican theory. Just as Kepler extended the astronomical side of the Copernican revolution into hitherto undreamed of directions, so too did Galileo transfigure Copernicus's crude adaptation of Aristotle. And in the case of both Kepler and Galileo powerful intuition was linked to crucial empirical observation. This was what distinguished their contribution to the scientific revolution from Descartes' much more purely philosophical approach.

Let us now see how Galileo's extension of Copernicus's notion brought him a long way towards the fundamental principle of science that is worthily named after him: the principle of Galilean invariance. We have seen that in the example of the ship Galileo nowhere stipulates the direction or speed of its motion; all that is required is that it should be moving at a uniform speed over the surface of the earth. This freedom was quite clearly suggested by Galileo's tabletop experiment, for which the 'sublimity' – and thus the resultant horizontal speed of the ball – can have any value. Thus, in discussing the motion of the ship and its parts, Galileo effectively introduced (without explicitly formulating the concept) a *class* of primordial motions – his example will be true if any uniform motion over the surface of the earth is a natural or 'primordial', motion, irrespective of its speed and direction. Thus, Galileo was *de facto* using all the key elements of the principle of Galilean invariance, though in a restricted form and also in a form only approximately true.

First, he recognized a *class of distinguished frames of reference* (frames moving at uniform speed on the surface of the earth), the most important property of these being that if phenomena are observed strictly within the frames it is impossible to establish the particular frame by such observations: without looking out of the cabin window, it is not possible to say whether the ship is under sail or not. Second, he attributed this 'unobservability of the common motion' to a physical principle, namely, to the fact that the common motion corresponds to a *primordial motion*, shared by all the parts of the ship. Equally important is his principle of the composition of motions, which permits different motions to be superimposed on each other without any mutual interference. This means that, if a collection of bodies share one common motion, and this happens to be primordial circular motion about the centre of the earth, they can themselves superimpose on the circular motion certain relative motions among themselves. This comes out particularly clearly in Galileo's discussion of why birds in flight do not suffer any 'inconveniences' from rotation of the earth, a point on which Sagredo expresses a touching concern for the hapless birds:[92]

If only the flying of birds didn't give me as much trouble as the difficulties raised by cannons and all the other experiments mentioned put together! These birds, which fly back and forth at will, turn about every which way, and (what is more important) remain suspended in the air for hours at a time – these, I say, stagger my imagination. Nor can I understand why with all their turning they do not lose their way on account of the motion of the earth . . . , which after all much exceeds that of their flight.

Salviati reassures him with these words:[93]

if you drop a dead bird and a live one from the top of a tower, the dead one will do the same as a stone; that is, it will follow first the general diurnal motion, and then the motion downward, being heavy. But as to the live bird, the diurnal motion always remaining in it, what is to prevent it from sending itself by the beating of its wings to whatever point of the compass it pleases? And such a new motion being its own, and not being shared by us, it must make itself noticeable. If the bird moves off toward the west in its flight, what is there to prevent it from returning once more to the tower by means of a similar beating of its wings? For after all, its leaving toward the west in flight was nothing but the subtraction of a single degree from, say, ten degrees of diurnal motion, so that nine degrees remain to it while it is flying. And if it alighted on the earth, the common ten would return to it; to this it could add one by flying toward the east, and with the eleven it could return to the tower.

As Segrè[94] comments, such conclusions, the fruit of Galileo's long rumination on the implications of terrestrial mobility, 'are among the most profound scientific insights ever achieved'. The principle of Galilean invariance is certainly the fourth of the baker's dozen associated with the study of terrestrial motions, though in this case Galileo must share the honour for it with Huygens, as we shall see in Chap. 9.

In view of its importance as the paradigm of the most powerful of all scientific principles, i.e., principles which state that in all circumstances certain positive effects will never be observed – *principles of impotence*, as Whittaker[95] calls them – we conclude this section with Galileo's own account of the 'nullity' of such experiments:[96]

For a final indication of the nullity of the experiments brought forth, this seems to me the place to show you a way to test them all very easily. Shut yourself up with some friend in the main cabin below decks on some large ship, and have with you there some flies, butterflies, and other small flying animals. Have a large bowl of water with some fish in it; hang up a bottle that empties drop by drop into a wide vessel beneath it. With the ship standing still, observe carefully how the little animals fly with equal speed to all sides of the cabin. The fish swim indifferently in all directions; the drops fall into the vessel beneath; and, in throwing something to your friend, you need throw it no more strongly in one direction than another, the distances being equal; jumping with your feet together, you pass equal spaces in every direction. When you have observed all these things carefully (though there is no doubt that when the ship is standing still everything must happen in this way), have the ship proceed with any speed you like, so long as the motion is uniform and not fluctuating this way and that. You will discover not the least change in all the effects named, nor could you tell from any of them whether the ship was moving or standing still. In jumping, you will pass on the floor the same spaces as before, nor will you make larger jumps toward the stern than toward the prow even though the ship is moving quite rapidly, despite the fact that during the time that you are in the air the floor under you will be going in a direction opposite to your jump. In throwing something to your companion, you will need no more force to get it to him whether he is in the direction of the bow or the stern, with yourself situated opposite. The droplets will fall as before into the vessel beneath without dropping toward the stern, although while the drops are in the air the ship runs many spans. The fish in their water will swim toward the front of their bowl with no more effort than toward the back, and will go with equal ease to bait placed anywhere around the edges of the bowl. Finally the butterflies and flies will continue their flights indifferently toward every side, nor will it ever happen that they are concentrated toward the stern, as if tired out from keeping up with the course of the ship, from which they will have been separated during long intervals by keeping themselves in the air. And if smoke is made by burning some incense, it will be seen going up in the form of a little cloud, remaining still and moving no more toward one side than the other. The cause of all these correspondences of effects is the fact that the ship's motion is common to all the things contained in it, and to the air also.

This passage brings us close to one of the central problems of the book. In modern terms, what is the origin of the distinguished class of *inertial frames of reference*? In Galileo's worldview, his residual Aristotelianism provided an explanation of sorts for the existence of a distinguished 'frame of reference' such as the ship's cabin. There is even an explanation of a class of such frames of reference. As already noted, they are tied into the contingent world. But when first Huygens and then Newton enlarged and altered the class of preferred frames, going over from uniform circular motions over the surface of the earth to uniform rectilinear motions in all three dimensions, they broke the last link with the Pythagorean cosmos, leaving us speeding through the enigmatic void. A crude return to primitive Aristotelianism is clearly impossible. The only alternative to the void is a more sophisticated – dynamical and physical – reconnection of the links between the here and now and the world at large. Kepler hinted at the direction we must go – even if erudite philosophers shake their heads in disapproval.

7.6. Galileo and absolute motion

The history of dynamics and especially the debate about the absolute or relative nature of motion demonstrates that certain ideas about the nature of things can become so ingrained in the mind as to defy almost completely any possibility of being dislodged. The beginning of such a tendency, specifically a belief in the independent and real existence of space, can be discerned in Galileo's writings. As explained earlier in this book, one of its aims is to lay bare the origin of our conceptions about space, time, and motion. Let us therefore consider what Galileo thought about these subjects.

We begin with one of the central questions of this book. How is motion to be described? To what is motion to be referred – space or other bodies? If we ask what Galileo thought about this subject, we find that in neither the *Dialogo* nor the *Discorsi* is there any general discussion of the nature of motion (such as is found in Descartes, Newton, and Leibniz). Galileo's beliefs, perhaps unconscious, have to be deduced from the manner in which he describes motion. Luckily, this can be done quite readily.

As one reads through the *Dialogo*, it becomes more and more apparent that, although he still retained a quite remarkably strong vestigial Aristotelianism in his concept of position, Galileo, as Maier (p. 48) quite rightly pointed out, instinctively regarded motion as taking place relative to space – which was of course the space of Euclidean geometry. Despite his profound remarks quoted above about the inoperative nature of motion common to a collection of bodies, the qualification 'as if it did not exist' is revealing and means in Galileo's mind precisely what it implies – the motion does actually exist. Whether he made his famous retort '*Eppur si muove*' ('And yet it moves'), or not, nothing could be more characteristic of Galileo's attitude: the mobility of the earth was for him an absolute and objective fact.

In this section I shall first review the evidence in Galileo's writings on which the conclusion of the foregoing paragaph is based, and then finally consider the explanation of his belief that motion takes place with respect to space rather than other bodies in the universe.

Right at the start of the *Dialogo*, in the review of Aristotelian physics and cosmology, there are several revealing comments – and omissions. Although Galileo comments on and emphasizes one of the most characteristic features of Aristotelian motionics, namely, that in it diversity of motions is an 'original principle', i.e., the elements are defined by their motions, he seems to have been completely uninterested in Aristotle's concern,evidently shared by Copernicus and Kepler, for an epistemologically sound definition of position by means of something material and visible. Statements like those of Copernicus in *De Revolutionibus* quoted in Chap. 5 and of Kepler in Chap. 6 are conspicuous by their absence throughout the whole of the *Dialogo*. If Galileo believed motion to be purely relative, he missed countless opportunities to say so. On the other hand, there was still a relational side to his thought, because he retained the concept of a Pythagorean cosmos, i.e., that the world[97] 'is of necessity most orderly, having its parts disposed in the highest and most perfect order among themselves'. But this must imply a strong degree of relationism, for the optimum order must clearly be expressed by the relative order of the bodies among themselves. It is also implicit in the concept of the primordial motions as Galileo defined them – towards the centre of the earth in a straight line in order to attain the 'proper place' and around the earth in a circle so as to 'remain there'. However, quite early on comes a passage which suggests that Galileo conceived of motion as taking place relative to space. In talking about the 'creation of order from chaos', Salviati describes:[98] 'primordial chaos, where vague substances wandered confusedly in disorder, to regulate which nature would very properly have used straight motions . . . But after their optimum distribution and arrangement it is impossible that there should remain in them natural inclinations to move any more in straight motions.'

But, one asks, how are straight motions defined? In a primordial chaos, in which everything is in a state of motion, straight lines cannot be defined operationally. Once the heaven of fixed stars or the solid frame of the earth is broken up, the parts wandering 'confusedly in disorder', there are no fixed points by means of which a straight motion can be defined. The concept can only be given meaning if Galileo had an underlying concept of motion as taking place with respect to space.

This passage could perhaps be dismissed as a mere poetic allegory but for the fact that in numerous places throughout the *Dialogo* Galileo indicates quite clearly that he regarded the state of rest as something totally different from motion – this is what gives the intensity to the whole of his passionate defence of Copernicanism. Particularly revealing is his attitude to the stars. Although he no longer thought of the stars as attached to a crystalline sphere but rather as distributed (in very great numbers) over a finite region at an immense distance from the sun, he still appears to have regarded them as fixed relative to one another and, more significantly, in a state of overall rest. Whereas Copernicus and Kepler say explicitly that the stars *define* the state of rest, Galileo says that *they are at rest*. For example[99] 'the fixed stars (which are so many suns) agree with our sun in enjoying perpetual rest'. This is a bald statement of fact and noticeably more emphatic than, for example, Copernicus's statement:[100] 'As a quality, moreover, immobility is deemed nobler and more divine than change and instability, which are therefore better suited to the earth than to the universe.'

It is moreover clear from numerous passages in the *Dialogo* that Galileo regarded the question of the earth's mobility as a quite definite either–or question. Either the earth moves or it does not. Speaking of the choice between the two, he says unambiguously[101] 'one of the arrangements must be true and the other false'. The reason for this is the difference between rest and motion, which are such that between them Galileo cannot imagine a greater dissimilarity:[102] 'Are not these two conclusions such that one must needs be true, and the other false? . . . moving eternally and being completely immovable are two very important conditions in nature, show the very greatest dissimilarity. . . . Eternal motion and permanent rest are such important events in nature and so very different from each other . . . it is impossible that one of two contradictory propositions should not be true and the other false.'

But such a standpoint only makes sense if there is some ultimate standard of rest or motion. If motion is purely relative, it is simply not possible to say that any particular body is at rest or in motion. There is only relative motion, and that belongs to the system of bodies considered together. Galileo must instinctively have regarded motion as taking place relative to space, otherwise his statement makes no sense.

The clearest evidence for Galileo's belief in the reality of motion with respect to space is to be found in his theory of the tides, of which he was very proud and not a little blind to its defects, as we have already mentioned in Chap. 6. He wrote a long letter[103] on his theory to Cardinal Orsino in 1616 and it also formed the subject of Day 4 of the *Dialogo*. In fact, it was actually Galileo's intention to name the whole work *Dialogue on the Tides* (*Dialogo del Flusso e Reflusso del Mare*), but he was made to change the title by Pope Urban VIII.[104] The gist of the theory is as follows.

Galileo observed, quite correctly, that if one of the barges carrying sweet (i.e., fresh) water to Venice happens to run aground, the abrupt deceleration of the barge causes the water to surge forward relative to the vessel. He therefore concluded that, if the earth's motion is accelerated or decelerated, the tides might be explained by such an effect. Galileo was completely convinced that the earth both rotates about its axis and moves in its orbit around the sun. Thus (Fig. 7.2), if the speed of the earth in its orbit is V and the speed on the surface of the earth due to its rotation alone is v, then the total speed at point A is $V + v$ but at B it is $V - v$. For the purposes of this discussion, the curvature of the earth's orbit can be ignored. Galileo compares the magnitudes of these speeds, $V + v$ and $V - v$, and notes that they are different, *ergo*, the motion of a point on the surface of the earth is accelerated. (It is important to note that Galileo did not have the Newtonian concepts of velocity and acceleration as vectors.) From this Galileo constructs a theory of the tides – the details of which need not concern us here – and indeed makes the tides the crowning piece of evidence in favour of the Copernican system and the earth's mobility. But, of course, as we now well know, it follows from Galilean invariance (of all things!) that the translational motion can have no influence at all. It can just as well be set equal to zero, and, since $|v| = |-v|$, the effect disappears.

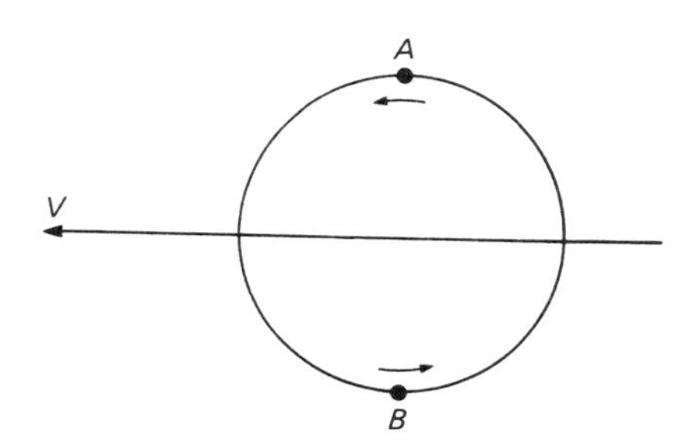

Fig. 7.2.

In view of the title of this study, it is particularly interesting that Galileo actually uses the words *absolute motion* several times in the discussion of the tides (but nowhere else in the *Dialogo*):[105] 'it must happen that in coupling the diurnal motion with the annual, there results an absolute motion of the parts of the surface which is at one time very much accelerated and at another retarded by the same amount.'

Although Galileo uses the expression *absolute motion*, he does not seem to feel that it is necessary to explain precisely what it means. With respect to what is this absolute motion? Mach comments:[106] 'It is noteworthy that Galileo in his theory of the tides treats the first dynamic problem of space without troubling himself about the new system of coordinates. In the most naive manner he considers the fixed stars as the new system of reference.'

This, like several of Mach's comments about Galileo, is rather misleading. It is certainly true that Galileo did not trouble himself about the system of coordinates. However, I do not think there is any warrant for saying that Galileo instinctively regarded the fixed stars as 'the new system of reference'. This would after all imply that the tides (an undoubted physical effect on the earth) were somehow causally related to the earth's speed relative to the distant stars, a notion entirely alien to Galileo, who strongly opposed Kepler's 'occult' suggestion that the tides were caused by the moon. No, the role of the stars is at most that of accidental spectators. The parts of the earth are accelerated and decelerated *relative to space*. That this same space happens to contain stars at rest relative to it is, from the point of view of the tides, a pure accident.

Wohlwill[107] calls Galileo's theory of the tides one of the most interesting mistakes in the whole history of science. It appears to me a good demonstration of how a conceptual way of thinking that, despite successful application, has no *ultimate* anchor in empirical observations can take hold of the imagination and lead one to an entirely false conception of the world. As this is all very relevant to the subject of this book, it is worth asking what it was that led Galileo, almost certainly unconsciously, to a position much more absolutist than Copernicus's. In particular, why was it that in thinking about the instantaneous configuration of the bodies in the world he still retained an Aristotelian cast of mind and was quite happy to use concepts such as *proper place* (which derive their meaning from a global and relational concept of position as typified in the Pythagorean cosmos) but as soon as his thoughts turned to motion he seems to have forgotten the world and instinctively to have assumed motion takes place in space?

The answer must lie in a combination of the syndrome to which Maier drew attention (p. 48) and Galileo's supreme belief in geometry and a Platonic view of the world that was much bolstered by his dynamical discoveries. It is possible, as Ptolemy, Copernicus, and Kepler seem to have done, to regard the theorems of geometry as tools to tell one about the disposition of objects relative to one another in space. But if one takes Euclidean geometry seriously, it is much more congenial to regard triangles, circles, and other geometrical figures as truly real and not as merely something formed by chance alignments of bodies as they move relative to one another. In Chap. 11 we shall see how Newton was quite explicit on this point. He conceived triangles and spheres as possessing just as real an existence in space as a statue that a sculptor could hew from a block of marble.

Also one should not forget the fact that Galileo was the creator of the science of rational motionics. His writings bear eloquent testimony to his own conceptual struggles to formulate what are now some of the most basic concepts of science. Reading in the *Dialogo* and the *Discorsi* the passages in which Salviati patiently and with great care explains to Simplicio and Sagredo concepts such as that of instantaneous speed and its continuous variation – concepts which, as we have seen, were developed only with difficulty by the Mertonians – one realizes how indispensable space must have been to Galileo (and them) as an underlying concept. Indeed, without it velocity as a unitary concept dissolves – a single velocity in space is replaced by a huge and completely indeterminate number of relations to other objects in the universe. Had Galileo stopped to consider that daunting prospect, he would surely never have made any progress at all. Thus, motionics rested on space, strengthening an already strong geometrical predisposition.

Geometrical figures were always seen as the paradigms of the Platonic Ideas or Forms, the eternal reality behind the deceptive and transient world. Galileo's natural inclination to order things according to the ideal of geometry – not an empirical science in his day but an expression of necessary truths – must have been greatly strengthened by his dynamical discoveries. Galileo must surely have believed that laws like those of perpetual and uniform circular motion and the odd-numbers rule were the reality behind an order perceived only imperfectly and approximately in the transient world. Of such perfection in themselves, how could they possibly be an expression of relations holding between bodies in this shifting contingent world?

Laws of such apparent geometrical perfection point to a transcendent world – beyond direct sense perception – governed by mathematics and geometry. Science in this view is not the ordering of sensual empirical facts that in and by themselves constitute the *totality of reality*. It is rather a journey by which the soul finds its way back laboriously, through the Socratic process of recollection, to its antenatal disembodied state, in which it can enjoy the 'direct contemplation of unbodied reality', i.e., the

Forms.[108] It is clear that for Galileo the world had just as much of its Renaissance sparkle as it did for Kepler – but was perhaps for that very reason to be distrusted.

The debate about the absolute or relative nature of motion is at root a debate about reality: is reality transcendent, beyond our ken, or is the world simply what it appears to be?

7.7. At the threshold of dynamics

As discussed in Chaps. 1 and 2, one of the most remarkable things about the discovery of dynamics was the extreme difference between the study of celestial and terrestrial motions. In the study of the latter, no real solid progress at all was made until the sudden explosion of Galileo onto the scene. Why that happened when it did – in precisely the decade that the astronomical part of our story was brought to its triumphant conclusion – is clearly a question that belongs to a study of far wider scope than this can be. What we can consider here briefly are those aspects of terrestrial motions which made them suitable for revealing aspects of dynamics which it would have been very difficult if not impossible to deduce from the astronomical results. Three things especially are to be noted.

First, we saw in Chap. 6 that one of Kepler's biggest problems in trying to comprehend planetary motions was that he was fixing his attention on mathematical relationships in observed motions at too high a level; he was not able to get down to the deeper, more fundamental, and more revealing level. His problem was much too complicated for that. The advantage which Galileo had over Kepler is highlighted by comparing Kepler's law $r = 1 + e \cos \theta$ (where θ is the eccentric anomaly) for the planet–sun distance and Galileo's odd-numbers law for the distance of descent of a freely falling body. Both laws were discovered empirically and, in fact, almost exactly contemporaneously. In many ways the actual discovery of Kepler's relationship was a far greater achievement than the finding of the odd-numbers rule. It certainly required a great deal more work and sophistication, to say nothing of the preparatory work of the earlier astronomers. Yet to Kepler it did not (and could not) yield any deep dynamical secret. It did not contain the time and could not, by itself, tell Kepler anything really useful. That could only come from a mathematical analysis made in conjunction with his other discoveries, an analysis moreover that was way above anything that either Kepler or Galileo could have attempted. In contrast Galileo found the simplest problem that exists in dynamics: one-dimensional and with constant acceleration. Yet for all its simplicity it was still nontrivial and, containing the time explicitly, was representatively characteristic of the basic structure of dynamics. This is why Galileo deserves great credit for realizing that the entrancing beauty of the odd-numbers rule was not the complete story and for finding the deeper and much simpler law that lay beneath it: the law of uniform increase of speed with time. From this, as we shall see, the full structure of dynamics was eventually built up. Kepler too had a sense for something deeper behind his law; it was just that he had no chance of finding it. Thus, despite all his labour, he was not granted the satisfaction which was Galileo's – that of experiencing 'just once the perfect understanding of one single thing'.

The second great advantage of terrestrial motions was the easy possibility they offered of varying the initial conditions and thus revealing the very characteristic structure we find in dynamics, in which universal laws are coupled with initial conditions, the actually realized motions being determined by the two in conjunction. This is another way in which the terrestrial motions made the deep structure of dynamics more evident, and we have already noted the first steps taken by Galileo in that direction. In contrast, the motions in astronomy are given as unique examples and there is no scope for varying the initial conditions. Only in one respect did astronomy offer an advantage of this kind – because several planets, each at different distances from the sun, are observed simultaneously, comparison of their motions reveals the distance dependence of the solar force which acts upon them. Kepler already had a good idea of this and set out to find it but with no hope of success given the difficulties he faced. Later, as we shall see in Chap. 10, the distance dependence of the solar force was the first useful information to be extracted from the astronomical data.

The third principal advantage of terrestrial motions was that they suggested much more readily the notion of rectilinear uniform motion as a basic norm than any of the phenomena in astronomy could have done. However, as we have seen, Galileo did not fully grasp this advantage, though it did play an important role in some further work of his to be discussed in Chap. 9 and was also used *de facto* in the projectile work. In the next chapter we shall see how Descartes clearly recognized what Galileo missed.

Important as all these advantages of the study of terrestrial motions were, Galileo's supreme achievement was simply that he made the study of motion as an *empirical and quantitative* subject a matter of intense importance and interest. Coupled with his two solid and tangible results – the law of free fall and the parabolic law for projectiles – this ensured that relatively soon a complete theory of dynamics would have to be found in one form or another. He was the Hipparchus of terrestrial motions. In a famous remark in the *Discorsi*, Galileo himself summarized the importance of his own work:[109] 'There have been opened up to this vast and most excellent science [of motion], of which my work is merely the beginning, ways and means by which other minds more acute than mine will explore its remote corners.' That exploration is still continuing.

If the inevitability of a way forward was now clear, it is much less certain whether the actual route that Newton took was preordained. The next chapter will show how Descartes introduced at least two twists into the story that by no means followed the logical path that one might have expected. It is at this point of the story that the quotation at the beginning of the book from Kepler's *Astronomia Nova* is shown to be even more apt than it was at the time he made it.

However, before we turn to that, it is fitting to close this chapter on Galileo with what can only be called a spiritual anticipation of one of the key ideas of Einstein's general relativity. It will by now be abundantly clear to the reader that one of the most decisive steps taken by Galileo was in his *decomposition* of motion. In Chap. 10 we shall see how this did, in fact, to a very large degree predetermine the form in which Newton discovered dynamics, especially after the intervention of Descartes. Yet there is one passage in the *Dialogo* from which it is perfectly clear that in his heart Galileo would have preferred a form of the theory of motion in which this does not occur. It is hard to imagine an approach more geometrical than the one Galileo actually created. Nevertheless, the passage we are about to consider reveals a hankering on Galileo's part for an even deeper and more harmonious geometrization of motion than either he or Newton achieved.

It comes in the passage in the *Dialogo* in which Galileo discusses the actual path taken by bodies when they are dropped from towers on the rotating earth. Although Galileo's basic principles are quite close to the truth, the mathematical details of the problem were well above the level at which Galileo could cope and his treatment contains both technical and conceptual flaws.* In fact, we shall see in Chap. 10 that the final breakthrough to dynamics occurred when Hooke prompted Newton into a precise mathematical study of this very problem. What is interesting in the present connection is that Galileo mistakenly believed when he wrote the *Dialogo* that the path in *space* which the falling body would follow would be a circle and that it would moreover move in its circular path with an exactly uniform speed (Galileo did not take into account the annual motion of the earth, only its diurnal rotation). This is what Salviati has to say:[111]

> if we consider the matter carefully, the body really moves in nothing other than a simple circular motion, just as when it rested on the tower it moved with a simple circular motion.
>
> The second is even prettier; it moves not one whit more nor less than if it had continued resting on the tower. . . .

* Galileo's mistakes in treating this problem gave rise to quite a long history of attempts to rectify them. Koyré followed them up in a well-known paper.[110] We shall omit this intermediate history and only return to the problem at the point at which it was attacked in earnest by Newton.

From this there follows a third marvel – that the true and real motion of the stone is never accelerated at all, but is always equable and uniform. . . . So we need not look for any other causes of acceleration or any other motions, for the moving body, whether remaining on the tower or falling, moves always in the same manner; that is, circularly, with the same rapidity, and with the same uniformity.

Clear evidence of Galileo's enthusiasm for these ideas is revealed in the fact that Sagredo is lost for words to praise them:[112] 'I tell you that I cannot find words to express the admiration they cause in me.' Stillman Drake, in his notes on his translation of the *Dialogo*, makes the following comment about Salviati's words 'even prettier':[113] 'This remark, though based on an erroneous demonstration, is particularly noteworthy for the light it throws on the deepest scientific predilections of the author. Galileo's attempt thus to discover an equivalence among "natural" motions is a philosophical anticipation of the concept of world lines in modern physics.'

Although the full significance of these words may be lost on the readers who have not yet learnt about the special and general theory of relativity – which cannot possibly be anticipated at this point – those who are familiar with Einstein's work will certainly recognize the truth of Drake's comment. If Newton's greatest achievement was to *decompose* motion in a gravitational field into an inertial part and a part caused by the gravitational force, Einstein's even greater achievement was to reunite the two parts (perhaps under the influence of Mach's remark (p. 54) to the effect that 'Nature does not begin with elements') into the single law of geodesic motion of bodies subject to only gravitational and inertial forces.

There can be no more fitting conclusion to this chapter on Galileo than Drake's observation of his spiritual affinity to Einstein. Perhaps one should rather say of Einstein to Galileo. But we should also not fail to note how much Galileo looked backwards as well as forwards: his commitment in the above passage to perfectly uniform circular motion was every bit as intense as Copernicus's and the Greeks'. Galileo is probably the most striking example in history of a great scientist poised between the very ancient and the truly modern.

8

Descartes and the new world

8.1. Introduction

'From the start of the fourteenth century the grandiose edifice of Peripatetic physics was doomed to destruction.' We encountered these words of Duhem in Chap. 4. The complete destruction was a long time coming, but when it finally came the tottering structure was brought down in spectacular fashion by Descartes (1596–1650).

The process began, as we saw, with 'a long series of partial transformations'. Initially, in the High Middle Ages, the transformations were mainly in attitudes of mind. With Copernicus, hard science began to make a real contribution. The famous supernova explosions of 1572 and 1604 shattered faith in the immutability of the heavens. Then came the telescope, a veritable Joshua's trumpet, and blasted gaping holes in the structure. But still it did not quite topple to its destruction. It survived, in fact, only because an alternative edifice was not immediately constructed in its place. Galileo was its last tenant. Like all tenants he was only too ready to complain to the landlord about the appalling state of repair of the old palace – without thinking about moving out into a new one. During the very years in which he wrote the *Dialogo*, which, as we have seen, was still permeated with the ancient idea of perfectly circular motion, Descartes was at work on plans for a new edifice.

Descartes is, perhaps, the most enigmatic of the figures who appear in this book. He played an important, indeed central, role in the discovery of dynamics, but this came about almost incidentally, not through intensive study of actual motions in their own right. His contribution was of a kind quite different from that of the astronomers and Galileo; it was much more purely philosophical and is to be compared with Aristotle's. Before we examine the details, it will be worth saying a few words about the man.

The son of a well-to-do lawyer and judge, René Descartes was educated at a famous Jesuit school in La Flèche in France. The Jesuits seem to have left a permanent mark on his personality. It is noteworthy that he presents a picture of someone fully prepared to topple Aristotle but very reluctant to essay a frontal encounter with the Church. He appears to have been devout, and died 'in the faith of his nurse', to whom he was devoted.[1] There is an interesting difference between Galileo's writings and Descartes'. No one reading the *Dialogo* could be in any doubt that Galileo believed in the earth's motion and wanted the intelligent reader to see through the subterfuges of his token impartiality in the presentation of both sides of the argument. In the case of Descartes' works published during his lifetime the reader remains in doubt. Descartes was in fact a secretive man and his proudly proclaimed motto was *bene vixit, bene qui latuit* (he has lived well who has hid well).[2] As Santillana says,[3] Descartes chose to wear a mask 'from the moment he stepped upon the stage'. In Sec. 8.6 we shall see why.

With moderate inherited wealth, Descartes resolved to live a life dedicated to the finding of a more satisfactory system of philosophy than the one he had learnt at school. After a decade of an almost nomadic existence he settled more or less permanently in Holland, though frequently changing his abode. He lived there from 1628 to 1649, when he was finally inveigled into moving to Stockholm to instruct the young Queen Christina of Sweden in philosophy. He died the following year, a victim of the Swedish winter.

For the development of his physical ideas, his early collaboration with the Dutchman Isaac Beeckman (1588-1637),[4] with whom he worked during the years 1619 to 1623, was very important. Like Descartes, Beeckman had a concept of inertial motion and should perhaps be given some credit for the idea. It was, however, Descartes, not Beeckman (who published nothing in his lifetime), who presented the idea of inertial motion, in a form quite close to that finally adopted by Newton, to the world with a great flourish and thereby influenced the course of further developments.

From about 1628 Descartes began systematic elaboration of his ideas. For our purposes his two most important works were his *Le Monde* (*The World*), which was more or less complete by 1633 and contains the key ideas of Cartesian physics, and his *Principles of Philosophy* (published in 1644). For reasons that will be explained in Sec. 8.6, *The World* was withheld from publication by Descartes and only appeared after his death in 1664. Descartes became famous in 1637 with the publication (anonymously, though the authorship soon became common knowledge) of his *Discours de la Méthode* (*Discourse on Method*). This work, to which the interested reader is recommended as the best introduction to his general philosophy, is held by many to mark the beginning of modern philosophy, and it was in it that Descartes expounded his famous method of Cartesian doubt, encapsulated in the saying *Cogito, ergo sum* (I think, therefore I am). The attitude of mind that informs this philosophy is of central relevance for the absolute/relative debate and will, in particular, be considered in Sec. 8.3. However, a few words are appropriate at this point.

Descartes' main concern was to find absolutely secure grounds of knowledge and to achieve these he took as his first principle:[5] 'That whoever is searching after truth must, once in his life, doubt all things; insofar as this is possible.' Practising such doubt, he concludes that he cannot even take the existence of the material world as an undoubted fact; for the sense perceptions that he has of it might simply be conjured up in his mind by a 'malignant demon'. But the one thing that he absolutely cannot doubt is his own existence as a *thinking* being, hence the famous saying, which is stated explicitly in his *Principles* in §I.7.

From the one thing in which he is absolutely confident, that he truly exists as a thinking subject, Descartes draws a conclusion that is characteristic of his whole philosophy – that the only ideas that one can trust when philosophizing are the *clear and distinct* notions formed by the mind. He arrives at this conclusion through one of the most important arguments in his system, namely, that an essentially good God must exist and that, being essentially good, God would not implant clear and distinct ideas in our minds if they were not truthful. Thus,[6] 'we never err when we assent to only things which are clearly and distinctly perceived'. On such a basis, Descartes develops a philosophical system that puts great emphasis on the clarity of concepts and rational thought. In this respect he greatly strengthened a somewhat similar tendency in Galileo, in whom however the empirical input is vastly more important, indeed essential for suggesting the key ideas.

Descartes was, in fact, sorely deceived as to his ability, using his own 'clear and distinct' ideas, to elaborate a comprehensive account of the material world. His detailed physics proved to be very largely useless. He nevertheless played an important role, for three main reasons: (1) First, the overall conception of the world, being based on a truly universal concept of matter and motion, was much more conducive to the final emergence of dynamics than the decrepit Aristotelianism it replaced; (2) his physics was in fact (despite the Cartesian principles) based, at least qualitatively, on empirical input, the great potential significance of which Descartes correctly realized, and this resulted in the recognition of the law of inertia as the first law of motion; (3) his basic physical scheme, which anticipated the structure of mature Newtonian dynamics in several important respects, prompted Huygens and Newton to study quantitatively two key phenomena that Galileo failed to treat adequately and these provided the final two of the 'baker's dozen' needed before the definitive synthesis of dynamics could occur.

Descartes was therefore instrumental in bringing the study of terrestrial motions to the point at which it was able to supply the hints that finally cracked the astronomical problem, which all this while, through three generations, awaited the final *coup de grâce*.

However, the way in which all this happened was not at all smooth. There were some very curious twists, as we shall see.

8.2. The new world

As we have seen, both Kepler and Galileo developed rather than completely broke an essentially Aristotelian concept of the world. They followed the tradition of the Hellenistic astronomers, transforming qualitative Aristotelianism through an infusion of quantitative empiricism. In the end, the new wine was, of course, too searching for the old bottle. Descartes' approach was quite different. In spirit he was a pre-Socratic in the mould of Leucippus and Democritus; he hankered after one or two simple clear concepts that, at a stroke, would explain the workings of the entire material world.

To gain a deeper understanding of what Descartes was trying to achieve and identify the sources of his inspiration, one can do no better than to read the admirably brief and eloquent opening chapters of *The World*, which, at the time of its writing, Descartes undoubtedly regarded as his *magnum opus*, even though he was very nervous about the reception it would get.[7] His aim was no more and no less than to give a rational explanation for all the phenomena of the material world. The work is written as a charming fable: Descartes says he will not attempt to explain the real world but will instead describe an imaginary world, a 'new world', about which the most important thing is that it is completely determined and described by an absolute minimum of properties. In fact, he puts into this imagined world nothing but *matter*, whose sole property is that it possesses extension, and, vitally important, *motion*. His assertion is that, provided this matter and motion satisfy (by God's ordinance) certain almost self-evident laws, such a world, *whatever* its initial condition, would of *necessity* evolve into a world indistinguishable from the one we observe around us. Hence his famous assertions: 'Give me extension and movement and I will reconstruct the world'[8] and 'The entire universe is a machine in which everything is made by figure and movement.'[9] Descartes was nothing if not ambitious. He justified his confidence in the following characteristic words:[10]

> But I shall be content with showing you that, besides the three laws that I have explained, I wish to suppose no others but those that most certainly follow from the eternal truths on which the mathematicians are wont to support their most certain and most evident demonstrations; the truths, I say, according to which God Himself has taught us He disposed all things in number, weight, and measure. The knowledge of these laws is so natural to our souls that we cannot but judge them infallible when we conceive them distinctly, nor doubt that, if God

had created many worlds, the laws would be as true in all of them as in this one. Thus, those who can examine sufficiently the consequences of these truths and of our rules will be able to know effects by their causes and (to explain myself in the language of the School) will be able to have demonstrations *a priori* of everything that can be produced in that new world.

It is significant that Descartes opens *The World* with a discussion of what, since Locke, has become known as the problem of the primary and secondary qualities. This problem was first raised with full clarity in the modern age by Galileo in his 'bestseller' *The Assayer* published in 1623. Galileo argued that there are certain fundamental properties of matter, above all extension and motion, which truly do exist in the external world and have a nature more or less as we perceive them. These are the *primary qualities*. With these are contrasted the *secondary qualities* such as heat, taste, and colour, which Galileo argued are merely sensations produced in the mind by the interaction between the external world and our sense organs and brain:[11]

To excite in us tastes, odors, and sounds I believe that nothing is required in external bodies except shapes, numbers, and slow or rapid movements. I think that if ears, tongues, and noses were removed, shapes and numbers and motions would remain, but not odors or tastes or sounds. The latter, I believe, are nothing more than names when separated from living beings.

It is with a similar discussion that Descartes opens *The World*. What was for Galileo a passing remark is for him the central concern. In *The World* and the later *Principles* he expounds the dream of the rationalists: to explain as many of the secondary qualities using as few primary qualities as possible.

Let us now briefly outline the basic scheme by which Descartes sought to achieve this. It is a curious amalgam of ancient atomism and ancient plenism with one or two new and, as events showed, fruitful ideas. Descartes had in common with the atomists the idea that the material world is composed of a single matter, which exists in different shapes and sizes. These pieces are assumed to be completely homogeneous; in fact, according to Descartes they have absolutely no other properties than extension - they are *purely mathematical* (or *geometrical*) figures. Unlike the atomists, who assumed that the atoms remain completely unchanged and can be neither broken up nor amalgamated into new pieces, Descartes assumed that the shapes of his individual pieces of matter were constantly being changed as a result of mutual collisions. In fact, he assumed that, whatever initial condition God might have chosen at the beginning of his 'new world', there would after a certain time develop a situation in which there exist only three basic types of pieces, *elements* as Descartes calls them. The first element consists of very small pieces that move with great rapidity and are associated with fire; the pieces of the second element are of intermediate size and spherical, having been ground into this shape like pebbles on the seashore. The pieces of the third element are gross and slow and make up the body of the earth. The programme, for it was a programme, of Cartesian physics was to show how *all* qualities of bodies[12] 'can be explained without the need of supposing for that purpose anything in their matter other than the *motion, size, shape, and arrangement* of its parts' (my italics). This extreme rationalist programme came to dominate scientific thought for well over half a century. As we shall see, its influence was initially beneficial for the discovery of dynamics but then more of a hindrance; it was also one of the factors that delayed the acceptance of Newton's theory of gravitation on the Continent for several decades.

In common with Aristotle, Descartes assumed the world to be a plenum – he did not permit any empty spaces anywhere. The reasons for this are not brought out very clearly in *The World*, but they evidently had to do with the explanations that Descartes provided for certain key phenomena, to which we shall shortly come. In the later *Principles* the absence of a vacuum is raised to a metaphysical principle of the very highest importance; as it plays an important part in Descartes' concept of motion we shall consider this point in later sections of this chapter.

The most distinctive, original, and valuable part of the Cartesian scheme (from the point of view of the discovery of dynamics) was the role played in it by motion. It will be recalled from Chap. 2 that one of Aristotle's most telling criticisms of the atomists was that Leucippus and Democritus had not stated with sufficient clarity what precise motions their atoms followed. This criticism, which was evidently a key factor in Aristotle's development of his alternative scheme, had little impact in antiquity. As we noted, in Lucretius's masterpiece *De Rerum Natura*, written about 300 years after Aristotle, the motions of the atoms are still remarkably vague despite the precision and clarity with which the poem is otherwise generally argued.

Whether consciously or not, Descartes took up Aristotle's challenge. He lived in the age in which the *zeitgeist* became reintoxicated with the heady pre-Socratic vision of the infinite universe. (Descartes was 4 when Giordano Bruno was executed, 14 when the *Starry Messenger* amazed the world.) Moreover, the Euclidean systemization of geometry had long since purged the atomists' void of its distinguished direction of eternal falling. In the ideal world of Euclid, close packed with points and infinite lines, this anthropomorphic relic was replaced by thoroughgoing directional democracy. Descartes grasped it. Two simple ideas were born and espoused with an eloquence that would never permit them to be forgotten again: first the idea that any piece of matter, once set in motion, would continue to move for ever with the speed initially imparted to it if it were not for the intervention of other matter and, second, the idea that this motion would be along a straight line in space, again if it were not for the intervention of other matter.

We shall look at the origin of these ideas, which were fused by Huygens and Newton into the modern law of inertia, and their significance for the discovery of dynamics in Secs. 8.4 and 8.5 and mention here only that crucial for their recognition and significance in Descartes' eyes was the long recognized but hitherto little studied phenomenon of *centrifugal force* (the name was coined by Huygens in 1673). The stone flung from the sling is by far the most important empirical input in Cartesian physics and figures prominently throughout *The World*. It clearly played an important role in suggesting to him the two components of his 'law of inertia'. Simultaneously, centrifugal force was destined to carry the main explicatory burden within his programme for the rational explanation of all the observed phenomena of nature.

We have here, in fact, the great difference between Descartes and Galileo. Galileo was interested in motion in its own right, as a deeply interesting subject of empirical study, Descartes only as a *means of explanation*. Descartes did not stand in awe before motion and reverently and patiently unravel its secrets as had Galileo (following the example set by the astronomers in their patient watching of the planets). As we shall see later, he was forced to think hard about the ontological nature of motion, but as far as his physics was concerned he took it more or less for granted, convinced he already possessed knowledge of the laws of motion.

He used motion rather than studied it. In a letter to Beaune, he said:[13] 'Although my entire physics is nothing but mechanical, nevertheless I have never examined questions which depend on measurement of velocity in detail.'

This is the first of several paradoxes we meet in Descartes: we owe the first outline of dynamics to Descartes' explicatory urge, not a study of motion *per se*.

There is no point in going into the details of Cartesian physics, which does rather tend to degenerate into 'hand waving' once Descartes leaves the exposition of his basic ideas, in which he achieved a lucidity that rivals Galileo's and which was undoubtedly also a factor in his great influence. It will however be helpful to give some more details of the overall scheme and also sketch the mechanisms that he proposed for some key phenomena. I follow loosely the account given in *The World*.

One of the most characteristic features of Descartes' physics is his *circle of motion*. Since the world is a plenum, matter can only leave one place if other matter comes in to fill the void. Depending on the disposition of the pieces of matter, this may involve quite an extended circular readjustment of matter. The circle must however always close because volume of matter is exactly conserved in Cartesian physics.

We may mention in passing that, viewed in the historical perspective of mechanical explanations of nature, the plenum, which Descartes needed primarily for his theory of light (see below), proved to be rather a redundant feature. For it had been required by Aristotle mainly to produce motion at a distance by direct transfer or 'push' (through pressure). However, this role of the plenum was largely obviated by Descartes' own 'law of inertia', i.e., rectilinear persistence of motion, and his rules of collision (to which we shall also come). The plenum was in the event a blessed nuisance – you had to have hordes of little pieces of matter squeezing through all the interstices in order to leave absolutely no gaps anywhere (Cartesian physics is rather like the rush hour on the London Underground) – and it was soon dropped by some of Descartes' most enthusiastic followers, notably Huygens. What they did, in fact, was transfer the Cartesian concept of persistence of motion to atomism. The associated acceptance of the vacuum was greatly helped by the work of Pascal, Torricelli, Boyle and others and by the famous vacuum demonstrations of Guericke in 1654.[14]

The notion of a circle of motion, which is, of course, quite different from the ancient circular motions of the astronomers, is central to one of the most distinctive features of the Cartesian cosmology – the theory of vortices. According to Descartes' 'laws of nature', to which we shall come in Sec. 8.4, God created the world with a certain initial amount of motion of the individual pieces of matter and then, through the laws of collision, ensured that this motion always remained in the world in an unchanged amount, being merely passed from certain pieces of matter to others (the details will be considered in Chap. 9).* Descartes posited that as a result of the combined influence of these factors – the collisions and the constancy of the total amount of motion – the world would, on a large scale, settle down into a system of huge vortices, in which the various particles swarmed around centres at which he placed the sun and the other stars. Thus, each vortex was assumed to be at least as large as the solar system. Figure 8.1 shows the diagram by means of which Descartes illustrated his vortex scheme (in *The World* he called each of the vortices a heaven, since the various planets were assumed to be carried around in them by the plenal fluid). The sun is at the centre S of one vortex. The

* It may be mentioned that the idea of conservation of motion and matter is another of those intuitive ideas like the uniform flow of time (discussed in Chaps. 2 and 3) that were originally assumed instinctively by philosophers and then justified by genuine discoveries based on sound empirical observations and theory. However, in all such cases the subsequent justifications demonstrated that what was initially assumed as a metaphysically obvious truth rests ultimately on sophisticated facts about the interconnection of different phenomena in the world, the clarification of which is very far from trivial. Before the end of this volume we shall meet another example of this kind in the concept of absolute space. The clarification of the basis of conservation laws by what is known as *Noether's theorem* will play an important role in Vol. 2.

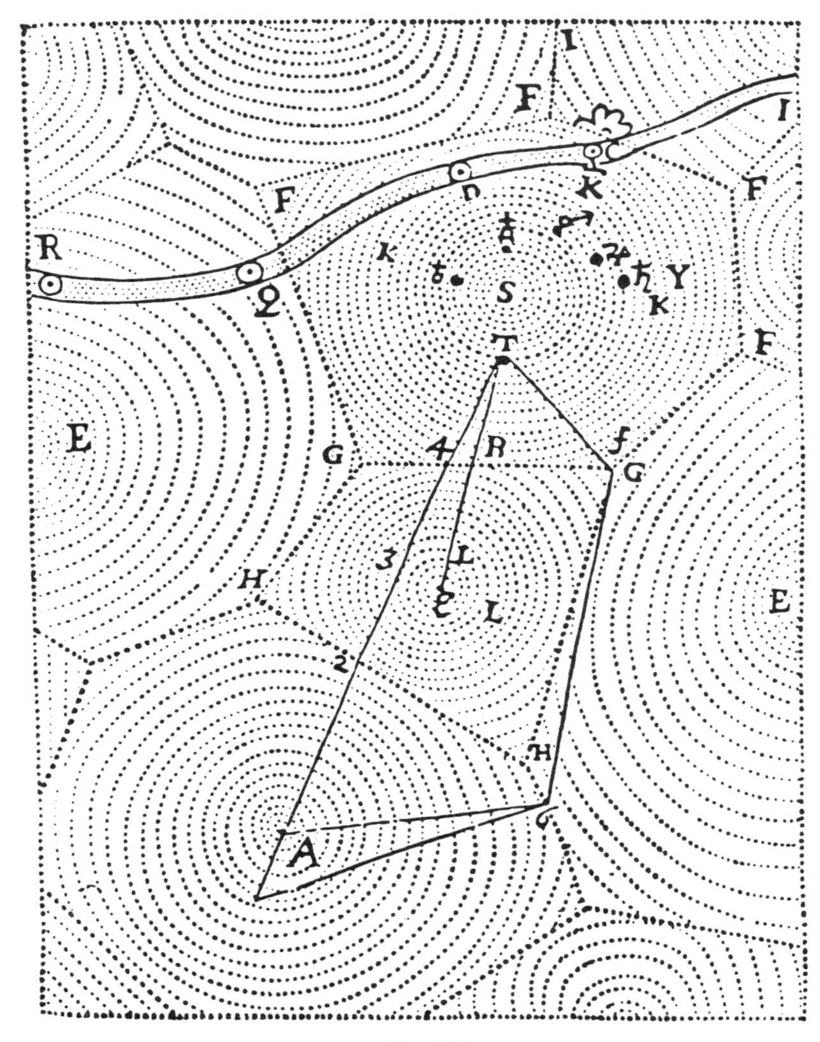

Fig. 8.1.

planets (indicated by their symbols; the earth is T) are carried around the sun in the circles as shown. Other stars, with planets assumed around them too, are at the centres of the other 'spiders' webs' (at the points marked E and A). The curious tube shown passing through the top of the solar system represents the path of a comet, which Descartes assumed to be more massive than a planet, so that it would have sufficient strength to plough through the current of one vortex and pass into another, unlike the planets, which, being lighter, would be carried with the vortex like twigs and feathers in a whirlpool.

It should be said that Descartes almost always talks as if his vortices are two-dimensional cylindrical formations. They must, however, be three-dimensional and spherical, a fact that presents all sorts of complications with which Descartes hardly begins to come to terms. In fact, the modern reader, coming to Descartes after a detailed reading of Kepler's *Astronomia Nova* and Galileo's great dialogues, cannot help feeling a sense of disappointment and also some surprise that Descartes had the influence that he did. Of course, Kepler's physical theory of the planetary motions, for example, was quite wide of the mark, but, as we have seen, it was developed on the basis of sound observations and was used as a valuable heuristic guide in the successful finding of more detailed results, which were expressed by very precise mathematical laws. In comparison the following passage from Descartes' *Principles* shows the level at which Descartes worked. It also shows that, for all his much-vaunted claim to be describing everything in nature by mathematics, in reality his physics very seldom advanced beyond pictorial analogy; the truth of his comment to Beaune is nowhere more evident than in this passage:[15]

let us assume that the matter of the heaven, in which the Planets are situated, unceasingly revolves, like a vortex having the Sun as its center, and that those of its parts which are close to the Sun move more quickly than those further away; and that all the Planets (among which we [shall from now on] include the Earth) always remain suspended among the same parts of this heavenly matter. For by that alone, and without any other devices, all their phenomena are very easily understood. Thus, if some straws [or other light bodies] are floating in the eddy of a river, where the water doubles back on itself and forms a vortex as it swirls: we can see that it carries them along and makes them move in circles with it. Further, we can often see that some of these straws rotate about their own centers, and that those which are closer to the center of the vortex which contains them complete their circle more rapidly than those which are further away from it. Finally, we see that, although these whirlpools always attempt a circular motion, they practically never describe perfect circles, but sometimes become too great in width or in length, [so that all the parts of the circumference whch they describe are not equidistant from the center]. Thus we can easily imagine that all the same things happen to the Planets; and this is all we need to explain all their remaining phenomena.

This may seem persuasive on first reading, but the minute one starts to think about the details of the motions of the planets that Kepler had demonstrated so conclusively, above all the fact that the centres of the planetary orbits are scattered around the sun with second foci at equal distance on the side further away from the sun and that the inclinations of the orbits are all different but fixed, the task of reproducing them by a vortex mechanism is immediately seen to be quite hopeless, as Newton was to point out with withering scorn in the *Principia*.[16]

We now consider briefly some of the phenomena that Descartes was most keen to explain. This will demonstrate the positive new part of his programme as compared with the original atomism – the much greater emphasis placed on motion and the anticipation of Newton's law of inertia. We have already seen that Descartes posited an inherent tendency of bodies to continue moving in a straight line if once set in motion. However, if at rest, they would remain so. To be set in motion, a certain force must be supplied - or rather the motion of a moving body must be transferred in a collision to the body at rest in order to set it in motion. It is most important that Cartesian physics allowed absolutely no other mechanism for the acquisition of motion. In a way, this was a useful rigour and subsequently played a fruitful role in the discovery of dynamics in much the same way as the notion of uniform circular motion had in astronomy (even to the extent of being abandoned at the crucial moment of synthesis). Its great merit was that the idea was amenable to mathematical treatment. However, very little of this mathematics was supplied by Descartes himself.

Let us first consider Descartes' explanation of hardness and liquidity.[17] He notes that if two small parts of matter are touching each other, 'some force is necessary to separate them'. Thus, if a body is composed of parts that are all at rest relative to each other, the mere fact of this will mean that it will not give way readily. Therefore,[18]

> to constitute the hardest body imaginable, I think it is enough if all the parts touch each other with no space remaining between any two and with none of them being in the act of moving. For what glue or cement can one imagine beyond that to hold them better one to the other?
>
> I think also that to constitute the most liquid body one could find, it is enough if all its smallest parts are moving away from one another in the most diverse ways and as quickly as possible, even though in that state they do not cease to be able to touch one another on all sides and to arrange themselves in as small a space as if they were without motion. Finally, I believe that every body more or less approaches these extremes, according as its parts are more or less in the act of moving away from one another. All the phenomena on which I cast my eye confirm me in this opinion.

The question of what 'glue' or 'cement' holds bodies together was one of several topics that dominated the natural philosophy of the seventeenth century – and helped to distract it from the precise quantitative study that Galileo had demonstrated could be so fruitful.

Next we come to light. Significantly, *The World* has an alternative title: in full it is *Le Monde ou Traité de la Lumière* (*The World or Treatise on Light*). Descartes' idea was as follows. According to his scheme (outlined in Chaps. 2, 8, 13 and 14 of *The World*) the material that collects at the centre of the vortices is of the finest type, i.e. is made up of the smallest particles, which move with the greatest rapidity (first element). In fact, it is these bodies that constitute the sun. The region between the sun and the planets is occupied by the larger (but still, of course, minute) spherical

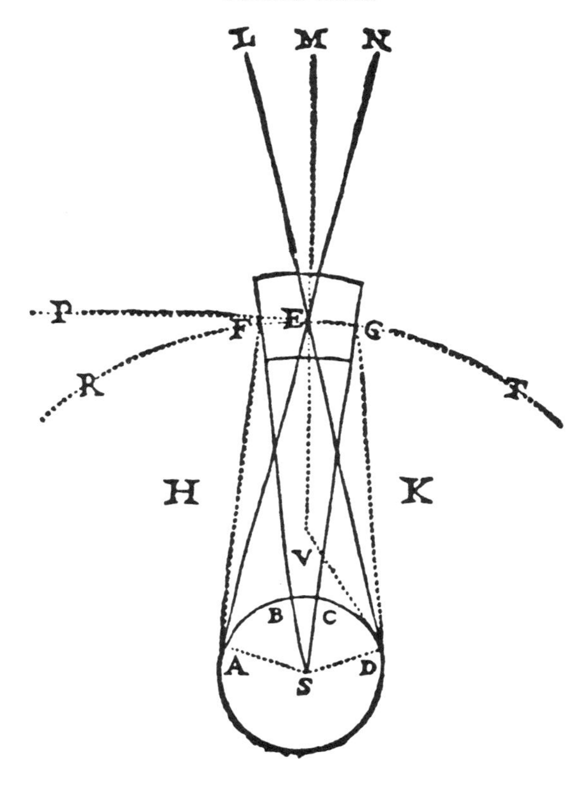

Fig. 8.2.

particles of the second element, which constitute the vortex, which in turn carries around the planets, which are made of the more bulky third element. Now by their inherent nature, all these rapidly moving particles have a tendency to carry on straight forwards in their motion. They are, however, constantly being prevented from doing this by the presence of the objects in the vortex further out from the centre than they are. This has two consequences: first, the individual particles are constantly being deflected into circular motion; second, the inner particles exert a constant pressure on the outer particles. According to Descartes, this pressure, which he assumes is transmitted instantaneously through the vortex (this is why he needs a plenum), is what the eye senses as the light of the sun (the pressure can also be 'reflected' by more solid objects like the moon and planets). Figure 8.2 is the diagram that Descartes used to explain how the pressure is transmitted from the body of the sun, with centre S, to the orbit of the earth at E. Descartes concludes his account of this explanation of light with the words:[19] 'Thus, if it were the eye of a man that was at the

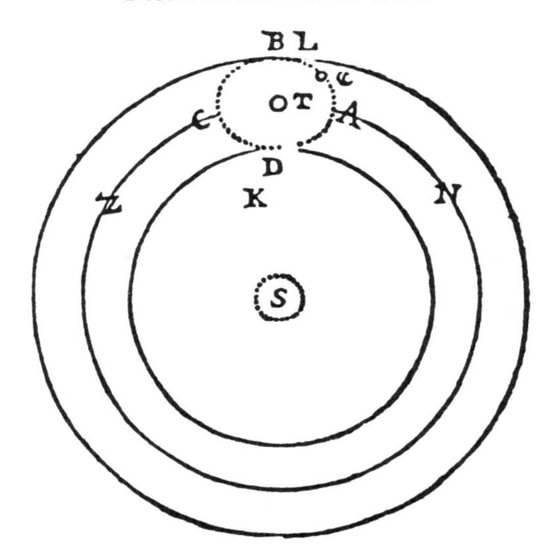

Fig. 8.3.

point E, it would actually be pushed, both the sun and by all the matter of the heaven between the lines AF and DG.'

Finally, we come to Descartes' account of the phenomenon of weight (Chap. 11 in *The World*). Descartes argued that not only is the earth carried around the sun in a large vortex (or heaven) but that also there is a small vortex or second heaven which surrounds the earth and rotates in the same sense as the main vortex. In his scheme, which is shown in Fig. 8.3, the moon is carried around at the outer edge of the second (little) heaven. Now the little heaven plays a crucial role in the explanation of weight, which Descartes says is as follows:[20]

> Now, however, I would like you to consider what the weight of this earth is; that is to say, what the force is that unites all its parts and that makes them all tend toward its center, each more or less according as it is more or less large and solid. That force is nothing other than, and consists in nothing other than, the fact that, since the parts of the small heaven surrounding it turn much faster than its parts about its center, they also tend to move away with more force from its center and consequently to push the parts of the earth back towards its center.

In other words, terrestrial gravity is produced by a kind of centrifuge effect; the smaller and faster particles of the second heaven have a greater tendency to recede from the centre of their rotation than do the particles of the third element, which forms the earth, and this causes bodies made up of the third element to be pushed back towards the centre of the earth.

Now in fact – and quite unbeknown to Descartes – this conception of the origin of weight was a great deal better than one might at first blush suppose. In the following chapter on Huygens we shall see why. It undoubtedly played a positive role in the discovery of the last of the 'baker's dozen' needed before the synthesis of dynamics could commence.

There is little point in going into any further detail of the mechanistic and pictorial accounts that Descartes provided for these and other phenomena. It is, however, well worth pointing out that, *almost* by accident, Descartes provided the one vital clue to the cracking of the planetary problem that completely escaped both Kepler and Galileo. Kepler came closest to the solution of the problem with his idea that the motion of the planets is produced by the combined influence of two forces – the azimuthal force produced by the rotating sun and a further magnetic force that alternately attracted the planets towards the sun or pushed them away again. Descartes changed the competition from one between *two forces* into the interplay of a tendency on the part of all matter to proceed forwards in a straight line and the intervention of a second factor, which causes the planets to be deflected into a continual circular motion. This was a crucial step; although he was wrong about the mechanism producing the deflection, he was right in the basic formulation. Within a decade or two of Descartes' death several talented men were coming to grips with the problem of the planets in the framework of a formulation that sooner or later was bound to yield the secret.

In Sec. 8.4 we shall look in more detail at the precise form in which Descartes formulated his fundamental principles. Before that we shall consider the relationship between Descartes' more general philosophy and some underlying aspects of the absolute/relative question. However, we are already in a position to understand how it was that Descartes came to perform the office of midwife to the birth of dynamics (in which capacity I contrast him to Galileo, the *creator* of motionics). Descartes' anticipation of the law of inertia was the child of his consuming desire to explain and the manner in which he understood explanation. The elements he originally allowed himself in his self-imposed rigour were just three: space, matter, motion. Nothing else was allowed to have explicatory power. Kepler, it will be recalled, proposed a mechanical explanation of the motion of the planets, but his entire clockwork was driven by the animalistic power of the rotating sun, which he simply had to invoke since his physics taught the inherent tendency of all motion to decay. Descartes banished such demiurges with contempt. But, since the world manifestly persists, he needs must have a perpetual source of explanation; the spring must not dry up up. *Ergo*, matter and motion must last forever. This is the origin of Descartes' twin principles of the conservation of matter and of motion. He could not do without them.

8.3. The Cartesian concept of substance and the divide between materialism and idealism

Descartes is famous, if not infamous, for the line he drew between mind and matter (which includes the human body). He recognized the existence of just two substances – thinking matter (*res cogitans*) and extended matter (*res extensa*). It was the latter that was to become the hallmark of materialism: dead inert matter informed (or rather governed) solely by laws of nature laid down by God from the beginning of time.

The concept of substance dominates not only Descartes' thought but also the Newtonian and post-Newtonian debate about the nature of space. It will therefore be as well to look at his concept more closely. He says:[21]

> However, corporeal substance and created mind, or thinking substance, can be understood from this common concept: that they are things which need only the participation of God in order to exist. Yet substance cannot be initially perceived solely by means of the fact that it is an existing thing, for this fact alone does not *per se* affect us; but we easily recognize substance from any attribute of it, by means of the common notion that nothingness has no attributes and no properties or qualities. For, from the fact that we perceive some attribute to be present, we [rightly] conclude that some existing thing, or substance, to which that attribute can belong, is also necessarily present.

This is fairly conventional philosophy, at least as regards the concept of substance. However, Descartes then introduces the distinctively Cartesian idea with which we are already familiar, namely, that 'each substance has one principal attribute, thought, for example, being that of mind, and *extension* that of body'[22] (my italics):

> And substance is indeed known by any attribute [of it]; but each substance has only one principal property which constitutes its nature and essence, and to which all the other properties are related. Thus, extension in length, breadth, and depth constitutes the nature of corporeal substance; and thought constitutes the nature of thinking substance. For everything else which can be attributed to body presupposes extension, and is only a certain mode [or dependence] of an extended thing; and similarly, all the properties which we find in mind are only diverse modes of thinking. Thus, for example, figure cannot be understood except in an extended thing, nor can motion, except in an extended space; nor can imagination, sensation, or will, except in a thinking substance. But on the contrary, extension can be understood without figure or motion; and thought without imagination or sensation, and so on; as is obvious to anyone who pays attention to these things.

As we have seen, most of the *Principles* is actually taken up by an extended discourse in which Descartes claims to show that the entire material world and all its phenomena can be understood more or less completely on the basis of this one single principal attribute of corporeal substance (in conjunction with motion). However, this is not the point which is to be made here but rather that Descartes, like many philosophers, could not conceive a world without substance – something that stands under and acts as support to the brightly coloured and variegated world that we actually perceive through our senses. As outlined in the passages above, substance itself is not directly accessible to perception but only the attributes, of which substance is the bearer.

The philosophical counter-attack to Cartesian–Newtonian materialism was mounted by Berkeley (1685–1753) when he pointed out[23] that the concept of substance is not derived through Cartesian introspection but empirically – from the observational fact of *persistent correlation of attributes*. We form the concept of body from the invariable correlation of qualitatively different deliverances of the various senses: a certain visual shape, taste, smell, and texture are invariably found together with the thing that we call an apple; the word *apple* is nothing but a convenient symbol to denote this remarkable *correlation of phenomena*. In this view, the attributes and their correlation are sufficient in themselves. There is no need to posit an underlying substance to carry them and explain the correlation. Substance is purportedly an explanation of correlation but is in reality, say the relationists, merely an alternative name for the correlation itself – a pleonasm.

In the discussion of Aristarchus and the application of trigonometry to astronomical problems, space was characterized as a safety net without which we feel that it is unsafe to embark on celestial acrobatics. Space is just one case of a general syndrome – one might call it the *substance reflex*. We instinctively look for support. At the deepest level, this is probably what explains the passionate rejection of Copernicanism in the first century after it was proposed – for it taught that *there is no ground under our feet*. Moreover, absolute space can be seen as a conceptual substitute for the ground that Copernicus so deftly pulled from under us. (And even he, as we have seen, felt he had to cling onto the walls of his medieval cosmology, i.e., the fixed stars.)

The standpoints of Descartes and Berkeley represent the two philosophical extremes between which the debate about the absolute or relative nature of motion moves. They are reflected in quite different attitudes to the scientific endeavour and different interpretations of what it is about. Let us start with materialism in the exteme form expressed in Descartes. For some reason or other, he thinks he can understand space and motion directly; that somehow the comprehension of motion is no different from the comprehension of numbers; therefore, like arithmetic, its basic laws are *a priori*, not dependent on experience; the same is true of extension, which he regards as the *essential* attribute of matter. Thus, extended matter and motion are lumped together with numbers and space and strictly separated from all the sensual (secondary) qualities.

This standpoint of Descartes leads to something of a paradoxical situation – for it was with *mental* conceptions that he constructed a *material* world. This is the whole point of his fable of the 'new world'.

In this connection, it is worth noting a very characteristic difference between Kepler, the empiricist, and Descartes, the rational philosopher, in the grounds they have for believing that the matter throughout the universe is one and the same. (The abolition of the Aristotelian distinction between quintessence and the four corruptible elements, i.e., between the heavens and the earth, was one of the major qualitative differences between the old and the new science.) In the *Astronomia Nova*,[24] Kepler's standpoint is that since the earth manifestly obeys *the same laws of motion* as the other planets one can reasonably conclude that it is made of the same matter. Descartes is able to find an *a priori* metaphysical argument[25] for the same conclusion, asserting that pure extension is the essence of corporeal substance and introspection of our mind reveals that pure extension is the same here as it is there!

It is also worth mentioning a very illuminating point made by Koyré. Writing of Descartes, he says:[26] 'The God of a philosopher and his world are correlated. Now Descartes' God, in contradistinction to most previous Gods, is not symbolized by the things he created; He does not express Himself in them. There is no analogy between God and the world; no *imagines* and *vestigia Dei in mundo*'. This goes a long way to explain Kepler's ecstatic empiricism and Descartes' rather contemptuous attitude to the contingent world. As we saw in Ptolemy, the inspiration that sustained the long and weary hours of observation came in significant part from the expectation that divinity itself would be revealed by all his labours. It was much the same with Kepler. This was the attitude that made him look on observation as an exciting voyage of exploration – exploration of the divine in which great surprises are to be expected. For how could man expect to be able to anticipate the divine?

It was a very deliberate and conscious decision on Descartes' part to withdraw the divine completely from the contingent world and reduce it to mere *res extensa* (the use of the passive *extensa* is deliberately symbolic; cf. Koyré[27]).

Two points should be made about this. First, by making the material world completely homogeneous and passive, Descartes simultaneously made it a lot less interesting. This may well explain why, despite what proved to be some very promising ideas, Descartes' actual physics yielded virtually no solid results. His ideas only became fruitful when Huygens and Newton added the empirical and quantitative aspect that is so characteristic of Galileo and Kepler.

The second point has to do with overall concepts and the difficulty of changing them. The very approach that Descartes adopted in identifying what he regarded as reliable concepts – involving a sounding of his conceptions, a careful sifting through them, and then the ruthless rejection of all those to which there appeared to attach the slightest doubt – of necessity meant that Descartes finished up with a set of concepts that do indeed seem beautifully 'clear and distinct'. One of the reasons why Descartes is important for the subject of the book is that some of the concepts which he identified in this manner became essentially the concepts that underlie the absolutist concept of motion. Their very clarity, coupled with the brilliant success of Newtonian dynamics, has meant that they have become deeply ingrained in the scientific consciousness and extremely difficult to dislodge. Moreover, they are tailor-made to fit the law of inertia in the form that it was formulated, first by Descartes, then by Newton. Perhaps the biggest hurdle that the advocates of relative motion and a conceptually different formulation of the law of inertia must overcome is that they are trying to persuade people whose deepest concepts were developed – in part, no doubt, unconsciously – for the precise purpose of giving expression to the Cartesian–Newtonian law of inertia.

Perhaps the most striking example of such concepts is provided by that of *Cartesian coordinates*. By introducing[28] the idea that the distance of a point from a line can be represented algebraically by a number, Descartes made geometry amenable to algebraic treatment and thereby made a most important contribution to the development of mathematics in the seventeenth century. The value of his method was greatly enhanced by Leibniz,[29] to whom the name Cartesian coordinates is due, when he pointed out the convenience of making the coordinate axes orthogonal. Leibniz thereby introduced for the plane orthogonal coordinate systems analogous to the spherical orthogonal systems that Ptolemy had introduced in astronomy. It should be emphasized that the ubiquitous use of Cartesian coordinates in problems of dynamics does much to disguise the problem of the ultimate invisibility of space. Because the conceptual axes are drawn on paper, the actual absence of reference marks in space can be conveniently forgotten.

Let us now briefly consider why the diametrically opposed philosophies of *idealism*, founded by Berkeley,[23] and the rather less extreme *phenomenalism* or positivism supported by Mach,[30] were able to play a positive role in counteracting the dangers inherent in the Cartesian approach. One of the important aspects of these philosophies is that they challenge the Cartesian idea that certain concepts or ideas are more fundamental than others. This is closely related to the distinction between primary and secondary qualities. As we have seen, Descartes argued that only the former – concepts such as motion, size, and extension – are truly comprehensible; only they can correspond to real features of the external world. All the other sensations we have of colours, sounds, etc. must be illusions created, in Descartes' view, by the[31] 'close and profound union of our mind with the body'. The secondary qualities are to be explained by the primary qualities. But according to the rival view, neither primary nor secondary qualities are intelligible in any ultimate sense – they are literally the *data* of human experience, the one no less than the other. The phenomenalists give up all attempt to 'construct the world' in the Cartesian manner. Instead, they concentrate attention on the sense perceptions and seek merely to *correlate the data*. Berkeley overcame the notorious dichotomy in Cartesian philosophy between mind and matter by denying the existence of matter altogether. He asserted that nothing exists except God and minds, in which God directly implants the sense perceptions. Mach took a view somewhat less extreme (and certainly less theological) but no less radical: since we have access to sense perceptions and nothing else, speculation about the world is futile. Science, in his view, consists of nothing more than establishing relations between the phenomena that are presented to our senses. It is inevitable that in such an approach much more attention is concentrated on the attempt to establish relationships between different observed phenomena, as opposed to the attempt to derive all such empirical phenomena from certain clear and distinct metaphysical first principles.

Ever since he first propounded the view in his *Mechanics*, Mach's philosophy has been contentious. Many scientists adopt a negative attitude to Mach's Principle because they associate it with phenomenalism and the denial of an external world. It would be quite inappropriate to enter into a long discussion here about the pros and cons of Machian positivism. The aim of the present section is more to inform the reader of the existence of the two main streams of thought that lie behind the absolute/relative debate. It is certainly not the case that a commitment to Mach's Principle necessarily involves adoption of thoroughgoing phenomenalism or idealism. Einstein is an example of a great scientist who took useful ideas from Mach without in any way fully accepting his system. In fact, systematic adherence to either one or other of the extreme viewpoints is almost certainly inimicable to the progress of science. Some of the defects of Cartesian physics have already been listed; it should also be pointed out that the approach that attempts to dispense totally with auxiliary concepts such as substance and deal solely with direct sense percepta has also had very little success hitherto in constructing viable theories. What we shall see in Vol. 2 is the way in which the relational standpoint led to very fruitful modifications to the rigidly Cartesian concepts of space and time that Newton adopted and thereby helped to lead Einstein to exciting new theories that have withstood experimental testing extraordinarily well. And before this volume is ended we shall see how the recognition that all good theory must ultimately be anchored in observation and observable things led to a very helpful clarification of the true basis of Newtonian dynamics.

We conclude this section with another of the several paradoxes that we find in Descartes. As we shall see in Secs. 8.6 to 8.8, Descartes was simultaneously the initiator of *both* the absolute and relative standpoints in the theory of motion. But it is a further irony that Descartes not only formulated the basic materialistic philosophy in which the concept of absolute space came to play such a prominent part but also stimulated development of the diametrically opposed philosophical school of idealism. For ultimately the Berkeleian and Machian solution to the conundrum of primary and secondary qualities sprang from the very same doubt that Descartes sowed when he asked so persistently and insistently whether the world really exists and whether we can ever get to know its essential nature if we have access to only our sense perceptions, not the world itself. Descartes and Berkeley merely reacted in different ways to the same problem which Descartes had posed in such acute form. Bertrand Russell points out[32] that there is 'in all philosophy derived from Descartes, a tendency to subjectivism, and to regarding matter as something only knowable, if at all, by inference from what is known of mind'.

Thus, the great divide in post-Cartesian philosophy – between materialism and idealism – is reflected, as we shall see, in the debate about absolute and relative motion, and both debates, the general philosophical one as well as the specific debate about motion, can be traced back to a common source in Descartes.

8.4. The stone that put the stars to flight

In this section we look in more detail at the origins of Descartes' most important contributions to the discovery of dynamics, namely, the outlining of its embryonic structure and the formulation of the principles, advanced as the first principles of the theory of motion, that were later fused into the law of inertia.

We begin by considering the empirical input in the principles that were later transformed into the law of inertia. In contrast to Galileo, who drew on both celestial and terrestrial motions in arriving at his notions of persistence of motion, Descartes does not seem to have had any direct inspiration from astronomy. The examples he gives to support his principles come exclusively from terrestrial motions. This in fact is a strength in Descartes, since his inertial motion is purely rectilinear; there is none of the curious mixture of linear and circular 'inertia' that we find in Galileo. Descartes presents two examples with admirable clarity:[33]

> For example, if a wheel is made to turn on its axle, even though its parts go around (because, being linked to one another, they cannot do otherwise), nevertheless their inclination is to go straight ahead, as appears clearly if perchance one of them is detached from the others. For, as soon as it is free, its motion ceases to be circular and continues in a straight line.

By the same token, when one whirls a stone in a sling, not only does it go straight out as soon as it leaves the sling, but in addition, throughout the time it is in the sling, it presses against the middle of the sling and causes the cord to stretch. It clearly shows thereby that it always has an inclination to go in a straight line and that it goes around only under constraint.

The stone flung from the sling is in fact the image that Descartes most frequently uses throughout *The World*. Figure 8.4 shows Descartes' first diagram in the book (it is repeated in the course of his explanation of the nature of light), about which Descartes says:[34] 'Suppose a stone is moving in a sling along the circle marked AB and you consider it precisely as it is at the instant it arrives at point A: you will readily find that it is in the act of moving (for it does not stop there) and of moving in a certain direction (that is, toward C), for it is in that direction that its action is directed in that instant.'

How or why Descartes came to the recognition of the importance of this law is difficult to say. However, the influence of Buridan and the medieval schoolmen is evident. The above quotations emphasize the *rectilinearity* in the law of inertia; the persistence of uniform motion (the other part of the law of inertia) was in fact, as we shall see, of rather greater importance to Descartes and in discussion of this aspect Descartes explicitly mentions the schoolmen. He points out[35] that they have a difficulty in explaining 'why a stone continues to move for some time after being out of the hand of him who threw it'. If there is any one point at which the decisive step to the basic structure of modern dynamics was taken, then surely in the simple way that Descartes turned this problem upside down. He remarks

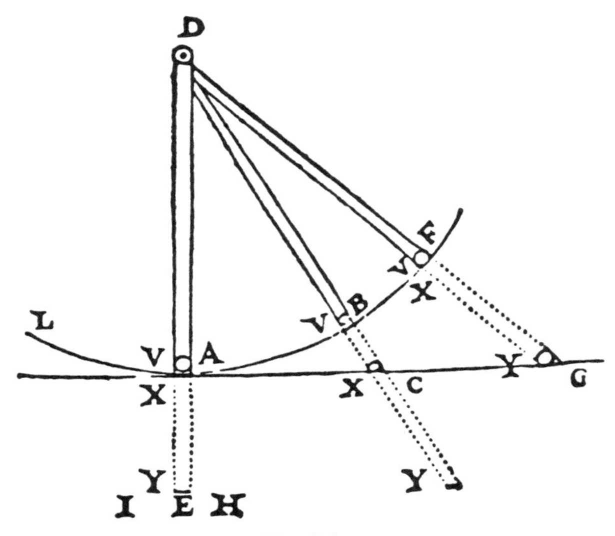

Fig. 8.4.

simply: 'We should ask, instead, why does the stone not continue to move forever?' This limpid question announces the transition from a dynamics that, to use modern terminology, is of first order in the time derivatives to one that is of second order. There were still innumerable difficulties and puzzles to solve, but the step had been taken.

After these comments about the empirical and historical origins of Descartes' concepts, let us go through his formal arguments a little more methodically. As pointed out in Sec. 8.2, the first outline of an embryonic form of modern dynamics grew out of Descartes' desire to explain all the phenomena of nature by motion. Very early in *The World*, before the above quotations and, significantly, in his discussion of liquidity and hardness, we find the first outline of dynamics in a form recognizably similar to that which it still has. Speaking of the motion of the elements of his matter, he says:[36]

I do not stop to seek the cause of their motion, for it is enough for me to think that they began to move as soon as the world began to exist. And that being the case, I find by my reasoning that it is impossible that their motions should ever cease or even that those motions should change in any way other than with regard to the subject in which they are present. That is to say, the virtue or power in a body to move itself can well pass wholly or partially to another body and thus no longer be in the first; but it cannot no longer exist in the world.

We have here first hints of the approach developed by Huygens and taken still further by Newton, namely, that it is not necessary to consider what causes motion but rather what changes it. Thus, motion in itself has a tendency to persist which is to be taken for granted; however, it can be changed by the interaction of matter with other matter. Thus, to formulate the theory of motion one must state the laws which say precisely how motion persists between interactions and also the laws which govern the interactions themselves.

This outline of the formalized structure of the science of motion was most important and original. Kepler made a first attempt in this direction, but his overall scheme was seriously flawed, for the reasons that we have already explained. In contrast, Galileo formulated no overall structure at all. The very notion of laws of motion was first clearly formulated by Descartes in Chap. 7 of *The World*, which has the very significant title 'On the laws of nature of this new world'. We find here the notion of a 'law of nature,' expressed for the first time in a form recognizably similar to the one it still has:[37]

Know, then, first that by 'nature' I do not here mean some deity or other sort of imaginary power. Rather, I use that word to signify matter itself, insofar as I consider it taken together with all the qualities that I have attributed to it, and under the condition that God continues to preserve it in the same way that He created it. For from that alone (i.e., that He continues thus to preserve it) it follows of necessity that there may be many changes in its parts that cannot, it seems to me, be properly attributed to the action of God (because that action does not change) and hence are to be attributed to nature. The rules according to which these changes take place I call the 'laws of nature'.

It is worth mentioning that the idea expressed here of God controlling the world 'at one remove', as it were, i.e., through 'laws of nature', is a characteristically medieval idea and was very widely accepted in the seventeenth century. Jaki,[38] among others, believes that it was a significant factor in the emergence of the scientific revolution from medieval Christian philosophy, for in such an intellectual framework one is led naturally to believe in a rationally ordered natural world – and, moreover, encouraged to study it by rational means.

Descartes actually set up three laws (or rules) of nature in *The World*. The first can be called a generalized law of inertia, according to which things have an innate tendency to stay in the same state (Descartes had a precise notion of *state*, which he used more or less in its modern scientific sense). The very significant step that he took was to include motion within this law:[39]

The first is that each individual part of matter always continues to remain in the same state unless collision with others forces it to change that state. That is to say, if the part has some size, it will never become smaller unless others divide it; if it is round or square, it will never change that shape without others forcing it to do so; if it is stopped in some place it will never depart from that place unless others chase it away; and if it has once begun to move, it will always continue with an equal force until others stop or retard it.

There is no one who does not believe that this same rule is observed in the old world with respect to size, shape, rest, and a thousand other like things. But from it the philosophers have exempted motion, which is, however, the thing I most expressly desire to include in it.

In neither *The World* nor the later *Principles* is Descartes particularly precise in his actual formulation of his first law as it applies to motion. What he means does, however, become abundantly clear in his applications, especially those given in the *Principles*. The essence of his first law is that the *quantity of motion* that a body has will remain constant unless it interacts with other bodies. This becomes rather clearer in Descartes' second rule in *The World*:[40]

I suppose as a second rule that, when one of these bodies pushes another, it cannot give the other any motion except by losing as much of its own at the same time; nor can it take away from the other body's motion unless its own is increased by as much. This rule, joined to the preceding, agrees quite well with all experiences in which we see one body begin or cease to move because it is pushed or stopped by some other.

Descartes formulates here, still very imprecisely, his idea that the *quantity of motion* is conserved in collisions. In *The Principles*, in which Descartes discusses the actual rules governing collisions[41] (which we shall discuss in Chap. 9), he makes it clear that by quantity of motion of a body he means the product of the *volume* of the body and its *speed*. Note that Descartes takes the volume, since he has no concept of mass such as Newton developed. Because he relies completely on geometrical concepts, he is forced to regard volume as the sole indicator of how much substance any given body contains. It is equally important to note that Descartes uses the *speed* and not the *velocity* of the body to obtain his quantity of motion. In technical terms, this means that his quantity of motion is a scalar, and not a vectorial, quantity. Cartesian physics lacks the strict directionality found in Newtonian dynamics. Thus, whereas Newton has a law of conservation of momentum, which, expressed in modern terms, involves the conservation of three quantities (the components of the momentum along each of the three axes), Descartes has conservation of just one quantity.

There is nevertheless a weakened form of the Newtonian directionality, which we have in fact already encountered in the earlier discussion of the sling. Descartes formulates it in *The World* in the form of his third law:[42]

> I will add as a third rule that, when a body is moving, even if its motion most often takes place along a curved line and (as has been said above) can never take place along any line that is not in some way circular, nevertheless each of its individual parts tends always to continue its motion along a straight line. And thus their action, i.e., the inclination they have to move, is different from their motion.

Although, as we have seen, the idea of rectilinear persistence of motion played a most suggestive and important leading role in Cartesian physics, which Descartes emphasized by the use of the special technical term *determination* (*determinatio*) to denote the direction of motion, as actually *used* by Descartes his third rule proved to be very weak, since it is not associated with any rigid conservation law. Thus, according to his scheme, if two bodies collide the sum of the quantities of their motion must be strictly conserved but, in contrast to Newtonian dynamics, there is no restriction imposed on the directions by a simple and universally valid conservation law. Not even the distribution of the total amount of motion between the two bodies is fixed. From the point of view of mature Newtonian dynamics, this was a serious flaw in the scheme. Descartes was always faced with the problem of finding further arguments that could be used to determine fully the outcome of particular collisions. As we shall see in Chap. 9, this led to all sorts of problems.

It will be evident from the above that in one obvious respect Descartes did not precisely formulate what was later to become Newton's law of inertia; for the persistence of the amount of motion and the rectilinearity of that motion were formulated as two separate rules. A more substantial difference is that conceptually his rules 1 and 2 (especially as formulated in *The World*) are rather strongly linked, more so than rules 1 and 3 (the two parts of the subsequent law of inertia). Indeed, one can say that the heart of Cartesian physics is his notion of the persistence of the amount of motion, understood as a scalar quantity. Rule 1 of *The World* expresses this persistence between interactions, rule 2 the persistence in interactions. In his later *Principles of Philosophy*, Descartes reformulated his scheme[43] in a form that brought it rather closer to the schemes that Huygens and Newton later developed. His rule 1 remained essentially unchanged and was formulated as law 1, while rule 3 of *The World* (expressing rectilinearity between collisions) was moved into second place as law 2. Rule 2 of *The World* became law 3 of the *Principles*; at the same time Descartes attempted to clarify its precise significance by supplying seven detailed and explicit rules which were supposed to describe what actually happened in collisions (to be considered in Chap. 9). This had the effect of making laws 1 and 2 rather more clearly into the laws that described the motion between collisions, while the remaining law and its explication by seven rules described the interactive part of his embryonic dynamics. This paved the way for the rather natural fusing of the two parts of the law of inertia. Simultaneously it highlighted the fact that the nontrivial part of dynamics resided in the interactions and thereby helped to concentrate attention on this important matter. Incidentally, the process of fusing of the two parts of the law of inertia can be seen actually occurring in Newton's early work, for in his earliest notes it is still formulated as two separate laws, a clear indication of the influence of Descartes.

Although Cartesian physics suffered from serious flaws, especially in the part dealing with the interactions (which was deduced by Descartes more or less purely *a priori* and, in contrast to the inertial part, contained only minimal empirical input), it is startling to find how modern Descartes was in certain key respects. It is not just that he swept away the last relics of the well-ordered Pythagorean cosmos that seemed to provide the support to Galileo's world view nor that he adopted a completely democratic attitude to motion, regarding all motions as essentially the same, so that the Aristotelian distinction between natural and violent motions was abolished. His new comprehensive views, so conducive to an ordered scientific approach, were of course extremely important and rapidly became an integral part of the general scientific outlook. No, the modernity of Descartes is even more striking. He clearly anticipated the overall view of the dynamical history of the universe that only crystallized several generations after Newton. As is well known, Laplace (1749–1827) epitomized the Newtonian revolution in physics by his concept of God the supermathematician who sees at an instant the positions and velocities of all the material bodies in the world and from this knowledge of the instantaneous state of the universe can immediately calculate the entire future evolution of the world using the laws of motion. Though Descartes' laws were different in detail from Newton's the overall Laplacian concept is already prominently present in *The World*.

The idea of the *instantaneous state*, so characteristic of modern physics, is clearly expressed: instantaneous position and speed are advanced as the basic elements of dynamics. Descartes supposes that at an initial time God creates pieces of matter in space, imparting to them diverse instantaneous velocities. From then on the laws of nature take over. The entire history of the material world is regarded as the evolution, subject to these laws of nature and nothing else, from the initial conditions that God in his wisdom happened to choose (for a slight qualification of this statement, see the footnote on p. 432).

Thus, Descartes' *The World* contains most of the key ideas of the scientific revolution: materialism – that all the variety of the visible world is to be explained by matter in motion (albeit often at a level that makes it invisible to our senses); determinism – that the whole world unfolds from an initial state in accordance with immutable laws (Aristotelian teleology, still lurking around in the contemporary *Dialogo* of Galileo, is banished and disappears without trace); isotropy and homogeneity of space and time – the laws of nature are exactly the same at all times and in all places.

Even today the main thrust of the scientific endeavour is essentially along the lines laid down by Descartes and this despite several major revolutions (Newtonian, relativistic, and quantum) to say nothing of significant shifts of direction such as occurred in the development of the field concept by Faraday and Maxwell.

One involuntarily starts when reading Descartes' conception of the fundamental problem of cosmology – how the presently observed order in the universe could have evolved from primordial chaos. Having introduced his concept of matter 'that has been so well stripped of all its forms and qualities that nothing more remains that can be clearly understood' and which we are to conceive 'as a real, perfectly solid body' and having explained how it can be divided into pieces, he continues:[44]

> Let us suppose in addition that God truly divides it into many such parts, some larger and some smaller, some of one shape and some of another, as it pleases us to imagine them. It is not that He thereby separates them from one another, so that there is some void in between them; rather, let us think that the entire distinction that He makes there consists in the diversity of the motions He gives to them. From the first instant that they are created, He makes some begin to move in one direction and others in another, some faster and others slower (or indeed, if you wish, not at all); thereafter, He makes them continue their motions according to the ordinary laws of nature. For God has so wondrously established these laws that, even if we suppose that He creates nothing more than what I have said, and even if He does not impose any order or proportion on it but makes of it the most confused and most disordered chaos that the poets could describe, the laws are

sufficient to make the parts of that chaos untangle themselves and arrange themselves in such right order that they will have the form of a most perfect world, in which one will be able to see not only light, but also all the other things, both general and particular, that appear in this true world.

The creation of order from chaos is, of course, one of the most venerable ideas of literature and philosophy and belongs to the dawn of history. The novelty in Descartes is in the precise manner in which he conceives it happening. In fact, this and other passages have an uncanny similarity to many an introduction to modern papers on inflationary cosmology, which aim to show that however the world began the laws of nature will have inexorably forced the universe into the general appearance it now presents.*

The main reason for this striking modernity of Descartes is quite simply that he was the first person to see clearly (or perhaps, rather, the first to say so unambiguously) that laws of motion are needed, not so much to describe motion itself, as to describe *change of motion*. In the theory of motion, the line that divides the modern world from the ancient was not like the Rubicon, a river that the world, Caesar-fashion, had to cross. It was a line written in French with a goose quill by a man who by the very act of writing it put himself on our side and left Kepler and Galileo on the other.

'Nous doit [sic] plutoſt demander purquoy elle ne continue pas toujours de ſe mouvoir?' ('We should ask, instead, why does the stone not continue to move forever?')

8.5. The discovery of inertial motion: Descartes and Galileo compared

We conclude this part of the chapter with a reflection on the genesis of the law of inertia, the foundation of dynamics which Immanuel Kant[46] went so far as to describe as providing the basis 'on which the very possibility of a true natural science rests'. How great are the respective contributions made by Galileo and Descartes to this momentous advance?

As regards the priority of discovering and recognizing the importance of the natural persistence of motion, there is no doubt that it belongs to Galileo. He had it almost before Descartes (born 1596) was out of his swaddling clothes and he published the essential idea (in the *Letters on Sunspots*) while Descartes was still at school at La Flèche. More interesting is the question: what led Descartes to inertia? Was he influenced, perhaps

* Even more uncanny (for a modern physicist at least) is the devout Descartes' anticipation[45] of the modern creationist argument: that actually God created the world only six thousand years ago but in a form and with laws of nature that make it indistinguishable from one that had evolved out of such chaos in accordance with the Cartesian scenario – he created it with the appearance of an evolutionary history. This is another point at which the reader is left wondering what Descartes really believed.

indirectly, by Galileo? By his major writings, certainly not; they were published too late to influence *The World*. Indirect influence by word of mouth is possible. As a young man, Descartes travelled widely in Holland and Germany and, by his own account, was an exhaustive interrogator of anyone possessed of original thought. Nevertheless it must be said that the whole manner of presentation of the subject in *The World* suggests that Descartes arrived at his idea independently, and through a quite different line of development. What he did essentially was put a new interpretation on the phenomenon to which the medievalists had drawn attention by their impetus theory.

A much more direct influence than Galileo's will certainly have been that exerted on Descartes by his Dutch physicist friend Isaac Beeckman, with whom, as we noted, he was in regular contact several years before he wrote *The World*. As early as 1614 Beeckman noted in the margin of his *Journal* that 'a stone thrown in vacuum does not come to rest' and that it therefore 'moves perpetually'.[47] Perhaps Beeckman should be credited with the formulation of the first law of motion and not Descartes. Whatever the truth, the outcome of the Beeckman–Descartes line of development was a set of interrelated ideas that owed little to Galileo.

Why did inertia remain with Galileo but a facet of motion but was made by Descartes into the very foundation of everything? This is surely due to not only the accident of dates of birth but also the differences of temperament. Galileo looked intently at nature, the great book of the world, in order to learn the language she spoke. When he had learnt the first few basic words he began to construct his own sentences. He worked outward, using confidently the elements he had found around him.

Descartes looked at nature in a way quite different from Galileo's: not to spy out one by one her veiled secrets but rather merely to find confirmation of laws he had already worked out in his own mind, or, in complicated matters, to find which of several schemes, all equally consistent with his basic laws, was actually realized in any particular case. A man with such ambitions just had to have a great principle. Descartes was lucky in lighting on one that proved very serviceable. It proved to be the diamond for which he was looking. But it was others, using techniques learnt from Galileo, who were destined to make it into the crown jewel.

In fact, the really important thing about the discovery of the law of inertia was not so much the finding of the law itself as the *demonstration of what could be done with it*. This achievement was almost exclusively Galileo's; there is nothing remotely approaching the derivation of the law of projectile motion or the resolution of the problems posed by the earth's mobility in Descartes' work. Descartes was, as it happens, very blind to the achievements of Galileo, about whom he made a much quoted disparaging remark: 'Without having considered the first causes of nature, he has only looked for the reasons for certain particular effects, and . . . he has thus built without foundation.'[48] Because of the accidents of time, Galileo had no opportunity to reply to this compliment, but of the two the nineteenth-century historian of science Whewell said:[49] 'If we were to compare Descartes with Galileo, we might say that of the mechanical truths which were easily obtainable at the beginning of the seventeenth century, Galileo took hold of as many, and Descartes as few as was well possible for a man of genius.'

Descartes simply failed to realize what Galileo was up to. For he automatically assumed that Galileo was engaged on the same enterprise as himself – the discovery of universal principles that 'explain' the world and all its myriad phenomena. He was quite wrong to say that Galileo 'looked for the reasons for certain particular effects'. Galileo did not 'look for reasons'; he looked for *characteristic phenomena*, the words of nature, and sought to reproduce them with quantitative exactitude – to tell us precisely *how they are*, not to beguile us with hidden causes of their coming to be. Then, having grasped a phenomenon here, he looked around to find the same one elsewhere – and succeeded.

On the question of foundations, Descartes was on slightly safer ground, though here too he failed to recognize that what he should have contrasted were the relative merits of two different approaches. The problem is: whence do we find secure foundations for science? Descartes exercised his mind to find all at once a foundation capable of supporting the entire universe. He would even have despised Atlas for carrying only the heavens and not the earth as well. How could Galileo's tabletop experiment mentality appear to Descartes as anything but puny? But Galileo had just as much awareness as Descartes for great issues, it was merely that his instinct told him to set about the problem in a quite different way. After all, the very thing that the Copernican revolution had questioned was foundations. Profoundly aware of the fact that we and the earth, cut loose from the moorings, were adrift on the sea, he looked about him for handy pieces of driftwood to make a raft. Being a very skilled craftsman, he chose and measured his wood with great care and added nothing to the raft before he knew that it fitted exactly. In this way he slowly extended the raft and, almost paradoxically, finished up with foundations of a sort of which Descartes never dreamed: secure, because taken straight from nature; and productive, because he also laid down workable ground rules for extending the raft – first measure carefully before trying to add anything, otherwise it is liable to fall off again.

On this surprising and quite unexpected foundation, put together in the hour of need following the cosmic shipwreck, the mighty edifice of science still rises, pitching it is true when the sea heaves, but with an extraordinary buoyancy before which Galileo's mentor, Archimedes, would have stood speechless, not even able to exclaim 'Eureka'.

But what of the grandiose foundations of the Jesuits' pupil? Of the half dozen or so tenets of Cartesian physics, so loudly proclaimed as the self-evident truths of nature by the proud rationalist, only two survived more or less intact the death of their promulgator. The stricture of Whewell is fair. To return to the language metaphor: Galileo took the trouble to learn a living language and thereby ensured that it is still spoken today, the *lingua franca* of the scientific revolution. Descartes, the man in a hurry, made up his own Esperanto and, as a physicist, deservedly suffered Zamenhof's fate – the basic idea was so original that everyone remembers it, but the language itself did not catch on.

Galileo's motionics, for all the oddity of its construction, was a functioning system and played a vital complementary role *vis-à-vis* Descartes' embryonic dynamics, revealing what in it was good, and what bad. Without Galileo, Huygens and Newton would never have found the sound grains in the heap of Cartesian chaff, nor how to make proper use of them, a trick that Descartes never learnt. But all this said, there is still no awareness in Galileo of the breathtaking change of direction that Descartes initiated. As Mach, a great admirer of Galileo, said somewhat grudgingly of Descartes:[50] 'The merit of having first *sought after* a more universal and fruitful point of view in mechanics, cannot be denied Descartes.' In fairness, he should have added: *'and found'*.

Perhaps even that is not the correct characterization. Some of the most important advances in physics occurred when a particular investigator effected a complete change of direction, as, for example, when Kepler introduced the idea that the heavenly bodies not merely dance around each other but actually cause each other's motions. First comes the idea, then realization. Descartes had one or two very good ideas but failed to bring any of them to realization. His work was the *Mysterium Cosmographicum* without the *Astronomia Nova*. More than any other figure in the history of dynamics, Descartes bequeathed to his successors much manifestly incompleted business. We shall therefore return to aspects of Cartesian physics in the chapters on Huygens and Newton: the law governing impacts of bodies, the study of centrifugal force, and, above all, the debate about the nature of motion. But this brings us now to the most surprising twist of all in the story of the discovery of dynamics.

8.6. The intervention of the Inquisition

By a curious irony, in 1633, Descartes was in the final stages of preparing *The World* for publication when the news broke of Galileo's condemnation by the Inquisition. Descartes was dismayed and wrote immediately to his friend Mersenne:[51]

In fact I had decided to send you my *World* as a New Year's gift; and no more than a fortnight ago I was still determined to send you at least a part of it, if the whole could not be transcribed in that space of time. But I shall tell you that, after I had someone enquire in Amsterdam and Leiden . . . whether Galileo's *World System* was available; . . . I was informed that it had indeed been printed, but that all the copies had been simultaneously burned in Rome and Galileo himself subjected to some sanction. This astonished me so much that I have more or less decided to burn all my papers, or at least to permit no one to see them. For I could not imagine that an Italian, especially one who is, so I hear, well-considered by the Pope; could have been condemned for anything other than the fact that he doubtless attempted to establish the Earth's motion. I well know that this view was formerly censured by some Cardinals, but I thought I had heard that it was being taught publicly, even in Rome; and I confess that if it is false, so are all the foundations of my Philosophy, since they clearly demonstrate this motion. And it is so connected to all the [other] parts of my *Treatise*, that I cannot omit it without rendering the remainder completely defective. But since I would not wish, for anything in the world, to write a discourse containing the slightest word which the Church might disapprove; I would, therefore, prefer to suppress it, rather than publish it in a mutilated version.

Implausible as it may seem, the letter quoted above is the unambiguous starting point of a process that culminated in Einstein's principle of general covariance and the creation of general relativity.

Luckily, a significant part of Descartes' *The World*, from which we have quoted extensively already, survived and was published posthumously in 1664. We are therefore able to compare it with his 'definitive' work, the *Principles of Philosophy*, published in 1644 (six years before his death). The comparison is most interesting: nearly all the key ideas of Cartesian physics are clearly expressed in *The World* and survive essentially unaltered in the *Principles*. However, the basic concept of motion undergoes radical revision. In the early *The World*, Descartes is no more concerned with the frame of reference than Galileo. Like him, he seems to have had an intuitive concept of motion as taking place in space, this space being to all intents and purposes much the same as Newton's absolute space – what I called earlier intuitive space, the void of the atomists without the distinguished direction of eternal falling. In contrast, the *Principles* pose the question of the referential basis of motion as a matter of prime importance. Space is said categorically not to exist and all motion is declared to be purely relative. On this basis, Descartes actually succeeds in finding a formal definition of motion according to which Copernicus's theory of the solar system is in essence correct and yet the earth is still maintained to be at rest! It has been suggested that Descartes may have formulated this particular formal definition with his tongue in his cheek (Koyré[52]), and indeed it is hard to conceive that he really believed it. Yet for all that he argues the general case for relativity of motion with considerable fervour. Unlike Galileo, he does not leave the reader obvious clues indicating that although he says one thing he means another. In fact, he asserted relativism so persuasively that Huygens, the greatest physicist of the second half of the seventeenth century except for Newton, adopted the position himself, clearly influenced to a great extent by the *Principles*.

The *Principles* is in fact a tangle of contradictions. The formal theory of motion is pure relativism but never settles down to any definite theory of relative motion. Descartes launches the acrobat on his way but gives no guidance as to the way the trapeze will swing. The formal theory of relative motion then comes to an abrupt end when Descartes announces his laws of nature. These are formulated as laws of motion but with complete disregard to the foregoing theory of relative motion. They are in essentials the same as in *The World* and make perfect sense in intuitive space, none at all in the relative world just described. Whether Descartes saw the discrepancy is difficult to tell. There is not a word of explanation nor any hint to the percipient reader which suggests that he did. Since Cartesian physics contains other major defects of the same order of magnitude of which the author was clearly unaware, it is at least possible that Descartes was fooled by his own casuistry. It is all very intriguing: that Descartes raised momentous issues is beyond question; that he raised them under pressure of the Inquisition is equally clear; but what in his heart he really believed eludes us.

What is not in doubt is the most remarkable of the paradoxes we meet in Descartes: that he was simultaneously the effective founder of the two diametrically opposed concepts of motion that are the subject of this book – both the absolute and the relative!

An achievement worthy of a philosopher. Let us now see how he did it.

8.7. Descartes' early conception of motion

It is quite clear that at the time Descartes wrote *The World* he instinctively conceived space as something that exists independently of matter. He has the concept of space as the *container* of matter. He describes[53] the world of his fable as a 'wholly new one, which I shall cause to unfold . . . in imaginary spaces. The philosophers tell us that these spaces are infinite, and they should very well be believed, since it is they themselves who have made the spaces so.' There is obvious irony here, but we have no reason to doubt that Descartes did at this stage actually believe in space. He talks of stopping in it at 'some fixed place'.[54] He says of the matter which he conceives that [55] 'each of its parts always occupies a part of that space'. Significantly, in view of Descartes' later passionate denial of the vacuum, in *The World* he grants[56] that there could be void in some parts of the world: 'if there can be a void anywhere, it ought to be in hard bodies'.

More evidence of the firm roots that the concept of Euclidean space as the container of matter had in Descartes' mind is provided by his discussion in Chap. 7 of *The World* of the various types of motion imagined by philosophers (by whom he primarily means Aristotle). We recall from the footnote on p. 48 in Chap. 1 that Aristotle employed the Greek word *kinesis* to mean not only motion in the modern sense but, more generally, any change such as from hot to cold, red to blue, and so forth. As we have seen, Aristotelian physics considered all forms of change. Aristotle did not adopt the atomists' programme of reducing all change to motion, though what we call motion was, at least at times, perceived as more fundamental than other changes – after all, the natural motions of each of the elements were regarded as their characteristic defining properties. Thus, what Aristotle opposed to atomism was a kind of qualitative dynamics of all change, not just motion. To use modern terms, it is as if one were to regard colour, heat, and taste as dynamical variables more or less on an equal footing with position and momentum.

It was this type of physics that Descartes was intent upon sweeping away in its entirety. He objects strongly to the idea of, say, change of colour being regarded as 'motion'. For him there is only one motion, motion in space. This comes out very clearly in the following passages. He is at pains to point out that the 'motion' philosophers speak of is 'very different from that which I conceive'. The text continues:[57]

They themselves avow that the nature of their motion is very little known. To render it in some way intelligible, they have still not been able to explain it more clearly than in these terms: *motus est actus entis in potentia, prout in potentia est*, which terms are for me so obscure that I am constrained to leave them here in their language, because I cannot interpret them. (And, in fact, the words, 'motion is the act of a being in potency, insofar as it is in potency,' are not clearer for being in [English].) On the contrary , the nature of the motion of which I mean to speak here is so easy to know that mathematicians themselves, who among all men studied most to conceive very distinctly the things they were considering, judged it simpler and more intelligible than their surfaces and their lines. So it appears from the fact that they explained the line by the motion of a point, and the surface by that of a line.

Note the explicit appeal that Descartes makes to the world of Euclid: space and motion in it is the source of his inspiration.

We continue with Descartes' argument:

The philosophers also suppose several motions that they think can be accomplished without any body's changing place, such as those they call *motus ad formam, motus ad calorem, motus ad quantitatem* ('motion with respect to form,' 'motion with respect to heat,' 'motion with respect to quantity'), and myriad others. As for me, I conceive of none except that which is easier to conceive of than the lines of mathematicians: the motion by which bodies pass from one place to another and successively occupy all the spaces in between.

It is clear from the final sentence of the quotation that Descartes initially drew his concept of motion from his concept of Euclidean space, which, at the time he wrote *The World,* he conceived to have an even more basic existence than matter; for he clearly imagined God placing matter *in space,* which is there waiting to receive the matter before it arrives. His immediate conception was of matter moving along straight lines in space. In Descartes' *The World* there is still a hierarchy of absolutes in the material world: space is the referent of matter. This is important, since in the later *Principles* Descartes denies the existence of space in the most categoric of terms – without in reality being able to dispense with it.

The above quotations, together with the whole tenor of *The World,* confirm that at the time it was written relativity of motion was the least of Descartes' concerns. He treats space and motion exactly in the way Maier (p. 48) describes – motion is relative to the empirical space of practical experience. It is noteworthy that although Descartes told Mersenne (in the letter quoted in Sec. 8.6) that the foundations of his entire philosophy depended on the earth's motion the problems he actually confronts are completely different from the ones that dominate the discussions of Copernicus, Kepler, and Galileo. For example, he nowhere considers the evidence of the earth's mobility and the fact that it must be deduced from observation of motions in the heavens. The precise details of the actual motions of the planets within the solar system – from which Kepler deduced such powerful arguments for heliocentricity – are of minimal interest to Descartes. Why then was the Copernican doctrine of such importance to Descartes in *The World*? The alternative title, *Treatise on Light,* provides the clue. The work is, in the first place, a theory of the nature of light. But the sun is the main source of light known to man and it was agreed by all in Descartes' time that the planets and moon merely reflect light. In Descartes' scheme the sun and planets were composed of quite different elements (of the first and third kinds, respectively). As Mahoney points out in his introduction to *The World,* Descartes was a Copernican above all because he needed the prime source of light at the centre of his vortex:[58] 'It would be hard to imagine a Ptolemaic sun of the first element swirling in a small, contained vortex in the fourth of seven orbits otherwise occupied by large bodies of the third element, all within a vortex of the second element rotating about a stationary earth of the third element.'

This comment is interesting, first, in showing how the Copernican revolution had come to influence conceptions of the world well outside the narrow confines of the interpretation of celestial motions from which it sprang, and, second, in showing how these secondary effects then reacted back on the main stream of the study of motion, providing the stimulus that led to the embryonic form of dynamics that we considered in the earlier sections of this chapter. These ideas were then developed within the mainstream by Huygens and Newton, who continued the tradition of the true study of motion in its own right developed by the astronomers and Galileo. Thus, the lesser of the unexpected twists to which I referred at the end of Chap. 7 was this indirect contribution to the basic structure of dynamics. The greater was the introduction of relativity in a quite new guise, which we now consider.

8.8. Descartes' revised concept of motion

We now come to the intriguing part of the story: the publication of the *Principles* (which Koyré[59] called a 'second edition' of *The World*). Compared with the charming freshness of *The World*, this is a somewhat tedious book. Descartes intended it as a comprehensive textbook of philosophy designed to replace Aristotle – and to a large degree succeeded. It is full of formal definitions. The immediate sources of his inspiration are less evident; in places he is obviously covering his tracks. The most striking difference from *The World* is the large portion given over to a formal theory of place and motion. Descartes insists on complete relativism and ironically adopts a position even more Aristotelian than the Stagirite's. Comparing *The World* with the *Principles*, we realize the lengths he was driven, in the intervening decade, to come to terms with Rome.

The all-embracing container space of the earlier work disappears without trace. This is reflected in a pronounced hardening of Descartes' commitment to plenism. In *The World* he was prepared to grant the possibility of a vacuum. It is now denied in terms stronger than any Aristotle used. The basic position of the plenists was that an empty container was a *physical* impossibility since the ambient medium would immediately press its inner walls together with great force. In contrast, Descartes now asserts that an empty container is a *logical* impossibility, that one can no more conceive such a thing than one can conceive valleys with the intervening mountains removed.[60] On the face of it, this is an admirable economy on Descartes' part: whereas his scheme previously contained three elements, matter, motion, space, it now contains only two. The redundant duplication of properties inherent in characterizing both space and matter by the same principal attribute, extension, is pointed out to good effect. Descartes insists that space, being nothing, simply cannot exist and is anyway superfluous. You do not need extended space to contain material extension. In his view matter *is* extension – it has no other attributes at all – and extension occupies itself; it has no need of anything to occupy.

This development from the earlier to the later work appears logical. In *The World* Descartes already seems close to remarking on the superfluity of space as a *container* of matter. His characteristic standpoint that matter is defined by extension and nothing else is already clearly formulated. It is in the new theory of place (needed once space has been banished) and motion that Descartes introduces quite new elements, unexceptionable perhaps in themselves but totally at variance with his laws of motion, which he retains unaltered.

Descartes' claim that his matter is characterized by extension and absolutely nothing else is not actually true; for it may also possess motion. Indeed, most matter must possess motion or otherwise his whole enterprise falls apart. Officially at least these are the only two elements of the Cartesian world. Before going on to discuss his theory of place and motion, we must, however, point out a flaw that I do not think Descartes anywhere addresses. *Per se*, extension is completely uniform. However, Descartes constantly assumes that it is meaningful to distinguish a piece of matter *here* from a *different* piece *there*. In fact, his extended matter is divided up into pieces of all different shapes and sizes. But if matter is nothing but pure uniform extension, what property is it that gives it this particulate structure?

There is here a very fundamental difficulty, inherent in the concept of Euclidean space just as much as in Descartes' pure material extension. If uniform extension is the sole essential property in the windswept Cartesian ontology, whence come the distinguishing attributes that enable one to speak of the piece of substance that is here rather than there? Descartes and all such rationalists who strive to make the world as simple and uniform as possible are always forced to fall back on some extra attribute, which merely serves to distinguish different pieces of substance that would otherwise be totally indistinguishable. But this extra attribute, meant to serve as nothing but a label, runs quite counter to the whole thrust of uniformization, which is supposed to have such reassuring explicatory powers. This is a problem that we encounter through the whole theory of motion and, indeed, any conceptual 'theory' which strives to make the world uniform. Most rationalists are less than candid (or clear) about this inherent difficulty, which will be discussed in Vol. 2 in connection with Leibniz's work.

Let us now consider the 'official' Cartesian theory of place and motion, bearing in mind that, strictly, it requires the introduction of at least one further attribute to permit unambiguous identification of different pieces of matter. (It might be argued that different pieces of the Cartesian matter could be identified by means of the second attribute that is allowed in the scheme – motion. However, this is not possible since Descartes' whole purpose is to *define* motion. The definition rests completely on the concept of relative transference of distinguishable pieces of matter. Descartes cannot save his concept of motion without at least one further attribute of matter.) The most important concept in the new theory is that of the *external place*, or *situation*, of some given piece of matter. This immediately plunges us into Cartesian relativism. (I use the world *relativism* to avoid the ambiguity inherent in relativity. Cartesian relativism is predominantly kinematic, with, however, as we shall see, some ill-defined dynamic elements.) In order to determine situation[61]

we must take into account some other bodies which we consider to be motionless: and, depending on which bodies we consider, we can say that the same thing simultaneously changes and does not change its place. Thus, when a ship is heading out to sea, a person seated in the stern always remains in one place as far as the parts of the ship are concerned, for he maintains the same situation in relation to them. But this same person is constantly changing his place as far as the shores are concerned, since he is constantly moving away from some and towards others.

This sort of problem is quite familiar from Aristotle. But now comes a highly important development. For Aristotle, position was relative but in the end determinate, the quintessential Catherine Wheel providing the ultimate frame of reference. Descartes must be given the credit for being the first to look a bleak prospect squarely in the face. For he continues[62]

Furthermore, if we think that the earth moves [and is rotating on its axis], and travels from the West toward the East exactly as far as the ship progresses from the East towards the West; we shall once again say that the person seated in the stern does not change his place: because of course we shall determine his place by certain supposedly motionless points in the heavens. Finally, if we think that no truly motionless points of this kind are found in the universe, as will later be shown to be probable; then, from that, we shall conclude that nothing has an enduring [fixed and determinate] place, except insofar as its place is determined in our minds.

The extraordinary sting in the tail, 'except insofar as its place is determined in our minds', was to dominate the discussion of the nature of motion for seventy years or more. It brought all thinkers up with a severe jerk. Absolute space was born of Newton's revulsion to such a preposterous notion. As will be seen later, Descartes himself clearly had a deep Freudian inability to believe in the doctrine he here proclaims so uncompromisingly. However, the lesson of Copernicanism was beginning to sink in. The great cosmic debate had raged for a century. Just when it appeared to be resolved in favour of Copernicus and Galileo – yes, the earth does move – Descartes put the whole debate in a totally different perspective. He questioned the law of the excluded middle, *tertium non datur*. From now on the question was no longer: does she move or does she not? It was transformed into: what is motion? This question, worthy of Pilate, was the *deus ex machina* that Descartes summoned up to pluck him from between the Copernican Scylla and the Roman Charybdis. It created the subject of much of the remainder of this volume and the next.

That the question was posed so late in the debate is not so much a mystery as it might at first seem. In a closed world with stable rim no great mental (as opposed to intuitive) adjustments were required to transfer motion from the sun to the earth. In spirit and in terms of the underlying attitude to motion, Copernican and above all Keplerian heliocentricity were remarkably Aristotelian. The *Cartesian* revolution, which, in contrast to the *Copernican* revolution, really did blow Aristotelian cosmology to smithereens, was actually based on two developments that were certainly fostered by the astronomical revolution but had nothing to do with its central thesis. These were the gradual advance of the idea that the universe is infinite and the notion that all matter in the universe is in mechanical interaction, so that even the supposedly fixed stars can be pushed about in the general plenum that transmits motion from one region to another. This is what Descartes is hinting at in the suggestion that there may be 'no truly motionless points' in the universe – the doubt that a generation later would nag Newton into the conviction that space truly exists.

If position is so relative, in the last resort mind-dependent, what hope do we have at all of defining *motion*? Descartes' treatment of this question, in which he veers from one position to another, seems to have induced sea sickness in his contemporary readers; today it largely evokes wry smiles because of the one or two places in which Descartes is rather obviously toadying to Rome. For all that, the impression of sincerity was convincing enough to persuade all the great figures of the seventeenth century that he was in earnest with his advocacy of relativism. The credit of having started the first real debate about the ultimate nature of motion cannot be denied Descartes.

Descartes begins with a definition of situation, or external place, that is startingly reminiscent of Aristotle, indeed a posthumous consultation for the destruction of his cosmos:[63]

> external place can be taken to be the surface which most closely surrounds the thing placed. It must be noticed that by 'surface' we do not understand here any part of the surrounding body, but only the boundary between the surrounding and surrounded bodies, which is simply a mode.

In this formal statement Descartes, literally by fiat, defines position by the immediately adjacent matter. But then he straight away admits a possible role of more distant matter. For he continues immediately:

> Or to put it another way, we understand by 'surface' the common surface, which is not a part of one body more than of the other, and which is thought to be always the same provided that it retains the same size and shape. For even if the whole surrounding body, with its surface, is changed; we do not on that account judge that the surrounded thing changes its place if it maintains the same situation

among those external bodies which we consider to be at rest. For example, if we suppose a boat to be driven in one direction by the flow of a river, and in the other by the wind, with perfectly equal force (so that it does not change its situation between the banks), anyone will easily believe that it remains in the same place, although all its surrounding surfaces change.

We are back to square one. All the formal discussion of motion and position in the *Principles* is a toing and froing between two extremes. One limit is definite, the other indefinite. The definite limit is the surface of the body under immediate consideration. The other limit is indefinite, because the Cartesian universe is indefinite (Descartes was curiously reluctant to say explicitly that the universe is infinite; he would only say it is larger than any given size we might be able to imagine, cf. Koyré[64]). Under such circumstances, Descartes' assertion that position is ultimately in the mind is not only plausible but true, certainly so if an independent space is denied and we seek to define position by the matter present in the universe at large. There is no court of final appeal. This, of course, is what Aristotle feared and his finite world so neatly obviated.

Descartes spends a lot of time on the definite limit, the immediate surface of the body considered. He was almost certainly consciously using it to please Rome; it is doubtful whether he fooled himself into believing that it provided a satisfactory concept of position (and hence motion) even if he wanted to. He was probably content to know that he had transformed the problem of the earth's motion out of all recognition and into a form that – theologically speaking at least – was far less threatening. Let us therefore first consider what purports to be his 'definite' definition of motion, the counterpart to his refurbished Aristotelian definition of position. He tells us what movement 'properly speaking is':[65]

If, however, we consider what should be understood by movement, according to the truth of the matter rather than in accordance with common usage (in order to attribute a determinate nature to it): we can say that it is *the transference of one part of matter or of one body, from the vicinity of those bodies immediately contiguous to it and considered as at rest, into the vicinity of [some] others.*

The virtue of this definition from Descartes' point of view was that according to his theory of the cosmic vortex, all the planets, the earth included, are carried around in a plenal fluid. Since the fluid carries the earth with it, there is no relative motion between the earth and *the immediately adjacent matter*. Thus, Descartes can assert, surely with very little internal conviction, that the earth, *pace* Copernicus, is at rest!

Absurd as this argument may appear, Newton's famous bucket experiment (to be discussed later though no doubt most readers are familiar with it, which I here assume) is evidence of the seriousness with which the seventeenth century took these arguments of Descartes. It is often wondered how Newton could have been so naive as to suppose that the thin wall of his containing bucket could have been the dynamical determinant of the contained water's motion. Anyone would surely realize that, if motion is relative, one must look further afield for the relevant matter. However, seen in the light of the above definition, it seems Newton was engaging in a direct dialogue with Descartes, since the bucket wall is, of course, the surface immediately surrounding the water.

We can probably thank Descartes' own lack of confidence in the above argument for a further argument that he introduced almost immediately and which undoubtedly marks the first appearance of a notion that, as we shall see in Vol. 2, leads on inexorably to the postulate of general covariance and some of the most basic ideas of general relativity.

In a series of comments on the formal definition given above, Descartes finally comes to a remark that seems to have made the biggest impact of all on his contemporaries. It is this:[66]

> Finally, I have stated that this transference is effected from the vicinity, not of any contiguous bodies, but only of *those which we consider to be at rest*. For the transference is reciprocal; and we cannot conceive of the body AB being transported from the vicinity of the body CD without also understanding that the body CD is transported from the vicinity of the body AB, and that exactly the same force and action is required for the one transference as for the other.

As we have reached a crucial point in Descartes' argument, let us pause a moment and review the various forms through which the problem of relativity had passed from the early discussions of Buridan and Oresme, bearing in mind that we are looking for the emergence of a clear distinction between kinematic relativity, on the one hand, and Galilean relativity, with its essentially dynamical element, on the other. In the case of kinematic relativity one must, moreover, distinguish several subaspects of the problem: (1) the ontological question, i.e., what *is* motion? It was primarily this question that Aristotle addressed, as did Descartes in the earlier quotations in this section; (2) irrespective of whatever concept we may have of motion, how and where can we find *evidence* that motion is actually taking place? This, of course, is the problem that features so prominently in the work of Copernicus, Kepler, and Galileo; (3) finally, there is the technical question: if motion is indeed relative, how is it possible to say that any particular body is moving in a straight line or any other definite trajectory? It is this last question that is crucial for the formulation of the law of inertia. It does not feature at all as a problem in the work of Copernicus and Kepler, since they both believed the fixed stars were truly fixed relative to one another; this was all they needed to have a well-defined concept of relative motion. Galileo, as we have seen, seems to have been curiously unconcerned about the problem. His main contribution was in showing how Galilean invariance, which he formulated only qualitatively, could provide an explanation for the nonobservance of dynamical effects of the earth's motion. He made no contribution to the aspects (1) and (3) listed above.

In contrast Descartes raised (1) very forcibly, was not really concerned about (2) and, as we shall see, seems completely to have failed to recognize the existence of (3) at all. He is a classic case of a philosopher using two different concepts of motion simultaneously; as a result, he introduced a great deal of confusion into the whole subject of relativity, confusion that we shall find difficult to untangle in the case of Huygens (Chap. 9) and Leibniz (in Vol. 2). Particularly significant in this connection, because of the influence that it had on Huygens, was his statement 'and we cannot conceive of the body AB being transported from the vicinity of the body CD without also understanding that the body CD is transported from the vicinity of the body AB, and that *exactly the same force and action is required for the one transference as for the other*' (my italics).

Although Descartes never satisfactorily explains what he means here by 'force and action' (such words, though frequently used by Descartes, do not properly fit into his 'official' scheme based solely on extension and motion), this bald assertion of *dynamical* relativism and complete reciprocity of motion goes far beyond the assertion of *optical* (or kinematic) relativism so familiar from the original arguments of Copernicus and especially Galileo.

What is significant about this passage is that it begins to introduce a dynamical element into the discussion, doing it moreover in a way that is not found in Galileo. He was concerned in the first place with the motion of single bodies. Salviati reassured Sagredo that *individual birds* would not lose their way relative to the rotating earth. But, as we shall see in the next quotation, Descartes considers the situation in which *two* bodies are simultaneously considered to be moving relative to the earth. This was very important for the development of dynamics, since it marked the beginning of the transition from single-particle motionics to two-particle and many-particle dynamics, to which Huygens made very significant contributions. But it simultaneously made the problem of relativity much more sophisticated and bewildering, so much so that neither Descartes nor Huygens was able to achieve clarity on the subject. Descartes did in fact look the Gorgon in the face but recoiled rather rapidly. He posed the problem rather well but baulked at giving any satisfactory answer, as we see from the following section of his *Principles* (together with the figure – Fig. 8.5 – that Descartes provided), in which he explains 'why the movement which separates two contiguous bodies is attributed to one rather than to the other':[67]

The main reason for this is that we do not think a body moves unless it moves as a whole, and thus we cannot understand that the whole earth moves just because

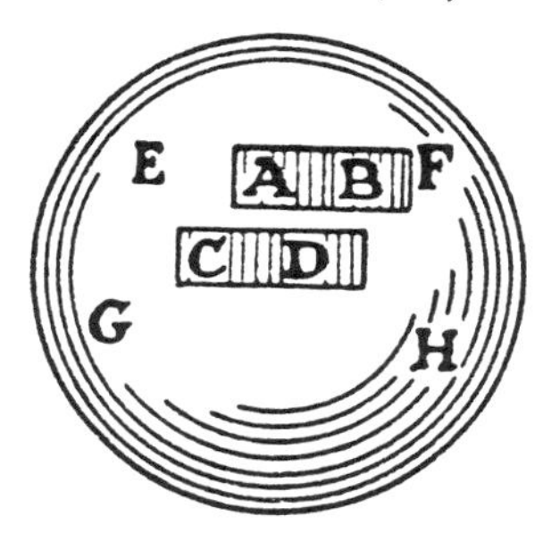

Fig. 8.5.

some of its parts are transported from the vicinity of some other smaller bodies which touch them; because we often notice around us many such transferences which are contrary to one another. For example, if the body EFGH is the earth, and if, upon its surface, the body AB is transported from E toward F at the same time as the body CD is transported from H toward G; then even though we know that the parts of the earth contiguous to the body AB are transported from B toward A, and that the action employed in this transference must be neither different in nature nor weaker in the parts of the earth than in the body AB; we do not on that account understand that the earth moves from B toward A, or from the East toward the West; because in view of the fact that those of its parts which touch the body CD are being similarly transported from C toward D, we would also have to understand that the earth moves in the opposite direction, i.e., from West to East; and these two statements contradict each other. Accordingly, lest we deviate too far from the customary manner of speaking, we shall say that the bodies AB and CD, and others like them, move; and not the earth. Meanwhile, however, we must remember that all the real and positive properties which are in moving bodies, and by virtue of which we say that they move, are also found in those contiguous to them, even though we consider the second group to be at rest.

Much of the confusion surrounding the use of the word relativity, which is still widespread today, can be traced back to the respective reactions of Huygens and Newton to these crucial sections of Descartes' *Principles*. Huygens picked up the dynamical hints that Descartes had given and forged out of them a greatly strengthened form of Galileo's invariance principle (the ship's cabin argument); on the other hand, he seems to have been almost as blind as Descartes to subaspect (3) that I listed above, and hence left the conceptual problems of kinematic relativity essentially unsolved. Newton, on the other hand, noted the glaring discrepancy in Descartes's treatment of relativity, and this led him to the concepts of absolute space and time. However, that is sufficient anticipation of their work. We must now return to Descartes.

There is not much point in following him through all his contortions, some of which do him little credit, and are often hard to follow, being based on incorrect physics. They exasperated Newton but eventually persuaded Huygens. It is, however, worth noting some passages in which Descartes uses arguments and concepts that have a Machian flavour. We shall see at the same time the use he made of the complete relativity of motion postulated above. It will be recalled that, kinematically, the Copernican and Tychonic systems are identical. Now it suits Descartes' purposes to argue that the Tychonic system actually fails to achieve the very thing it was designed to do, namely, ensure the immobility of the earth. According to it, the earth is supposed to be at rest with the sun circling it and the planets (except the earth, of course) in turn circling the sun. To make his point, Descartes takes it for granted that the entire cosmos is filled with his plenal fluid, which, he says, must also sweep around the earth, since it carries all the other celestial bodies. There is thus a relative motion between the earth and the plenal fluid touching it. At this point he invokes the argument from above. The situation is entirely reciprocal he says. Any assertion to the effect that the plenal fluid moves automatically entails the conclusion that the earth moves just as well. Descartes actually goes further. There is, after all, much more plenal fluid, which stretches away indefinitely, than there is earth. On the other hand, the actual relative motion of separation between the fluid and the earth, which occurs only where they touch, is given and therefore seems very insignificant when considered against the great bulk of the plenal fluid as compared with the earth. On this score, one ought to say that it is the earth which moves, not the remainder of the universe.[68]

This is not very far removed from the Machian idea that the amount of motion of a given body is to be measured by the amount of motion it has relative to *all* the other matter in the universe. There are several passages in the *Principles* in which Descartes argues in such vein, as, for example, in the passage quoted below. However, he never attempts to knit them together into a well-defined relational theory of motion. Nor could he, even if he had in his heart believed in his professed relativism. He faced two problems; one of his own making, the other probably insuperable. The one of his own making was the absurd notion that motion is determined solely relative to contiguous matter; the insuperable problem was in the indefinite outer boundary of the Cartesian universe. Both problems are highlighted in the following passage, in which Descartes explains 'how there can be innumerable diverse movements in the same body':[69]

Each individual body has only one movement which is peculiar to it, since it is understood to move away from only a certain number of bodies contiguous to it and which are considered at rest; nevertheless, it can also participate in innumerable other movements, inasmuch as it is a part of other bodies which have other movements. For example, if a sailor travelling on board his ship is wearing a watch; although the wheels of his watch will have only a single movement peculiar to them, [it is certain that] they will also participate in that of the voyaging sailor, for they and he together form one body [which is transported as a unit]; they will also participate in the movement of the ship tossing on the ocean, and in that of the ocean itself, [because they follow its currents]; and, finally, in that of the earth, if [one supposes that] the entire earth is moved, [because they form one body with it]. All of these movements will indeed be in the wheels of the watch; but because we do not ordinarily conceive of so many movements at one time, and because we cannot even know all [those in which the wheels of the watch participate]; it will suffice for us to consider in each body the one movement which is peculiar to it [and of which we can have certain knowledge].

This passage shows clearly the difficulty of the problem Descartes raised. In a democratic theory of relative motion, which ought to follow from the situation Descartes describes at the beginning of this passage, and in which each piece of matter in the universe should do its own little bit to define the motion of the given piece under consideration, you have to know when you can stop counting the votes. But Descartes resolutely refused to draw any parliamentary boundaries for the hapless returning officer. In fact, he rather obviously abused the relativity of motion that he asserted and, more or less as it pleased him, chose whatever reference matter to define motion that would give him the answer that suited his purposes. In a way, Descartes got his deserts when he was accused of gerrymandering.

The real lesson to be learnt is that Descartes faced a far greater problem than the Inquisition. The hurdle which a relational theory of motion must take, and one wonders if it ever can, is infinity. With an indefinite, or, worse, infinite, constituency, democracy of motion becomes a mockery. Before the infinite even Einstein admitted defeat. But we anticipate.

The remainder of this particular story is quite quickly told. After the comprehensive statement of his principles, philosophical and physical, Descartes turns to the causes of motion and the laws it must satisfy. We see immediately that his conversion to relativism was not much more than skin deep; for, as we noted in Sec. 8.4, the laws of motion formulated in the *Principles* are in their essentials identical to those formulated (but never published) more than a decade earlier in *The World*. We rediscover all the familiar elements, above all the two principles that Newton, following Descartes' lead, was later to make into the first law of motion. Both of these principles presuppose of course that motion is quite definite; Descartes talks blithely of uniform motion in a straight line, extraordinarily without a single word of explanation for his *non sequitur*. Having cast the unfortunate reader loose on the shifting sea of relativism, so completely devoid of stable landmarks, he appears of a sudden to have made a miraculous return to *terra firma*. This part of the *Principles* clearly moves in the conceptual world that Descartes inhabited before the Inquisition gave him such a rude shock. It is the world of intuitive space, of straight lines that are citizens of that other democracy we mentioned earlier, Euclidean directional democracy, a different kind of democracy that can function without constituency boundaries.

We find again too the idea, entirely meaningless without a definite space to which it is referred, that God created a definite quantity of motion at the first instant of the world and ordained his laws of nature in such a way as to ensure that this given quantity of motion is preserved exactly constant in all subsequent collisions between matter. Descartes fails to note that this is meaningless without some quite definite frame of reference, the very existence of which he had been at such pains to deny.

One could multiply the examples but there is no point, especially since a particularly revealing manifestation of the absolute nature of Descartes' treatment of motion will come to light when we consider Descartes' seven rules of collisions in the following chapter.

This then is the confused state in which Descartes' *Principles* left the theory of motion in the middle of the seventeenth century. On the one hand, he advanced laws that cry out for an absolute space in order that they may be formulated at all; but on the other he succeeded in making the second half of the seventeenth century far more acutely aware of the dilemma that motion poses than the first half had been despite all the drama of the Copernican debate. Which of these two great strands of thought would gain the upper hand in the sequel? The choice could not be long delayed.

At the end of *Love's Labour's Lost*, Don Adriano de Armado tells the audience: 'You, that way; we, this way.' They could almost have been Descartes' parting words as he set off on his ill-fated journey to Sweden and comparatively early death in 1650: across the Channel, Newton opted for absolute space: on the Continent, Huygens and Leibniz, after vacillation and not a little confusion, followed Cartesian relativism.

9

Huygens: relativity and centrifugal force

9.1 Introduction

If Descartes exerted a stronger influence on the development of dynamics than the quality of his work on motion as such might have led one to expect, with Huygens it was the other way round. In accordance with the tally promised in the Introduction, all but $2\frac{1}{2}$ of the 'baker's dozen' of insights needed before the synthesis of dynamics could commence were to hand when Descartes died in 1650. The remainder were all obtained by Huygens, who almost certainly had sufficient mathematical skill to put them together and claim the prize of immortal fame that actually went to Newton. But it did not happen that way. Huygens' results were published in only mutilated form in his lifetime; a decade and more after he had discovered them and before any of them had been published Newton rediscovered nearly all of them. Then, more than another decade later, fortunate circumstances prodded Newton into the work that finally led to the creation of dynamics. During all this time Huygens was held back from taking the final step to dynamics by *an attitude of mind*. It was, in fact, the very same attitude of mind that led him to his great discoveries in the first place but which then became a hindrance. The hero and villain of this story was likewise one and the same person – Descartes.

Christiaan Huygens (1629–95) was the son of the wealthy and distinguished Dutch diplomat and poet Constantijn Huygens. He was truly fortunate in his birth and never had any real financial worries in his life, though his health was never good. Like several other of the great figures of the scientific revolution (including Newton and Leibniz) he never married. He was able to devote his entire life to science and through brilliant studies in several different fields rapidly established himself as the leading scientist on the Continent. He was a very considerable, if somewhat old-fashioned, mathematician (in mathematics he did not quite achieve the significant results of Descartes, Fermat, Newton, and Leibniz), a superb astronomer (he not only made important observations, including the discovery of a moon of Saturn and the establishment of the interpretation of that planet's rings, but also designed micrometers for measuring small angles with a telescope and, with his brother, Constantijn, developed great skill in grinding lenses), and a brilliant inventor (his invention of the pendulum clock will be discussed shortly). Besides his exceptionally important work in dynamics, which will be the subject of this chapter, he did brilliant work in optics, in which he excelled even Newton. His *Traité de la Lumière*,[1] which has been translated into English,[2] is one of the great works in the history of science and contains the formulation of the famous wave-front principle that is named after him.

Huygens was in his prime in the age in which science became, if not 'big-business', at least a major intellectual activity. It developed into a formalized undertaking and specialized journals were created specifically for the publication of scientific results. The scientific revolution acquired a momentum that it has never since lost. The Royal Society was founded in London in 1660 (first charter 1662) and Huygens became a member following a visit to London in 1663. On a second visit in 1689 he met Newton. In 1666 Huygens became one of the founding members of the French Academy of Sciences (Académie Royale des Sciences), of which he was to become the most distinguished member. As a member he was awarded a generous stipend and had an apartment in the Bibliothèque Royale. He lived mainly in Paris during the period 1666 to 1681. For the history of mathematics, physics, and philosophy one of the most important events of this period was the visit of Leibniz to Paris during the years 1672 to 1676, when the young Leibniz met Huygens and learnt from him much that was of great value in both mathematics and physics. This encounter was of decisive importance for the subject of the present study and its consequences will concern us both in this volume (Chap. 12) and again in Vol. 2.

The key to both the strengths and weaknesses of Huygens' approach to dynamics are to be found in his mechanical philosophy of nature. He read Descartes' *Principles of Philosophy* at the impressionable age of 15 or 16. Descartes was in fact a friend of Huygens' father and an occasional visitor to the family home in Holland, where he met and was impressed by the young Christiaan. Huygens was carried away by the idea of a mechanical explanation of the phenomena of nature along the lines proposed by Descartes. In accordance with this programme, all phenomena must be given a microscopic interpretation in terms of invisible particles of matter, the only elements allowed in this explicatory enterprise being, as we have seen, the size, shape, and motion of the particles.

In his *Treatise on Light*, Huygens says that 'it is not possible to doubt that light consists in the motion of certain matter'; he speaks of the 'true philosophy in which the grounds of all natural effects are derived from mechanical causes' and says that if this approach is not adopted one must 'completely give up all hope of ever understanding anything in physics'. It must be emphasized that *mechanical* is used here in the sense of direct physical contact. Like Galileo and Descartes, Huygens rejected the concept of attractive forces acting over distances. He completely accepted Descartes' basic position, namely that the motion in the world was put into it at the initial instant and has since been regulated by the action of the law of inertia (not yet called such) and the rules of impact of bodies. His main difference from Descartes (apart from being a vastly superior physicist) was in following Gassendi's lead* in reviving the ancient atomistic theory of particles moving in a void.

Huygens took the mechanistic programme very seriously and treated the problems that it posed with rigorous mathematics rather than Cartesian pictorial analogy. There is quite a detailed account of this work in the article on him in the *Dictionary of Scientific Biography*.[5] Huygens' work in this field anticipates the great work of the second half of the nineteenth century done on the kinetic theory of gases by Maxwell, Boltzmann, and Gibbs. By his introduction of mathematics and at least some sound physical principles he transformed atomism from qualitative philosophy into a genuine science, though in retrospect we can see that his attempts were premature and too ambitious. In the end, the main consequence of his work was that it provided the final elements needed for the creation of the definitive structure of dynamics. The description of this work will take up the bulk of the present chapter.

We conclude this introduction with a brief mention of Huygens' invention of the pendulum clock, which revolutionized the science of time keeping and played a significant role in the final emergence of dynamics. It will be recalled from the early chapters of this book that one of the most serious hindrances to the discovery of dynamics was the absence of suitable clocks. As in so many other things, it was Galileo who made the decisive breakthrough with his observation that the period of a

* The French priest Pierre Gassendi (1592–1655), whose main achievement was the revival of ancient atomism, is an important peripheral figure in our subject. As already noted, he performed important services in observing the transit of Mercury across the sun in 1631 and the experiment in which a heavy weight was dropped from the mast of a ship under sail in order to test Galileo's ideas about motion (1640). Like Descartes and Beeckman, with both of whom he maintained contact, he developed the idea of inertial motion and, in fact, published a correct form of the law just before Descartes published his much more influential *Principles of Philosophy*. Wohlwill[3] believes that Gassendi's work would by itself have ensured the establishment of the law of inertia in its correct form rather than the Galilean form with 'circular inertia'. It is quite possible that his ideas about space and time – which Gassendi taught existed independently of their content and provided the general frame of any knowledge of reality – influenced Newton's views. Newton undoubtedly read Gassendi, who was one of the first to adopt such universal concepts of space and time. The reader wishing for more information about Gassendi is referred in the first place to the article on him in the *Dictionary of Scientific Biography*.[4]

pendulum is independent of the amplitude of its oscillation (this is the discovery that Galileo is supposed to have made while watching the swinging of a great chandelier in the cathedral at Pisa, his own heart beat serving as the clock). Attempts were quite soon made to exploit this property for time-keeping purposes, and these showed that the period is independent of the amplitude only if the amplitude is small. Huygens made both technical and theoretical contributions of the first order of importance. He designed (in 1657) and patented (in 1658) a brilliantly successful pendulum clock, which transformed time-keeping. He also worked on the theoretical problem of designing a pendulum in which the period is independent of the amplitude. In 1659 he showed that if the path of the bob of a pendulum follows the curve known as a cycloid the period will be completely independent of the amplitude. His demonstration of how the pendulum must be designed in order to achieve this led him to the mathematical theory of evolutes of curves.

This work was important for the burgeoning science of dynamics for three reasons especially. First, by showing how the period T of a pendulum, its length l and what is now known as g, i.e., the acceleration of free fall, are related by the famous formula $T = 2\pi\sqrt{(l/g)}$ (which Huygens derived in his *Horologium Oscillatorum* (1673)[6]), Huygens provided a method for accurate measurement of g. Together with the accurate measurement of the radius of the earth, which occurred somewhat later, this datum was essential for linking the strength of terrestrial gravity to the strength of the force that keeps the moon in orbit and played a major role in establishing Newton's theory of universal gravitation. Second, in 1671 Richer was sent by the French Academy of Sciences to Cayenne on a mission having several aims* and he found that at the latitude of Cayenne (5°N) the pendulum clock that he had taken with him went slower than at Paris by about $2\frac{1}{2}$ minutes in a day.[7] This was direct evidence that the strength of gravity near the equator was less than at Paris; as we shall see later, this observation was of great importance for providing direct dynamical evidence of the earth's rotation and oblateness (in turn a consequence of its rotation). Third, it was found[8] that clocks taken to the tops of mountains went slower than when at the bottoms of the mountains. This was the clearest possible evidence that the strength of gravity decreases with increasing distance from the centre of the earth and provided important support for the idea that its strength falls off as $1/r^2$ with increasing distance r from the body producing the gravitational attraction. This result was also important in leading to the distinction

* The most important was to make observations of Mars at identical times as an observer at Paris to enable an accurate value for the parallax of Mars to be obtained. This was the first determination of a distance within the solar system made with really good accuracy and led to a dramatic enlargement of the dimensions of the solar system.[7]

between *weight* and *mass*, which was another most important step in the clarification of the fundamental dynamical concepts and will be discussed in Chap. 12.

When we come to discuss Newton's concept of time, in Chap. 11, we shall find confirmation of the importance of this aspect of Huygens' work. For the clarification of the time concept it was important in providing a second reliable motion by means of which the 'universal and uniform' flow of time (understood in the sense explained in Sec. 3.15) can be directly measured. And whereas the earth's rotation was a unique motion the pendulum provided in principle as many motions as one might reasonably want. Huygens can be said to share with the Hellenistic astronomers the credit for the practical clarification of the way in which the passage of time is manifested in the world. Nor must Galileo's contribution be forgotten. It is also worth mentioning that the invention of the pendulum clock is one of several examples of how trade and general economic development played an important part in the discovery of dynamics, since one of the great stimuli to the development of accurate chronometry was the need to determine longitude at sea, clearly a matter of prime importance for a seafaring nation such as Holland.

We now turn to the fundamental contributions that Huygens made to the discovery of dynamics. It must not, however, be thought that little or no work could be done on dynamics before its definitive structure had been established by Newton. Quite the contrary; many of the formulas and results that students of dynamics must learn when they begin the subject were discovered by Galileo and Huygens before Newton published the *Principia* in 1687.

9.2. Collisions and relativity: general comments

In their book *Gravitation*, Misner, Thorne, and Wheeler include a quotation from *The Sacred Wood*, in which T. S. Eliot remarks that when a new work of art is created something happens simultaneously to all the works of art which preceded it:[9] 'The existing monuments form an ideal order among themselves, which is modified by the introduction of the new (the really new) work of art among them.' Nowhere in physics does this quotation come to mind more forcibly than when one reads Huygens' works on dynamics with the benefit of hindsight illuminated by Einstein's genius.

We shall be concerned primarily in this chapter with Huygens' two posthumous works *De Motu Corporum ex Percussione* (*On the Motion of Bodies in Collisions*)[10] and *De Vi Centrifuga* (*On Centrifugal Force*).[11] The ideas and techniques that Einstein used to create the special theory of relativity develop directly those of the first of these two works, while *De Vi Centrifuga* anticipates several key ideas of general relativity. In particular, Huygens comes close to formulating Einstein's principle of equivalence – that locally it is impossible to distinguish between inertial and gravitational forces.

It was shown in the previous chapter how Descartes had brought the problem of motion to a head. Arguing on the basis of the 'pure light of reason', he had in particular stated a number of laws of impact. One of the great advances achieved by Huygens was in correcting the mistakes that Descartes had made in his rules of impact. Huygens made his discoveries during the 1650s but did not publish his results until 1669, when he responded to a general invitation issued by the Royal Society of London in 1668 for scientists to submit the correct laws of impact. He published the same laws at the same time in French in the *Journal des Sçavans* (18 March 1669).[12] His publication (in Latin) in the *Philosophical Transactions*[13] of the Royal Society was preceded by those of Wallis, who gave the laws of inelastic collisions, and Christopher Wren,* who, like Huygens, treated elastic collisions and gave essentially the same rules (without, however, any general principles of their derivation). Unfortunately, the beautiful arguments by which Huygens arrived at many of his conclusions had to await the posthumous publication of *De Motu Corporum* in 1703.

The primary interest of *De Motu Corporum* for this book is that it contained the first clear statement of the restricted principle of relativity for mechanical phenomena. Galileo had of course come extremely close to stating this principle, but Huygens went beyond him in several important respects. First, he applied the principle quite generally to all (inertial) systems moving uniformly with respect to each other in any direction, i.e., Galileo's (actually incorrect) restriction to uniform motions over the surface of the earth was lifted; second, whereas Galileo only considered the motion of *single* bodies, Huygens (clearly stimulated by Descartes' rules of impact) considered the *interaction* of bodies; by treating systems of bodies he drew attention to possible new results that can be extracted from Galileo's principle, thereby making explicit what in Galileo is only implicit; third, he combined the principle of relativity with a further principle that historically was to prove to be of as great significance as Galileo's principle – the principle of the conservation of energy; by means of these two principles the problem of elastic collisions of two bodies can in fact be *completely solved*; fourth, he formalized much more clearly than Galileo the concept of frames of reference and considered how the expression of a law of motion stated in one frame of reference must be transformed when one and the same phenomenon is expressed in different frames of reference. He was the first practitioner of transforma-

* Wren (1632–1723) had already had a very distinguished career in science before turning to architecture at the age of 30. He was Savilian professor of astronomy at Oxford from 1661 to 1673. There are numerous references to his work as a geometer in Newton's *Principia*.

tion laws. It is in the combination of transformation laws and the selection of certain simple principles or empirical facts from which far reaching conclusions are then drawn that he so closely resembles Einstein (one should, of course, word that too the other way round!). *De Motu Corporum* is a pre-run of the 1905 paper that created special relativity. In fact, the editors of Huygens' works comment that if the principle of relativity should carry anyone's name, then it should be Huygens'.[14]

De Motu Corporum is a beautiful exercise in logical deduction from certain simple premises. This was, of course, what Descartes had attempted; unfortunately, he only seldom selected sound premises. There is another important difference: Descartes' 'deductions' were mostly pictorial. He posited, as it were, certain principles underlying nature, which he regarded as a mechanism, and was then content to explain in qualitative terms how these principles operated in any particular part of the mechanism one chose to examine. Huygens' deductions are on an altogether different basis – strictly mathematical and simultaneously subjected to empirical verification. The failure of Descartes and the success of Huygens were an important lesson for science and emphasized the significance of Galileo's work (which was grasped by Huygens but escaped Descartes): unless the natural scientist is prepared to accept the rigour of mathematical discipline, coupled with accurate observation, the essence of the laws of physics will elude him. The vision will be blurred to such an extent that the key points carrying the explicatory burden, i.e., the actual laws of nature, will disappear entirely. The sharpness of focus is far more important than Descartes intuitively imagined. It is not the case that with poor focus you 'see' the same laws but not quite so precisely – you simply do not see the laws at all. There is a world of difference between the hunch that the sun attracts the earth and the precise statement which gives the strength of the force and states further that there is exact proportionality between the force and the *acceleration*. Of course, Descartes wanted to go in this direction but was far too sloppy in application (and absurdly ambitious in what he attempted to explain). Practising a self-discipline that Descartes completely lacked, Huygens demonstrated a paradox: by taking a seemingly much shorter step, but one that is in exactly the right direction, you end up by going much further. As Neil Armstrong remarked:[15] 'A short step for a man, a giant leap for mankind'.

9.3. Descartes' theory of collisions

There is no clearer way to contrast Descartes and Huygens than through comparison of the principles by which they sought to determine the laws of collision between bodies, so we begin by considering Descartes' work, which is described in Part II of his *Principles*.[16]

We must first say something about the mass, weight, and bulk (volume) of bodies. Descartes assumed that the dynamical properties of bodies are determined by their *volume* (and speed of course). He did not have the Newtonian concept of *mass* and was convinced that *weight* could be explained as a manifestation of bulk, speed, and centrifugal force. In treating collinear collisions of two bodies, Descartes assumed accordingly that a collision of two bodies, 1 and 2, is characterized by the velocities of each body before the collision, u_1 and u_2, and after it, v_1 and v_2 (we measure velocities from left to right in the usual convention, i.e., positive u_1 corresponds to motion from left to right). In addition, each body is assumed, in the formally stated rules of collision at least, to have a volume, or bulk, that remains unchanged in the collision. It is convenient to denote the bulks by m_1 and m_2, since they play a role closely analogous to mass in Newtonian physics.

The collision problem can then be posed as follows. Given m_1 and m_2 together with the initial velocities u_1 and u_2, to find the post-collision velocities v_1 and v_2. Since there are *two* unknowns, the solution of the problem requires knowledge of two equations relating the six quantities (four of which are known). Descartes had no difficulty in finding one such condition. His whole natural philosophy is based on the *a priori* idea that the *quantity of motion*, defined as the bulk times the speed (NB: speed, not velocity) remains constant through all vicissitudes. Thus, one condition can be written down immediately

$$m_1|u_1| + m_2|u_2| = m_1|v_1| + m_2|v_2|. \tag{9.1}$$

There are two serious difficulties in the Cartesian theory of impacts. The first is that the relation (9.1) is in fact quite false for general collisions; the second is the missing second condition. In an attempt to provide it, Descartes imagines that collisions are rather like two men having a fight.[17] Each is endowed with a certain pre-collision 'force to move or to resist movement'. Then, according to Descartes, to find the outcome of any given collision, 'it is only necessary to calculate how much force to move or to resist movement there is in each body; and to accept as a certainty that the one which is stronger will always produce its effect'.

This principle is a lot less clear than it seems. The power of a moving body to move can be measured by $m|u|$, the characteristic quantity that appears in Eq. (9.1). But what about the power of a body at rest to resist movement? Guided by intuition, Descartes had the feeling that bodies at rest do resist motion. But where can he find a quantitative measure of this resistance? There is no speed with which he can multiply its bulk m. Descartes overcomes this difficulty by measuring its resistance to motion as the speed it *would acquire*, after the collision, if set in motion. I do not propose to go into the detailed arguments that Descartes employed in the attempt to supply his missing condition. Most modern physicists would see them as *ad hoc*; they were certainly very largely wrong. The main reason for this is undoubtedly that Descartes made no serious attempt at all to deduce or confirm his rules of impact by observation. Instead he relied almost exclusively on pure thought. This is why his detailed physics generally seems of such a poor quality compared with that of Galileo and Huygens. Nevertheless, he did make two enduring and important contributions to the problem of collisions, namely, the clear posing of the problem of *describing them mathematically* and his idea of determining their outcome by positing the conservation of a certain quantity (even if his own choice proved to be wrong). (Gabbey[18] has pointed out that this fruitful idea of Descartes was probably in large part introduced in reaction to a problem in Beeckman's work, in which motion could be lost in collisions, so that the world would eventually run down. For a detailed discussion of Descartes' work and the way in which it influenced Newton, the reader is referred to a further paper of Gabbey.[19]) The most notorious mistake which Descartes made was in his rule 4, which asserts that a body C at rest could never be set in motion by a smaller body B (even if only slightly smaller). Instead, B would have to spring back with equal speed in the opposite direction (in order for Eq. (9.1) to be satisfied). Of the seven rules that Descartes proposed, only one, the first, was correct.*

It is worth summarizing the rules which Descartes gives for collisions of identical bodies. A general collision of two such bodies can be represented formally by

$$u_1,\ u_2 \rightarrow v_1,\ v_2. \tag{9.2}$$

Descartes gives rules (Nos. 1, 3, 6 in his enumeration) which state the outcome of such collisions between equal bodies for different values of u_1 and u_2. He starts with rule 1, the only one of his seven that is in fact correct (for what are called *elastic collisions*; see below), which states that if two equal bodies approach each other with exactly equal but opposite speeds the 'contest of strength' between them must necessarily result in a draw (since they are clearly 'equally strong'). Since (9.1) must hold always, the

* The hopelessly ambitious nature of Descartes' physics is revealed by the fact that these seven rules of impacts, together with his form of the law of inertia, were the only explicit rules which he gave in the *Principles* to account for the entire phenomena of nature. But his physics envisaged much more than collisions in which the individual bodies remain intact. He also relied on collisions to *change the shapes* of the pieces of matter; for he argued that as a result of such collisions the various pieces of matter will, whatever shapes God may have given them initially, be worn away until they approximate to one of three basic shapes (with corresponding sizes too). We see that there are in fact many unknowns on the right-hand side of Eq. (9.1), since m_1 and m_2 must in this more general scheme actually represent all possible sizes and shapes of the collision products (which will also necessarily be more than two). Since he did not even begin to discuss this problem, Descartes' claim to be able to 'reconstruct the world' from 'extension and movement' was mathematically equivalent to trying to find the values of infinitely many unknowns from a few linear equations.

only possible outcome of the collision is that the two spring back with reversed but undiminished speeds:

$$u,\ -u \rightarrow -u,\ u. \tag{9.3}$$

In rule 3, Descartes supposes that one body approaches the collision with a slightly greater speed than the other. The initial condition is thus $u + \frac{1}{2}a,\ -(u - \frac{1}{2}a)$, where $u \gg a > 0$. In this case the 'contest of strength' will clearly be decided in favour of the faster body, which therefore 'gets its way', which means that it continues on its way, transferring, however, some of its motion to the other in order to ensure that (9.1) is satisfied:

$$u + \tfrac{1}{2}a,\ -(u - \tfrac{1}{2}a) \rightarrow u,\ u. \tag{9.4}$$

Finally, in the case when one of the bodies is at rest and the other has speed $2u$, Descartes asserts (rule 6) that the outcome will be

$$2u,\ 0 \rightarrow -\tfrac{3}{2}u,\ \tfrac{1}{2}u, \tag{9.5}$$

i.e., a rule quite different from the limit obtained from (9.4) when $a \rightarrow 2$.

It is revealing to consider the *relative* speeds of the bodies before and after the collisions in each of the three cases. The relative speeds before the collision are in all cases $2u$ (the particular initial speeds were chosen to achieve this), but after the collision they are $2u$, 0, and $2u$.

However, we now recall Descartes' assertion that all motion is relative. If that is the case, one might suppose that the frame of reference in which any collision is observed should not make any difference to its objective outcome. But, as we have noted, the relative speeds before the collisions are the same in all three cases. Let us therefore go over in the case of collisions (9.3) and (9.4) to frames of reference in uniform motion relative to the frame in which the rules are specified by Descartes and such that in the new frames one of the bodies is at rest. Then the three collisions are observed in such a frame of reference as

$$2u,\ 0 \rightarrow 0,\ 2u, \tag{9.3$'$}$$

$$2u,\ 0 \rightarrow 2u - \tfrac{1}{2}a,\ 2u - \tfrac{1}{2}a, \tag{9.4$'$}$$

$$2u,\ 0 \rightarrow -\tfrac{3}{2}u,\ \tfrac{1}{2}u. \tag{9.5}$$

These results have a rather startling implication, of which Descartes himself seems to have been completely unaware (at least, there is nothing in the *Principles* to suggest that he was aware of it),* namely that there is a

* It is true that Descartes qualifies his rules with the important provision that they are supposed to hold for collisions in the absence of any contiguous bodies and that 'the application of these rules is difficult, because each body is always surrounded by many contiguous ones'.[20] However, it is hard to believe that the differences between the outcomes of Eqs. (9.3′), (9.4′), and (9.5) would not show up in the world in the most obvious manner. We may also note how unphysical was Descartes' instinct: the $a \rightarrow 0$ limit in (9.4) is not at all the same as (9.3), so that initial conditions differing infinitesimally lead to totally different outcomes. This observation was the basis of a devastating criticism of Descartes' rules made by Leibniz.[21]

criterion of absolute rest. According to Eqs. (9.3′), (9.4′), and (9.5), it is not possible to predict the outcome of a collision between equal bodies if only the relative speed of the two is known. The outcomes in the three cases are entirely different. However, in all cases, once the collision has been observed it is immediately possible to say what transformation is necessary in order to go over to Descartes' original frame of reference. The frame in which collisions unfold as in Eqs. (9.3), (9.4), and (9.5) therefore provides the criterion of absolute rest.

It is quite clear that Descartes formulated his rules of impact using his 'pre-Inquisition' concept of space, which is an essential element in 'gauging' the respective strengths of the two bodies in their ensuring 'contest of strength' and defining the quantity of motion.

It is worth mentioning here an aspect of scientific theories which will figure prominently in later discussions in Vol. 2 of the respective merits of the approaches that regard motion as absolute or relative. It is this: by what criterion does one say that a theory is a good one? An empiricist will say that the job of the scientist is to find out how the world *is*, not how he would like it to be. Thus, empirical truth is the prime criterion of such an approach. Taking this a bit further, we may say that an investigator will have 'learnt to read nature's book' if what he sees around him at any one instant enables him to *predict the future*. Descartes' prime concern was rather different: as we have seen, his main aim was to *understand* the world. By understand, he meant that he could form a clear picture in his mind of what is happening. Seen in this light, Descartes would not have been perturbed by the fact that the left-hand sides of Eqs. (9.3′), (9.4′), and (9.5) are identical but the collision outcomes are all quite different. For he is able to 'understand' these collisions in the conceptual framework in which he works (in reality, absolute space), which gives meaning to the concept of the 'strengths' with which two bodies enter collisions. If challenged explicitly on the matter, Descartes would no doubt have expressed himself happy to sacrifice the ability to predict the future from observed initial data for the sake of a clear and rational explanation of what is happening. We shall see in Vol. 2 how Poincaré put the absolute/relative debate in a particularly illuminating light by examining the matter from the point of view of the predictability of the future. For the moment we will only say that Cartesian rationalism is not synonymous with predictive power – nor a true understanding of how the world really works.

But it is now time to pass from Descartes to Huygens and see the transformation he wrought in Cartesian physics. In the history of dynamics there are few more beautiful pieces of work than Huygens' replacement of Cartesian guesswork by the rigorous derivation of the laws of impact from general principles resting on a secure empirical basis.

9.4. Huygens' theory of collisions

Huygens opens *De Motu Corporum* with two fundamental propositions, or assumptions. The first is as follows:

> When once a body has been set in motion, it will, if nothing opposes it, continue that motion with the same speed in a straight line.

We note how simply the law of inertia has been fused out of the two parts that it possessed in Descartes' *Principles*. So far as I know, this is the first clear statement of the law formulated: (*a*) as a single law, (*b*) with the clear recognition of it as the most basic law of motion. (Although *De Motu Corporum* was not published in his lifetime, Huygens repeated much the same assumption at the very beginning of the second part of his *Horologium Oscillatorum* published in 1673, where he also gives a very clear statement of Galileo's principle of composition of inertial and gravitational motion.)

Whereas Huygens' first assumption follows more or less directly from Descartes, in his second he breaks new ground. He states the principle of relativity in a form very little different from the one in which it is still used today:

> The motion of bodies and their speeds, uniform or nonuniform, must be understood as relative to other bodies that are regarded as being at rest even if these together with the others partake in a further common motion. Thus, when two bodies collide but have in addition a further common uniform motion, they impart to each other impulses that, viewed by one that also partakes in the common uniform motion, are exactly the same as if the uniform motion common to all were not present.
>
> Thus we say that if the occupant of a boat travelling with uniform speed lets two equal spheres approach each other with speeds that are equal relative to the occupant and the parts of the ship then as a result of the collision each must spring back with speeds that, relative to the occupant, are exactly the same as if the occupant let the same spheres collide with the same speed when the ship is not travelling or he were on land.

We shall look at the consequences that Huygens extracts from these principles in a moment. It should, however, be emphasized that there is in these two principles of Huygens, both of which reveal the strong influence of Descartes, a contradiction, or at least a difficulty, just as great as in Descartes. If motion is relative to other bodies, which bodies are they that define the uniform motion in a straight line which, according to Huygens' first principle, is the basis of his entire mechanics? Neither in its formulation nor anywhere else in *De Motu Corporum ex Percussione* does Huygens offer any clarification. This was probably very wise on his part. Had he stopped at that point and attempted to unravel all the mysteries of motion, he would probably have made no progress at all. As it was, doubts on this score may well have been one of the reasons why he did not publish *De Motu Corporum* in his lifetime (the long delay in publishing even the results was also probably due to a reluctance to challenge in public the prevailing orthodox Cartesianism). If the first principle stands totally without visible support, the principle of relativity is formulated with words that seem to tie it in to the real contingent world but in a curiously vague way. In expressing the principle in words, Huygens anticipates the photographer's art of vignetting: the central point is in clear focus, but the picture shades off gradually into nothing. There is no supporting background, the picture hangs literally in the air. We see a boat and there is talk of a river bank, but there it ends.

It is a delicate balancing act in the limbo created by Descartes' final demolition of the cosmos. Huygens appears to be searching for the true ocean through which the Galilean galley courses but does not find it. The opening statement seems like an acceptance of Cartesian relativism ('The motion of bodies . . . must be understood as relative to other bodies that *are regarded as being at rest*') but some other more stable background is immediately implied by the 'further common uniform motion'. But where is the court of final appeal that this common uniform motion implies, that can define such a mathematically precise concept as uniform rectilinear motion? Huygens answers with an example: the local surface of the earth. He clearly cannot have believed the earth to have been the terminal referent of motion. If questioned, he would presumably have answered that the surface of the earth serves his purpose since it is, at least to a good enough approximation, itself moving uniformly. But when questioned how the concept of uniform motion tallies with the relativism he espouses in his opening statement, he would probably have sighed and admitted defeat.

He clearly had what Mach,[22] in speaking of Newton, called the *tact* of the great natural scientist. He could sense the immense but as yet untapped potential in Galileo's cabin cameo and knew that *uniformity* of the relative motion between two such cabins was an essential prerequisite for the obtaining of correct results. But he was also very open to the intuitive force of Cartesian relativism, and this leads him to use words and phrases which suggest (though certainly not by any explicit proposition or theorem) that the Galileo–Huygens principle of relativity is somehow a necessary consequence of Cartesian relativism. It is most important to realize that, despite the identity of expression, the actual use which Huygens made of the principle of relativity gave to the word *relativity* a significance that is quite absent in Descartes. For a start, in the latter's *Principles*, there is nothing resembling Galileo's principle. As we have seen, Descartes had very little interest in the questions that were of central concern to Galileo. The clearest indication that Huygens' relativity marks a completely new departure compared with Cartesian relativism is the word *uniform*, a crucial restriction, which occurs nowhere in Descartes' general discussion of the true nature of motion (though it does of course appear in his formulation of the laws of motion, which, as we have seen, were formulated in a totally different conceptual framework).

Another point which must be made here is that the Galileo–Huygens principle of relativity (which has a very precise empirical content) simply cannot be deduced as a necessary consequence of Cartesian relativism (which, when accepted in its most extreme form, actually denies the possibility of an objective theory of motion by asserting that place and motion are ultimately determined 'only in our minds'). Even if all motions are truly relative, one could very easily imagine that motion with respect to the universe as a whole would show up in local physics, as was pointed out in Sec. 1.2. It is quite easy to construct models in which this is the case. It is, in fact, very difficult to establish precisely what Huygens did understand by relativity. The words he uses seem to suggest he meant Cartesian relativism, i.e., kinematic relativity, but his actual principle gives expression to Galilean relativity, which, as we have pointed out several times, is not the same thing at all. We shall return to this question a little later and also in Chap. 12, when we compare the positions of Newton and Huygens.

Let us now return to Huygens' derivation of the laws of impact. Having praised Huygens so highly earlier, I should for all that admit that the overall presentation of elastic collisions on a straight line as given in *De Motu Corporum* is not quite as clear as it might be, with subsidiary assumptions being added as the exposition proceeds. As we shall see later, the main reason for this seems to have been Huygens' desire to derive as many of his results as possible by rational arguments from principles that would command general assent (rather in the manner of Euclid's axioms). Huygens too had a strongly rationalistic approach, but he took care to choose principles in agreement with empirical fact. For the sake of clarity, I shall break loose somewhat from Huygens' detailed exposition and merely concentrate on the most important principles he uses, showing how they are sufficient to solve the two-body collision problem in its entirety. In Sec. 9.5, I shall come back to the question of the precise route by which Huygens appears to have obtained his first results, as this is of no small interest.

We begin by considering the positive use that Huygens makes of his principle of relativity. It is a fascinating exercise: he can be observed feeling his way forward, careful step by careful step, to the laws of transformation between coordinate systems that are distinguished by being what we now call inertial. Perhaps it is precisely because he failed to achieve clarity on the status of these systems (and therefore does not say explicitly they are distinguished) that he moves so very cautiously. Nevertheless, his instinct did not fail him, and he separates cleanly the two essential elements – the purely kinematic aspect of the problem, which consists of establishing how an event described in one frame of reference will appear when viewed from a different frame of reference, and the exploitation of the physical element of the relativity principle, the empirical statement that collisions as observed on the boat are unaffected by whatever uniform velocity the boat may have relative to the bank.

He supposes a man in a boat on a river performing collision experiments and a man on the bank who has two tasks, the first of which is merely that of a spectator watching these collisions. This is the purely kinematic aspect of the situation. As before, we measure velocities from left to right, so that c denotes a speed of the boat as observed by the man on the bank from left to right and $-c$ the same speed but in the opposite direction. However, for the moment we concentrate on the man in the boat, in which he lets two hard (by which is meant, in modern terms, perfectly elastic) balls collide. Let their velocities before the collision be u_1 and u_2 and after it v_1 and v_2. Then the collision can be denoted symbolically by

$$u_1, u_2 \rightarrow v_1, v_2. \tag{9.6}$$

Now if at the same time the boat is moving with uniform speed c relative to the man on the bank, then from the bank the collision (9.6) will be observed as

$$u_1 + c, u_2 + c \rightarrow v_1 + c, v_2 + c. \tag{9.7}$$

For the moment this is a purely kinematic relationship. Huygens now wishes to make it clear, however, that in reality the collision must not be thought of as taking place *in the boat*. It is not tied to the boat, it could in fact be regarded just as well as taking place *on the bank*. He makes this point as follows. He supposes that the man on the boat effects the collision by standing on its deck with outstretched arms holding in his hands strings by which the balls are suspended. He then moves his hands with the uniform velocities u_1, u_2 relative to the deck, thereby causing the balls to collide with the same velocities. Huygens now arranges that as the boat comes past the spectator on the bank, he joins his hands with the man on the boat, so that they both hold the ends of the string. The man on the bank moves his hands at the speeds which ensure that they travel exactly together with the hands of the man in the boat, i.e., $u_1 + c$ and $u_2 + c$ relative to the bank. Now the balls clearly cannot 'know' which of the two men is causing them to collide. Huygens illustrates his argument with the characteristic picture shown in Fig. 9.1. In fact, he repeats such arguments many times, and, to save the artist redrawing the full picture, just leaves the hands of the two men as they lock together, as in Fig. 9.2.

It is now that the physical element is introduced. The collision (9.7) is a collision that actually takes place in the spectator's frame of reference. It has initial velocities $u_1 + c$, $u_2 + c$. But suppose that the man on the boat

Fig. 9.1.

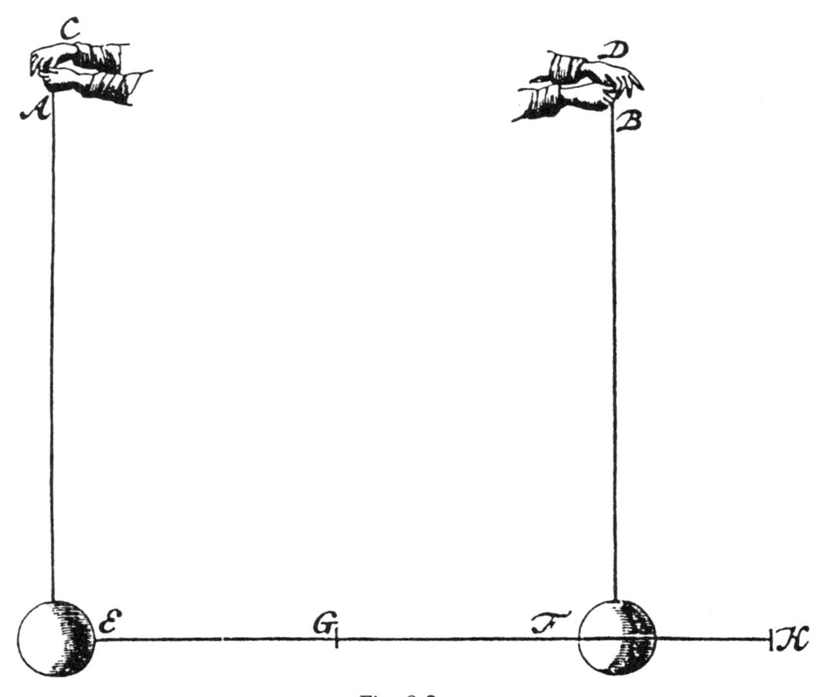

Fig. 9.2.

arranges a collision in which the initial velocities *relative to the boat* are $u_1 + c$, $u_2 + c$. What will be the outcome then? It is at this point that Cartesian guesswork gives way to physical principle. By the Galileo–Huygens principle, the final velocities relative to the boat will be just as they are in Eq. (9.7), despite the fact that the new collision with initial velocities in the boat $u_1 + c$, $u_2 + c$ is a quite different collision. Thus, by means of the principle of relativity, knowledge of the outcome of just *one* collision suffices to *predict* the outcome of a complete one-parameter family of collisions, i.e., given Eq. (9.6) we can immediately deduce (9.7), in which c can have any value.

Huygens illustrates this general rule by considering the one rule of impact that Descartes succeeded in getting right, the one which states that if two equal elastic bodies approach each other with equal but opposite speeds they will rebound from the collision with the speeds reversed but unchanged in magnitude:

$$u, -u \rightarrow -u, u. \tag{9.8}$$

Choosing now c in (9.7) to be equal to u, we immediately deduce a new collision:

$$2u, 0 \rightarrow 0, 2u. \tag{9.9}$$

The main point of interest about (9.9) is that it is completely different from the rule that Descartes gave for a collision with initial velocities $2u$, 0. We recall that according to his rule 6, the collision should be

$$2u, 0 \rightarrow -\tfrac{3}{2}u, \tfrac{1}{2}u.$$

Although Huygens' correct result (9.9) should have sounded the death knell of *a priori* Cartesian physics, Descartes' overall scheme continued to lead a protected life, dominating the overall pattern of thought long after it had been shown to suffer from the most serious defects. There is a parallel here between Descartes' scheme and the one it was intended to replace, i.e., Aristotle's. It was a case of the curate's egg: good in parts. Both suffered spectacular failures but both also had triumphs (the unique centre of Aristotelian cosmology was lost, but the planets were indeed found to move in nearly perfect circles; the rules of impact were wrong but the law of inertia right). But the remarkable staying powers of both these schemes cannot be explained alone by their partial successes. Much rather, they must have made a very direct appeal to people's intuition, giving eloquent expression to instinctively held beliefs about the nature of the world. That Cartesianism survived for so long, many years after the appearance of Newton's *Principia*, shows clearly how reluctant is the human mind to throw off conceptions that have once taken root.

Impressive as is the generation of the one-parameter family of collisions (9.7) from the single collision (9.6), the principle of relativity alone does not suffice to solve the collision problem. Having already done enough to earn one Nobel Prize, Huygens then goes on to do the work for a second.

To solve the general problem Huygens must somehow or other deal with the problem of the collision of *unequal* bodies. In the case of equal bodies it was, of course, comparatively obvious to take the example of their approaching each other with equal speeds and then springing back with undiminished speeds. Granted this, then the relativity principle solves his problem for all collisions between equal bodies. Following along this line of attack (in which we have so far followed Huygens faithfully) genuine further progress can only be made by making a definite assumption about how at least one actual collision between *unequal* bodies unfolds; only then can the relativity principle be used to generate yet further collisions. It is at this point that a valuable point of principle will be obscured if we follow Huygens too closely, so I shall go over to a more modern approach and put the really important new idea that Huygens introduced in the forefront of the discussion.

We must begin by saying something about masses. In *De Motu Corporum*, Huygens in fact speaks of bodies being merely larger or smaller. In the statement of the laws of impact sent to the Royal Society and the *Journal des Sçavans*[12] he says expressly that the bodies are assumed to be made of the same material or that their magnitude is to be estimated by their weight. Thus, he effectively made *weight* the dynamically crucial factor in collisions (along with the relative velocity – see below). Since weight is proportional to mass and Huygens did not consider a situation in which the bodies are moved into different gravitational fields he was effectively operating with the Newtonian mass concept though not clearly recognized as something distinct from weight. We shall therefore continue to use the symbol m to denote the 'size' of the bodies, identifying m with the mass.

What Huygens did in effect was discover a partial form of the law of conservation of energy. This again is a fascinating combination of elements taken from Descartes and Galileo. For a start, the stimulus to the passage from single-particle motionics to two-particle dynamics comes from Descartes rather than Galileo; so too does the idea that the details of collisions are governed by a general conservation law. However, Huygens realizes that the conserved quantity is not Descartes's quantity of motion but something rather different. He hit on the correct quantity by pondering that other great law of Galilean motionics: the law of free fall.

Huygens notes that according to Galileo a body that falls freely through a height h acquires a velocity u whose square is proportional to h, namely, $u^2 = 2gh$. Moreover, it will acquire exactly the same velocity if it descends down any inclined or even curved planes (assumed perfectly smooth, of course) through the same height. If it is then deflected into the horizontal

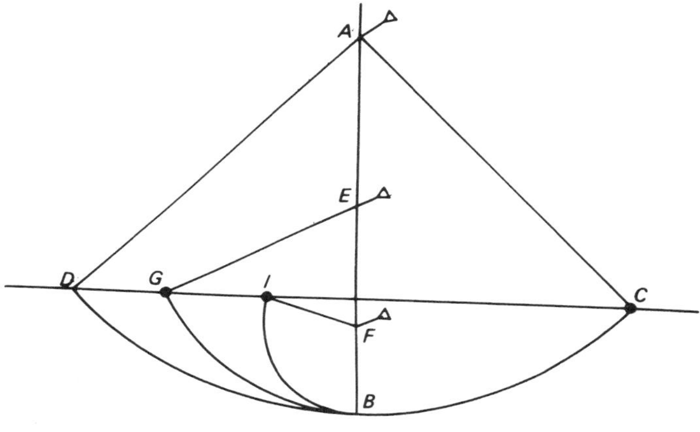

Fig. 9.3.

plane it will retain that velocity for as long as one wishes but can be made to ascend again to exactly the height h from which it originally descended.

To highlight Huygens' debt to Galileo it is here worth quoting a famous passage from Day 3 of the *Discorsi*, together with Galileo's deeply suggestive diagram (Fig. 9.3). Here is the passage:[23]

> Imagine this page to represent a vertical wall, with a nail driven into it; and from the nail let there be suspended a lead bullet of one or two ounces by means of a fine vertical thread, AB, say from four to six feet long, on this wall draw a horizontal line DC, at right angles to the vertical thread AB, which hangs about two finger-breadths in front of the wall. Now bring the thread AB with the attached ball into the position AC and set it free; first it will be observed to descend along the arc CBD, to pass the point B, and to travel along the arc BD, till it almost reaches the horizontal CD, a slight shortage being caused by the resistance of the air and the string; from this we may rightly infer that the ball in its descent through the arc CB acquired a momentum [*impeto*] on reaching B, which was just sufficient to carry it through a similar arc BD to the same height. Having repeated this experiment many times, let us now drive a nail into the wall close to the perpendicular AB, say at E or F, so that it projects out some five or six finger-breadths in order that the thread, again carrying the bullet through the arc CB, may strike upon the nail E when the bullet reaches B, and thus compel it to traverse the arc BG, described about E as center. From this we can see what can be done by the same momentum [*impeto*] which previously starting at the same point B carried the same body through the arc BD to the horizontal CD. Now, gentlemen, you will observe with pleasure that the ball swings to the point G in the horizontal, and you would see the same thing happen if the obstacle were placed at some lower point, say at F, about which the ball would describe the arc BI, the rise of the ball always terminating exactly on the line CD.

Huygens develops such considerations in a very original way. He supposes that *two* bodies, say of mass m_1 and m_2, are let fall from heights

h_1 and h_2, deflected into the horizontal with the acquired velocities $u_i = \sqrt{(2gh_i)}$, $i = 1, 2$, allowed to collide, and then to ascend inclined planes to heights $\bar{h}_1$ and $\bar{h}_2$.

In principle, $\bar{h}_i \neq h_i$ since the speeds will be changed by the collision. As the second fundamental principle introduced in *De Motu Corporum*, Huygens requires that as a result of the collision the *centre of gravity of the system of two bodies should not be raised*. We see here the influence of the science of statics and the experience gained from simple mechanical systems for raising weights which had suggested rather strongly that it is never possible to get work done for nothing. Huygens' principle just formulated extended such ideas by assuming that merely through the collision of bodies one should not be able to get the centre of gravity of a system of bodies to rise spontaneously.

When, in 1673, Huygens published in Paris his masterpiece *Horologium Oscillatorum*,[6] he again used this principle to solve a famous problem known as the *centre-of-oscillation problem* (for a discussion of this see Ref. 5 or Mach's *Mechanics*[24]). Huygens pointed out that if through any purely mechanical process involving gravity it were possible for the centre of gravity of a system of bodies to rise spontaneously then it would be possible to build a perpetual motion machine. This principle, like the entire work, caused a considerable stir. The idea of deducing the outcome of experiments in physics by means of arguments which deny the possibility of perpetual motion machines has proved to be one of the most powerful tools of theoretical physics. It is another of these *impotence* principles to which attention was drawn earlier near the end of the chapter on Galileo. Such principles played a central role in the discovery of the laws of thermodynamics and quite recently helped to reveal a most surprising connection between thermodynamics, gravity, and quantum field theory (Hawking's theory of the evaporation of black holes[25]). Huygens therefore has the distinction of being the first physicist to have made systematic use of 'impotence' principles. It is one of the reasons why, in certain respects, he seems so remarkably modern and a precursor of Einstein.*

* Some time after I had written the above I came across some very interesting comments of Martin Klein in an introduction he wrote to the recently published English translation of Mach's book on thermodynamics (*Principles of the Theory of Heat* (Vienna Circle Collection, Vol. 17), Reidel, Dordrecht (1986)). He points out that Mach, through his *Mechanics* (which Einstein read avidly), may well have introduced the young Einstein to several typically Huygensian techniques, above all the use of the relativity principle. In particular, Mach gives an account in the *Mechanics* of Huygens' use of it in his work on collisions. Klein believes that Huygens was the only person to make *positive* use of the principle of relativity before Einstein employed it to such dramatic effect in his special theory of relativity, and he thinks it possible that Einstein was therefore directly influenced by Huygens through his reading of Mach (*loc. cit*, note 50, p. 419). This would obviously explain why Huygens' work reminds one of Einstein's so often.

But we must now get back to the collision problem. This is the point at which we part company from Huygens' treatment, since it is simpler to assume that after the collision the centre of gravity reascends to exactly the same height that it had before the collision, a result that Huygens enforces by other assumptions (we shall discuss these shortly). The centre of gravity before the collision is at height $H = (m_1h_1 + m_2h_2)/(m_1 + m_2)$; substituting Galileo's law of free fall, according to which $u_i = \surd(2gh_i)$, $i = 1, 2$, we obtain

$$H = \frac{m_1h_1 + m_2h_2}{m_1 + m_2} = \frac{1}{2g}\left(\frac{m_1u_1^2 + m_2u_2^2}{m_1 + m_2}\right). \tag{9.10}$$

Writing down the corresponding relation for the height $\bar{H}$ of the centre of gravity after the collision (when the velocities are v_1 and v_2) and setting $H = \bar{H}$, we obtain immediately

$$m_1u_1^2 + m_2u_2^2 = m_1v_1^2 + m_2v_2^2. \tag{9.11}$$

Thus, in any elastic collision (9.11) must always hold. But now consider (9.11) in some particular case and in some particular frame of reference. The same collision can be observed in a frame of reference moving with uniform velocity c relative to the original frame. In the new frame, it appears as the collision

$$u_1 + c,\ u_2 + c \rightarrow v_1 + c,\ v_2 + c. \tag{9.12}$$

By the relativity principle, this will be the collision that actually takes place when $u_1 + c$ and $u_2 + c$ are the initial velocities. But by the energy principle the law corresponding to (9.11) will also hold for the new collision (9.12). Thus,

$$m_1(u_1 + c)^2 + m_2(u_2 + c)^2 = m_1(v_1 + c)^2 + m_2(v_2 + c)^2. \tag{9.13}$$

Multiplying this out and substracting (9.11), we obtain

$$m_1u_1 + m_2u_2 = m_1v_1 + m_2v_2. \tag{9.14}$$

The two relations (9.11) and (9.14) are then sufficient to determine the outcome of any elastic collision with given m_1, m_2, u_1, u_2. (The formal two-fold ambiguity of the solution due to the fact that (9.13) is a quadratic equation merely corresponds to the possibility of an unphysical collision, in which the particles would have to pass through each other rather than spring back from each other, as in the physical solution, which is the one that must be chosen.)

It must be admitted that something of the marvellous attaches to these results, especially when one considers them in the light of the historical development and all those long centuries in which philosophers speculated about motion but had in truth only the vaguest notions of what in reality happens. One of the ironies in the discovery of dynamics is that as long as men stood on what they regarded as firm ground (the earth before Copernicus) they in fact had no firm foundation at all to their speculations about motion, but very soon after that most disconcerting of all events, when Copernicus took the ground from under men's feet, secure foundations for the theory of motion were in fact identified (in the Galileo–Huygens principle of relativity) in the very process of adjustment to the shock and the re-establishment of bearings in the new environment. We exchanged the earth for a whole family of rafts travelling uniformly with respect to each other. We only found our footing when we took to these providentially provided life-rafts – inertial frames of reference, as they came to be known. The oddity was – and is – that the rafts have a well-defined relationship to each other (that of uniform translational motion) but do not in themselves appear to rest on any foundation. They are simply there, and we know not whither they are carried by what appears to be a mysterious *cosmic drift*.

Although *De Motu Corporum ex Percussione* does not use the above derivation of the laws of elastic impact, and the law (9.14) – the law of conservation of momentum – is not expressly stated there as a law, the brief communication that Huygens sent in 1669 to the Royal Society and the *Journal des Sçavans*[12] included very clear statements of three of the most fundamental laws of nature: the conservation of momentum, the conservation of (kinetic) energy in collisions, and the centre-of-gravity (centre-of-mass) law. In stating the first of these, Huygens, though he does not mention Descartes by name, explicitly states that his (Descartes') fundamental law, that the *quantity of motion* is always conserved, is not correct. We recall that this law states that

$$m_1|u_1| + m_2|u_2| = m_1|v_1| + m_2|v_2|. \tag{9.14'}$$

According to Huygens' rule 5 communicated to the *Journal des Sçavans*,

> The quantity of motion that two bodies possess may either increase or decrease as a result of their collision; but the quantity towards the same side, the quantity of motion in the contrary direction having been subtracted, always remains the same.

This is, of course, the momentum conservation law (9.14). The energy conservation law is formulated as rule 6:

> The sum of the products of the size of each hard body multiplied by the square of its velocity is always the same before and after their collision.

Huygens concludes by saying that he has noted an 'admirable law of Nature', according to which the common centre of gravity of two or three or any number of bodies always advances uniformly in the same direction in a straight line before and after a collision.

In terms of awarding insights that belong to the 'baker's dozen', both the principle of relativity itself and the momentum conservation law (which was rediscovered by Newton) were vitally important, and for these Huygens scores $1\frac{1}{2}$ points. In the longer term, the energy conservation theorem was perhaps even more important but Huygens himself never made much out of it and it played no significant part in Newton's work. It was made into something of great import by Leibniz;[26] this is something we shall consider in Vol. 2.

9.5. Collisions in the centre-of-mass frame

For a variety of reasons, it will be worth our while considering collisions in what is called the *centre-of-mass frame*. According to the energy conservation law, we have in a general frame (i.e., for arbitrary speed of Huygens' barge)

$$m_1u_1^2 + m_2u_2^2 = m_1v_1^2 + m_2v_2^2. \tag{9.15}$$

But we can always choose the frame of reference in such a way that, for body 1, say, the magnitude of its pre-collision velocity is equal to the magnitude of its post-collision velocity, i.e., $|u_1| = |v_1|$. But then Eq. (9.15) implies that the same is true for the second body, i.e., $|u_2| = |v_2|$.

Then the momentum conservation law reduces to

$$m_1u_1 + m_2u_2 = -m_1u_1 - m_2u_2,$$

from which it follows that

$$u_1/u_2 = -m_2/m_1, \tag{9.16}$$

i.e., the magnitudes of the velocities of the bodies are in the inverse ratio of their masses. Now suppose we place the origin at the centre of mass of the two bodies, and let the bodies be at x_1 and x_2. Then by the definition of the centre of mass $x_1m_1 = -x_2m_2$ and

$$x_1/x_2 = -m_2/m_1. \tag{9.17}$$

Comparison of Eq. (9.17) with (9.16) shows that the two bodies approach their common centre of mass with velocities of magnitude in inverse proportion to their masses, collide at it, and rebound with their velocities reversed but undiminished.

We note an interesting fact; if collisions are studied in the centre-of-mass frame, Descartes' fundamental law of motion, according to which his *quantity of motion* is conserved, is correct, i.e., Eq. (9.14′) holds.

It seems that this result, which is not true in a general frame, was an important factor in Huygens' discovery of the correct laws of impact, as is pointed out by Westfall.[27] For in the earliest unpublished papers of Huygens we find that he started by considering the collision case covered by Descartes' rule 1 and what in *De Motu Corporum* he called his Assumption 2, according to which two equal bodies that are perfectly hard and move towards each other with equal speed in opposite directions will after colliding spring away from each other with no loss of speed.[28] Then comes a very interesting comment. He says that[29] 'the force of collision is the same whether A is at rest and B moves with a given speed, or A moves to the right and B to the left with half that speed.' Although not identical, this passage immediately brings to mind §II.29 of Descartes' *Principles*, in which it was stated that 'we cannot conceive of the body AB being transported from the vicinity of the body CD without also understanding that the body CD is transported from the vicinity of the body AB, and that exactly the same *force and action* [my italics] is required for the one transference as for the other.' It seems that consideration of the relativity implicit in this statement and similar statements of Descartes led Huygens to realize that a complete family of collisions can be generated in the case of equal bodies from his first assumption (Assumption 2, i.e., Descartes' rule 1) by the Galileo relativity principle. That was the first breakthrough. But how was he to treat collisions of unequal bodies? In notes made at the end of his life on the problem of absolute and relative motion Huygens mentions §II.29 of Descartes' *Principles* and says[30] that it is basically correct 'except where he says that the same force and action are required to transport AB from the vicinity of CD as to transport CD from the vicinity of AB, which is true if AB is equal to CD but not otherwise'. Somehow or other, possibly through experiment, Huygens found quite soon the key to the problem of generalizing his Assumption 2 in *De Motu Corporum*, for in Axiom 3 of a further paper of 1652 he says:[31] 'If a larger body A strikes a smaller body B, but the velocity of B is to the velocity of A reciprocally as the magnitude A to B, then each will rebound with the same speed with which it came.' This is immediately followed by the triumphant comment: 'If this is granted, everything can be demonstrated. Descartes is forced to grant it however.' Huygens is quite right; for, in conjunction with the relativity principle, this result, which in *De Motu Corporum* (in the proof of his Proposition 7) he says is in excellent agreement with experiments, completely solves the problem of elastic collisions. However, for some reason he appears to have been loath to take this empirical but not intuitively obvious fact as one of the cornerstones of his theory of impacts and commented already at the time that 'it must be seen whether it can be demonstrated from principles that are better known'. This appears to be the origin of the somewhat unsatisfactory (and yet most fruitful – because it led to the energy conservation law) proliferation of hypotheses employed in the later *De Motu Corporum*, in which only the relativity principle and the law of inertia are advanced as truly fundamental principles. The desire for 'principles that are better known' seems to reflect a Cartesian hankering for *a priori* certitude, which in turn is reflected in a certain obscurity, already noted, in key formulations of principles: one cannot be sure whether Huygens is putting them forward as empirical facts, hypotheses, or *a priori* truths. This applies particularly to his formulation of the relativity principle.

An important point to note, frequently stressed by Huygens, is that the relative velocity with which bodies approach each other in elastic collisions is equal to the relative velocity with which they spring apart. It is also worth mentioning another point which Leibniz,[32] who learnt about collisions from Huygens, emphasizes strongly, namely, that the *effect* of a collision, by which he means the amount elastic bodies contract when they collide, depends only on the relative velocity of the collision – another manifestation of the Galileo–Huygens relativity principle. All of these results tend to strengthen the notion that a collision between two bodies is really something that can be regarded as quite detached from the world. In this view, the 'collision-in-itself' is the collision as it unfolds in the centre-of-mass system. We can then imagine the 'collision-in-itself' being 'towed' past us (on one of Huygens' Dutch barges) at any uniform speed we like to choose. In such a view, the overall speed of the 'towing' is a thing purely in the mind of the observer and we get quite close to Cartesian relativism (though the 'thing' which is 'towed past' us is very much a fact of experience). As noted above, when Huygens speaks of the relativity of motion it is difficult to say what precisely he does mean (and we shall see in Chap. 12 that he was on one point at least seriously confused); the account of collisions just given is perhaps the closest we can get to what he thought. It certainly goes in the direction of a further comment that Descartes made in the crucial (but enigmatic) §II.29 of his *Principles*, which is as follows: 'Thus, if we wish to attribute to movement a nature which is absolutely its own, without referring it to any other thing; then when two immediately contiguous bodies are transported, one in one direction and the other in another, and are thereby separated from each other; we should say that there is as much movement in the one as in the other.' The key to Huygensian relativity seems to be that the latter part of this statement is corrected and made precise by the centre-of-mass formulation for collisions of unequal bodies but the idea of detachment from the rest of the universe is accepted.

Deferring now until Chap. 12 further discussion of what Huygens understood by relativity, it is worth pointing out here the significant development which Huygens' mechanics represents as compared with Galileo's motionics. As we saw, Galileo had, as it were, an atomic theory of *motions*, according to which each general motion is decomposed into 'atomic' elements. In contrast, the 'atoms' in Huygens' mechanics are *collisions*: the world consists of nothing but particles of matter which travel on straight lines in uniform motion except when involved in collisions. All the nontrivial dynamics is carried in the collisions, which, in accordance with the above discussion, are to be regarded as autonomous units. We have already pointed out how seriously Huygens took the mechanistic programme, even going so far as proper calculations to show how microscopic vortical particles, colliding with bodies, can give rise to the phenomena of gravity on the earth along the lines outlined by Descartes (see, for example, Ref. 5). The importance of the mechanistic philosophy as a stimulus to his pioneering work on collisions can be seen in a preface that he intended for *De Motu Corporum*:[33] 'For if the whole of nature consists of certain particles from the rapid motion of which all the diversity of things arises, and by the extremely rapid impulse of which light is propagated and spreads through the immense spaces of the heavens in a moment of time, as many philosophers deem probable, this examination [of nature] will seem to be helped no small amount if the true laws by which motion is transferred from body to body be made known.'

Finally, it should be emphasized that Huygens assumed perfect hardness, i.e., elasticity of collisions, in his *De Motu Corporum* and for the atoms of his mechanistic philosophy. This is why the bodies spring back with undiminished speed and kinetic energy is conserved. We shall consider the more general case of only partly elastic collisions in the next chapter.

9.6. The enigma of relativity

Huygens' work brings us to a point at which we can for the first time formulate clearly the problem that this study addresses. All the elements of the enigma are present in at least embryonic form. There is first of all one fact which comes before all theorizing and detailed observation: any observation of the real world is relative. You cannot put one end of a ruler opposite a mark that you can see, the other opposite a mark that you cannot see, and make an observation that tells you anything about the world. Nothing can ever get around this fact of existence. Relativity (in the wide sense of the word) is a fact of life. But alongside this extremely general fact are two *observational facts* of a very specific nature. The first is the fact of inertial motion. As we have seen, Huygens seems to have accepted it without question. After the publication of Descartes' *Principles* in 1644 it seems to have been accepted quite widely. For example, the English philosopher Henry More, whose ideas on space are supposed to have influenced Newton and who will be considered in Chap. 11, referred to it in 1652 as[34] 'that prime *Mechanicall* law of motion persisting in a straight line'. The difficulty was in saying in precisely what it consisted. It all seemed very much more reassuring in Galileo's writings, in which the phenomenon was described within the definite framework of a recognizably Aristotelian cosmos, albeit one that was being stretched and readjusted to its very limit. In fact, if Galileo had attempted to show systematically how not only the diurnal rotation of the earth but also its motion around the sun can escape detection, he would probably have been forced to admit that the tottering structure was past saving. In the end, the job of demolition was left to Descartes. Huygens then had no alternative but to say farewell to the Galilean galley sailing over the blue seas of the earth under the all embracing canopy of a well-ordered cosmos and push out the life-rafts into the universal ocean, in which they are undoubtedly carried by a cosmic drift as powerful as any Gulf Stream but with two particularly odd features: (1) there is nothing immediately apparent which shows whence the stream comes and what determines its course; (2) the stream flows simultaneously in all directions and at all velocities, like a multiple infinity of airport conveyor belts carrying passengers from every possible point in every possible direction at every possible speed – but always at a strictly uniform speed.

The attentive reader may have wondered why I showed the seemingly superfluous Fig. 9.2 a few pages earlier. The reason is that there is something very symbolic about those disembodied hands and the lifeless balls dangling from them. They illustrate in graphic form two conflicting processes that make up the whole enigma we are trying to comprehend. Ideally, for greatest effect, Fig. 9.1 should be preceded by a picture depicting the way in which Galileo had his first intimations of the law of inertia. Cast our minds back barely more than 60 years to Galileo's Pisan tract, *De Motu*.[35] At that time, he clearly conceived motion as being determined in all its aspects by the cosmos, the one unique frame of reference that gives meaning to motion, the structure that defines the loci of the teleological goals. Thus, as 'Fig. 9.0', the reader is asked to picture a ball rolling round the earth, whose centre coincides with the centre of the Aristotelian cosmos. The complete universe is shown. Without it the picture loses all its meaning. The celestial spheres and the centre of the universe are every bit as important for understanding the motion of that ball as is the geography of Bunyan's world for the pilgrim's progress.

Figure 9.1 shows how the concept of motion has changed in two generations. Unable to fathom the cosmic drift, Huygens has been forced to vignette away not only the universal framework but also the earth. But even Fig. 9.1 shows too much. The mysterious conveyor belts are after all completely invisible. Lange[36] commented that, conceptually speaking, the main effect of the discovery of dynamics was that the concept of motion became *completely disengaged from the contingent world*.

It was surely only with an eye to economizing on printing costs that Huygens dispensed with everything but the hands in Fig. 9.2. Yet what could illustrate more perfectly the termination of the process described by Lange? As formulated by Huygens, the law of inertia, the first principle of motion, stands naked before us with no visible support or relation to anything contingent, seemingly in complete contradiction to the all embracing relativity that we noted at the beginning of this discussion.

But now we cast our eyes down to the colliding balls. Here, a completely different tendency is at work, indeed, a diametrically opposed one. For, as we have seen, one of the most important results of *De Motu Corporum* is the *law of relative velocities*, according to which, in any elastic collision between two bodies of any mass, the relative velocity of the bodies before impact is equal in magnitude to the relative velocity after impact.

In the law of conservation of the magnitude of the relative velocities we have a fact in perfect accord with the idea that only relative quantities should determine motion. Huygens' result is embryonic but characteristic in the sense that in the more general case when bodies interact through finite-range forces (as opposed to pure contact forces as in the collisions we have been considering) the outcome, i.e., the change in the velocities brought about by the interaction, is always completely determined by purely relative quantities (the relative speeds and the successive relative distances). Thus, the mutual accelerations which bodies impart to each other are determined quite explicitly, literally before our eyes, by relative quantities we can grasp with the greatest ease. But what governs the motion before and after the impact eludes us.

These mysterious results came to light because Huygens, like Kepler and Galileo before him, succeeded in decomposing observed motions into components. Perfectly clean decomposition seems to be the hallmark of some of the greatest of the contributions to dynamics: Kepler's finding of the point at which to peel the earth's motion from the Martian motion, Galileo's analysis of projectile motion into the inertial and gravitational components, and now, in Huygens, the clean separation of the effect of interaction of two bodies from what, for want of a better expression, I have called their cosmic drift.

Huygens' achievement leads us straight to the enigma that gave rise to the present book: we are in the curious position of being able to say unambiguously what causes the cosmic drift to be changed but not what governs the drift itself. The two parts of the Galileo–Huygens principle of relativity seem to be saying almost contradictory things: that there is a manifest and palpable cause of the one but not the other.

When Huygens died, about 40 years after completing *De Motu Corporum ex Percussione*, he was still impaled on the horns of this dilemma: *is motion absolute or relative*? The work to be described in the next two sections helped to make the dilemma especially acute.

9.7. Centrifugal force: the work done prior to Huygens

Centrifugal force, or the tendency of bodies swung in a circle to recede from the centre, played an extremely important role in the arguments about the rotation of the earth, in the discovery of the law of inertia, and in the discovery of the law of universal gravitation; it played a key role in the emergence of the precise concept of *force* and, thus, of fully-fledged dynamics; and it continues to play a central role in the discussion about the absolute or relative nature of motion. The expression *centrifugal force* (*vis centrifuga*) itself is due to Huygens, who was the first person to derive the correct expression for its magnitude.

As with so many topics in the discovery of dynamics, the discussion of centrifugal force begins with Copernicus and *De Revolutionibus*. It will be recalled that Copernicus had been misled by a poor translation of the *Almagest*, and he incorrectly reported Ptolemy as having said that things which undergo abrupt rotation seem likely 'to fly apart unless they are held together by some bond'. This was therefore perceived by Copernicus as an argument against the rotation of the earth. His first counter to this argument was to suggest that if the earth rotates then such motion will certainly be *natural*. As we have seen, this concept of a natural circular motion of the earth and its parts was central to the emergence of Galileo's concept of inertia.

Copernicus, however, was able to produce a further argument to counter the supposed Ptolemaic argument. If Ptolemy was concerned for the earth, how much more he should fear for the heavens:[37]

> But why does he not feel this apprehension even more for the universe, whose motion must be the swifter, the bigger the heavens are than the earth? Or have the heavens become immense because the indescribable violence of their motion drives them away from the center? Would they also fall apart if they came to a halt? Were this reasoning sound, surely the size of the heavens would likewise grow to infinity. For the higher they are driven by the power of their motion, the faster that motion will be, since the circumference of which it must make the circuit in the period of twenty-four hours is constantly expanding; and, in turn, as the velocity of the motion mounts, the vastness of the heavens is enlarged. In this way the speed will increase the size, and the size the speed, to infinity.

Quaint as this vision of a mad, runaway dash to infinity may appear, it does succeed in capturing some of the true nature of centrifugal force. It is worth quoting a further comment of Copernicus:[38]

> But beyond the heavens there is said to be no body, no space, no void, absolutely nothing, so that there is nowhere the heavens can go. In that case it is really astonishing if something can be held in check by nothing.

The next major development in this question comes in Day 2 of Galileo's *Dialogo*,[39] in which the question is raised of whether the earth's rotation might not cause bodies on its surface to be flung off into space. This passage in the *Dialogo* is remarkable in two respects: as a glaring logical inconsistency on Galileo's part but also as one of the nearest misses in the history of science. The passage in question comes immediately after the lengthy section in which Galileo shows why the stone dropped from a high tower is observed to fall vertically downward and is not left behind by the earth's rotation. He had, of course, refuted such a suggestion by his Copernican concept of the 'ineradicable' tendency of all parts of the earth to circle with it, to partake 'naturally' of the earth's rotation.[40]

On the basis of this argument, Galileo could just as easily have argued that all objects lying on the surface of the earth participate equally in the diurnal revolution and that therefore the problem does not arise – it would be quite contrary to their inborn nature to fly off the earth. However, Galileo here appears to have felt there to be great strength in the argument of the stone leaving the sling. Moreover, in view of his commitment to circular inertia, it is remarkable that at this point he clarifies the phenomenon of the sling and makes it quite clear that when the stone leaves the sling it departs along the tangent to the circle in which it is being swung, commencing on this motion at the point of release and continuing with a uniform motion along the straight line defined by the tangent.* This was already an important clarification compared with Copernicus and the somewhat confused idea that such bodies move radially outwards while simultaneously continuing in the circular motion.

One reads the relevant passage in the *Dialogo* with bated breath, for Galileo seems to be on the point of discovering how satellites could circle the earth. On the one hand, he grants the tendency of bodies on the earth to fly off along the tangent; on the other, he relies on the tendency of the weight of the bodies to draw them back down to earth. Although for only a single initial velocity and direction, the satellite problem is here correctly formulated. The excitement and interest of the passage comes from the fact that Galileo had long been in possession of all the elements needed for its solution. He appeals to a figure (given in simplified form in Fig. 9.4) in which the circle represents the locus (i.e., path in space) of a fixed point on the surface of the earth, while the horizontal tangent AB represents the path that would be taken by a body subject to no gravity if released at A with horizontal velocity equal to the surface rotation velocity of the earth. In a certain time, the point A will move to D on the circumference of the circle, while a body at A in the absence of gravity would have travelled along the horizontal line AB a distance equal to the arc AD, taking it to a point E just short of the intersection of OD with AB. In the limit when the arc AD is very short, ED measures the distance that accelerated motion in free fall must be able to carry the body downward in the same time if it is not to rise up from the earth, as argued by the anti-Copernicans. Galileo now invokes his as yet unpublished results from the *Discorsi* and asserts, without proof but quite correctly, that the speed with which a body falls freely increases linearly with the time. It is perhaps unfortunate that at this point Galileo uses a single figure to represent two quite different

* In the chapter on Galileo, we noted that the simultaneous existence of two basically different 'laws of inertia' in his writings probably reflects the distinction, not yet overcome in Galileo's mind, between natural and violent motions.

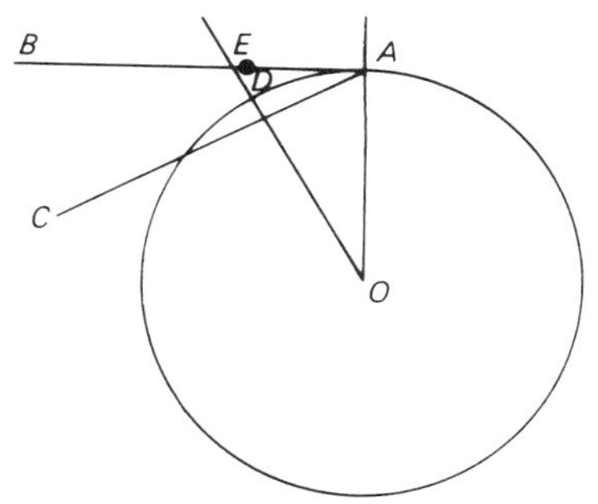

Fig. 9.4. Simplified version of Galileo's diagram explaining (incorrectly) why gravity should always overcome the centrifugal tendency of bodies resting on the surface of the earth however rapid the diurnal rotation.

things. In modern terminology, he takes AB as axis to represent not only the horizontal distance but also the *time* elapsed from the instant of release of the body at A. This time is also measured from A, increasing towards B. Simultaneously, he takes AO to be an axis measuring the *vertical speed* acquired in free fall. He takes the straight line AC to represent the speed acquired in free fall with some particular acceleration as a function of the time. This is very confusing, because the line AC appears to represent a *path* in space but in fact represents a *speed* as a function of time. Now comes the egregious mistake. Galileo argues that since the straight line AC is always inside the circular locus of the point on the surface of the earth (this is true whatever the magnitude of the acceleration), the tendency to fall will always overcome the tendency to fly off. The mistake is, of course, that it is not the *velocity* acquired in given time that is important, but the *distance* d that is travelled in the given time, and that is not a linear but a quadratic function of the time, $d = \frac{1}{2}at^2$, as in fact Galileo well knew. That is what he is famous for. Galileo seems to have been confused by the fact that he had used a single figure to represent distance, time, and velocity simultaneously. Thus, he somehow imagined that, merely because the line AC is superimposed on the circle representing the earth, it represents position in space. He was no doubt also anxious not to give away as yet too many of his results on the law of free fall. Whatever the reason, he inadvertently got himself into a Zeno type paradox. Ironically, it is precisely after this false demonstration that Galileo praised the power of geometry, making his famous comment that 'trying to deal with physical problems without geometry is attempting the impossible'.

Perhaps one should say that, without the calculus – many ideas of which are implicit in Galileo's argument (which I have given in a very simplified form) – it was extremely understandable that Galileo got in a tangle. But it is fascinating to consider what might have been had Galileo spotted his mistake. As we shall see, it was left to Huygens and Newton to rectify the mistake.

Incidentally, this passage in the *Dialogo* contains such a clear formulation of not only the correct law of inertia but also the problem of centrifugal force that it might well have been sufficient on its own, without the intervention of Descartes, to lead Huygens and Newton (who both studied it carefully) to the central insights from which dynamics was synthesized. Without the work of Kepler the *Principia* is utterly inconceivable; without Galileo's barely conceivable. But although Descartes clearly did influence developments quite strongly, Huygens and Newton might have found sufficient stimulus to their work from other sources. If there is a point at which Descartes unambiguously exerted a positive influence that could not have come from any other source, then it was in the stimulus he gave to the study of collisions. In other cases (except the absolute/relative question, which I am not considering here), his intervention tended to highlight problems and ideas already present in one form or another rather than introduce them.

A typical example of this is the case of centrifugal force. The credit for having put the problem of its accurate mathematical description at the centre of attention in the study of motion is undoubtedly Descartes'. We have seen that in his *The World*, exactly contemporary with Galileo's *Dialogo* as regards its writing, Descartes made the breakthrough to very nearly the modern conception of inertia and its relation to the phenomenon of centrifugal force. (By modern I mean post-Newton but pre-

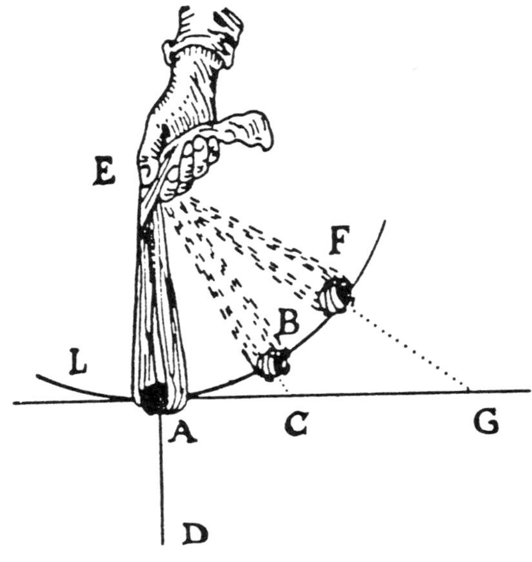

Fig. 9.5.

Einstein.) Moreover, we have also seen that precisely this phenomenon was chosen by Descartes to play the central explicatory role in the framework of his physical conceptions.

Thus, the concept of centrifugal force, or endeavour from the centre (*conatus a centro*), as Descartes called it, occupies a prominent position in *The World* and the *Principles*. Several sections of Part III of the latter[41] (§§56 to 59) are devoted to a formal discussion of the phenomenon, in which, for example, Descartes considers the force with which a stone swung in a sling (Fig. 9.5) will strive to recede from the centre E. The main interest of this discussion is that Descartes poses the problem as one of prime importance but fails to treat it adequately, despite the fact that the heading to §59 even announces that he will determine 'how great the force of this striving is'. However, when he actually comes to the hurdle, the horse refuses, and the discussion ends with a purely qualitative statement: 'since what makes the rope taut is nothing other than the force by which the stone strives to recede from the centre of its movement, we can judge the quantity of this force by the tension.'

9.8. Huygens' treatment of centrifugal force

It is from this qualitative treatment of Descartes, rather than Galileo's quantitative but incorrect treatment, that Huygens took his point of departure. His treatment of centrifugal force illustrates, perhaps more clearly than anything else in the literature, the transition from embryonic Cartesian dynamics to fully fledged dynamics. It shows us what Descartes lacked and Galilean motionics was able to provide – a well-defined concept of force, though not, it is true, one that Galileo used. In fact, both Galileo and Descartes frequently use the word force but always with a vague undefined meaning. Huygens saw that a precise definition of force could not be obtained unless Cartesian dynamics was complemented with an awareness of something that it completely lacked: acceleration. He realized that through acceleration it was possible to form a precise concept of force.

Huygens solved the problem of centrifugal force in the year 1659 as a by-product of his attempt to find a mechanical explanation of gravity, in which he followed Descartes' suggestion that terrestrial gravity is to be explained mechanically by a kind of centrifuge effect of the cosmic vortex. His results were written up in the form of a manuscript entitled *De Vi Centrifuga*[11] (*On Centrifugal Force*), which, however, was only published posthumously, together with *De Motu Corporum ex Percussione*, in 1703. However, when Huygens published his *Horologium Oscillatorum* in 1673 he added at the end, without proofs, a number of the main theorems on centrifugal force. Their publication was an important development in the history of dynamics, since they made it possible to attack the problem of the planetary motions quantitatively, and this led rapidly to a precise clarification of the essence of the problem, as we shall see in Chap. 10.

The stimulus to Huygens' work in this field is apparent in the opening sentence of *De Vi Centrifuga*: 'Weight is a striving to downward motion.' Huygens' concern with finding a mechanical explanation for gravity had, from his point of view, both a negative and a positive effect. The negative one was that he tried to perfect the Cartesian theory of vortices as the explanation of the planetary motions and terrestrial gravity and therefore did not attempt to see if gravity could explain not only the falling of apples but also the motion of the moon and planets. He persisted in the vain search for an explanation of gravity, whereas Newton found an explanation for the structure of the world in terms of gravity. The positive effect was that he was led to look closely at Galileo's work on free fall. After all, he wanted to explain Galileo's results and, unlike Descartes, he was not content with purely qualitative arguments. He wanted to get exact quantitative results. This must have undoubtedly impressed upon him that the essence of terrestrial gravity is uniform *acceleration* and from this came his great breakthrough: that the dynamically significant quantity is acceleration and that the only serviceable definition of force is by means of acceleration. It is slightly ironic that Huygens came to this insight, since he was more at home in the Cartesian conceptual world of uniform motion interrupted by instantaneous changes of velocity in collisions; changes of speed brought about by gradual acceleration do not belong to the basic concepts of Cartesian or Huygensian physics. This is another demonstration of the strength of Galilean motionics; like Keplerian celestial motionics, once the true *facts* had been discovered, they of necessity imposed the introduction of appropriate concepts despite a climate of philosophical opinion that was unfavourable.

Thus, like *De Motu Corporum ex Percussione*, Huygens' *De Vi Centrifuga* draws heavily on Galileo's work.

Huygens first of all reviews Galileo's theory of free fall and descent down inclined planes, putting particular emphasis on the fact that uniform acceleration is the essence of free fall. He then turns to consider the 'tug' that a weight exerts when suspended vertically on a string (Fig. 9.6) or else is prevented from rolling down an inclined plane (Fig. 9.7). He says that we are to regard this tug as giving an indication of what the body is trying to do; the 'tug' is a measure of what the body would do if no longer constrained by the string. He recalls that according to Galileo a body, if released, begins to descend in accordance with the odd-numbers rule, passing through the successive distances 1, 3, 5, 7, . . . in successive units of time. Thus, the total distance traversed increases as the square of the elapsed time: 1, 4, 9, 16, . . . However, he points out that it would be incorrect to attempt to measure the tug by the distance that the released body falls in a given time, say, one second. He illustrates his point by Fig. 9.8, in which the suspended ball touches a smooth curved wall at C, at

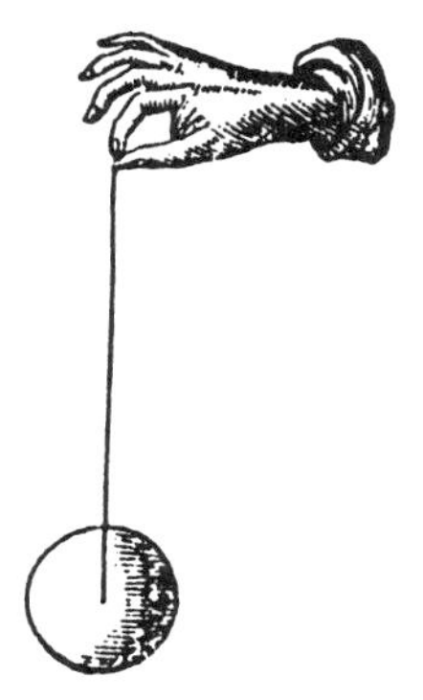

Fig. 9.6.

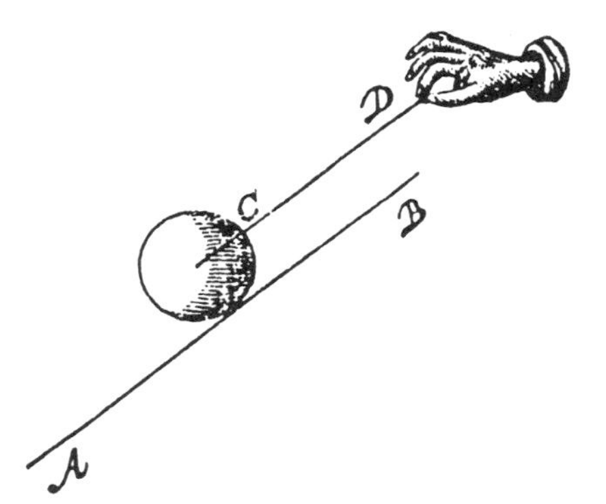

Fig. 9.7.

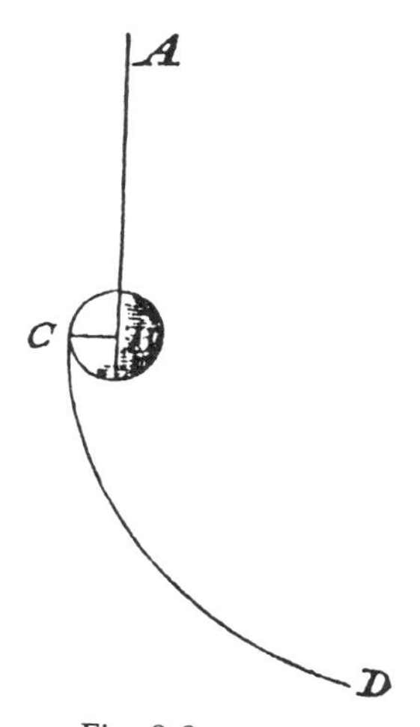

Fig. 9.8.

which point the wall is vertical. In this case, as we know from direct experience, the tug is just the same as if the wall were perfectly plumb and did not curve below the point C. But if the ball in Fig. 9.8 is released, *it is only at the initial instant* that its motion is identical to that of vertical free fall; after the initial release the ball does not, because of the constraint imposed by the curved wall, fall so rapidly.

Huygens thus arrives at his most important conclusion: that if we want to find a feature of the motion actually realized when the body is released that corresponds *exactly* to the tug it exerts before the moment of release, then the only feature which serves our purpose is the *instantaneous acceleration at the moment of release along the direction of the tug*. Huygens' actual words are: 'It is therefore clear that when we wish to determine the force we must consider, not what happens during a length of time after the body has been released, but rather what happens in an arbitrarily small amount of time at the beginning of the motion.'

The concept of *force* had been born.

Force is something which causes the body on which it acts to be accelerated in a given direction. The strength of the force can be measured in two ways: by the 'tug' it exerts if the body is constrained or by the instantaneous acceleration which results on release.

Huygens immediately generalizes the concept beyond the immediate phenomenon from which it was deduced, thus converting another of his small carefully measured steps into a giant leap. He asserts that whenever we experience a tug, no matter what its origin, there is a constraint imposed on the body that tugs and that release of the constraint will be immediately followed by *acceleration* of the body in the direction of the tug. The acceleration will be directly proportional to the tug.

Although Huygens did not write down such a formula, we can express his conclusion by the equation

$$a = cT,$$

where a is the acceleration, c is a constant of proportionality, and T is the tug.

After these preparations, Huygens says: 'Let us now see how great is the striving that bodies swung round on a string or attached to a wheel have to recede from the centre of the circle.' This is illustrated in Fig. 9.9, in which the circle BG with centre A represents a wheel which rotates in the horizontal plane. Huygens supposes that on the wheel is fixed a small ball. When it reaches the point B, it is released. Following Descartes closely (though without quoting him), Huygens asserts that the ball, when released, will move away along the tangent to the circle at B, moving along the straight line BH with uniform speed. Here again we see how he accepts the law of inertia without reservation or comment. Simultaneously he thereby tacitly assumes a frame of reference in which

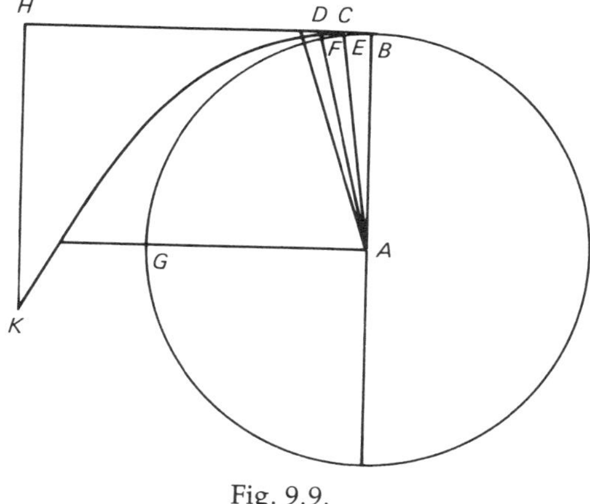

Fig. 9.9.

it holds and in which the basic calculations are made. These are initially purely kinematic. The wheel is imagined to be very large, so that a man (attached firmly) may sit on the wheel as it rotates. It is supposed that the man restrains the ball from flying away before it reaches the point B by a short string held in his hand, the other end of which is attached to the ball. Huygens comments that the tension that the man feels in the string is clearly exactly the same as if the string were attached to the axle A of the wheel. Later, Huygens demonstrates the identical effect of the centrifugal tug and the effect of gravity by supposing that the tug in the string needed to prevent the ball flying away tangentially is supplied by letting the other end of the string hang down vertically at A with a suitable weight at its end. This is a beautiful illustration of Einstein's famous equivalence principle – in their observable effects, centrifugal force and gravity are locally indistinguishable. This is a point which Huygens is very keen to make, with similar but not quite identical intention to Einstein two and a half centuries later.

The problem that Huygens sets himself is to describe the motion of the ball, after it has been let loose at B, from the point of view of the man sitting on the wheel. Thus, he goes over in effect to a coordinate system rotating with the wheel. Let us first of all consider the kinematics of the situation in the nonrotating frame. This is shown in enlarged scale in Fig. 9.10. In a short time interval, the man is carried to point E, while the ball, travelling inertially along the tangent is carried to K, the distance BK being equal to arc BE. In the next time interval of equal length, the man is carried to F, the ball to L, and so forth. Now comes the decisive point. Huygens asks: what is the ratio of the lengths of the arcs EK, FL, MN? These are the successive arcs along which the body appears to travel as seen from the man. The answer is that if the time intervals are taken

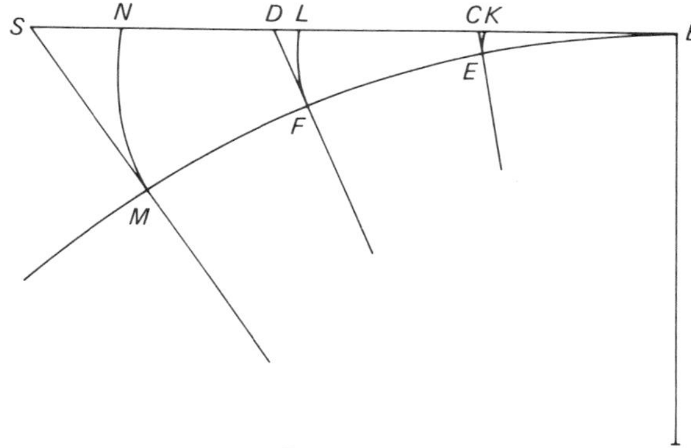

Fig. 9.10.

sufficiently short, the lengths of the arcs EK, FL, MN increase as 1, 4, 9, 16 . . ., i.e., exactly as in Galileo's law of free fall. In fact, as far as the man on the wheel is concerned, the motion that the ball executes is kinematically the same as the initial motion of the ball when released in the situation shown in Fig. 9.8. For as seen from the man's point of view the resultant curve followed by the ball is as shown in Fig. 9.11. This curve, BRS, is found as follows. Suppose the wheel has turned quarter of a circle. Then if r is the radius of the wheel, the ball will have travelled in Figs. 9.9 and 9.10 the distance $\frac{1}{2}\pi r$ along BH. But in the rotating frame of the man (Fig. 9.11) the ball will be at R, where the distance NR is $\frac{1}{2}\pi r$. The curve BRS is in fact found by 'unwinding' a string wrapped round a fixed circle. The curve BRS can be described simply by attaching a pencil to the end of the string at B (which can pass once round the circle, for example, and have its other end attached at B) and pulling it away from the circle, keeping it taut the whole time as the curve BRS is described. The important thing about this curve from Huygens' point of view – and he takes some trouble to prove it – is that at B it touches the radius AB, i.e., AB is the tangent to BRS at B. Thus, as far as the motion *in the immediate vicinity* of B is concerned, the situation is identical to the one shown in Fig. 9.8. It is just as if there were a force of gravity directed along AB. Moreover, Huygens can readily establish how fast the acceleration actually takes place – this is a purely kinematic calculation given the radius and rate of rotation of the wheel – so that the magnitude of the centrifugal force can be immediately calculated.

In view of the great importance of the equivalence principle in Vol. 2, when we come to consider how Einstein attempted to understand the force of inertia, it is worth quoting in full the following passage from Huygens, which shows just how close he did come to formulating the fundamental principle of general relativity. We should not, however, be

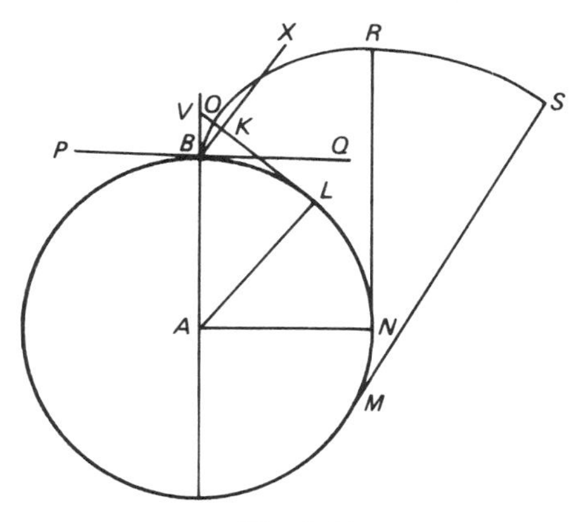

Fig. 9.11.

misled into attributing too much to Huygens. Three geniuses addressed themselves to centrifugal force: Huygens, Newton, and Einstein. Huygens because he was looking for an explanation of gravity in terms of inertia, Einstein because he was looking for an explanation of inertia in terms of gravity (so far as I know, unaware of what Huygens had done 250 years earlier), and Newton because he realized centrifugal force was an effect that would enable him, through Kepler's Third Law, to determine the distance dependence of the strength of the force that keeps the planets circling the sun. The biblical story of David slaying Goliath with a sling is a very apt metaphor for the part that the sling played in the history of dynamics. This simple effect, potent in the hands of the deft who know the point at which to strike, brought down much more than just the hero of the Philistines. But let us have Huygens' words; although the least of the three latterday Davids, he was the first:[11]

Thus, we must regard the intervals EK, FL, MN as if they grew from unity as squares: 1, 4, 9, 16, etc. And thus the striving on the rotating wheel of the attached ball is just the same as if it wished to move away along the continuation of the straight line that joins it to the centre, and indeed in an accelerated motion in which in equal intervals of time it traverses distances that grow successively as the numbers 1, 3, 5, 7, etc. It is sufficient that this progression holds at the beginning; for later the ball may travel in accordance with other proportions or laws of motion. That has no bearing on the striving that is present before the motion commences. But this striving is very similar to that which weights suspended on a string have to falling. From this we shall also conclude that the centrifugal forces of unequal bodies which are however rotated around equal circles with equal speeds bear to each other the same proportion as their weights, or their bulks. For as all heavy bodies strive to fall with the same speed and acceleration, and as further this striving has a greater power the larger the bodies are, the same effect must hold for bodies fleeing from a centre, whose striving is, as we have shown, entirely analogous to the striving that is caused by gravity.

The seeming closeness of Huygens to Einstein's equivalence principle is in part an illusion created by the fact that Huygens' work was done before the discovery of the full scheme of dynamics by Newton. The illusion itself casts interesting light on the discovery of dynamics and the important part played by centrifugal force in clarifying the concept of force, without which dynamics could not have appeared. The point is that centrifugal force was the *second phenomenon* discovered in nature in which acceleration according to Galileo's odd-numbers rule was observed. Prior to the discovery of dynamics proper by Newton and the resulting extension of the force concept to vastly more phenomena than just these two effects, the similarity between the two effects of necessity seemed far more striking than after the discovery of dynamics. That the similarity could be noted at all was due to the fact that all bodies are equally subject to the two effects, i.e., free fall under gravity and centrifugal force. This is the equivalence principle proper, but is not what primarily struck Huygens, who saw the main similarity in the fulfilment, *at least initially*, of Galileo's odd-numbers rule. The italicized words emphasize a further important point about centrifugal force – although similar to the law of free fall, it describes a situation that is not identical: the similarity only holds at the initial instant. This began the process of generalization from Galileo's initial discovery to the full generality of Newtonian dynamics, which occurred when Newton had fully grasped the fact that the defining characteristic of physical forces is that they generate an accelerated motion that *at its initial instant* is described by Galileo's odd-numbers rule.

One more comment on this point before we return to Huygens' derivation of the actual formulas for centrifugal force. Einstein is greatly admired for having noted something very simple, well known for at least two centuries, which had escaped the attention of everyone else. It is what we have already mentioned several times – the equivalence principle: locally a gravitational field is indistinguishable from the 'apparent forces' generated by inertia on the passage to an accelerated frame of reference. But we now see that both Huygens and Newton had also been forcibly struck by exactly the same physical phenomenon two decades or more before Newton wrote his *Principia*. We see here again the ability of great physicists to spot *characteristic phenomena* in nature and exploit them to great advantage. In this century Einstein was able to use the equivalence principle to overthrow many of the principles of Newton's dynamics and his law of universal gravitation, replacing them by the general theory of relativity. But two and a half centuries earlier, Huygens and Newton, struck by exactly the same similarity as Einstein, used the very same equivalence principle to discover the concept of force, the universal theory of gravitation, and the general principles of dynamics – all things that were drastically modified by Einstein when, in greatly changed circumstances, he 'rediscovered' the equivalence principle. But now back to Huygens.

Having clarified to his complete satisfaction the conceptual aspect of the problem, Huygens could now proceed to evaluate the magnitude of the centrifugal force as a function of the radius and angular velocity of the circular motion. He had already established that the centrifugal force was proportional to the weight, or bulk, of the rotated body. The rest was pure kinematics.

There is no point in going through all Huygens' theorems, though we shall give the first two, since these already contain all the important results (the proportionality to m, the 'mass', having already been established). His Theorem 1 states:

> If two equal bodies pass around unequal circles in equal times, the ratio of the centrifugal force on the larger circle to that on the smaller is equal to the ratio of the circumferences, or the diameters.

This result follows directly from Fig. 9.12. The distance s travelled in time t in uniform motion with acceleration a is $s = \frac{1}{2}at^2$. Thus, the acceleration, which is the measure of the centrifugal force, is proportional to s. It is evident that AEGC in Fig. 9.12 is similar to ADFB. Therefore, the distances of recession from the centre, EG and DF, bear the same proportion as the diameters or the radii: EG/DF = AC/AB. From this it follows that f, the centrifugal force, is proportional to the radius r:

$$f \propto r. \qquad (9.17)$$

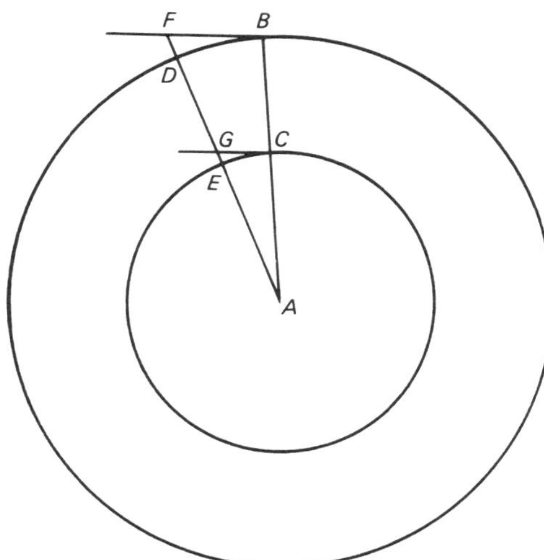

Fig. 9.12.

Theorem 2 states:

> If two equal bodies move around the same circle or wheels with unequal speeds, though uniformly, the ratio of the centrifugal force of the faster to the slower is equal to the ratio of the squares of the velocities.

This is illustrated in Fig. 9.13. The ratio of the speeds, v_1 and v_2, is as BD is to BC. But the distances of recession, FD and EC, are in the proportion of the *squares* of BD and BC (in the limit of very small arcs). It follows immediately that the centrifugal force is proportional to the square of the velocity:

$$f \propto v^2.$$

Thus, with these two theorems, Huygens has all the essential relations for centrifugal force, which are summarized in the well-known equations

$$f = m\omega^2 r = mv^2/r,$$

where $\omega = v/r$ is the angular velocity.

In the two theorems quoted above, Huygens only established the proportionality $f \propto m\omega^2 r = mv^2/r$. He fixed the constant of proportionality by considering a problem clearly stimulated by Galileo's discussion of the capability of centrifugal force to fling bodies from the surface of a rotating earth. He in effect posed this question: how fast must the earth rotate if the resulting centrifugal force (at, say, the equator, where the effect is largest) is to be exactly equal to the force of gravity (for once the centrifugal force becomes greater, bodies will rise up from the surface of

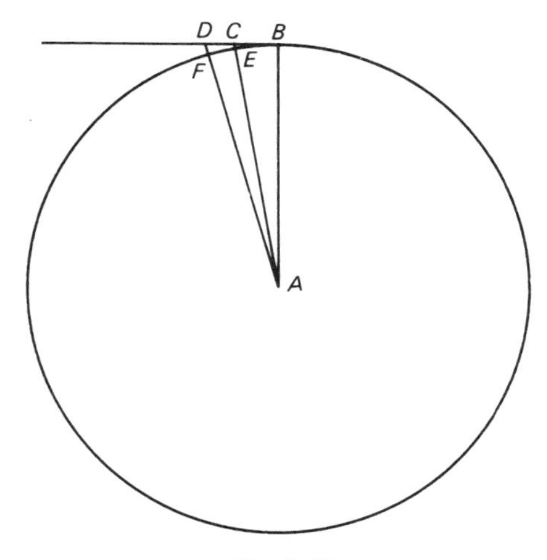

Fig. 9.13.

the earth). He gave the answer in his Theorem 5:

> If a body rotates in a circle with the speed that it would acquire by falling under gravity a distance equal to a quarter of the diameter of the circle, then the gravity of the body is equal to its centrifugal force, i.e., it pulls on the cord that restrains it just as strongly as if it were suspended by it.

By means of this theorem, together with the known strength of gravity and the rotational speed of the earth, Huygens, like Newton a few years later, was able to conclude that the centrifugal tendency of bodies on the surface of the earth is very small compared to the force of gravity.[42] We shall return to this question in the chapter on Newton.

It should be said that Huygens' coining of the expression *centrifugal force* was a little unfortunate and is something of a misnomer that has caused and still causes a great deal of confusion. The present writer is almost certainly not the only person whose physics teacher at school never tired of complaining: 'There is no such thing as centrifugal force!' The problem with Huygens' expression is its seeming to imply that centrifugal force is a force that acts on a body which is moving in a circle, whereas it is in fact the *reaction* exerted by such a body on the body which is constraining it to move in a circle. It will be convenient to defer the detailed discussion of this point to Chap. 10 and here merely make two comments about Huygens' derivation of the expression for centrifugal force.

The first is that it is unambiguously derived from the law of inertia. The tendency to inertial motion in a straight line is the underlying phenomenon from which all springs. There is no suggestion at all in Huygens' exposition that circular motion exists as some kind of natural motion on an equal footing with rectilinear inertial motion. Centrifugal force is something observed when a body is prevented from executing the motion to which it has a natural tendency. Moreover, it is clear that *any* constraint which forces a body to deviate from its natural rectilinear motion will produce an analogous tension in the string or push in the body which is deflecting it. The impression that a body forced to move in a circle is striving to get away from one specific point, namely, the centre of the circle, is a confusing illusion caused by the fact that the curve on which the body is forced to move is a circle rather than a completely general curve. If Huygens' analysis is repeated for motion along such an arbitrary curve, it will be found that the force corresponding to the centrifugal force of circular motion is directed towards the instantaneous centre of curvature at the point in question and its magnitude is governed by the instantaneous radius of curvature. Thus, the appearance of striving from one particular point is created in the case of purely circular motion by the fact that the successive centres of curvature all coincide because of the very particular shape of the constraining curve.

The second comment is to point out a significant extension in his work on centrifugal force of Huygens' technique of examining the same phenomenon in different frames of reference. We have already seen that in his treatment of the relativity principle Huygens pioneered the technique of transformation laws by showing how a single phenomenon could be viewed in different coordinate systems. In *De Motu Corporum*, the transformations were restricted to transitions between frames of reference in uniform motion relative to each other. In *De Vi Centrifuga*, Huygens went a step further and viewed phenomena from frames in a state of uniform rotation relative to the original frame of reference. This revealed an interesting fact that Einstein was to exploit and significantly generalize many years later. Namely, that if one makes a restriction to considering phenomena over only infinitesimal distances and times the basic concepts of dynamics – above all force – are generally covariant, i.e., the rules for defining and using these concepts are the same in any frame of reference.

Let us look at this for the case of force. Of course, if we define the force which acts on a body as its instantaneous acceleration (multiplied by its mass) in a given frame of reference, such a concept is trivially covariant, i.e., the same in all frames of reference. It gains nontrivial content through the physical relationship that holds between 'tug' and the purely kinematically defined acceleration. Namely, suppose we observe a body to have particular instantaneous acceleration in a quite arbitrary frame of reference. Huygens' analysis shows that if in that frame of reference we hold the body fixed and prevent the acceleration occurring then we can only do this by applying a 'tug' proportional to the instantaneous acceleration which the body exhibits when unconstrained. Moreover, the tug must be applied along the line of the instantaneous acceleration. Thus, in Huygens' example, the man sitting on the rotating wheel who constrains the body to move with him in the circle feels in his hand a tug indistinguishable from the tug exerted by a suspended body in a nonrotating frame. An elastic string is extended in just the same manner and a strain gauge will measure the same force in the two cases. The balancing of centrifugal force by a weight suspended at the centre of the circle, the possibility of which is explicitly pointed out by Huygens, is the most graphic illustration of this equivalence. Thus, it is the universal validity of the relationship between the 'tug' and the acceleration – in a rotating or nonrotating frame – that is the nontrivial essence of the covariance.

The use of accelerated frames of reference is another of the ways in which Huygens looks forward to Einstein, leapfrogging Newton, as it were. Incidentally, the concentration on infinitesimals and a purely local treatment is a point that should be especially emphasized. It does much to create the impression that the laws of motion (more generally the laws of physics) are a purely local matter, quite divorced from the universe at large. We have already made this point in the previous sections. It cannot be emphasized too strongly that the appearance of 'disengagement' is an illusion created by the very success of dynamics. As we have said, Huygens was the first person to succeed in doing for terrestrial motions what Kepler had done just over fifty years earlier for celestial motions: he succeeded in separating *absolutely cleanly* two components of motion – the uniform inertial motion, the mysterious cosmic drift common to all bodies, and the local deviations from it. By remaining quietly discreet about the inertial component, he inevitably concentrated attention on the local physics, and this created the impression that it is quite divorced from the universe at large. But this ignores the fact that it is only the cosmic drift that, so to speak, enables us to build the raft on which the local physics unfolds (this statement will be elaborated on in Chap. 12). Just as Kepler succeeded in bringing the planetary motions around the sun into absolutely sharp focus by learning how to subtract the earth's motion cleanly, so too did Huygens and Newton, building on the hints they inherited from Galileo and Descartes, learn how to identify those components of the overall motion of bodies that have *an identifiable local cause*. The *rationality* of motion only appeared after this clean separation of the two components had been achieved.

9.9. Why Huygens failed to win the greatest prize

We conclude this chapter by considering briefly how it was that Huygens created pretty well all the concepts needed for a universal dynamics but yet did not take the decisive step (which was left to Newton). Leaving aside fortuitous events such as the occurrence of suitable stimuli (crucial in the case of Newton, as we shall see), the main factor was undoubtedly Huygens' dogmatic attachment to contact-mechanical physics. He just could not countenance the concept of attractions between bodies at spatially separated points. Huygens' standpoint is very thoroughly documented by Westfall,[43] from whom I quote some of the most telling points.

A decisive event was the meeting in 1669 at the Académie in Paris to discuss the nature of gravity. Roberval and Frenicle argued that gravity 'is caused by an attraction between terrestrial bodies and the earth, a mutual attraction of like bodies for like' (almost identical to Kepler's viewpoint expressed 60 years earlier), but Huygens would have none of it. His reply opened with this uncompromising statement:[44]

> To discover a cause of weight that is intelligible, it is necessary to investigate how weight can come about while assuming the existence only of bodies made of one common matter in which one admits no quality or inclination to approach each other but solely different sizes, figures, and motions. . . .

Nearly twenty years later, on being advised of the forthcoming publication of Newton's *Principia*, he commented that he looked forward to seeing it and did not care that Newton was not a Cartesian 'as long as he does not propose any suppositions like that of attraction to us'.[45]

There are few more revealing passages in Huygens' writings than the following letter, which he wrote towards the end of his life and which is quoted by Westfall:[46]

> M. Descartes had found the way to have his conjectures and fictions taken for truths. And to those who read his *Principles of Philosophy* something happened like that which happens to those who read novels which please and make the same impression as true stories. The novelty of the images of his little particles and vortices are most agreeable. When I read the book of *Principles* the first time, it seemed to me that everything proceeded perfectly; and when I found some difficulty, I believed it was my fault in not fully understanding his thought. I was only fifteen or sixteen years old. But since then, having discovered in it from time to time things that are obviously false and others that are very improbable, I have rid myself entirely of the prepossession I had conceived, and I now find almost nothing in all his physics that I can accept as true, nor in his metaphysics and his meteorology.

As Westfall says, Huygens was cruelly deceived in thinking he had thrown off the influence of Descartes. He had shaken off all the *details* of Cartesian philosophy yet remained completely ensnared by the general mechanical philosophy to which he had been introduced at the impressionable age of 15 or 16.

The disaster from Huygens' point of view was that it was precisely in the mathematical treatment of attraction that his own concept of force, derived so elegantly while Newton was still a schoolboy at Grantham, came into its own and enabled Newton to make the final breakthrough to dynamics and the recognition of the universal role played by accelerations in all motions.

Perhaps the cruellest evidence of just how near Huygens got to the greatest prize is to be seen in the fact that Newton consciously inverted Huygens' own expression *centrifugal force* when looking for a suitable name to describe any force, but specifically a gravitational force, that tends toward a given centre, coining the expression *centripetal force*.

The point is that in accordance with Huygens' own treatment of centrifugal force the planets must, as they orbit around the sun, strive to recede from it with a force given by Huygens' formula. Since the planets do not fly off along the instantaneous tangent to their orbits, there must be some force that counteracts this tendency. Huygens, like Descartes, believed that this force was itself produced by the centrifugal tendencies of vortical particles, and his main effort was devoted to finding the precise mechanism by which the counteracting force could be produced. But Newton, like several other people in England, took seriously the idea that the force which keeps the planets in their orbits emanates directly from the sun; instead of trying to find the mechanism that counteracts the centrifugal tendency, he used the known strength of that tendency to deduce the strength of the putative force that emanates from the sun. As we shall see in the next chapter, this then led him on to the discovery of universal gravitation. This discovery, besides its immense significance in its own right, served simultaneously to demonstrate that all the parts required to constitute a full and consistent system of dynamics were now to hand. Huygens had all the parts in his hands but came nowhere near using them to full effect. He created the tools but, having exhausted the fruitfulness of the contact-mechanical philosophy, only put them to work on relatively minor problems.

This chapter should end with a worthy appreciation of Huygens' overall achievement. He made some of the most important discoveries in science – he was the first to reveal the true power of the Galilean relativity principle, he found the first form of the law of conservation of energy, and he found the correct definition of force. But, as we have pointed out, the very stimulus that led him to make his greatest discoveries, including, we must not forget, the wave principle named after him (also, it may be noted, another key phenomenon used by Einstein and derived directly from Huygens), prevented him from exploiting them to the full. The desire to put solid bone under the flesh of Descartes' contact-mechanical philosophy of nature was both his strength and his weakness. His tact and sureness of touch took him to the very brink of discovering the universal phenomenon at the heart of dynamics: the Second Law of Motion. As Westfall says:[47] 'Huygens possessed the conceptual equipment that such a dynamics required. His mechanical philosophy of nature effected its early abortion.'

Descartes gave him the stimulus to reach the brink but, like the man sitting on the rotating wheel, held back Huygens from breaking loose on the flight to the greatest scientific discovery of all times. The metaphor for Christiaan Huygens comes from falconry: he is the bird imprinted on his earth-bound falconer, Descartes. He had fashioned himself wings with which he might have flown to unimagined heights but was restrained by a string that temperament and circumstances never gave him cause to break. His clarification of the centrifugal phenomenon and the elucidation of the concept of force had an elegance that surpassed Newton's, and anticipated his by several years.

If Huygens was a falcon content to remain on the perch having brought home the sleekest hares ever caught on the Lord's estate, Newton was the soaring eagle with an eye to catch the moon and the very stars. Breaking the Cartesian tether impetuously, he set forth into uncharted regions – whither we now must follow.

10

Newton I: The discovery of dynamics

10.1. Introduction

Isaac Newton was born on Christmas Day 1642 according to the old calendar and 4 January 1643 according to the new style. The difference between the calendars is often exploited to make a nice symbolic point – that Newton was born in the year, 1642, that Galileo died. As was noted in Chap. 7, in his *Discorsi* Galileo predicted that:[1] 'there have been opened up to this vast and most excellent science [of motion], of which my work is merely the beginning, ways and means by which other minds more acute than mine will explore its remote corners.' Newton's was the mind destined to fulfil this confident – and accurate – prophecy.

Newton was a so-called posthumous child – his father, a yeoman farmer, died three months before he was born in the hamlet of Woolsthorpe in Lincolnshire. He barely survived his birth, but in fact lived to the ripe age of 84, dying in 1727, fêted as the greatest scientific genius of all time. He was notorious for obsessive secrecy and neurotic distrust, a character trait that has been attributed to the fact that his mother remarried when he was only two, and for nine years, until her second husband – an unsympathetic but prosperous minister of the church – died, the young Isaac was left in the care of his grandmother. His hatred for his stepfather is well documented. For this and other details of his life the reader is referred, for example, to Westfall's relatively recent biography *Never at Rest*.[2]

Newton's mother initially wished him to take over the management of what was now a quite considerable estate but his academic promise and manifest unsuitability for such work resulted in his being sent to school at Grantham, where he was noted for the brilliance of the mechanical devices that he constructed. Newton, like Huygens, combined brilliant intellectual gifts with superb craftsmanship and design skills. In the summer of 1661 Newton matriculated at Trinity College, Cambridge, where his well-to-do mother gave him a very meagre allowance, which initially he was forced to augment by performing menial tasks for his fellow students.

At that time Cambridge, like many other universities in Europe, was still largely in the grip of the fossilized Aristotelianism immortalized by Galileo's Simplicio. However, as Gascoigne has pointed out,[3] it was by no means completely closed to new influences and within a few years Newton began to assimilate the ideas of the scientific revolution, which he obtained above all from Descartes' *Principles of Philosophy* and *La Géometrie*, from Galileo's *Dialogo*, and from Gassendi's writings. Another major influence was the chemist Robert Boyle (1627–1691). It appears that he was also influenced, especially in his ideas on space and time, by the Cambridge Platonist Henry More[4] (1614–1687), who also introduced him to alchemy, which was to take up a great part of Newton's energies over many years.

On the basis of his readings and in almost complete isolation from other scientists, Newton embarked, at the age of about 22, on independent studies. As Westfall writes:[5] 'The first blossoms of his genius flowered in private, observed silently by his own eyes alone in the years 1664 to 1666, his *anni mirabiles*.' Towards the end of this period, in 1665, the university was closed by the famous plague, which caused Newton to return to his home, where in the garden, as he asserted in later years, he 'began to think of gravity extending to the orb of the Moon.'[6] There are quite extensive records of the unpublished studies that Newton did during these early years, which included brilliant pioneering work in mathematics (working out the essentials of the calculus), optics (including the famous experiments on the spectral decomposition of light and the invention of the reflecting telescope that is named after him), and dynamics. His brilliance in mathematics came to the attention of Isaac Barrow, who had recently become the first Lucasian professor of mathematics. Aiming at even higher things – he soon became Master of Trinity College[7] – Barrow resigned the professorship and recommended Newton, who had been made a fellow of Trinity in 1667, as his successor. Newton obtained the post in 1669, which, if not a sinecure, at least did not involve too much work.

Newton came to national and even international note primarily through his work on optics, though some knowledge of his work on mathematics was spread through correspondence and actually reached Leibniz. This was later to lead to the acrimonious controversy as to the priority in the discovery of the calculus and the (unjustified) charge of plagiarism made against Leibniz.[8] It was Newton's reflecting telescope that first brought him recognition and election to the Royal Society. This encouraged him to submit a paper on optics in 1672, which was reasonably well received but led in the next two or three years to a bitter dispute between him and Hooke. Newton was so mortified by the experience that he more or less withdrew from public life and devoted himself with increasing energy to alchemy and the study of the early Christian Church, to which he was led in an attempt to prove that the doctrine of the Trinity was a fabrication of some of the early church fathers (Newton was a Unitarian, a fact which he kept secret).

Of his early work in dynamics, recorded in his famous *Waste Book* and other unpublished papers, nothing leaked out and he seems to have more or less laid aside the studies in which he had effectively caught up with Huygens' work described in the previous chapter (which was also unpublished). It was really outside events and the quickening interest in problems of motion that eventually brought Newton to the centre of the stage. This growing interest had found reflection in the general invitation issued in 1668 by the Royal Society for submission of the correct laws of impact. The outcome was the response of Wallis, Wren, and Huygens already mentioned in the previous chapter. It is worth quoting here an extract of a letter written at that time (May 1669) by Oldenburg, the indefatigable secretary of the Royal Society during its first decade and a half and editor of the first purely scientific journal in the world (*Philosophical Transactions*) to Hieronymo Lobo; Westfall gives it very appropriately opposite the contents page of his book *Force in Newton's Physics*:[9]

> Our Society is now particularly busy in investigating and understanding Nature and the laws of motion more thoroughly than has been done heretofore . . . . Since Nature will remain unknown so long as motion remains unknown, diligent examination of it is the more incumbent upon philosophers . . . .

Even more significant was the publication in 1673 of Huygens' *Horologium Oscillatorum*, a copy of which was presented to Newton. In his letter to Oldenburg acknowledging the gift he made a favourable comment on Huygens' theorems on centrifugal force. More directly relevant was the fact that it quickened the pace of the attack on the problem of the dynamical treatment of the planetary problem, on which Hooke especially had been developing promising ideas. In particular, he proposed explicitly that it was a force directed towards the sun that kept the planets in their orbits. This was a most significant refinement of the original Cartesian formulation of the planetary problem. In 1679 Hooke, who had just been appointed to succeed Oldenburg as secretary of the Royal Society, wrote to Newton in conciliatory vein, asking him among other things for an opinion of his theory. This correspondence stimulated Newton into making what was probably the single most important discovery in the history of physics – the demonstration from Kepler's Laws that the planets must be attracted to the sun by a force whose strength decreases as the square of the distance from the sun. The synthesis of dynamics – in its essentials – was brought about by this work.

For the first time the extraordinary power of the tools forged by Galileo, Huygens, and the youthful Newton was fully revealed.

If there was now no going back, the way forward was still comparatively leisurely and Newton, who let out no hint of his discovery, worked only intermittently if at all on dynamics in the following years, though it is evident that his ideas were developing apace. The final stimulus again came from outside. For several years not only Hooke but also Halley and Wren had been pondering the problem of the planets and slowly closing in on the correct solution without however having the necessary mathematical ability to crack the nut. In particular, both Halley and Wren had had the idea of using Kepler's Third Law in conjunction with Huygens' formula for centrifugal force to demonstrate that the force of attraction towards the sun must fall off as the square of the distance from the sun. They did not know that Newton himself had had a very similar idea nearly twenty years earlier, at the end of his *anni mirabiles* period, after his own rediscovery of Huygens' results. What neither Hooke, nor Halley, nor Wren could prove was that such a force must give rise to an elliptical orbit with the sun at one focus. When the three men met in January 1684 at a meeting of the Royal Society Hooke's claim that he could demonstrate all three laws of planetary motion was met with scepticism by the other two.[10] Wren offered a book worth forty shillings as a prize for the correct demonstration, but his two-month limit was exceeded without the answer being forthcoming. Later in the same year Halley travelled to Cambridge, where he put the problem to Newton, only to be told that he, Newton, had already solved the problem but could not, at the moment, put his hand on his proof. However, he promised Halley to provide it. Thus began the train of events which led eventually to the publication in 1687 of the *Principia*, a date which more than any other can be taken as the definite commencement of the scientific age.

The chapter which now follows will look in more detail at the development of Newton's ideas and attempt to identify the crucial clarifications of the most important concepts in the synthesis of dynamics. In the following chapter we shall specifically look at the reasons which led Newton to introduce his concepts of absolute space, time, and motion. The final chapter will round off Vol. 1 by examining the final conceptual clarification of some of Newton's most important concepts. This work, which only occurred in the second half of the nineteenth century, is important simultaneously as marking the conclusion of this great enterprise that began more than two and a half millennia ago in ancient Greece (and Babylon) but also in preparing the ground for Vol. 2, in which we trace the reaction to Newton's ideas and the part this reaction played in the creation of general relativity.

10.2 A comment on the significance of Newton's early work

A somewhat difficult question, especially relevant in the context of the problem of the origin of Newton's ideas about space, time, and motion, is that of the relationship of his early work on dynamics to that of the mature *Principia*. My own view is that by the end of the 1660s Newton had worked out – but tested to only a very limited degree – all the basic principles of his mature dynamics except the notion of forces acting at a distance towards definite centres. If he did have the idea in the plague year of gravity as a universal force acting over great distances, the surviving unpublished papers reveal little evidence of it. What they do show is the development of a conceptual framework into which such a notion can be introduced very naturally and without difficulty. In fact, as will be argued later in this chapter, it seems to me that there was a certain inevitability about the form that dynamics took and the way in which its development unfolded after 1650. The reason for this is to be sought in the preparatory work done by Kepler, Galileo, and Descartes and in certain facts of the natural world, above all the manifestations of the gravitational force. Quite strong support for this view is to be found in the fact that Huygens and Newton, working in complete isolation but from the same sources and the same facts, arrived, *grosso modo*, at identical results in the penultimate stage before the final synthesis. (There are nevertheless one or two residual differences of considerable interest to which attention will be drawn.) Moreover, with the benefit of hindsight, we can see that the route attempted by Huygens to go beyond his first great insights was doomed to failure. Newton was extremely fortunate in being prodded into the only direction that was capable of leading to success. For, at that time, only the astronomical data contained the facts capable of justifying fully the generic concept of force.

Thus, I see the early period as providing the elaboration of the necessary concepts; the later period as revealing the power latent but as yet unrecognized in the early work. This results in an evaluation of Newton's development that differs in some respects from other accounts, for example Westfall's,[11a] especially with regard to the part played by relativity (in both forms, kinematic and Galilean), and Whiteside's[11b] (with regard to Newton's attitude to the law of inertia). The significance of Hooke's intervention is also interpreted somewhat differently. The reader is therefore asked to bear in mind that the account that follows might not be accepted by all historians of science. In fact, since the relevant documents are all published and readily available in libraries (they are listed in Ref. 12), the reader is strongly encouraged to consult the originals and make up his or her own mind. After this caveat, let us now begin with Newton's early clarification of the force concept.

10.3. Three types of force

The most distinctive feature of Newton's dynamics is the concept of *force*, or, rather, *forces*, since Newton actually employed three related but distinct concepts of force. (It should be noted that in the seventeenth century and for quite a long time afterwards the word *force* (*vis*) was used with many different meanings, often rather imprecise.) The clarification of these three different forces, the relationships they bear to one another, the manner in which they are to be applied, and the extraordinary range of their application is very largely the history of the final stages of the discovery of dynamics.

It is ironic that the word *dynamics* was coined by Leibniz[13] to characterize a *fourth* concept of force. This was his famous *vis viva*, the modern kinetic energy (apart from a factor $\frac{1}{2}$). But energy is the one concept that hardly features in Newton's dynamics, and he never introduced it as a special concept. The word *dynamics* was transferred from its original narrow Leibnizian meaning to its present meaning during the first half of the eighteenth century and was made standard by d'Alembert in his *Traité de Dynamique* (Paris, 1743). The subsequent development of dynamics was to show that energy is, in fact, the most important concept of the four, which is posthumous consolation of a sort to Leibniz for the excessively rough treatment he got at Newton's hands over the question of the discovery of the calculus.

Two of the three force concepts that Newton employed developed out of his work on collisions; the third, out of his work on centrifugal force, or *conatus a centro* (endeavour from the centre), as, following Descartes, he initially called it. We begin with collisions.

10.4 Collisions

Newton's solution of the collision problem came about a decade after Huygens'; it belongs to his *anni mirabiles* period (1664–6) and is all the more remarkable for having been done in complete isolation in Cambridge, which at that time was, as has been pointed out, very much an intellectual backwater. In fact, Newton was only in physical isolation. Through books he had access to Galileo, Descartes, and Gassendi – and that was more than enough stimulus.

Newton's early work on collisions (or *reflections*, as he called them) survives in what he called his *Waste Book** (see Herivel[12a]), which contains the complete solution to the problem of collisions, both elastic and

* The *Waste Book* was in fact a large notebook that Newton inherited from his hated stepfather, Barnabas Smith, in which the rector of North Witham had entered a few theological jottings. As a boy Newton had threatened Smith and his mother 'to burne them and the house over them' (Ref. 2, p. 53).

inelastic. Newton's work on collisions laid the foundations of his mature dynamics. We shall see how, in conjunction with his work on centrifugal force, done at much the same time, it was able to provide for him all the really basic concepts that he needed. All that was lacking was the final stimulus and, perhaps, one crucial suggestion; these came from Hooke and Halley a decade and a half later.

Of particular interest are the differences in approach between Newton and Huygens. From the very beginning, the latter's work was dominated by the relativity principle. Though the relationship between the Galilean relativity that Huygens actually used and the relativism which he found in Descartes may not have been clear to Huygens, his use of the principle was completely sound and enabled him to avoid all the problems in which Descartes had got so bogged down in trying to determine the outcome of collisions by estimating the respective forces of the bodies involved in collisions. After an early use of the force concept, which, as we have seen, may have helped him to his initial breakthrough, Huygens completely eschewed the approach via 'forces' and sought instead universal rules that govern the outcome of all collisions. We have seen how well he succeeded.

In contrast, Newton retained a prominent role for the concept of force but went far beyond Descartes in two important respects: he arrived at a precise quantitative concept of force and he realized that the appearance of force is subject to a precise reciprocal phenomenon – forces never appear in isolation but always in pairs at least. This insight led eventually to the formulation of his famous Third Law. One of the main differences in emphasis between Newton and Huygens as regards the basic principles of dynamics is to be seen in the prominence that Newton gives to the Third Law, whereas the relativity principle is put in the forefront by Huygens.

The entries in the *Waste Book* take the form of a series of propositions or axioms. The formal arrangement of these propositions is not so strictly logical as in Huygens' *De Motu Corporum* or in Newton's own *Principia*, published more than twenty years later. A distinction is not clearly drawn between assumptions, axioms (or laws), empirical observations, and deductions from adopted axioms. Although more advanced than mere jottings, the entries in the *Waste Book* were clearly still in a stage prior to that in which they could have been published as a formal tract. For all that the key insights can be very readily identified – though not the manner by which Newton arrived at them, which appears to have been by rational thought rather than through experiment.

The general framework in which the problem is approached seems to be clearly due to Descartes. Everything is based upon the idea that the natural state for a body is to continue in a straight line with uniform motion (or else to remain at rest). Collisions are events that deflect a body from one such state into another. Note the important point, emphasized in the previous chapter, that the very concept of a collision is made possible by the clear definition of inertial motion. The *quantification* of collisions is not possible without the well-defined pre-collision and post-collision states with respect to which they are themselves defined. This led Newton to one of his two most important insights: that whatever the ontological status of force might prove to be, its *quantitative measure* must be through *change in inertial motion*. Moreover, it was equally clear to Newton that speed (or, rather change of speed) alone is not a sufficient characterization of 'force'. Clearly, more force is needed to change the speed of a large body than a small. At this stage Newton did not possess a clear concept of mass (as we shall see in Chap. 12, it is, in fact, doubtful whether he ever did); like Descartes, he obviously thought that the only distinction between bodies was one of size. (Size here means the volume actually occupied by matter, which is itself taken to have uniform density if free from voids and interstices.) Newton thus identified force with change in speed multiplied by the *bulk* of the body. He arrived at a concept very close to Descartes' quantity of motion. It is, however, evident from the earliest entries in the *Waste Book* that Newton recognized the importance of the *direction* of motion. The first worked examples[14] demonstrate that Newton knew it is the *directed quantity of motion* (which Newton usually simply calls *motion* – it is synonymous with the modern *momentum* if the bulk is identified with the mass) that is conserved in collisions, not Descartes' directionless quantity of motion. He makes this point explicitly in an early definition, which employs the Cartesian expression *determination*:[15] 'Those Quantitys [i.e., bodies] are said to have the same determination of their motion which move the same way, and those have divers which move divers ways.'

To understand Newton's conceptions of force we must understand his conception of inertial motion. The evidence from his notebooks and unpublished tracts on dynamics suggest that to a late date he conceived the phenomenon of inertial motion somewhat after the manner of medieval impetus theory. That is, a body in motion has in it a power or force that maintains it in uniform motion in a straight line unless it is affected by some external body or force.

It seems probable, on the basis of the available evidence, that for a long period Newton regarded this force, which, following his own usage, I shall call the *inherent force* (Newton also used the expression *innate force*), as having a quite definite and unique value. Now this cannot be the case unless motion is conceived as taking place with respect to some quite definite frame of reference. It will be argued in Chap. 11 that Newton never wavered in his belief in what he later came to call absolute space. Initially it was no doubt purely instinctive; later it hardened into a formalized conviction. All we need to know at this stage is that for

Newton objects moved through a space that was as real for him as the green felt of a snooker table. Thus, the *direction* and *speed* of rectilinear motion were perfectly well-defined quantities in Newton's mind.

However, what distinguished the earliest work of Newton from Descartes was his awareness that speed and direction of a physical body cannot be considered in isolation. Speed by itself is not the measure of the inherent force. It is not a scalar quantity but inescapably a directed, vectorial quantity. For one-dimensional collision problems this is evident from the definition given a little earlier; more remarkably it is implicit in the work on two-dimensional problems that establishes the centre-of-mass theorem,[16] which will be discussed a little later.

This is intimately tied up with his second use of the word force, by which he means a capacity to *change motion*. Newton called this *motive force*. The first point to make is that the motion which is changed is the inertial motion of a body which it has when free from external disturbances; it is the motion characterized by the inherent force. The second point to make is that the change of motion can take place *in any direction*. Newton seems to have conceived of changes in inertial motion being, as it were, the addition or subtraction of 'bits of motion' to the already existing inertial motion, i.e., as increments or decrements to the body's inherent force. But since these increments or decrements are essentially *directed* quantities, it is not possible to add them to something that does not have the same essential nature, from which it follows that the inherent force is a vectorial quantity.

It must be emphasized that such a standpoint is not explicitly verbalized in the early notebooks and manuscripts. But the worked examples suggest that Newton worked instinctively on such a basis. It has startling consequences for anyone who approaches the problem of motion with the naive intuition that Descartes very largely employed, regarding the *speed* of motion as the all-important concept. Consider, for example, the case of uniform circular motion, in which the direction of motion is constantly changing because of an applied force but the total speed remains the same: a 'force' appears to be acting without achieving any effect – the speed is not changed (in modern terms we say that the force 'does no work'). The same thing can be seen in the use of the parallelogram of forces (or rather motions), which is certainly one of the most important discoveries in dynamics and is what distinguishes most unambiguously the Galilean and post-Galilean study of motion from what preceded it. In the Newtonian view, a body before the application of a force has a motion characterized by a definite speed and direction, say by the vector **AB** (Fig. 10.1). When the force is applied (say instantaneously in a collision), the vector **AB** has added to it the vector **AC**. The resultant motion is then **AD**. But the length of **AD** depends entirely on the direction of **AC** relative to **AB**. Changes of motion are more sophisticated than the

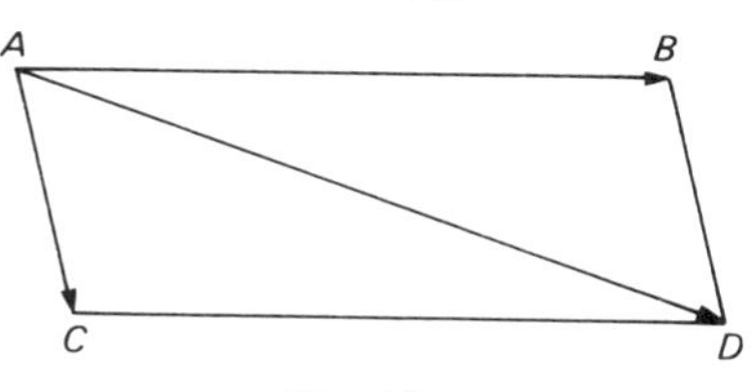

Fig. 10.1.

simple addition or subtraction of lengths. Although Galileo is not mentioned on the subject in the *Waste Book*, Newton explicitly states the rule of composition of motions by means of the parallelogram of motions.[17]

I said that Newton employed three different force concepts. It would in fact be rather more accurate to say that he employed two: that of the inherent force and that of the motive force, the latter being of two kinds – impulsive (acting instantaneously) and continuous (changing the speed and direction continuously). In Newton's mind there was no essential difference between the two kinds of motive force. According as it suited his purposes, he either regarded an impulsive motive force as the integrated effect of an intense continuous force acting over a very short period of time or, much more commonly, imagined a continuous force as the outcome of innumerable infinitesimal impulsive forces applied successively at very short time intervals. Throughout his entire work on dynamics Newton switched promiscuously from the one to the other of these two kinds of motive force without exhibiting the slightest discomfiture.

Let us now see how Newton clarified the idea that force can be regarded not only as something that resides within a body but also as something external to the body, something that causes a change in the body:[18]

> Hence it appeares how and why amongst bodys move[d] some require a more potent or efficacious cause others a lesse to hinder or helpe their velocity. And the power of this cause is usually called force. And as this cause useth or applieth its power or force to hinder or change the perseverance of bodys in theire state, it is said to Indeavour to change their perseverance.

It was in developing the concept of force along the lines of this 'Indeavour' that Newton arrived at what was probably his most important original insight in the basic structure of dynamics, his Third Law. The next quotation clearly reveals Newton's dual perspective of force:[19]

> Now if the bodys *a* and *b* meete one another the cause which hindereth the progression of *a* is the power which *b* hath to persever in its velocity or state and is usually called the force of the body *b* and as the body *b* useth or applyeth this force to stop the progression of *a* it is said to Indeavour to hinder the progression of *a* which endeavour in body [*b*] is performed by pressure and by the same reason the body *b* may bee said to endeavor to helpe the motion of *a* if it should apply its force to move it forward; soe that it is evident what the Force and indeavor in bodys are.

As yet, this is purely qualitative. The real breakthrough, the insight that solved the problem of collisions literally at a stroke, was in Newton's enunciation of the rules that govern the application of 'Indeavour'. He formulates rules that govern the *changes in motion* (*motion*, again, understood in the Newtonian sense, i.e., the modern momentum) in any collision. Three brief propositions (which between them contain the bulk of the content of his subsequent Second and Third Laws) solve the problem of collisions:[20]

119. If *r* presse *p* towards *w* then *p* presseth *r* towards *v* [i.e., in the opposite direction; see Fig. 10.2]. Tis evident without explication.
120. A body must move that way which it is pressed.
121. If 2 bodys *p* and *r* meet the one the other, the resistance in both is the same for soe much as *p* presseth upon *r* so much *r* presseth on *p*. And therefore they must both suffer an equall mutation in their motion.

In essence, the statement 'both suffer an equall mutation in their motion' solves the problem of collisions in Newton's approach. It is the condition of conservation of momentum which we have already met in Huygens' work. It is, of course, only a single condition and we know that two conditions are needed to determine the outcome of a one-dimensional collision uniquely. It is, however, a condition that is valid for *all* collisions. In contrast, the condition of conservation of the kinetic energies holds only in the case of elastic collisions. In this sense, Newton's principle gives the general solution to the collision problem to the extent that one exists. He was moreover completely clear as to the essential facts of elasticity and the part it plays in collisions. Earlier in the *Waste Book* he had treated elastic collisions. For example:[21]

If two equall and equally swift bodys (*d* and *c*) meete one another they shall bee reflected, so as to move as swiftly frome one another after the reflection as they did to one another before it. For first suppose the sphaericall bodys *e*, *f* [Fig. 10.3] to have a springing or elastic force soe that meeting one another they will relent and be pressed into a sphaeroidicall figure, and in that moment in which there is a period put to theire motion towards one another theire figure will be the most sphaeroidical and theire pression one upon the other is at the greatest, and if the

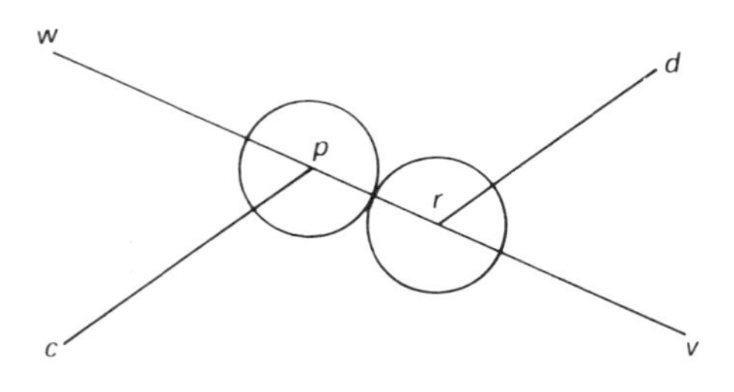

Fig. 10.2. Reproduced from: J. W. Herivel, *The Background to Newton's 'Principia'*, Clarendon Press, Oxford (1965).

endeavour to restore theire sphaericall figure bee as much vigorous and forcible as theire pressure upon one another was to destroy it they will gain as much motion from one another after their parting as they had towards one another before theire reflection.

There then follows a similar argument for perfectly hard bodies and it ends with the same conclusion, namely, that the bodies 'shall move from one another as much as they did towards one another before theire reflection'. This is the special case of the energy conservation law for the elastic collision of two equal bodies and, when it holds, completes the solution of the collision problem.

It is remarkable that Newton immediately generalizes this result to the case when the bodies in the collision have not only unequal sizes but also unequal speeds (or, *celerities*, to use Newton's favourite word):[22]

There is the same reason when unequall and unequally moved bodys reflect, that they should separate from one another with as much motion as they came together.

The entire proof of this striking generalization (which is Huygens' law of the equality of the pre-collision and post-collision relative speeds) that Newton gives is in the 'There is the same reason'.

The picture of collisions that emerges from the *Waste Book* is that of mutually induced changes of motion. So much as body a changes the motion (i.e., momentum) of body b, just so much body b changes the motion of body a. This is the only universally valid rule governing collisions. The actual outcome of any particular collision depends on the elasticity of the bodies involved in the collision. If they are perfectly elastic, they will rebound from each other without any loss of *relative* speed; but if they are imperfectly elastic the relative speed v will be reduced by the collision by a factor k, the coefficient of restitution, which takes values in the interval $0 \leq k \leq 1$, with $k = 1$ corresponding to perfect elasticity and $k = 0$ to complete inelasticity, when the bodies stick together after the collision. In a passage in the *Principia* that we shall shortly quote, Newton reports the experimental fact that k is a property of the two bodies and has, to the accuracy of the measurements that he made, a value that is independent of the relative speed with which the two bodies collide.

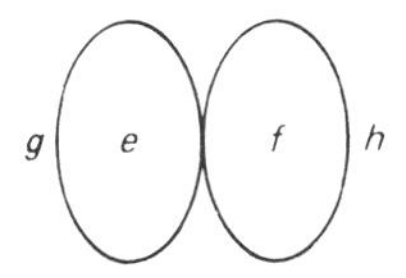

Fig. 10.3. Reproduced from: J. W. Herivel, *The Background to Newton's 'Principia'*, Clarendon Press, Oxford (1965).

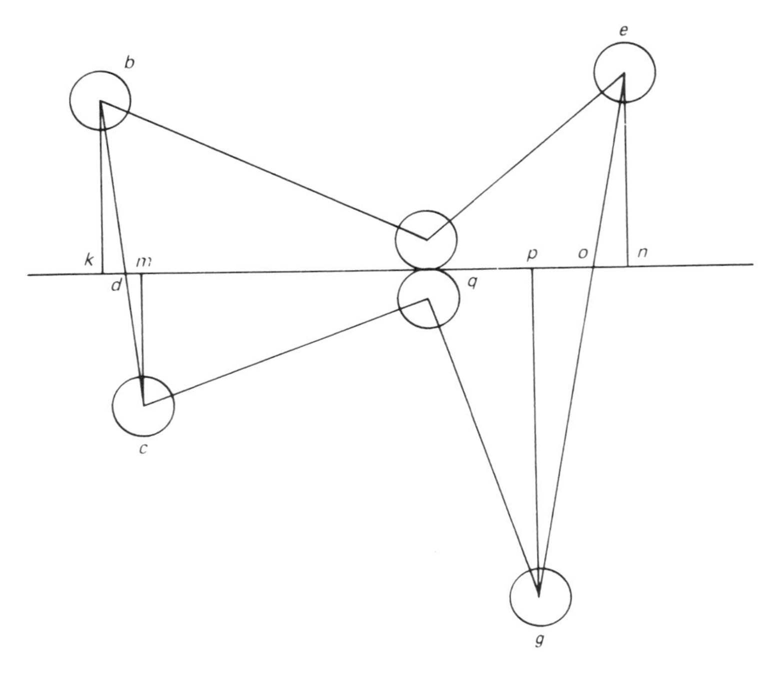

Fig. 10.4. Reproduced from: J. W. Herivel, *The Background to Newton's 'Principia'*, Clarendon Press, Oxford (1965).

Newton's fundamental rule of the 'equall mutation of the motion' led him to the conclusion that the centre of mass (in the *Waste Book*, Newton calls it the centre of motion) of the bodies continues in a state of rest or of uniform motion in a straight line during the entire period of the collision. This is the result, which, as we saw in the previous chapter, Huygens described as 'an admirable law of Nature'. The centre-of-mass theorem appears as an important result in the *Waste Book*,[23] and it will be worth saying a little about it.

Newton first of all considers the motion of two bodies that move inertially and do not collide (he considers both the case when they move in a common plane and also when this is not the case) and shows that in all cases the motion of their centre of mass is along a straight line and with uniform speed. He then shows that, in accordance with the rules of collision formulated above, the motion of the centre of mass, for elastic or inelastic collision, is unaffected by the collision, continuing to proceed along the same straight line and with the same uniform speed as before. This study culminates with the two following propositions, which I give together with the respective figures (Figs. 10.4 and 10.5).

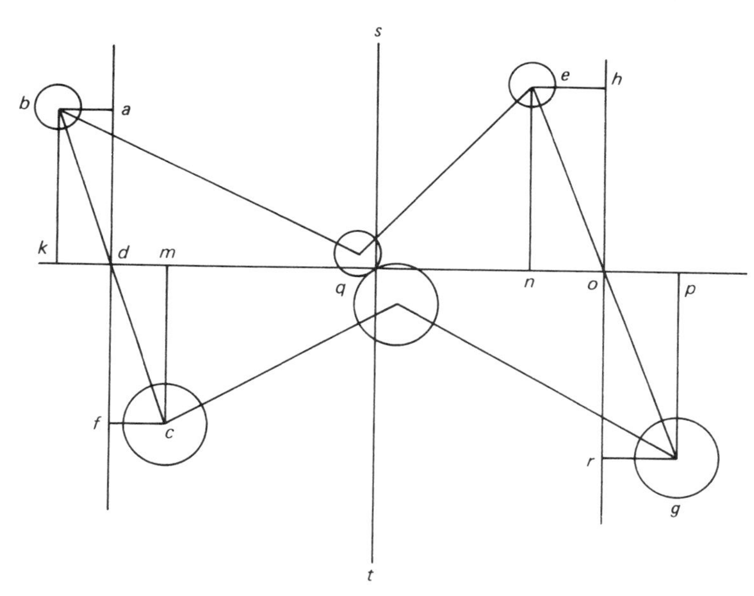

Fig. 10.5. Reproduced from: J. W. Herivel, *The Background to Newton's 'Principia'*, Clarendon Press, Oxford (1965).

31. If two bodys (*b* and *c*) meete and reflect one another at *q* [Fig. 10.4] their center of motion shall bee in the same line (*kp*) after reflection in which it was before it.

32. If the bodys (*b* and *c*) reflect at *q* [Fig. 10.5], to *e* and *g*, and the centers of their motion describe the line *kdop* the velocity of that center (*o*) after reflection shall bee equall to the velocity of that center (*d*) before reflection.

To give the flavour of Newton's proofs, which use all of his subsequent three laws of motion in embryonic form, together with the law of composition of velocities (note that the bodies in Figs. 10.4 and 10.5 have a common motion from left to right as well as towards each other in the top–bottom direction), I give the proof of the first of these propositions (recall that throughout Newton's *motion* is, subject to the proviso that bulk is identified with mass, the modern *momentum*):

For the motion of *b* towards *d* the center of their motion is equall to the motion of *c* towards *d* [by an earlier proposition] then drawing $bk \perp kp$, and $cm \perp kp$, then $cd{:}bd{::}cm{:}bk$, therefore the bodys *b* and *c* have equall motion towards the points *k* and *m*, that is towards the line *kp*. And at their reflection so much as (*c*) presseth (*b*) from the line *kp*; so much (*b*) presseth (*c*) from it (ax. 121). Wherefore they must have equall motion from the line *kp* after reflection, that is drawing $gp \perp kp$ and $ne \perp kp$, (*e*) and (*g*) have equall motions towards (*n*), (*p*), then drawing the line *eog* tis $ne{:}eo{::}gp{:}go$. Therefore *e* and *g* have equall motion from the point *o* which [by the earlier proposition] must therefore be the center of motion of the bodys (*b*) and (*c*) when they are in the places *g* and *e*, and it is in the line *kp*.

In treating thus the motion in two dimensions Newton achieved a sophistication that exceeded Huygens' work described in the previous chapter.

Examined in the light of the attempts by Descartes and others (but not Huygens) to solve the problem of collisions, Newton's solution is above all remarkable for the way in which it passes from vague and muddled anthropomorphic concepts of force to precisely defined quantitative rules which simply state what happens in collisions. The notion that the outcome of a collision is to be predicted by examining which of the colliding particles has the greatest 'force' and assuming that the particle will 'get its way' in the ensuing contest of strength is replaced by the bald statement that in an actual collision 'soe much as *p* presseth upon *r* so much *r* presseth on *p*', so that 'they must both suffer an equall mutation in their motion'.

How simple is the solution when at last it is found! The entire Newtonian success rests on the recognition of the *equality* of the action and the reaction. In his article 'Action and reaction before Newton', J. L. Russell concluded:[24] 'There was nothing original in the principle that every action involves a reaction. This had been asserted by so many people as an accepted fact that it must surely have been common knowledge. The principle of equality was, however, another matter.' We see again that dynamics was created little step by little step, each of which transformed a vague qualitative awareness of some phenomenon into a precise quantitative statement. Particularly striking in the present case is the generality of the phenomenon which Newton succeeded in quantifying at the age of about 23 or 24: his rule applies in all collisions. Years later, when formulating his Third Law, he generalized the rule still further to encompass all interactions, not merely collisions. It was the key to much that was best in his dynamics. It is interesting to note that Newton himself, as is clear from the way in which he presented the laws of motion in the *Principia*, regarded this law, his *Lex Tertia*, as the only one in which he himself made an original contribution to the foundation of dynamics. In the draft *De motu*,[25] which immediately preceded the *Principia*, he says of the first two laws that 'they are now widely accepted' and in the *Principia*[26] implies that they were clearly recognized and employed as such by Galileo when he demonstrated the parabolic motion of projectiles.

How Newton arrived at his rule of the 'equall mutation of the motion' I do not know. Simon Schaffer has suggested to me that it may have come from a deep and close reading of Descartes' rules of collision, whose manifest conflict with empirical fact (which was widely recognized) clearly indicated the need for new principles radically different from those employed by Descartes. Gabbey too (Ref. 19, Chap. 9) believes the influence of Descartes was decisive. Whatever, the truth, I think we simply have to accept the fact that Newton arrived at the rule as an original and primitive insight and very soon realized its significance and the fact that it more or less completely solved the problem of collisions. Perhaps, like that other great discovery in the history of dynamics, the odd-numbers rule in free fall, it was first recognized in one of those table-top experiments that Descartes disdained. Newton was, after all, not only a superb theoretician. He was also a great experimentalist. However, as already pointed out, the presentation in the *Waste Book* does not suggest an experimental origin. Indeed, Newton appears rather to approach the problem in a Cartesian and rationalistic frame of mind. It may be that he had not yet fully developed his characteristic emphasis on the need to attack the secrets of nature through empirical phenomena. It was this that was to set him apart from so many of his contemporaries, including Huygens, who, as we have seen, retained throughout his life a predeliction for a Cartesian explicative approach.

It is worth pointing out in this connection a significant difference of presentation of these matters between the treatment given in the *Waste Book* and the *Principia*. In the former, the most important results are simply presented as axioms, rather in the manner in which Huygens introduced his most important ideas in the form of propositions. But in the *Principia* Newton takes a lot of trouble to produce evidence for the Third Law in the form of empirical results on collisions. There is a beautiful account of an experiment he did that refined a famous demonstration of the collision laws that Wren performed before the Royal Society. Wren's arrangement is shown in Fig. 10.6. Pendula of different weights are suspended from the points C and D and allowed to collide with various speeds, which can be calculated readily by Galileo's law of descent (this is a fine example of indirect overcoming of the difficulty of measuring speeds, the problem that held back the development of dynamics for so long). The speeds that they acquire as a result of the collision can be equally calculated from the height to which they reascend.

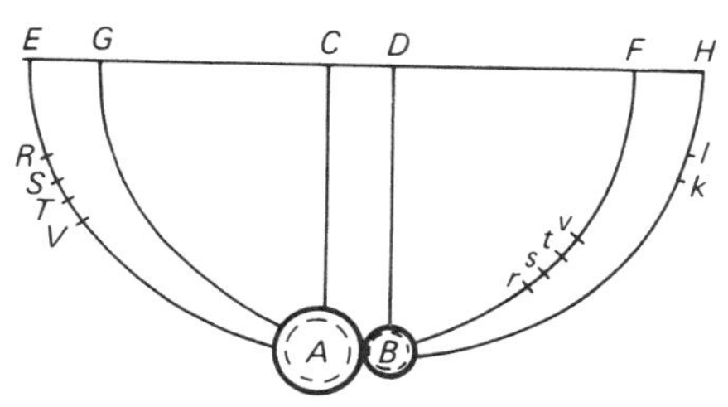

Fig. 10.6.

Newton comments on this experiment:[27] 'to bring this experiment to an accurate agreement with the theory, we are to have due regard as well to the resistance of the air as to the elastic force of the concurring bodies.' There then follows a detailed account of the elegant way in which he did this; by taking into account the air resistance 'everything may be subjected to experiment, in the same manner as if we were really placed *in vacuo*'. (It is worth noting in passing how firmly Galileo's basic principles had become established by the time Newton was writing the *Principia*, i.e., the point of view that there are *ideal laws of motion* and that these are to be deduced from the observed motions by careful elimination of extraneous disturbances. (Note also that the vacuum has become established, in Newton's mind at least – there were still plenty of plenists on the Continent – as the ideal medium in which motion can be studied.) Newton's summary of the experiment was as follows: 'Thus trying the thing with pendulums of 10 feet, in unequal as well as equal bodies, and making the bodies to concur after a descent through large spaces, as of 8, 12, or 16 feet, I found always, without an error of 3 inches, that when the bodies concurred together directly, equal changes towards the contrary parts were produced in their motions and, of consequence, that the action and reaction were always equal.'

He went to considerable trouble to do the experiment with a wide range of bodies and not just the perfectly hard or elastic ones of the theory of Wren and Huygens and found:[28]

the bodies to return one from the other with a relative velocity, which is in a given ratio to that relative velocity with which they met. This I tried in balls of wool, made up tightly, and strongly compressed. For, first, by letting go the pendulous bodies, and measuring their reflection, I determined the quantity of their elastic force; and then, according to this force, estimated the reflections that ought to happen in other cases of impact. And with this computation other experiments made afterwards did accordingly agree; the balls always receding one from the other with a relative velocity, which was to the relative velocity with which they met as about 5 to 9. Balls of steel returned with almost the same velocity; those of cork with a velocity something less; but in balls of glass the proportion was as about 15 to 16. And thus the third Law, so far as it regards percussions and reflections, is proved by a theory exactly agreeing with experience.

With this we conclude the discussion of Newton's work on collisions.* Though the end results for elastic collisions are the same as Huygens obtained, the approach seems strikingly different. It is in the prominent

* Newton's early papers also contain some striking work on collisions of *extended rotating bodies*. I omit the discussion of this work, which is reproduced and discussed in Herivel's book,[12a] only because I have attempted throughout this book to concentrate on the emergence of those dozen or so elements that proved to be sufficient for the ultimate synthesis of dynamics. This is the story of how that happened, not of everything that occurred along the way.

role accorded to the relativity principle that Huygens' work differs most from Newton's, for in neither the *Waste Book* nor the *Principia* does the relativity principle appear as one of the principles from which the results on collisions are deduced. In fact, so far as I have been able to discover, the Galileo–Huygens relativity principle is not explicitly stated in any of Newton's early unpublished work. As we shall see, it only makes its first appearance in the period immediately preceding the writing of the *Principia*. This absence of an explicit reference to the relativity principle may be a significant clue to the development of Newton's thinking about the relative or absolute nature of motion and we shall return to this point later. It should also be noted that for inelastic collisions the results that can be deduced from Huygens' and Newton's most fundamental principles are not identical. Newton's Third Law says that momentum is conserved in every collision; the principle of relativity says nothing definite about the outcome of any collision – it merely shows one how to generate a whole family of collision outcomes from a given collision that is known to occur. The connection between the principle of relativity and the laws of mature Newtonian dynamics will be discussed later.

Thus far we have met two of the force concepts that Newton employed, the inherent force and the impulsive motive force, which changes the inherent force in discrete (i.e., finite) vectorial amounts in collisions. We now come to the third force, the force which changes motion continuously. Like Huygens, Newton discovered this force by solving the problem posed but not solved by Descartes – the problem of finding a quantitative measure of centrifugal force, the *conatus a centro*.

10.5. Centrifugal force: the paradigm of a continuously acting force

The discovery of dynamics is simultaneously the story of how men learnt, from careful observation of phenomena, to project numbers into nature. It is the story of the *quantification of nature*. One can look at this as the progressive discovery of rulers – the word used in a generalized sense – for measuring off numbers encoded in the seemingly perfect ordering of the world that Pythagoras sensed as the harmony of the cosmos.

The first ruler was the ordinary ruler which discovered 'ordinary' geometry, the quotes here being used merely to distinguish conventional three-dimensional geometry from the four-dimensional geometry of motion discovered by Galileo and Newton. We can usefully recall the stages that led to the very high level of sophistication of quantification that mature dynamics represents. The key step was the transition from the ruler that measures distances to the ruler that measures speeds. This comes in a beautiful passage in Galileo's *Discorsi*,[29] in which he speaks of the need for a measure of speed and shows where it can be found. He is discussing two of his most important results: first, that when a body free of all resistance falls a definite height it acquires a definite speed, this speed being exactly the same for all bodies; second, the fact that if such a body is deflected into the horizontal and can continue to move freely its speed will remain unaltered. Thus, an object let fall from a definite height and then deflected into the horizontal becomes a kind of ruler, a unit of speed. The key point, strongly emphasized by Galileo, is that two well attested physical phenomena, the law of free fall and the law of inertia, make it possible to convert a purely geometrical ruler into a 'ruler of speeds'. In the previous section we saw a very fine example of the utility of this ruler in Wren's collision experiment with pendula.

Galileo actually suggests that as unit of uniform speed one should take the height of a javelin, i.e., about six feet. As we noted in Chap. 7 it is very suggestive that, in his discussion of parabolic motion, Galileo supposes that a body is let fall from a certain height H, deflected into the horizontal, and then allowed to fly off the table top on which the experiment is being done. By varying H, you thereby vary the horizontal component of the generated uniform inertial motion and parabolas of different curvatures are obtained as a result. Considering the far-reaching consequences of the concepts which he was introducing, the name Galileo gave to the height H was very apt. As we recall, he called it the *sublimity*. We have seen in both Chaps. 7 and 9 how fruitful was the introduction of these ideas. It was the sublimity that converted three-dimensional geometry into four dimensions and simultaneously combined what was best in Plato and Aristotle. Platonic clarity finally dispersed the mists that had so long obscured the magnificent mountain range which Aristotle had sensed.

What was sublime about the work of Huygens and Newton in solving Descartes' problem of finding the quantitative measure of *conatus a centro* was that it took this process one stage further. They showed how Galileo's ruler, the ruler of uniform motion, could itself be used to measure a derivative concept one storey higher, the force of motion, or the acceleration.

It is the *principle* by which this was done with which we are mostly concerned; for this principle, tested on the smooth round nut of centrifugal force, finally cracked the world apart.

The principle, of which Newton never loses sight, is that the quantitative measure of force is *the amount by which it can change inertial motion in a given time*, motion being understood as the product of bulk and velocity. In tackling the problem of centrifugal force, the main problem that Newton faced was finding a unit in terms of which he could measure the centrifugal force. We can illustrate the device he employed by considering first the simpler example of a body (of mass m) moving along a straight line with initial speed v. A collision imparts to it a velocity increment Δv. Then, according to Newton, the force applied to the body, the impulsive *motive force*, is $m\Delta v$. The inherent force of the body before the collision was mv. Then the motive force can be measured in terms of the inherent force by the ratio: $m\Delta v/mv = \Delta v/v$. From the point of view of mature Newtonian dynamics, there are two problems with such a definition: (1) the ratio is a scalar quantity and is meaningful only in the case of one-dimensional motion; if Δv and v are not collinear, as in the general case, we cannot in this way find a meaningful measure of the motive force itself but only its scalar magnitude, measured by $|\Delta v|/|v|$; however, this is sufficient for Newton's purposes; (2) because of the Galileo–Huygens relativity principle, the velocity v by means of which the inherent force is defined is not a well-defined quantity; in fact, by going over to a different frame of reference moving relative to the first with velocity w, one can make the resultant velocity $v' = v - w$ (and also its scalar magnitude) take absolutely any value. This is, in fact, the root of one of the problems that flaw the logical consistency of Newton's famous Scholium in the *Principia* on absolute and relative motion, and we shall have more to say about it in the next chapter. I suspect that even when he became aware of the problem Newton never ceased to doubt that there was just one unique distinguished frame of reference, the one that is at rest relative to absolute space, in the existence of which he appears to have believed with an unshakeable tenacity. At the time of his early work on centrifugal force he may well have been unaware of the difficulty. Whatever the truth, he used a method that is open to criticism and yet managed to obtain a result that is quite correct. In modern terminology, he used a noncovariant method to obtain a correct covariant result.

Some way into the *Waste Book*,[30] we find what seem to have been Newton's first thoughts on the problem. As Newton's understanding of the precise nature of centrifugal force has been the topic of much discussion,[11a,11b,12a] let us look closely at what Newton has to say in general qualitative terms before we consider his solution to the problem of quantifying the *conatus a centro*. The question at issue is this: when a body moves in a circle (or some other closed curve), what are the basic dynamical elements which determine the body's motion? Prior to his work on the *Principia*, Newton has not left us many passages in his documents from which his attitude to this question can be deduced, but the passage which now follows is one (Newton's figure is reproduced in Fig. 10.7):

20. If a sphaere *oc* move within the concave shaeicall of cilindricall surface of the body *edf* circularly about the center *m*, it shall press upon the body *def* for when it is in *c* (supposeing the circle *bhc* to be described by its center of motion and the line *cg* a tangent to that circle at *o*) [it] moves towards *g* or the determination of its motion is towards *c* therefore if at that moment the body *efd* should cease to check it it would continually move in the line *cg* (ax, 1, 2) obliqly from the center *m*, but if the body *def* oppose it selfe to this endeavour keeping it equidistant from *m*, that

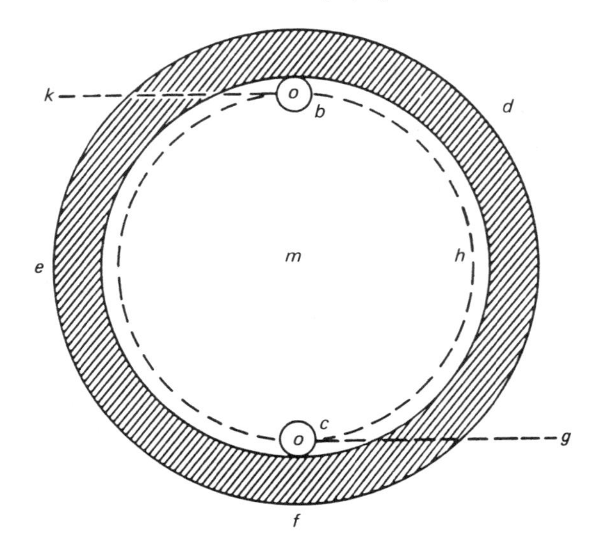

Fig. 10.7. Reproduced from: J. W. Herivel, *The Background to Newton's 'Principia'*, Clarendon Press, Oxford (1965).

is done by a continued checking or reflection of it from the tangent line in every point of the circle *cbh*, but the body *edf* cannot check and curbe the determination of the body *co* unless they continually presse upon one another . . . .
21. Hence it appears that all bodys moved circularly have an endeavour from the center about which they move, otherwise the body *oc* would not continuusly presse upon *edf*.

It is passages like the conclusion drawn in Newton's §21 which lead Herivel,[12a] Westfall,[11a] and Whiteside[11b] to conclude that, because Newton speaks of 'an endeavour from the center' he believed in the real existence of a force, acting radially from the centre and serving as one of the factors that determine the motion of the body. This conclusion is based in part on an apparent similarity between the terminology employed by Newton and Borelli[31] in his theory of the motion of the planets and the satellites of Jupiter, which was published in 1666, i.e., at much the same time that Newton was attacking the problem of centrifugal force. However, the similarity of terminology is, in my opinion, misleading. To demonstrate this, it is necessary to spell out the details of Borelli's theory of orbital motion around celestial bodies, large portions of which have been translated by Koyré in his study of the astronomical revolution.[32]

The most important feature of Borelli's scheme, which is a hybrid of Cartesian and Keplerian elements, is that the motion of an orbiting body is the outcome of the interplay of *three forces* that act simultaneously on the body. First, the orbiting body is held to have a natural tendency, akin to gravity, to move towards the central body about which it is orbiting. In the absence of other forces, it would simply fall into the central body. However, just as in Kepler's theory of planetary motion, the central body is assumed to rotate together with 'rays' that emanate radially from it, and these 'rays', rotating with the central body, impart a circular motion to the body. Now according to Borelli this 'circular motion confers on the moving body an *impetus* to move away from the centre of revolution, as we know from experience by spinning a wheel, or whirling a sling, whereby a stone acquires the *impetus* to move away from the centre of revolution'.[33] According to Borelli, the motion of the orbiting body is governed by these *three* forces; the 'rays' act as flails carrying the body in the transverse direction while the radial motion is governed by the balance (or, rather, imbalance in the case of noncircular motion) between the other two forces.

Although it must be admitted that in places Borelli is difficult to follow and occasionally seems to have at least some inkling of a more correct understanding of things, it is quite clear that the Borellian conception of the *impetus* from the centre being generated by a circular driving force is quite definitely fundamentally wrong. It is now worth considering in some detail what Descartes actually said on the subject. We shall see that he is much nearer to the truth but still exhibits a degree of confusion. The following passage is from §57 in Part III of his *Principles*, which we know Newton read closely (Descartes' figure has already been reproduced as Fig. 9.5 (p. 482), to which the reader is referred):

Now, inasmuch as it often happens that several different causes act simultaneously against the same body and some impede the effect of others; depending on whether we consider the former or the latter, we can say that this body strives or tends to move in [several] different directions at the same time. For example, the stone A, when rotated in the sling EA around E, definitely tends from A toward B, if all the combined causes which determine its movement from A to B are considered simultaneously; because it is in fact thus transported. But we can also say that, in accordance with the law of motion explained previously, the same stone tends toward C when it is at point A, if we consider only the force of its own movement [and agitation]; assuming AC to be a straight line which is tangent to the circle at point A. For [it is certain that], if this stone (coming from L) were to emerge from the sling when it reached point A, it would go toward C and not toward B; and the sling, though it impedes this effect, does not impede the striving [toward C].

Thus far Descartes' discussion, although only qualitative, is perfectly correct. The most important point is that Descartes, unlike Borelli, has the correct concept of inertia. In his scheme there is evidently no need for 'flails' to maintain the circular motion. On the basis of this passage (and several others in Descartes) one would say that in his mind there are only two factors determining the body's motion: rectilinear inertia and the deflecting effect of the sling. However, Descartes concludes the passage with this comment:

Finally, if instead of considering all the force of its motion, we pay attention to only one part of it, the effect of which is hindered by the sling; and if we distinguish this from the other part of the force, which achieves its effect; we shall say that the stone, when at point A, strives only [to move] towards D, or that it only attempts to recede from the center E along the straight line EAD.

Before discussing what appears to be a flagrant contradiction between this passage and the part immediately preceding it, let me make a remark that may be helpful. We recall that Descartes was not primarily interested in studying motion *per se*. He was far more interested in using motion to provide mechanical explanations for things like weight and light. In fact, by far the most important element in his explicative scheme was precisely the *conatus a centro*. This approach to the question is very noticeable in the passage we have just quoted. Descartes says explicitly that he is going to pay attention 'to only part' of all the force that determines the motion. It was the 'part' in which he was particularly interested. I believe that much of the confusion surrounding the meaning attached to the term *centrifugal force* by different authors in the seventeenth century can be attributed to this distortion of the dynamics of the problem. Although it is perfectly obvious that the young Newton was much more directly concerned with fundamental dynamical questions than Descartes, it is equally clear that until quite late in his development he undoubtedly had considerable interest in mechanical type explanations of gravity, for example. It was only in the extraordinary burst of concentrated work that led to the *Principia* that the fundamental dynamical aspect came completely to the forefront and Newton shook off the entanglement of mechanical explication. We should never forget the irony of the discovery of dynamics – the uncovering of the basic structure was an almost inadvertent by-product of Descartes' desire for explanation. And the reason why Huygens failed where Newton ultimately succeeded is that the Dutchman never did abandon explication as the prime aim of the exercise. This can therefore explain why in the early Newton we find frequent references to *conatus a centro* (and then later, after 1673, when Huygens coined the expression, to *centrifugal force*) without this necessarily meaning that Newton would employ the concept in a fundamental analysis of the forces that determine motion. What I mean by this will become clearer shortly.

But now to Descartes' inconsistency. In the first part of his §57 he says very emphatically that the stone, by virtue of 'the force of its own movement', will definitely travel along the instantaneous tangent. How then are we to understand his curious assertion in the final passage that the force of the stone's motion contains a component that is purely radial and outwards? Chapter 13 in *The World* (which was not available to Newton) casts some light on this question, for there Descartes invites us to 'imagine the inclination this stone has to move from A toward C as if it were composed of two other inclinations, of which one were to turn along the circle AB and the other to rise straight up along the line EAD' (I have changed Descartes' lettering to match Fig. 9.5). This suggests that, in the framework of his basic scheme, Descartes regarded *conatus a centro* as a kind of mathematical fiction, or, perhaps, as a concept that was not the most appropriate way of approaching the analysis of the overall factors that govern the stone's motion but was still of the greatest interest for his own explicatory purposes.

Quite what Newton made of Descartes' §57 we can only guess. Although the final part of §57 leads into two further paragraphs in which Descartes attempts to quantify the radial striving by means of a quite inappropriate analogy, to which neither Huygens' nor Newton's quantitative treatment of centrifugal force bears the remotest resemblance, Newton's very first qualitative discussion of the problem as quoted above does at least superficially parallel Descartes' §57. Let us now look at Newton's words more closely.

On the basis of Newton's §20 (p. 517–18), considered for the moment without reference to his §21, it is clear that his conception of circular motion is quite different from Borelli's (there is no trace of Borelli's 'flails' maintaining the circular motion). The primacy of inertial motion is absolutely clear, in agreement with the prominent role accorded to it in Newton's treatment of collision problems elsewhere in the *Waste Book*. If it were not for the deflecting effect of the cylindrical surface, the body 'would continually move in the line *cg*', the tangent at *c* to the circle in which it is moving. According to Newton's account, the circular motion is the outcome of just *two* dynamical factors. The first is its rectilinear inertia, the second is the 'reflection of it from the tangent line in every point of the circle'. Newton comments further, in the ellipsis at the end of §20, that the situation is just the same when a stone is whirled in a sling, the deflection in this case being produced by the tension in the string.

Given that in Newton's mind circular motion is the outcome of just two dynamical factors, and that these are the perfectly correct ones in the light of mature Newtonian dynamics, how then are we to understand Newton's expression 'endeavour from the center'? It is here necessary to point out that although the body's motion is governed by only two forces there is a third force at work, namely, the *force of reaction* of the body exerted on the body which deflects it. Note that Newton says the body and the cylindrical surface 'continually presse upon one another'. We have here too the germ of his Third Law, though not expressed nearly so clearly as in the work on collisions. Now this manifestation of the Third Law has a very surprising aspect. Although the fact that the body and the deflecting surface are constantly being brought into mutual contact by the body's having an inertial tendency to move forward uniformly and rectilinearly, i.e., 'obliqly from the center' as Newton correctly says, the acceleration (in the Newtonian sense) of the body from its rectilinear inertial path is purely radial, towards the centre. Thus, in accordance with mature Newtonian dynamics, the force with which the cylindrical surface acts on the body, which is parallel to the vectorial acceleration, is purely radial. Therefore, the reaction force, with which the body acts on the deflecting surface, is also purely radial but in the opposite direction, i.e., exactly away from the centre. This is the devilishly tricky and confusing part of the story; for the tension in the string or the pressure on the cylindrical surface really *are purely radial* even though the body's primal tendency is to move rectilinearly forward, i.e., 'obliqly from the center'. We have here the explanation for Huygens' apt yet confusing expression *centrifugal force*.

When physics teachers say 'there is no such thing as centrifugal force' but merely a tendency to move forward in a straight line, this is only part of the story; for the minute this tendency is thwarted, a reaction force comes into play and this force is both perfectly real and directed radially outward.

By the time Newton came to write the *Principia* he clearly understood this, for he completes one of the proofs of the formula for centrifugal force (it derives from the *Waste Book* work we are about to describe) with these words:[34] 'This is the centrifugal force, with which the body impels the circle; and to which the contrary force, wherewith the circle continually repels the body towards the centre, is equal.' The absence of such a statement in the *Waste Book* at the corresponding place in the proof suggests that Newton had not yet achieved such clarity in the 1660s. I find confirmation of this as yet imperfect sensing of the importance of the Third Law in circular motion in some very interesting definitions that Newton gives in a paper, written around 1670 and called *De gravitatione* (Ref. 12b, p. 121ff) (this paper will concern us greatly in Chap. 11). However, as this would take us into a rather complex discussion, I do not propose to take this matter any further here and will instead merely state my conclusions.

They are that one must distinguish carefully between Newton's instinctive understanding of the basic principles that govern motion and his use of the expression 'endeavour from the center' which, given its prominent employment by Descartes and Huygens, he could hardly avoid employing himself. I suspect that by it he meant one of two things, both of which can be taken from his very earliest comments in the *Waste Book* quoted above. We have already noted how in his §20 Newton says that the primal inertial tendency of the body strives to carry it 'obliqly from the center'; moreover he refers to this striving as an 'endeavour'. Thus, one possible meaning of 'endeavour from the center' could be as a convenient shorthand for this inertial tendency; if Newton did use the expression in this sense (and I believe there are several examples, to which we shall come, from the period after 1679 in which he did), then the use was perfectly correct in the light of mature Newtonian dynamics. However, in his §21 above, Newton unquestionably says that 'all bodys moved circularly have an endeavour from the center about which they move, otherwise the body *oc* would not continuously presse upon *edf*.' What does Newton mean here? He could just possibly mean oblique endeavour from the centre, as above, which would result in pressure on *edf*. However, I think more probably he does mean a radial endeavour, because, as we know, the pressure exerted on *edf* is radial, not oblique (this is what is so mysterious about the effect and made it so difficult to comprehend). However, that granted, I can find no evidence to suggest that Newton (or Huygens for that matter) ever regarded the radial endeavour as a fundamental factor determining the body's motion in the Borellian sense. Such a notion is totally lacking from the discussion in §20, in which Newton does approach the problem from the point of view of fundamental dynamics. Thus, even if Newton was not *consciously* aware until much later of the fact that centrifugal force is not a force that governs the motion of the considered body but a reaction force exerted by that body as a consequence of its being deflected from its primal inertial motion, I believe he knew instinctively perfectly well how to tackle problems of circular or more general motion.

In fact, the very clearest evidence that Newton did not regard centrifugal force as a real force in the sense of Borelli is to be found in his first proof, to which we now come, of the quantitative expression for centrifugal force.

Therefore, after this rather lengthy but, I feel, necessary introduction, let us now see how Newton succeeded in quantifying the 'endeavour from the center'. He uses what at the first sight seems, compared with Huygens' approach, to be a very crude, rough and ready, approach. Nevertheless, guided by a very sound instinct, he eventually succeeded in transforming it into what, from the point of view of the discovery of dynamics (which is to be contrasted with the discovery of the law of universal gravitation), proved to be his most important discovery – the recognition of the dynamical significance of Kepler's Second Law of planetary motion (the area law). Newton's first attempt comes immediately after the qualitative discussion considered above.

Suppose a body moves within a circular rim with speed v. When it passes a given point A, its velocity vector points in one direction. After half a circuit, at the diametrically opposite point B, the velocity vector has been exactly reversed. Thus, the force with which the rim acts on the body has been capable of exactly reversing the velocity. Newton's first conclusion was that the integrated effect of the force which the rim exerts in half a circuit is equal to twice the inherent force of the body. For the instantaneous speed of the body at any moment is v, and this measures the inherent force. But between A and B a velocity of this magnitude has been completely reversed. Newton therefore concluded that the force which had been applied was $2 \times v = 2v$, i.e., twice the inherent force.

But a correction he made shows how aware he was of the vectorial nature of force. Let us consider what actually happens in circular motion. At any instant, the rim is constantly deflecting the instantaneous velocity. Moreover the deflection is always exactly perpendicular to the instantaneous velocity; there is an instantaneous acceleration a. Of course, as the body proceeds around the rim, the direction of the instantaneous velocity is constantly changing, and with it the direction (but not the magnitude) of the instantaneous acceleration. Newton's idea was to get a measure of this acceleration by supposing it were *to act continuously in a straight line, after the manner of gravity*. Such a conceptual *rectification* of the acceleration is not, of course, what happens as the body passes around the rim, because the acceleration acts in different directions. Nevertheless, it still succeeds in completely reversing the motion in half a circuit. Newton therefore corrected his original conclusion, making an inequality out of an equality. His final conclusion was that [35] 'the whole force by which a body . . . indevours from the centre . . . in halfe a revolution is more than double to the force which is able to generate or destroy its motion, that is to the force with which it is moved'.

Newton's next attack on the problem is, in fact, the first major dynamical entry in the *Waste Book*,[36] though there is no doubt that chronologically it was a later entry. Newton continues the logic of the approach in the passage we have just considered. He imagines a ball constrained by a globe or ring to move within a circle. As a first step, he considered a body moving along the four sides of a square, being reflected elastically at the four corners by a circumscribed ring (Fig. 10.8).

When the body strikes the circle at b, the change in the normal component of its velocity is $2v \sin(\pi/4)$. Now suppose, as Newton did, that the square is made into a regular polygon of N sides, with N large. Let θ be the angle between a side of the polygon and the tangent to the circle at the point at which the side meets the polygon. Then the change in the normal component is $2v \sin\theta$ and this is $\approx 2v\theta$ when N is large and θ is therefore small. Measured from the centre of the circle, the angle turned through between two collisions, ϕ, is 2θ, i.e., the change is proportional to the angle turned through: $\delta v = v\phi$. In the limit $N \to \infty$ we therefore conclude that if all these little increments were to act in a straight line 'like the force of gravity'[37] the speed acquired would simply be $v\phi$, where ϕ is the arc through which the body has revolved; thus when $\phi = 1$, i.e., the body has travelled through one radian, the acceleration would have generated speed v. This is precisely Newton's conclusion:

If the ball *b* revolves about the center *n* the force by which it endeavours from the center *n* would beget soe much motion in a body as there is in *b* in the time that the body *b* moves the length of the semidiamiter *bn*.

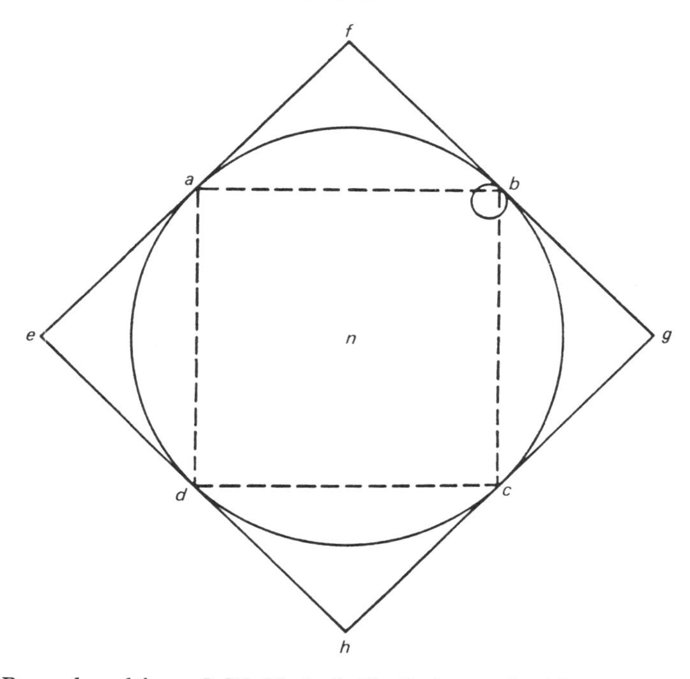

Fig. 10.8. Reproduced from: J. W. Herivel, *The Background to Newton's 'Principia'*, Clarendon Press, Oxford (1965).

Let us now consider Newton's derivation of the endeavour from the centre by polygonal approximation to a circle in the light of the earlier discussion of the status in Newton's mind of that endeavour. The fact is that such an endeavour appears nowhere in the entire derivation. The motion in the polygon is made up of rectilinear sections of pure inertial motion interrupted by reflections, at which an impressed force changes the inertial motion. Indeed, at the crucial point in the proof Newton uses the expression 'the force of all the reflections is to the force of the bodys motion'. Thus, there are only two forces which govern the motion, inertia and the force of reflection produced by the rim. Newton uses perfectly correct dynamical arguments to obtain the quantitative expression for the 'endeavour from the center'. He clearly cannot have regarded that endeavour as playing any part in determining the motion of the body. (It is inconceivable that Newton should have imagined a total transformation of the dynamical determining factors in the limit when the polygon becomes a perfect circle.) Nowhere more clearly than in this proof do we see the need to distinguish between Newton's use of a potentially confusing expression and his instinctive dynamical practice. Whatever Newton's words might suggest, his deeds do not reveal circular motion as ultimately determined by a balance between an impressed force towards the centre and a Borellian type centrifugal force. The moment he attacks the problem in earnest he correctly decomposes it into its two dynamically correct components: inertial motion modified by a supervening force.

To conclude this discussion of Newton's treatment of centrifugal force, we must also mention Newton's celebrated paper 'On circular motion', which is believed to post-date the *Waste Book* entries, but was certainly completed before 1669. This piece of work (written in Latin) is almost identical in its central point to Huygens', as is immediately evident from the figure (Fig. 10.9) and the opening proposition.[38]

The endeavour from the centre of a body A revolving in a circle AD towards D is of such a magnitude that in the time [corresponding to movement through] AD (which I set very small) it would carry it away from the circumference to a distance DB: since it would cover that distance in that time if only it were to move freely along the tangent without hindrance to its endeavour.

Note again that though Newton does indeed speak of endeavour from the centre the origin of the effect in inertial motion is explicit in the 'move freely along the tangent without hindrance to its endeavour', from which it is quite clear that Newton saw the unimpeded endeavour as being 'along the tangent'. We shall return to this paper in Sec. 10.6, where we consider its applications.

In this section I have wanted to emphasize the importance of centrifugal force in clarifying the basic concept of force and thereby the basic structure of dynamics. This conceptual significance is independent of the specific type of force which acts in any particular case.

Highly significant in this connection is the comment that Newton makes at the end of the entry in Folio 1 of the *Waste Book* and which shows his concern to find a quantitative measure of force valid under circumstances more general than the special case of circular motion:[39]

If the body *b* moved in an Ellipsis that its force in each point (if its motion in that point bee given) [will?] bee found by a tangent circle of Equall crookednesse with that point of the Ellipsis.

Quite apart from the suggestive evidence this provides of very early interest on Newton's part in the planetary problem,* the significant expression here is 'a tangent circle of Equall crookednesse with that point of the Ellipsis'. It shows how clearly Newton had understood the insight which Huygens had reached about seven years earlier – that when a body

* Even earlier evidence for Newton's interest in the problem is one of the first things he apparently wrote at all on the subject of dynamics, in a notebook that antedates even the *Waste Book*:[40] 'Note that the mean distances of the primary Planets from the Sunne are in *sesquialiter* [i.e. $\frac{3}{2}$ power] proportion to the periods of their revolutions in time.' How indicative this reference to Kepler's Third Law is of Newton's awareness of the importance of the quantitative aspect of motion and how clearly it distinguishes him from Descartes with his pictorial and qualitative explicatory approach!

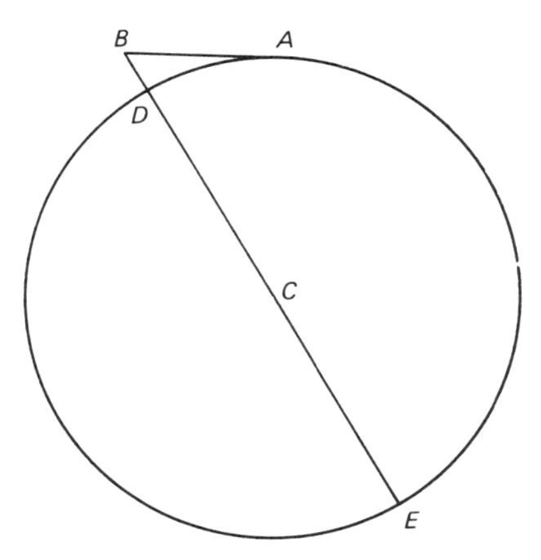

Fig. 10.9. Reproduced from: J. W. Herivel, *The Background to Newton's 'Principia'*, Clarendon Press, Oxford (1965).

moves uniformly in a circle it is being continuously drawn away from the instantaneous tangent with increments in the motion perpendicular to the tangent that increase *in direct proportion to the time,* i.e., exactly as in Galileo's law of free fall. But whereas Huygens' instinctive reaction was to exploit this identity of the phenomena in order to find a mechanical explanation of gravity in terms of a Cartesian centrifuge, i.e., to find confirmation for his preconception of the nature of gravity, Newton seems primarily to have been concerned to make absolutely sure he had a failsafe method of measuring the presence of a force from its characteristic manifestation. Having this method secure, he can then turn to the task of establishing the actual nature of the forces present and operating in nature. We see already in the young Newton the most valuable talent of the mason, the ability to lay absolutely secure foundations. We see the early intimation of that superb one sentence characterization in its Preface (written in 1686) of what the *Principia* is all about:[41] 'For the whole burden of philosophy seems to consist in this – from the phenomena of motions to investigate the forces of nature, and then from these forces to demonstrate the other phenomena.'

But, if this is to be done, the very first task is to make absolutely sure that you know the precise manner in which a force is revealed. This is why the work on centrifugal force was so important for the clarification of the force concept as a key element of dynamics. And it all developed out of the overall structure of incipient dynamics as laid down by Descartes coupled with Galileo's law of free fall, as we see again in the opening sentence of Newton's proof in his paper 'On circular motion': 'Now since this endeavour, provided it were to act in a straight line in the manner of gravity, would impel bodies through distances which are as the square of the times . . .'.

Some of the greatest discoveries in theoretical physics were made by men who had the ability to spot one particular phenomenon in nature and whose instinct told them that this was but one manifestation of a universal phenomenon. They had the ability to fathom the way nature works from the merest of hints. With good reason does Newton say in his Rules of Reasoning in Philosophy prefaced to Book III of the *Principia* 'nor are we to recede from the analogy of Nature, which is wont to be simple, and always consonant to itself'.

I am not suggesting that Newton was already consciously embarked on a systematic programme leading him inexorably to the *Principia*. He seems to have worked in a much too eclectic and desultory fashion for that. But he was testing the implications of Galileo's assured quantitative results when fitted into the overall framework supplied by Descartes.

10.6. Newton's early applications of the formula for centrifugal force

In the paper 'On circular motion', Newton finally caught up with Huygens' sophistication in handling the problem of centrifugal force and then promptly outclassed him in his applications. It will be worth looking in some detail at what he did. In essence, what Newton had found was that the acceleration towards the centre in circular motion is V^2/R, where V is the speed of the circular motion and R the radius of the circle. Now the time of a complete revolution in a circle is $2\pi R/V$, so that in this time a body subject to the acceleration V^2/R will, by the standard formula $s = \frac{1}{2}at^2$, traverse the distance

$$d = \frac{1}{2}\frac{V^2}{R}\left(\frac{2\pi R}{V}\right)^2 = 2\pi^2 R$$

This result underlies all the applications that Newton makes; for he considers objects rotating in circles of known radii at known speeds and compares the distances that will be generated in a given time by the corresponding accelerations. His first application was to complete the job begun by Galileo but vitiated by the remarkable mistake mentioned in Chap. 9. Newton did this by calculating the centrifugal force that acts on a body on the surface of the earth at the equator. Crucial here was the value that he adopted for the radius of the earth – not so much for his first calculation but for one that was to follow it. In fact, the figures that Newton's employs in this paper and also the calculations done on the so-called Vellum Manuscript[42] (which dates from the same period) demonstrate unambiguously that Newton took his value for the radius of the earth from Galileo's *Dialogo*, in the very part in which Galileo discusses the problem of bodies being thrown off the earth by its rotation.[43] As a result, Newton used a value that was only about 82% of the correct radius. He concluded that the centrifugal acceleration at the equator would[44] 'in a periodic day . . . impel a heavy body through $19\frac{3}{4}$ terrestrial semidiameters'. Newton then points out that such an acceleration would in one second move the body through $\frac{5}{9}$ inches.

He now compares this result with the effect of gravity, which he says 'moves heavy bodies down about 16 feet in one second'. This corresponds to an acceleration due to gravity of 32 feet/second^2, a figure with a very respectable accuracy. It will be recalled from Sec. 9.1 that accurate determination of g, the acceleration due to gravity, was difficult prior to Huygens' discovery of the formula $P = 2\pi\sqrt{(l/g)}$ for the period of a simple pendulum (published in 1673 in the *Horologium Oscillatorum*). The Vellum Manuscript (see Ref. 42) reveals that Newton found an alternative method of using pendula to determine g accurately. It was based on two important results relating to simple and conical pendula. In a conical pendulum the bob moves in a circle with the string maintaining a fixed angle θ to the vertical. Considering the problem of finding the equilibrium angle θ in a frame of reference rotating with the bob (in which there is a genuine centrifugal force), we find that three forces act on a bob of unit mass: the gravity force (vertically downwards), the tension T in the string, and the centrifugal force V^2/R (radially outwards) (Fig. 10.10). In the steady situation in which $\theta =$ const the radial component of T must balance the centrifugal force and the vertical component must balance the gravity force. It immediately follows from this, by elimination of T, that the ratio of the gravity force to the centrifugal force is equal to the tangent of $90° - \theta$, i.e., the ratio is tan φ, where $\varphi = 90° - \theta$. This was Newton's first result. Now let l be the length of the conical pendulum, R be the radius of the

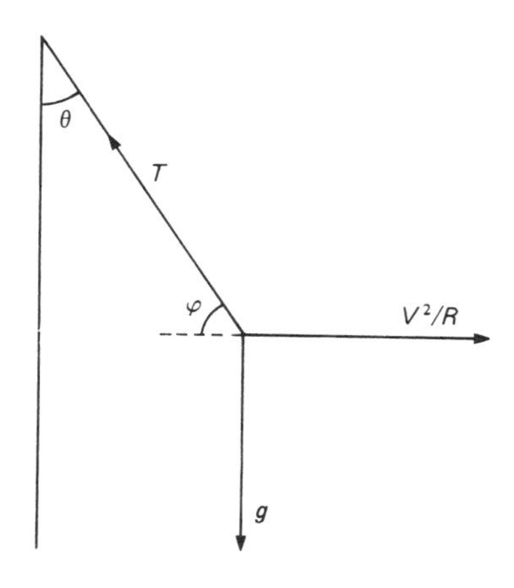

Fig. 10.10.

circle in which it moves, and V be the speed of its rotational motion. Then $g/(V^2/R) = \tan\varphi$. But $R = l\cos\varphi$, from which it follows that $V = \cos\varphi\sqrt{(gl/\sin\varphi)}$. The bob moves in a circle of circumference $2\pi l\cos\varphi$ and its period is therefore $P = 2\pi l\cos\varphi/V = 2\pi\sqrt{[(l/g)\sin\varphi]}$.

But we know from Huygens' formula for a simple pendulum (Sec. 9.1), that this is the period of a pendulum of length $l\sin\varphi$. In fact, the second result that Newton used was precisely this: that the time of complete revolution of a conical pendulum is equal to the period of oscillation (to and fro) of a simple pendulum in which the length is equal to the depth of the 'centre' of the conical pendulum below the point of support. Through this equivalence, i.e., by finding accurately the period of a simple pendulum, Newton implicitly used Huygens' formula and was able to obtain a good value for g, greatly improving the accuracy of the figures given for free fall by Galileo.

The upshot of this beautiful work was that Newton concluded[45] 'the force of gravity is of such a magnitude that it moves heavy bodies down about 16 feet in one second, that is about 350 times further in the same time than the endeavour from the centre [would move them], and thus the force of gravity is many times greater than what would prevent the rotation of the earth from causing bodies to recede from it and rise into the air.' A few years earlier Huygens had reached essentially the same conclusion, namely that the ratio was 265,[46] a value that purely fortuitously was much closer to the correct value of just over 288.[47]

These almost identical conclusions of Huygens and Newton demonstrate graphically how extraordinarily close Galileo himself had come to results which were at the very threshold of mature dynamics. We shall shortly see another example. They demonstrate equally clearly the decisive role in the discovery of dynamics played, on the one hand, by the Copernician revolution, which literally forced this problem into a position right at the top of the agenda, and, on the other, by Galileo's discovery of the law of free fall. In fact, as noted in Chap. 9 (p. 482), this work shows that Galileo's work alone would probably have led to the discovery of dynamics in some form or other without the contribution of Descartes. Herivel calls Newton's calculation of the centrifugal effect of the earth's rotation 'his first great practical discovery in dynamics'.[48]

But this was a problem in which fateful error followed fateful error. Galileo's had already meant that about three decades elapsed between the publication of the *Dialogo* and the correct demonstration that bodies would not be thrown off the earth by its rotation. Now, through no fault of either, Galileo's incorrect value for the radius of the earth led Newton to make a second error, which almost certainly delayed the discovery of the law of universal gravitation by nearly another twenty years. This is what happened. Where Huygens had been content to consider terrestrial applications of his formula for centrifugal force, Newton, the 'impetuous eagle', turned his eye to the moon and planets. He attempted to establish, first, the magnitude of the centrifugal acceleration corresponding to the moon's motion around the earth. Crucial here is the *distance of the moon* – the centrifugal force is $\omega^2 r$, where ω is the known angular velocity of the moon in its motion around the earth and r is the radius of the moon's orbit. From the astronomers Newton knew that r was, on average, about sixty times the radius of the earth. The ratio was known with good accuracy; the problem was in the value Newton used for the earth's radius. Because he took that about 18% too small he automatically obtained a value for the radius of the moon's orbit that was too small by the same amount. Using this incorrect value, he first concluded that 'the endeavour of the surface of the earth at the equator is about 12½ times greater than the endeavour of the moon to recede from the centre of the earth.' This result, being based on correct angular velocities and a correct *ratio* of the radii, is perfectly satisfactory. But Newton went further. He compared his correct determination of the acceleration of free fall on the earth with the moon's endeavour, for which he had, of course, obtained a value that was significantly too small. The fateful conclusion was 'the force of gravity [as at the surface of the earth] is 4000 and more times greater than the endeavour of the moon to recede from the centre of the earth.' In fact, Newton's '4000 and more' should have been 4375, whereas the correct result, which it is hard not to believe Newton was hoping to obtain, is 3600. It is interesting to note that the place at which Newton says '4000 and more' is the only point in the entire paper which does not give an exact figure. It almost looks as if he could not bring himself to face up to the brutal truth. We shall come back to this result in a moment.

Newton gives two further astronomical applications. The first is a rather curious comment related to the question of why the moon always presents the same face to the earth. He says: 'And if the moon's endeavour from the earth is the cause of her always presenting the same face to the earth, the endeavour of the lunar and terrestrial system to recede from the sun ought to be less than the endeavour of the moon to recede from the earth, otherwise the moon would look to the sun rather than to the earth.' It should be noted that at that time the sun's distance from the earth was still very inaccurately known (as we saw in Chap. 9, the first precise parallax determination of a distance – from which all distances followed automatically – within the solar system was made five years later in 1671–2). On the basis of this hypothesis Newton concluded (incorrectly, since his hypothesis was not sound, though it was certainly intriguing) that the ratio of the earth–moon distance to the earth–sun distance must be greater than 559.5/100 000.

The second application was, in the long term, of far greater significance. It is the first documented use of Kepler's Third Law to compare the strengths with which the centrifugal tendencies of the planets will cause them to recede from the sun. Newton says: 'Finally since in the primary planets the cubes of their distances from the sun are reciprocally as the squares of the numbers of revolutions in a given time: the endeavours of receding from the sun will be reciprocally as the squares of the distances from the sun.' Perhaps it will be slightly clearer to express this in symbols. An equivalent formulation of Kepler's Third Law is that the velocity of a planet in its orbit is proportional to $1/\sqrt{R}$; i.e., $V \propto 1/\sqrt{R}$. But its centrifugal tendency F is V^2/R, so that $F \propto 1/R^2$.

We can now appreciate the significance of the unfortunate mistake Newton made in the case of the moon. Had he used the correct radius of the earth* he would have found that the strength of the moon's endeavour away from the earth was reduced compared with the force of gravity on the surface of the earth by exactly the square of the ratio of the radius of the earth to the radius of the moon's orbit around the earth. In conjunction with the very striking result obtained for the planets, this must surely have made it seem almost inescapable that the moon is restrained from flying off from the earth by precisely the same force of gravity that causes apples to fall to the earth and that a precisely analogous force of gravity exerted by the sun is what keeps the planets in orbit around the great luminary. For the rate of decrease of the gravitational force would in both cases be as $1/R^2$.

The paper 'On circular motion' that we have been discussing is the primary piece of evidence in the thorny question of when precisely Newton did get his idea of universal gravitation. As is well known, when Halley presented the *Principia* to the Royal Society in 1686 on Newton's behalf, Hooke claimed that Newton had been guilty of plagiarism – that he, Newton, had used without acknowledgement his own idea that the planets are kept in orbit around the sun by a force that is[50] 'in a duplicate proportion to the Distance from the Center Reciprocall'.

This question is treated very fully by Herivel,[51] so I will give only a brief summary of some of the key points. The first is the curious fact that in the paper 'On circular motion' Newton does not anywhere mention the idea of gravity as being the force that restrains either the moon or the planets; he only compares the endeavours away from the centre. Of course, he may have had publication of the paper in mind and might not have wanted to give away valuable ideas. A second point is that the paper 'On circular motion' does not quite tally with the accounts that Newton gave several years after the publication of the *Principia*, i.e., after the unpleasant controversy with Hooke. There are three different versions of this post-*Principia* account, all more or less agreeing in their essentials. I quote

* Had Newton taken the trouble, quite accurate values of the radius of the earth were available and had been determined by Snel (1617)[49a] and Norwood (1635).[49b] Years later Newton used the value published by Picard in his *Mesure de la Terre* (1671).[49c]

the account from Whiston, who recalls how Newton told him (in 1694) that when in Woolsthorpe at the time of the plague he had had the idea that the earth's gravity might extend to the moon and how he had put the idea to the test:[52]

Upon Sir Isaac's First Trial, when he took a Degree of a great Circle on the Earth's Surface, whence a Degree at the Distance of the Moon was to be determined also, to be 60 measured Miles only, according to the gross Measures then in Use. He was, in some Degree, disappointed, and the Power that restrained the Moon in her Orbit, measured by the versed Sines of that Orbit, appeared not to be quite the same that was to be expected, had it been the Power of Gravity alone, by which the Moon was there influenc'd. Upon this Disappointment, which made Sir Isaac suspect that this Power was partly that of gravity, and partly that of Cartesius's Vortices, he threw aside the Paper of his Calculation and went to other Studies.

Two points about this are to be noted: (1) In 'On circular motion' Newton used Galileo's value for the radius of the earth, which does not match with what is said here, though the error is much the same; (2) Whiston's account implies an explicit test was made, which is not evident from 'On circular motion'. If any early document of Newton's were to come to light in which the idea of gravitational attraction is explicitly mentioned, then clearly the matter would be settled beyond all doubt. However, none has to date and the chances of this seem slim. Moreover, at the time of the correspondence in 1686 with Halley over Hooke's claim, Newton referred to a paper that he had written 'above 15 yeares ago & to ye best of my memory was writ 18 or 19 years ago'. From what he writes to Halley[53] it is virtually certain that this paper is the paper 'On circular motion'. What this means is that in 1686 Newton was unable to put his hands on any other documentary evidence that proved his point beyond any gainsaying (but, of course, it could well be true that 'he threw aside the Paper'). Thus, there remains a suspicion that Newton was not entirely frank in the matter.

Against this must be set the plausibility of the apple story. It has an authentic ring about it. Let us have the story again – it is surely worth the telling. According to Stukeley's account of his discussions in April 1726 with Newton:[54] 'After dinner, the weather being warm, we went into the garden and drank thea, under the shade of some apple trees, only he and myself. Amidst other discourse, he told me, he was just in the same situation, as when formerly, the notion of gravitation came into his mind. It was occasion'd by the fall of an apple, as he sat in a contemplative mood.'

According to Pemberton:[55] 'The first thoughts, which gave rise to his *Principia*, he had, when he retired from Cambridge in 1666 on account of the plague. As he sat alone in a garden, he fell into a speculation on the power of gravity: that as this power is not found sensibly diminished at the remotest distance from the center of the earth . . . that this power must extend much further than was usually thought; why not as high as the moon, said he to himself?'

And according to Newton himself:[56] 'And in the same year [1665 or 1666] I began to think of gravity extending to the orb of the Moon, and having found out how to estimate the force . . . compared the force requisite to keep the Moon in her Orb with the force of gravity at the surface of the earth, and found them to answer pretty nearly.'

On this question the jury is still out. For myself I believe Newton; it would be sad indeed if he had fabricated the perfect and poetic little story of the apple merely as a sordid means to make his case against Hooke seem plausible. However, it must be emphasized that there is a world of difference between gravity understood in a vague sense of attraction towards some body and the precise notion of universal gravitation, according to which each and every particle of matter attracts every other particle of matter with a precisely defined force that has a strength proportional to the mass of the attracting body and decreases inversely in proportion to the square of the distance from the attracting mass. For example, it is well known that Newton, like most of his contemporaries, dabbled in various mechanical explanations for gravity. As Wilson[57] has pointed out very clearly, such mechanical explanations can generate a force that impels apples and the moon to the earth and the planets to the sun. However, there is no universality in such mechanisms; they are specific to the centres to which the mechanical mechanisms propel the bodies; moreover, there is no law of action and reaction, so that the moon can be impelled towards the earth without the earth being simultaneously impelled towards the moon. Thus, it is easy to believe, as I do, that around 1666 Newton seriously considered the possibility that terrestrial gravity could extend to the moon and govern its motion. But this does not mean that at that time he had even the remotest inkling of the theory of universal gravitation. Indeed, as we shall see later, there is a great deal of evidence which suggests that *universal* gravitation did not occur to him as a serious possibility until very late indeed. But one thing is certain: by about 1669 at the latest Newton knew how to calculate the strengths of centrifugal endeavours. He therefore knew in principle the strength of the force that must counter the centrifugal tendency – whatever may have been the nature of that force. He had completed some of the most important preparatory work for what would prove to be his greatest feat.

10.7. The development of Newtonian dynamics

It was argued in Sec. 10.2 that the main significance of Newton's early work on dynamics lay in his clarification of general principles rather than in the specific results he obtained, important as these were. In particular, we have now seen how Newton developed and tested different concepts of force. Throughout his life, Newton employed all three of these concepts, using the same word force (qualified if he felt it necessary to distinguish between them by adjectives such as *inherent, impressed,* etc.) to denote them. Among the three, the *practical* importance of inherent force was quite different from that of the other two. *De facto* it came to signify for Newton the phenomenon it was meant to explain, i.e., he regarded the inherent force as something akin to an animistic force that kept any given body following its inertial motion. But that was its sole purpose; it was a cause invented to explain a phenomenon Newton felt unable to accept at its face value, as a simple datum beyond explanation. During the composition of the *Principia* Newton at least partially shook off this relic of medievalism, which survived well into the seventeenth century, but the concept was never completely exorcised. We shall return to this important conceptual matter in Chap. 12, since it is closely related to the concept of mass and Einstein's interpretation of what Mach's Principle should achieve.

On the general question of Newton's development, the unpublished early work on dynamics and the *Principia* reveal to my mind remarkably little evidence of any really significant shift in Newton's attitude to the most basic elements of dynamics. He used essentially the same three force concepts in all his work up to and including the *Principia* and moreover used them throughout with almost complete mastery. (I am referring here to Newton's mastery of the physics – the rigour of his mathematics is another matter.) It is true that the words used to describe the concepts changed, and that was significant for the *exposition* of the work. But I do not think it made much difference to the mental processes by which Newton arrived at his great results or the essential content of the concepts.

In fact, the early and mature Newtonian dynamics are not distinguished by any fundamental change in the basic concepts but rather by an *extension* of their application. This occurred in two directions: (1) in the early Newtonian dynamics, all forces were assumed to be *contact* forces; the first great extension of the *Principia* was in the consideration of forces acting over great distances – it was the transition from direct contact to action at a distance; (2) the second great extension was that the early Newtonian dynamics dealt with problems *with very special initial conditions,* whereas in the *Principia* Newton made the breakthrough to being able to handle problems with *arbitrary initial conditions.** The story of the *Principia* is the story of how the mature Newton was prodded by Hooke

* It is above all the absence in Newton's early work of problems with general initial conditions and nonconstant forces that makes it difficult to establish the extent to which he fully comprehended the significance and range of the principles he employed, or, for example, to determine precisely what he understood by endeavour from the centre. What is certain is that the later work joins on seamlessly to the earlier and there is no point at which a principle adopted earlier is abandoned or refuted.

and Halley into applying exactly the principles that the young Newton had developed on a restricted set of problems to a much wider set. It was, as already suggested, the transition from cracking a few nuts to cracking the world. But he used the same nutcrackers.

He had to use the same nutcrackers because their design had been almost, though not quite completely, predetermined for him by Galileo and Descartes. From Galileo he had the parallelogram rule for *physical* motions, i.e., the rule that when a body executes a compound motion under the influence of two simultaneously acting causes the resultant motion is the diagonal of the parallelogram of the motions corresponding to the two causes acting separately. Equally important, from Galileo he had the law of free fall with the vitally important fact that *the velocity increases in proportion to the time* (and not in proportion to the distance traversed, as had hitherto been widely supposed). This gave him the paradigm of *physical force* as something that changes the motion by adding equal increments of motion in equal increments of time. It is not possible to over-emphasize the importance of this result; together with the integrated form of this law (i.e., that the distance increases as the square of the time), it was the foundation of most of Newton's great applications of dynamics. Very important here was his early realization, in which he was anticipated by Huygens, that in the case of a force that varies either in space or time the infinitesimal increments of motion are always proportional to the infinitesimal interval of time over which the force is applied (and that they are in the direction in which the force is applied). It was by systematic application of this principle (in conjunction with the parallelogram rule) that Newton performed his greatest feat: the demonstration from Kepler's laws of the inverse square law of gravitation.

From Descartes Newton had the significantly improved (as compared with Galileo) concept of inertial motion.* This provided him with the benchmark from which to operate. From Descartes he also had the heightened awareness, to a large degree absent in Galileo, that the primary task of dynamics is to study how bodies deflect each other from their respective inertial motions. Rather more insistently than Galileo, Descartes also provided the stimulus to attack the problem of centrifugal force. He also demolished the residual Galilean vestiges of the Aristotelian–Pythagorean cosmos and made motion truly universal.

If Galileo and Descartes each provided one arm of the nutcrackers, what was Newton's contribution? The linchpin, of course: *Lex Tertia* was the insight that linked the two arms together. Together with the law of inertia and the embryonic Second Law, it supplied Newton with an

* Despite what Westfall[11a] and Whiteside[11b] suggest, I can find no evidence at all which indicates that Newton ever doubted the universal validity of rectilinear inertia (reverting, for example, to Galilean circular inertia for problems on cosmic scales). Whiteside's arguments have been countered by Herivel.[58]

essentally complete set of rules for calculating the motions of bodies. For the first time since Copernicus set the earth in motion, a coherent scheme for calculating the motions of bodies was beginning to appear. Before Copernicus – and even for a while after him – cosmology had supplied a fixed framework. The early astronomers saw their task as the description of the motion of individual bodies within that fixed framework. But Newton was confronted with the task of describing the celestial acrobatics implicit in Aristarchus's reaching out by trigonometry to sense the distance of the sun. And although the correct way to formulate the law of inertia would yet present – and still does present – severe conceptual difficulties, the other two laws, already moderately well clarified, supplied the general rules governing the way bodies moving through space deflect one another from their respective inertial motions.

If in 1666 Newton already had such a good understanding of the basic elements of dynamics and an apparent interest in the problem of the elliptical planetary orbits – the problem that more than any other brought on the full flowering of Newton's genius and the unshakeable foundation of his mature dynamics – why did he not solve that problem then?

I think there were at least two reasons. Probably the most important was that in 1666 Newton lacked the stimulus to attack the problem in its full generality. For although the subsequent history was to show how basically sound his early work was, Newton could not have known this at the time. This is closely connected with the question of the date at which Newton had the concept of universal gravitation and the related notion of forces acting at a distance towards a definite centre rather than through direct contact. We have already noted that if Newton did have the idea that the force which keeps the planets in their orbits is the same force as is responsible for terrestrial gravity it was only the merest glimpse of the fully developed concept, which quite definitely did not appear before the early 1680s. In fact, the full significance of what he was finding probably only dawned on Newton in the early and mid 1680s as he unearthed more and more evidence of the universality of gravity and the quite extraordinary diversity of problems that this opened up to mathematical analysis. It was only then that he grasped the power of the method which he had forged nearly twenty years earlier. One could say that Newton developed a theory of forces in the 1660s without being aware that they actually existed in nature!

Of course, Newton was perfectly aware of the existence of forces of impact, but that may well have exhausted in his mind the spectrum of possible forces. He was probably still very much under the spell of the Cartesian world view, i.e., the mechanical picture of the world in which the motion of macroscopic bodies is explained by the shapes, motions, and collisions of microscopic bodies. But such a philosophy does not lend itself to a great deal of mathematical analysis. Newton had solved the problem of collisions and there did not seem to be all that much left to do. He undoubtedly understood that a force of some kind kept the planets from flying away from the sun in rectilinear inertial motion but may well have supposed, like Descartes, that it was supplied by collisions with particles in a Cartesian vortex, as the quotation earlier from Whiston suggests. But Cartesian physics was all so qualitative; trying to subject it to mathematical analysis was a bit like trying to pin a jelly to a wall. When Newton discovered it more than 15 years later with all its ramifications, universal gravitation had a great liberating effect. It suddenly created a host of problems amenable to exact solution. Mathematical physics as a major growth industry – still growing today in a most impressive fashion – dates from the discovery of universal gravitation. But back in the 1660s Newton probably lacked the stimulus and also the clues as to how he should proceed. He stood on Darien but could not see the Pacific because his vision was befogged by Cartesian physics.

It should also be borne in mind that dynamics was only one of Newton's many interests, which included optics, mathematics, alchemy, and theology; with the last two especially he became particularly absorbed.[2] Moreover, before he made his great discoveries in dynamics he could not know just how significant that subject would prove to be. It is a profound mistake to attempt to evaluate Newton's priorities with a modern awareness of the relative significance of the various subjects he studied.

A quite different but perhaps almost as important problem was that he had not yet quite learnt *how to deal with time* and the *variability* of the planets' motions. This is a most important technical matter, which we will consider in Sec. 10.9.

Thus, having mastered pretty well all the basic elements of dynamics, Newton more or less laid the subject on one side for more than a decade. If he had not meanwhile made something of a name for himself by his discoveries in mathematics and optics (see, for example, Westfall[2]), his brilliant early work in dynamics would probably have come to nothing. But powerful forces were at work. Great insights once achieved have a way of working upon receptive minds. As noted in Sec. 10.1, vitally important steps on the road to dynamics were the publication in 1669 of the correct laws of impact by Wallis, Wren, and Huygens and the publication in 1673 by Huygens as an appendix to his *Horologium Oscillatorum* of the correct formulas for centrifugal force. It was this in particular that got things moving again. Several people, most notably Wren and Halley, had the idea of using Huygens' formula in conjunction with Kepler's Third Law to do what Newton had done already several years earlier: to find the strength of the force with which the planets attempt to recede from the sun. But what is only implicit in the early Newton paper now becomes quite explicit. The search is on for the force that *attracts* the planets and thus balances the centrifugal tendency.

10.8. The Hooke–Newton correspondence of 1679

Even before Huygens published the formulas for centrifugal force, Hooke (1635-1702) had achieved a considerable degree of clarity as to the qualitative dynamical elements involved in the problem of orbital motion. From about 1664 he had been pondering the problem in a basically Cartesian framework, and in May 1666 presented a paper to the Royal Society on the subject. In it, he said:[59]

I have often wondered why the planets should move about the sun according to Copernicus's suggestion, being not included in any solid orbs . . . nor tied to it, as their center, by any visible strings; and neither depart from it beyond such a degree, nor yet move in a strait line, as all bodies that have but one single impulse, ought to do . . . But all celestial bodies, being regular solid bodies, and moved in a fluid, and yet moved in a circular or elliptical lines, and not strait, must have some other cause, besides the first impressed impulse, that must bend their motion into that curve.

Hooke illustrated these suggestive comments, which have the effect of highlighting the specific problem of the planets within a generally Cartesian approach, with a notable demonstration employing a conical pendulum. He quite correctly sensed that such a pendulum does mimic the planetary situation and reveals the two essential elements of the problem: the inertial tendency of a body to continue in the direction of a motion imparted to it at any instant and a deflecting force towards a fixed centre (produced in the case of the pendulum by the horizontal projection of the tension in the string). By varying the initial conditions, Hooke was able to generate circular and elliptical orbits and thus show qualitatively how orbits something like the planetary orbits could be generated by the interaction of inertia and a deflecting force. Moreover, by hanging a little supplementary pendulum below the bob of the main pendulum he was able to reproduce qualitatively the motion of the moon around the earth as the earth orbits the sun. These simple but beautiful experiments of Hooke represent the first clear formulation of the planetary problem in qualitatively correct terms. Indeed, the analogy would be perfect but for the fact that the deflecting force towards the centre increases with the distance from the centre instead of decreasing with the square of the distance. For this reason it is not possible to obtain elliptical orbits with the force centre at one of the foci – it is always at the centre of the ellipse. Nevertheless, Hooke must clearly be given the credit for *posing the problem* for the first time in the correct form – always the first step to finding the solution. Newton *may* have had the idea at about the same time but no firm evidence of this survives beyond the tantalizing two or three sentences in 'On circular motion'.

We have already mentioned another stimulus to the solution of the problem which also occurred in 1666. It was the publication by Borelli

(1608–1679) of his book *Theoricae Mediceorum Planetarum ex Causis Physicis Deductae*, ostensibly a physical theory of the motion of the moons of Jupiter (which, following Galileo, he called the Medicean planets) but in fact simultaneously a theory of the motion of the planets in the Copernican framework (which, to avoid offending the Inquisition, he did not want to emphasize). As we have seen, Borelli's theory is an interesting mixture of Keplerian ideas (Kepler's influence is already manifest in the title of the book, which recalls the subtitle of the *Astronomia Nova*) with the important new idea that the planets are subject to a natural tendency towards the sun. Borelli is above all important in insisting, like Kepler and Hooke, that celestial motions are governed by essentially the same physical causes as terrestrial motions. For, despite widespread knowledge of the ideas of Kepler and Descartes, there were still eminent astronomers in the mid seventeenth century who held to the Ptolemaic and Copernican notion that celestial motions were purely geometrokinetic and subject to quite different rules from the terrestrial motions. And there were still several who held to the Tychonic cosmology, i.e., who assumed an earth at rest around which the sun revolved with the planets in turn revolving around the sun. These questions have been very well discussed by Russell[60] and Wilson.[57] Borelli's work, for which the reader is referred to Koyré,[32] is particularly interesting on account of his frank questioning of why the celestial bodies neither fall into the respective central bodies nor depart ever further and further away from them. However, as regards the dynamical formulation of the problem, Borelli's scheme is manifestly inferior to Hooke's, since it requires three forces instead of two and the law of inertia is at the very best only most obscurely understood.

Thus, from about 1666 the idea of some force attracting the planets towards the sun began to gain quite wide currency. In the next three or four years Hooke seems to have developed his ideas with rather more precision and in 1670 gave a lecture at Gresham College in London on the problem of proving the rotation of the earth, at the end of which he returned to the problem of the planetary motions. This lecture was published in 1674 in his *Attempt to Prove the Motion of the Earth*. I reproduce in its entirety the final section of the booklet devoted to this problem, since apart from the greater precision with which the problem is posed the passage also contains a remarkable anticipation of some aspects at least of the theory of universal gravitation. Here it is:[61]

At which time also I shall explain a System of the World differing in many particulars from any yet known, answering in all things to the common Rules of Mechanical Motions: This depends upon three Suppositions. First, That all Coelestial Bodies whatsoever, have an attraction or gravitating power towards their own Centers, whereby they attract not only their own parts, and keep them from flying from them, as we may observe the Earth to do, but that they do also attract all the other Coelestial Bodies that are within the sphere of their activity;

> and consequently that not only the Sun and Moon have an influence upon the body and motion of the Earth, and the Earth upon them, but that ☿ also ♀, ♂, ♃, and ☊ [Mercury, Venus, Mars, Jupiter, and Saturn] by their attractive powers, have a considerable influence upon its motion as in the same manner the corresponding attractive power of the Earth hath a considerable influence upon every one of their motions also. The second supposition is this, That all bodies whatsoever that are put into a direct and simple motion, will so continue to move forward in a streight line, till they are by some other effectual powers deflected and bent into a Motion, describing a Circle, Ellipsis, or some other more compounded Curve Line. The third supposition is, That these attractive powers are so much the more powerful in operating, by how much the nearer the body wrought upon is to their own Centers. Now what these several degrees are I have not yet experimentally verified; but it is a notion, which if fully prosecuted as it ought to be, will mightily assist the Astronomer to reduce all the Coelestial Motions to a certain rule, which I doubt will never be done true without it. He that understands the nature of the Circular Pendulum and Circular Motion, will easily understand the whole ground of this Principle, and will know where to find direction in Nature for the true stating thereof. This I only hint at present to such as have ability and opportunity of prosecuting this Inquiry, and are not wanting of Industry for observing and calculating, wishing heartily such may be found, having my self many other things in hand which I would first compleat and therefore cannot so well attend it. But this I durst promise the Undertaker, that he will find all the great Motions of the World to be influenced by this Principle, and that the true understanding thereof will be the true perfection of Astronomy.

Within a decade Hooke's confident prediction in the final sentence was to be triumphantly vindicated – and again, just as he predicted, by someone else, not himself. The saddest part of the story is that Hooke found that person, badgered him into making the greatest discovery ever in science, and yet got nothing but pain and misery as reward. As long as historians study the discovery of dynamics they will have to consider the relative merits of Hooke and Newton in the making of this key discovery which crowned the two-millennial undertaking. Perhaps the best that one can do is relate the facts as they are known. And it is certainly true, as Westfall points out,[62] that the first clear and correct statement of the dynamic elements of orbital motion on record is due to Hooke.

However, despite some further suggestive evidence[50] it is not clear whether Hooke should be credited with the precise notion that Newton eventually formulated, namely, that *each piece of matter* always exerts and is subject to gravitational attraction. Above Hooke merely speaks of mutual attraction *to the centres* of the various bodies; in his *Cometa* of 1679 he says that comets can lose the gravitating principle they have when entire if their parts become 'confounded or jumbled' (Ref. 57, p. 152). Also, although Hooke mentions the moon above (and in Ref. 50) he does not say explicitly that it is kept in its orbit by exactly the same gravitational attraction as makes objects fall to the earth. Still less is there any suggestion of a comparison of the strength of gravity on the surface of the earth and the strength of the force needed to keep the moon in its orbit. This crucial insight seems to have been Newton's and Newton's alone. Nobody else seems to have grasped with full clarity and awareness the essential point that there must be complete identity between the falling of apples in country gardens and the motion of the moon – that it too falls towards the earth.

The final stage in the gradual clarification of the planetary problem came, as pointed out in Sec. 10.1, with the publication in 1673 by Huygens of the correct formula for centrifugal force. This enabled both Wren and Halley to resolve the one point that Hooke left undecided in his Gresham College lecture of 1670 – the verification of the 'several degrees' of 'these attractive powers'. It appears that they passed on to Hooke, who was a weak mathematician,* the result of their own application of Huygens' formula to Kepler's Third Law, for by 1679 Hooke was sure in his own mind that the strength of the attractive power decreased as the square of the distance.

After this introduction, let us now look at the most important exchange in the Hooke–Newton correspondence.

This was initiated by Hooke on 24 November 1679. He wrote to Newton in his capacity as Secretary to the Royal Society and hoped that Newton would continue his 'former favours to the Society by communicating what shall occur to you that is philosophicall'. Hooke continued:[64] 'For my own part I shall take it as a great favour if you shall please to communicate by Letter your objections against any hypothesis or opinion of mine, And particularly if you will let me know your thoughts of that of compounding the celestiall motions of the planetts of a direct motion by the tangent & an attractive motion towards the centrall body.'

Newton responded on 28 November somewhat evasively, saying that he had little inclination at the present to philosophy. Of Hooke's specific proposal he said that he did not recall hearing of 'your Hypotheses of compounding ye celestial motions of ye Planets, of a direct motion by the tangt to ye curve'. However, he did propose to Hooke a possible test of the rotation of the earth based on the fact that a body at the top of a high tower will actually have a faster rotation speed than one at its bottom, so that, if it is dropped, then 'outrunning ye parts of ye earth [it] will shoot forward to ye east side of the perpendicular . . . quite contrary to ye opinion of ye vulgar'.

Newton's description of his proposal takes us straight back to the problems that figure so prominently in Galileo's *Dialogo*:[65] 'Suppose then

* For an evaluation of Hooke and his many other remarkable achievements the reader is referred to Westfall's article on him in the *Dictionary of Scientific Biography*.[63] He was a brilliant but rather sad figure, characterized by fertility of imagination and a powerful intuition but not a capacity to elaborate the full potential of his ideas. One of his greatest achievements was the invention and development in 1658/9 for Robert Boyle of the air pump in its modern form.

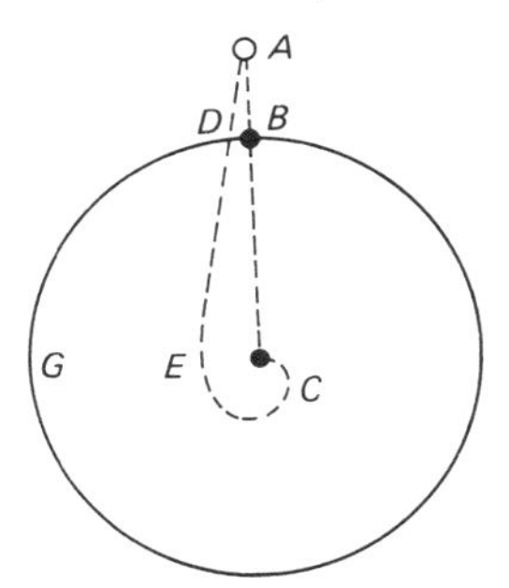

Fig. 10.11.

BDG represents the Globe of ye Earth carried round once a day about its center C from west to east according to ye order of ye letters BDG; & let A be a heavy body suspended in the Air & moving round with the earth so as perpetually to hang over ye same point thereof B. Then imagin this body B let fall & it's gravity will give it a new motion towards ye center of ye Earth without diminishing ye old one from west to east . . .'. Newton's figure, which we may have to thank for the final synthesis of dynamics, is reproduced in Fig. 10.11.

Hooke replied on 9 December, agreeing with Newton's conclusion, but he then ventured to suggest that Newton had made a mistake with the shape of the curve he had communicated in his letter. Westfall has pointed out how much Hooke's correction appears to have mortified Newton:[66] more than thirty years later he referred to the mistake as 'a negligent stroke with his pen'. Hooke's comment was as follows:[67] 'But as to the curve Line which you seem to suppose it to Desend by (though that was not then at all Discoursed of) Vizt a kind of spirall [Fig. 10.12] which after sume few revolutions Leave it in the Center of the Earth my theory

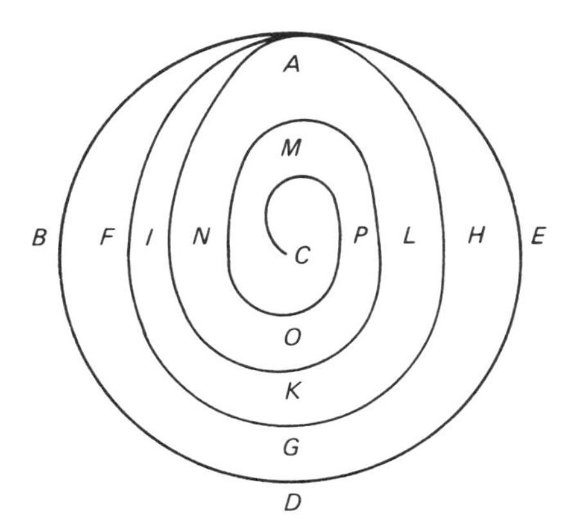

Fig. 10.12.

of circular motion makes me suppose it would be very differing and nothing att all akin to a spirall but rather a kind Elleptueid.'

Newton replied to this letter on 13 December, granting that he had made a mistake:[68] 'And also that if its gravity be supposed uniform it will not descend in a spiral to ye very center but circulate wth an alternate ascent & descent made by it's *vis centrifuga* & gravity alternately overballancing one another.'

Westfall[69] and Whiteside[11b] see Newton's use here of the notion of the *vis centrifuga* and gravity alternately 'overballancing' one another as evidence that he was still confused about the correct formulation of the dynamical elements of the problem, treating it in Borellian fashion. However, Newton is, in fact, merely agreeing with Hooke's objection, and his own formulation of the problem in distinctly Galilean terms on 28 November had not revealed any confusion. The point I should like to make is that as soon as Newton fixes his attention on gravity as the cause of the body's deflection from rectilinear inertial motion his account is perfectly clear and derives straight from Galileo. This comes out especially in the passage that immediately follows the last quotation. It is the second clear example in the pre-*Principia* documents where Newton outlines in qualitative terms the basic dynamical elements involved in orbital motion. There are again just two: rectilinear inertia and gravity, which causes the deflection. Borellian confusion is quite absent. Here is the passage (my italics):

Yet I imagin ye body will not describe an Ellipsœid but rather such a figure as is represented by AFOGHIKL &c. [Fig. 10.13] Suppose A ye body, C ye center of ye earth, ABDE quartered with perpendicular diameters AD, BE, wch cut ye said curve in F and G; AM ye tangent in wch ye body moved before it began to fall & GN a line drawn parallel to ye tangent. When ye body descending through ye earth (supposed pervious) arrives at G, the determination of its motion shall not be towards N but towards ye coast between N & D. *For ye motion of ye body at G is compounded of ye motion it had at A towards M, & of all ye innumerable converging motions successively generated by ye impresses of gravity in every moment* of it's passage

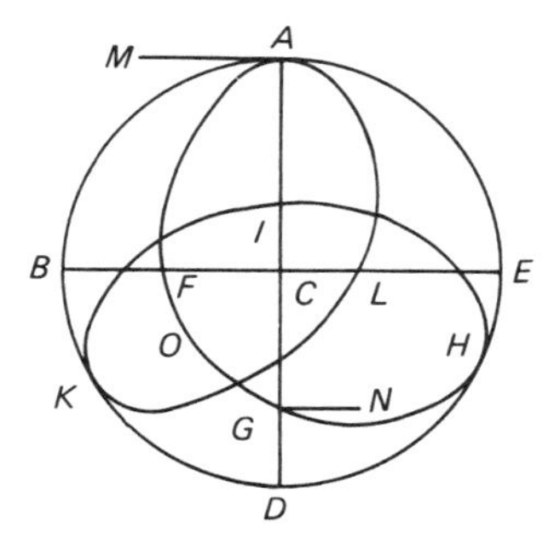

Fig. 10.13.

from A to G: The motion from A to M being in a parallel to GN inclines not ye body to verge from ye line GN. The innumerable & infinitly little motions (for I here consider motion according to ye method of indivisibles) continually generated by gravity in its passage from A to F incline it to verge from GD towards D, & ye like motions generated in its passage from F to G incline it to verge from GN towards C. *But these motions are proportional to ye time they are generated in*, & the time of passing from A to F (by reason of ye longer journey & slower motion) is greater then ye time of passing from F to G. And therefore ye motions generated in AF shall exceed those generated in FG . . . . Thus I conceive it would be if gravity were ye same at all distances from ye center. But if it be supposed greater nearer ye center . . . .

Although not without certain minor technical flaws, this passage demonstrates a complete mastery of the basic principles. Note in particular the parts I have italicized. There is not a trace here of any mysterious or imperfectly understood concept of *vis centrifuga*. It is simply an elaboration of the words Newton had used on 28 November when he posed the problem in the first place.

Let me emphasize again the point about this being pure Galileo. If it were to be translated into Renaissance Italian and inserted into the *Dialogo*, it could pass for Salviati himself. Nowhere in Newton's writings does one see more clearly than here the extent to which Galileo predetermined the structure of dynamics. We have already noted that much of Galileo's motionics survived the transition to dynamics unscathed. In this letter of Newton to Hooke we see how the transition occurred. In terms of mathematical operations, Newton was doing *exactly the same* as Galileo. And we know that Newton studied closely the relevant passage in the *Dialogo*, since he used Galileo's numerical data in the paper we discussed earlier (and corrected that egregious mistake of Galileo). No wonder Newton credited Galileo with the discovery of the first two laws of motion.[70] The basic dynamic elements of this particular problem are all there in Galileo; the real difference is in the *metaphysical framework* in which they are conceived to operate. Here the shift is huge; Descartes and the passage of time (the one as ruthless as the other) had done their work. The one really genuine conceptual innovation on Newton's part, referred to at the end of the above passage, is in the supposition that the force of gravity could vary with position. In the work to be described in the next section we shall see how Newton's far greater mastery of mathematics enabled him to handle this significant generalization of Galileo's ideas.

In the absence of firm documentary evidence, and in view of a certain coyness on Newton's part in his correspondence with Halley about the part played by Hooke,[71] it does seem that Hooke's innovation was the insistence on the reality of the attractive force to a centre. Whether or not Newton had it as a clear idea before the Hooke–Newton correspondence is a moot point; that Hooke had it is beyond question. But most important of all was the fact that his persistent probing and testing of the reclusive and crusty Newton finally forced him to take such forces seriously. Indeed, later in the same letter of 13 December Newton said: 'Your acute Letter having put me upon considering thus far ye species of this curve, I might add something about its description by points *quam proximè*'.

But what he did in the way of determining the curve as accurately as possible Newton kept to himself. Luckily it appears that the fruit of this correspondence, which was as beautiful as anything Newton (or anyone else for that matter) ever did, may well have survived. We have reached the point at which the work of Kepler and the astronomers at long last receives the attention it deserves.

10.9. The area law, Newton's treatment of time, and the solution to the Kepler problem

Near the end of Sec. 10.7 I said that in the 1660s Newton had not yet quite learnt how to deal with time and the variability of the planets' motions. The reason why he and Huygens were able to solve the problem of centrifugal force was two-fold: (1) the simplicity of circular as compared with elliptical geometry; in particular, the acceleration is always exactly towards the centre of the circle, so that the deviation from the tangent is always a direct measure of the force acting. (2) The fact that uniform circular motion *is its own clock*. For the uniform circulation of the body gives a direct measure of time in the very geometry of the problem under consideration. *Time is simply measured by the distance traversed along the perimeter of the circle by the body.* This can be seen particularly clearly by referring back to Fig. 10.9 and the text which opens Newton's paper 'On circular motion'. Note in particular the expression 'in the time [corresponding to movement through] AD'. In fact, the interpolation here, 'corresponding to movement through', is Herivel's. Newton simply says, 'in the time AD (*in tempore AD*)'. This embedding of time into spatial geometry is one of the most characteristic features of Newton's dynamical techniques; I suspect he may have learnt it from Galileo's treatment of the earth's rotation (see the discussion in the previous chapter, p. 480). (If he did, he certainly never let it trip him up in the way it did Galileo.)

The proper treatment of time is so all important in dynamics because *forces generate motion in direct proportion to time*. This was the key insight Newton won from Galileo's work. Thus, to measure the strength of the force, it is necessary to know the deflection *in unit time*. This is what makes uniform circular motion comparatively easy to treat, because the time elapsed is always directly proportional to the distance traversed around the circle; the deflection in unit time becomes the deflection in unit distance.

In his comment (p. 526) about the force acting when a body moves in an ellipse, Newton took only the first step towards solving the problem of determining the strength of the force. First, the statement as he made it presupposes that the force acts at *right angles* to the instantaneous tangent. Second, the deviation from the tangent can only serve as a measure of the force if one knows the time taken to traverse unit distance along the direction of the tangent. But in the elliptical motion of the planets equal distances along the perimeter are not traversed in equal times. It was the solution to this problem that Newton lacked in 1666; it was a technical difficulty associated with specific problem solving rather than a question of one of the basic principles of dynamics.

It was here that Kepler, 75 years after his first two great discoveries, at long last came into his own (alas with very stingy recognition on Newton's part). For not only did his laws force people to confront the problem of elliptical orbits described about the sun at one focus. His other laws of planetary motion gave Newton the absolutely essential hints without which the problem would certainly not have been solved in the seventeenth century. The part played by Kepler's Third Law in making it possible, through the formula for centrifugal force, to determine the distance dependence of gravity is well known and has already been emphasized. What is much less widely appreciated is that Kepler's area law played an equally if not even more important role. For it was through this law that Newton learnt, first, how to show that the planets must be subject to a central force and, second, how to master the *variability* of orbital speed. Ptolemy's great discovery and Kepler's intuitive feel for its vital importance finally bore the fruit they deserved. Newton learnt how to prove the existence of long-range central forces and how to embed time into space in problems involving variable speed. The geometrization of motion that Galileo had always known was possible was finally completed. Kepler's area law was the crucial and final link that extended the dimensions of geometry by one, from three to four.

Where and when Newton first learnt about Kepler's area law have been discussed by Whiteside,[11b] Russell,[60] and Wilson.[57] In his answer to Hooke's request for a comment on his [Hooke's] theory of motion, Newton wrote:[72] 'But how ye Orbits of all ye Primary Planets but ☿ [Mercury] can be reduced to so many concentric circles through each of wch ye Planet moves equal spaces in equal times (for that's ye Hypothesis if I mistake not your description) I do not yet understand.' In this sentence Newton appears to exhibit a rather surprising degree of ignorance about astronomy, for the planetary orbits are neither circular nor concentric. The one thing he appears to have got right is the area law. In fact, it is very well attested that from an early date Newton was familiar with both Kepler's First and Third Laws. The situation with regard to the Second Law is not nearly so clear. Although it was comparatively well known to astronomers (as Russell[60] has shown, contrary to a quite widely held view) and also to mathematicians (on account of Kepler's famous request to them for assistance in solving his problem (p. 307)), many astronomers persisted, as we noted in Chap. 6, in the use of equant mechanisms because they were so much easier for calculations and gave such remarkably good results. (Kepler's great achievement, of course, was his recognition that they were not quite good enough.) Neither of the astronomical sources that Newton is known to have used repeatedly in the 1660s gives the area law. However, Wilson[57] argues, rather persuasively, that Newton probably learnt about it early in the 1670s and at the latest by 1676. For all that, the area law was not nearly so readily accepted as the First Law, and Newton himself experimented with an equant mechanism as late as the spring of 1769, as Whiteside has pointed out.[11b]

Whatever the truth, Newton's recognition of the dynamical significance of the area law was undoubtedly one of the great turning points in the history of science, and it is widely assumed that this occurred very soon after (if not during) his 1679/80 corresponding with Hooke. Newton himself said as much ('I found now that whatsoever was the law of the forces wch kept the Planets in their Orbs, the areas described by a Radius drawn from them to the Sun would be proportional to the times in wch they were described'[73]) and the claim cannot be reasonably denied.[74]

Papers that might correspond to this very piece of work have been published by Herivel,[75] and they are what we shall now discuss, even though there has been considerable controversy about their date ever since Herivel (later supported by Westfall) suggested that they could represent Newton's first solution of the Kepler problem (see Ref. 129, pp. 108–17 and Ref. 2, p. 387, note 145). Without wishing to enter the lists, it does seem to me that on internal grounds Herivel and Westfall have at the least made a quite good case for a 1679/80 dating of these papers. In discussing them now I do not wish to commit myself to such a dating. It is merely that Newton must – and on this all are agreed – have proved some such results around the beginning of 1680. The results discussed in this section must at the least correspond broadly to the most important insights that Newton then gained, and they are what I am interested in getting across.

One of the most interesting things about these papers is that Newton does not start by assuming the inverse square law and then demonstrate the motion that follows. He goes the opposite way, following the direction already taken in the early papers, namely, he uses the observed facts to deduce the nature and strength of the force which must be acting.

There are some points which should be made in this connection. Given an entirely arbitrary motion in some curve, one can always take the tangent to that curve at some given point P (Fig. 10.14) in the motion and say that the particle, having at that point a given instantaneous velocity, would by the action of the law of inertia alone be carried in unit time to the

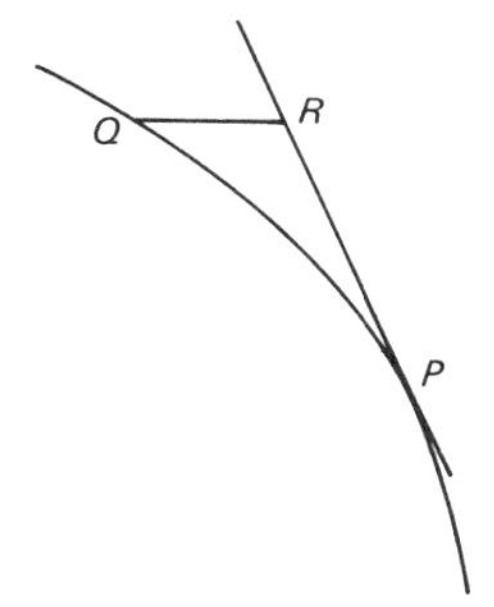

Fig. 10.14.

point R. Suppose however it is actually taken to the point Q. One may then say that the 'extra' motion which generates the deviation RQ is due to the action of a force. In the limit when the unit time goes to zero, R and Q approach P and RPQ becomes an infinitesimal triangle, half of a parallelogram in which PQ is the diagonal. Newton's assumption is that the distance RQ, which increases as the square of the time, is proportional to the strength of the force acting and that the force acts along RQ. (There are, of course, some tricky mathematical questions related to this limit-based definition, to which I shall return later.) For an arbitrary motion, there will be nothing significant or striking about the 'force' that is obtained. Moreover, one could easily choose some other definition of force. What Newton actually discovered was that the force, defined in this specific manner, turns out to have very remarkable properties. In the case of the planetary problem, it is always found to point exactly towards the sun and to have a magnitude inversely proportional to the square of the distance to the sun. This was the remarkable truth that Newton discovered hidden below the more superficial mathematical relationships of Kepler's laws.

A further point which should be especially emphasized is the universality of the force phenomenon. It is often said that, through the inverse square law, Newton was able to demonstrate the commonality (universality) of terrestrial gravity and the force which keeps the moon in its orbit. But even deeper than this quantitative linking of the strengths and of greater significance for the basic structure of dynamics is the fact that force generates motion in proportion to time. This establishes a very far-reaching universality of the force phenomenon and is what makes the quantitative treatment of free fall by Galileo and then centrifugal force by Huygens and Newton such significant milestones in the discovery of dynamics. It should also be emphasized that the unambiguous identification from empirical observations of the force of attraction of the planets towards the sun was simultaneously the justification for using the concept of inertial motion and extending the decomposition that Galileo had achieved in the projectile problem to problems involving vastly greater distances. Thus, the solution of the Kepler problem provided three things at once: Newton's First and Second Laws and the specific law of gravitational force. The crucial test comparing the strength of terrestrial gravity with the force of attraction of the moon was therefore only the final link – though it was of course vitally important in that it closed the circle and showed that the whole held together.

After these introductory comments, let us now look at this marvellous piece of work, the synthesis of all that was best in Kepler and Galileo – and Descartes and Hooke to give them their due. Newton begins by stating the hypotheses that he is going to use:

Hypoth. 1 Bodies move uniformly in straight lines unless so far as they are retarded by the resistance of the Medium or disturbed by some other force.
Hyp. 2 The alteration of motion is ever proportional to the force by which it is altered.
Hyp. 3 Motions imprest in two different lines, if those lines be taken in proportion to the motions and completed into a parallelogram, compose a motion whereby the diagonal of the Parallelogram shall be described in the same time in which the sides thereof would have been described by those compounding motions apart. The motions AB [Fig. 10.15] and AC compound the motion AD.

It is interesting to note that the parallelogram rule is stated here as an independent hypothesis, and not, as in the *Principia*, as a consequence of the First and Second Law. This is one of the arguments advanced by Herivel and Westfall for a pre-*Principia* dating of these papers.

All the three hypotheses had appeared in one form or another in the *Waste Book*, though the parallelogram rule is stated with rather more precision. The new element is the use that Newton makes of them. Proposition 1 breaks totally new ground in the application of these principles; Newton lays bare the dynamical origin of what we have called Kepler's Zeroth Law (that the planets move in invariable planes that all intersect the sun) and his Second Law (the area law):

Prop. 1 If a body move *in vacuo* and be continually attracted toward[s] an immoveable center, it shall constantly move in one and the same plane, and in that plane describe equal areas in equall times.

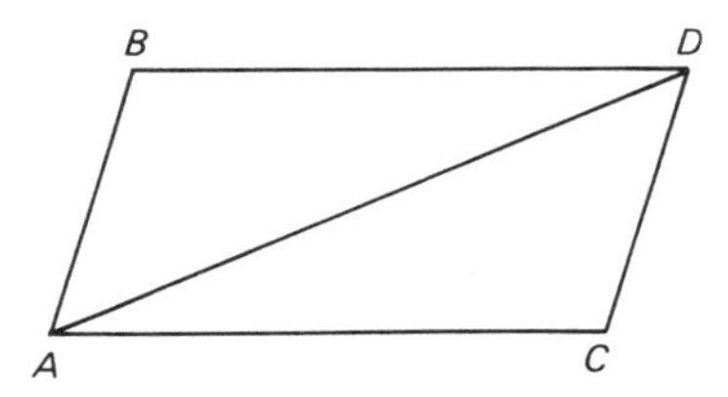

Fig. 10.15. Reproduced from: J. W. Herivel, *The Background to Newton's 'Principia'*, Clarendon Press, Oxford (1965).

Newton's proof is so simple and so crucial to the further applications that I give it in full even though it has already appeared in paraphrase in Chap. 1. It is also beautiful testimony to the central role in dynamics of Galileo's rule of compounding physical motions; note too how beautifully it develops out of Newton's *Waste Book* polygonal proof of the formula for centrifugal force, as Herivel has pointed out to me:

> *Proof.* Let A [Fig. 10.16] be the center towards which the body is attracted, and suppose the attraction acts not continually but by discontinued impressions made at equal intervalls of time which intervalls we will consider as physical moments. Let BC be the right line in which it begins to move from B and which it describes with uniform motion in the first physical moment before the attraction makes its first impression upon it. At C let it be attracted towards the center A by one impuls or impression of force, and let CD be the line in which it shall move after that impuls. Produce BC to I so that CI be equall to BC and draw ID parallel to CA and the point D in which it cuts CD shall be the place of the body at the end of the second moment. And because the bases BC CI of the triangles ABC, ACI are equal those two triangles shall be equal. Also because the triangles ACI, ACD stand upon the same base AC and between two parallels they shall be equall. And therefore the triangle ACD described in the second moment shall be equal to the triangle ABC described in the first moment. And by the same reason if the body at . . . . Suppose now that the moments of time be diminished in length and encreased in number *in infinitum*, so that the impulses or impressions of the attraction may become continuall and that the line BCDEFG by the infinite number and infinite littleness of its sides BC, CD, DE etc. may become a curve one: and the body by the continual attraction shall describe areas of this Curve ABE, AEG, ABG etc. proportionall to the times in which they are described.

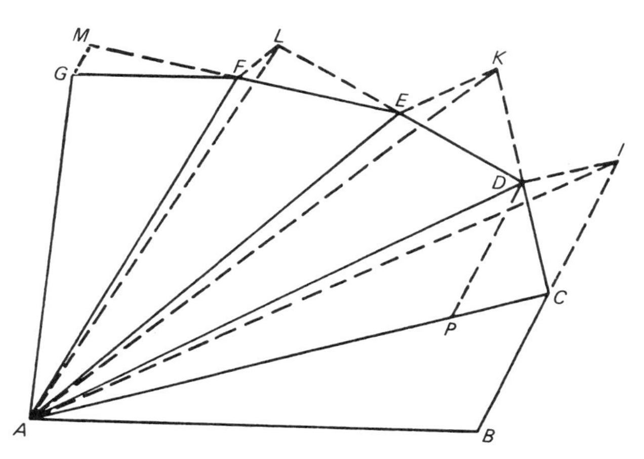

Fig. 10.16. Reproduced from: J. W. Herivel, *The Background to Newton's 'Principia'*, Clarendon Press, Oxford (1965).

So here we have (leaving aside questions relating to the rigour of Newton's proof) the truly dramatic turn which would have so amazed Kepler. Within the framework in which he had worked, the significance of the area law had seemed to be quite clear: a solar force pushed the planets around in their orbits and its strength decreased simply as the distance. But now Newton had changed almost everything: forces govern accelerations, not speeds; the area law says nothing at all about the distance dependence of the strength of the force; and the force is directed straight towards the sun, not at right angles to the radius vector from the sun to the planet. In one crucial respect however Kepler was completely vindicated: the sun was unambiguously implicated in the motion of the planets. He, of course, also supplied all the mathematical relationships in a form marvellously suited to Newton's analysis. Nowhere more than in the case of Newton's interpretation of the area law does one see the extraordinary consequences that can flow from a really precise mathematical description of motion. The equant device was already amazingly good but could not possibly have given Newton the insight that he got from the area law and its decisive shifting of the key point from the void to the occupied focus.

But how the problem was now transformed! Note, in particular, the power of the law of inertia and the way in which it ensured, for central forces, that each planet would remain on a fixed plane for ever. Many of the particular Keplerian mechanisms on which so much thought and ingenuity had been lavished were made redundant at a stroke. Just as Copernicus and Kepler had cleared out the proliferation of Ptolemaic epicycles, so Newton now cleared out the Keplerian mechanisms and special forces. Conversely, we see what powerful empirical support for the validity of the law of inertia the astronomical results did provide.

The dynamical interpretation of the area law was obviously the key piece of evidence that forced Newton to take seriously the idea of a force of attraction directed towards the sun. From now on the mathematics would dictate the physical interpretation. Simultaneously, Newton's results showed how the various elements of his embryonic but very largely untested dynamics could be fitted together to give a decidedly nontrivial result. The synthesis had begun. Finally, half of the problem of interpreting the motion of the planets was now completed: they must be attracted towards the sun by a force (defined in the Newtonian sense) of some as yet undetermined strength. Moreover, as we shall now see, the solution of the first half of the problem was of considerable assistance in the solution of the second half, the determination of the actual strength. To this we now turn.

The second proposition is:

Prop. 2 If a body be attracted towards either focus of an Ellipsis and the quantity of the attraction be such as suffices to make the body revolve in the circumference of the Ellipsis; the attraction at the two ends of the Ellipsis shall be reciprocally as the squares of the body in those ends from that focus.

Once again, I give the proof in full, since it shows how Newton solved the problem of *measuring force* by the deviation from the instantaneous tangent (the line of the inertial motion that would arise if the deflecting force were to be 'switched off' abruptly) and how he used Kepler's area law to get the correct measure of time at each point of the orbit. Here is the proof:

Proof. Let AECD be the Ellipsis [Fig. 10.17], A, C its two ends or vertices, F that focus towards which the body is attracted, and AFE, CFD areas which the body with a ray drawn from that focus to its center, describes at both ends in equal times: and those areas by the foregoing Proposition must be equal because proportionall to the times: that is the rectangle $\frac{1}{2}$AF × AE and $\frac{1}{2}$FC × DC must be equal supposing the arches AE and CD to be so very short that they may be taken for right lines and therefore AE is to CD as FC to FA. Suppose now that AM and CN are tangents to the Ellipsis at its two ends A and C and that EM and DN are perpendiculars let fall from the points E and D upon those tangents: and because the Ellipsis is alike crooked at both ends those perpendiculars EM and DN will be to one another as the squares of the arches AE and CD, and therefore EM is to DN as FC^q [i.e., FC^2 in modern notation] to FA^q. Now in the times that the body by means of the attraction moves in the arches AE and CD from A to E and from C to D it would without attraction move in the tangents from A to M and from C to N. Tis by the force of the attractions that the bodies are drawn out of the tangents from M to E and from N to D and therefore the attractions are as these distances ME and ND, that is the attraction at the end of the Ellipsis A is to the attraction of the other end of the Ellipsis C as ME to ND and by consequence as FC^q to FA^q. W.W. to be dem.

Most of the conceptual points that are important about this proof have already been made. We see how naturally it follows on from the earlier work, in particular how Newton's very early remark (p. 526) about measuring the strength of the force in elliptic motion by a 'tangent circle of Equall crookednesse' can now bear fruit since he has in the meantime

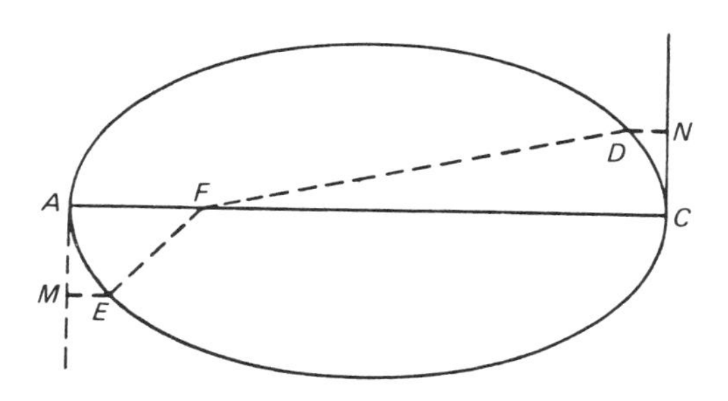

Fig. 10.17. Reproduced from: J. W. Herivel, *The Background to Newton's 'Principia'*, Clarendon Press, Oxford (1965).

found the missing link – the embedding of time in space in the case of variable motion. We note also that other characteristic feature of all Newton's work: his promiscuous use of the two different force concepts – instantaneous impulses in Proposition 1 (motive force) and continuous acceleration in Proposition 2 (accelerative force). In the proof of Proposition 3, the statement of which we now give, he returns once more to discrete impulses followed by a passage to the continuous limit:

Prop. 3. If a body be attracted towards either focus of any Ellipsis and by that attraction be made to revolve in the Perimeter of the Ellipsis: the attraction shall be reciprocally as the square of the distance of the body from that focus of the Ellipsis.

I shall not give the complete proof of this proposition, since it is full of purely technical matters relating to the geometry of ellipses. Nevertheless explicit quotation at a couple of crucial points will bring before our eyes the inexorable application of the Newtonian nutcrackers, applied now to the elliptical nut. The following passage illustrates the ideas of rectilinear inertial motion along the tangent, the action of the supervening force (with the motion it generates being compounded by the parallelogram superposition rule with the inertial motion), and the deviation from the tangent in unit time, *in the direction of the force* (which Newton knows from his first proposition), as the measure of the strength of the force acting. Note that in Newton's figure (Fig. 10.18) PX is meant to be infinitesimally small and the attracting focus is at F. Here are Newton's words:

Let P be the place of the body in the Ellipsis at any moment of time and PX the tangent in which the body would move uniformly were it not attracted and X the

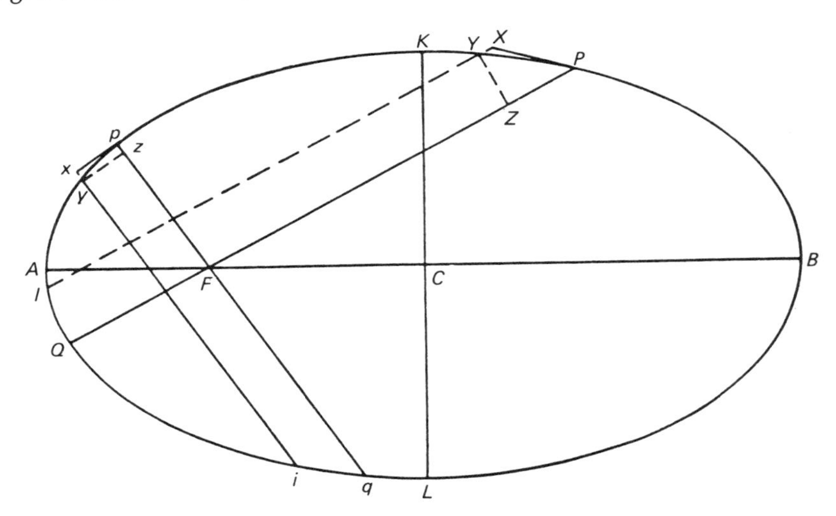

Fig. 10.18. Reproduced from: J. W. Herivel, *The Background to Newton's 'Principia'*, Clarendon Press, Oxford (1965).

place in that tangent at which it would arrive in any given part of time and Y the place in the perimeter of the Ellipsis at which the body doth arrive in the same time by means of the attraction. Let us suppose the time to be divided into equal parts and that those parts are very little ones so that they may be considered as physical moments and that the attraction acts not continually but by intervalls once in the beginning of every physical moment and let the first action be upon the body in P, the next upon it in Y and so on perpetually, so that the body may move from P to Y in the chord of the arch PY and from Y to its next place in the Ellipsis in the chord of the next arch and so on for ever. And because the attraction in P is made towards F and diverts the body from the tangent PX into the chord PY so that in the end of the first physical moment it be not found in the place X where it would have been without attraction but in Y being by the force of attraction in P translated from X to Y: *the line XY generated by the force of the attraction in P must be proportional to that force and parallel to its direction that is parallel to PF* [my italics].

The other main conceptual point is the application of the area law, which has already been used in the crucial assertion that the force acts in the direction of PF, to obtain a measure of time. Newton imagines exactly the same process at p, the time taken to travel the distance px being equal to that to traverse PX. Newton then comments: 'And because the lines PY py are by the revolving body described in equal times, the areas of the triangles PYF pyF must be equal by the first Proposition.' From this point on the problem is pure geometry and exploits very ingeniously quite a considerable number of geometrical properties of the ellipse, some of which were well known, others of which Newton appears to have derived himself. Newton combines all these to obtain an *exact* relationship that holds for any line XI drawn parallel to PQ, which passes through the focus F. This states that the distance XY is equal to

$$\frac{AB \cdot PQ \cdot YZ^2}{XI \cdot KL^2},$$

where YZ is the perpendicular from Y onto PQ. There is a similar relation for xy. Now in the expression for XY, KL and AB are fixed, and in the limit in which Y approaches P the ratio PQ/XI tends to unity. Newton then envisages a limiting process in which Y tends to P and simultaneously y tends to p but in such a way that the ratio of the areas of the two triangles PYF and pyF remains exactly equal to unity, i.e., $PF \cdot YZ = pF \cdot yz$. But through the above exact geometrical relationship this immediately determines the ratio XY/xy of the force-induced deflections in equal infinitesimal times. The ratio is immediately found to be pF^2/PF^2. Newton completes his solution of the Kepler problem with the lapidary words: 'And therefore the attraction in P will be to the attraction in p as pF^q to PF^q, that is reciprocally as the squares of the distances of the revolving bodies from the focus of the Ellipsis.'

We have in the work described in this section the very core of Newton's greatest contribution to dynamics. In what does the essence of this work consist? It consists in the adaptation of the key elements of Galileo's work on projectiles, which involved the compounding of two motions that admit this with especial ease on account of two significant simplifications, namely, the orthogonality of the two motions and the fact that one is constant and the other is uniformly accelerated, to the much more difficult situation in which the motions are not orthogonal and the acceleration is not uniform. Newton achieved the generalization by 'infinitesimalizing' Galileo's procedure, that is, he broke the motion up into a succession of very small stretches, in each of which he applied the Galilean technique. We note that in the process he introduced a certain imbalance not present in Galileo, for whom we recall that the two motions in the projectile problem were treated as primordial atomic motions on an equal footing. But in Newton's treatment, following the adoption of the law of inertia, the two compounded elements are no longer on the same footing: one is the instantaneous inertial motion, the other is the motion generated by the external force in the infinitesimal time corresponding to the considered stretch. It is this imbalance that creates the conceptual difficulties in Newtonian dynamics as it was originally formulated; for it is always *motions* that are compounded but Newton speaks of an innate force that maintains the inertial motion but an impressed force that generates the other motion which is compounded with the instantaneous motion. The two forces are therefore heterogeneous – one maintains motion, the other creates it.

We note also that the imbalance is reflected in the mathematics; for the two sides of the 'Galilean parallelogram' are of different orders in terms of infinitesimal analysis: if the infinitesimal arc PY along the motion in Fig. 10.18 is taken to be of order ε, then PX is of order ε but XY is of order ε^2. We shall come back to this point briefly later in the chapter and will merely say here that Newton generally managed to get the physical results he needed despite quite tricky problems in the mathematics.

10.10. The genesis of the *Principia*: Ulysses draws forth Achilles

It is remarkable that even the successful solution of the Kepler problem did not stir Newton to seek publication or even, it seems, to follow up his ideas at all energetically. Over six years elapsed between the Hooke–Newton correspondence and the publication of the *Principia*. In this section we shall briefly recount this part of the story.

One of the most important developments was the appearance of a magnificent comet in early November 1680 and its reappearance after sunset in December. The nature of comets had been hotly disputed for more than a century. Brahe's assertion (which, incidentally, Galileo[76] refused to accept) that comets pass clean through the putative orbs of the planets was a very important factor in concentrating minds on the problem of the nature of motion of both the planets and comets. For Kepler, as we saw, it was the decisive piece of evidence in ruling out all crude mechanical contrivances as the explanation of planetary motions. Kepler himself believed that comets, which do not remain in the ecliptic, were bodies subject to laws quite different from those of the planets. He believed that they passed right through the solar system, travelling along straight lines. This, in fact, was the opinion held by the majority of astronomers – and Newton – at the time when the great comet of 1680/81 appeared.

It was this event which again caused astronomy to play a significant role in the development of dynamics. The main protagonist, apart from Newton, was John Flamsteed (1646–1716), who was appointed the first Astronomer Royal in 1675, a post he held until his death. He compiled a famous star catalogue, by far the most accurate hitherto. In accordance with Kepler's theory of comets, the majority of astronomers assumed that the comet of 1680/81 was not one but two. Flamsteed secured for himself a minor but important role in the history of dynamics by espousing the idea that there was only a single comet. He developed a theory according to which it approached the sun, being attracted initially by a magnetic force of the sun before being repelled again.

According to his theory, the comet passed *between* the earth and the sun. Flamsteed worked quite hard on his theory and on the interpretation of his observations and wrote them up in letters to Halley[77] and arranged that they should be brought to Newton's attention through an intermediary. Now Newton himself had become extremely interested in the comet, and he took up Flamsteed's request for comments on the developed theory with considerably more enthusiasm than the similar request from Hooke. He recounted[78] how he had questioned people about the comet and reported the observation of 'one of our Fellows, Dr Babington', who around the 22 or 23 November 'between 5 & 6 saw the tayle of ye Comet shoot over Kings College Chappel from east to west. Twas a frosty morning & a very clear & starry sky . . . . The tayl ran from one end of ye chappel to ye other . . . .' Newton supplied a little sketch

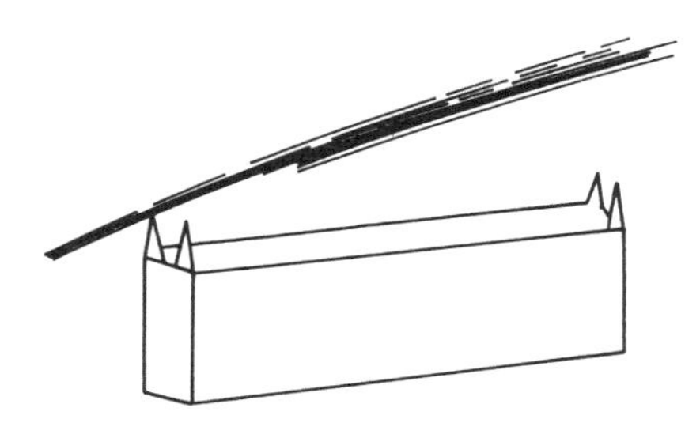

Fig. 10.19.

(Fig. 10.19). In a later letter,[79] he reported his own extensive observations of the comet from January to March 1681.

The main point of interest in the ensuing correspondence is that Newton quite strongly rejected Flamsteed's idea of a single comet. He argued that there were two and clearly seems to have believed that each had travelled on rectilinear paths. As Westfall points out,[80] only a year earlier Newton had solved the problem of orbital motion for a planet circling the sun. At this stage he does not appear to have believed comets were subject to the same laws. Thus 'the letter allows us to measure crudely the progress of his thought toward the concept of universal gravitation.' He did not yet think gravitational attraction applied equally to all bodies in the solar system. On the other hand, his letters to Flamsteed (including a draft that was not actually sent[81]) suggest that he was at least toying with the idea. For in his first letter, speaking of Flamsteed's proposal that the comet passed between the earth and the sun, he says: 'The only way to releive this difficulty in my judgmt is to suppose ye Comet to have gone not between ye ☉ [the sun] & Earth but to have fetched a compass about ye ☉ . . . .' In the same breath he says he 'can easily allow an attractive power in ye ☉ whereby the Planets are kept in their courses about him from going away in tangent lines', and, in fact, his objection to Flamsteed's proposal at this point seems to concentrate on its being *magnetical*, with first attraction and then repulsion, an idea which Newton strongly attacks. In the unsent draft he is more specific:[82] 'But all these difficulties may be avoyded by supposing ye comet . . . to have been attracted all ye time of its motion . . . & thereby to have been as much retarded in his recess as accelerated in his access & by this continuall attraction to have been made to fetch a compass about the sun in ye line ABKDF, the *vis centrifuga* at C overpow'ring the attraction & forcing the Comet there notwithstanding the attraction, to begin to recede from ye sun.' (Incidentally, we note here Newton's continuing use of the concept of centrifugal force despite the fact that he had in the meanwhile solved the Kepler problem – and indeed in the previous quotation had spoken of the planets 'going away in tangent lines'. I see this as further evidence that Newton used the expression as a shorthand for a phenomenon whose true dynamical significance he had always understood.)

Newton's interest in comets was reawakened by the appearance in 1682 of the comet that has since been named after Halley. Work that he did at that time suggests that it was at about this period that he worked seriously on the hypothesis of curved orbits for comets.[83] By late December 1684 his notions of universal gravitation had clearly become very much more precise and he was engaged upon thorough mathematical treatment of the problem that had already been apparent to Kepler – the small deviations of the planets from the exact Keplerian laws. These are due very largely to the superposition of the gravitational attraction of the planets on top of the dominant force of the sun. Kepler had been concerned with the slow secular perturbations. Newton, in contrast, was concerned with the perturbation of Jupiter on the motion of Saturn and asked Flamsteed[84] 'if you ever observed Saturn to err considerably from Keplers tables about ye time of his conjunction with Jupiter'. He explained to Flamsteed that Saturn 'so oft as he is in conjunction with Jupiter ought (by reason of Jupiters action upon him) to run beyond his orbit about one or two of ye suns semidiameters'. A couple of weeks later he wrote again and announced:[85] 'I do intend to determin ye lines described by ye Comets of 1664 & 1680 according to ye principles of motion observed by ye planets.' Finally, in September 1685 he wrote to Flamsteed that[86] 'I have not yet computed ye orbit of a comet but am now going about it.' He admitted that 'taking that of 1680 into fresh consideration, it seems very probable that those of November & December were ye same comet.' Flamsteed was suitably gratified and answered:[87] 'I am heartily glad that you have the Theory of comets under your consideration: we have hitherto onely groped out the lines of their Motions.' Flamsteed was alas another of the men with whom Newton was later to have a furious row.[88]

The theory of comets, especially the comet of 1680/81, figures prominently in the *Principia*. In the first edition (1687), Newton computed in Book III a parabolic path for the 1680/81 comet. In 1695, Halley worked out an elliptic orbit with very large eccentricity, proposing it as an alternative to Newton's parabolic orbit. Newton then collaborated with Halley[89] on the details of cometary orbits, and in the second (1713) and third (1726) editions of the *Principia* gave tables for both parabolic and elliptical orbits.

For the development of dynamics, Newton's work on comets was important for at least two reasons. First, in extending the range of application of a particular force; from being a force *specific* to the planets, the inverse square law of attraction became a force that applied to all bodies within the solar system. Equally important, it forced Newton to attack dynamical problems with ever more general initial conditions. From circles in the *Waste Book* he progressed through the correspondence with Hooke to the elliptical motion of the planets and then on, through the comets, to parabolic and hyperbolic orbits. This demonstrated clearly that the various different shapes of the orbits reflect the possibility of setting a body in motion at the initial time with arbitrary direction and speed relative to the attracting centre. Newton was thus brought to grasp the *overall structure* of the science of dynamics and realize the full potential of the techniques he had developed.

At the beginning of this chapter we quoted Galileo's remark that he had opened up the 'vast and most excellent' science of motion. Elsewhere in the *Discorsi* he spoke of the 'numerous and wonderful results which in future years will command the attention of other minds'.[90] How true he was, even if it took the assistance of a great comet to rivet Newton's mind to the problem. But even then, in the early 1680s, Newton's attention remained desultory. Just as he himself said, he really was the boy on the seashore picking up one pretty pebble after the other while the great ocean of truth lay all undiscovered before him.

But then came the decisive turn – the journey made by 'the most acute and universally learned Mr *Edmund Halley*' (as Newton terms him in the *Principia*[91]) in August 1684 to Cambridge. After this, things would never be the same again.

As the reader will know from Sec. 10.1 (if not before), Halley came with a specific request: what would be the curve described by a planet if subject to a force of attraction by the sun inversely proportional to the square of its distance? Newton replied it would be an ellipse and promised to supply the proof. In November the delighted Halley received a short treatise called *De motu corporum in gyrum* ('On the motion of bodies in an orbit').[92] This reproduces the essentials of the great results discussed in Sec. 10.9 but adds significantly more.

There are several points of considerable interest about this paper. First, it opens with the definition of the concept that more than any other dominates and distinguishes the *Principia* – the concept of a centripetal force:[93] 'I call centripetal that force by which a body is impelled or drawn towards any point which is regarded as a centre [of force].' This immediately leads to a significant clarification of the expression of the law of centrifugal force. Thus:[94] 'The centripetal forces continuously pull back the bodies from the tangents [which they would follow by the action of their innate forces alone] to the circumferences.' Second, it already has the flavour, so characteristic of the definitive *Principia*, of being a general mathematical treatise on motion. Newton delights in stating theorems, establishing corollaries, and posing problems in the formal manner so beloved by mathematicians ever since Euclid. He discourses on the practical significance of his formal results in extended scholia (a *scholium* was originally an ancient exegetical note or comment upon a passage in a Greek or Latin author [*OED*] and was used by Newton and other mathematicians to illustrate or amplify points of interest). Particularly interesting in the light of the earlier comment about the significance of the work on comets in extending the scope of dynamics is the following problem:[95]

> Given that the centripetal force is inversely proportional to the square of the distance, and knowing the magnitude of that force, required to find the ellipse which a body describes when projected from a given point with given velocity in a given straight line.

Thus, we have here the first formulation of a general initial-value problem. (Newton in fact comments on his proof that 'this is the way when the figure is an ellipse. But it can happen [if the speed has the correct value] that the body moves in a parabola or a hyperbola.') There is also an explicit reference to the possibility of defining the 'orbits of comets'. (This proved to be a very difficult question and delayed Newton for two months in the autumn of 1685 until he had found a 'good method'.[96]). A final point of interest about the paper is its inclusion of two propositions on the motion of bodies in resisting media, a further indication of the generality at which he was now aiming.

Halley immediately recognized the importance of the work and went post-haste to Cambridge to discuss it with Newton and get him to have it entered in the register of the Royal Society to ensure Newton's priority. This was done by Halley when he reported on the matter to the Royal Society on 10 December 1684. Meanwhile Newton's life had been transformed, and he had started work on one of the most astonishing labours of intellectual man: a comprehensive treatise on motion, the aim of which was to show how the entire gamut of observed motions – both terrestrial and celestial – could be deduced from a mere handful of general principles formulated in a mathematically rigorous framework. As Westfall comments:[97] 'The problem had seized Newton and would not let him go.' For about 18 months, until the spring of 1686, he worked on the project with incredible energy. His famulus of the time, Humphrey Newton, reported his rapt concentration and how he would often completely forget to eat the food prepared for him.

Within about six months Newton had come to grips with and in principle solved the major tasks that confronted him. On the one hand there was the need to order the conceptual framework. He had to identify the basic principles on which the complete system should rest. A later reworking of the original *De motu corporum in gyrum* and various other papers that predate the *Principia*, which are also reproduced by Herivel,[98] provide evidence of this. There were two sides to this work – the selection of the best empirical evidence for the adopted laws and concepts and the finding of the best words to express these concepts. Then there were numerous specific problems that must be treated within the framework of the chosen concepts and laws. To avoid repetition, this work will be described together with the account given of the *Principia* in the following section. Here we shall merely complete the account of the genesis and publication of the *Principia*.

In the autumn of 1685 Halley returned to Cambridge, where he was shown the work that Newton had so far done. In the year since his first visit, Newton's early solution of the planetary problem had been transformed into a complete theory of universal gravitation. From this point on, Halley devoted all his efforts to ensuring the publication of Newton's work and to preparing the scientific community, both national and international, for the appearance of the impending masterpiece. This mission was to involve him in a great deal of delicate diplomacy and not a little risk and cost (Halley paid for the printing of the *Principia*).[99] In later years Halley would refer to himself with justifiable pride as [100] 'the Ulysses who produced this Achilles'. He undoubtedly needed all the skills of the wily Ithacan.

Towards the end of April 1686 a first version of the work, with its full title *Philosophiae Naturalis Principia Mathematica* (*Mathematical Principles of Natural Philosophy*) was presented to the Royal Society and on 19 May a resolution was passed that it should 'be printed forthwith'.[101] Halley immediately wrote to Newton with the news that his 'Incomparable treatise' had been presented to the Society and that they had ordered its printing. Halley then had to broach another most unfortunate matter:[102]

> There is one thing more that I ought to informe you of, viz, that Mr Hook has some pretensions upon the invention of ye rule of the decrease of Gravity, being reciprocally as the squares of the distances from the Center. He sais you had the notion from him, though he owns the Demonstration of the Curves generated therby to be wholly your own; how much of this is so, you know best, as likewise what you have to do in this matter, only Mr Hook seems to expect you should make some mention of him, in the preface, which, it is possible, you may see reason to praefix. I must beg your pardon that it is I, that send you this account, but I thought it my duty to let you know it, that so you may act accordingly; being in myself fully satisfied, that nothing but the greatest Candour imaginable, is to be expected from a person, who of all men has the least need to borrow reputation.

Given the circumstances of Hooke's knowledge, i.e., his complete ignorance of what Newton had already achieved, this request was not unreasonable. Moreover, in one of the final letters of the correspondence of 1679/80 Hooke had explicitly told Newton that[103] 'my supposition is that the Attraction always is in a duplicate proportion to the Distance from the Center Reciprocall', and in yet another letter had told Newton[104] that it 'now remaines to know the proprietys of a curve Line (not circular nor concentricall)' subject to a central attraction of the kind he envisaged and that he was sure 'you will easily find out what that Curve must be'. With the benefit of the access to Newton's unpublished papers we can see clearly that the suggestion of the inverse square law was the least of Hooke's services. Hooke's supreme service was to make Newton take the concept of central attraction seriously in the consideration of the planetary problem. Once Descartes had transformed the conceptual approach to the problem and put what was later to become the law of inertia in the forefront, the concept of a deflecting force just had to appear in any problem not involving uniform rectilinear motion. As we have seen, it did very naturally appear in the work of both Huygens and Newton. It had to from the very general logic of the situation. What Hooke did was propose a *specific force* responsible for the deflection. He argued for an attractive force towards a centre. *This* was the greatest contrast between Hooke, on the one hand, and, on the other, Descartes, Huygens, and the explicit worked examples which have survived from the early Newton. For in all these cases the deflecting force was a *contact force*, supplied through a string in the case of Descartes and Huygens and the circular rim in the case of Newton.

After an initially restrained response to Halley's letter, Newton wrote again to the long-suffering man and unleashed the full fury of his venom against Hooke. This is again a point at which I would recommend the reader to consult the readily available originals (Ref. 12c, pp. 431–47). Any such story always loses immediacy in retelling. Apart from the rather cruel light that the exchange casts on Newton's troubled personality – and the favourable light in which Halley is put – the main interest of the exchange is in showing how the problem of the planetary motions had been brought clearly into focus by the mid and late 1670s. Hooke, Halley, and Wren were all hovering around the prize. The vital importance of Huygens' input of the correct formula for centrifugal force is clearly established. Newton commented:[105] 'ye honour of doing it in this is due to Huygenius.'

From its first crude outline in Descartes the problem had passed through the qualitative stage of Borelli and Hooke and was now posed with quantitative exactitude. It awaited only a genius commensurate to the task. The solution of the problem was not in itself a surprise. What was breathtaking, despite the accuracy of Hooke's prediction, was the sudden appearance of the theory of universal gravitation with all its ramifications fully explored. Halley had asked for the solution to a specific problem and had received it in generous measure. The attraction of the planets to the sun was now established beyond doubt. But in the first draft of *De motu corporum in gyrum* there was as yet no hint of a *universal* theory of gravitation. What changed the concept of the world so completely and utterly was Newton's demonstration, in the minutest detail, that each and every little piece of matter exerts a force on every other piece of matter in the universe, the force it exerts being proportional to its mass and universely proportional to the square of the distance to the attracted body.

Wilson has argued,[57] rather convincingly in my mind, that Newton's rather vague notions of attraction were transformed into the precise – and vastly more far reaching – theory of universal gravitation very late indeed, in the period between the two drafts of *De motu* in the autumn of 1684. This would tally with the letters to Flamsteed mentioned earlier. Newton's two great contributions to dynamics were the establishment of its overall structure and the specific discovery of the law of universal gravitation. As regards the general structure of dynamics, I differ from Westfall[11a] and Whiteside[11b] in seeing most of that as having been settled in the 1660s. But it took the recognition of the dynamical significance of the area law in 1679/80 to transform the general framework into a powerful formalism capable of solving nontrivial dynamical problems.

The same recognition simultaneously concentrated attention on centripetal forces – indeed it showed that they exist in a very real mathematically definable sense – and further applications must then have led, towards the end of 1684, to the comparatively sudden (and almost miraculous) appearance of universal gravitation. Westfall and Whiteside place the bulk of both major discoveries comparatively late, I do that for only one.

It is worth emphasizing that the establishment of the theory of universal gravitational attraction was a revolution that overthrew a revolution that at the time – just forty years earlier – had seemed just as breathtaking, convincing, and comprehensive: Descartes' dream of explaining the entire material world by a single primitive substance differentiated solely by figure, magnitude, and motion. At a stroke Newton replaced this Cartesian world – which was a pure figment of the imagination – with a scheme that did as much, but of the three Cartesian pillars, retained only one (motion), the other two (figure and magnitude) being replaced by mass and gravitational force. And this scheme worked. The first explicit intimation of what was to come is found in a significantly extended draft of *De motu corporum in gyrum*, which probably dates from the end of 1684 or the beginning of 1685. At the end of an important scholium on centripetal forces comes the passage which shows that Newton, using the accurate value of the radius of the earth that Picard, in another of the important studies sponsored by the Académie Royale des Sciences, had published in 1671 in his *Mesure de la Terre*,[106] had definitely repeated – and now with success – his early *experimentum crucis*:[107]

> For certainly gravity is one kind of centripetal force: and my calculations reveal that the centripetal force by which our Moon is held in her monthly motion about the Earth is to the force of gravity at the surface of the Earth very nearly as the reciprocal of the square of the distance [of the Moon] from the centre of the Earth.

And thus the great idea burst suddenly upon the amazed world. As Athena from the head of Zeus, so universal gravitation from Newton's. It appeared all at once in the *Principia*, fully clothed and armed. The defence against snipers was impregnable: almost all conceivable evidence was marshalled and the theoretical elaboration was done with severe mathematical rigour and meticulous detail.

But we anticipate a little. Halley had still to compose the tempers of Hooke and Newton – the end was an admission by Newton to Halley[108] in a letter of 14 July 1666 that Hooke's 'Letters occasioned my finding the method of determining Figures, wch when I tried in ye Ellipsis, I threw the calculation by . . . for about 5 yeares' and a solitary and rather curmudgeonly remark by Newton well into the body of the *Principia* that if periodic times in circular orbits are as the $\frac{3}{2}$th powers of the radii then the centripetal forces will be inversely as the squares of the radii 'as Sir

PHILOSOPHIÆ
NATURALIS
PRINCIPIA
MATHEMATICA.

Autore *J S. NEWTON*, *Trin. Coll. Cantab. Soc.* Matheseos Professore *Lucasiano*, & Societatis Regalis Sodali.

IMPRIMATUR.
S. PEPYS, *Reg. Soc.* PRÆSES.
Julii 5. 1686.

LONDINI,

Jussu *Societatis Regiæ* ac Typis *Josephi Streater*. Prostat apud plures Bibliopolas. *Anno* MDCLXXXVII.

Fig. 10.20.

Christopher Wren, Dr Hooke, and Dr Halley have severally observed'.[109] That matter having been patched over as well as Newton's prickly temperament permitted, printing of the great work could proceed and was completed on 5 July 1687. It had already been granted an 'imprimatur' a year earlier by everyone's favourite diarist – Samuel Pepys, the then president of the Royal Society. The title page is reproduced as Fig. 10.20.

10.11. The *Principia*: its structure, fundamental concepts and most important results

The *Principia* is a very substantial work – Motte's English translation published by the University of California Press runs to nearly 550 pages. The aim of this section is to give a brief survey of its contents and clarify the emergence of some of the concepts that appear in it. However, two major topics are deferred until the following chapters – Newton's concept of mass and the entire discussion of the nature of space, time, and motion.

The book begins with an ode in Latin by Edmond Halley, which does, in fact, give quite a good summary of its contents. It opens with the line 'Lo, for your gaze, the pattern of the skies!' and promises the reader 'the Laws which God, Framing the universe, set not aside But made the fixed foundations of his work'. It will reveal 'The force that turns the farthest orb' and show how the planets, as they move 'through the boundless void' are sped 'in motionless ellipses'. Echoing the spirit that informed Lucretius's *De Rerum Natura*, which was written primarily to dispel fear of capricious gods, Halley assures the reader 'Now we know The sharply veering ways of comets, once A source of dread, nor longer do we quail'. The 'silver moon' is also to yield the secrets of her 'travel with unequal steps, As if she scorned to suit her pace to numbers – Till now made clear to no astronomer'. The reader will learn why 'the Seasons go and then return' and 'The Hours move ever forward on their way' and 'How roaming Cynthia bestirs the tides'. The optimistic spirit of the age finds expression too: through 'reason's light' the 'clouds of ignorance' are 'Dispelled at last by science'. Halley concludes the ode with ecstatic praise of Newton, saying 'Nearer the gods no mortal may approach'.

The book itself opens with a preface by Newton written in 1686. Right at the start, Newton makes the rather remarkable claim that geometry is but part of universal mechanics:

> To practical mechanics all the manual arts belong, from which mechanics took its name. But as artificers do not work with perfect accuracy, it comes to pass that mechanics is so distinguished from geometry that what is perfectly accurate is called geometrical; what is less so, is called mechanical. However, the errors are not in the art, but in the artificers. He that works with less accuracy is an imperfect mechanic; and if any could work with perfect accuracy, he would be the most perfect mechanic of all, for the description of right lines and circles, upon which

geometry is founded, belongs to mechanics. Geometry does not teach us to draw these lines, but requires them to be drawn, for it requires that the learner should first be taught to describe these accurately before he enters upon geometry, then it shows how by these operations problems may be solved. To describe right lines and circles are problems, but not geometrical problems. The solution of these problems is required from mechanics, and by geometry the use of them, when so solved, is shown; and it is the glory of geometry that from those few principles, brought from without, it is able to produce so many things. Therefore geometry is founded in mechanical practice, and is nothing but that part of universal mechanics which accurately proposes and demonstrates the art of measuring.

The *Principia* demonstrates how faithfully Newton followed Galileo's lead in seeing the essence of mechanics in geometry. The emphasis throughout the work on the geometrical aspect of motion, both in the ordinary three dimensions (shape of orbits) as well as in four dimensions, is most pronounced, but the passage quoted here is particularly interesting in seeming to anticipate the recognition of Riemann and Einstein that the explanation for geometry is to be sought in dynamics. This is a question that will figure prominently in Vol. 2. Meanwhile it may be noted in anticipation of the following chapter that the strongly geometrical aspect of Galilean–Newtonian dynamics was a major factor in the emergence of the concept of absolute space.

The *Principia* proper starts with definitions of the fundamental concepts. Significantly, pride of place is given to the mass concept. On the one hand this emphasizes the materialistic philosophy that underlies the whole of the *Principia*. On the other it highlights the fact that this crucially important concept had hitherto received remarkably little attention – not that Newton's treatment can be regarded as entirely satisfactory. However, this is a topic that is to be deferred until Chap. 12, so I will only mention here that already in the gloss of his definition of mass, Newton draws attention to the distinction between *mass* and *weight*, saying that these are proportional to each other 'as I have found by experiments on pendulums, very accurately made'.

Newton's second definition is that of what is now called *momentum*: 'The quantity of motion is the measure of the same, arising from the velocity and quantity of matter conjointly'. Although the 'quantity of matter' had never been properly defined – and, as we shall see in Chapter 12, its definition still defeated Newton – the momentum concept, albeit one of the most fundamental, is perhaps the least innovative of Newton's contributions to dynamics. As we saw in Chap. 4, Buridan already had a clear intuitive grasp of the notion. Nevertheless, as already emphasized more than once, we should be on our guard against the mistake of not paying due attention to the precise clarification of concepts that have long been intuitively anticipated. When they are finally put on a secure empirical basis, they are almost always found to have subtle and far-reaching implications that were by no means appreciated in the early intuitive period. In the case of momentum this applies especially to the way in which it appears in Newton's Second Law.

Newton's third definition, like the second, is already familiar in its essence from his early work and in its content does not differ much from the statement of his First Law:

The *vis insita*, or innate force of matter, is a power of resisting, by which every body, as much as in it lies, continues in its present state, whether it be of rest, or of moving uniformly forwards in a right line.

Like the mass concept, this definition will be discussed in more detail in Chap. 12. It is only worth mentioning here that the explicit inclusion of the state of rest as a state of the body on a dynamically equal footing with a state of uniform motion is a comparatively late development. In the treatise *De motu* (both versions) the persistence of the state of rest is not mentioned. I do not see this as evidence of a profound reorientation of Newton's concept of motion, i.e., as the sudden recognition that rest and motion have the same ontological status. I see it rather as the simple recognition that his earlier definitions were not quite complete. The addition did not add anything significantly new to the methods by which Newton solved specific problems – always a good criterion for establishing if something important has happened.

The remaining definitions (IV–VIII) all deal with impressed forces, i.e., the forces that change inertial motion. Newton begins with a general definition:

An impressed force is an action exerted upon a body, in order to change its state, either of rest, or of uniform motion in a right line.

After this Newton defines a centripetal force as a particular case of a general impressed force:

A centripetal force is that by which bodies are drawn or impelled, or any way tend, towards a point as to a centre.

It is interesting to note that the general definition of impressed force is more or less identical to Newton's early formulations in the *Waste Book*. We see how the successful application of the specialized concept of a centripetal force demonstrated the fruitfulness of the early notions of force and eventually led to their advancement to the forefront of Newton's scheme. It is well worth quoting here part of Newton's amplification of the concept of impressed force and how it continually acts to change the inertial motion. A great deal of the story of dynamics is contained in the two hundred words or so in which Newton, before ever he postulates his laws of motion, illustrates the action of force and inertial motion by explaining how artificial satellites could circle the earth:

A projectile, if it was not for the force of gravity, would not deviate towards the earth, but would go off from it in a right line, and that with an uniform motion, if the resistance of the air was taken away. It is by its gravity that it is drawn aside continually from its rectilinear course, and made to deviate towards the earth, more or less, according to the force of its gravity, and the velocity of its motion. The less its gravity is, or the quantity of its matter, or the greater the velocity with which it is projected, the less will it deviate from a rectilinear course, and the farther it will go. If a leaden ball, projected from the top of a mountain by the force of gunpowder, with a given velocity, and in a direction parallel to the horizon, is carried in a curved line to the distance of two miles before it falls to the ground; the same, if the resistance of the air were taken away, with a double or decuple velocity, would fly twice or ten times as far. And by increasing the velocity, we may at pleasure increase the distance to which it might be projected, and diminish the curvature of the line which it might describe, till at last it should fall at the distance of 10, 30, or 90 degrees, or even might go quite round the whole earth before it falls; or lastly, so that it might never fall to the earth, but go forwards into the celestial spaces, and proceed in its motion *in infinitum*.

Comparison of this passage with the account Galileo gave of the reasons why bodies are not thrown off the earth by the action of centrifugal force (Chap. 9, p. 480) shows just how close he was to the entire story of dynamics. His mathematics might just have sufficed to solve the satellite problem for circular motion at least but he lacked the overall metaphysics – and made that remarkable slip.

The remaining definitions at the beginning of the *Principia* are amplifications of the force concept over which we may pass. Then follows the famous Scholium on absolute and relative space, time, and motion. This will be discussed in the next two chapters, so we now turn directly to the next part of the *Principia*, which is the formulation of the laws of motion and the deduction of some very general consequences from them (Corollaries I–VI). It is here necessary to say something about the preliminary stages through which Newton's formulations of his laws (and definitions) passed before taking their definitive form in the *Principia*. In this connection it is especially interesting to compare the first and the revised form of *De motu*. If we leave out of consideration the part of the tract that has to do with motion in a resisting medium, the first version of *De motu* rests on two definitions and three hypotheses. The two definitions are of centripetal force, which we have already given, and of innate force, which is worth giving explicitly, especially for its metaphysical overtones and distinct echo of the medieval Buridan:[110] 'And I call that the force of a body or the force innate in a body by reason of which it endeavours to persist in its motion along a straight line.' The hypotheses are as follows (Hypothesis 1, which is omitted, merely states that for certain specified propositions which follow the resistance of the medium is zero):[111]

Hypothesis 2. Every body under the sole action of its innate force moves uniformly in a straight line to infinity unless anything extraneous hinders it.
Hypothesis 3. A body is carried in a given time under the combined action of [two]

forces so far as it would be carried by the forces acting separately in succession in equal times.

*Hypothesis 4.** The space described by a body at the beginning of its motion under the action of any centripetal force is proportional to the square of the time.

These three hypotheses, coupled with the definition of centripetal force, are truly admirable in expressing the way in which dynamics grew out of Galileo's work – indeed how it was created by the application of rigorous mathematics resting on just three hypotheses suggested by empirical observations. The first of the hypotheses reflects the promotion brought about by Descartes of the law of inertia to the status of the first and most fundamental law. (I say 'brought about' because Descartes himself certainly did not conceive the law of inertia in the way Newton did – his service was to bring it to the forefront, after which the law was honed into its definitive form by Huygens, Newton, and others.) The second is straight from Galileo and the third reflects the 'infinitesimalizing' of Galileo's law of free fall noted at the end of Sec. 10.9 (and, indeed, in Huygens' derivation of the formula for centrifugal force). This is the assumption that made it possible to break down the orbital problem into a string of beads in which each bead is a miniature Galilean projectile problem. Incidentally, the title of Newton's pre-*Principia* treatise – 'On the motion of bodies in an orbit' – is eloquent evidence of the fact that it was precisely the solution of the Kepler problem that marked the maturation of dynamics as a nearly complete science. This realization is what slowly dawned on Newton, and it was what determined an almost explosive growth in his aims.

Between the revised and extended version of *De motu* a significant change takes place. Newton drops the word *hypothesis* and employs instead *law*. In the *Principia* the laws of motion are called the Axioms, or Laws of Motion. This reflects Newton's desire to present dynamics in as rigorously mathematical form as possible. It is not, of course, that he regards the laws of motion in the way Descartes did, as truths self-evident to the sufficiently trained and luminous intellect. Far from it; Newton is especially keen to identify the empirical basis of the laws. But it is with him a point of honour to reduce the number of axioms to a bare minimum. This in fact led him to obscure somewhat the empirical springs of his science and to attempt to derive more from his laws than can in truth be done. This tendency is already manifest in the revised version of *De motu*; the laws of motion are in fact changed almost out of recognition. Only the law of inertia remains essentially the same:[112] 'By its innate force alone a body will always proceed uniformly in a straight line provided nothing hinders it.' In place of the other two hypotheses we now have

* This hypothesis was listed by Newton but was not actually written down. However, it was clearly intended by Newton (see Ref. 2, p. 413, note 33).

Law 2. The change in the state of movement or rest [of a body] is proportional to the impressed force and takes place along the straight line in which that force is impressed.

Law 3. The relative motions of two bodies contained in a given space are the same whether the space in question rests or moves perpetually and uniformly in a straight line without circular motion.

Law 4. By the mutual actions between bodies the common centre of gravity does not change its state of motion or rest. It follows from Law 2.

Law 2 is of course Newton's Second Law. The Second Law presents a problem in an interpretational study such as is attempted here. One can certainly say that the discovery and explication by Newton of the dynamics that it, in conjunction with the Third Law, represents constitute the heart of Newton's contribution to dynamics. The Third Law being – for all its importance – relatively simple to grasp and formulate, Newton's outstanding achievement is certainly to be seen in the elaboration of the content of the Second Law and in the recognition of its universality. But in one sense Newton obscured rather than clarified the situation with his actual formulation of the Second Law.

One can look at the matter this way. The motions in the world are given by nature. The task of dynamics is to attempt to find laws that govern them. The ground rules of the overall strategy for finding these laws had crystallized within a very few years of Descartes' publishing of his *Principles of Philosophy*. This is reflected by the elevation of the law of inertia to the first law of motion[113] – 'that prime *Mechanicall* law of motion persisting in a straight line'. This had the effect of more or less *predetermining the remaining laws of motion*. For, the motions being given and the law of inertia having been adopted, the remaining laws were forced to appear – as the complement, so to speak, of the law of inertia in the domain of observed motions. But if the overall strategy was predetermined (unconsciously, of course, but with hindsight we can see that it was), the actual implementation was nevertheless a supremely difficult task. The most important step was what Newton, using the preparatory work done by Galileo on the simplest of compound motions, so successfully accomplished in the orbital problem.

As the overall structure of dynamics took shape, Newton correctly sensed that the First Law must be complemented by the laws which describe how bodies interact and mutually deflect each other from their inertial motions as posited by the First Law. If the *Principia* is considered in its totality – definitions, laws, and explications thereof in highly nontrivial examples – there is no question but that Newton succeeded brilliantly. What one can fault is the manner in which the most important empirical facts – grasped by Galileo, used by Newton in his solution of the Kepler problem, and still formulated (in the form of hypotheses 3 and 4) as primal laws in the first draft of *De motu* – are demoted to consequences of the allegedly more fundamental Second Law.

For in the revised draft we find that the laws posited above are immediately followed by two lemmas:

Lemma 1. A body acted on simultaneously by [two] forces describes the diagonal of a parallelogram in the same time as it would the sides if the forces acted separately.
Lemma 2. The distance a body describes from the beginning of its motion under the action of any force whatsoever is in the duplicate ratio of the time.

But nothing that Newton has formulated hitherto – in his definitions and laws – permits him to deduce these results. The truth is much rather that these empirical facts are what suggested in the first place the structure of laws and definitions from which they are now supposedly deduced. It is not that Newton's laws and definitions are wrong but merely that they are too skeletal – they are not fleshed out with sufficient empirical content. As Mach was careful to emphasize in later editions of his *Mechanics,*[114] such comments on Newton's ordering of his definitions and axioms in no way diminishes his achievements or one's admiration of them. There is however much to be gained – both from the point of view of intellectual satisfaction as well as for a proper appreciation of the points at which changes might be made in the Newtonian scheme – in laying bare the empirical springs of the great science that took shape in the *Principia*.

The unconscious process of obscuring the empirical basis of dynamics is taken even slightly further in the *Principia*. Lemma 1 above reappears as Corollary I to the laws of motion,[115] but the vital Lemma 2 does not appear until quite some way into Book I, where it is formulated as the tenth[116] of a series of decidedly mathematical lemmas. Moreover, the result is in fact obtained – as it only can be – from Newton's implicit assumption of the actual manner in which forces act.

In this way Newton obscured the truth that is the real heart of his dynamical scheme – that the force is proportional to the instantaneous acceleration. This was the result that both Huygens and Newton deduced by their infinitesimalization of Galileo's law of free fall. The modern reformulation of Newton's Second Law in the form $\mathbf{F} = m\mathbf{a}$ (for the case of constant mass) has restored the original insight of Huygens and Newton to the prominence it deserves.

We must now note the appearance – so far as I know for the first time in Newton's writings – of the law of Galilean relativity (Law 3). We see that it appears together with the centre-of-mass theorem (Law 4), which is stated cryptically to be a consequence of Law 2. This latter is a clear indication that Newton must have come to realize that an important element was still lacking from his dynamics, namely, his Third Law, for his Law 2 is quite insufficient in itself to derive the centre-of-mass theorem. There must in fact have been two or perhaps three distinct stages in Newton's advance to the comprehensive treatment of motions that we find in the *Principia*. The first is the one we have treated at length and which was certainly by far the most difficult from the technical point of view – the mastering of the Kepler problem. This stage remains restricted to what can be called single-particle dynamics in the force field of a fixed attractive centre. The other two stages correspond to the recognition of the full import of the concept of universal gravitation and the further recognition that all changes in motion are brought about by the interaction of at least two bodies and that if body A changes the state of body B then body B will simultaneously change the state of body A. Wilson[57] has emphasized the intimate connection between the final emergence of universal gravitation and the appearance for the first time in the pre-*Principia* papers of the embryonic Third Law (which we shall shortly discuss). Attractions are now mutual and truly universal, and what in the *Waste Book* applied only to collisions is now extended to all interactions. Further, in parallel with this progressive encompassing of the full spectrum of physical possibilities now being opened up to analysis, we have Newton's growing desire, already noted, to create a general mathematical formalism adequate for the proper expression of these great insights.

The two drafts of *De motu* reveal the dramatic extensions in the process of taking place. In the first draft we see the bridgehead firmly established. In the second we see the troops beginning to spread out in all directions to occupy the rich territory now made defenceless by the fact that the greatest problem has been comprehensively cracked. We have already noted Newton's announcement of the successful moon test; as the implications begin to sink in, Newton realizes more and more places at which the concept of universal gravitation can and must be tested. The planetary problem is immediately seen to be considerably more sophisticated than the problem of a single fixed force centre controlling the motion of a single planet. All the bodies in the solar system must interact with one another and produce small perturbations from Kepler's laws – hence the questions sent at this time to Flamsteed about the perturbations of Saturn by Jupiter. Newton grasps the fact that in truth the planetary motions are fearfully complicated but he yet sees through to the essence of the situation: Kepler's laws hold to a first approximation and in addition there is one beautifully simple consequence of the laws of motion – despite the eternal tangle of the exact planetary orbits there is a distinguished point in the solar system that has a very special motion. It is the centre of mass of the complete system of planets and the sun.

This insight, the first clear recognition of how everything in the solar system hangs together, is summarized in a notable passage in the revised version of *De motu*. Newton now understands how the natural world works in its gross details and announces it to the human world in these words:[117]

Moreover the whole space of the planetary heavens either rests (as is commonly believed) or moves uniformly in a straight line, and hence the communal centre of gravity of the planets either rests or moves along with it. In both cases the relative motions of the planets are the same, and their common centre of gravity rests in relation to the whole of space, and so can certainly be taken for the still centre of the whole planetary system. Hence truly the Copernican system is proved *a priori*. For if the common centre of gravity is calculated for any position of the planets it either falls in the body of the Sun or will always be very close to it. By reason of this deviation of the Sun from the centre of gravity the centripetal force does not always tend to that immobile centre, and hence the planets neither move exactly in ellipses nor revolve twice in the same orbit. So that there are as many orbits to a planet as it has revolutions, as in the motion of the Moon, and the orbit of any one planet depends on the combined motion of all the planets, not to mention the action of all these on each other. But to consider simultaneously all these causes of motion and to define these motions by exact laws allowing of convenient calculation exceeds, unless I am mistaken, the force of the entire human intellect.

We must now consider Newton's standpoint with regard to the law of Galilean relativity and the centre-of-mass theorem. As we see, both appear as laws in the revised version of *De motu*, though the latter is said to follow from Law 2 (Newton's Second Law). At this stage Newton's Third Law does not yet appear, though, as we have seen, it was very clearly formulated for impacts in the *Waste Book*. Does this mean that Newton recognized Galilean relativity as an empirical fact of the first importance, took it as a law, and then deduced his Third Law from it by means of arguments along the lines of those used by Huygens in his work on collisions? This seems to me most unlikely, and there does not appear to be any evidence to support such a derivation. Moreover, the *Waste Book* and the definitive treatment in the *Principia* suggest quite the opposite – that the Third Law was the primal insight and that Newton attempted to derive both Galilean relativity and the centre-of-mass theorem from it and the other laws.

I therefore see the laws as formulated in the revised version of *De motu* as corresponding to the stage at which Newton had recognized the importance of Galilean relativity but had not yet seen his way to a derivation of it from more primitive laws. As for his Third Law, its omission at this stage may simply have been an oversight. Particularly interesting in this connection are the *Drafts of Definitions and Laws of Motion* which Herivel reproduces.[118] He concludes that they are slightly later in date than the revised version of *De motu*, and this certainly matches the logical development that they represent. Leaving aside the question of motion in a resisting medium (which will be considered later), we find that there are now *five* fundamental laws of motion though it is clearly indicated that one at least can be seen as a consequence of the others. The first two are in essence Newton's First and Second Laws. But for the first time we now find the Third Law stated explicitly. Here are the three remaining laws:[119]

3. As much as any body acts on another so much does it experience in reaction [this is followed by several well chosen empirical confirmations].
4. The relative motion of bodies enclosed in a given space is the same whether that space rests absolutely or moves perpetually and uniformly in a straight line without circular motion. For example, the motions of objects in a ship are the same whether the ship is at rest or moves uniformly in a straight line.
5. The common centre of gravity of [a number of] bodies does not change its state of rest or motion by reason of the mutual actions of the bodies. This law and the two above mutually confirm each other.

As in the revised version of *De motu*, these laws are followed by the same Lemma 1 and Lemma 2 as above.

The final stage of this process is what we find in the *Principia*. The fundamental laws of motion have been pared down to just three:[120]

Law I. Every body continues in its state of rest, or of uniform motion in a right line, unless it is compelled to change that state by forces impressed upon it.
Law II. The change of motion is proportional to the motive force impressed; and is made in the direction of the right line in which that force is impressed.
Law III. To every action there is always opposed an equal reaction: or, the mutual actions of two bodies upon each other are always equal, and directed to contrary parts.

The laws of motion are then followed by the six corollaries already mentioned, which are themselves followed by a scholium in which Newton discusses the empirical evidence for the Third Law of Motion. As already pointed out, Newton seems to have regarded the first two laws as so well established as to need absolutely no justification, and he states baldly (but decidedly anachronistically) that Galileo used the first two laws when he discovered the law of descent and the projectile law. On the other hand, he devotes several pages to the empirical justification of the Third Law, some of which we have already considered. As this does not involve any further points of particular interest, we turn immediately to the Corollaries.

The first is Lemma 1 of the revised version of *De motu*, i.e., Galileo's law of the composition of two motions produced by the simultaneous action of two forces, by which Newton usually understands, as before, the innate force and an impressed force. Corollary II is quite new though conceptually is closely related to Corollary I. It deals with the composition and decomposition of forces in statics and by implication shows how statics is subsumed in the more comprehensive science of the laws of motion ('for on what has been said depends the whole doctrine of mechanics variously demonstrated by different authors'[121]).

Corollary III states the law of conservation of momentum:[122]

The quantity of motion, which is obtained by taking the sum of the motions directed towards the same parts, and the difference of those that are directed to contrary parts, suffers no change from the action of bodies among themselves.

Newton in fact proves this corollary only for the case of collisions. The proof shows the dominant role played by the Third Law in his overall scheme. It begins thus: 'For action and its opposite reaction are equal, by Law III, and therefore, by Law II, they produce in the motions equal changes towards opposite parts. Therefore if the motions are directed towards the same parts, whatever is added to the motion of the preceding body will be subtracted from the motion of that which follows; so that the sum will be the same as before. If the bodies meet, with contrary motions, there will be an equal deduction from the motions of both; and therefore the difference of the motions directed towards opposite parts will remain the same.' Then follows some numerical examples, after which Newton concludes with a remark that, on the one hand, takes us back to the *Waste Book*, in which he had solved two-dimensional problems so effortlessly, but, on the other, demonstrates what a complete science dynamics had now become:[123]

But if the bodies are either not spherical, or, moving in different right lines, impinge obliquely one upon the other, and their motions after reflection are required, in those cases we are first to determine the position of the plane that touches the bodies in the point of impact, then the motion of each body (by Cor. II) is to be resolved into two, one perpendicular to that plane, and the other parallel to it. This done, because the bodies act upon each other in the direction of a line perpendicular to this plane, the parallel motions are to be retained the same after reflection as before; and to the perpendicular motions we are to assign equal changes towards the contrary parts; in such manner that the sum of the conspiring and the difference of the contrary motions may remain the same as before.

Corollary IV is stated as follows:[124]

The common centre of gravity of two or more bodies does not alter its state of motion or rest by the actions of the bodies among themselves; and therefore the common centre of gravity of all bodies acting upon each other (excluding external actions and impediments) is either at rest, or moves uniformly in a right line.

The proof follows closely the one with which we are familiar from the *Waste Book*. The main difference is that Newton considers the case of an arbitrary number of bodies and not just two. He first shows that if the bodies do not interact at all the common centre of gravity must move uniformly and in a straight line. That interaction has no effect on this result is proved by the definition of the centre of gravity and the Third

Law: 'Moreover, in a system of two bodies acting upon each other, since the distances between their centres and the common centre of gravity of both are reciprocally as the bodies [i.e., the masses], the relative motions of those bodies, whether of approaching to or of receding from that centre, will be equal among themselves. Therefore since the changes which happen to motions are equal and directed to contrary parts, the common centre of those bodies, by their mutual action between themselves, is neither accelerated nor retarded, nor suffers any change as to its state of motion or rest.' This result is extended to a system of more than two bodies by the assumption that all interactions within a system reduce to interactions between pairs of bodies:[125] 'But in such a system all the actions of the bodies among themselves either happen between two bodies, or are composed of actions interchanged between some two bodies; and therefore they do never produce any alteration in the common centre of all as to its state of motion or rest.' It is worth noting that at this point Newton makes an assumption which goes beyond his three laws of motion, making in effect a restriction on the manner in which the forces of nature act. It is particularly interesting to note that he does the same in the proof of his famous Corollary V:[126]

> The motions of bodies included in a given space are the same among themselves, whether that space is at rest, or moves uniformly forwards in a right line without any circular motion.

The proof of this key proposition, which I give in full, is a great deal shorter than the proofs given of the two previous propositions (the italics are mine):

> For the differences of the motions tending towards the same parts, and the sums of those that tend towards contrary parts, are, at first (by supposition), in both cases the same; *and it is from those sums and differences that the collisions and impulses do arise with which the bodies impinge one upon another.* Wherefore (by Law II), the effects of those collisions will be equal in both cases; and therefore the mutual motions of the bodies among themselves in the one case will remain equal to the motions of the bodies among themselves in the other. A clear proof of this we have from the experiment of a ship; where all motions happen after the same manner, whether the ship is at rest, or is carried uniformly forwards in a right line.

The part I have italicized is especially interesting for the discussion which follows in Chap. 11. The first point to make is that it is a major assumption, quite unrelated to Newton's three laws of motion, about the manner in which interactions take place (we note that once again Newton considers only changes produced by collisions). He has of course his Third Law, but this only says that the action and reaction are equal in magnitude and opposite in direction. As pointed out in Sec . 1.2, there is, however, no reason whatever why the *strength* of interaction (the impulses) between two bodies (to consider the simplest case) should be the same when their centre of mass moves through absolute space with a uniform velocity as when it is at rest. If there were such a dependence, the Third Law would still ensure the validity of Corollaries III and IV but Corollary V would not hold. In anticipation of the following chapter let it merely be noted here that despite his firm insistence on the reality of absolute space and absolute motion it is striking that Newton appears to have assumed instinctively that interactions are *purely relative*.

We conclude this discussion of the corollaries with Corollary VI, which is of a quite unrelated nature and seems hardly to warrant the exalted status given it immediately following the laws of motion. It is:

> If bodies, moved in any manner among themselves, are urged in the direction of parallel lines by equal accelerative forces, they will all continue to move among themselves, after the same manner as if they had not been urged by these forces.

Newton presumably includes it because, to a good first approximation, it describes the situation that obtains for the earth–moon system in the gravitational field of the sun and especially (to a very good accuracy) for Jupiter and its system of moons. However, it is interesting to note that this corollary anticipates Einstein's equivalence principle; for as most readers will know, Einstein made the fact that all bodies fall with equal acceleration in a given gravitational field into the physical basis of his general theory of relativity, in which he succeeded in reformulating Newton's theory of gravitation in a conceptually quite different framework. We see here again something noted already more than once – that all three great pioneers in dynamics (Galileo, Huygens, and Newton) often came remarkably close to some of Einstein's deepest insights. This is one of the especial fascinations of dynamics – how one and the same empirical fact can be noted and used in totally different ways by great scientists living centuries apart.

The main body of the *Principia* is divided into three Books. Books I and II are basically of a formal mathematical nature with Book I treating the case of motion when there is no resistance while Book II treats motion in resisting media. These two books are essentially the mathematical theory of the motion of bodies with mass moved in accordance with Newton's three laws and interacting in accordance with definite forces. This work, an almost incredible *tour de force* that did for the theory of motion what Euclid had done for geometry, is perfectly characterized in Newton's Preface by the words[127] 'rational mechanics will be the science of motions resulting from any forces whatsoever, and of the forces required to produce any motions, accurately proposed and demonstrated'. Newton makes a point of stressing that he offers the work 'as the mathematical principles of philosophy'. The overall logic of his book is to posit formally the concept of force in conjunction with his laws and the definition of mass, derive the mathematical consequences that flow from them, and then demonstrate that the existence of certain forces with mathematically well-defined properties, above all forces of gravity, is proved by the phenomena of nature. Hence his remark that we have already quoted: 'For the whole burden of philosophy seems to consist in this – from the phenomena of motions to investigate the forces of nature, and then from these forces to demonstrate the other phenomena.' The climax of the work is thus in Book III, in which Newton, as he announces in the Preface, gives 'an example of this [the finding of the forces and the demonstration of the other phenomena] in the explication of the System of the World; for by the propositions mathematically demonstrated in the former Books [I and II], in the third I derive from the celestial phenomena the forces of gravity with which bodies tend to the sun and the several planets. Then from these forces, by other propositions which are also mathematical, I deduce the motions of the planets, the comets, the moon, and the sea.'

To do justice to the *Principia* I should have to write another chapter at least as long as the present one. But as this book is about the discovery of dynamics and the problem of whether motion is absolute or relative (and not about dynamics itself) I shall excuse myself with an account of the crucial sections on the orbital problem and brief summary of some of the most important of Newton's results and applications – and the now familiar exhortation to the reader to take up the *Principia* and read for himself (or herself). It must be said that the *Principia* was and is a rather daunting book – at the time it was published on account of the novelty and difficulty of the subject for Newton's contemporaries, in the modern age on account of the now archaic mathematical techniques and means of expression that Newton employs. Moreover, in the first two books Newton does often get carried away and tends to explore anything which seems capable of analysis. He loves to add one corollary to another after his propositions. Indeed, Newton himself, when he opens Book III, comforts the reader about the first two books:[128] 'Not that I would advise anyone to the previous study of every Proposition of those Books; for they abound with such as might cost too much time, even to readers of good mathematical learning. It is enough if one carefully reads the Definitions, the Laws of Motion, and the first three sections of the first Book.' (One wonders quite why Newton gives this advice only when the reader has completed reading what he is here advised not to read!) In fact, the reader with a sound knowledge of dynamics and its applications can readily derive a great deal of interest from the *Principia,* since the propositions and their physical significance are often immediately apparent to the modern reader. Very often the most rewarding parts are to be found in the scholia and general comments. One of the chief points of interest is simply the extraordinary range of topics – both mathematical and physical – that Newton covers and the sheer comprehensiveness of his treatment. One is reminded of Carlyle's aphorism on Frederick the Great to the effect that what makes genius is, in the first place, a 'transcendent capacity for taking trouble'. It is as if Schrödinger, having discovered his wave equation, had then gone on, single-handed, to obtain all that flood of results in the quantum mechanics of atomic systems in which so many physicists had a part in this century at the end of the 'twenties and beginning of the 'thirties. And Newton did it all in about 18 months to two years.

Let me therefore now just mention some of the most salient points, especially in the part on orbital problems, and the most interesting of the applications.* We have already anticipated the most important tendencies. The crucial role of Kepler's area law in both the final breakthrough to mature dynamics and the clarification of the concept of force is reflected in the fact that Newton begins the exposition proper with the derivation of this law. It is the content of his very first proposition: 'The areas which revolving bodies describe by radii drawn to an immovable centre of force do lie in the same immovable planes, and are proportional to the times in which they are described.' Just as Kepler learnt to move freely in his mind's eye through the solar system when once he had firmly established the halving of the eccentricity of the earth's orbit, so too did Newton learn to treat the most general of orbital problems when once he had grasped the dynamical significance of the area law. And thus the second great synthesis (the creation of dynamics) was added to the first (Kepler's discovery of the laws of planetary motions).

Newton concludes his first few propositions on the subject of the area law with the following scholium, which shows precisely how he uses the phenomena of motions, described with mathematical exactitude, to draw conclusions of far reaching physical significance. He uses mathematics to study nature in a manner of which Descartes never even dreamed:

> Since the equable description of areas indicates that there is a centre to which tends that force by which the body is most affected, and by which it is drawn back from its rectilinear motion, and retained in its orbit, why may we not be allowed, in the following discourse, to use the equable description of areas as an indication of a centre, about which all circular motion is performed in free spaces?

* My treatment of the detailed mathematical aspects will, I am afraid, be very sketchy. An especially interesting question relating to the period of development up to and including the *Principia* is that of the relative importance, *vis à vis* the physical insights, of Newton's growing ability to cope with the mathematics needed to solve the orbital problems. D. T. Whiteside, who has devoted many years to the study of Newton's mathematics, has asserted that[129a] 'the continuous growth during the period 1664–84 of Newton's expertise with the various orders of the infinitely small was a significant conditioning factor on the effective expression and forceful persuance of his dynamical researches.' I am not always able to follow all of Whiteside's arguments, but he is surely right that this is an important point, and it belongs to a study more comprehensive than the present one can attempt to be. Readers interested in these questions can begin by consulting Whiteside's works listed in Ref. 129.

There are then several propositions which follow up this line of development. It will here be worth describing in general terms the method employed by Newton to solve orbital problems, since this reveals particularly clearly the central role played in Newton's work by the astronomical results – and also the way in which Newton makes extensive use of calculus-type techniques without employing formal equations of the infinitesimal calculus. Newton's point of departure, indeed the basis of his entire work, in the crucial early sections of the *Principia*, which lay the foundations for his later demonstration of the law of universal gravitation, is that one knows from observation that a body is moved in a given curve at a given (position-dependent) speed such that the radius vector from a given point to the moving body sweeps out equal areas in equal times. Note that for a given curve (which may but need not be closed) and for any given point it is always possible to prescribe a law of motion in the curve such that equal areas are swept out in equal times about the given point. (There may of course be singular situations in which the area law is violated at certain points, but for a point strictly within a closed, everywhere concave, curve this does not occur.) In contrast, for an arbitrarily specified law of motion it is a highly nontrivial result that there does exist a point from which the radius vector to the moving body sweeps out equal areas in equal times. In fact, one of Newton's first results[130] is to show how knowledge of the velocity at three points in the orbit suffice in such a case to determine the position of the point about which equal areas are described and from that to find the velocity at any other point in the orbit. Newton's construction is shown in Fig. 10.21. The velocities are given at P, Q, and R; the perpendiculars AP, BQ, and CR at these points to the tangents to the curves (PT, TQV, RV) have heights that are inversely proportional to the speeds at P, Q, and R. The line AD is parallel to PT, DBE to TQV, and CE to RV. Newton shows that the point S is the force centre about which the equal areas are described. The basis of his construction is a corollary of the area law, namely, that the height of the perpendicular (the dashed line, my addition to Newton's figure) dropped from S onto, say, the tangent PT is

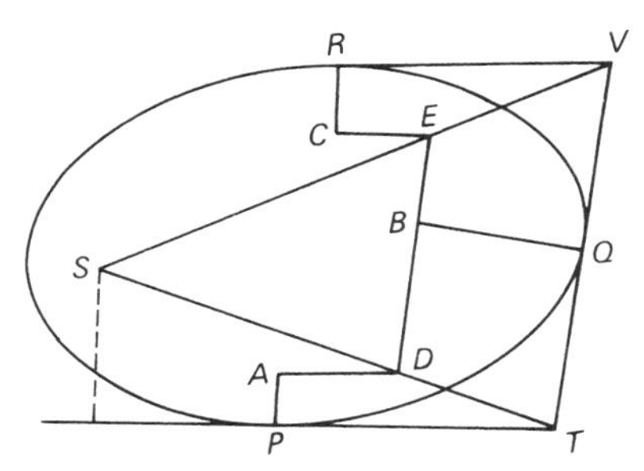

Fig. 10.21.

inversely proportional to the speed at P. For in an infinitesimal time dt the body will traverse a space vdt in the direction of the tangent at P, v being the instantaneous velocity at that point. Let h be the height of the perpendicular dropped onto PT. Then the area swept out by the radius vector SP in time dt is $\frac{1}{2}hv$dt, but this must be constant by the area law, from which it follows that h is inversely proportional to v. (A particular delight of the *Principia* is the remarkable number of important dynamical results that Newton succeeds in deducing from the formula for the area of a triangle in terms of its base and height.)

Just as in the solution of the Kepler problem, but now for the perfectly general case of bodies moved in arbitrary curves by a centripetal force towards a fixed centre at some given point, i.e., in such a way that the area law holds with respect to that point, Newton then proceeds to solve the problem of determining the *distance dependence* of the force that produces such motion. The elements of his method are illustrated in Fig. 10.22, in which the force centre is at S and the continuous curve PW is the curve in which the body is moved. At a certain time the body will be at P and its instantaneous velocity will be along the tangent PT. The dashed arc is the circle, with centre at C, that approximates the orbital curve at the point P. One can think of it as being obtained by taking an orthogonal Cartesian coordinate system with origin at P, x-axis along PT and y-axis along the perpendicular PC and then finding the Taylor expansion of the function $y(x)$ that describes the orbital curve. By the choice of the coordinate system, the constant and linear terms of this expansion will be zero and the first nonvanishing term will be the quadratic term, which, in the neighbourhood of P, represents the circle that best approximates the orbital curve at P. This curve is important, since it measures the instantaneous deviation in space of the orbit from the instantaneous tangent. In

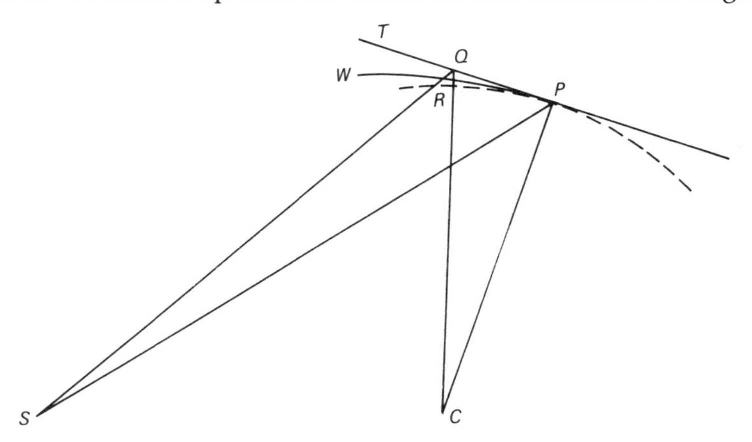

Fig. 10.22.

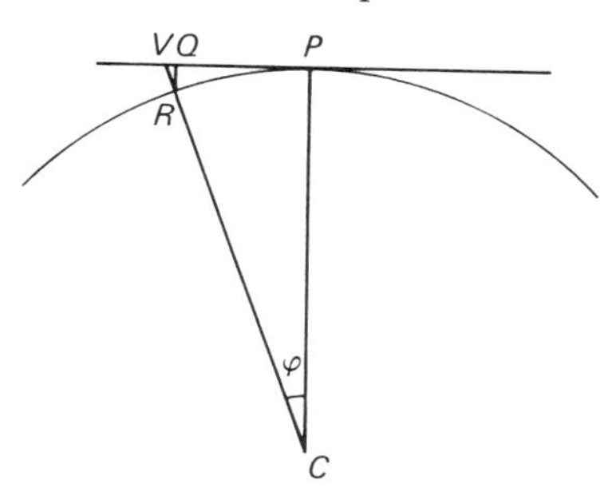

Fig. 10.23.

the case of circular motion with force centre at the centre of the circle, when S, the force centre, coincides with C, the instantaneous centre of curvature, this deviation is all that one needs to determine the force acting (Fig. 10.23). (If a body revolves in a circle and turns through the small angle φ, then the deviation QR from the tangent (which is equal to VR in the limit $\varphi \rightarrow 0$) is proportional to $1 - \cos \varphi$, or the versed sine of φ. This trigonometric function, which is seldom used today, appears frequently in the *Principia* on account of this fact.) In the general case of nonuniform motion in a closed curve about some arbitrary point that acts as a centre of a centripetal force (Fig. 10.22), there are two complications in the way of finding the deviation in unit time from the instantaneous velocity (and hence finding the force acting): the nonuniformity of the motion and the fact that the force centre does not coincide with the instantaneous centre of curvature. In principle Newton overcame these difficulties by using the area law (and the known geometry of the orbital curve and known position of the centre from which equal areas are swept out) to determine how much the radius vector SR (R is the point on the *continuous curve* of the actual motion reached at the given time) advances from SP in time dt. In the limit when this tends to zero, the deviation QR from the instantaneous inertial motion at P is measured by the deviation of the instantaneous circle of curvature at P from the instantaneous tangent PT, this deviation, which is measured by the versed sine of the arc of the circle of curvature (i.e., the versed sine of angle QCP), being multiplied by a suitable trigonometric factor to take account of the fact that RQ is inclined at an angle to QC, which coincides with CP in the limit when Q tends to P. The important point here is that, from the area law, Newton knows that the force is along QR towards S (in the limit QR becomes parallel to PS) and this tells him by how much the actual deviation induced by the force is greater than the simple perpendicular separation between Q and the curve of the orbit (which is the force-induced deviation in the case of circular motion). We see again how the area law simultaneously solves two separate problems and reduces the complete orbital problem to one of pure geometry, albeit the geometry of limit increments of curves rather than synthetic geometry in the spirit of the ancients.

To summarize: Newton's solution of the orbital problem consisted of three main components. First, and most important of all, was the clear understanding of what needed to be calculated. He needed the insight that the area law solved the problem of determining the time over which the deviation occurs and simultaneously the problem of determining the direction in which it acts (without both of which the strength of the force, measured by an infinitesimal quantity of *second* order, cannot be found). Finally, he needed a high degree of geometrical competence to do the actual computations for the case of a general orbit, when S and C do not coincide. Difficult as this problem may be, it is nevertheless purely technical once the conceptual points have been clarified.

Let us take this opportunity to note just how drastic and sophisticated is the Newtonian division of the physically given motions of the planets in their orbits into inertial motion and force-induced deviation from it. At each new point of the orbit, the inertial motion with respect to which the deviation is measured is a different one and advanced mathematics is necessary to make the decomposition. Inertial motion pure and simple is never present but is always a mathematical construct. Nevertheless, the result of the analysis is unambiguous and certainly provides very good grounds for the notion that bodies have an inherent tendency to uniform rectilinear motion in the absence of deflecting forces. Inertial motion is demonstrated by mathematics from the empirically observed motions by 'subtracting' the effect of the sun in accordance with a well-defined and physically plausible prescription. Newton's work clearly established the notion of inertial motion, and with it his idea of absolute space, on a vastly more secure basis than it ever had prior to this work. Because this dramatic discovery, the full import of which can readily escape a modern scientist, does indeed provide the basis of dynamics and with it most of modern science, it clearly warrants careful examination, which we shall attempt to give in Chaps. 11 and 12.

Having developed his general technique, Newton used it to determine the centripetal force that must act in the case of bodies that move in various curves about particular points. His first important result with physical applications concerns the case of elliptical motion when the centre of force is at the *centre* of the ellipse. Newton shows[131] that in this case the force must increase in proportion to the distance from the attracting centre. After that he moves onto the more difficult problem of the centre of force being at one focus and establishes the inverse square law for this case. He shows that the same is true for bodies moved in the other conic sections.

As a very simple illustration, which Newton does not in fact give, let us see how the inverse square law follows *approximately* from the small-

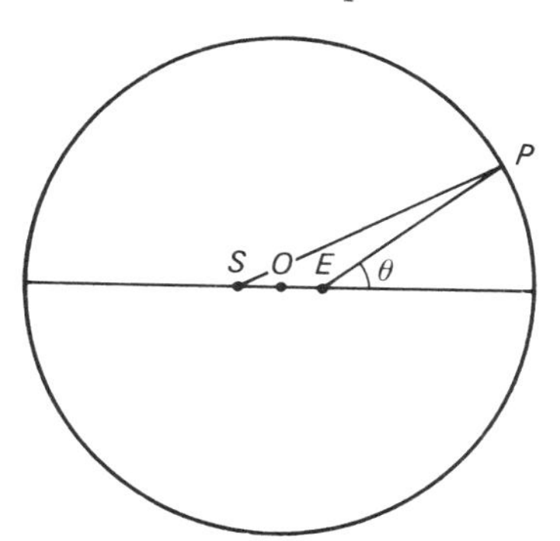

Fig. 10.24.

eccentricity limit of Kepler's laws. Thus, suppose a circle (Fig. 10.24) and a point E eccentric from the centre at the small distance er, i.e., $e \ll r$, where r is the radius of the circle. Let a ray rotate about E with uniform angular velocity and let the body be situated at all times at the point P at which the spoke cuts the circle. Then we know that to a good approximation equal areas are swept out by the radius vector SP, where S is the point at equal distance from the centre O of the circle opposite to E. To first order in e, the length of EP is $r(1 - e \cos \theta)$ and the distance travelled along the circumference in the time corresponding to an increase in θ by $d\theta$ is $r(1 - e \cos \theta)\, d\theta$. The radial deflection of the body, i.e., the deviation from the tangent is proportional to the square of this distance, i.e., to first order it is proportional to $r(1 - 2e \cos \theta)$. This is the radial deflection towards the centre. But for small e, this is also, to order e, the deflection towards S. On the other hand, the distance of P from S is, again to first order in e, equal to $r(1 + e \cos \theta)$. Since $1/(1 + e \cos \theta)^2 \approx 1 - 2e \cos \theta$, we see that the radial deflection towards S is, to the given accuracy, exactly what corresponds to the inverse square law. This example uses the area law nontrivially but still corresponds, in the adopted approximation, to trivial geometry (with the deflection towards the centre of curvature equal to the deflection towards the force centre).

It is appropriate to mention here the beautiful explanation that Newton gives for the phenomenon known as the *libration* of the moon. As is well known, the moon always presents the same face to the earth. However, this is only approximately true; in the course of a month the terrestrial observer is permitted to see a little bit of the 'other side' of the moon. The largest effect, the *libration in longitude* (which amounts to about 8°) comes about for the following reason. The moon, like most celestial bodies, rotates about its axis. In the early part of this book it was repeatedly emphasized that the rotation of such bodies is to a high degree uniform, this being basically a consequence of the conservation of angular momentum. Now it so happens that the period of rotation of the moon is exactly equal to its period of revolution around the earth; this is why we always see the same face of the moon. Moreover, as we know from Chap. 3, the moon has a relatively large eccentricity and therefore moves with pronounced nonuniformity in its orbit. Since its axial rotation is uniform, this is what allows us to see around the edge of the moon. The particularly beautiful point, which Newton makes, is that since the apparent motion of the moon appears almost exactly uniform when observed from the *void focus* of the moon's orbit (this is Ptolemy's equant phenomenon), the moon will always present more or less exactly the same face towards that focus. In Newton's words:[132] 'the same face of the moon will be always nearly turned to the upper focus of its orbit; but, as the situation of that focus requires, will deviate a little to one side and to the other from the earth in the lower focus.' Thus, the moon, gazing sempiternally on the void focus of its orbit, is a perpetual monument to Ptolemy's great discovery of the enigmatic equant in the days before Kepler had begun the work that eventually identified the physical cause of the celestial motions and shifted the point of interest from the void to the occupied focus – to the centre of force that finally made sense of the celestial ballet. And this explanation of the moon's libration in longitude* simultaneously reminds us of the crucial importance for theoretical astronomy of the uniformity of the rotation of celestial bodies (providing the 'ticks' of the astronomical clock) and the equant phenomenon.

It is in fact doubly appropriate to recall Ptolemy at this point, since the manner in which Newton presents the orbital problem in the *Principia* reveals very clearly the astronomical antecedents of dynamics and emphasizes the *continuity* of the development all the way from Hipparchus and the very first attempt to provide a rational explanation for the nonuniformity of the solar motion through Ptolemy, Copernicus, Kepler, and right on to Newton. The central problems throughout the entire period were those of finding the geometrical orbit and its position in space relative to the observer and of finding an algorithmic prescription for the law of advance of the celestial body on its orbit with the passage of time (as measured by the rotation of the earth). This latter problem was solved first by Hipparchus with the simple assumption of uniform motion, then by Ptolemy with his equant prescription, and then improved again by Kepler with his area law. It was this algorithm, progressively refined, that made astronomy work. It was therefore especially appropriate that Newton then transformed the very same algorithm into a powerful tool

* The *libration in latitude* is also a consequence of the conservation of angular momentum, since the moon's rotation axis is not perpendicular to the plane of its orbit but tipped at an angle of $6\frac{1}{2}°$, which permits us to see over first one and then the other pole. There was quite a long and interesting pre-history to Newton's explanation of the moon's libration in longitude which Gabbey has traced.[133]

for solving dynamical problems, taking advantage of Kepler's great work to encode information about time in the spatial geometry. And because of the close connection between the area law and some of the most significant advances in astronomy, the dynamics of the orbital problem as treated by Newton in the *Principia* still retains a great deal of the geometrokineticism of ancient astronomy. The spirit of Hipparchus and Ptolemy still breathes in the geometrical constructions that Newton employs, even though dynamics is on the point of progressing to a language and techniques in which the influence of the early astronomy can hardly be traced at all. But in the *Principia* one can see explicitly how Newton merely continued the tradition started by Kepler of introducing modifications in astronomical practice that were suggested by terrestrial physics and then found fruitful application in the heavens. And, whereas Kepler had to rely on frankly very poor terrestrial physics, Newton could use three thoroughly good results: Galileo's law of free fall and the law of composition of motions, and the law of inertia. Thus it was that the combination of these three elements with Kepler's three laws yielded, in the manner just described explicitly, the two great Newtonian discoveries – the law of universal gravitation (using Kepler's First and Third Laws) and the realization (brought about by the use of the Second Law) that central forces do truly exist and that the laws of dynamics were now in principle complete, i.e., that dynamics had been discovered.

In modern terms, the significance of the area law is that it represents what is known as a *first integral* of the orbital problem. As already pointed out (p. 413 fn), first integrals (conserved quantities) and the dynamical symmetries with which they are associated will play an increasingly central role in our discussion in Vol. 2. The area law is not the only first integral that played an important part in the emergence of dynamics. The conservation of momentum and energy must also be mentioned, especially the former, since it was crucial to the solution of the impact problem, which, on the one hand, quickened the interest in dynamical problems, and, on the other, gave Newton the clue to his Third Law.

Finally, these considerations show that dynamics was not discovered by a systematic programme of painstaking observation followed by theoretical interpretation designed to solve a well-defined problem such as happened in the case of Brahe and Kepler's attack on the problem of the planets. It was much rather the case that Newton gradually came to the realization that he was in principle capable of solving *any* specific problem of motion that was put or occurred to him. But if that was the case it must follow that, without being fully seized of the fact, he possessed all the elements of dynamics. Hence the dramatic extension of the ambit of his investigations and the transformation of the framework in which he conceived them that we find intimated in the progression from the first to the revised version of *De motu corporum in gyrum* and brought to fruition in the *Principia* in such a short space of time.

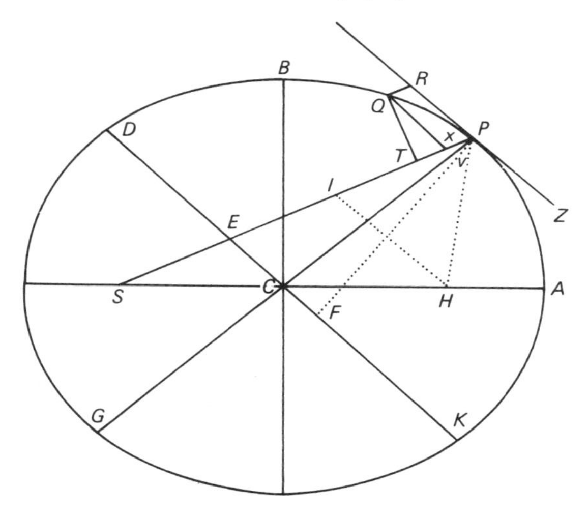

Fig. 10.25.

Let us end this discussion of the orbital problem by reproducing (Fig. 10.25) from the *Principia* the diagram which Newton used to illustrate his deduction of the inverse square law from Kepler's First and Second Laws. This is also very appropriate because it illustrates in one diagram the really essential contributions that went into the discovery of dynamics: the synthetic geometry of the ancients, Kepler's first two laws, and Newton's infinitesimalization of Galileo's projectile work. Simultaneously it shows the old mathematical techniques being pushed to their very limit and presages the introduction of the infinitesimal calculus on a systematic basis. The planet is at P, the sun at the focus S, and RQ is parallel to PS. The contribution of synthetic geometry is in giving Newton exact relationships between the distances RQ, QT (the perpendicular from Q onto PS) and fixed elements of the ellipse (in this particular proof, the semimajor and semiminor axes and chord of the ellipse through one of the foci perpendicular to the major axis – the *latus rectum*). Galileo's contribution (somewhat strengthened by Descartes) is represented by the tangent PR and the notion it defines of the instantaneous inertial motion at P, the deflection RQ, its compounding with PR by the parallelogram rule, and the growth of the length of RQ quadratically with infinitesimal displacement of R along the tangent and hence with time (the infinitesimalization of the law of free fall). Kepler is of course represented by the complete ellipse, the focal position of the sun, and Newton's use of the area of the triangle PQS to measure the lapse of time. All the essentials are put together in Newton's remark that the strength of the force acting at P is, in the limit when Q goes to P, directly proportional to the distance RQ and inversely proportional to the square of the area PS·QT, which measures the area of the triangle PQS. This measure, which Newton regularly used, is simply the application of Newton's Second Law in the form that the instantaneous acceleration a is proportional to the force f and Galileo's law for the distance s of free fall in time t under acceleration a: $s = \frac{1}{2} at^2$, from which there follows $f \propto s/t^2$, and, translated into the geometrical terms of Fig. 10.25, $f \propto RQ/(QT^2 \cdot PS^2)$. All that is then needed is the exact relationship from synthetic geometry valid for arbitrary position of Q and the recovery from it of the inverse square force law as an exact limit of infinitesimal calculus when Q tends to P.

So much for the general approach to the orbital problem and its particular application to the Kepler problem. We now mention briefly some of the many other topics covered in the *Principia*, starting with the perturbation of orbits. In Sec. 9 of Book I ('Motion of bodies in movable orbits; and the motion of the apsides') Newton showed that if a body can be made to revolve in a closed fixed orbit by a certain centripetal force to a given centre then, by the addition of a force towards that same centre whose strength decreases inversely as the cube of the distance to that fixed centre, one can cause the original fixed orbit to precess in its plane about the fixed centre while the body moves along the moving curve in accordance with the same law as it followed in the fixed orbit. Newton used this work in the *Principia* in his studies of the difficult problem of the perturbations of the moon's orbit, in which he had only partial success.[134] (Some years after the publication of the *Principia* Newton complained to Halley about his work on the moon that[135] 'his head ached from studying this problem'; it apparently 'kept him awake' and 'he would think of it no more'.)

The *Principia* also devotes considerable space to the solution of problems with arbitrary initial conditions. We have already quoted an example from the early draft *De motu*. Throughout the *Principia* we find many more. It was through propositions such as these that the concept of *physical determinism* became so firmly established in the scientific mind. What was already implicit in Descartes' embryonic form of dynamics is here made explicit. Laplace is already waiting in the wings! However, as we shall see in Vol. 2, there is a most interesting twist to this question of determinism in dynamics. Poincaré was the first to point out clearly that Newtonian dynamics is not as deterministic as one might imagine; in fact, the precise extent to which it falls short of ideal determinism is the most accurate characterization one can obtain of the difference between absolute and relative theories of motion. So as not to leave the reader completely in the dark as to the basis of this assertion, let me merely say that Newtonian dynamics is fully deterministic if the initial positions and initial velocities in *absolute space* (or a frame moving uniformly through it) are specified. However, specification of the initial *relative positions* and *initial relative velocities* (which are all that are observable) does not quite suffice to determine the future evolution of a system of particles. For more details the reader is referred to Ref. 24 (second reference) in the Introduction.

Another extremely important result in Book I is Newton's proof of his celebrated theorem (or rather theorems) of potential theory (as it is now called). In accordance with the notions of universal gravity that he had developed, the falling apple is attracted by each and every piece of matter in the earth. Each piece of matter exerts a force proportional to its mass and inversely proportional to the square of the distance to the attracted body. At a great distance from the earth it is evident that the integrated effect of all the pieces will be to produce a force that decreases as the square of the distance from the centre of the earth. But what will happen at short distances from the surface of the earth? In his celebrated moon test, Newton had, it seems, instinctively assumed that the $1/r^2$ dependence holds exactly all the way from the orbit of the moon to the surface of the earth. But in the light of the theory of universal gravitation this is by no means obvious – indeed it would appear to be somewhat miraculous that the effects of all the pieces of matter in the earth should combine together to give exactly a $1/r^2$ dependence just outside the earth.

Newton seems to have become aware of this problem as he worked on the *Principia*, and he eventually succeeded in proving a result (among several others to do with this problem) of which he was justly proud and used in his correspondence in 1686 with Halley[12c] to bolster his claims *vis-à-vis* Hooke as to the priority of finding the $1/r^2$ law. His result was this: if the density distribution of any body is spherically symmetric, then the gravitational attraction to that body will follow the $1/r^2$ law exactly all the way from infinity to the surface of the body. Within the body it will follow a quite different law, the strength at any point at distance r from the centre being the same as the force that would be exerted by taking all the mass within the radius r (but not the mass outside the sphere), and supposing it concentrated at the centre. In the case of a sphere of uniform density this results in a force that increases from the centre of the sphere in direct proportion to the radius until the surface is reached and then falls off as $1/r^2$. By this result, which Newton called[136] 'very remarkable', it was then possible to bring the moon test into very impressive agreement with the theory of universal gravitation. Perhaps more than anything else in the *Principia* these results of Newton demonstrate the remarkable power of the mathematics that was now being unlocked in support of physical investigation.

A final major topic in Book I that we must mention is the mutual interaction of bodies. This is where the Third Law, suggested initially by Newton's work on collisions, came into its own for the case of attractions acting over great distances within the solar system. This side of Newton's work has already been mentioned in connection with its first appearance in *De motu* (p. 573–4). In the *Principia* we find the first hints given there expanded into one of the most important sections of Book I; it carries the title 'The motions of bodies tending to each other with centripetal forces'. We cannot go into the details of this section but it will be worth quoting Newton's introduction to it, first because it will give the reader the flavour of Newton's magisterial style (as translated into English by Motte, of course), which is manifested especially when he passes from one broad topic to another, and second because of its bearing on the absolute/relative debate. Here are Newton's words:[137]

> I have hitherto been treating of the attractions of bodies towards an immovable centre; though very probably there is no such thing existent in nature. For attractions are made towards bodies, and the actions of the bodies attracted and attracting are always reciprocal and equal, by Law III; so that if there are two bodies, neither the attracted nor the attracting body is truly at rest, but both (by Cor. IV of the Laws of Motion), being as it were mutually attracted, revolve about a common centre of gravity. And if there be more bodies, which either are attracted by one body, which is attracted by them again, or which all attract each other mutually, these bodies will be so moved among themselves, that their common centre of gravity will either be at rest, or move uniformly forwards in a right line.

Turning now to Book II, our discussion here must be very brief indeed even though it contains a wealth of fascinating results. The stimulus to the writing of this book seems to have come from Newton's concern to establish the nature of the medium through which the planets move or, alternatively – if Cartesian vortex theory is correct – by which the planets are carried. As we noted in the discussion of *De motu*, the earliest of the drafts that finally developed into the *Principia* already adumbrates the problem of resistance. In a scholium included in the revised version of *De motu* the thrust of Newton's thinking is clearly revealed:[138]

> Thus far I have considered the motion of bodies in nonresisting media; so that I may determine the motion of celestial bodies in ether . . . . Now ether penetrates freely but does not offer sensible resistance. . . . Comets . . . are carried with immense speed indifferently in all parts of our heavens yet do not lose their tail nor the vapour surrounding their heads [by having them] impeded or torn away by the resistance of the ether. And the planets actually have now persisted in their motion for thousands of years, so far are they from experiencing any resistance.

My impression from reading Book II is that its primary aim was to create a framework of theoretical continuum mechanics by means of which Newton could conclusively prove his own conviction that the planets move through an ether completely free of resistance and simultaneously demolish comprehensively Descartes' vortex theory. However, if the work did have this initial stimulus, it certainly acquired a life of its own and produced many results of considerable interest and value for general physics quite independent of Newton's opposition to Descartes and the Cartesians. One need only mention his attempt at theoretical derivation of the speed of sound[139] on the basis of the elastic force and density of the medium, which, though flawed (he obtained as speed in air 979 feet per second, while the experimental value is about 1142 feet per second), was regarded by Laplace, who completed his theory, as a[140] 'monumente de son génie'.

The main topics that Newton considers are the motion of bodies that are resisted in accordance with two basic laws (both suggested to him by experiment): in proportion to their velocity through the medium and in proportion to the square of their velocity. He reports extensive experiments made with pendula to establish the actual resistance of air. He considers circular motion of bodies in resisting media. Then follow sections on the density and compression of fluids and hydrostatics. He considers the motion of fluids and the resistance they offer to bodies moving through them. Then comes the section on the propagation of motion (pulses) through a fluid in which he attempted to derive the speed of sound. Finally, he devotes a section to the circular motion of fluids, at the end of which he says:[141] 'I have endeavored in this Proposition to investigate the properties of vortices, that I might find whether the celestial phenomena can be explained by them.' His main conclusion is that Kepler's Third Law presents innumerable difficulties for any vortex theory and throws out the taunt:[142] 'Let philosophers then see how that phenomenon of the $\frac{3}{2}$th power can be accounted for by vortices.' In a further scholium he finds even more difficulties, and concludes Book II with these words: 'The hypothesis of vortices is utterly irreconcilable with astronomical phenomena, and rather serves to perplex than explain the heavenly motions. How these motions are performed in free spaces without vortices, may be understood by the first Book; and I shall now more fully treat of it in the following Book.'

Thus we come to Book III, *The System of the World*, the triumphant conclusion of the *Principia* and the part that made such a tremendous impact on his contemporary Englishmen (although Newton's towering intellect was also rapidly recognized on the Continent, particularly by Huygens and Leibniz, acceptance of his universal theory of gravitation was more tardy. Shortly after Newton died in 1727 Voltaire became very active and effective in propagandizing his work). Newton has worked out in detail in Books I and II the general theory and applications of the twin concepts of force and inertial motion. The burden of Book III is announced in the famous sentence:[143] 'It remains that, from the same principles, I now demonstrate the frame of the System of the World.' Stating several carefully selected phenomena (which are mostly taken from celestial phenomena – one of them will be discussed in the next chapter), Newton demonstrates by means of his mathematics that *the phenomena reveal the existence of a specific but universal force in nature: the force of gravity*. He makes good the promise in the Preface: 'from the phenomena of motions to investigate the forces of nature, and then from these forces to demonstrate the other phenomena'. The existence of the gravitational force and its manifold consequences are demonstrated with impressive thoroughness. Pride of place is given to the famous moon test, now made explicitly with an accurate radius of the earth[144] 'as the *French* have found by mensuration'. Newton then moves on to a most important topic – the demonstration of the *universality* of gravity. For the primary astronomical phenomena establish essentially only that the celestial bodies move around the attracting force centre subject to a $1/r^2$ force while the moon test demonstrates the identity of this force and the force of gravity on the surface of the earth. How do we get from these results, impressive as they are, to the notion that each piece of matter attracts with a force that is proportional to its mass and equally is attracted in a given gravitational field with a force that again is proportional to its mass?

Newton begins this task by drawing attention to another of Galileo's great discoveries – one indeed that is so well known we did not bother to mention it in Chap. 7: the fact that[145] 'all sorts of heavy bodies (allowance being made for the inequality of retardation which they suffer from a small power of resistance of the air) descend to the earth *from equal heights* in equal times.' Newton remarks that the 'equality of times we may distinguish to a great accuracy, by the help of pendulums'. He describes another of his simple but beautiful experiments, displaying the rare sense of the great natural scientists for what is crucially important in the seemingly simplest of all phenomena:[146]

> I tried experiments with gold, silver, lead, glass, sand, common salt, wood, water, and wheat. I provided two wooden boxes, round and equal: I filled the one with wood, and suspended an equal weight of gold (as exactly as I could) in the centre of oscillation of the other. The boxes, hanging by equal threads of 11 feet, made a couple of pendulums perfectly equal in weight and figure, and equally receiving the resistance of the air. And, placing the one by the other, I observed them to play together forwards and backwards, for a long time, with equal vibrations. And therefore the quantity of matter in the gold (by Cor. I and VI, Prop. XXIV, Book II) was to the quantity of matter in the wood as the action of the motive force (or *vis motrix*) upon all the gold to the action of the same upon all the wood; that is, as the weight of the one to the weight of the other: and the like happened in the other bodies. By these experiments, in bodies of the same weight, I could manifestly have discovered a difference of matter less than the thousandth part of the whole, had any such been.

Here too we see Newton, busy with his experiments in his rooms at Trinity, brushing shoulders, as it were, with Einstein. Intent on the experiment that gives such powerful support to his own lucid vision of the world and the forces that shape it, he cannot really have had any inkling of how the very same result (confirmed with much greater accuracy in the meanwhile) that he reports here would be used by Einstein to create a totally different theory of gravitation – and a totally different view of the world. Words cannot do justice to the marvel of the thing – the magnitude of what each was able to extract from the one simple experimental fact and the utter transformation wrought by Einstein in the seemingly impregnable worldview of his illustrious predecessor. Using such simple ideas suggested by such simple experiments, let me thus eschew words and leave the reader, who will surely be aware of the experimental basis of Einstein's theory of general relativity* (people who are not aware of that are not likely to pick up a book such as this), to ponder these things.

Now back to Book III. By his pendulum experiments Newton showed that all terrestrial matter is attracted to the earth by a force proportional to its mass. He then points out that the celestial motions prove that this property must also hold for the material of which the planets and satellites are made:[147] 'That the weights [i.e., gravitational attractions] of Jupiter and of his satellites towards the sun are proportional to the several quantities of their matter, appears from the exceedingly regular motions of the satellites. For if some of those bodies were more strongly attracted to the sun in proportion to the quantity of their matter than others, the motions of the satellites would be disturbed by that inequality of attraction.' Thus, what the pendula establish on the earth the celestial bodies confirm in the heavens. Taken together, Newton's empirical proofs for what is now called the equivalence principle – the principle that all bodies fall equally fast in a given gravitational field – are extremely persuasive.

It is much harder to find direct experimental proof for the other half of Newton's central proposition, namely, that bodies are not only attracted but also attract in proportion to their mass and that the force of attraction of any celestial body is the sum of the attractions of each of its individual parts. The main arguments in the *Principia* are to a high degree theoretical and rely on the Third Law and Newton's demonstration that the force of attraction outside a spherical body with radially symmetric mass distribution decreases exactly as $1/r^2$, where r is the distance to the centre of the body. Newton uses this result, in conjunction with the Third Law, to argue from the known $1/r^2$ dependence outside the earth to the universality of the attraction exerted by all parts of the earth:[148] 'Since all the parts of any planet A gravitate towards any other planet B . . . and to every action corresponds an equal reaction; therefore the planet B will, on the other hand, gravitate towards all parts of the planet A.'

Earlier in this chapter I mentioned the growing ability of natural scientists to 'project numbers into nature'. By the development of

* It will, of course, be explained in Vol. 2.

trigonometry and theories of the motion of celestial bodies the astronomers showed how distances between objects far removed from the earth could nevertheless be measured. In the *Principia* Newton takes this process a dramatic step further – in effect he is able to weigh (or, rather, measure) the mass of distant objects. For since the gravitational attraction of any body is proportional to its mass, and all bodies such as satellites of planets 'fall' equally fast in the field of the parent planet, it is a simple matter to deduce, from the observed orbital elements of the satellites, the mass of the attracting object (relative to the mass of a chosen reference object). In this way Newton concluded[149] that if the sun has by definition unit mass then the masses of Jupiter, Saturn, and the earth are 1/1067, 1/3021, and 1/169 282, respectively. (Newton appears to have made an error with a factor 2 in his value for the earth, which is about two times larger than it should be.)

The concept of the gravitational force established a completely new paradigm (in Kuhn's sense[150]) for our conception of the world. The *Principia* contains numerous hints that the gravitational force is almost certainly not the only such force in nature. Newton was however very loathe to speculate in public about things he could not demonstrate mathematically and summarized his standpoint in the Preface with the following words:[151]

> I wish we could derive the rest of the phenomena of Nature by the same kind of reasoning from mechanical principles, for I am induced by many reasons to suspect that they may all depend upon certain forces by which the particles of bodies, by some causes hitherto unknown, are either mutually impelled towards one another, and cohere in regular figures, or are repelled and recede from one another. These forces being unknown, philosophers have hitherto attempted the search of Nature in vain; but I hope the principles here laid down will afford some light either to this or some truer method of philosophy.

If, according to Whitehead's famous saying,[152] 'The safest general characterization of the European philosophical tradition is that it consists of a series of footnotes to Plato', one might with even more justice say that physics since Newton has consisted of the elaboration of these two sentences. It is however worth pointing out that there are significant differences of emphasis between the first edition of the *Principia*, published in 1687, and the two later editions (1713 and 1726). In the first edition, Newton seemed much more prepared to conceive force as truly existent, as an irreducible metaphysical element of the world. This was certainly how he was interpreted (and criticized) by his contemporaries, especially Huygens and Leibniz. In the second edition of the *Principia*, Newton went to some pains to emphasize that he had an open mind about forces – that their existence is inferred from the phenomena by mathematical rules but that they could still have an explanation in terms of unseen mechanisms. His famous dictum: *Hypotheses non fingo* ('I frame no hypotheses') only appeared in the remarkable General Scholium added at the end of *The System of the World* in the second edition of 1713.[153]

At this point we really must come to an end of this very incomplete survey of applications of dynamics given in the *Principia* – there are numerous topics that here have been barely mentioned or even completely omitted – for example, Newton's explanation of the precession of the equinoxes and the tides, to say nothing of the extensive work in Book III on comets and the theory of the moon. One further topic – the oblateness of the earth – will be considered in Chap. 11.

What is perhaps most marvellous about the whole book is the way it grew out of those two definitions and three hypotheses – about 120 words in all – that open the first version of *De motu corporum in gyrum*. Newton may have reformulated them to a considerable extent, but they are in essence, augmented only by the mass concept and the Third Law, the secure basis of the entire work. Delicate little beads they seem yet what a weight of explication they carry.

Lest the reader should get the impression from the above account that the *Principia* is a flawless masterpiece, it should be said that close study of it by modern scholars reveals many places in which it can be faulted. For example, Newton was quite capable of using dubious arguments and of fudging results to make his results look better than they actually were. In particular, his treatment of the more difficult topics such as the motion of the moon and the precession of the equinoxes left much to be desired. These were problems that were to absorb the energies of the great mathematicians of the eighteenth century; in several cases Newton's work merely represents a promising first step. Nevertheless, in the breadth and coherence of its vision and in the demonstration of what mathematics coupled to accurate observation could achieve the *Principia* was truly remarkable. Readers wishing to learn more about the book may like to consult I. B. Cohen's *Introduction to Newton's 'Principia'*.[154]

Let us now look forward to the next chapter and the main concern of this book – the question of the nature of motion. The man who first posed the question in acute form appears rarely in the *Principia* but he lurks behind many a passage whose significance would escape the modern reader. In fact, throughout the *Principia* we detect the influence, generally implicit rather than explicit, of Descartes. The title alone is eloquent testimony to that.[155] Descartes had entitled his major work *The Principles of Philosophy* (*Principia Philosophiae*), announcing thereby his intention to supplant Aristotle. By also employing the word *Principia* in the title of his work, Newton staked out a claim to be taken as seriously as Descartes. Also striking is the way in which Newton emulated Descartes in setting up laws or axioms of motion (Descartes actually called his *laws of nature*). But very interesting too is the implied rebuke to Descartes (for his overweening ambition) in using the title *The Principles of Philosophy*.

Newton's natural caution, based on his recognition of the *empirical* origin of our knowledge, is reflected in his more restricted title: *The Mathematical Principles of Natural Philosophy*.

If Halley played the part of Ulysses in luring Newton out of his lair in Cambridge, Descartes was the Hector whom Newton came forth to slay. The extent to which the *Principia* is a polemic against Descartes should never be forgotten, least of all in the Scholium on absolute and relative motion.

Which brings us now to the central conflict.

11

Newton II: absolute or relative motion?

11.1 General introduction: the period up to Newton

The famous Scholium in the opening pages of the *Principia* on absolute and relative motion together with the whole subsequent development of the debate, including the discovery of general relativity, constitutes one of the most ironic and curious chapters in the history of science. I will go so far as to say that if Descartes had finished *The World* a few months earlier, and hence published it before the Inquisition condemned Galileo, Newton might never have felt the need to formalize his views about space and time in such an outspoken manner – and then general relativity might never have been created. For the Scholium was Newton's response to Descartes' squirming before the Inquisition. And if Newton had omitted the Scholium, with its categorical and ultimately untenable endorsement of absolute space, would Leibniz, Berkeley, and Mach have been stimulated to oppose the concept so vigorously? Or to defend the relativity of motion with such conviction as to call forth a second Achilles every bit as great as Newton?

Let us start this discussion with generalities and then move on to particulars. Broadly speaking, the concept of motion employed by the really major figures in the history of science was relational up to Galileo. The main reasons for this appear to have been, first, the intellectual dominance of Aristotle, second, the fact that, alone among the disciplines, astronomy developed as a quantitative empirical science and third, related to the second, the unchanging aspect of the heavens. The need to relate all motions to the distant sphere of the stars, which are assumed to be relatively fixed, is most pronounced in both Copernicus and Kepler.

However, as we have seen, there also existed from the earliest times an alternative concept of space which married the intuitive space of practical experience with the formalized mathematical space of the Greek geometers. Very often this concept coexisted, perhaps unconsciously, with an 'official' relational concept of position in the mind of one and the same person – Aristotle is a striking example of this. This alternative space-based view seems to have gathered strength in the second half of the sixteenth century. Several Italian philosophers and also Giordano Bruno played an important part in this process, which is well described by Jammer[1] in his book on concepts of space. Also recommended are Koyré's book *From the Closed World to the Infinite Universe*[2] and Grant's *Much Ado About Nothing*,[3] in which the conclusions differ somewhat from Koyré's. In fact, one could see the evolution of the concepts of space rather well in dialectical terms with thesis calling forth antithesis, which is in turn replaced by a modification of the original thesis. For we saw how the atomists advanced the thesis of the void, which was admittedly very meagrely developed but nevertheless was sufficient to bring out strongly the Aristotelian antithesis, in which position is determined by matter and the existence of space is denied. But meanwhile the perceived difficulties of Aristotle's scheme, above all the problem of what exists outside its outermost sphere, were sufficient to keep the alternative viewpoint alive and it gained strength steadily through the sixteenth century. Sometime around 1630, give or take a decade, the space-based concept of position and motion seems to have moved into the ascendancy, perhaps unconsciously at first but then explicitly, as in Gassendi (see the footnote on p. 453).

A host of factors seems to have stimulated the transition from the relational to the absolute standpoint. Perhaps the deepest reason of all was psychological need – even the greatest and most daring minds could not conceive a world without some sort of solid foundation. Space had to be more or less invented when the material frame started to fall apart. For although both Copernicus and Kepler had a relational concept of motion their frame of reference was solid and unchanging. An equally important fact was that the serious study of motion as a universal discipline commenced exactly at the time of the astronomical revolution. Confidence in the earth as a frame of reference for describing motion was obviously seriously undermined by the realization that the earth itself already had at least a two-fold motion – rotation about an axis and motion around the sun – to say nothing of the motion of the earth's axis corresponding to the precession of the equinoxes. Among the most striking examples of the way in which the idea of a stable foundation of the world held a tenacious grip on the mind are Galileo's firm conviction that the sun, despite its manifest rotation, is in a state of absolute translational rest (along with the stars) and Newton's extraordinary assertion (to which we shall come later) that the centre of mass of the solar system is the centre of the world and also at rest. Equally important in the emergence of the concept of absolute space (and time) was the thrust to geometrization of motion, which is extremely pronounced in the Galileo–Descartes–Newton triad, and the associated reassertion of Platonic concepts. Closely related to this was the revival of the ancient doctrine of atomism and the idea that material particles moved through a void. Thus, the desire for a solid foundation went hand in hand with the search for a geometrized container of the world. Absolute space could be seen as the reified and simultaneously Platonized (i.e., geometrized) void.

The most interesting person as regards his underlying concept of motion is Galileo. He is quite clearly a transitional figure. The idea of the well-ordered cosmos is still very strong in him, and he clearly arrived at the concept of inertial motion in an explictly Aristotelian – and hence relational – context. But his theory of the tides and numerous revealing tell-tale passages in his writings show that he was, deep down, an absolutist. Highly significant in this connection is his use of the expression *absolute motion* in the justification of his theory of the tides. Prior to Newton, this is the only use of the expression (employed in the Newtonian sense) that I have come across.

The inescapable need for a concept more or less identical to Newton's absolute space comes out more clearly in Descartes' *The World* than anywhere else. This is, of course, highly ironic in view of his later advocacy of relativism (and complete denial of space), but, as we have seen, the relativism of his *Principles* is but froth on the surface of a deep underlying absolutism. The central aim of Descartes' whole philosophy – the explication of the totality of physical phenomena by universal motion of matter, treated geometrically and quantitatively (in principle, at least) – made that inevitable. His central concept, that of the conserved quantity of motion (understood in the technical sense of bulk times speed) together with the rectilinearity of undisturbed primordial motion (watched over meticulously by God to ensure that there is never loss of the least portion of either matter or motion), is simply meaningless without a quite definite frame of reference in which to prescribe it. But Descartes was far more radical in his conceptions than either Copernicus or Kepler. *The World* set in motion not merely the earth but every last particle of the universe. Descartes was truly the creator of the restless universe. But, driven by an explicatory urge he could not tame, he unconsciously created an underlying world of perfect rest to carry the storm he had summoned forth. That is another of the ironies of the discovery of dynamics – the extent to which the man who prided himself on the clarity of his thought appears to have been totally unaware (even in *The World*) of the fact that he not only set the entire universe into restless motion but simultaneously created, as its foundation, the epitome of rest. For, quite unconsciously, he embedded the motion, as it were, in a perfectly translucent block of conceptual glass, extending from infinity to infinity, through which the parts of the universe move – and he did this despite his perfectly sincere espousal in the *Principles* of the idea (which does not seem to have had anything to do with the Inquisition) that the extended nature of matter makes the concept of an independent space redundant.

In the letter to Mersenne (quoted in Sec. 8.6), Descartes wrote that if the doctrine of the earth's motion was false 'so are all the foundations of my philosophy, since they clearly demonstrate this motion'. Descartes was quite right, even though reasons other than the nature of space and motion were probably uppermost in his mind at the time he wrote. The decade of tortuous trimming by Descartes that followed the Inquisition's attack on Galileo did not change the reality of the foundations of his philosophy one jot. All it did was add a highly confusing gloss. But below the curious and contradictory scratchings on the surface the block of glass remained intact.

It will now be worth saying something about the period, in which more than 40 years elapsed, between the publication of Descartes' *Principles of Philosophy* and Newton's *Principia*. Two major forms of the theory of motion developed, just as they had in antiquity – one based on the idea of a plenum and the other on the revival of the atomic theory by Gassendi, which occurred, as we have noted, at roughly the same time as Descartes published his *Principles*. Simultaneously, more and more people came to accept the idea of an infinite universe and to feel an increasing need for a more clearly defined concept of space. Koyré's book[2] is especially recommended as a review of this process.

One aspect of the development worthy of note is the persistent tendency of thinkers throughout the ages to seek perfection. Aristotle thought he had found it in quintessence and its perfectly circular motion. It is interesting to note that as the concept of quintessence fell apart and evidence of manifest impermanence and imperfection of the heavens accumulated from astronomy – beginning with the two famous supernovae of 1572 and 1604 and then becoming a flood following the discovery of the telescope – all the divine and admirable attributes of quintessence appear to have been transferred to the concept of uniform motion through perfectly uniform space. The Cambridge neo-Platonist Henry More, who had corresponded with Descartes and is believed to have had quite a significant influence on Newton,[4] is full of the perfections of space, which he compared – and even identified – with God's. What is also interesting is that the Newtonian concept of space that today we grasp with such ease also had to be evolved into explicit consciousness as a precise concept. Koyré comments:[5] 'Absolute space is infinite, immovable, homogeneous, indivisible and unique. These are very important properties which Spinoza and Malebranche discovered almost at the same time as More . . .'.

Moreover, clarification of these concepts was to quite an extent stimulated by difficulties perceived in Descartes' nominal relativity of motion.

One sees here the impact of the infinitization of the concept of space and the associated conceptual breakup of the sphere of the fixed stars, in both of which Descartes played a major part. So long as the world was finite and definite, it seemed almost natural to define position by the fixed stars. But if matter everywhere wanders around in vague confusion, all points of reference seem completely lost. According to More:[6] 'the Cartesian definition of motion is repugnant to all the faculties of the soul, the sense, the imagination and the reason.'

Commenting on his difficulties, Koyré says:[7] 'He feels that when bodies move, even if we consider them as moving in respect to each other, something happens, at least to one of them, that is unilateral and not reciprocal: it *really* moves, that is, changes its place, its internal *locus*. It is in respect to this "place" that motion has to be conceived and not in respect to any other.'

Particularly illuminating in the light of the remarks about substance in the chapter on Descartes (Sec. 8.3) are the arguments by which More, who accepts the reality of extension and also the void (following the lead of the ancient atomists), seeks to demonstrate the real existence of space:[8]

> a real attribute of any subject can never be found anywhere but where some real subject supports it. But extension is a real attribute of a real subject (namely matter), which [attribute] however, is found elsewhere [namely there where no matter is present], and which is independent of our imagination. Indeed we are unable not to conceive that a certain immobile extension pervading everything in infinity has always existed and will exist in all eternity (whether we think about it or do not think about it), and [that it is] nevertheless really distinct from matter.
>
> It is therefore necessary that, because it is a real attribute, some real subject support this extension. This argumentation is so solid that there is none that could be stronger. For if this one fails, we shall not be able to conclude with any certainty the existence in nature of any real subject whatever. Indeed, in this case, it would be possible for real attributes to be present without there being any real subject or substance to support them.

Commenting on this argument of More, Koyré sums it up rather nicely as follows:[9] 'Attributes imply substances. They do not wander alone, free and unattached, in the world. They cannot exist without support, like the grin of the Cheshire cat, for this would mean that they would be attributes *of nothing*.'* Thus, the whole tendency of the reaction to Descartes' 'official line' – that all motion is relative – was to accept that extension must have a substance to support it, but that this support is not matter but space – indeed, absolute space. From this point on, such arguments form a central element in the absolute/relative debate.

*It goes without saying that the relationist would look elsewhere in *Alice* for the correct characterization of attributes – to Tweedledum and Tweedledee: the important thing is the persistent correlation, the fact that they always appear *together*.

One of the ironies about the law of inertia is the tremendous success it achieved despite having such seemingly dubious foundations. It must have corresponded to some deep instinct of the seventeenth century mind; how else can one explain the fact that both Galileo and Descartes were instinctively ready to speak about motion taking place in a straight line and seem to have sensed no feeling that any sort of epistemological clarification is called for? The law of inertia was used in practice for many years before the real dilemma it posed became fully apparent. After Descartes' *Principles*, and despite its crass incompatibility with Cartesian relativism, it seems very soon to have established itself as the foundation of the burgeoning science of mechanics. Galileo had already demonstrated the immense practical value of such a law without, however, elevating it to any particular pre-eminence. Descartes showed how attractive it was to take the law, admittedly still formulated in two parts, as the very foundation of the theory of motion. Evidence of its complete triumph in this role and rapid adoption can be seen in its acceptance by Huygens and in the reference to it by More (p. 476) as 'that prime *Mechanicall* law of motion persisting in a straight line'.

One can see in a way how it happened. There is first of all the instinctive prejudice pointed out by Maier (p. 48). But clearly very important was also the thrust toward mathematization of motion; this sits very uneasily with a thoroughgoing relativity, or at least it does if one selects a particular body and wishes to describe *its motion*. Such an approach was inevitable in the early days of the science. It is probable that the millennial history of astronomy here left its mark. The fixed stars having exhibited absolutely no motion among themselves throughout the entire history of astronomy, the motion of each individual planet against the background of the stars was something with an almost concrete reality. Astronomy always led the way. Thus, as astronomy, the problem of the planets finally solved, moved onto new problems, motionics picked up its cast-off swaddling clothes and likewise sought to track bodies. Initially, the earth served as a more or less adequate frame, but both Galileo (theory of the tides) and Descartes (vortex theory of planetary motion) advanced into territory in which the earth was clearly perceived as inadequate, not to say irrelevant. But this was happening at just the time that the sphere of the fixed stars was itself being dissolved (in the mind at least – Halley's discovery of the proper motions of the stars was still nearly a century ahead). Motionics inherited the idea of quantitative tracking but without the backcloth on which to do it. There was nothing for it: space had to fill the breach.

Reassurance must have come from the fact that locally at least it always seemed to make sense to speak of motion taking place along a straight line. In this way, the immediate terrestrial environment was extrapolated to the entire universe. Relationists could see this as a late form of the flat-earth fallacy. Motion when observed locally and in the absence of disturbances appears to be straight, uniform, and quite independent of the bodies in the universe. The law of inertia was simply extrapolation of this local straightness and uniformity without limit, just as extension of the apparent local flatness of the ground gave rise to the flat earth.

Closely related to this was the phenomenon of centrifugal force and the role played in the discovery of the law of inertia by the stone whirled in a sling. For although the empirical reality of a stone, sling, and a boy David whirling it is that the overall process can be decomposed into the motions of all the individual parts involved relative to all the other observable objects in the universe, hardly anyone is likely to conceive it in that way. Attention is automatically concentrated on the circular path of the stone prior to its release, and all the attendant details are abstracted away. The circle remains suspended in our mind's eye and we instinctively embed it and the tangent to it in space, without which an ideal (mathematical) circle cannot be conceived.

It is worth pointing out that the very possibility of this happening is actually a consequence of the law of inertia itself. This, in fact, was Galileo's great insight – the explanation of why we observe no trace of the earth's rotation. As he noted, the explanation for the stability of our immediate environment is that all the bodies around us share in the same basic motion – with respect to what is immaterial provided only we all share that one common motion. Thus, boy David can just as well stand upon the deck of a ship coursing over the seas and whirl his sling with as deadly effect as he did in the desert before Goliath.

The very success of the Galilean explanations for the nonmanifestation of the earth's rotation has an uncomfortable converse. The exquisite cameo acted out in the cabin of the Galilean galley shows us butterflies and candles, not the story of how the cameo itself comes to be. We who sit and have our existence in these pockets of local stability, these havens of order coursing through the restless ruins of the Pythagorean cosmos that Descartes laid waste, are singularly ill-placed to conclude from the local order observed in a particular local frame the law that determines the motion of the frame itself. It is a bit like trying in mathematics to deduce a function from the first term of its Taylor expansion. Of course it then looks uniform and linear – and 'clear and distinct'.

There is a curious confirmation of this in the mistake made by Mach, mentioned in Sec. 7.4 on p. 378. As we saw in the chapter on Galileo, the existence of the distinguished frames of reference in the original formulation of 'Galilean invariance' had a 'Machian' explanation of sorts: they were the frames moving over the surface of the earth, never leaving their 'proper place' within the perfectly ordered cosmos on which Galileo laid such stress. But in the *Discorsi* Galileo, while still retaining his concept of circular inertia, talked exclusively about phenomena 'on the tangent plane' to the earth, i.e., phenomena in which the curvature of the earth could be ignored. This led Mach to assume automatically that Galileo was speaking of rectilinear inertia in space when in fact he had in mind circular inertia about the earth.

In thus misunderstanding Galileo, Mach, who at least partially granted his error[10] when it was pointed out by Wohlwill,[11] was a victim of the flat-earth fallacy. The moral of the story is clear: extrapolate at your peril.

To conclude this section: it is amusing to reflect that, at the end of the debate initiated by Descartes' espousal of relativism, More (and others) had developed, and Newton took over, concepts of space and motion that did not differ in any essentials from the intuitive concepts Descartes accepted unquestioningly at the time he wrote *The World*. Thus, 20 years after the publication of the *Principles of Philosophy*, the concept of motion was apparently back where it had been before the dramatic intervention of the Inquisition. That this is true in essence is confirmed by the fact that throughout the entire diversion through the relational phase the first law of motion remained quite unchanged. In the 'prime Mechanicall law of motion' the uniformity and rectilinearity – concepts that are meaningless in Cartesian relativism – were never questioned.

The importance of the diversion was that it sowed the seed of doubt, even though germination occurred only centuries later.

11.2. Newton: general comments

And so we come to Newton. The surviving manuscripts permit, I believe, a reasonably accurate reconstruction of the development of Newton's standpoint in dynamics. His detailed early study of Descartes' *Principles* is well documented; some early acquaintanceship with Galileo is beyond doubt.[12] Indeed, one finds it hard to believe that he did not 'devour' the *Dialogo* at much the same time as the *Principles*. He also studied Gassendi, and some of his early formulations of the concepts to be used in studying motion suggest an influence (at least in wording) from that quarter (see Westfall[13]).

I suggest that his initial interest was a practical one rather than an immediate concern with the foundational concepts. This is the picture that emerges from the dynamical entries in the *Waste Book*, in which he solves a succession of practical problems one after another. At that stage he seems to have worked a bit like Galileo, making just so many premises as sufficed to solve the problem in hand. He solved most of them using just four principles: that the primary dynamical quantity is motion (understood in the technical sense as the product of bulk and directed speed), that bodies move uniformly in straight lines unless disturbed, that collisions are governed by the law of the 'equall mutation of the motion', and, finally, that force causes and is measured by the deviation from inertial motion.

It was *almost** inevitable that once Newton felt obliged, under the pressure of Cartesian relativism, to clarify his conceptual notions, the concepts of absolute space and time would emerge. This was dictated by the simple fact that Newton, like Galileo, the one mentor he freely acknowledged in dynamics, solved all his problems *as geometrical problems on a piece of paper*. The elements with which he worked were right lines, circles, and ellipses. Where else can they be described but in Euclidean space and what is the sheet of paper but its two-dimensional reification? Just as characteristic was the way in which he embedded time into space and thereby made it amenable to geometrization. In his proof of the area law, the same straight line that represents the motion in space of the body simultaneously represents the uniform flow of time. *Tempus* is a right line.

No one who has successfully solved great problems will lightly discard the most basic of the concepts that made the solution possible. Very revealing in this connection is the review that Einstein made of his own theory, general relativity, when he began to suspect that it did not implement Mach's Principle. He commented[14] that the theory 'rests on three principal points of view': (a) the relativity principle, (b) the equivalence principle, (c) Mach's Principle. Now 'principle (b) was the point of departure of the whole theory and entailed the setting up of principle (a); it certainly cannot be abandoned if the basic ideas of the theoretical system are to be retained'. Thus it was that Einstein, when it came to the crunch, was fully prepared to jettison Mach's Principle rather than abandon 'the basic ideas of the theoretical system'.

If Einstein was prepared to abandon one of his most cherished ideas in preference to tinkering with the elements of a working dynamical system, how much less enthusiasm will Newton have had for flirting with relativism, a doctrine that does not seem to have played any part in his early solutions of dynamical problems?

To substantiate this last assertion, that Cartesian relativism played no role in Newton's discoveries or his basic approach to problem solving, it should first of all be repeated that there is nothing in Cartesian relativism that remotely resembles the extremely precise relativity principle that plays such a large part in Huygens' mechanics. The relativity principle has a precise empirical content that clearly entirely escaped Descartes; by his own account, he never bothered himself with the detailed study of motion that would have been necessary to discover it. It must also be emphasized again that inertial motion of undisturbed bodies is by no means synonymous with the relativity principle. Inertial motion is, of course, Galileo invariant; that was indeed the entire point of its introduction by Galileo to explain the nonobservation of effects of the earth's motion. Although implicit in the cabin cameo that Galileo

* This qualification is necessary if for no other reason than that Huygens' development was not identical to Newton's. However, as we shall see in Chap. 12, Huygens was confused.

describes, there is no *a priori* reason why interactions should possess the same Galilean invariance as inertial motion. They might well be lacking that precise invariance, which was in fact what Descartes' intuition led him to believe. (Descartes' laws of nature are the same everywhere and at all times but are not invariant on the transition to frames of reference in uniform motion relative to one another.) As it happens, the interactions in Newtonian dynamics turn out to have an invariance group much larger than the Galilean transformations. Indeed, the central problem of Mach's Principle is to try and understand how pure inertial motion and the interactions as observed in nature can come to have such different invariance groups. This will be one of the main topics in Vol. 2.

We can come in at this same group of questions from a different tack. One of the keys to understanding a great part of what Newton has to say in the Scholium on absolute, space, time, and motion is the fateful expression in Descartes' *Principles*, §II.29, already mentioned in the chapter on Descartes, in which he asserts that 'transference is effected from the vicinity, not of any contiguous bodies, but only of *those which we consider to be at rest*' (Descartes' italics). Now although there are several rather contradictory things about Descartes' 'official' relativism (i.e., all that formal theory of the 'true' nature of motion which he invented for the sake of the Inquisition), one rather reasonable interpretation of this statement is that we can elect to describe motion by means of a frame of reference in a completely arbitrary state of motion, i.e., we can choose to study the motion of some particular body by referring it to some other body which, since its choice is entirely up to us, can therefore be in an arbitrary state of motion. It is evident from the Scholium (as we shall see) that this is precisely how Newton interpreted Descartes. Although neither Newton nor Descartes was in a position to express it thus in modern terms, one could say that implicit in Descartes' assertion was the idea that the laws of motion are invariant with respect to a group of transformations containing infinitely many parameters; for a body moving arbitrarily can be described by a function of the time represented by a Taylor expansion with infinitely many parameters (coefficients).

Now as regards the dynamical scheme which Newton put together and used with such dramatic effect, it is invariant, with respect, so to speak, to only the zeroth and first parameters of this Taylor expansion – with respect to arbitrary displacements in space (arbitrary relocation of the origin and arbitrary reorientation of the coordinate axes) and with respect to transformation to another frame in a state of uniform rectilinear motion with respect to the first. This state of affairs can be contrasted with the two quite different schemes that Newton found in Descartes' *Principles*. On the one hand, the 'official' relativism designed for the Inquisition corresponds to invariance with respect to infinitely many parameters, while the concrete concept of motion and interactions that Descartes actually used (and Newton instinctively adopted in so far as it did not blatantly contradict reality, as in collision theory) has invariance only with respect to the zeroth parameter; there is not even Galilean invariance.

But what did Newton think about Galilean invariance? The evidence of what he wrote, published and unpublished, suggests he only became fully aware of it comparatively late and that he never saw it as one of the most fundamental features of the world. As we saw in Chap. 10, it appeared briefly for the first time as a law, only to be rapidly (and invalidly) transformed into a corollary. The contrast with Huygens' prominent (but as yet unpublished) use of the relativity principle in his work on collisions is very striking. It is also interesting to compare Newton's formulation of the relativity principle with Huygens' in Chap. 9. Newton clearly implies (p. 571) that the state of rest has a real meaning ('whether the space in question rests'), whereas Huygens (p. 462) uses vaguer expressions to enunciate the principle ('has some further common motion').

A point worth making is that Galilean invariance did not completely undermine the entire edifice of Newton's conceptual scheme. After all, he persuaded himself that it was in fact a simple consequence of his laws of motion (though his very brief proof is itself intriguing in revealing that he did automatically believe the *forces of interaction* to be purely relative and directly derivable from the relative configuration of the matter in the world, putting them thus in a singular contrast to inertial motion, which is totally disengaged from the material contents of the world). The relativity principle is undoubtedly compatible with Newton's scheme, and thus Newton could have argued that motion through absolute space is perfectly real even though the specific structure of the laws of motion make it impossible to observe an overall uniform motion.

Whatever the truth on this particular point, I believe that throughout his life Newton believed instinctively in the reality of motion through space and that in the Scholium, and also in the much earlier and unpublished *De gravitatione*, which we shall discuss shortly, he aimed in essence to show the nonsensicality of the notion that the laws of motion are invariant with respect to a transformation group containing infinitely many parameters (to use this anachronistic but convenient mode of expression). To the extent that relativism contains infinitely many such parameters, while Newtonian dynamics contains only the three corresponding to uniform motions in three mutually perpendicular directions (ignoring here the parameters corresponding to relocation of the origin and reorientation of the axes), Newton was clearly far nearer the truth than Descartes. Moreover, his great grasp of dynamics held him back from asserting anything that could definitely be said to be categorically wrong; thus, he did not repeat a mistake such as Galileo had made with the tides. Nevertheless, the failure to face up to the fact of

Galilean relativity is one of the defects of the Scholium; before we are finished we shall find two more. But it is now time to consider *De gravitatione*.

11.3. Newton's early discussion of motion and *De gravitatione*

The dynamical entries in the *Waste Book* contain few explicit statements by Newton on the subject of the basic framework within which he conceived motion. However, as has been argued at some length, he clearly worked instinctively in an intuitive (absolute) space, just like Galileo and the pre-Inquisition Descartes. The first really clear statement of his overall position comes in the remarkable *Laws of Motion* paper, which appears to be only slightly later in date than the *Waste Book* entries. The opening of this paper is as follows:[15]

The Laws of Motion
How solitary bodys are moved

Sect. I. There is an uniform extension space or expansion continued every way with out bounds: in which all bodys are each in severall parts of it: which parts of space possessed and adequately filled by them are their places. And their passing out of one place or part of space into another, through all the intermediate space is their motion. Which motion is done with more or lesse velocity accordingly as tis done through more or lesse space in equal times or through equall spaces in more or lesse time. But the motion it selfe and the force to persevere in that motion is more or lesse accordingly as the factus of the bodys bulk into its velocity is more or lesse. And that force is equivalent to that motion which it is able to beget or destroy.

The opening sentences describe almost exactly the framework in which Descartes operates in *The World* when he is formulating his laws of motion. Particularly striking is the more or less complete identity between Newton's sentence beginning 'And their passing . . .' and the key sentence in *The World* (see p. 438) in which Descartes speaks of 'motion by which bodies pass from one place to another and successively occupy all the spaces in between'. It was, of course, precisely this concept that Descartes later rejected in the *Principles* when discussing the formal theory of motion, but nevertheless continued to use implicitly in the formulation of his laws of motion. For understanding the thrust of much of what Newton has to say about space and motion, it is important to note that Descartes explicitly rejects the central thesis of *The World* in the *Principles*, where in §II.24 he condemns the idea that movement 'as commonly interpreted, is nothing other than *the action by which some body travels from one place to another*' (Descartes' italics). The expression 'as commonly understood', which is also rendered[16] 'in the vulgar sense', is almost certainly the reason why Newton employs the expression 'the common people' in the opening of the Scholium in the *Principia*. For he is intent, not without a certain malicious pleasure, on turning the tables on Descartes and demonstrating that the opinion which Descartes describes as vulgar or common is the one that must be accepted as true.

It is therefore most important to distinguish between the two Cartesian incarnations: the instinctive Descartes of *The World*, still present implicitly in all that part of the *Principles* that deals specifically with the laws of motion, and the Descartes who espoused relativism to please Rome. It is important to bear in mind that for a reader such as Newton Descartes' *Principles* provided no internal clue as to the origin of the crass contradiction between the two concepts of motion – one explicit and riddled with obvious difficulties, the other implicit and at least self-consistent – that exist side-by-side in the work.

And it was the second concept that Newton instinctively adopted. From the earliest pages of the *Waste Book* through to the completion of the *Principia*, all of Newton's work on dynamics was based on the concept, which he must have taken from Descartes, of uniform rectilinear motion of undisturbed bodies. The main question at issue is: uniform and rectilinear with respect to what? There is, however, a less important issue, with which we deal first – the *conception* that Newton had of the *cause* of the uniform rectilinear motion. Here there does seem to have been some movement on his part. For as we have seen the young Newton appears to have *conceived* this motion very much in the manner of medieval impetus theory, whereas the mature Newton moved more to the Cartesian standpoint of regarding uniform rectilinear motion as a natural *state* which will persist of its own accord without the intervention of any external or internal agency. However, this did not alter any of the mathematical consequences of Newton's theory. It did not touch its structure and was purely cosmetic. The mathematics – and hence the objective content – of an existing theory is indifferent to the metaphysics through which it is interpreted (though a change in the metaphysics may later induce a real change in the theory).

Now to the real question: with respect to what is the uniform rectilinear motion defined? Galileo, the pre-Inquisition Descartes, and Newton all had the same instinctive response – it is with respect to space: absolute space. If such a space is granted, the formulation of the law of inertia presents no problem. This is still true even when allowance is made for the Galileo–Huygens relativity principle. The unique absolute space is then replaced by a family of spaces, in each of which any given undisturbed body moves uniformly and rectilinearly. Thus, in Newton's spaced-based concept of motion, the only ambiguity comes from the existence of an equivalent family of frames of reference, in *all of which* the motion of any particular undisturbed body is uniform and rectilinear. If its motion is straight and uniform in one such frame of reference, *it is straight and uniform in all the other frames of the family*. This is why Galileo–Huygens relativity was something of an embarrassment to Newton but in no way threatened to undermine the essential structure of his dynamics.

The situation is quite different if we take Cartesian relativism seriously. For according to Descartes we can elect to describe motion *by any bodies we care to choose*. But this brings in a quite disastrous ambiguity. For a motion that happens to be straight and uniform with respect to one particular set of reference bodies will certainly not be so with respect to others whose selection is arbitrary. Whereas the law of inertia based on space is invariant on the transition between the Galileo–Huygens equivalent frames, a freedom which is characterized by *three parameters*, there is not the remotest chance of making the law when based on material bodies invariant in an analogous manner because there is no restriction imposed on the bodies. The freedom in this case is characterized by *infinitely* many parameters. It was this latter suggestion, the post-Inquisition incarnation of Descartes (so incongrously juxtaposed in the *Principles* with the pre-Inquisition incarnation), which Newton attacked so violently; and he attacked it essentially from the standpoint of the pre-Inquisition Descartes.

Unlike Westfall,[17] I do not see much evidence for significant change on Newton's part with regard to the most fundamental concepts of his dynamics (that is, apart from his wholehearted adoption of centripetal force from about 1680). From the start the law of inertia was the basis of his dynamics and continued adherence to that, coupled with the challenge to attack the problem of the planets, enforced the appearance of the specific centripetal force concept. Unlike Huygens he does not seem ever to have taken thoroughgoing relativism as a serious proposition for the basis of a dynamical theory. He was from the start instinctively and deeply opposed to the infinite number of parameters with respect to which dynamics would be invariant that is implicit in Cartesian relativism. It seems to me that there was a growing awareness that three parameters (to accommodate Galilean invariance) would have to be granted, but the invective against Descartes was all against the proposition that infinitely many should be granted. And as regards the status of inherent force and the law of inertia, the change here was also comparatively slight, the mature Newton coming to recognize that inherent force serves not much purpose between interactions and is really only manifested as a resistance to change of the inertial state.*

* I should like to take this opportunity to express especial thanks to Prof. Westfall for extended correspondence on the differences noted above and in Chap. 10. It seems to me that the most substantial point of disagreement remaining between us relates to the final sentence of the above paragraph and similar comments earlier. Westfall believes the change to which I refer was highly important because 'Newton could not fit circular motion into his dynamics satisfactorily until he made the switch, because in circular motion the continual exercise of impressed force produced no change in the internal force bearing the body forward'. As explained in Sec. 10.5, I believe that Newton's treatment of the problem of circular motion was in essence correct from the beginning, so that no significant change occurred.

So now to the paper *De gravitatione et aequipondio fluidorum*, named by its opening words, which suggest that it will be about weight and buoyancy but which was broken off before it got anywhere near a detailed discussion of these topics. What survives is actually a violent attack on Descartes' post-Inquisition theory of space, matter, and motion. Unlike the Scholium in the *Principia*, in which Newton did not deign to mention Descartes' name, employing instead a coded language (which has helped to confuse the issue), *De gravitatione* attacks Descartes head-on and at times goes into detailed discussion of individual articles of the *Principles*. The paper, which was discovered comparatively recently (1940s), is certainly not earlier than 1668 but certainly not much after that date either. As the title indicates, it was written in Latin. The extracts which follow are taken from the translation in Hall and Hall.[18]

Newton commences with some definitions that could easily have come from *The World*:

Definitions

The terms *quantity, duration* and *space* are too well known to be susceptible of definition by other words.

Def. 1. Place is a part of space which something fills evenly.
Def. 2. Body is that which fills place.
Def. 3. Rest is remaining in the same place.
Def. 4. Motion is change of place.

After a few amplifications, Newton turns his guns squarely on Descartes:

For the rest, when I suppose in these definitions that space is distinct from body, and when I determine that motion is with respect to the parts of that space, and not with respect to the position of neighbouring bodies, lest this should be taken as being gratuitously contrary to the Cartesians, I shall venture to dispose of his fictions.

After a summary of the most important propositions from Part II of Descartes' *Principles*, Newton launches the attack:

Indeed, not only do its absurd consequences convince us how confused and incongruous with reason this doctrine is, but Descartes by contradicting himself seems to acknowledge the fact. For he says that speaking properly and according to philosophical sense the Earth and the other Planets do not move, and that he who declares it to be moved because of its translation with respect to the fixed stars speaks without reason and only in the vulgar fashion (Part III, Art. 26, 27, 28, 29). Yet later he attributes to the Earth and Planets a tendency to recede from the Sun as from a centre about which they are revolved, by which they are balanced at their [due] distances from the Sun by a similar tendency of the gyrating vortex (Part III, Art. 140). What then? Is this tendency to be derived from the (according to Descartes) true and philosophical rest of the planets, or rather from [their] common and non-philosophical motion?

We have here already the germ of the famous distinction made in the Scholium between absolute and relative rotation.

Newton is merciless in exposing the flagrant inconsistency of Descartes' arguments (Newton seems not at all to have realized what had got Descartes into his contortions). Turning to Descartes' discussion of comets, Newton says:

The philosopher is hardly consistent who uses as the basis of Philosophy the motion of the vulgar which he had rejected a little before, and now rejects that motion as fit for nothing which alone was formerly said to be true and philosophical, according to the nature of things. And since the whirling of the comet around the Sun in his philosophic sense does not cause a tendency to recede from the centre, which a gyration in the vulgar sense can do, surely motion in the vulgar sense should be acknowledged, rather than the philosophical.

Then follows quite a lot more in the same vein. We come after a while to a passage that expresses Newton's deep conviction in the reality of motion:

For the motions that really are in any body, are really natural motions, and thus motions in the philosophical sense and according to the truth of things, even though he contends that they are motions in the vulgar sense only.

Newton now turns his wrath on the idea that the immediately contiguous bodies play any significant role in defining or determining the motion of a body:

But besides this, from its consequences we may see how absurd is this doctrine of Descartes. And first, just as he contends with heat that the Earth does not move because it is not translated from the neighbourhood of the contiguous aether, so from the same principles it follows that the internal particles of hard bodies, while they are not translated from the neighbourhood of immediately contiguous particles, do not have motion in the strict sense, but move only by participating in the motion of the external particles: it rather appears that the interior parts of the external particles do not move with their own motion because they are not translated from the neighbourhood of the internal parts: and thus that only the external surface of each body moves with its own motion and that the whole internal substance, that is the whole of the body, moves through participation in the motion of the external surface. The fundamental definition of motion errs, therefore, that attributes to bodies that which only belongs to surfaces, and which denies that there can be any body at all which has a motion peculiar to itself.

This passage looks forward to the famous bucket experiment. It shows, as hinted in the comment on pp. 444–5, that a main aim of the bucket experiment was to discredit the notion that motion is determined by *contiguous* matter. This was, in fact, one way in which Descartes' contortions resulted in a distortion of the presentation in the Scholium. What I mean by this is that Newton did not really have a worthy opponent with whom to contend when he wrote the Scholium. As Berkeley,

Leibniz, and above all Mach were to demonstrate later, a far more cogent case for the relational nature of motion could be made than the confused arguments Descartes had advanced. Descartes was such an easy target to hit, Newton was not fully extended.

But to continue with Newton. He has demonstrated the absurdity of determining motion by contiguous bodies. He now considers the difficulties involved in referring motion to distant bodies:

Secondly, if we regard only Art. 25 of Part II, each body has not merely a unique proper motion but innumerable ones, provided that those things are said to be moved properly and according to the truth of things of which the whole is properly moved. And that is because he understands by the body whose motion he defines all that which is translated together, and yet this may consist of parts having other motions among themselves: suppose a vortex together with all the Planets, or a ship along with everything within it floating in the sea, or a man walking in a ship together with the things he carries with him, or the wheel of a clock together with its constituent metallic particles. For unless you say that the motion of the whole aggregate cannot be considered as proper motion and as belonging to the parts according to the truth of things, it will have to be admitted that all these motions of the wheels of the clock, of the man, of the ship, and of the vortex are truly and philosophically speaking in the particles of the wheels.

We now come to the heart of Newton's contentions. I give the following passage in full and then make some comments:

From both of these consequences it appears further that no one motion can be said to be true, absolute and proper in preference to others, but that all, whether with respect to contiguous bodies or remote ones, are equally philosophical – than which nothing more absurd can be imagined. For unless it is conceded that there can be a single physical motion of any body, and that the rest of its changes of relation and position with respect to other bodies are so many external designations, it follows that the Earth (for example) endeavours to recede from the centre of the Sun on account of a motion relative to the fixed stars, and endeavours the less to recede on account of a lesser motion relative to Saturn and the aetherial orb in which it is carried, and still less relative to Jupiter and the swirling aether which occasions its orbit, and also less relative to Mars and its aetherial orb, and much less relative to other orbs of aetherial matter which, although not bearing planets, are closer to the annual orbit of the Earth; and indeed relative to its own orb it has no endeavour, because it does not move in it. Since all these endeavours and non-endeavours cannot absolutely agree, it is rather to be said that only the motion which causes the Earth to endeavour to recede from the Sun is to be declared the Earth's natural and absolute motion. Its translations relative to external bodies are but external designations.

The first comment to be made is to draw attention to Newton's passionate belief that there must be 'one motion' that is 'true, absolute and proper'. I think this desire is closely related to comments I made in an earlier chapter (p. 401) about the difficulties which Galileo had in formulating concepts of motion. Motionics and dynamics started out as a study of the motion of *individual bodies* (note that Newton's *Laws of Motion* paper carried the subtitle 'How solitary bodys are moved'). Both Galileo and Newton were looking for laws that describe *individual motions of individual bodies*. This formulation of the problem more or less necessitated the introduction of absolute space, especially once the overriding importance of inertial motion had been recognized.

A second comment is that the above passage is remarkable for anticipating and rejecting Mach's idea that rotation relative to distant matter rather than space might in some way generate the centrifugal forces we observe locally. This sort of argument is not, in fact, repeated in the Scholium but it is interesting to note that Newton at least considered the possibility. It should however be noted that Newton's arguments are formal rather than physical. What I mean by this is that if local centrifugal force were to be generated by some physical mechanism there is no reason to suppose that all the 'endeavours and non-endeavours' would have to agree absolutely; for each rotating planet and orb would make their own contribution and the net effect would be obtained by integrating all such contributions. It is clear that such a manner of thinking was alien to Newton. In this respect he was completely geometrokinetic. He could just about contemplate the idea that inertial motion might be *defined* relative to other matter but had no sense of a *physical determination*. In fact, in *De gravitatione* he still clearly inhabited the world of the prevailing orthodoxy – the mechanical philosophy in which the only interaction between bodies is through direct contact.

The final comment is a warning that the 'translations' in the final sentence (and which also come again later) are not on any account to be confused with the specifically rectilinear translatory motions of the type encountered in Galileo–Huygens transformations, i.e., in this sentence Newton is not contrasting uniform rectilinear motion with circular motion; he just means any change in position relative to external bodies.

If the passage just quoted did not find a reflection in the Scholium, the ideas that immediately follow figure there very prominently, albeit in a rather different form. If motion is relative to some particular chosen reference body, it is only necessary to apply to the reference body a force, and thereby move it, for an apparent motion to be generated in the body under study although it is clear that nothing has happened to it. For example:

> It follows from the Cartesian doctrine that motion can be generated where there is no force acting. For example, if God should suddenly cause the spinning of our vortex to stop, without applying any force to the Earth which could stop it at the same time, Descartes would say that the Earth is moving in a philosophical sense (on account of its translation from the neighbourhood of the contiguous fluid), whereas before he said it was resting, in the same philosophical sense.

Newton gives several more examples in similar vein. This part of his onslaught terminates in a passage which shows how clearly he perceived the crass contradiction between the two Cartesian incarnations: i.e., how the 'official line' of the post-Inquisition *Principles* is quite incompatible with the pre-Inquisition laws of motion as they are nevertheless presented in the *Principles*. Newton points out that Descartes' own concepts of motion make a mockery of the Cartesian law of inertia, the 'prime Mechanicall law of motion':

> Lastly, that the absurdity of this position may be disclosed in full measure, I say that thence it follows that a moving body has no determinate velocity and no definite line in which it moves. And, what is worse, that the velocity of a body moving without resistance cannot be said to be uniform, nor the line said to be straight in which its motion is accomplished. On the contrary, there cannot be motion since there can be no motion without a certain velocity and determination.

Let us now look at the arguments Newton gives to support his contention that the Cartesian laws of motion contradict Cartesian relativism. They also show how deeply Newton was influenced by an insight which he almost certainly got from Descartes, or at least from the prevailing mechanical orthodoxy, according to which all matter is in motion. I am referring to the words in the following passage which I have italicized:

> But that this may be clear, it is first of all to be shown that when a certain motion is finished it is impossible, according to Descartes, to assign a place in which the body was at the beginning of the motion; it cannot be said whence the body moved. And the reason is that according to Descartes the place cannot be defined or assigned except by the position of the surrounding bodies, and after the completion of a certain motion the position of the surrounding bodies no longer stays the same as it was before. For example, if the place of the planet Jupiter a year ago be sought, by what reason, I ask, can the Cartesian philosopher define it? Not by the positions of the particles of the fluid matter, for the positions of these particles have greatly changed since a year ago. Nor can he define it by the positions of the Sun and fixed stars. For the unequal influx of subtle matter through the poles of the vortices towards the central stars (Part III, Art. 104), the undulation (Art. 114), inflation (Art. 111) and absorption of the vortices, and other more true causes, such as the rotation of the Sun and stars around their own centres, the generation of spots, and the passage of comets through the heavens, change both the magnitude and positions of the stars so much that perhaps they are only adequate to designate the place sought with an error of several miles; and still less can the place be accurately defined and determined by their help, as a Geometer would require. *Truly there are no bodies in the world whose relative positions remain unchanged with the passage of time*, and certainly none which do not move in the Cartesian sense; that is, which are neither transported from the vicinity of contiguous bodies nor are parts of other bodies so transferred. And thus there is no basis from which we can at the present pick out a place which was in the past, or say that such a place is any longer discoverable in nature. For since, according

to Descartes, place is nothing but the surface of surrounding bodies or position among some other more distant bodies, it is impossible (according to his doctrine) that it should exist in nature any longer than those bodies maintain the same positions from which he takes the individual designation. And so, reasoning as in the question of Jupiter's position a year ago, it is clear that if one follows Cartesian doctrine, not even God himself could define the past position of any moving body accurately and geometrically now that a fresh state of things prevails, since in fact, due to the changed positions of the bodies, the place does not exist in nature any longer.

Now as it is impossible to pick out the place in which a motion began (that is, the beginning of the space passed over), for this place no longer exists after the motion is completed, so the space passed over, having no beginning, can have no length; and hence, since velocity depends upon the distance passed over in a given time, it follows that the moving body can have no velocity, just as I wished to prove at first. Moreover, what was said of the beginning of the space passed over should be applied to all intermediate points too; and thus as the space has no beginning nor intermediate parts it follows that there was no space passed over and thus no determinate motion, which was my second point. It follows indubitably that Cartesian motion is not motion, for it has no velocity, no definition, and there is no space or distance traversed by it.

I have given this passage in full since it is another of the parts of *De gravitatione* that do not really reappear in the Scholium, which is often very condensed and does not go fully into the difficulties which Newton perceived in Cartesian relativism. It shows, once and for all, that relativism and a meaningful statement of the law of inertia as formulated by Descartes himself are simply incompatible.

Newton concludes this part of the discussion with the statement of the inescapable need for absolute space:

So it is necessary that the definition of places, and hence of local motion, be referred to some motionless thing such as extension alone or space in so far as it is seen to be truly distinct from bodies.

Without such a concept, he could see no way of making sense of Descartes' law of inertia, nor, importantly, all the numerous significant results in dynamics that he already had to his own credit.

11.4 *De gravitatione*: Newton's discussion of space and body

Having arrived at his fundamental conclusion that motion must 'be referred to some motionless thing such as extension alone or space', Newton proceeds to analyze the concepts of extension and space. He feels it is incumbent on him to point out, once again, the error of Descartes, this time in asserting that we cannot form a concept of *spatial* extension that is in any way different from the concept of *material* extension. He announces that he will explain what 'extension and body are, and how they differ from each other'. For he considers it 'most important to overthrow [that philosophy] as regards extension, in order to lay truer foundations of the mechanical sciences'.

Newton begins by declining to define extension as either 'substance or accident or else nothing at all'. For 'it has its own manner of existence which fits neither substance nor accidents'. Newton lists several reasons why it is difficult to conceive extension as a substance and then comes to this very characteristic passage:

we can clearly conceive extension existing without any subject, as when we may imagine spaces outside the world or places empty of body, and we believe [extension] to exist wherever we imagine there are no bodies, and we cannot believe that it would perish with the body if God should annihilate a body, it follows that [extension] does not exist as an accident inherent in some subject. And hence it is not an accident. And much less may it be said to be nothing, since it is rather something, than an accident, and approaches more nearly to the nature of substance. There is no idea of nothing, nor has nothing any properties, but *we have an exceptionally clear idea of extension*, abstracting the dispositions and properties of a body so that there remains only the uniform and unlimited stretching out of space in length, breadth and depth. And furthermore, many of its properties are associated with this idea; these I shall now enumerate not only to show that it is something, but what it is.

The italics in this passage are mine. The words so indicated characterize Newton's deepest conviction. The properties that Newton lists as belonging to extension show how remarkably 'substantial' it appeared to him. It really is a perfectly uniform and translucent block of glass extending from infinity to infinity and has all the properties of such a block of glass except the glass! As we see in the following extracts:

1. In all directions, space can be distinguished into parts whose common limits we usually call surfaces; and these surfaces can be distinguished in all directions into parts whose common limits we usually call lines; and again these lines can be distinguished in all directions into parts which we call points. And hence surfaces do not have depth, nor lines breadth, nor points dimension, unless you say that coterminous spaces penetrate each other as far as the depth of the surface between them, namely what I have said to be the boundary of both or the common limit; and the same applies to lines and points. Furthermore spaces are everywhere contiguous to spaces, and extension is everywhere placed next to extension, and so there are everywhere common boundaries to contiguous parts; that is, there are everywhere surfaces acting as a boundary to solids on this side and that; and everywhere lines in which parts of the surfaces touch each other; and everywhere points in which the continuous parts of lines are joined together. And hence there are everywhere all kinds of figures, everywhere spheres, cubes, triangles, straight lines, everywhere circular, elliptical, parabolical and all other kinds of figures, and those of all shapes and sizes, even though they are not disclosed to sight. For the material delineation of any figure is not a new production of that figure with respect to space, but only a corporeal representation of it, so that what was formerly insensible in space now appears to the senses to exist. . . .

2. Space extends infinitely in all directions. For we cannot imagine any limit anywhere without at the same time imagining that there is space beyond it. And hence all straight lines, paraboloids, hyperboloids, and all cones and cylinders and other figures of the same kind continue to infinity and are bounded nowhere. . . .

3. The parts of space are motionless. If they moved, it would have to be said either that the motion of each part is a translation from the vicinity of other contiguous parts, as Descartes defined the motion of bodies; and that this is absurd has been sufficiently shown; or that it is a translation out of space into space, that is out of itself, unless perhaps it is said that two spaces everywhere coincide, a moving one and a motionless one. Moreover the immobility of space will be best exemplified by duration. For just as the parts of duration derive their individuality from their order, so that (for example) if yesterday could change places with today and become the later of the two, it would lose its individuality and would no longer be yesterday, but today; so the parts of space derive their character from their positions, so that if any two could change their positions, they would change their character at the same time and each would be converted numerically into the other. The parts of duration and space are only understood to be the same as they really are because of their mutual order and position; nor do they have any hint of individuality apart from that order and position, which consequently cannot be altered.

4. Space is a disposition of being *qua* being. No being exists or can exist which is not related to space in some way. God is everywhere, created minds are somewhere, and body is in the space that it occupies; and whatever is neither everywhere nor anywhere does not exist. And hence it follows that space is an effect arising from the first existence of being, because when any being is postulated, space is postulated. And the same may be asserted of duration: for certainly both are dispositions of being or attributes according to which we denominate quantitatively the presence and duration of any existing individual thing. So the quantity of the existence of God was eternal, in relation to duration, and infinite in relation to the space in which he is present; and the quantity of the existence of a created thing was as great, in relation to duration, as the duration since the beginning of its existence, and in relation to the size of its presence as great as the space belonging to it.

Moreover, lest anyone should for this reason imagine God to be like a body, extended and made of divisible parts, it should be known that spaces themselves are not actually divisible, and furthermore, that any being has a manner proper to itself of being in spaces. For thus there is a very different relationship between space and body, and space and duration. For we do not ascribe various durations to the different parts of space, but say that all endure together. The moment of duration is the same at Rome and at London, on the Earth and on the stars, and throughout all the heavens. And just as we understand any moment of duration to be diffused throughout all spaces, according to its kind, without any thought of its parts, so it is no more contradictory that Mind also, according to its kind, can be diffused through space without any thought of its parts.

5. The positions, distances and local motions of bodies are to be referred to the parts of space. And this appears from the properties of space enumerated as 1. and 4. above, and will be more manifest if you conceive that there are vacuities scattered between the particles, or if you pay heed to what I have formerly said

about motion. To that it may be further added that in space there is no force of any kind which might impede or assist or in any way change the motions of bodies. And hence projectiles describe straight lines with a uniform motion unless they meet with an impediment from some other source. But more of this later.

6. Lastly, space is eternal in duration and immutable in nature, and this because it is the emanent effect of an eternal and immutable being. If ever space had not existed, God at that time would have been nowhere; and hence he either created space later (in which he was not himself), or else, which is not less repugnant to reason, he created his own ubiquity. Next, although we can possibly imagine that there is nothing in space, yet we cannot think that space does not exist, just as we cannot think that there is no duration, even though it would be possible to suppose that nothing whatever endures. This is manifest from the spaces beyond the world, which we must suppose to exist (since we imagine the world to be finite), although they are neither revealed to us by God, nor known from the senses, nor does their existence depend upon that of the spaces within the world. But it is usually believed that these spaces are nothing; yet indeed they are true spaces. Although space may be empty of body, nevertheless it is not in itself a void; and *something* is there, because spaces are there, although nothing more than that. Yet in truth it must be acknowledged that space is no more space where the world is, than where no world is, unless perchance you say that when God created the world in this space he at the same time created space in itself, or that if God should annihilate the world in this space, he would also annihilate the space in it. Whatever has more reality in one space than in another space must belong to body rather than to space; the same thing will appear more clearly if we lay aside that puerile and jejune prejudice according to which extension is inherent in bodies like an accident in a subject without which it cannot actually exist.

Note here particularly the reference in 5. to the law of inertia. From the moment Newton encountered the concept of inertial motion (presumably in Descartes or Gassendi) he does not seem to have ever wavered from his belief in space and inertial motion within it as the prime law of nature. To that extent his debt to Descartes (in his pre-Inquisition incarnation) was very great. It is also worth noting that at this stage at least Newton seems to have believed in a finite world in an infinite space. This is quite interesting, since, as we shall see in Vol. 2, a completely relational theory of motion can be constructed without too much difficulty for a world conceived in this manner.

Otherwise I would just like to ask the reader to let these fairly extensive passages from Newton work upon the mind. Newton expresses himself with such marvellous clarity, it is a privilege to be able to step within his mind and examine the workings of the intellect that, more than any other, shaped the scientific view of the world. But there is a further purpose in giving these passages at length. Concepts change. In barely more than a century we have passed from a completely matter-based concept of motion, as found in Copernicus and Kepler, to the space-based concept of Newton. The pendulum will swing again. For all his pre-eminence,

Newton does not represent the central position of the pendulum, the point of equilibrium. He is, in fact, at one of the extremes.

Another remarkably interesting thing about Newton is the extent to which, like the two other giants of dynamics, Galileo and Einstein, he combined a very strongly aprioristic and rationalistic case of mind with an almost equally strong empiricism. Descartes was, of course, the supreme advocate of the rationalistic approach; he was convinced, as we have seen, that all the essential concepts which we need to form a picture of the world are implanted in the mind by God. Now, as regards the concepts of space and time, Newton was every bit as rationalistic as Descartes. He really did have complete confidence that his mind contained a perfect and already formed concept of space and extension. As he said: 'We have an exceptionally clear idea of extension.' But at the same time he had an equally remarkable faith in empiricism. On several occasions in the *Principia* he makes it clear that by intently examining nature as manifested to us in the phenomena we can learn really fundamental things about the way nature works. I have already mentioned his one-sentence characterization of what the *Principia* is all about. Newton's faith in the power of empiricism finds its most remarkable expression in a remark that he made at the end of his General Scholium on the nature of God which he added to the second edition of the *Principia*, published in 1713. After the discussion of the nature of God, Newton concludes with the sentence (added[19] as an afterthought as the book was going through the press):[20] 'And thus much concerning God; *to discourse of whom from the appearances of things*, does certainly belong to Natural Philosophy' (my italics). Thus, empiricism can even tell us things about the nature of God. This is a most revealing and characteristic remark.

The division which Newton makes is clear and interesting. The concepts of space and time belong to the rationalistic side of his thinking. But as regards the nature and properties of *matter* he had a remarkably open mind. One of his main charges against Descartes was that in claiming to know *a priori* the essential properties of matter he was almost blasphemously playing the role of God – that he was limiting the powers of God to create matter with whatever essential properties he might care to choose. Whereas Newton had no hesitation in saying that God had no alternative to creating matter in space and time, he took a decidedly positivistic attitude about matter and asserted that its inner secrets were hidden from us. In the case of matter, all we could do was study its behaviour. Newton begins his discussion on this subject with the following comment, which it will be seen stands in a most marked contrast to the certainty of his views about space:

Now that extension has been described, it remains to give an explanation of the nature of body. Of this, however, the explanation must be more uncertain, for it does not exist necessarily but by divine will, because it is hardly given to us to know the limits of the divine power, that is to say whether matter could be created in one way only, or whether there are several ways by which different beings similar to bodies could be produced.

To demonstrate how easily we could be misled, Newton gives the following example:

it must be agreed that God, by the sole action of thinking and willing, can prevent a body from penetrating any space defined by certain limits.

If he should exercise this power, and cause some space projecting above the Earth, like a mountain or any other body, to be impervious to bodies and thus stop or reflect light and all impinging things, it seems impossible that we should not consider this space to be truly body from the evidence of our senses (which constitute our sole judges in this matter); for it will be tangible on account of its impenetrability, and visible, opaque and coloured on account of the reflection of light, and it will resonate when struck because the adjacent air will be moved by the blow.

Thus we may imagine that there are empty spaces scattered through the world, one of which, defined by certain limits, happens by divine power to be impervious to bodies, and *ex hypothesi* it is manifest that this would resist the motions of the bodies and perhaps reflect them, and assume all the properties of a corporeal particle, except that it will be motionless. If we may further imagine that that impenetrability is not always maintained in the same part of space but can be transferred hither and thither according to certain laws, yet so that the amount and shape of that impenetrable space are not changed, there will be no property of body which this does not possess. It would have shape, be tangible and mobile, and be capable of reflecting and being reflected, and no less constitute a part of the structure of things than any other corpuscle, and I do not see that it would not equally operate upon our minds and in turn be operated upon, because it is nothing more than the product of the divine mind realized in a definite quantity of space. For it is certain that God can stimulate our perception by his own will, and thence apply such power to the effects of his will.

In the same way if several spaces of this kind should be impervious to bodies and to each other, they would all sustain the vicissitudes of corpuscles and exhibit the same phenomena. And so if all this world were constituted of this kind of being, it would seem hardly any different.

This whole discussion continues a lot further but the above passage already gives a sufficient flavour. It will be seen that Newton takes a decidedly positivistic approach, very much in the spirit of Mach. It is almost pointless for us to speculate about the inner nature of matter. All we can do is correlate the deliverances of our senses – 'which constitute our sole judges in this matter'. We have a clear anticipation of Newton's famous *hypotheses non fingo*[21] ('I frame no hypotheses') in his discussion of the nature of gravitation.

Westfall[22] believes, correctly in my opinion, that this willingness of Newton to take the evidence of the phenomena at their face value and seek to describe them by mathematical formulae, as opposed to attempting to come to some definitive comprehension of matter in mechanical terms deduced from metaphysical principles (as Descartes was trying to do), was a major factor in enabling Newton, rather than Huygens, to discover the law of universal gravitation. It is certainly very striking that when Newton was prodded into action by Hooke he seems instinctively to have felt that the correct course of action was to use the empirically known facts (Kepler's Laws) to deduce the nature of the forces rather than make a guess about the nature of the world and then seek confirmation of it in the phenomena (as was Huygens' approach). Of course, there is abundant evidence to show that Newton speculated unceasingly about possible mechanisms that would explain gravity and the other phenomena of nature. Nevertheless, as regards matter and our ability to learn about it, he was extremely realistic, sober, and positivistic.

Which makes his *a priori* attitudes to space and time all the more remarkable.

11.5. The Scholium on absolute space, time, and motion

Newton's Scholium at the beginning of the *Principia*, immediately after his formal definitions and before the statement of his laws of motion, is of such transcendent importance in any discussion of the problem of absolute and relative motion that it needs to be stated in full, despite the fact that it is one of the most quoted passages from the whole history of physics. Let us therefore start by having the text[23] in full and then comments. For convenience of identification, I have identified the individual paragraphs (after item IV) by prefixing [a], [b], [c], etc.

SCHOLIUM

Hitherto I have laid down the definitions of such words as are less known, and explained the sense in which I would have them to be understood in the following discourse. I do not define time, space, place, and motion, as being well known to all. Only I must observe, that the common people conceive those quantities under no other notions but from the relation they bear to sensible objects. And thence arise certain prejudices, for the removing of which it will be convenient to distinguish them into absolute and relative, true and apparent, mathematical and common.

I. Absolute, true, and mathematical time, of itself, and from its own nature, flows equably without relation to anything external, and by another name is called duration: relative, apparent, and common time, is some sensible and external (whether accurate or unequable) measure of duration by the means of motion, which is commonly used instead of true time; such as an hour, a day, a month, a year.

II. Absolute space, in its own nature, without relation to anything external, remains always similar and immovable. Relative space is some movable dimension or measure of the absolute spaces; which our senses determine by its position to bodies, and which is commonly taken for immovable space; such is the dimension of a subterraneous, an aerial, or celestial space, determined by its position in respect of the earth. Absolute and relative space are the same in figure and magnitude; but they do not remain always numerically the same. For if the earth, for instance, moves, a space of our air, which relatively and in respect of the earth remains always the same, will at one time be one part of the absolute space into which the air passes; at another time it will be another part of the same, and so, absolutely understood, it will be continually changed.

III. Place is a part of space which a body takes up, and is according to the space, either absolute or relative. I say, a part of space; not the situation, nor the external surface of the body. For the places of equal solids are always equal; but their surfaces, by reason of their dissimilar figures, are often unequal. Positions properly have no quantity, nor are they so much the places themselves, as the properties of places. The motion of the whole is the same with the sum of the motions of the parts; that is, the translation of the whole, out of its place, is the same thing with the sum of the translations of the parts out of their places; and therefore the place of the whole is the same as the sum of the places of the parts, and for that reason, it is internal, and in the whole body.

IV. Absolute motion is the translation of a body from one absolute place into another; and relative motion, the translation from one relative place into another. Thus in a ship under sail, the relative place of a body is that part of the ship which the body possesses; or that part of the cavity which the body fills, and which therefore moves together with the ship: and relative rest is the continuance of the body in the same part of the ship, or of its cavity. But real, absolute rest, is the continuance of the body in the same part of that immovable space, in which the ship itself, its cavity, and all that it contains, is moved. Wherefore, if the earth is really at rest, the body, which relatively rests in the ship, will really and absolutely move with the same velocity which the ship has on the earth. But if the earth also moves, the true and absolute motion of the body will arise, partly from the true motion of the earth, in immovable space, partly from the relative motion of the ship on the earth; and if the body moves also relatively in the ship, its true motion will arise, partly from the true motion of the earth, in immovable space, and partly from the relative motions as well of the ship on the earth, as of the body in the ship; and from these relative motions will arise the relative motion of the body on the earth. As if that part of the earth, where the ship is, was truly moved towards the east, with a velocity of 10 010 parts; while the ship itself, with a fresh gale, and full sails, is carried towards the west, with a velocity expressed by 10 of those parts; but a sailor walks in the ship towards the east, with 1 part of the said velocity; then the sailor will be moved truly in immovable space towards the east, with a velocity of 10 001 parts, and relatively on the earth towards the west, with a velocity of 9 of those parts.

[a] Absolute time, in astronomy, is distinguished from relative, by the equation or correction of the apparent time. For the natural days are truly unequal, though they are commonly considered as equal, and used for a measure of time; astronomers correct this inequality that they may measure the celestial motions by a more accurate time. It may be, that there is no such thing as an equable motion, whereby time may be accurately measured. All motions may be accelerated and retarded, but the flowing of absolute time is not liable to any change. The duration or perseverance of the existence of things remains the same, whether the motions

are swift or slow, or none at all: and therefore this duration ought to be distinguished from what are only sensible measures thereof; and from which we deduce it, by means of the astronomical equation. The necessity of this equation, for determining the times of a phenomenon, is evinced as well from the experiments of the pendulum clock, as by eclipses of the satellites of Jupiter.

[b] As the order of the parts of time is immutable, so also is the order of the parts of space. Suppose those parts to be moved out of their places, and they will be moved (if the expression may be allowed) out of themselves. For times and spaces are, as it were, the places as well of themselves as of all other things. All things are placed in time as to order of succession; and in space as to order of situation. It is from their essence or nature that they are places; and that the primary places of things should be movable, is absurd. These are therefore the absolute places; and translations out of those places, are the only absolute motions.

[c] But because the parts of space cannot be seen, or distinguished from one another by our senses, therefore in their stead we use sensible measures of them. For from the positions and distances of things from any body considered as immovable, we define all places; and then with respect to such places, we estimate all motions, considering bodies as transferred from some of those places into others. And so, instead of absolute places and motions, we use relative ones; and that without any inconvenience in common affairs; but in philosophical disquisitions, we ought to abstract from our senses, and consider things themselves, distinct from what are only sensible measures of them. For it may be that there is no body really at rest, to which the places and motions of others may be referred.

[d] But we may distinguish rest and motion, absolute and relative, one from the other by their properties, causes, and effects. It is a property of rest, that bodies really at rest do rest in respect to one another. And therefore as it is possible, that in the remote regions of the fixed stars, or perhaps far beyond them, there may be some body absolutely at rest; but impossible to know, from the position of bodies to one another in our regions, whether any of these do keep the same position to that remote body, it follows that absolute rest cannot be determined from the position of bodies in our regions.

[e] It is a property of motion, that the parts, which retain given positions to their wholes, do partake of the motions of those wholes. For all the parts of revolving bodies endeavor to recede from the axis of motion; and the impetus of bodies moving forwards arises from the joint impetus of all the parts. Therefore, if surrounding bodies are moved, those that are relatively at rest within them will partake of their motion. Upon which account, the true and absolute motion of a body cannot be determined by the translation of it from those which only seem to rest; for the external bodies ought not only to appear at rest, but to be really at rest. For otherwise, all included bodies, besides their translation from near the surrounding ones, partake likewise of their true motions; and though that translation were not made, they would not be really at rest, but only seem to be so. For the surrounding bodies stand in the like relation to the surrounded as the exterior part of a whole does to the interior, or as the shell does to the kernel; but if the shell moves, the kernel will also move, as being part of the whole, without any removal from near the shell.

[f] A property, near akin to the preceding, is this, that if a place is moved, whatever is placed therein moves along with it; and therefore a body, which is moved from a place in motion, partakes also of the motion of its place. Upon which account, all motions, from places in motion, are no other than parts of entire and absolute motions; and every entire motion is composed of the motion of the body out of its first place, and the motion of this place out of its place; and so on, until we come to some immovable place, as in the before-mentioned example of the sailor. Wherefore, entire and absolute motions can be no otherwise determined than by immovable places; and for that reason I did before refer those absolute motions to immovable places, but relative ones to movable places. Now no other places are immovable but those that, from infinity to infinity, do all retain the same given position one to another; and upon this account must ever remain unmoved; and do thereby constitute immovable space.

[g] The causes by which true and relative motions are distinguished, one from the other, are the forces impressed upon bodies to generate motion. True motion is neither generated nor altered, but by some force impressed upon the body moved; but relative motion may be generated or altered without any force impressed upon the body. For it is sufficient only to impress some force on other bodies with which the former is compared, that by their giving way, that relation may be changed, in which the relative rest or motion of this other body did consist. Again, true motion suffers always some change from any force impressed upon the moving body; but relative motion does not necessarily undergo any change by such forces. For if the same forces are likewise impressed on those other bodies, with which the comparison is made, that the relative position may be preserved, then that condition will be preserved in which the relative motion consists. And therefore any relative motion may be changed when the true motion remains unaltered, and the relative may be preserved when the true suffers some change. Thus, true motion by no means consists in such relations.

[h] The effects which distinguish absolute from relative motion are, the forces of receding from the axis of circular motion. For there are no such forces in a circular motion purely relative, but in a true and absolute circular motion, they are greater or less, according to the quantity of the motion. If a vessel, hung by a long cord, is so often turned about that the cord is strongly twisted, then filled with water, and held at rest together with the water; thereupon, by the sudden action of another force, it is whirled about the contrary way, and while the cord is untwisting itself, the vessel continues for some time in this motion; the surface of the water will at first be plain, as before the vessel began to move; but after that, the vessel, by gradually communicating its motion to the water, will make it begin sensibly to revolve, and recede by little and little from the middle, and ascend to the sides of the vessel, forming itself into a concave figure (as I have experienced), and the swifter the motion becomes, the higher will the water rise, till at last, performing its revolutions in the same times with the vessel, it becomes relatively at rest in it. This ascent of the water shows its endeavor to recede from the axis of its motion; and the true and absolute circular motion of the water, which is here directly contrary to the relative, becomes known, and may be measured by this endeavor. At first, when the relative motion of the water in the vessel was greatest, it produced no endeavor to recede from the axis; the water showed no

tendency to the circumference, nor any ascent towards the sides of the vessel, but remained of a plain surface, and therefore its true circular motion had not yet begun. But afterwards, when the relative motion of the water had decreased, the ascent thereof towards the sides of the vessel proved its endeavor to recede from the axis; and this endeavor showed the real circular motion of the water continually increasing, till it had acquired its greatest quantity, when the water rested relatively in the vessel. And therefore this endeavor does not depend upon any translation of the water in respect of the ambient bodies, nor can true circular motion be defined by such translation. There is only one real circular motion of any one revolving body, corresponding to only one power of endeavoring to recede from its axis of motion, as its proper and adequate effect; but relative motions, in one and the same body, are innumerable, according to the various relations it bears to external bodies, and, like other relations, are altogether destitute of any real effect, any otherwise than they may perhaps partake of that one only true motion. And therefore in their system who suppose that our heavens, revolving below the sphere of the fixed stars, carry the planets along with them; the several parts of those heavens, and the planets, which are indeed relatively at rest in their heavens, do yet really move. For they change their position one to another (which never happens to bodies truly at rest), and being carried together with their heavens, partake of their motions, and as parts of revolving wholes, endeavor to recede from the axis of their motions.

[i] Wherefore relative quantities are not the quantities themselves, whose names they bear, but those sensible measures of them (either accurate or inaccurate), which are commonly used instead of the measured quantities themselves. And if the meaning of words is to be determined by their use, then by the names time, space, place, and motion, their [sensible] measures are properly to be understood; and the expression will be unusual, and purely mathematical, if the measured quantities themselves are meant. Upon which account, they do strain the sacred writings, who there interpret those words for the measured quantities. Nor do those less defile the purity of mathematical and philosophical truths, who confound real quantities with their relations and sensible measures.

[j] It is indeed a matter of great difficulty to discover, and effectually to distinguish, the true motions of particular bodies from the apparent; because the parts of that immovable space, in which those motions are performed, do by no means come under the observation of our senses. Yet the thing is not altogether desperate; for we have some arguments to guide us, partly from the apparent motions, which are the differences of the true motions; partly from the forces, which are the causes and effects of the true motions. For instance, if two globes, kept at a given distance one from the other by means of a cord that connects them, were revolved about their common centre of gravity, we might, from the tension of the cord, discover the endeavor of the globes to recede from the axis of their motion, and from thence we might compute the quantity of their circular motions. And then if any equal forces should be impressed at once on the alternate faces of the globes to augment or diminish their circular motions, from the increase or decrease of the tension of the cord, we might infer the increment or decrement of their motions; and thence would be found on what faces those forces ought to be impressed, that the motions of the globes might be most augmented; that is, we

might discover their hindmost faces, or those which, in the circular motion, do follow. But the faces which follow being known, and consequently the opposite ones that precede, we should likewise know the determination of their motions. And thus we might find both the quantity and the determination of this circular motion, even in an immense vacuum, where there was nothing external or sensible with which the globes could be compared. But now, if in that space some remote bodies were placed that kept always a given position one to another, as the fixed stars do in our regions, we could not indeed determine from the relative translation of the globes among those bodies, whether the motion did belong to the globes or to the bodies. But if we observed the cord, and found that its tension was that very tension which the motions of the globes required, we might conclude the motion to be in the globes, and the bodies to be at rest; and then, lastly, from the translation of the globes among the bodies, we should find the determination of their motions. But how we are to obtain the true motions from their causes, effects, and apparent differences, and the converse, shall be explained more at large in the following treatise. For to this end it was that I composed it.

11.6. Comments on the Scholium

About a quarter of the way into Book I of the *Principia*, there is a magical moment. Newton has just solved his great problem: *If a body revolves in an ellipse; it is required to find the law of the centripetal force tending to the focus of the ellipse*. He then comments that, as in a previous problem involving an ellipse,[24] 'with the same brevity with which we reduced [that] problem to the parabola, and hyperbola, we might do the like here; but because of the *dignity of the Problem* . . ., I shall confirm the other cases by particular demonstrations' (my italics).

It is *dignity* that distinguishes the Scholium too (as indeed the whole of the *Principia*). The summit of Galileo's ambition in the *Dialogo* was to show that the phenomenon of the tides revealed the existence of an alternately accelerated and retarded absolute motion, that from the tides the reality of absolute motion of the earth could be proved. In the Scholium Newton is explicitly seeking to demonstrate, through phenomena, the transcendent basis of all motion. But he aims even higher; the boy from Grantham has set his eye on the *anatomy of God*. Not that this is stated explicitly in the 1687 edition of the *Principia*: but in Query 28 appended to his *Opticks* in 1706, Newton came clean about his aspirations and wrote:[25]

And these things being rightly dispatch'd, does it not appear from Phænomena that there is a Being incorporeal, living, intelligent, omnipresent, who in infinite Space, as it were in his Sensory, sees the things themselves intimately, and throughly perceives them, and comprehends them wholly by their immediate presence to himself: Of which things the Images only carried through the Organs of Sense into our little Sensoriums, are there seen and beheld by that which in us perceives and thinks. And though every true Step made in this Philosophy brings us not immediately to the Knowledge of the first Cause, yet it brings us nearer to it, and on that account is to be highly valued.

As we shall see in Vol. 2, Newton's concept of absolute space as the *sensory* (or *sensorium*) *of God* was an important factor in prompting Leibniz to initiate the famous Leibniz–Clarke correspondence, which was itself important in helping to revive the idea of the relational nature of motion in the second half of the nineteenth century. But again we anticipate; we must come back to the Scholium, in which Newton set out to ensure that 'these things' are 'rightly dispatch'd'. No one was ever better qualified than he to see that they were.

Even at the rather more modest level of finding the true basis of motion, the Scholium is shot through with grandeur. In the spirit of what it attempts it is natural philosophy at its very best. From careful observation and analysis, Newton is seeking [26] 'to derive two or three general Principles of Motion from Phænomena'. More clearly than ever before, the essential difficulty of the undertaking is laid bare, particularly in the opening sentence of the final paragraph: 'It is indeed a matter of great difficulty to discover, and effectually to distinguish, the true motions of particular bodies from the apparent; because the parts of that immovable space, in which those motions are performed, do by no means come under the observation of our senses.'

Seen in terms of his polemic with Descartes (whom we note that Newton does not now even mention), the Scholium was an almost complete success – as I already indicated, more or less as an infinity of parameters is to three. But as a demonstration of the anatomy of God it failed; as a demonstration that the existence of an absolute immovable space follows of necessity from the observed phenomena it failed; and as the most accurate characterization of what precisely the *Principia* is able to tell us about the observed world it also failed. In attempting to deduce the unseen from the seen, Newton overshot the mark. But the attempt remains one of the glories of the quest for the absolute, the ground of being. Newton's failures have a lustre which many another man's successes lack.

The successes of the Scholium speak for themselves. Within half a century or so Newtonian dynamics conquered the world. Men came to accept his concepts of absolute space and time – and they worked brilliantly. Descartes' confused notions were completely forgotten. You have to look quite hard to find the faults in the Scholium. If we spend time on that rather than the successes, it will not be through any meanness of spirit. For to find a flaw in Newton is to strike gold. So comprehensive was his genius, it appeared to open all doors into nature, to leave nothing really major to discover. Life after Newton seemed a mere walking through the garden into which his genius had directed us.

But the flaws in the Scholium point the way down little narrow paths that Newton, his eyes fixed on the contemplation of God's majesty, failed to note. Followed far enough, they lead to doors that, like the one Copernicus opened, take us into worlds of which Newton never dreamed – nor any man for that matter.

The flaws in the Scholium are three.

First, although purified of the extreme polemicism of *De gravitatione*, of which it is evidently an improved version, it is nevertheless distorted by being aimed too specifically at Descartes, above all his Aristotelian idea that place and motion are defined *by the immediately contiguous bodies* which we happen to choose as reference bodies.

Second, the physical arguments that Newton invokes to prove his point are not taken from the full generality of the dynamics which he expounds in so masterly a fashion in the body of the *Principia*. Instead he reverts to the restricted dynamics of his early work; the only dynamical problem he considers is that of perfectly circular motion. This gives the impression that the all-important distinction is between rotation and absence of rotation, whereas in reality the decisive distinction is between unaccelerated and accelerated motion (a very special example of which is circular motion). Newton was probably ill-served by the fact that when he came to write the Scholium he was able to fall back on his fully elaborated draft of *De gravitatione*, written almost twenty years earlier and long before that dramatic extension of the applications of his dynamics into which he was prodded by Hooke, the comet of 1680/81, and Halley.

Third, throughout the Scholium Newton persistently fails to acknowledge the existence of one of the most important results of his own dynamics, the famous Corollary V to his Laws of Motion. It is this corollary that, in the Newtonian scheme, gives expression to the Galileo–Huygens relativity principle. As we have seen, Newton states it as follows:[27]

> The motions of bodies included in a given space are the same among themselves, whether that space is at rest, or moves uniformly forwards in a right line without any circular motion.

In the Scholium, Newton adopts a severely empirico-inductive approach. The existence of absolute space, or rather absolute motion, is to be demonstrated from the phenomena as interpreted by Newton's laws. But Corollary V, alleged (according to Newton's somewhat mistaken demonstration) to be one of the most direct consequences of the laws, demolishes the basis of any claim that the *unique* speed and direction of absolute motion can be determined from the phenomena, quite counter to what is implied in the Scholium.

It is not possible to believe Newton was unaware of this uncomfortable fact. Indeed the internal evidence of *De gravitatione* and the Scholium (which is much more carefully worded and goes out of its way to emphasize the difficulty of the problem) suggests that full realization of the import of the relativity principle came between the composition of these two tracts. It is hard to avoid the conclusion that Newton was committed to the concept of absolute space to such an extent that he simply could not bring himself to look unpalatable evidence squarely in the face. Lofty though it is in its aspirations, the Scholium is marked by a certain disingenuousness. For all that, especially for reasons that will become apparent in the next chapter, I would still say the Scholium was far more right than wrong. Cartesian relativism was thoroughly routed and the idea that motion is relational rather than absolute was reborn in the second half of the nineteenth century shorn of its most manifest Cartesian absurdities.

The Scholium was Newton's attempt to interpret the content of his dynamics by identifying the referential basis of motion. It was his attempt to explain visible motion in terms of invisible space and time. The final clarification of the immense achievement of Newtonian dynamics only came about two centuries later when Neumann, Mach and Lange showed that the true understanding of what Galileo and Newton had achieved required one to interpret visible motion, not in terms of invisible space, but in terms of visible matter. They achieved the final conceptual clarification which put Newtonian dynamics in its true perspective. It was not, as Newton believed, a distant intimation of the relationship between the material world and God, but something rather closer to home, though not, I think, any the less wonderful for that: the intimate and unbelievably delicate and precise bond of matter to matter, the fine and subtle net that permeates the palpable created world.

Thus, the evaluation in depth of at least the latter parts of the Scholium depends on this post-Newtonian conceptual clarification of Newton's dynamics; it therefore belongs to the next chapter. We conclude this chapter by a discussion of all but the final paragraphs of the Scholium and those aspects of the remainder of the *Principia* which touch on the subject matter of the Scholium and are not dependent on the further clarification.

The early paragraphs up to and including III speak for themselves. We are already familiar with the basic concepts from *De gravitatione*, though it is true that the words absolute space, time, and motion used to denote them are new. (I presume Newton got the idea of using the word absolute from Galileo's use of the expression *absolute motion* in the *Dialogo*.) It is worth making one comment about IV and noting that the existence of myriad different relative motions, which this highlights, was already clearly recognized by Aristotle, in whom we encountered similar passages. We see how little the basic problem has changed. There is, however, a dramatic change of attitude. No less than More and Newton, Aristotle sought to find a concept of motion according to which any given body has just one true motion. But his instinctive concept of motion was matter based, and he sought the solution to his problem through the self-contained closed and finite universe. In contrast, Newton's concept of motion was instinctively space based and he embraced an infinite space (if not an infinite world). It is interesting to note a correlation between the concepts of motion and of the divinity. The divine in Aristotle is vastly less powerful than Newton's God. The Aristotelian universe was uncreated and eternal and the role of the divine was reduced to supplying the inspiration that kept the quintessence turning in its perfect circle. But Newton's God not only created the world but also ruled over it as a lord over his servants.[28]

With the paragraph identified as [a] we come to the part of the Scholium in which Newton is on strongest ground. The significance of this paragraph has already been anticipated in Sec. 3.15, where we discussed the equation of time (as it later became known – Newton refers to it in the Scholium as 'the astronomical equation'). We recall that the accuracy of lunar theory had made it necessary for Ptolemy to distinguish between the 'time' defined by the natural days (passage of the sun across the meridian) and the 'time' that governed the motion of the celestial bodies (sidereal time, defined by the passage of a given star across the meridian). As we saw, the existence of this more or less uniquely defined time did not strike Ptolemy as particularly remarkable, since he found that it marched exactly in step with the diurnal rotation of the stars, i.e., in step with what, in the prevailing cosmology, he regarded as the most fundamental motion of all – the rotation of the universe. Nevertheless, the very fact that the distinguished 'time' was found by Ptolemy to be *common to all the motions* of the individual bodies would have permitted him to reconstruct the same 'time' even if it had not been instantiated by the diurnal motion of the 'fixed' stars. For, having demonstrated the pervasive 'marching-together' of the law-governed celestial motions, Ptolemy could in principle have dispensed with the concrete realization. And this is the point, forced upon him by the intervening change in cosmological conceptions, which Newton is making. For, post-Copernicus, no one could identify the rotation of the earth with the eternal circling of the very frame of things; nor, post-Descartes, could Newton any longer hang onto the Aristotelian reflex – that uniform motions do exist. The new awareness, child of Copernican cosmology and the Cartesian philosophy of universal motion of matter, is expressed in Newton's 'It may be, that there is no such thing as an equable motion, whereby time may be accurately measured.' But the objective 'marching-together' of phenomena that permitted the ancient astronomers to devise their schemes and predict the future appearance of the heavens subsisted as an empirical reality just as much in Newton's as in Ptolemy's day. Deprived of Ptolemy's option of attributing, at least unconsciously, the remarkable interconnection of things to the[29] 'revolution of the universe', Newton looks instead to 'the flowing of absolute time'. This is another classic example of the substance reflex mentioned in Sec. 8.3: persistent correlations must have some carrier.

Newton's canonization of intuition in the form of absolute time to play the part of this carrier had the effect of freezing the concept of time. Not until the late nineteenth century did the idea gain ground that the proper task of interpretation was precise characterization of the empirical correlations and not the conceptualizing of a metaphysical framework. This will be discussed in Chap. 12. But we may note already here that Newton did correctly grasp the essence of the facts as they were known in his time – as we saw, they had been known since Ptolemy – and his metaphysical concept of absolute time is, for practical purposes, identical to the astronomers' ephemeris time (see p. 181), as we see by his reference to the astronomers' correction of the natural days 'that they may measure the celestial motions by a more accurate time'.

Newton was, in fact, familiar with practical calculations involving the equation of time. In his correspondence with Flamsteed about the 1680/81 comet, Newton enquired[30] whether the observations were made using equated time or only the 'time by ye Sun's course'. In 1672 Flamsteed himself had made the first modern study of the equation of time.[31] In the revised version of *De motu* Newton says explicitly[32] 'absolute time . . . is that whose equation astronomers investigate.' In the triad of Newton's absolutes – space, time, and motion – absolute time is the only one that corresponds closely to empirical practice. (By this I mean that the manner in which astronomers actually determine ephemeris time corresponds closely to what Newton says on the subject of time in the Scholium, whereas the method used to determine position, and hence motion, is based on a principle quite alien to the spirit of the Scholium, as we shall see in Chap. 12.) It is true that, along with the other two, Newton's absolute time was completely overthrown by the revolution of the special theory of relativity, but that introduced entirely new considerations unrelated to the questions under consideration here.

To summarize for the moment Newton's discussion of time: it was based on a sound empirical foundation, laid first by Ptolemy and strengthened by more recent discoveries (note the references to 'the experiments of the pendulum clock' and 'eclipses of the satellites of Jupiter'), which demonstrated the possibility of introducing an essentially unique time parameter with respect to which innumerable different motions could simultaneously be made to obey basically the same laws of motion. Although the rotation of the earth relative to the stars still provided, and would continue to provide for about another two hundred years, the only really reliable clock, Newton correctly foresaw the need for its replacement by ephemeris time.

Finally, we see again in the *practical* measurement of time how objective interconnections of phenomena, once clearly recognized, survive unscathed radical adjustment of the overarching metaphysical framework in which they are conceived. The situation with regard to time is just the same as the fate of Galileo's motionics – the conceptual framework proved transitory but many of the concrete details and techniques were admirably durable: used in their proper place they will probably last forever.

Paragraph [b] is treated in much greater length in *De gravitatione*, but it is quite clear that Newton's concept of absolute space and time had not changed one jot in the intervening 15–20 years. It is worth recalling here a sentence in the earlier work not reproduced in the Scholium, since it casts light on the somewhat curious second sentence of [b]: 'Suppose those parts to be moved out of their places, and they will be moved (if the expression may be allowed) out of themselves.' In *De gravitatione* (item (3) on the properties of space, p. 619) Newton said 'the parts of duration derive their individuality from their order' and follows this with the quaint comment about yesterday changing places with today. It is evidently a similar thought that Newton has in mind in the Scholium, but he does not express himself here so clearly. In the Scholium, he also omits the very explicit statement (p. 619) of the complete identity of the parts of space: 'The parts of duration and space are only understood to be the same as they really are because of their mutual order and position; nor do they have any hint of individuality apart from that order and position, which consequently cannot be altered.' This extreme insistence on uniformity, with the denial of any genuine distinguishing attribute apart from mere numerical ordering, has already been commented upon in Sec. 8.3. It was the absence of 'any hint of individuality' that Leibniz found so unconvincing in Newton's absolute space and attacked with considerable success in the Leibniz–Clarke correspondence (Vol. 2).

Paragraph [c] is primarily of interest on account of its final sentence and the comment that there may be 'no body really at rest, to which the places and motions of others may be referred'. This recognition will remain as a permanent tribute to Descartes, for he really was the first to bring to the fore the profound difficulties implicit in the break up of the sphere of the fixed stars. (We may mention in passing that the title of Westfall's biography of Newton, *Never at Rest*, perfectly characterizes the conceptual picture of the universe with which Newton was grappling in the Scholium, though in fact the words are taken from a letter in which Newton is evidently describing his own restless mind:[33] 'he that is able to reason nimbly and judiciously about figure, force, and motion, is never at rest till he gets over every rub.')

Paragraph [d] is a bit confused but certainly helps to emphasize the dignity (or rather magnitude) of the problem which Newton is attacking.

The idea that 'in the remote regions of the fixed stars, or perhaps far beyond them, there may be some body absolutely at rest' anticipates perhaps Neumann's concept of the mysterious *Body Alpha*, which we shall encounter in the next chapter. The final comment ('that absolute rest cannot be determined from the position of bodies in our regions') seems almost a give-away line but, in fact, announces a very profound shift. Newton renounces once and for all the attempt to define rest kinematically (by the mere relation to external bodies) and anticipates the shift in the second half of the Scholium to the search for a *dynamical criterion of rest*. He very nearly succeeds: of all the infinite terms in the Taylor expansion of a body's motion, Newton in essence (if not in clear conceptualization) succeeded in catching all but the first.

The following paragraph, [e], provides the clearest evidence in the Scholium of Newton's desire to discredit Descartes and, by the same token, it demonstrates the distorting effect that Descartes and the Inquisition exerted on the Scholium. What modern reader could possibly realize that the statement 'if the shell moves, the kernel will also move' is just a late reverberation of the momentous events in 1633 when Galileo was forced to his knees by the Inquisition and the offended pride of a vain pope? It will be recalled that Descartes, with an eye to his assertion that the earth does not move, had produced a definition of motion according to which the kernel of the nut cannot be said to move, the reason for this being that its immediately contiguous containing surface, the shell, has no motion *relative to the kernel*. Henry More had already pointed out absurd consequences such as these in his discussion of Descartes' definition of place.[34]

One of the more intriguing aspects of the *Principia* is the seriousness and thoroughness with which Newton set about testing and comprehensively demolishing Cartesian tenets. The bucket experiment has already been mentioned, as have the arguments against Descartes scattered throughout the *Principia*, especially in the Scholium. A little gem of this kind is the statement near the beginning of paragraph [e] to the effect that 'the impetus of bodies moving forwards arises from the joint impetus of all the parts'. This is a beautiful dynamical argument against Descartes' absurd notion that the kernel does not move even though the shell does (the kernel has no 'true philosophical motion'). Newton is simply commenting upon the fact that if you fill a box with lead and let the box strike another object, the outcomes of the collision will be completely different in the presence and absence of the lead even though the lead has no 'true philosophical motion', being at rest relative to its immediately contiguous container. The lead may be 'philosophically at rest' but provides striking physical evidence in a collision that it does have a motion just as real as its container's. That Newton felt obliged to make this seemingly obvious point shows, on the one hand, how powerful a grip

Descartes had taken on the seventeenth-century mind but equally how baffling the problem of motion was and how carefully Newton felt each piece of evidence for the reality of motion should be weighed. With that tact which Mach so admired in the great natural philosophers, Newton found all the evidence; it would have been almost superhuman for him to have synthesized into the bargain a conceptually impregnable interpretation of the evidence.

To summarize the discussion of paragraph [e]: Newton concentrates his fire exclusively on the notion that contiguous bodies are to be used to define motion. It is a remarkable fact that (so far as I have been able to establish) Newton never once in his published work mentions the possibility that, unlike local bodies, the *totality* of bodies in the material universe (which Newton seems to have been prepared to believe is finite, in contrast to space) might provide a frame of reference appropriate for defining the motion of one particular body within the universe. This is in very striking contrast to, for example, Copernicus and Kepler. I suspect there were three reasons for this: (1) Newton's *a priori* rationalistic concept of space, which he was very loath to abandon; (2) excessive concentration on the demolition of Descartes' specific definition of motion by means of local bodies; (3) an acute awareness, for which Descartes must take the credit, that all the bodies in the universe are most probably in a state of motion relative to each other. Conceptually, as we have noted several times, it was easy for Copernicus and Kepler to define motion relative to the ultimate frame of what they believed were truly fixed stars. It is impossible to say whether Newton seriously considered the possibility of using all the bodies in the universe to define motion. Even if he did, he may well have concluded that the bodies being 'never at rest' the technical difficulties were insuperable and dismissed the thought forthwith.

We pass rapidly over paragraph [f], which repeats very largely the arguments from *De gravitatione* in which Newton showed that motion relative to more distant objects could not be used to define motion satisfactorily (we repeat that in neither the Scholium nor the early work did Newton consider the totality of all relative motions). It is worth noting that this is the only place in the Scholium in which Newton says explicitly that absolute space is infinite ('from infinity to infinity').

Paragraph [g] shows how fateful for the discussion of motion was Descartes' extraordinary claim in the *Principles* that motion is defined by those contiguous bodies[35] *'which we consider to be at rest'*. We have already seen how Huygens used an almost identical phrase (p. 462) in his *De Motu Corporum ex Percussione*, though in a much more precise context than Descartes ever contemplated. It was probably this Cartesian phrase more than any other which made Newton see red. It threatened to take all the objectivity out of the world and leave a mere mishmash that man could interpret at his pleasure (as indeed Descartes did when it served his purposes). No wonder Newton accused Descartes of impious atheism. In Newton's eye, the glory of the world was the structure of its frame. Descartes threatened to remove all the props and leave a confused porridge into which he could stir his arbitrary mischief.

I said earlier that Newton instinctively interpreted Descartes to mean that the motion of any given body contained infinitely many arbitrary parameters. This comes out clearly in Newton's sentence 'it is sufficient only to impress some force on other bodies with which the former is compared, that by their giving way, that relation may be changed, in which the relative rest or motion of this other body did consist.' Since the force applied to the 'other bodies' can obviously be quite arbitrary, in particular, time dependent, it is clearly possible to introduce infinitely many arbitrary parameters in this way. This was the Cartesian relativism to which Newton objected so violently.

To make the point once more: Descartes caused Newton to misdirect his fire. The contiguity of the bodies of reference; the implication that they could be moved arbitrarily; and the further implication that motion was to be determined by some finite (but unspecified) number of such bodies – all these targets, not difficult to hit, diverted Newton's attention from the real threat to his attempt to establish space as the referential basis of motion. He missed the points of genuine danger. We shall see in the next chapter that a small but highly significant modification of the prescription for selecting the bodies of reference draws the sting of all Newton's arguments against using material bodies to provide the referential basis of his dynamics and actually makes Newton's achievement all the more dramatic, impressive, and epistemologically clear. And in Vol. 2 we shall show that if the material universe in its entirety is assumed to supply the frame of reference for describing its own motion one can not only provide an epistemologically unexceptionable framework for describing motion but also give plausible dynamical explanations for the most mysterious of all phenomena in motion – the law of inertia itself.

To come now towards the conclusion of the present discussion of the Scholium, we pass over the crucial paragraph [h] (with its appended dig in paragraph [i] at Descartes, who is the evident butt of the shaft directed at those who 'defile the purity of mathematical and philosophical truths') and most of the equally crucial final paragraph; for these arguments from the absolute nature of circular motion are best deferred to the next chapter. We just note at the very end of the Scholium that the Achilles whom Halley, the Ulysses of the seventeenth century, had drawn out from his lair in Cambridge had a heel every bit as vulnerable as Homer's petulant hero. Having shown that the speed of circular motion can be established entirely independently of the 'remote bodies' placed in that space, i.e., that the speed of circular motion is in some true sense absolute (the precise sense will be made clear in the next chapter), Newton apparently wants to complete his demonstration that the absolute motion can be completely determined. For he says that if the tension in the cord is precisely what is to be expected on the supposition that the globes rotate and the distant bodies rest, then we can conclude 'the bodies to be at rest; and then, lastly, from the translation of the globes among the bodies, we should find the determination of their motions'. Newton here seems to imply that we can determine the instantaneous direction of motion of the globes in absolute space. Two things need to be noted in this connection: (1) Newton has only shown (within the framework of his theory) that the distant bodies do not rotate; he cannot from that deduce that they are in a complete state of rest (the system as a whole may have a uniform translational motion in absolute space). Therefore, his claim falls to the ground. (2) To complete the determination of motion, Newton seems to need distant bodies such as the stars, which are assumed to be at rest. So the final conclusion of the Scholium appears in truth to be an admission of defeat: it is not possible to determine the direction of motion without reference to other bodies! We shall find a much more striking and unambiguous confirmation of this in the next section.

Finally, a comment that looks forward to Chap. 12 and the brief discussion there of Huygens' standpoint with regard to this question. In *De gravitatione*, as we have seen, Newton spelt out in detail what he evidently regarded as the most manifest absurdity of Descartes' scheme: the assertion, on the one hand, of complete relativity of motion, and, on the other, the formulation of his laws of motion, which say bodies have an innate tendency to persist in uniform rectilinear motion with uniform speed. In the early unpublished work Newton showed explicitly how such a notion is simply irreconcilable with general Cartesian relativism. Now although this consideration is clearly implicit in much of what Newton says in the Scholium, especially in paragraph [g], it is not actually stated anywhere there explicitly as a bald fact. We see the same process at work as we noted with Newton's axiomatization of dynamics. In the desire to present the entire matter in exalted terms worthy of the great mathematicians Newton has a tendency to omit the explicit statement of some important details. In this case the clear statement of the non-sensicality of the Cartesian standpoint may have been omitted because it seemed so self-evident to Newton. Alternatively, he may simply have wanted to avoid any direct discussion of Descartes, regarding that as below his dignity. It is interesting to note in this connection that 'the Cartesians' are in fact mentioned explicitly in the set of pre-*Principia* definitions and laws (which Herivel dates somewhat later than the two versions of *De motu*). Included among them is an evident precursor of the Scholium, with which it is worth ending this section, since it stands in much the same relation to the Scholium as does the complete set of pre-*Principia* documents of 1684/85 to the complete *Principia*. In the preparatory documents we find the most important results and ideas listed precisely and sparely; in the *Principia* we meet them all again but greatly amplified. The passage in question is perhaps the most compelling of Newton's statements on the need for absolute space. Here it is:[36]

> The motion of a body is its translation from one place to another, and is consequently either absolute or relative according to the kind of place. But absolute motion is in fact distinguished from relative in circular motions by the endeavour to recede from the centre, which in an entirely relative circular motion is zero, but in a circular motion relative to bodies at rest may be very large, as in the celestial bodies which the Cartesians believe to be at rest, although they endeavour to recede from the sun. The fact that this endeavour [from the centre of circular motion] is certain and determinate argues some certain and determinate quantity of real motion in individual bodies in no wise dependent on the relations [between the bodies] which are innumerable and make up as many relative motions. For example, that motion and rest absolutely speaking do not depend on the situation and relation of bodies between themselves is evident from the fact that these are never changed except by force impressed on the body moved or at rest, and are always changed after [the action of] such a force; but the relative [motion and rest of a body] can be changed by forces impressed only on other bodies to which the relation belongs, and is not changed by a force impressed on both so that their relative situation is preserved.

Even here there is no explicit statement of the irreconcilability of rectilinear uniform motion and Cartesian relativism. It is, of course, the most immediate consequence of what is said and one might ask: Is it necessary to state the obvious? In fact, had Newton done so in the Scholium much confusion might have been avoided – as we shall see in Chap. 12.

11.7. The absolute/relative problem in the remainder of the *Principia*

The Scholium ends with stirring words: 'But how we are to obtain the true motions from their causes, effects, and apparent differences, and the converse, shall be explained more at large in the following treatise. For to this end it was that I composed it.'

We read through the *Principia* with eager anticipation but have to put it down at the end virtually no wiser on this key question. Newton does not return to it specifically anywhere in the book. What he does do, of course, is present the science of dynamics with the extraordinary thoroughness and clarity that we described in Chap. 10. This in itself is powerful evidence in support of his overall conception, which is shown to be internally consistent and in agreement with observation. He provides implicit rather than explicit support for the notions of absolute space and time. Almost any of the topics that he considers has some bearing on the absolute-relative question, but one in particular warrants special mention:

the oblateness of the planets, including the earth, produced by their rotation. Although Newton does not point this out, the oblateness provides evidence for the reality of rotation every bit as convincing as the bucket experiment and the thought experiment with the two globes attached to each other with a string.

Newton treats the oblateness at some length in Book III of the *Principia*, beginning his discussion with Proposition XVIII, which states that 'the axes of the planets are less than the diameters drawn perpendicular to the axes'. For[37] 'the equal gravitation of the parts on all sides would give a spherical figure to the planets, if it was not for their diurnal revolution in a circle. By that circular motion it comes to pass that the parts receding from the axis endeavor to ascend about the equator; and therefore if the matter is in a fluid state, by its ascent towards the equator it will enlarge the diameters there, and by its descent towards the poles it will shorten the axis.' Newton points out that this is confirmed by observations of Jupiter, which is both observed to rotate rapidly (the period is just under 10 hours) and to be sensibly flattened at the poles.

In the case of the earth, Newton made quite detailed calculations with a view to interpreting the pendulum observations made by Richer in 1672 in Cayenne (see Sec. 9.1) and other similar observations, including one by Halley in 1677 at St Helena (also near the equator). The period of pendula of given length at a location near the equator is changed compared with European latitudes by the rotation of the earth through two factors: first, the rotation causes the oblateness, so that points at sea level at the equator and in Europe are in gravitational fields of different strengths. But, in addition, the rotation also produces a centrifugal effect, as both Huygens and Newton showed, and this results in a stronger effective reduction of gravity at the equator than in Europe. Using a very simplified fluid model of the earth Newton predicted, on the basis of the earth's radius and rotation speed, that the ratio of the earth's equatorial diameter to its pole-to-pole diameter should be 230/229, i.e., the relative compression should be 1/230, which Newton found[38] to agree very well with Richer's observations. In fact, Newton was a bit optimistic in his evaluation of Richer's accuracy ('whose diligence and care seems to have been wanting to the other observers'). The correct value[39] of the relative compression is actually about 1/298.

The accurate measurement of the earth's oblateness became the subject of a major scientific undertaking in the first half of the eighteenth century. Maupertuis and Clairaut made geodetic measurements in Lapland (while La Condamine and others travelled to the equatorial regions of South America). Comparison of these observations with ones made in France confirmed Newton's prediction (and disproved a theory of Cassini, according to whom the pole-to-pole diameter was greater). This led Voltaire to make some innocent fun of Maupertuis, pointing out that he had only confirmed in the boring Lapland what Newton had known without leaving home:[40]

> Vous avez confirmé dans les lieux pleins d'ennui
> Ce que Newton connut sans sortir de chez lui.

As we shall see in Vol. 2, the oblateness of rotating bodies has been used more than once to bring home in the most graphic fashion the mystery of rotational motion.

But it is now time to turn to two of the most curious features of Book III, which show more clearly than anything else that Newton was attempting too much in the *Principia*. The point is that the entire abstract theory of orbits described by bodies moving under different centripetal forces is worked out in Book I under the assumption that absolute space exists. But when, in Book III, the theory now complete, Newton announces with justified pride that 'it remains that, from the same principles, I now demonstrate the frame of the System of the World', we find that, *de facto*, Newton refers all motions to the centre of mass of the solar system, assumed to be at rest, and makes in addition the assumption that the distant stars are at rest. It is this assumption which enables Newton to fix directions.

Both these assumptions are very interesting and reveal the difficulties Newton faced in trying to achieve an unconditional victory over Cartesian relativism. Let us begin with the part played by the fixed stars. Near the start of Book III, Newton lists several key phenomena from which he then proceeds to deduce the nature of the forces which must act between the bodies in the solar system. As given in the second and third editions, Phenomenon I, for example, is as follows (the italics are mine):[41]

> That the circumjovial planets, by radii drawn to Jupiter's centre, describe areas proportional to the times of description; and that their periodic times, *the fixed stars being at rest*, are as the $\frac{3}{2}$th power of their distances from its centre.

Now the interesting thing about the italicized words is that they did not appear in the first edition of the *Principia*. A similar addition is made in the later editions to the analogous Phenomenon IV, which relates to the primary planets in their motion about the sun. The significance of the proviso is that the astronomers measure *sidereal periods*, i.e., they measure the length of time an object in the heavens requires to complete a revolution against the backcloth of the stars. It is against this backcloth that Kepler's law of equal areas swept out in equal times is found to hold. On the other hand, all Newton's mathematical calculations were made in absolute space. Thus, the link between observation and theory can only be made under the additional assumption that the fixed stars are at rest. Thus, this proviso, like the revealing (but possibly ambiguous) sentence at the end of the Scholium, according to which the instantaneous determination of circular motion is determined by reference to 'remote bodies', shows that Newton could not quite dispense with the distant stars. In this later example no ambiguity is possible. Newton needs the stars. About this, two comments should be made. The first is that Newton's theory was in fact better than he seems to have realized. It is a remarkable fact, only fully appreciated long after Newton's death, that directions in 'absolute space' can be unambiguously determined without reference to any distant stars (the quotation marks are used to indicate that the 'absolute space' is defined in a way different from Newton's). This will be discussed in the next chapter. The other comment is that the *Principia* completely lacks any suggestion that inertial forces, specifically centrifugal forces, could in any way be causally determined by the distant stars. As we have seen, *De gravitatione* did at least pose the possibility of some sort of connection (formal rather than physical) even if only as a rhetorical device to demonstrate its absurdity. In the *Principia* there is no hint at all in such a direction. The fact that the distant stars do not appear to be in motion in absolute space is recognized only fleetingly in the *Principia* and is treated as a fortuitous coincidence.

So much for the part played by the distant stars in the *Principia*. We now consider that other inconvenient difficulty about which Newton was so coy – the Galileo–Huygens relativity principle, as expressed in his Corollary V to the laws of motion. There are numerous references to this corollary throughout the *Principia*, indeed it is often prominently invoked, so there is no question of Newton deliberately suppressing awkward evidence. However, there is nowhere any admission that it makes rather a nonsense of the idea of an unambiguously defined motion in absolute space. Nevertheless, Newton seems to have felt obliged to establish a criterion of absolute rest and he did this in a rather curious way. I follow the presentation given in the later editions of the *Principia* (in the first edition the arguments were presented in a slightly different way but were in essence identical).

In Book III (*The System of the World*), Newton first of all establishes from the phenomena the facts about gravity and, equally important, the grounds for asserting that interplanetary space is a vacuum. Then follows a remarkable hypothesis:[42]

HYPOTHESIS I. That the centre of the system of the world is immovable.

Newton's comment on this hypothesis is exceedingly brief: 'This is acknowledged by all, while some contend that the earth, others that the sun, is fixed in that centre.'

An important point to be established is what precisely Newton meant by 'the System of the World', which is, in fact, the title of Book III of the *Principia*. Does Newton mean by it the entire universe or merely the solar system? In fact, comparison of the relevant passage in Book III with the analogous passage quoted in the final section of Chap. 10 (p. 574) shows clearly that Newton means the latter. For in the earlier passage he speaks of 'the whole space of the planetary heavens' being either at rest or in a state of uniform motion in a straight line, so that 'the communal centre of gravity of the planets either rests or moves along with it'. What is particularly interesting is that in *De motu* Newton seemed to have been happy to accept this state of affairs as a bald fact (he made no attempt to distinguish between the two possibilities) whereas in the *Principia* he appears to have felt it necessary to insist on a state of rest for the communal centre of gravity. This leads him to put forward a decidedly incongruous idea and suggests that he still had one foot in the old cosmology. This follows from the proposition which immediately follows Hypothesis I and in which Newton asserts that '*the common centre of gravity of the earth, the sun, and all planets, is immovable*'.

The proof is exceedingly brief:

> For (by cor. iv of the Laws) that centre either is at rest, or moves uniformly forwards in a right line; but if that centre moved, the centre of the world would move also, against the Hypothesis.

Newton's arguments are an extraordinary mixture of sound science and a residual geocentrism that verges on superstition. In the proof to the following proposition (*That the sun is agitated by a continual motion, but never recedes far from the common centre of gravity of all the planets*), he shows convincingly that there is only one truly distinguished point in the solar system, the centre of gravity of the complete solar system, and that, according to his principles, this is either at rest or in uniform rectilinear motion. But the second possibility is simply dismissed as inconceivable – on the ground, 'acknowledged by all', that the centre of the system of the world is immovable. Just as revealing is the comment Newton makes in his earlier version of *The System of the World*, in which, speaking of the idea that the centre of gravity of the solar system moves, he says:[43] 'But this is an hypothesis hardly to be admitted; and, therefore, setting it aside, that common centre will be quiescent.'

I find this almost the most intriguing passage in the entire *Principia*. It is the last vestige of geocentrism. Newton, like Galileo before him, simply cannot bring himself to believe in the absence of rest. There has just got to be a point at rest. Copernicus put the earth in motion; Galileo, as an article of faith, transferred the state of rest to the sun; even though he knew full well that the sun rotated in a rather mundane mechanical sort of a way, he nevertheless held fast to the idea that the sun as a whole enjoyed a state of divine rest. It was no different with Newton. His laws of motion and gravity enforced the recognition that the sun moves and permitted the existence of just one point that could, at least in principle, be at rest: the common centre of gravity of the solar system. It became Newton's last refuge. A plague on Corollary V: a truly still point was more important than the democracy of frames of reference imposed by the laws of motion.

This point was also the last vestige of the perfectly ordered cosmos, the belief that the observed architecture of the world is the direct handiwork of God. Where Descartes looked forward to Laplace and the modern inflationary cosmologists, Newton looked back to Kepler and beyond him to Pythagoras. Newton was pretty well certain that the stars were other suns with attendant planetary systems, but he had no doubt that the overall structure of the universe was the conscious design of God. Yet again he stood in direct opposition to Descartes. In the famous General Scholium[44] added to the second edition of the *Principia* in 1713, Newton argued that the great regularity observed in the solar system with all the primary planets and their satellites rotating in the same sense argued the conscious design of God: 'It is not to be conceived that mere mechanical causes could give birth to so many regular motions. . . . This most beautiful system of the sun, planets, and comets, could only proceed from the counsel and dominion of an intelligent and powerful Being.' In Question 31 of the *Opticks* he said explicitly of the origin of the world that one could not pretend that[45] 'it might arise out of a Chaos by the mere Laws of Nature' and again that the 'wonderful Uniformity in the Planetary System must be allowed the Effect of Choice'. These are all barbs clearly directed against Descartes.

For Kepler the sun was not at all on the same footing as the stars. It was a great globe, the image of God the Father, animating the planetary system and illuminating the vast magical cavern formed by the system of the fixed stars. In the teeth of all the evidence that his own colossal industry had uncovered, Newton clung to the last remnant of this cosmology. Whether he liked it or not, the walls of the cavern were demolished, the soul of the sun was killed, and mechanical order reigned almost supreme; but the still point, the centre of the world that really no longer existed, remained. How hard it was to wrench the human mind from the concept of perfect rest!

Keynes[46] did have a point when he called Newton 'the last of the magicians'.*

* There is a huge literature on the topics covered in this and the following chapter, much of which is rather inconclusive in being too philosophically rather than physically oriented; in addition much of it does not take account of the relatively recently discovered *De gravitatione*. This is not the case with Stein's paper.[47] Earman's paper[48] does a good job of establishing the strength of Newton's position and can serve as an introduction to the debate in the literature. The penultimate sentence of paragraph [i] on p. 627 has been altered from Ref. 23 on the basis of Ref. 47 (p. 184).

12

Post-Newtonian conceptual clarification of Newtonian dynamics

12.1 Introduction

As we have seen, Newton combined strongly rationalistic, *a priori* views about space and time with a markedly empirical approach to the nature of bodies and the interactions between them. This obscured the extent to which the whole of Newtonian dynamics could be put on an empirical foundation. The final chapter, therefore, in the discovery of dynamics should for completeness include the significant clarification achieved in the second half of the nineteenth century with regard to the foundations of Newtonian dynamics. This clarification did not in any way change the content of the science, but was valuable in several ways, of which three in particular may be mentioned. First, in its own right, in that it provided a coherent and intellectually satisfying account of how all the essential features of dynamics can be identified empirically in phenomena. In this respect, it merely completed what Newton had begun in such masterly fashion, particularly in his identification of forces. Second, by identifying very clearly the basic elements from which empirical dynamics can be constructed, it simultaneously showed that there is nothing *a priori* or sacrosanct about these elements. It showed that these elements depend on nature and not metaphysical necessity, as Newton in part believed, and that their selection is suggested ultimately by observation and could therefore be modified in the light of further observation. What observation could make, it could also unmake. Just as important, by showing how the theory could be built up on the basis of judiciously chosen facts of experience, it suggested that a quite different theoretical structure might be obtained by taking different facts of experience as the foundation. Then quite new theories could result after extension of the range of application. Third, the empirical definitions of *inertial system* by Lange, of *equality of time intervals* by Neumann, and of *mass* by Mach provided paradigms of the *operational definition* of the basic concepts of dynamics in terms of observable objects and processes. In themselves they were of academic and epistemological rather than practical value; they did not significantly change our knowledge or understanding of the real contingent world. However, within a few decades, Einstein was to fashion the operational definition of simultaneity very much after the manner of these paradigms and was thereby able to bring about a revolution every bit as great as Copernicus's. Thus, in completing the conceptual clarification of the old science, Lange, Mach and Neumann simultaneously started to lay the foundations of the new.

By jumping forward two centuries we shall unfortunately do some violence to the historical development. Many things very relevant to the absolute/relative question occurred during these two centuries. However, it does appear to me that it will not be possible to break the story into two volumes in any other way that makes sense conceptually. For in Vol. 2 we shall need to consider the organic development of several entirely new strands – analytical mechanics and the calculus of variations, the rise of field theory, and the overthrow of Euclidean geometry, to say nothing of special relativity – and it would not be appropriate to make a start on them in this volume even though many important parts of the development occurred before the work to be described in this chapter. On the other hand, what now follows does join on conceptually in a very natural manner and even serves to put the pre-Newtonian and Newtonian achievement in more striking perspective. In any case, the second half of this story, to which so many rich conceptual strands contribute, is better arranged according to the concepts than the chronology. So we now bid farewell to the miraculous century that opened with the great discoveries of Kepler and Galileo and closed with the towering achievement of the *Principia* and pass straight to a very different age, an age in which the intellectual climate had changed out of recognition and no one would dream of treating physics as an organic part of theology. This change too had its impact on the way the concepts of space, time, and motion were approached – as we shall see.

12.2. Neumann and *Body Alpha*

We begin with Carl Neumann's habilitation address 'On the principles of the Galilean–Newtonian theory', which was given at the University of Leipzig in 1869 and published as a little booklet a year later.[1] This address, in which Neumann, who is mainly noted for his work in mathematics, introduced his famous *Body Alpha*, marks the beginning of the critical re-examination of the basic concepts of dynamics which helped to prepare the ground for the revolutions at the beginning of this century.

Neumann begins with an illuminating discussion of what the word *explanation* means in science. He points out that it by no means implies that a complete explanation of all phenomena is provided, so that nothing mysterious and unexplained remains. No, explanation in empirical science is something very different: it is the reduction of innumerable phenomena to a few basic phenomena which in themselves are completely *inexplicable*. They are irrational or arbitrary, in the sense that no reason at all can be given why they are as they are and not otherwise. He illustrates this by the classical example of Galileo's law of parabolic motion, which, as we saw, Galileo 'explained' by showing how it could be decomposed into a horizontal inertial component and a vertical motion governed by the attraction toward the centre of the earth. The existence of each of the two separate components and the fact that they can be combined in accordance with a rule which states that each component continues to exist independently of the presence of the other are observational facts for which no explanation is forthcoming. That Galileo's discovery nevertheless has great explicatory power stems from the fact that the *initial conditions* of the parabolic motion may be extremely varied and yet the subsequent motion can always be understood in terms of just those two same components, which are always present. It is the *universality* of the components, the fact that they are invariably found in all projectile motions, that is the real measure of their worth.

The task of science, then, is to reduce the multifarious phenomena of nature to as few inexplicable but universal *basic concepts* as possible. Moreover, it will always be an important task of the natural scientist to identify and define as clearly as possible these foundational elements, i.e., to lay bare their empirical basis. Only in this way will the scientist be aware of the true basis of his work and fully aware of the changes that may be necessary in its foundations. As Neumann put it at the end of his address:

> However exalted and complete a theory may appear before us, we shall always be forced to give a most precise account of its principles. We must always keep before our eyes the fact that these principles are *arbitrary* and hence *mobile*; we must at every instant be able to see the consequences that a change of these principles would have on the complete structure of the theory, so that we are in a position to allow such a change to take place in good time – so that we can (in a word) protect the theory from *petrification*, from *rigidification*, which would be nothing but *harmful* and a *hindrance* to the progress of science.

It is in such a frame of mind that Neumann approaches the question of the *content* of the law of inertia. He objects strongly to the statement that a body subject to no external disturbance moves in a *straight line*, because we do not know 'what is to be meant by motion in *a straight line*'. For 'a motion that is *straight* when observed from the earth will appear *crooked* when observed from the sun. In brief: every motion that is *straight* with respect to *one* celestial body will appear *crooked* when referred to *every other* celestial body.' This, of course, is exactly the point that Newton was making at length against Descartes in *De gravitatione* and rather more succinctly (but also more obscurely) in the Scholium.

The solution which Neumann proposes to this problem is at first blush startling. He says that 'there must be a special body in the universe which serves us as the basis of our judgement, with respect to which all motions are to be referred'. He arrives at this conclusion because 'careful examination of the theoretical framework of Newtonian dynamics' clearly reveals its foundation, namely, that 'all actual or even conceivable motions in the universe are to be referred to *one and the same* body'.

The body itself is most mysterious: '*Where* this body is situated and the grounds for according a single body such a distinguished, indeed sovereign position – these are questions to which we receive *no* answer.' Unperturbed by these mysteries, Neumann states formally:

As *first principle* of the Galileo–Newton theory one should take the statement that at some unknown place in the universe there is an unknown body, indeed, an *absolutely rigid* body, a body whose shape and size are unchanged for all time.

Neumann proposes to call this the *Body Alpha* and says that the *second principle* of the theory is that the motion of any material point left to itself (i.e., subject to no disturbance) is rectilinear with respect to *Body Alpha*.

If Newton's concept of absolute space is remarkable, *Body Alpha* seems nothing short of fantastic. I have to admit that when I had got this far in Neumann's paper I began to wonder if he had given his address on the 1st of April (it was actually given on 3 November). However, it becomes clear from the later discussion that Neumann deliberately intended such an effect. He in fact uses the concept of *Body Alpha* in somewhat the same way (though not for the identical purpose) that Brecht uses masks in his plays to achieve his *verfremdungseffekt* (alienation). He asks if it is really necessary to bring in this principle that is so 'strange and alien (*sonderbar und befremdlich*)' and answers his question by saying that it is in order to bring out the extremely special nature of absolute motion in Newton's theory. For 'the character, the true essence of so-called absolute motion is that all changes of position are referred to *one and the same object*, which, moreover, is spatially extended and unchanging but otherwise not further particularized'. Neumann in fact admits that by introducing *Body Alpha* he is only giving expression with other words to what is ordinarily meant by absolute motion in absolute space. ('*Das dürften nur andere Worte für dieselbe Sache sein*.') The purpose of *Body Alpha* is to shock us into an awareness of how very special is the motion that Newton's First Law describes.

Just how special can be demonstrated in the following manner. Imagine a system of n material point particles, which may either be taken to represent the entire system of bodies in the universe or else a small subsystem of bodies sufficiently far removed from other bodies that it can be regarded as an isolated system not subject to any external forces. Let the masses of the particles be m_i, $i = 1, \ldots, n$, and suppose they interact in accordance with Newton's law of gravitation. Neumann poses the question that Newton raised at the end of the Scholium but did not discuss or solve in full generality in the *Principia*: given the observable (relative) positions of the bodies at successive times, to find how *Body Alpha*, i.e., absolute space, is oriented with respect to the successive instantaneous configurations of the particles. If we imagine the n-body problem solved in accordance with Newtonian dynamics, i.e., 'in absolute space', the general solution will contain $7n$ constants; these correspond to $3n$ initial coordinate components, $3n$ velocity components, and the n masses m_i. This solution gives the positions of the particles with respect to *Body Alpha* or absolute space. However, given the *absolute* positions, one can calculate how the relative configuration of the n point masses must vary. By comparing the actually observed relative configuration with the general solution, it is in principle possible to determine the unknown constants, that is, to deduce from the observed relative motions the corresponding absolute motions. However, the solution to the problem is not unique, for rather obvious reasons. The initial position of the coordinate origin and the initial orientation of the coordinate axes is entirely arbitary. Moreover, in accordance with Newton's Corollary V, the coordinate system as a whole is only determined up to a uniform rectilinear motion of the origin. Thus, the attempt to fix the coordinate system uniquely by observing the relative motions over a time interval sufficient to extract all the salient dynamical data about the system (the 'length' of this time interval will be discussed in Vol. 2 – it has great relevance to the absolute/relative debate) necessarily fails; all one can do is determine a *family* of coordinate systems, this family being characterized by nine arbitrary parameters (three fix the origin at the initial time, three the orientation of the axes, and three the velocity vector of the uniform translational motion allowed by Corollary V). Nevertheless, what can be fixed is still very striking. First, the n masses m_i can all be determined uniquely (in principle at least; one of them must, of course, be taken as the unit of mass). Once the masses m_i have been determined, the centre of mass of the n-body system is uniquely determined by the *relative* positions of the bodies. This is a point worth emphasizing, because it highlights the curious interplay between relative and absolute concepts in Newtonian dynamics. In any of the absolute frames of reference, the centre of mass is found to move rectilinearly with constant speed. The freedom in the absolute coordinate systems can be usefully restricted by taking the origin to coincide with the centre of mass. When this has been done, only *three* arbitrary parameters remain, the ones corresponding to the *orientation* of the coordinate axes.

It is remarkable that, within the framework of Newtonian dynamics, there exists a natural way to fix the orientation *almost uniquely* (for a given dynamically isolated system of bodies). As Neumann points out, this is made possible by a famous theorem discovered in the eighteenth century by Laplace: the theorem of the *invariable plane*, which is related to the conservation of angular momentum and which holds whenever purely central forces (centripetal forces in Newton's terminology) act. At this point we should perhaps review briefly the law of conservation of angular momentum, which can be stated as follows. In any absolute coordinate system (i.e., in what is now called an inertial coordinate system), let the instantaneous position of body i be given by the radius vector $\mathbf{r}_i$ and its instantaneous velocity by $\dot{\mathbf{r}}_i$ (the dot denoting differentiation with respect to time). Let $\mathbf{p}_i = m_i\dot{\mathbf{r}}_i$, i.e. $\mathbf{p}_i$ is the *momentum* of particle i. For the system of n bodies we form the *angular momentum vector*

$$\mathbf{M} = \sum_{i=1}^{n} \mathbf{r}_i \times \mathbf{p}_i. \qquad (12.1)$$

If the system as a whole is isolated (as we assume), then the vector $\mathbf{M}$ is different in different absolute coordinate systems, but the different $\mathbf{M}$s are each separately conserved, i.e., remain fixed in magnitude and direction, in the respective coordinate systems. It is worth giving explicitly the rules for finding $\mathbf{M}$ when the absolute coordinate system is changed. Suppose we go over to a 'primed' coordinate system at rest relative to the first and such that the origins are separated by the constant vector $\mathbf{a}$, so that

$$\mathbf{r}_i = \mathbf{r}'_i + \mathbf{a}.$$

Then $\mathbf{M}'$ and $\mathbf{M}$ are related by

$$\mathbf{M} = \mathbf{M}' + \mathbf{a} \times \mathbf{P}, \qquad (12.2)$$

where $\mathbf{P} = \Sigma_i \mathbf{P}_i$ is the total momentum of the system (the same in both systems, since they are relatively at rest). Note that in accordance with Eq. (12.2) the angular momentum is independent of the position of the origin if $\mathbf{P} = 0$, i.e., if the system as a whole is at rest in the chosen coordinate systems. Particularly interesting is the decomposition of the angular momentum into a component that can be called its *intrinsic* angular momentum and a component that is due to the motion of the system as a whole. Namely, let $\mathbf{R}$ be the position vector of the centre of mass of the system in an arbitrary coordinate system and $\mathbf{P}$ be the total momentum of the system in that same coordinate system. Further, let $\mathbf{M}'$ be the angular momentum vector in the coordinate system with origin at the centre of mass and axes parallel to the original coordinate system. Then the angular momentum $\mathbf{M}$ in the original system is

$$\mathbf{M} = \mathbf{M}' + \mathbf{R} \times \mathbf{P}, \qquad (12.3)$$

i.e., **M** breaks up into two distinct terms, one corresponding to the angular momentum in the rest frame and one due to the motion of the system as a whole.

Now for a given isolated system **M**′ defines a unique and fixed direction in absolute space. There is therefore a coordinate system that is more or less uniquely associated with an isolated system of bodies. It is the one in which the origin is placed at the centre of mass and one of the coordinate axes (by convention normally taken to be the z-axis of a rectangular system of Cartesian coordinate axes x, y, z) is taken along the direction of **M**′, so that $\mathbf{M}' = (0, 0, M)$. This direction is fixed in absolute space and the xy plane is Laplace's *invariable plane*.

This represents the maximum extent to which the laws of Newtonian dynamics can uniquely distinguish a coordinate system when applied to a given dynamically isolated system. In general, as we have seen, the laws of motion on their own leave nine parameters free. By going to the centre-of-mass system one can eliminate six of these parameters; by taking the z-axis of the Cartesian coordinate system along **M**′, two of the remaining three arbitrary parameters can be fixed. But in general it is not possible to fix the last free parameter: the azimuthal angle of the x- and y-axes about the z-axis remains free. However, it is important to note that, although the *orientation* of the x- and y-axes cannot be fixed, the axes themselves cannot rotate in absolute space; once the orientation has been fixed at the initial time it is then fixed for all times. Thus, by carefully noting the observable *relative* motions one can (in principle at least) construct coordinate axes that are 'non-rotating in absolute space'.

The remarkable nature of Newtonian dynamics which Neumann was trying to emphasize by introducing the concept of *Body Alpha* can now be illustrated by a thought experiment (in which we ignore corrections that general relativity introduces). Suppose there are two isolated material systems containing n and n' bodies respectively anywhere in the universe and separated by a very great distance. Imagine teams of astronomers sent by spaceship to each of the systems and instructed to observe the relative motions within their respective systems, without looking at any other bodies in the universe. They are to carry out the construction of the distinguished coordinate systems in the manner outlined above. Each will find their own respective system. That they can find one at all is remarkable enough. Even more remarkable is the result of comparing the two coordinate systems. Let them be N and N'. Then according to the laws of Newtonian dynamics it will be found that N and N' are related in an almost miraculous way. For, relative to N say, the origin of N' will be found to be either at rest or moving uniformly in a straight line. Moreover, the axes of N' will in general be inclined to those of N but *there will be absolutely no change in the relative orientation with time*. The axes are rigidly locked relative to each other.

Although Neumann did not give this specific example, it was above all this remarkable *rigidity* of the orientations that he wanted to emphasize by introducing the concept of *Body Alpha*.

It is worth spelling out what exactly happens in the Neumann process. It is what astronomers call *materialization of the frame of reference*. The Newtonian concept of absolute space is of no value whatsoever unless the absolute frame of reference can be explicitly linked to observable matter; this linking is what is meant by *materialization*. One can suppose the two teams of astronomers in the thought experiment described above taking 'snapshots' of the instantaneous configurations of their respective systems of point particles. These show the instantaneous *relative configurations*, which are all that are observable (we recall that, throughout the present volume, the velocity of light is assumed to be infinite). Once the astronomers have obtained enough snapshots of their system, they will be able to deduce the masses of the bodies and, for each snapshot, the position of the distinguished coordinate system (unique up to the fixing, once and for all, of the azimuthal angle) relative to the material frame provided by the corresponding instantaneous relative configuration of the bodies in the system. They can, so to speak, 'paint' the otherwise invisible axes of absolute space onto the snapshot. The actual observations of how the axes of systems N and N' are related to each other must always be made via the observable configurations, i.e., the snapshots.

There is no doubt that, at the ultimate level, all observation is relative. The thought experiment we have been describing is merely a graphic way of making this explicit. What therefore is the justification for introducing absolute space, or, to use Neumann's expression, *Body Alpha*? In what sense is it real? Neumann has the following comment:

In a purely mathematical investigation involving several variables simultaneously in which the relationship between these variables is to be represented in as clear a manner as possible, it is often expedient or even necessary to introduce an intermediate variable and then specify the relationship which each of the given variables has to this intermediate quantity. We find something similar in the physical theories. In order to get an overview of the connection between different phenomena presented simultaneously, it is often expedient to introduce a merely conceptual process, a merely conceptual substance, that, so to speak, represents an intermediate principle, a central point, from which the individual phenomena can be reached in different directions. The individual phenomena are linked to each other in this manner, in that each is related to the central point. Such is the role played by the luminous ether in the theory of optical phenomena, and the electric fluid in the theory of electric phenomena; and our *Body Alpha* plays a similar role in the general theory of motion.

We shall come back to these comments of Neumann in Vol. 2. But it is interesting to note that within half a century of his writing this passage all three specific examples which he mentioned – luminous ether, electric fluid, and *Body Alpha* – had taken less than ceremonious exits from the scene.

If the concept of *Body Alpha* encapsulated superbly the *rigidity* of the Newtonian frames of reference, it was not at all felicitous as a characterization of the Galilean invariance of dynamics. Neumann was forced to admit that *Body Alpha* was subject to a certain lack of definiteness – any particular *Body Alpha* can always be replaced by another provided only the new body is in a state of uniform rectilinear motion relative to the first. One of the most interesting phenomena to emerge from an historical survey of the discovery and development of physics in general and dynamics in particular is the persistent refusal of even the greatest minds to take Galileo's great principle at its face value – to accept that there simply is no such thing as absolute rest. We have seen how Newton tried to chain down the 'system of the world'. Neumann too was no exception. He clearly was unhappy about the whole family of *Bodies Alpha* and advanced the conjecture that if the totality of bodies in the universe is considered (which therefore, by implication, is assumed to be finite) the principal axes of the inertia tensor of the complete universe will be found to maintain a fixed orientation with respect to the *Body Alpha* that has its origin at the centre of mass of the universe. This would then in some sense at least be clearly distinguished and provide a genuine *material embodiment* of his 'central point'.

This is a somewhat startling suggestion, since the inertia tensor is a concept that normally only has a meaningful definition for a rigid body (see, for example, Landau and Lifshitz[2]); for a system of free particles, the inertia tensor can only be defined at each successive instant and will, even in quite simple cases, e.g. purely inertial motion, change. The principal axes could only maintain a fixed orientation relative to *Body Alpha* for highly exceptional initial conditions. The fact that Neumann was prepared to entertain conjecture of this kind betrays a hankering for the pre-Copernican verities of solid ground under the feet.

Brecht put masks on his actors to drum into the heads of the spectators that what they watch is not 'reality', but a presentation of human beings in circumstances different from their own. Neumann invented *Body Alpha* for just the opposite purpose, to say, 'This *is* reality and see how extraordinary it is.' He ended up by achieving Brecht's effect. Confronted so clearly with the full implications of absolute space, the late nineteenth century came to believe reality could not be so bizarre. By emphasizing so graphically the extraordinary nature of absolute space, Neumann hastened its demise, an eventuality that he did, in fact, anticipate as a remote possibility. We shall come back to this too in Vol. 2.

12.3. Lange and the concept of inertial systems

The publication in 1870 of Neumann's habilitation address was followed two years later by Ernst Mach's slim book *History and Root of the Principle of the Conservation of Energy*.[3] Much more forcibly than Neumann, Mach questioned the basis of Newtonian dynamics and came out quite openly with the suggestion that the concept of absolute space was not merely epistemological nonsense but could also be physically wrong. He implied that a theory of motion based on relational ideas might well lead to observational predictions at variance with Newtonian theory. Between them, these two publications stimulated a discussion about the nature of motion that, more than a century later, has still not abated. It is not the purpose of the present chapter to go into a discussion of Mach's arguments for a relational theory of motion. This chapter is concerned instead with the clarification of the conceptual basis of Newtonian dynamics, accepted as in essence correct, that was stimulated by Neumann and Mach. Within a decade the subject had become very topical, and important contributions to the debate were made by Streintz[4] and above all Lange.[5] It was Lange who coined the expression *inertial system*, which has since become standard. In 1886, Lange published the book to which reference was made in Chap. 9 (p. 477); its title can be rendered in English as *Historical Development of the Concept of Motion and its Probable Outcome*.[6] It contains much useful information but is unfortunately not easy to obtain. This section is devoted to Lange's formulation of the law of inertia.

Lange got his idea for formulating the concept of inertial system from a proposal that Neumann made in his habilitation address concerning time. His solution to the problem of defining time presents what appears at the first encounter to be an alternative to the concept of ephemeris time (p. 181 and p. 182fn). However, we shall see that in fact there is no effective difference. Neumann comments that the reference in Newton's First Law to the uniformity of inertial motion has no concrete meaning unless we know what is meant by 'equally long time intervals'. Neumann points out that *de facto* the rotation of the earth provided the *practical* unit of time in his day, but that it would be manifestly absurd to take the rotation of the earth as providing the ultimate time scale; for it was already certain in his time that the period of rotation of the earth could not be constant in accordance with the then known laws of physics (for example, tidal interaction of the moon and earth must cause a gradual slowing down of the earth's rotation). To make the earth a perfect time keeper, by definition, would be absurd; one could just as well take Mars or any other rotating celestial body. Any such arbitary definition comes up against all the objections which Newton found in Descartes' relativism.

Neumann got round the apparently insuperable difficulties which Newton had seen in Cartesian relativism in a most interesting and illuminating manner. What he did in essence was to change in a subtle and significant way the seemingly absurd statement Descartes made in a closely related context, namely, that motion is to be defined by bodies *'which we consider to be at rest'*. More than anything else, it was the apparently arbitrary nature of this proposal (which Descartes manifestly exploited and abused) that so aroused Newton's wrath. Motion cannot be made to depend on human whim! The subtlety of Neumann's modification was in providing objective criteria, suggested by the laws of motion themselves, for the selection of bodies with respect to which motion is to be determined. The key element in Neumann's concept is that of a material point *subject to no forces* (in German *sich selbst überlassen*, literally, *left to itself*). For brevity we can call such an object *force-free*. The essence of Neumann's definition of equal intervals of time, which Lange modified to define an inertial system, is that *motion and time are to be defined relative to force-free bodies*. For the moment we shall not consider the criterion used to establish that any particular body is force-free; in fact, as we shall see shortly, the original force-free concept of Neumann and Lange needs to be modified in a significant way before it can acquire experimental utility.

The use of force-free bodies to define motion and time can be (and is) called *dynamic*. Conceptually, the dynamic method is quite sophisticated since the most fundamental physical processes are defined in terms of themselves. The processes define their own means of definition!

Let us see what this means in concrete terms by examining Neumann's definition of equal intervals of time, which is the simplest and most straightforward example we shall meet. Neumann observes that a statement like 'A force-free material point travels equal distances in equal times' is empty of content even if you can identify a force-free body and know how to measure equal distances. But the situation is entirely different as soon as we consider a case in which we simultaneously observe *two or more* force-free bodies. For in this case the motion of one of them can, *by convention*, be taken as defining equal units of time (the times required to pass through equal distances). What this convention does is *materialize a clock*. Then, with the concretely realized clock, one can make a nontrivial experiment. Does a second force-free body travel equal distances in equal times as defined by the first force-free body? The answer is, of course, yes, or it is at least to the accuracy with which Newtonian dynamics holds and the experiment is properly performed.

Neumann states the part of Newton's First Law relating to the uniformity of the motion as follows:

Two material points, each force-free, move in such a manner that equal path distances of the one always correspond to equal path distances of the other.

Note that this definition is in agreement with the ancient requirement, noted in Chap. 6 in connection with Kepler's treatment of orbital speed (p. 305), that only like quantities should be compared. For in accordance with Newmann's prescription, a studied motion is not measured in terms of some abstract time but in terms of another actual motion. It is also worth noting that this concept of the effective meaning of equal units of time is quite close to Newton's own practice (though not his metaphysics). It will be recalled how Newton embedded time in space in the problem of centrifugal force by using the actual motion of the body in space to provide simultaneously the measure of time. Here is one further typical example, in which Newton is discussing Galileo's law of the parabolic motion of projectiles:[7]

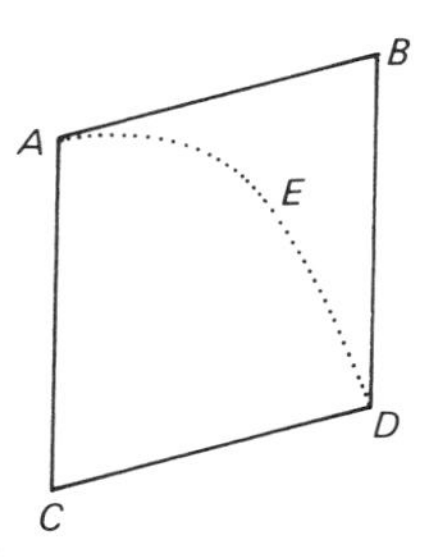

Fig. 12.1.

If the body A by its motion of projection alone could describe in a given time the right line AB, and with its motion of falling alone could describe in the same time the altitude AC; complete the parallelogram ABCD, and the body by that compounded motion will at the end of the time be found in the place D; and the curved line AED, which that body describes, will be a parabola, to which the right line AB will be a tangent at A; and whose ordinate BD will be as the square of the line AB.

It will be noted how Newton effectively uses the length of AB as the measure of the time. As emphasized in Chap. 10, Newton was very conscious of the fact that the *geometrization* of motion, the extension of ordinary three-dimensional geometry into the fourth dimension, depends crucially and explicitly on the law of inertia, for this is what *metrizes* the fourth dimension. However, Newton had no inclination to follow the direction which Neumann here opens up; for his whole instinct, like Galileo's in the question of the tides, was to interpret dynamics as the explanation of the seen in terms of the unseen, ultimately God, whereas the thrust of Neumann's definition is to the reinterpretation of dynamics, not so much as an explicative, but as a *correlative* science, as establishing the relationships between the seen and the seen.

There is no alternative to Neumann's approach. However much Newton may expostulate, the reality of dynamics is that it relates matter to matter, not matter to space and time. What the whole debate is really about is not whether motion is relative to matter or to space but the clarification of the precise nature of the connection which dynamics establishes between matter and matter (I use *matter* here in a generalized sense, as something which is unambiguously observable). In Newtonian dynamics the relationship between matter and matter appears highly bizarre; so bizarre that the case for absolute space acquires a certain cogency. Newtonian dynamics relates matter to matter but in a manner that seems *independent* of matter. For we have seen how it is possible, by careful observation of matter, to 'paint in' coordinate axes on each successive relative configuration of the observable masses in the universe; it is quite impossible to 'paint in' the axes without using the observable matter but, when this is done, there is absolutely nothing about the relative configuration of the observable matter to suggest why the axes, when painted in, move as they do relative to the observable matter. The rules of connection seem curiously disjunct from what they connect. Watching the dancer, we have learned the rules of the dance but they are not at all tailored to the dancer – who must dance, quite literally, in a straightjacket which hangs there seemingly in nothing. Absolute space and time (as determined empirically by careful observation of the seen) are to a high degree *irrational* – as we shall see in Vol. 2 when we come to compare Newtonian with more rational Machian forms of dynamics.

After this general reflection and before making the linkup between Neumann's 'time' and ephemeris time, let us now turn to Lange. His definition of inertial system is taken directly from the paradigm provided by Neumann's *operational* definition of uniformity of motion. The basis of Lange's proposal, given in Ref. 5, is a purely mathematical fact, which I shall present in a slightly simplified form. Imagine three straight lines x, y, z which meet at one point and are fixed rigidly relative to each other, i.e., the angles between them are fixed. Then take any nondegenerate plane triangle with vertices a, b, c and attempt to place vertex a on line x, b on line y, and c on line z. It is trivially obvious that the manoeuvre will succeed for at least two vertices but not necessarily for the third (put mathematically, the solutions to the equations that give expression to this problem sometimes have real and sometimes imaginary roots; the construction succeeds when the roots are real). For example, it is easily seen that if the three lines are mutually perpendicular and all the angles of the triangle acute, the construction is readily achieved. (Put a on line x, b on line y and lay the triangle in the plane defined by x and y. Drop the perpendicular from c onto ab and move a and b along x and y until the extension of this perpendicular passes through the point of intersection of x, y and z. Hold a and b fixed and then swing c up until it rests on line z.) If the construction succeeds for one particular set of lines and one particular triangle then there exists a whole set of triangles obtained from the given one by continuous variation of the triangle's sides for which the construction still succeeds (the triangle can be continuously varied in any way until the roots corresponding to the problem become imaginary).

It follows from this that for three point particles *moved arbitrarily* (subject to continuity) it is always possible to find fixed straight lines which meet at a point and on which the triangle can be fixed in the manner described above for a certain finite time interval (until the corresponding roots become imaginary). But this means we can always find a coordinate system such that three point particles moved in an entirely arbitrary manner will, for a finite time at least, move along straight lines emanating from one point.

Therefore, for three or fewer bodies we see that the rectilinearity of Newton's First Law has no physical content at all; for the construction goes through without any restriction at all on the bodies (there is no need for them to be force-free). There is here an exact parallel with Neumann's definition of uniform flow of time (for one body uniformity has no significance). Lange's operational definition of an inertial system is then as follows. At some instant, three force-free material particles are projected from a single point. At some later instant, choose arbitrarily a point O which lies outside the plane formed by the instantaneous triangle of the three bodies and join this point by straight lines to the three bodies. A three-dimensional corner is thereby constructed. Now take any three rectangular Cartesian axes with respect to which the three-dimensional corner occupies a completely fixed position. Then an inertial system is realized by these three Cartesian axes if the three-dimensional corner (and with it the Cartesian axes) is moved in such a way that the three bodies are always situated on the three corresponding edges of the corner. This requirement fixes the position of the corner and axes uniquely. As Lange emphasizes, this definition has virtually no physical content. For a finite time at least the same construction could be carried through for arbitrarily moved bodies.

According to Lange, Newton's First Law must then be formulated as follows. Given an inertial system realized in the manner described, any fourth force-free particle will be found to move rectilinearly in that system. Clearly, this is a highly nontrivial result.

There are several points worth making about this definition, of which the basic principle rather than any particular detail is important. In the chapter on Huygens, I introduced the concept of cosmic drift to describe inertial motion. Newton made a valiant effort to define and explain this cosmic drift in terms of transcendental, or metaphysical, concepts (absolute space and time). Neumann and Lange rejected this approach as being conceptually unsatisfying and experimentally valueless. Instead of attempting to put a foundation under the cosmic drift, *they made it the foundation*. For the important point is that the reference system, both in its spatial and its temporal part, is, in principle at least, constructed explicitly from material particles that are allowed to follow the cosmic drift. This is the twist to the theory of motion that was announced in Chap. 1. In our intuition we conceive that space and time exist, providing the framework within which motion takes place. Space and time define motion. But the truth is the other way round. For in the Neumann–Lange scheme the space and time coordinates are explicitly constructed from the motions. It is on them that the universal scaffolding of space and time are erected.

We have thus a curious situation in which a basic phenomenon (motion) is defined by a foundation which it simultaneously creates. This is characteristic of all operational definitions and is probably what makes them as difficult for many people as it was for Peter to accept that divine invitation to walk on water.

At root, the reluctance to accept operational definitions as the ultimate basis of physics is a manifestation of what I called the substance syndrome. It springs from the desire to have ground under one's feet. Confronted with the existence of something, we attempt to explain that something in terms of something else. But is this not a perpetual delusion? Perhaps the reader will excuse a digression into poetry, which may make my point better: that our understanding of the world rests ultimately on what we see (vision being taken here to embrace all the senses and other aids to observation), that universal foundations for science exist only in the patterns we perceive in the seen. For, when all is said and done, everything of which we are aware remains perpetually incomprehensible. Nothing is comprehensible except nothing. The nonexistent world is the only rational world we can fully understand. Exposed to the incomprehensible, we search for substance and support. The mist and clouds slowly disperse, but Atlas stands not there before our eyes supporting the world. We never get behind the spectacle of the senses. Pull aside the gaudy curtain and see what is behind! That is the cry. But the curtain is all. We do not have the option of leaving Plato's cave. Whatever brought us and the world into existence worked after the manner of Shakespeare's poet:[8]

> And, as imagination bodies forth
> The forms of things unknown, the poet's pen
> Turns them to shapes, and gives to airy nothing
> A local habitation and a name.

And though the whole 'grows to something of great constancy' the world we confront is nevertheless a *baseless fabric* and could always return from whence it came:[9]

> And, like the baseless fabric of this vision,
> The cloud-capp'd towers, the gorgeous palaces,
> The solemn temples, the great globe itself,
> Yea, all which it inherit, shall dissolve
> And, like this insubstantial pageant faded,
> Leave not a rack behind.

It is futile to look for a metaphysical substrate. The challenge is to trace and put our finger on the universal and unifying pattern. Newton was looking for the *loom* on which God wove the gorgeous fabric of the world. He mistook the woof and the warp for the loom.

To come back to more prosaic matters. The Neumann–Lange approach shifts the problem from the identification of space and time to the identification of force-free bodies. Is there not a major problem lurking behind the idea of force-free bodies? How else can a force-free body be recognized other than by uniform rectilinearity of its motion? Are we not confronted with a vicious circle?

This is the point at which to explain the modification of the force-free concept which must be made before it can be employed in practice. This leads us to the concept of a reference body whose deviation from the force-free state can be calculated from the known forces acting on it. Position determination with respect to such a body is then a two-step process involving observation of the position of the studied body relative to the reference body, whose own position is corrected by calculating how much it must deviate from a force-free trajectory. In this way the position of the studied body is related to an 'ideal' reference body. The ideal reference body is immaterial, a mathematical construct, but is nevertheless obtained by means of a perfectly real body, with which it is not possible to dispense. As this process too is conceptually quite sophisticated, it is worth pointing out explicitly how it relies on the fundamental structure of Newtonian dynamics. According to the theory, the real body would move inertially but for the interactions, which cause deflections from purely inertial motion. Now the important point about the deflections, to which attention has been drawn several times, is that they have *identifiable causes*. Thus, examining all the potential sources causing deflection from inertial motion, we can in principle calculate what the deflection of any given body from inertial motion must be. From the observed motion and the calculated correction we can then get back to the 'pure' inertial motion and hence determine an inertial frame of reference. In principle, the determination of ephemeris time (until its replacement by atomic time) was done in the same way, i.e., it was deduced from the observed motion of a selected reference body whose motion was calculated theoretically with allowance for all known perturbations. (Ephemeris, or Newtonian, time was officially defined by the apparent motion of the sun, as pointed out in Chap. 3, though the practical determination was by the apparent motion of the moon.) Thus, the idea of Lange and Neumann of 'plugging in' to ultimate standards supplied by force-free bodies is basically sound but in reality involves a much more comprehensive exercise in order to separate the force-free behaviour from the disturbances.

This somewhat roundabout way of arriving at a dynamically relevant spatiotemporal frame of reference is quite unavoidable and is forced upon scientists, above all astronomers, by the facts of life. A point to be especially emphasized is that the use of some material reference body can never be avoided. The reason for this is precisely the fact that inertial motion, which I have called cosmic drift, is not in any way tied to the objects that can be observed in our neighbourhood. There is absolutely no way of revealing the cosmic drift other than by letting some material body flow with it. To be of genuine practical value, our spatiotemporal reference system must 'tap' the dynamic behaviour of actual bodies. The system of reference bodies then provides the frame in which the motion of other bodies is described.

Before we examine this technique in a specific example, we should mention an important point of principle. All matter in the universe is subject to gravity – that is what universal gravitation means. This is a very serious threat to Lange's definition, for it puts a question mark over the basis of his entire construction. A force-free body simply does not exist, since all matter resides in the gravitational field created by matter in the universe. Even in the pre-Einstein world this fact was recognized as a serious restriction on the practical utility of Lange's definition, though at that time it could still be argued that for an almost perfectly isolated system such as the solar system appeared to be the gravitational field could be calculated from the observed matter distribution and hence taken into account in the manner just described. Nevertheless, it is hardly satisfactory to base the most important definitions of dynamics on a fortunate contingent accident of the way in which matter happens to be distributed in our immediate neighbourhood. I have not been able to discover who first pointed out this difficulty in Lange's definition. I have seen it attributed to Mach, but I have not been able to locate anything to that effect in his writings. Certainly the problem was clear to Russell in 1903 when he wrote *The Principles of Mathematics*.[10] In it he discussed a proposal made by Streintz[4] that in practical terms is essentially the same as Lange's and relies on the force-free concept, motion being referred to fundamental bodies or axes which do not rotate and are independent of all outside influences.* Russell comments: 'If motion means motion relative to fundamental bodies (and if not, their introduction is no gain from a logical point of view), then the law of gravitation becomes strictly meaningless if taken to be universal – a view which seems impossible to defend. The theory requires that there should be matter not subject to any forces, and this is denied by the law of gravitation.'

Russell left his comment at that. Einstein, anxious to retain the sound core of Lange and Streintz's idea, transformed it by one of the most subtle strokes in science. Giving up the attempt to distinguish conceptually between inertia and gravity, he altered the definition of force-free; he still defined motion relative to distinguished bodies, but stipulated that these be subject *to no forces of noninertial and nongravitational origin*. This is the story of the equivalence principle and Einstein's rediscovery of what Huygens had regarded as such a promising basis for explaining gravity – the fact that, in their observed manifestations, inertial forces are remarkably like gravity. We shall come to this in Vol. 2.

12.4. Determination of the earth's polar motion from satellite observations

Lange's formal definition of an inertial system might appear at first glance the sort of pedantry in which academics delight. Quite the opposite – in the present age of high precision space technology with the Strategic Defense Initiative (Star Wars) looming on the horizon it has become a matter of life and death. For both modern scientific and defence requirements, the ability to define position and motion conceptually and measure them experimentally have become tasks of the highest importance. Lange's *dynamic* definition of position has come into its own in our

* Streintz (conceptually) materialized 'nonrotating axes' by means of force-free gyroscopes. In practice, this is one of the easiest ways of obtaining nonrotating axes. In fact, gyroscopes give one of the most graphic ways of visualizing the empirical reality that underlies the concept of the family of dynamically equivalent inertial frames of reference. A triplet of mutually perpendicular force-free gyroscopes at a common point defines the origin and axes of one inertial coordinate system. Any other such triplet defines another. The axes of any one triplet of gyroscopes remain rigidly locked relative to each other and to any other triplet but the triplets themselves are in states of arbitrary uniform rectilinear motion relative to each other. It is worth noting that the planets, spinning around their axes, implement a system of such gyroscopes in space to a good approximation. Although the planets move around the sun in different orbits and the spatial configuration of the planets is constantly changing, their rotation axes maintain fixed angles relative to each other except for a very slow precession. This also illustrates the point about the possibility of calculating the corrections to force-free motion, since the slow precession of the planetary axes (giving rise in the earth's case to the precession of the equinoxes) can be calculated as a perturbation from the observed matter distribution in the solar system (cf. p. 249).

day. In Vol. 2, I hope to devote an entire chapter to the fascinating and rapidly developing subject of the various different frames of reference in current use in astronomy, space technology, and geophysics. The present section is a brief anticipation of that chapter, inserted here to show just how relevant Lange's definition is to modern problems. We shall consider the problem of determining the position of the rotation axis of the earth. It will also emphasize the following point. One can often find in the text-book literature Lange's concept of an inertial system defined simply by the statement that it is a frame of reference in which force-free bodies move rectilinearly (or, more generally, the laws of motion as formulated by Newton are found to hold). This, of course, is quite correct but it nevertheless misses most of the subtlety of Lange's definition, which has the virtue of emphasizing that the concept of an inertial system *has no practical value unless materialized by actual matter in motion*. The example we consider highlights this point most graphically.

One of the first major scientific discoveries in history was that of the precession of the equinoxes by Hipparchus sometime after 135 BC. As we saw in the chapters on Ptolemy and Copernicus, this discovery was simultaneously the first great frame-of-reference crisis. As soon as one thing starts moving with respect to another, problems of the definition of motion inevitably arise. In fact, determination of the constant of precession, which is an essential step in the determination of an inertial frame of reference for description of motion in the solar system, is one of the great classical problems of astronomy. In advance of the chapter promised for Vol. 2, readers can find a clear semipopular introduction to this topic in a paper by Clemence.[11] Accurate determination of astronomical frames of reference, including the fundamental dynamical frame of reference, is important for a great many purposes but this is apparently 'very hard to get over to the astronomical community at large, who tend to take the reference frame for granted', as C. A. Murray has pointed out to me.

We shall not discuss here this very basic question but will consider rather a particular problem chosen for its relevance to topics central to the absolute/relative debate. The earth rotates about an axis. That axis points instantaneously in a certain direction relative to the distant stars. It also cuts the surface of the earth at a certain point. With the passage of time both the position to which the rotation axis of the earth points on the celestial sphere as well as the position at which it passes through the surface of the earth vary. The later motion is typically around an irregular spiral curve of about 15 m diameter on average. The magnitude of the motion oscillates because there are two major components in this motion, one annual and the other with a component of about 430 days, and these interfere, the maximal effect occurring when the two components are in phase. The component with period around 430 days is the famous

Chandler wobble. The two components are of equal order of magnitude, having amplitudes of about 3 m and 4.5 m respectively.

How are such quantities measured? And, more importantly, with respect to what are they measured? The classical method, by which the effect was discovered in the 1890s by Chandler, an American geodesist, was by astronomical observations. In principle, astronomical observations can tell us two things: the point on the celestial sphere at which it is cut by the rotation axis of the earth (this is the point on the sky around which the stars describe the circles shown in the photograph at the front of the book) and the angular distance of this polar point from the zenith as observed from a particular observatory on the surface of the earth. Motion of the rotation axis 'in space' shows up as motion of the polar point on the celestial sphere, while motion of the polar axis relative to the earth – its 'wobble' about the rotation axis, which causes the rotation axis to puncture the earth's surface at a different point – is revealed by a change in the apparent distance from the zenith of the polar point.

A distance of three metres on the surface of the earth corresponds to about a tenth of a second of arc, and classical astronomical observations were just accurate enough for Chandler to make his discovery. A special programme of observations with dedicated instruments was started at the beginning of this century in order to monitor continuously the polar motion. Firstly under the auspices of the International Latitude Service, and subsequently the International Polar Motion Service and the Bureau International de l'Heure, observations have enabled the pole to be tracked with a statistical accuracy of the order of a hundreth of a second of arc. An account of this work together with the various subtle problems which arise in the study of the polar motion can be found in an article by Murray.[12] Although some improvements in accuracy were achieved by classical astronomical methods in recent decades, the accuracy of these methods nevertheless fall just short of what is needed for a really detailed study. One problem, always present in ground-based observations, is the effect of the earth's atmosphere. Another is extremely relevant to the discussion in hand and is much more fundamental, touching the very concept of motion. I am referring to the *proper motions* of the stars. As Halley was the first to show (p. 252), the stars are in continual relative motion; the important point here is that the magnitude of the effect of the proper motions as observed from the earth is as large as the effect of the Chandler wobble. There are, for example, around two hundred stars with a proper motion of about 1 second of arc per annum.

We have reached the level of accuracy at which the use of the stars as markers of position becomes decidedly questionable. The proper motion of the stars is, in fact, a fundamental problem in all basic frame-of-reference work, especially that on the determination of the constant of precession.[11] As the authors of a recent article in the *Scientific American* put it:[13] 'Measuring the exact position of a point on the earth by referring to the visible stars is rather like being on a boat and trying to gauge one's position by observing the positions of other boats in a large, disorderly fleet.' This takes us straight back to the problems of relational position determination with which Newton was grappling in *De gravitatione*.

In fact, at this point it is worth quoting part of a passage which Mach wrote about two decades before Chandler discovered his wobble:[3]

> if we wish to apply the law of inertia in an earthquake, the terrestrial points of reference would leave us in the lurch, and, convinced of their uselessness, we would grope after celestial ones. But, with these better ones, the same thing would happen as soon as the stars showed movements which were very noticeable. When the variations of the positions of the fixed stars with respect to one another cannot be disregarded, the laying down of a system of co-ordinates has reached an end. It ceases to be immaterial whether we take this or that star as point of reference; and we can no longer reduce these systems to one another. We ask for the first time which star we are to choose, and in this case easily see that the stars cannot be treated indifferently, but that because we can give preference to none, the influence of all must be taken into consideration.

This passage is quoted now, not for the sake of the final sentence, in which Mach is adumbrating what later became known as Mach's Principle, but rather to emphasize the seeming impasse reached. Have we in fact reached some fundamental limit? Is it meaningless to talk of polar motion of the earth to an accuracy significantly better than that achieved by Chandler?

One way round the difficulty is to look for reference objects vastly further away than the stars, which have therefore much smaller proper motions. This can in fact be done by means of quasars, the extraordinarily powerful sources of radiation at great cosmological distances, and the technique of Very-Long-Baseline Interferometry (VLBI) (see Ref. 13). While this is at present the most satisfactory method for studying the polar motion, it does not in principle differ from the classical astronomical methods and can be called kinematic rather than dynamic. There is, however, a quite different method based on observation of artificial satellites of the earth. This is a genuinely dynamic method and a beautiful application of the idea which underlies Lange's definition of an inertial system.

A satellite launched into orbit around the earth is, if sufficiently massive and high enough not to be seriously affected by the earth's residual atmosphere, an excellent example of a body whose deviation from inertial motion can be accurately calculated and used in the modified form of Lange's dynamical realization of an inertial frame. It is, of course, subject to the gravitational field of the earth, which keeps it in its approximately elliptical orbit. It is, however, possible to model the effect of the earth and thereby take into account the extent to which it deflects such artificial satellites from their inertial paths. The residual atmospheric resistance can also be calculated. When this has been done, the satellites effectively define an inertial frame of reference, with respect to which the motion of the almost rigid earth can be measured. The orbits of the satellites weave, as it were, a kind of basket framework around the earth. By measuring the distance from stations on the earth to the various satellites (this can be done either by laser ranging to the satellites or by measuring the Doppler shift of radio signals transmitted at a precisely defined frequency by the satellites) it is possible to determine the change in orientation of the earth relative to the local inertial frame as materialized by the artificial satellites. In this way it is possible to track not only the motion of the rotation axis but also, of course, the rotation of the earth itself. This is intimately related to the problem of time-keeping.

This is not the place to go into further details. The aim here is merely to point out the conceptual and practical progress that has been made. Descartes posed the great question: what is motion? His own answer seemed hopelessly confused. He insisted that motion must be measured relative to material bodies, but this appeared to open up a whole can of worms. It seemed to deny the possibility of universal as opposed to a particular definition of motion – for a different motion will be obtained for each particular reference body that is chosen. Without universality, scientific treatment is impossible and must be replaced by mere description. The importance of Lange's proposal is that it very neatly deals with the problem of the particularity of relational definition of motion. Descartes was correct to emphasize that we must always measure motion relative to material bodies. The breakthrough to a scientifically significant relational definition of motion came with the recognition that the reference body only acquires utility when it is allowed, so to speak, *to follow the cosmic drift*. Because the cosmic drift is universal, each particular reference body which is allowed to follow it is, by virtue of the universality of its motion, *departicularized*. Of course, the departicularization is not complete, since the cosmic drift is only defined up to a constant rectilinear motion. However, this does not detract from the scientific utility of the motion which is defined in this manner. As the earth wobbles in the cage spanned by the network of orbiting satellites, its motion *relative to the universal drift* is changed in a way that is directly determinable. The wobble is what is of scientific interest, and it is perfectly amenable to universal definition even though any actual determination of the wobble must always be made through particular means.

Of course, the recognition that the existence of the cosmic drift makes possible a universal definition of motion (modulo the always undetermined uniform rectilinear component) leaves unanswered the question: whence comes the universal drift? This is the question that Mach's

Principle sets out to answer and will obviously concern us greatly in Vol. 2.

To conclude this section, let us look briefly at some of the remarkable evidence that there really is a universal drift – that the determinations of motion by different and particular frame-of-reference materializations give a universal result. It is this coincidence among the particulars, the commonality among the diverse, which provides the reassurance that science is a meaningful undertaking. It is simultaneously the phenomenon that ultimately is responsible for creating the impression that there is solid ground under our feet. It is no coincidence that the overthrow by Copernicus of the tenaciously held belief in the fixed earth eventually led to at least partial clarification of the physical factors responsible for creating this remarkably powerful illusion.

The polar motion and rotation of the earth is currently determinable in four essentially different ways. In the last resort, all of them rest on materialization of a frame of reference that realizes inertial motion, i.e., partakes of the cosmic drift. But the diversity of the materializations is remarkable. Artificial satellites of the earth come just about as close as is

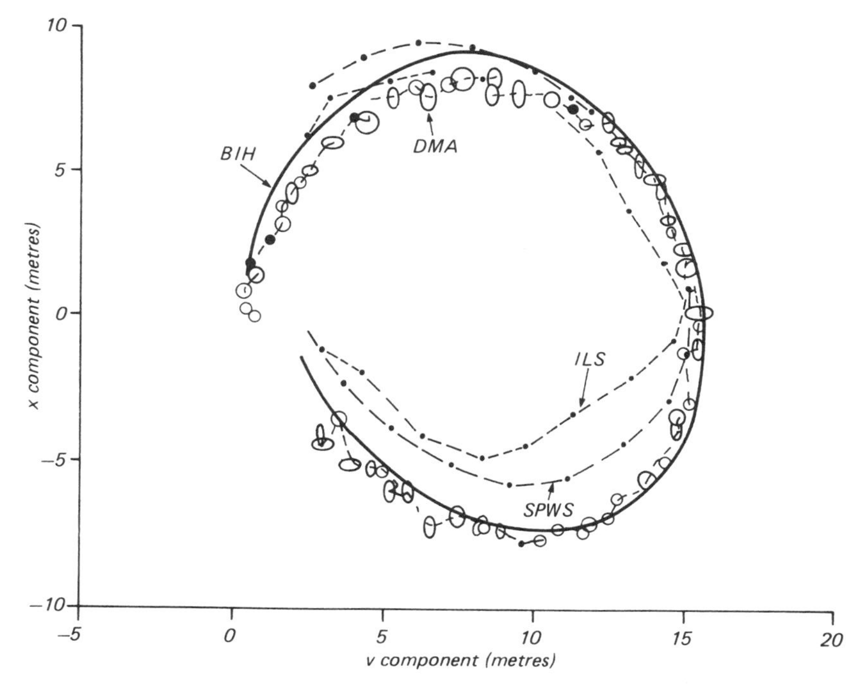

Fig. 12.2. Pole path for 1977. (Taken from C. Oesterwinter's article in: *Time and the Earth's Rotation* (eds D. D. McCarthy and J. D. Pilkington), IAU (1979), pp. 263–78.)

humanly possible to realizing Descartes' suggestion that motion is 'transference . . . from . . . bodies immediately contiguous'. Other frames of reference are materialized by the moon, the stars, and, finally, by the most distant objects observable in the universe – the distant quasars and radio sources. Figure 12.2 shows plots of the polar motion of the earth as determined by some of these different materializations. The remarkable agreement between the determinations is the coincidence that this study is all about: the coincidence between the earth's wobbles determined by the almost immediately contiguous satellites and those determined by the extragalactic sources at what for us is effectively the edge of space and time.

Mach spoke often of the profound interconnection of things. Figure 12.2 is a striking example of this.

12.5. Back to the Scholium

It was asserted in the previous chapter that the Scholium suffered from three main defects. The first, that it was distorted by being too specifically directed against the Aristotelian and Cartesian notion of position determination by immediately contiguous bodies, has already been adequately considered. We shall be here mainly concerned with the other defects – the concentration in the Scholium on a very special form of motion (circular) and the suppression of Galilean invariance. We shall also make good the claim in that chapter that in one respect at least Newton was on stronger ground than he appreciated.

This is in fact a good point at which to start. Right at the end of the Scholium Newton is forced to take a step which threatens to undermine the whole basis of the main contention of the Scholium – that there are dynamical criteria which enable us to deduce the true (absolute) motions from the observed relative motions (which Newton asserts, at the end of the Scholium, is the reason why he wrote the *Principia*: 'For to this end it was that I composed it [the *Principia*]'). It will be recalled that in the final paragraph [j] of the Scholium Newton imagined two globes, attached to each other by a cord, and rotating in space very far from any other bodies. He based his argument for the reality of absolute motion on the fact that, using the formula for the strength of centrifugal force, it is perfectly possible to establish the 'quantity of the circular motion' and also, through observation of the effect of application of forces to different faces of the globes, its determination (i.e., whether it is clockwise or anticlockwise when observed along the rotation axis from one side or other of the plane in which the globes move). However, to determine the actual direction of the globes at any instant, Newton is forced to use the distant bodies: 'from the translation of the globes among the bodies, we should find the determination of their motions'. We have seen too that in discussing the motions observed within the solar system Newton felt obliged to make the assumption that the distant stars are at rest.

This creates the impression that, as a dynamical theory, Newton's scheme is somehow incomplete – that, for example, the n-body problem can only be fully studied if there is a background of very distant stars to provide the ultimate frame of reference. However, Neumann's discussion of the n-body problem, summarized above, shows that, in principle, this is not necessary. By observing an isolated system for sufficiently long, it is possible to construct a coordinate system in which Newton's laws hold. This coordinate system is explicitly constructed from the observable relative distances and relative velocities and at any given instant has a perfectly definite and determinable position relative to the n observed bodies. In this respect, Newton's dynamics was even more impressive than he seems to have realized (or at least implied). On the other hand, this explicit construction of 'absolute axes' makes rather clear the extent to which the construction is not unique but subject to the arbitrariness associated with Galilean invariance.

Now to the distortion introduced by Newton's self-imposed limitation to circular motion. This is what brings the most serious confusion into the Scholium – and has mesmerized many people since then. The essential point of Newtonian dynamics is that, by its internal structure, it defines a family of frames of reference in which acceleration can be determined uniquely (acceleration is 'absolute' but velocity is not). Thus what Newton should have been doing was demonstrate the remarkable consequence of this ability to determine acceleration 'absolutely'. However, the limitation to circular motion led him to chase a will-o'-the-wisp; for what Newton really wanted to demonstrate was that velocity in absolute space can be determined, not acceleration. Frustrated by Corollary V in his attempt to determine rectilinear speed, he put all his hopes on showing how *circular speed* could be determined. But this seriously distorts the real content of Newtonian dynamics and makes the difference between rectilinear and circular motion appear excessively mysterious. A much truer picture emerges when one considers that circular motion is just *a very special case* of accelerated motion. It then becomes clear that the ability to determine a definite speed of circular motion is a consequence of the very special type of acceleration and is not really anything to do with the basic structure of dynamics, which is about the difference between accelerated and unaccelerated motion.

This can be illustrated as follows. Suppose two spaceships in empty space far from all gravitating matter. One is totally without power and therefore moves inertially; the other has power and is equipped with a modern inertial guidance device such as those used in aircraft for inertial navigation. The device is an accelerometer. If the powerless craft has on it a set of three mutually perpendicular gyroscopes, these define an inertial system in which the powerless craft is at the origin. The powered craft can now fly in a quite arbitrary way, accelerating and changing direction in all three dimensions just as often as the pilot wishes. The remarkable thing is that without using any observations apart from the readings of the accelerometer the pilot of the powered craft can in principle always immediately determine where he is in the inertial coordinate system defined by the gyroscopes on the unpowered craft. He has no need to look at either that craft or the distant stars. The gyroscopes within his own accelerometer are perfectly sufficient for the task. In particular, after an arbitrary journey, the pilot can return the second craft to the first along any route he cares to choose and will know exactly when he will meet up again. Indeed, there is not even any need for the first craft to remain inertial. They can both go off on arbitrary journeys and then agree to meet up again at a prearranged position in their original inertial frame of reference.

This fact emphasizes another remarkable point: there is no need for the frame of reference to be physically materialized for all time. Materialization at the beginning of the exercise is quite sufficient. Essentially, all the materialization is needed for is to ensure that at the start the two space craft are both 'plugged into' the same cosmic drift (in the conveyor belt analogy from Chapter 9 (p. 477), they must both be on the same 'conveyor belt' at the start of their manoeuvres; by means of the accelerometers they can then always get back to the origin on the 'conveyor belt'). In Vol. 2 we shall find a precise mathematical characterization of the remarkably small degree of 'extraneous materialization' that Newtonian dynamics requires to become a complete self-contained scheme.

Now in all the manifold possible journeys, with arbitrarily varying velocities, the velocity of the powered craft *in the inertial frame of reference materialized by the gyroscopes of the other craft* is always a well-defined quantity. So is the position. Both velocity and position are determined by integration. In the Scholium, Newton created a very misleading impression by considering a very special case. All such journeys would have served to prove his point that acceleration is absolute (and, equally remarkable, that within any definitely materialized inertial frame of reference both position and velocity can be determined by purely dynamical means), but he wanted to prove velocity is absolute and, probably confused to some degree, chose highly singular motions among the infinitude that are possible. By selecting uniform circular motions, he created the spurious impression that he had achieved direct contact with the absolute.

But one may ask, what precisely is the speed of circular motion that Newton claimed he was able to determine absolutely? Certainly not the speed in absolute space, since Newton can hardly have believed that either the bucket experiment or the one with the two globes would have any different outcome in a frame of reference in uniform rectilinear motion relative to the original frame in which the experiment is performed. In fact, Newton's persistent refusal to mention this fact – and his having overlooked it is barely credible – is the clearest evidence of disingenuousness on his part though there does also seem to be an element of straightforward misunderstanding. On the one hand Newton was trying to prove more than is admissible (that velocity is absolute), while, on the other, his self-imposed limitation obscured the truly remarkable power and universality of his dynamics. The instantaneous tension in the cord joining Newton's two globes is determined by their masses and instantaneous mutual accelerations. Only in the artificial case of constant circular motion can one pass from the accelerations to a uniquely defined circular speed. Newton's result is an artefact of special conditions, not a characteristic consequence of fundamental dynamics.

The root cause of his trouble seems to have been the attempt to dispense totally with matter when defining motion. Wherever specific applications of dynamics are made, the inescapability of a definite materialized frame of reference is unavoidable. Without it all equations are void of content. With it, there is no limit to what dynamics can achieve. (The ability to guide intercontinental ballistic missiles onto targets thousands of miles away by inertial guidance with frightening accuracy is one of the most striking illustrations of this.) The temptation to dispense with the particular materialization is very great. It seems rather odd that it is always necessary to have some matter defining the frame of reference. The very fact that, in principle, any matter will do helps to create the impression that the materialization of the frame of reference is superfluous. Lange, in particular, must be given the credit for seeing that this is not the case. He supported his case with a particularly apposite quotation from Mach's first published book:[3] 'Because a piece of paper money need not necessarily be funded by a definite piece of money, we must not think that it need not be funded at all.'

In this inflationary age, in which the funding of paper money by gold has been abandoned, this remark may make its point even better than it did in 1872!

Two small points to end this section. Readers familiar with the history of the discovery of quantum mechanics will recall that Schrödinger was initially of the firm belief that his wave function had a real physical meaning, representing a mass or charge density. He attempted to prove this by showing that the wave function has no spreading in time. Unfortunately, he showed this for the very special example of the harmonic oscillator, which is the one dynamical system for which there is no spreading. By choosing such a special example, Schrödinger seriously distorted the true content of his own theory.[14] Newton's restriction to uniform circular motion in the Scholium had a rather similar effect. Indeed one could argue that the relevant passages in the Scholium represent the only serious conceptual confusion in the entire *Principia*.

The other point concerns the thought experiment with the space craft. The reader may have noticed the crucial restriction 'far from all gravitating matter'. As soon as the experiment is attempted in nonconstant gravitational fields, the situation is completely changed. The second craft will then not be able to find its way back to the first without additional information about the real contingent world around it. The discussion of this significant complexity is deferred to Vol. 2.

12.6. Huygens and absolute motion

At this point it is appropriate to consider briefly once more the difficult question of what precisely were Huygens' views about the fundamental nature of motion. We shall see that his views changed, but that he never appears to have arrived at a coherent standpoint. The most extensive discussion of this question of which I am aware is to be found in Vol. 16 of the *Oeuvres Complètes* of Huygens, where the editors give several relevant texts together with a useful commentary.[15] Schouten has also published and discussed some of the more important texts.[16]

We can start this discussion by noting that throughout his life Huygens does not appear ever to have questioned the validity of the law of inertia as the most fundamental law of motion. More importantly, he never seems to have had more than passing doubts about the manner in which it should be formulated. We have already encountered his formulation in the work on collisions. Essentially the same formulation occurs in the opening of the second part of the *Horologium Oscillatorum*, in which he says that if it were not for gravity and air resistance 'every body that has acquired a certain motion would continue to move forward with the same speed in a straight line'.[17] As has been emphasized several times, such a statement is meaningless unless the motion in question is referred to some quite definite frame of reference. Huygens was at least partly aware of this problem, as is evident from an unpublished note, which may date from 1668, in which he remarked that 'the motion of a body can at the same time be truly uniform and truly accelerated if its motion is referred to other different bodies.'[18] As we saw in Chap. 11, this problem was the core of Newton's objection to Cartesian relativism; it was expressed very clearly in *De gravitatione* but unfortunately not so explicitly in the Scholium. In Huygens' case it seems never to have dominated his thought. In fact, in the period after the publication of the *Principia* this difficulty seems to have escaped him completely; for we find that on the one hand he asserts the complete relativity of motion with increasing conviction but yet continues to suppose that undisturbed bodies travel in straight lines with uniform speed. There is therefore a serious contradiction at the heart of Huygens' thought on this subject.

Let us now briefly review the main stages through which Huygens' ideas passed. At a very early stage in his life he became totally convinced of the validity of the Galileo relativity principle. This development has been well documented by Gabbey.[19] The conviction in the validity of the Galileo relativity principle never left Huygens throughout his entire life, and we have seen in Chap. 9 the brilliant use he made of the principle.

A new stage in his thought occurred after he had found the correct formulas for centrifugal force in 1659. His reaction to his discovery seems to have been exactly the same as was Newton's several years later – because the strength of the centrifugal force is mv^2/r, the circular speed of a stone whirled around on the end of a string can be deduced from the tension in the string without reference to any external bodies. Huygens therefore concluded that although rectilinear uniform motion is quite undetectable through dynamical phenomena, circular motion does truly exist and can be measured in various ways. This appears to have been his opinion for about a quarter of a century, or even longer. Particularly interesting in this connection are his reflections, dating from around 1686–87, on the cause of the oblateness of the earth, which, at that time, Huygens believed could not be due to rotation relative to the fixed stars but must be a consequence of a motion of rotation 'pure and simple'.[20]

The final stage in Huygens' development was brought about by the publication of Newton's *Principia* in 1687. This had a great impact on Huygens and stimulated him to publish his own ideas on the nature of gravity.[21] It appears that Newton's outspoken advocacy of absolute space and absolute motion also prompted Huygens to reconsider his views on the nature of motion and he came round to the view that all motion is relative. This process is especially well documented in an exchange of letters between Leibniz and Huygens which took place in 1694, the year before the latter's death. The correspondence has been published by Gerhardt.[22] In a letter of 29 May 1694 Huygens wrote that 'there is no real but only relative motion. I hold this to be quite certain, without wishing here to dwell on the arguments and experiments of Mr Newton. I see that he is in error and am curious whether he will not withdraw his opinion in the new edition of this book . . . Descartes did not understand enough of this matter.' In his reply of June 1694, Leibniz commented: 'I seem to remember that you yourself, Sir, were once of the same opinion as Mr Newton with regard to circular motion.' Leibniz is here referring to a discussion which he had had with Huygens in Paris more than twenty years earlier. Huygens answered on 24 August 1694 as follows: 'With regard to absolute and relative motion, I do admire how you have remembered that with regard to circular motion I was in earlier time of the same opinion as Mr Newton. That is correct, and it is only two or three years ago that I found a position closer to the truth.'

Huygens did not elaborate on this point except to say that he did not at all agree with Leibniz's idea, according to which 'among several bodies moved relative to each other each does have a definite degree of motion or true force'. In thus rejecting Leibniz's thesis (which will be discussed further in Vol. 2), Huygens committed himself to complete relativity of motion. How did he reconcile this with the facts of circular motion?

To answer this question, we must turn to his unpublished papers written in the last years of his life. The most revealing of these is perhaps the one reproduced as IV in Schouten's paper.[16] It is also published in the *Oeuvres Complètes* (Vol. 16, pp. 232–3). The date is uncertain but is certainly late. The paper contains several sentences which reveal how deeply Huygens sensed the pure relativity of motion; he anticipates the sentiments that we shall find in Vol. 2 in Berkeley's and Mach's writings. The paper begins with the statement that 'motion between bodies is only relative'. He points out that in the case of a single body (alone in the universe) it is quite impossible to conceive any difference between motion and rest of such a body. Very characteristic is the following question: 'If space extends infinitely in all directions, what is then the definition of a place or of rest?' By way of answer he continues: 'It is true people say that the fixed stars of the Copernican system may be at rest. Well, suppose they are all at rest relative to each other; but, taken all together with respect to what other body are they at rest or in what respect do they differ from bodies that move very rapidly in a certain direction? Therefore, one can neither say that a body is at rest in infinite space nor that it moves in that space, and thus rest and motion are only relative.'

However, when we come to Huygens' explanation of the apparently absolute nature of circular motion we find a deep confusion. He says that 'a rotational motion is a relative motion of the parts, which are driven in different directions but are held together by a string or a connection.' He continues: 'But can one say two bodies move relative to one another if their separation remains the same? This is perfectly possible provided an increase in the separation is prevented. In fact, on the circumference there exists opposite relative motion.' This rather remarkable statement is explained by Huygens in a figure (Fig. 12.3) and a footnote, in which he says: 'Let body A move along the straight line or rod AB and body C move along the straight line CD, which has the same direction. When A arrives at B and C at D the two bodies are certainly moved in opposite directions; however their separation changes very little and almost not at all. This is what happens in the case of rotating connected bodies.'

It seems almost certain that this is the argument which Huygens had in mind when he wrote to Leibniz and asserted that he had a counter to

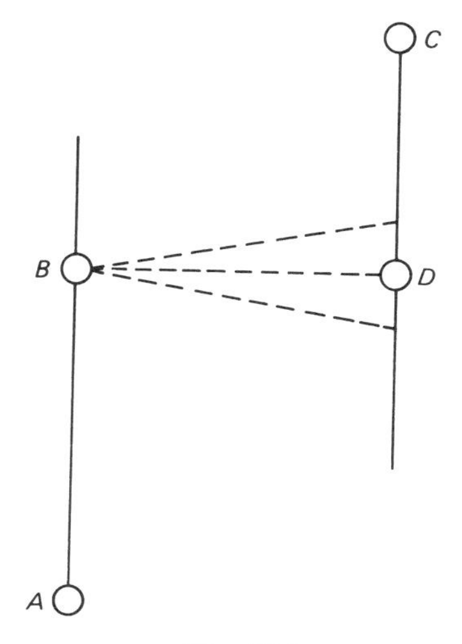

Fig. 12.3.

Newton's argument for absolute motion. Huygens' argument is perhaps the most striking example in the entire literature on the absolute/relative debate of the tyrannical grip exerted on the mind by the intuitive notion of space. With the very breath with which he asserts the relativity of motion Huygens uses kinematic concepts that cannot be formulated without some underlying concept of space. For with respect to what are the lines AB and CD straight?

Huygens clearly believed very deeply in the complete relativity of motion but it is evident that, unlike Newton, he did not at all grasp the full implications of such a doctrine. His use throughout his life of the law of inertia in an essentially Cartesian form demonstrates that. He understood very well Galileo's principle of relativity for uniform motions and was therefore correctly unimpressed by Newton's argument that each body has a *unique* absolute motion. And because Newton did not spell out explicitly and unambiguously in the *Principia* the inherent contradiction between complete Cartesian relativism and the law of inertia Huygens was not alerted to his own confusion. Out of this muddle came eventually a powerful stimulus to the search for an alternative, completely relational theory of motion. Huygens' espousal of complete relativity became known publicly, especially in the nineteenth century following the publication of his correspondence with Leibniz, but the confusion underlying this espousal did not see the light of day until much later.

12.7. Mach's operational definition of dynamical mass

Before concluding this volume, we consider the other important clarification of Newtonian concepts that occurred in the nineteenth century: Mach's definition of mass. This is particularly important in view of the part played by the mass concept in Einstein's understanding of Mach's Principle.

It is interesting that at the end of his habilitation address Neumann said that, the concept of motion having been clarified by the introduction of *Body Alpha* and that of uniform flow of time by means of the time scale defined by inertial motion, the clarification of the concepts of Galilean–Newtonian dynamics had been reduced to a completely straightforward exercise. In his view, the only tricky point which remained concerned the definition of mass, into the discussion of which he declined to enter 'as it would take us too far to concern ourselves further with these matters'. As it happens Mach had already given a lecture on precisely this topic a year before Neumann's habilitation, and he published the essence of his proposal four years later in 1872 as an appendix to his book *History and Root of the Principle of the Conservation of Energy*.[3] Quite apart from its conceptual importance in the context of Vol. 2, it is well worth looking at Mach's operational definition of mass in some detail since intellectually it is a most satisfying piece of work and prior to Einstein's definition of simultaneity is the most striking example of how the interpretation of phenomena well known can be completely changed by an appropriately chosen operational definition. Of course, as noted at the beginning of this chapter, Mach's definition of mass merely resulted in a reinterpretation of already known physics, whereas Einstein's definition of simultaneity opened the door into a world as unexpected as the one inadvertently discovered by Copernicus when trying to save uniformity of motion in the heavens. What Mach's definition shares with Einstein's is deftness and the way it puts familiar things in a totally different light.

It will be worth looking briefly at the development of the mass concept in a fairly long historical perspective. The interesting question is: Why did the *physical* mass concept take so long to emerge? The reasons for the tardy development of physical concepts in general and the mass concept in particular have already been noted in the early chapters: paucity of suitable phenomena and the alluring prospect of explanation by means of geometry. We have seen how the geometrism that preceded Aristotle threatened to retard the study of motion by providing what seemed to be an exceptionally powerful explicatory tool. It was just the same in the seventeenth century. As regards explication of the world, *shape* and *size* promised to achieve so much there seemed little need to develop any further concepts. It took the discovery of dynamics to show that nothing could be further from the truth. What moved Aristotle more than anything else in the direction of genuine dynamics was probably a gut awareness that three-dimensional geometry by itself could never explain why motion occurs at all. Motion is something genuinely new. Thus, where the atomists and Plato had a concept of matter determined almost exclusively by extension and shape, Aristotle felt the need to introduce attributes of matter that related directly to motion. Thus was born the concept of *occult quality* – an intrinsic property by which a body's motion is directly determined; occult because there was nothing in the appearance of the body to tell us that it possessed such a quality. The intellectual battle which raged in the seventeenth and early eighteenth centuries was all about such concepts, with the pure mechanists like Descartes and Huygens asserting that everything must be explained by means of the only comprehensible principles available to man: shape, size, and motion. When Newton introduced universal gravity, the concept was immediately denounced as an occult quality. It is ironic that the establishment in the *Principia* of the correct principles of dynamics passed almost without comment – an earth-shaking event if ever there was one, yet it aroused remarkably little interest compared with what was seen as the pernicious reintroduction of Aristotelian occult qualities. In this connection, the reader is particularly recommended the reading of Cotes' Preface to the second edition of the *Principia*;[23] it captures the flavour, intensity, and bitterness of the debate – and is also excellent philosophy of science.

Cotes and Newton won the debate and thereby ensured that the concepts of active and passive gravitational charge (as they are now sometimes called) became the first universally accepted *and clearly defined* physical quantities to be associated with bodies. The evidence provided by the phenomena and marshalled with such mastery by Newton was overwhelming: whatever may be the ultimate mechanism behind it, the possibility of associating an operationally quantifiable *attractivity* (as it may be called) in matter was simply undeniable. The empiricopositivistic approach to this question adopted by Newton and Cotes is remarkable and anticipates the attitude that Mach brought to the whole field of science: the phenomena manifestly permit the definition and determination of gravitational attractivity; therefore, let us accept it is a fact and get on with the extremely interesting job of showing just how many and diverse phenomena in the world can be explained by this one single property. Further speculation on its origin is at the moment out of place; it may be either an ultimate property irreducible to anything else or we may eventually find an explanation of it in terms of something even more primitive.

With the full glare of the spotlight focussed on gravity, the history of the concept of mass is most curious: it became established as a genuinely physical (as opposed to a purely geometrical) concept without anyone properly noting the fact. After more than two hundred years of brilliantly successful use, people suddenly realized that the mass concept presented a decided enigma. Neumann clearly recognized the existence of the problem, though he failed to provide the answer.

For our purposes the natural starting point for the discussion of mass is Kepler, though the reader is recommended the reading of Jammer's book on the mass concept[24] for the earlier history. It is particularly worth noting how philosophers struggled for centuries, indeed millennia, in a more or less fruitless attempt to characterize the essence of matter. The final appearance of a satisfactory definition through *dynamical behaviour* is really most remarkable and surprising when seen in the light of this philosophical endeavour. What is interesting about Kepler is the way in which his changed conception of motion led very naturally to the introduction of genuine physical concepts. As we have seen, the great early astronomers made no attempt to explain the celestial motions; they merely sought to describe them. Kepler's search for a cause of the motions led him to introduce first of all the concept of force. His instinct, prompted by the behaviour of the planets, then told him that individual bodies must be characterized by some quantity which measures its resistance to the applied force. In this way he introduced the idea of a physical quantity intrinsic to each and every body, one moreover that is present in each and every part of any given body, so that the total 'laziness' of a body is the sum of the 'lazinesses' of each of its parts. It is worth emphasizing that this notion of the sum of the individual parts producing the integrated effect is one of the most characteristic features of the transition from medieval teleology to a modern scientific viewpoint. In this respect, Kepler was much more modern than Galileo and perhaps even than Hooke.

It is a measure of Kepler's instinctive feel for the needs of dynamical theory that it is only necessary for one to replace the concept of laziness with respect to *motion* by laziness with respect to *change in motion* for one to arrive at precisely Newton's concept of inertial mass. Kepler's major share in the development of this concept is eloquently attested by the fact that Newton was not ashamed to take over Kepler's name to denote the correctly identified property. As he said, it may, 'by a most significant name, be called inertia'. (For an interesting discussion of how Newton came across the concept and his generally very unfair attitude to Kepler, the reader is recommended Cohen's paper.[25])

This is the point at which to make an important remark. It relates to the two different uses made of the word *inertia*. There has in fact been a subtle and potentially confusing shift in the commonly accepted meaning of this word. Both Kepler and Newton used the word for the capacity of matter to resist changes in its state. With Kepler this is quite explicit: he merely took over the Latin word for *laziness*. For Kepler it was quite clearly what we now call a scalar quantity, a number characterizing an intrinsic property of matter. When Newton first introduced the word into his own vocabulary, in *De gravitatione*,[26] he followed Kepler very closely. He said: '*Inertia* is force within a body, lest its state should be easily changed by an external exciting force.' Note that inertia is a resistance to change of state. What state is not specified, though it is obvious Newton is thinking in terms of change from one state of motion to another rather than the simple Keplerian idea of the transition from rest to motion. The point which needs to be made is that the concept of resistance to change of state is by no means the same thing as the state. Indeed the transition from Kepler to Newton illustrates this point perfectly: the concept of laziness survived but the state (or rather states) changed. *A priori*, there is absolutely no reason why the states between which a resistance barrier must be overcome are states of uniform rectilinear motion. They could be states of uniform circular motion or states of uniformly accelerated rectilinear motion – or indeed any unambiguously defined state of motion.

By the time Newton came to write the *Principia*, the states between which the resistance is manifested were made quite explicit. Here is Newton's Definition III and part of his gloss on it; this shows how closely he followed Kepler:

The vis insita, *or innate force of matter, is a power of resisting, by which every body, as much as in it lies, continues in its present state, whether it be of rest, or of moving uniformly forwards in a right line.*

This force is always proportional to the body whose force it is and differs nothing from the inactivity of the mass, but in our manner of conceiving it. A body, from the inert nature of matter, is not without difficulty put out of its state of rest or motion. Upon which account, this *vis insita* may, by a most significant name, be called inertia (*vis inertiæ*) or force of inactivity.

Throughout the *Principia*, Newton consistently uses words like *vis insita*, or *vis inertiae* to refer to this resistance. There is no suggestion at all in the *Principia* that the word *inertia* should be used to describe the *state*. It is the resistance to the change of state, and its magnitude is characterized by what is now called the *inertial mass*, the *m* which appears in the modern statement of Newton's Second Law in the form

$$m\mathbf{a} = \mathbf{F}.$$

Long after Newton was dead, the connotation of the word *inertia* was to a considerable degree transferred from the *resistance* to the *state*. What Newton called *Lex Prima*, the First Law, and stated in the form

Every body continues in its state of rest, or of uniform motion in a right line, unless it is compelled to change that state by forces impressed upon it

became known as the *Law of Inertia*. When the *Oxford English Dictionary* was compiled in the late nineteenth and early twentieth centuries, *inertia* was defined as 'that property of matter by virtue of which it continues in its existing state, whether of rest or of uniform motion in a straight line, unless that state is altered by an external force'. This is a very major shift of meaning and one that has passed almost unnoticed. It is interesting that *OED* quotes (in Latin) Newton's definition of *vis insita* given above (including the words *potentia resistendi*) but does not give any definition expressing the idea of a quantitative resistance to the change of state. In fact, among the dictionaries I have consulted Wallenquist's *The Penguin Dictionary of Astronomy* is alone in adding a second meaning:[27] 'the resistance to . . . change of state'.

It would clearly be quite inappropriate to attempt to reverse this shift of usage. An adequate distinction can be made by using *inertial mass* to refer to the resistance and *inertial motion* to refer to the state. Further evidence for the fact that the two concepts are totally different and need to be clearly distinguished is the circumstance of their origin. Kepler never really had an inkling of inertial motion. He in no way contributed to the recognition that it is a fundamental element of dynamics; that was entirely the service of Galileo and Descartes. They, in contrast, had no part in introducing the Keplerian concept of laziness, though Descartes used somewhat similar ideas and served as a useful means of communicating the concept from Kepler to Newton (see Cohen[25]). Perhaps the distinction could be made clear by referring to *Keplerian inertia* and *Galilean* (or *Cartesian*) *inertia*. This would certainly emphasize the immense utility of both concepts and do some sort of justice to Kepler, whose major role in developing the science of dynamics has been most regrettably underestimated. His contribution was by no means restricted to discovering the laws of planetary motion, crucial as that was.

More important than all this is that Einstein apparently failed to distinguish between the Keplerian and Galilean–Cartesian concepts. This will be a major concern in Vol. 2.

The remainder of this section is devoted to the curious history of Keplerian inertia – the manner in which the concept of inertial mass became established and clarified. This is the story of the breaking of the tyranny of geometry.

We saw in Chap 6 that Kepler appears to have drawn an explicit distinction between the amount of matter, or inertia, in a body and the volume which the body occupies. In this he was exceptional. So far as I can make out, nearly all the great natural philosophers of the seventeenth century simply identified the amount of matter in a body with the volume it occupied. None of them seems to have felt that there was anything more to matter than extension and shape. This is explicit in Descartes but seems to have been widely and unquestioningly accepted. Two main mechanisms were recognized for the explanation of the great diversity of mechanical behaviour, above all weight, observed in matter. Within the atomic conception, denser bodies were such by virtue of the fact that the atoms within them were more closely packed together. There was no suggestion that the individual atoms might have different intrinsic densities: they were all chips off one monolithic, perfectly uniform block. Very characteristic is the example, quoted by Westfall,[28] of Gassendi, who 'employed the image of a bushel of wheat which can be shaken to cause the kernels to fit more closely together'. Such a mechanism was not, of course, open to the plenists like Descartes and Leibniz. They had to rely on the differing ability of matter to give way to other matter, this difference being ultimately explained by the shapes, sizes, and above all the motions of the various microscopic bodies of which macroscopic bodies were supposed to be composed.

The words most commonly used to describe quantity of matter reflect this instinctive idea: *corpus* (body), *moles* (mass or bulk), and *massa* (lump, mass). When Newton first identified the modern momentum (mass times velocity) as the most fundamental dynamical concept, it seems that he regarded the mass as simply the volume of the matter. This comes out in a passage in the *Waste Book* in which Newton discusses collisions between unequal bodies. He first discusses the case of equal bodies and then the case when one is twice the other. He represents this by a figure in which one body is simply shown as having *twice* the volume of the other.[29]

We now come to the *Principia* and the definition of mass which opens the book. First, a general comment: my impression, after reading through the *Principia* while watching out for the way in which Newton used the word mass and its synonyms, is that Newton had simultaneously two quite distinct and unrelated concepts of mass without being fully aware of the fact. In many cases it appears that his mass concept was just the same as in the *Waste Book*, i.e., that in the last resort what counts is the (close-packed) volume of matter, conceived as chips hewn by God from a perfectly uniform block. Such an interpretation is supported by the gloss (quoted below) on Definition I, in which Newton says that he will refer to the quantity of matter 'under the name of body or mass' ('*sub nomine Corporis vel Massae*'). This use of the word *body* has a strong connotation of volume, which would be entirely consistent with Newton's early work in the *Waste Book*, as well as with the dominant concept of his age.

Alongside this concept, there are every now and then hints of a completely different way in which mass could be defined. The clearest example comes in Book III of the *Principia* (Prop. VI, Cor IV) where Newton says: 'By bodies of the same density, I mean those, whose inertias are in the proportion of their bulks.' It is very remarkable that in the early *De gravitatione* Newton gave an explicit definition of density along exactly these lines:[30] 'Bodies are denser when their inertia is more intense, and rarer when it is more remiss.' This is clearly a totally different approach to the problem and amounts to a *dynamic* definition of mass: the amount of mass is measured by the resistance it offers to a given force. We shall see that this is a clear anticipation of Mach's definition of mass.

There is nothing implausible about the suggestion that Newton had simultaneously two quite different mass concepts. He worked perfectly happily with no less than three different concepts of force and never once put a foot wrong. So far as I can understand his mind, he had such a sure and instinctive grasp of the way to solve problems in dynamics that he never had to take particular care over his definitions. By this, I do not want to suggest that he gave only cursory consideration to the formulation of his definitions and axioms; he certainly did spend time on them. What I mean is that successful solution of problems did not depend for Newton on complete prior and explicit clarification in words of his concepts. He somehow knew what needed to be done and did it. Moreover, his terminology lagged behind his insights; he used the old words to describe phenomena and concepts for which he had meanwhile acquired a much clearer dynamical understanding. This could well be what happened to the concept of mass.

Seen in this light, let us now look at Newton's definition of mass. On the face of it, the most important book in the history of science opens with one of the crassest of vicious circles:

The quantity of matter is the measure of the same, arising from its density and bulk conjointly.

Of this definition, Mach made a famous and much quoted criticism:[31] 'With regard to the concept of "mass", it is to be observed that the formulation of Newton, which defines mass to be the quantity of matter of a body as measured by the product of its volume and density, is unfortunate. As we can only define density as the mass of unit volume, the circle is manifest.' Clearly, everything hinges on the sense in which Newton understood the word *density*. If when he wrote this definition Newton had in mind density as defined above by means of inertia, then the problem disappears completely. But if that is the case it is most remarkable that Newton did not make that clear; apart from anything else, density used in such a sense would be entirely novel, as indeed is confirmed by the fact that in the quoted corollary Newton recognized the need to explain what he meant by density in that particular context. Moreover, Newton's gloss on Definition I seems to suggest that he had in mind the older usage from the *Waste Book*:

Thus air of a double density, in a double space, is quadruple in quantity; in a triple space, sextuple in quantity. The same thing is to be understood of snow, and fine dust or powders, that are condensed by compression or liquefaction, and of all bodies that are by any causes whatever differently condensed. I have no regard in this place to a medium, if any such there is, that freely pervades the interstices between the parts of bodies. It is this quantity that I mean hereafter everywhere under the name of body or mass. And the same is known by the weight of each body, for it is proportional to the weight, as I have found by experiments on pendulums, very accurately made, which shall be shown hereafter.

While this passage is far from decisive, it does appear to me consistent with the idea that in Newton's mind higher density is a consequence of closer packing of pieces of matter that in themselves are made of identical matter.

However, precisely what Newton meant is not all that important; indeed, examination of Jammer's survey[32] of what the large number of interpreters have made of Newton's definition and gloss persuades me that the rather meagre reward is not worth the effort. I suspect that the gloss is a vestige of the mechanical origin of dynamics, a last remnant of Descartes' London-Underground-in-the-rush-hour concept of the world.

What is clear from reading Newton's documents dating from the period just before the *Principia* is that the discovery of the law of universal gravitation was of decisive importance for his recognition of the need for a concept of mass distinct from weight. It was undoubtedly this insight which led to the promotion of the mass concept to pride of place at the head of the definitions in the *Principia*. It was forced upon Newton by his realization that the gravitational force (and hence weight) decreases with distance from the attracting body. In the gloss on his Definition VII he says 'the force of gravity is greater in valleys, less on tops of exceeding high mountains.'

At much the same time as Newton recognized clearly the need for the mass concept there also occurred the shift in his conception of the nature of the *vis insita* that was mentioned in Chap. 10 (p. 535). This process has been well documented by Herivel.[33] As was pointed out in Chap. 10, the young Newton conceived the force of inertia rather in the manner of Buridan's impetus theory, a force that keeps a body moving along at uniform speed on its Cartesian straight line through absolute space. But as he worked on the preparation of the *Principia* this original medievalism (which never adversely affected his ability to solve problems) gave way to the more mature (and, in fact, Cartesian) realization that resistance to change in motion is to be distinguished from the motion that takes place between interactions between bodies. On p. 679 we gave the start of Newton's gloss on his Definition III of *vis insita*, or *vis inertiae*. The gloss continues with these words:

But a body only exerts this force when another force, impressed upon it, endeavors to change its condition; and the exercise of this force may be considered as both resistance and impulse; it is resistance so far as the body, for maintaining its present state, opposes the force impressed; it is impulse so far as the body, by not easily giving way to the impressed force of another, endeavors to change the state of that other.

In conjunction with Newton's instinctive comment from near the end of the *Principia*, quoted earlier in this chapter, to the effect that density of mass is to be determined by inertia, this passage shows clearly that Newton had come to distinguish, if not yet completely clearly, between the state of inertia (uniform rectilinear motion) and the resistance to change in that state.

In the post-Newtonian period the acceptance of the mass concept as an independent physical concept seems also to have been not a little helped by Newton's introduction of the Latin synonym *massa* alongside *corpus* or *quantitas*, with their Cartesian mechanical overtones. Mass liberated itself little by little from its predynamical geometrical origin and came to be understood as that 'something' in matter which measures its inertia. It was transformed from meaning volume into a substance, the Atlas-type prop to support the observable attribute of resistance to acceleration. Like absolute space it became a metaphysical concept – something we fancy we can conceive clearly in our mind, though we are very hard pressed when called upon to explain it in unambiguous terms. We find ourselves going round in vicious circles like Newton.

In the mid-1860s, when he was in his late twenties, Mach set out to clear out all the metaphysical cobwebs from the kitchen of physics. Totally committed to empiricism, he began his mature work in the decade after Riemann completed the overthrow of *a priori* geometry. A few decades earlier Gauss had performed one of the most deeply symbolic experiments in history: he had made a serious attempt to see if there were 180° in a triangle. This was indeed an age to challenge all the old preconceptions.

Mach's view of Newton was that he had instinctively grasped most if not all the essentials of dynamics but that his presentation of the subject was deficient in failing to make clear what were the key observable phenomena on which the most important concepts of dynamics rested. Newtonian mechanics was the culmination of a long historical process in which metaphysics and speculative mechanics after the manner of Descartes had vied with empirical observation for the upper hand. Both strands were very pronounced in Newton himself. In the final product, empiricism got the upper hand but metaphysics fought a strong rearguard action; it appeared to make a greater contribution than it in fact did, mainly because Newton continued to use old words for new insights.

More clearly than any of his contemporaries, Mach realized that any successfully functioning scientific theory or discipline must in the last resort rest on experimentally observable phenomena. For otherwise it could not make meaningful and nontrivial statements about observed phenomena. In particular, most of the key concepts in science are those which permit scientists to associate definite numbers with the concepts. Examination of ways in which these numbers are actually determined would then give one a clue to the phenomena which provide the ultimate justification for the concepts themselves. In a sense, the method of determination of the value of say, the mass, will then be simultaneously the definition of the mass concept. We have already seen examples of this in Neumann's definition of inertial time scale and Lange's definition of inertial system (all position observations are made relative to observable matter, therefore an inertial system must be conceptually materialized, to match the experimental practice). Mach developed this technique into a systematic art – the art of the *operational definition* of the fundamental concepts of physics.

Mach's idea was to build up the concepts of dynamics systematically, starting with the most basic geometrical elements and progressing from them to the more sophisticated dynamical concepts. Although well aware of the revolution wrought by Gauss and Riemann in the conceptual foundations of geometry, Mach did not set himself the hopelessly ambitious task of sorting out geometry as well. Instead, he accepted distance measurement as given and also, for the purposes of the clarification of the concepts of mass and force, that clocks exist and that the distant stars define a frame of reference with respect to which all motions are to be defined. Then the first observational fact on which dynamics rests is the law of inertia: In the frame of reference defined for practical purposes by the stars, bodies are usually observed to travel in straight lines with uniform speed. So far, there is little difference here from Newton; it is on the transition from velocities to accelerations that Mach parts company from Newton, who, it will be recalled, made *force* the primary concept of dynamics. However, even Newton was clear that the existence of forces must be deduced from the observed accelerations they induce in bodies. Developing his systematic approach, Mach insists that the observable accelerations should come first. When this is done, it transpires that the definition of mass (and force) cannot be separated from the content of Newton's Third Law. For the observed facts are as follows: Bodies normally travel on straight lines relative to the stars but, under suitable conditions, *two or more bodies can mutually accelerate each other.* Mach's great insight was that it is the very special law which governs the manner in which this mutual acceleration takes place which makes possible the meaningful definition of mass.

Let us consider the observationally simplest case of interaction between bodies: collision of two bodies, 1 and 2, on a straight line. Experience soon shows that the mere size of the individual bodies is not at all decisive for the outcome of a collision between 1 and 2. The essential information extracted from a collision is the change in velocity that each body undergoes as a result of the collision. Let the changes be δv_1 and δv_2. Even if collisions between only two bodies are considered, an interesting observational fact emerges: if the precollision velocities are varied, the mutations δv_1 and δv_2 vary too, but remarkably the ratio $\delta v_1/\delta v_2$ is (provided we consider only nonrelativistic speeds) always the same and equal to some constant negative number, which we shall call $-C_{12}$ ($C_{12} > 0$). Suppose we now take a third body, 3, and perform collision experiments between 1 and 3 and 2 and 3. We obtain corresponding constants C_{12} and C_{23}. Now the really striking fact is that C_{12}, C_{13}, and C_{23} are found to satisfy a very particular relationship:

$$C_{12}C_{23} = C_{13}. \tag{12.4}$$

A priori, there is absolutely no reason why the individual constants C_{ij} should exist at all, nor why a relation such as Eq. (12.4) should hold between them. The point which Mach makes is that it is the existence of such relationships *which make possible the introduction of the mass concept*. Namely, the outcome of all collisions, elastic and inelastic, can be characterized by associating with each body i an intrinsic number m_i. These numbers are such that $C_{ij} = m_i/m_j$. In this representation, all the numbers m_i, the *masses*, are determined only up to a common constant factor, which is fixed by taking one of the bodies, say 1, as the *unit of mass*.

Mach had a wonderfully uncompromising attitude to the basic concepts of physics. Time and again he insisted that there is absolutely no *a priori* (merely logical) reason why the phenomena unfold in the way they do. Before we make collision experiments, there is no way we can predict how bodies will behave in experiments. We cannot form an idea in our minds of what bodies are like and on that basis predict in advance what the outcome of collisions will be. This was the route chosen by Descartes – with disastrous results. As Westfall says,[34] Newton's law of the 'equal mutation of the motions' was a result that the seventeenth century did not expect at all. And, as the above analysis shows, the one phenomenon leads simultaneously to the definition of mass *and* the formulation of the Third Law.

It is a further fact of experience that the same quantities m_i are obtained when the bodies do not interact instantaneously, as in collisions, but continuously, over finite times, as in the case of the planets. In this case it is the mutually induced *accelerations* which are found to satisfy a relation like Eq. (12.4), namely, if a_1 and a_2 are the accelerations which bodies 1 and 2 have induced in them by their mutual interaction, then defining

$$a_1/a_2 = -C_{12},$$

we again find that Eq. (12.4) holds.

We can conclude this discussion of the definition of mass with the following comment of Mach:[35] 'All uneasiness will vanish when once we have made clear to ourselves that in the concept of mass no theory of any kind whatever is contained but simply a fact of experience.' It is not only the mass concept that is deduced from the observation of collisions and continuous interaction of bodies. Forces too are to be obtained in a similar manner. In Mach's reformulation of Newtonian dynamics, Newton's Second Law ceases to be a law and becomes instead the *definition of force*:

$$\mathbf{F} = m\mathbf{a}. \tag{12.5}$$

Here, the acceleration **a** is a directly observed kinematic quantity, while the mass m is also an observable quantity, deduced in interaction experiments from the observed accelerations. The use of *facts of experience* to define concepts which are in turn used to summarize huge bodies of empirical facts terminates in Mach's presentation of dynamics with the statement of the conditions under which forces are manifested. Here again, the phenomena are remarkably particular. A given configuration of bodies is always found to give rise to definite and calculable forces, the most famous example of which is, of course, Newton's law of universal gravitation.

What then is the essence of dynamics? It is in the *recognition of universal correlations in the observed behaviour of bodies*. The basis of it all, the ground on which the correlations are observed, is the possibility of measuring distances and times and the existence of bodies that can be recognized at different times (without the ability to say that such and such a body is the same as the one we were observing a moment ago, we should make very little progress in dynamics: this question will be considered in Vol. 2). Using these basic possibilities of observation, we establish the characteristic way in which bodies accelerate themselves from one inertial motion into another. The very specific and completely universal manner in which the mutual accelerations always occur permits the introduction of the *mass concept*, the most important concept in Newtonian dynamics and the one which Newton, making a break with all previous tradition, very characteristically and significantly placed at the very beginning of the *Principia* even if he did not in the event succeed in providing a satisfactory definition.

We can summarize this part of the discussion with the following characteristic passage from Mach:[36]

If we pass in review the period in which the development of dynamics fell . . . its main result will be found to be the perception, that bodies mutually determine in each other *accelerations* dependent on definite spatial and material circumstances, and that there are *masses*. The reason the perception of these facts was embodied in so great a number of principles is wholly an historical one; the perception was not reached at once, but slowly and by degrees. In reality only *one* great fact was established. Different pairs of bodies determine, independently of each other, and mutually, in themselves, pairs of accelerations, whose terms exhibit a constant ratio, the criterion and characteristic of each pair.

Incidentally, this passage shows clearly how unjust are the criticisms very often made of Mach that he did not believe in theories and wanted to reduce everything to mere description – to replace a scientific theory of the world by 'mere geographical description'. Such an opinion represents a serious misunderstanding of what Mach was trying to do. He was bitterly opposed to speculative, explicatory theories in the manner of Descartes and all the seventeenth century mechanists and atomists. What Mach wanted was the raw truth, the facts of experience. There is a thrill, with which no amount of speculation can compare, in knowing how the world actually *is*. Even today, 350 years after they were written, Galileo's words 'but to just what extent this acceleration occurs has not yet been announced' produce a tingle of excitement in the reader. Hence Mach's great admiration for Galileo:[37] 'Galileo did not supply us with a *theory* of the falling of bodies, but investigated and established, wholly without preformed ideas, the *actual facts* of falling.'

Equally typical is the sentence with which he (Mach) ended his first published book:[3] 'The object of natural science is the connection of phenomena; but the theories are like dry leaves which fall away when they have long ceased to be the lungs of the tree of science.'

Critics of Mach should not be misled by his pejorative use of the word *theory*, the predominant meaning of which has changed quite significantly since Mach's time (indeed, in part due to his influence). Mach used *theory* in the sense of 'an explanation . . . of a group of facts or phenomena', whereas it is nowadays used much more in the sense of 'a statement of . . . general laws, principles' (*OED*). Mach condemned Maxwell's derivation of the laws of electromagnetism from underlying mechanical models of the ether, but once they had been purified of the props which Maxwell's imagination originally needed and had been transformed into statements of the mere facts governing electromagnetic interactions Mach gave them his enthusiastic assent. Thus, Mach had absolutely no quarrel with what is today known as the Maxwell–Faraday theory of electromagnetism. The whole difference between Mach's concept of physics and 'descriptive geography' is in the *universality* of the connections between the phenomena.

Mach's clarification of the mass concept more or less completes the story of the discovery of dynamics and the clarification of its empirical basis.

As the mass concept will later play a central role in the discussion of Einstein's work, let us end this section with an explicit warning that Mach made on the subject:[38]

The concept of mass when reached in the manner just developed renders unnecessary the special enunciation of the principle of reaction [the Third Law]. In the concept of mass and the principle of reaction, . . ., the same fact is *twice* formulated; which is redundant. If two masses 1 and 2 act on each other, our very definition of mass asserts that they impart to each other contrary accelerations which are to each other, respectively, as 2 : 1. . . . As soon therefore as we, our attention being drawn to the fact by experience, have *perceived* in bodies the existence of a special property determinative of accelerations, our task with regard to it ends with the recognition and unequivocal designation of this *fact*. Beyond the recognition of this fact we shall not get, and every venture beyond it will only be productive of obscurity.

The last sentence of this quotation proved to be remarkably prophetic. Mach felt that his definition of mass had removed all obscurity about the concept that made its first hesitant and somewhat muddled appearance in Kepler's 'laziness'. There is a very important point to be made about the mass concept: the concept itself identifies an intrinsic property of individual bodies, i.e., it makes possible the association of a unique number, the mass m_i, with each body in the universe. However, the phenomenon through which this is made possible is one that involves at least two bodies – because the essence of the phenomenon is mutual acceleration of bodies. This mutuality of the acceleration makes it clear that the original idea of Kepler which Newton developed, namely that of the *resistance* of a body to a change of its state, is a misleading way of looking at the phenomenon, and one moreover that is based on rather direct subjective experience. When we push a body, we feel a resistance; we naturally think the essence of the phenomenon is resistance to pushing. But what we perceive as resistance is in reality the body pushing us, something it must do in accordance with the law of action and reaction as soon as we push the body. Thus, the essence of the mass phenomenon is not resistance *per se* but mutuality of acceleration.

It is necessary to make this point because, as pointed out in the Introduction, Einstein regarded an integral part of Mach's Principle as consisting of explaining 'resistance to acceleration'. He seems to have regarded this resistance as somehow a resistance against acceleration with respect to the universe as a whole, with the implication that the resistance phenomenon would disappear altogether if there were no masses in the universe. But this completely ignores the essence of the phenomenon, which is mutuality of the acceleration of at least two bodies. Einstein can hardly have believed that Newton's Third Law (or its generalization in relativistic physics) could somehow cease to hold if masses were progressively removed from the universe. And as long as the Third Law continues to hold, the concept of a quantity *intrinsic* to each and every body in the universe, the *inertial mass*, will survive unscathed. It must, or otherwise no content could be given to the Third Law.

Einstein did not heed Mach's final words quoted above. The almost total confusion surrounding the history of Mach's Principle in this century is ample confirmation of Mach's prediction made in 1883: 'and every venture beyond it will only be productive of obscurity'.

12.8. Synoptic overview of the discovery of dynamics

We have reached halfway. The Pythagorean cosmos has been dismantled, the world atomized. Laws of nature now govern its each and every part. The world's soul, or the universal spirit, has been killed.

Volume 2 will show how the Pythagorean cosmos is reassembled – and in the process transformed out of recognition.

The concept of motion that emerges at this halfway station can be characterized as *semi-interactive*. The earliest laws of motion, employed in astronomy, were completely noninteractive. Both practically and conceptually, geometrokinetic laws are the easiest to formulate. Given a fixed background, the concept of uniform straight or circular motion is not a difficult one. But as science matured, the observations became more detailed, the pattern more complicated. The first geometrokinetic laws were essential because without them the detail of the pattern could never have been put in such clear relief. Only when the gross structure has been defined does the fine structure come into focus. Nowhere is this more clearly demonstrated than in Kepler's work. The key to all his success was getting the orbit of the earth right. Once the equantized circle of the earth's orbit had been properly placed, Mars at last acquired a definite orbit. The fine details of this orbit carried the hints that led to the demise of the geometrokinetic scheme that revealed them in the first place. Kepler transformed the glorious lantern that God had hung at an advantageous point to illuminate the mysterious cavity of the world into the motor which drove the planets. The concept of motion was completely changed; where bodies previously *defined* each other's motion they now *determined* it. Newton completed what Kepler began.

But a geometrokinetic element remained. Galileo pulled it out of the heavens and put it straight into his motionics as one of the elements of the parabolic motion of projectiles. The net effect of Descartes' intervention was merely to straighten a geometrokinetic element that Galileo had imported from the closed circle-oriented world of Aristotle. More than anything else this Cartesian straightening of circular Galilean inertia brought the problem of motion to the centre of the stage. Galileo's circles were defined relative to the matter in the cosmos. What defined the straight lines with which Descartes replaced them? Descartes never realized he had created this problem. He initiated the real debate about the ultimate nature of motion quite unaware of the fact it was his own rectilinear motion rather than the Inquisition which brought the issue to a head.

Newton's solution to the problem, absolute space and time, left dynamics in a semi-interactive state. It divided the original geometrokinetic law of perfectly uniform circular celestial motions into two: the law of inertia, which remained geometrokinetic (though space-based instead of matter-based), and the law of universal gravitation, which epitomized the totally new interactive concept of motion. With interactive motion came necessarily the introduction of physical quantities – mass, electric charge, gravitational charge, etc. – without which it is not possible to realize a rational theory of interactive motion; they are needed to measure the capacity of bodies to generate motion and the response they exhibit to externally applied forces.

It was only in the nineteenth century that the startling consequences of a semi-interactive dynamics became fully apparent. Under the pressure of applications, which of necessity are matter-based and not space-based (you cannot make an observation of how something observable moves relative to something that is unobservable) the factual nature of absolute space and time were identified: a family of immaterial *inertial systems* whose location, orientation, and motion must ultimately be deduced from the observed motion of matter and, as it were, 'painted in' on that matter. What makes the family of inertial systems so mysterious is not so much its specific structure (which is odd enough – rotationally absolutely rigid but slightly undetermined translationally) but rather the manner in which it is 'suspended' in a decidedly arbitrary manner in the midst of the observable matter. The nature of this curiously disjunct suspension of the invisible inertial systems in the observable matter will be spelt out in more detail in Vol. 2. We have already been introduced to what is the most important thing about this 'suspension' – that it is *arbitrary*. This is the price which Newton pays for the only semi-interactive nature of his dynamics: ultimately, it leads to a loss of predictive power, as will be shown in Vol. 2. There is a nice way of summarizing the Copernican–Newtonian revolution. It eliminated the Aristotelian gulf between the heavens and the earth but created a gulf between the seen (matter) and the unseen (inertial systems). But every gulf we perceive in the world is more probably of our own conceptual making than a real feature of the world.

It seems that it is only very slowly that genuinely new ideas develop. There was much early criticism of Newtonian dynamics, and absolute space and time were sensed to be the outlandish creatures that the nineteenth century proved them to be. But nearly 200 years elapsed between the publication of the *Principia* and the first suggestion of the idea that for the first time offered a genuine prospect of bridging the gulf opened up between the seen and the unseen: Mach proposed the elimination of the cut between matter and the inertial systems. Interactive gravity represented all that was best about the scientific revolution: there was no hope of improvement in trying to turn back the clock from interactionism to geometrokineticism. The way forward for Mach was to complete the work that Kepler and Newton had begun and make dynamics completely interactive. Only such an approach could solve satisfactorily the problem of the frame of reference. Volume 2 will be basically devoted to the implementation of this idea. We can close this part of the discussion with Mach's clearest statement of the ideal of a *seamless dynamics*:[39]

> The natural investigator must feel the need of further insight – of knowledge of the *immediate* connections, say, of the masses of the universe. There will hover before him as an ideal an insight into the principles of the whole matter, from which accelerated and inertial motions result in the *same* way. The progress from Kepler's discovery to Newton's law of gravitation, and the impetus given by this to the finding of a physical understanding of the attraction in the manner in which electrical actions at a distance have been treated, may here serve as a model.

Raine[40] quoted the first two sentences of this passage, calling them 'a clear, if distant, presentiment of a general theory of relativity'. When I looked up the original quotation, having not looked at the passage in question for several years, I was delighted to find the third sentence, which had previously made little or no impact on me. But, meanwhile, a year immersed in the pre-Newtonian study of motion had shaped my awareness of the problem and the historical development on exactly those lines – Kepler introduced the idea of interaction, Newton implemented it only incompletely.

To conclude this first half of our study, it is worth drawing attention to some important points that so far have hardly been emphasized but which must certainly influence the way in which we attack the problem of motion.

Perhaps the most important point to make in this connection is the overriding importance of *distance*. It is on the possibility of measuring distance that ultimately the whole of dynamics rests. All the higher concepts of dynamics – velocity, acceleration, mass, charge, etc. – are built up from the possibility of measuring distance and observing the motion of bodies. Examination of the writings of Descartes and Newton reveals no awareness of the potential problems of an uncritical acceptance of the concept of distance. Both men clearly saw extension as something existing in its own right with properties that simply could not be otherwise than as they, following Euclid, conceived them. As Newton said in *De gravitatione*:[41] 'We have an exceptionally clear idea of extension.' The overthrow of the simple Euclidean/Cartesian verities by Gauss and Riemann is one of the most important factors that we shall have to take into account in the attempt to create that seamless, fully interactive dynamics which is our goal. It will complicate our task enormously but will also lend the problem a grandeur and dignity of which not even Newton dreamed. It will not be simply a question of interaction between matter and matter but also of interaction between matter and space.

Even deeper than this problem of interaction between space and matter, which Riemann posed and Einstein cracked, is the question of the origin of quantitative measures of distance altogether. Everything here is still shrouded in mystery; in Vol. 2 we shall consider some of the issues.

It should also be pointed out that hitherto time has played only a relatively small part in our considerations. For millennia time was the Cinderella of physics – always taken for granted and completely upstaged by space and geometry. This was mainly, as we have seen, because the empirical manifestations of time, discovered by the astronomers, were, despite being highly nontrivial, nevertheless largely in accord with prescientific intuition. Einstein was the Prince Charming who rescued Cinders from her unjust neglect – not that the poets had failed to pay due tribute. Perhaps the most remarkable aspect of the dramatic entry of time onto the centre of the stage was the carriage in which Einstein brought Cinderella to the ball – the Galileo–Huygens principle of relativity, that venerable first fruit of the Copernican revolution. As we have seen, Newton attempted, not quite correctly, to obtain this principle as a consequence of his three laws of motion. The relativity principle, which played little or no part in the discovery of Newtonian dynamics, appeared to be demoted. But the whirligig of time brings in its revenges. Huygens' instinct, earlier than Newton's, may not have opened up the high road to prerelativistic dynamics but in the end it proved to be the surer. The relativity principle was found to be stronger than originally appreciated, Newton's laws weaker. The shift in the relative importance of space and time has been most pronounced in this century, so much so that many people now regard the central problem in the quantization of general relativity as the elucidation of time – 'in reconciling the diametrically opposite ways in which relativity and quantum mechanics view the concept of time' (Kuchař[42]). Physicists of the twentieth century have caught up with the enigma that has always fascinated poets, none more than the late Elizabethans writing just at the time Kepler and Galileo laid the foundations of modern dynamics.

Nor must we forget in this connection to mention the finite velocity of light, which plays such a role in relativity. If the relativity principle was the carriage in which Cinders came to the ball, light was the horse that drew the carriage. The light of the world: yes, we see the world by light but Einstein put a deeper significance into that 'seeing' than ever man could possibly have imagined. Yet again were we disabused of the comforting notion that we have a basically correct picture of what the world is like and merely need our eyes to tell us what are the objects in it and how they are disposed. Along with the 'dynamization' of geometry (an apt coining as we shall see), the relativization of time is a second major factor that immensely complicates the implementation of the Machian programme for the unified treatment of motion. And if several people (Clifford, and not least Mach himself) anticipated Einstein in seeing that geometry might need to be dynamized, no one had even an inkling of the relativization of time. Truly a bolt from the bluest of skies. We still lie spreadeagled on the ground.

This coda to the discovery of dynamics should come to an end. How do we end on the right note? Let me stick my neck out. The acid test of any account of motion should surely be: can it answer comprehensibly and unambiguously the question which started the process that endowed the word revolution with its present primary connotation: does the earth move? Is it true, as is so often asserted, that the Copernican and the Ptolemaic representations are equally valid at the most fundamental level and that therefore the question of whether the earth moves or not is ultimately a question of convention? We are now in a position to answer this question *within the context of the level of understanding achieved by Newton, as clarified by Lange's introduction of the concept of an inertial system.*

We must first insist that motion must be defined relative to something which is definite and observable. Copernicus chose the stars and was aware that at least in part this amounted on his part to a *conventional* definition of motion ('since the heavens, which enclose and provide the setting for everything, constitute the space common to all things, it is not at first blush clear why motion should not be attributed rather to the enclosed than to the enclosing, to the thing located in space rather than to the framework of space'[43]). As regards the question of whether the earth or the heavens rotates, the difference between Copernicus and Ptolemy is indeed convention as long as we take no account of the *dynamical* discoveries that completed the Copernican revolution. However, as regards the second motion that Copernicus proposed (the annual as opposed to the diurnal motion) there is an unambiguous difference between Ptolemy and Copernicus, for Ptolemy taught that the earth is at rest relative to the stars, i.e., the distance from the centre of the earth to the stars does not change. Thus, if motion is defined as motion relative to the matter of the universe as a whole, Copernicus is without doubt correct about his second motion and Ptolemy simply wrong. The question of the diurnal motion remains tantalizingly open.

When the laws of Newtonian dynamics, reformulated in terms of inertial systems, are taken into account, the problem of the earth's motion takes on a *totally new complexion*. Whereas before there existed only kinematic definitions of motion (by means of identified bodies), there now appears, for the first time in the history of human thought, a second (and seemingly totally distinct) possibility of defining motion: relative to the *family of inertial systems*. It is important to be clear (in the context of Newtonian dynamics) precisely what this family is. Unlike the individual inertial systems, the family of inertial systems is a unique entity. Operationally it can only be defined by means of the totality of the matter which exists in the universe. It is identified in an ongoing process by careful observation and analysis of the matter in the universe (in principle, all matter has to be taken into account). It is itself completely invisible and must be 'painted onto' the constantly changing configuration of the contingently observable matter in the universe. It is 'suspended' on matter in a curious way – a rigid straightjacket whose existence is inferred by the rules of the dance. The discovery of this second means of defining motion was totally unexpected, as was the very curious feature of this definition of motion – that it is not motion itself that is defined but only acceleration. In fact, with hindsight we can now recognize the intimate connection between this mysterious feature of dynamics and the fact that for so long the idea that the earth moves could be dismissed as fancy; for there is no dynamical definition of motion *per se,* only a definition of acceleration – and the effects of the earth's dynamical acceleration were below the detection level as far as everyday phenomena were concerned. Thus it was that Copernicus's cutting of the earth's moorings led very rapidly to the flushing out of what is perhaps the deepest and most enigmatic of the principles of physics – the Galileo–Huygens principle of relativity. This is what Galileo grasped: if the earth moves, the principle has to hold.

Consideration of the different ways in which the two definitions of motion – kinematic and dynamic – leave open considerable ambiguity whenever we attempt to say precisely what is the motion of any particular body emphasizes just how remarkable is the appearance of the second, dynamical method. As regards the kinematic definition, Descartes was absolutely right – there are as many different motions as there are reference bodies. Only in the case of a finite universe containing a limited number of bodies does any sort of well-defined concept of motion – relative to the totality of the bodies – appear. In contrast, the dynamical definition of motion is completely free of this infinitude of ambiguity in which Descartes took refuge from the Inquisition – though the price which Newton had to pay (and paid most grudgingly) for this certainty was Corollary V: the final pinning down of motion (which Galileo hoped to do by the tides) is just not possible. Thus, in accordance with the dynamic definition of motion the earth does truly move, and it does so in any of the allowed inertial frames of reference. But a mystery still remains since the frame of reference is not quite unique.

Was Mach right to say *one* great fact was discovered? It really should be two. While recognizing the value of Lange's concept of inertial system, Mach did not perhaps give it sufficient due. Newton was quite right to give the exalted names absolute space and time – and sensorium of God – to the mysterious entity of whose existence he was completely certain even if precisely the right way of describing it eluded him. Seen in the perspective of the history of man's growing comprehension of the universe this was truly an extraordinary discovery: a new, totally distinct, and well-defined method of determining motion – the first dramatic *dénouement*. Who are we to criticize Newton, the 'last of the Sumerians and Babylonians',[44] for seeing in the as yet imperfectly comprehended dynamical frame of reference, invisible but all-powerful, the demidivine structure of the world? And how mysteriously it hangs there, suspended on the fleeting contingent world!

The Newtonian concepts of absolute space and time are the most striking examples of concepts grasped intuitively but imperfectly in the pre-scientific age and then shown to have a genuine basis in phenomena. As in all such cases, the fact that they seem so close to the original notions has tended to hide the remarkable nature of their discovery, the deep sophistication of the procedure by which their empirical existence is demonstrated, and the truth that the reality when found is always more wonderful and subtle than its anticipation in metaphysical and intuitive concepts. Newton's absolute space was anticipated, the family of inertial frames of reference and the manner in which they are empirically determined were not. We shall have to wait until Vol. 2 to see how absolute time, also anticipated so long before Newton, required similar modification. However, the main conclusion to be drawn from the two and a half millennia that this volume has covered can already be stated: the dynamical frame of reference, and not the ground under our feet or a sphere of fixed stars in the heavens, is the true backbone of the world.

But what is its origin?

ABBREVIATIONS FOR WORKS QUOTED FREQUENTLY IN THE REFERENCES

Almagest: *Ptolemy's Almagest*, transl. and annotated by G. J. Toomer, Duckworth, London (1984) (The paginations are from this translation; the H numbers give the page in Heiberg's Greek text published by Teubner and used by Toomer (HI and HII for volumes 1 & 2).)

Astr. Nova: Kepler's *Astronomia Nova*, in: *Gesammelte Werke*, Vol. 3, ed. M. Caspar, C. H. Beck'sche Verlagsbuchhandlung, Munich (1937); I also used Caspar's translation into German: *Neue Astronomie*, R. Oldenbourg, Munich (1929).

Clagett: M. Clagett, *The Science of Mechanics in the Middle Ages*, The University of Wisconsin Press, Madison (1979).

De Caelo: Aristotle's *De Caelo* (*On the Heavens*); all quotations are from: *Aristotle VI: On the Heavens* (transl. by W. K. C. Guthrie), Loeb Classical Library, Heinemann, London (1971).

De Rev: Copernicus's *De Revolutionibus Orbium Coelestium* (1543); all quotations are from: *Copernicus: On the Revolutions* (transl. by E. Rosen), Macmillan, London (1978).

Dialogo: Galileo's *Dialogo Sopra i due Massimi Sistemi del Mondo Tolemaico e Copernicano*, Florence (1632); in: *Le Opere di Galileo Galilei*, Vol. 7, Florence (1897); all quotations are from the translation: *Dialogue Concerning the Two Chief World Systems – Ptolemaic and Copernican*, transl. by S. Drake, University of California Press, Berkeley (1953).

Dict. Sci. Biog.: *Dictionary of Scientific Biography*, Vols. 1, . . ., Charles Scribner's Sons, New York (1971).

Discorsi: Galileo's *Discorsi e Dimostrazioni Matematiche Intorno a due Nouve Scienze*, Elsevier, Leiden (1638); *Le Opere di Galilei*, Vol. 8, Florence (1898); all quotations are from: *Two New Sciences*, transl. by H. Crew and A. de Salvio, Macmillan (1914), reprinted by Dover, New York.

Epitome: Kepler's *Epitome Astronomiae Copernicanae* (*Epitome of Copernican Astronomy*) (1618, 1620) in: *Gesammelte Werke*, Vol. 7, ed. M. Caspar, C. H. Beck'sche Verlagsbuchhandlung, Munich (1953); all quotations are from the translation of Book IV in: *Great Books of the Western World*, ed. R. M. Hutchins (1952), Vol. 16.

Ges. Werke: *Johannes Kepler Gesammelte Werke*, ed. M. Caspar *et al.*, C.H. Beck'sche Verlagsbuchhandlung, Munich (1938 . . .).

HAMA: O. Neugebauer, *A History of Ancient Mathematical Astronomy*, Vols. 1–3, Springer-Verlag, Berlin (1975).

Mechanics: E. Mach, *Die Mechanik in ihrer Entwicklung Historisch-Kritisch Dargestellt* (1883); all quotations are from the translation: *The Science of Mechanics: A Critical and Historical Account of Its Development*, transl. of 9th German edition by T. J. McCormack, Open Court, La Salle, Ill. (1960).

Opere: *Le Opere di Galileo Galilei*, Vols. 1–20, Florence (1890–1909).

Oeuvres: *Oeuvres de Descartes*, 12 vols., eds. C. Adam and P. Tannery, Léopold Cerf, Paris (1897–1910).

Oeuvres Complètes: *Oeuvres Complètes de Christiaan Huygens*, 22 vols., Martinus Nijhoff, The Hague (1888–1950).

Physics: Aristotle's *Physics*; all quotations are from: *Aristotle IV: Physics, Books I-IV* (transl. by P. H. Wicksteed and F. M. Cornford), Loeb Classical Library, Heinemann (1980).

Principia: Newton's *Philosophiae Naturalis Principia Mathematica* (1687); all quotations are from: *Sir Isaac Newton's Mathematical Principles of Natural Philosophy and his System of the World*, Motte's translation of 1729 revised by F. Cajori, University of California Press, Berkeley (1962).

Principles: Descartes' *Principia Philosophiae* (1644) in: *Oeuvres de Descartes*, Vol. 8, eds. C. Adam and P. Tannery, Paris (1905); all quotations are from: *Principles of Philosophy*, transl. by V. R. Miller and R. P. Miller, Reidel, Dordrecht (1983).

Westfall (1971): R. S. Westfall, *The Concept of Force in Newton's Physics: The Science of Dynamics in the Seventeenth Century*, Macdonald, London (1971).

Westfall (1980): R. S. Westfall, *Never at Rest: A Biography of Isaac Newton*, Cambridge University Press, Cambridge (1980).

World: Descartes' *Le Monde* (1664), in: *Oeuvres de Descartes*, Vol. 11, eds. C. Adam and P. Tannery; partial translation in: *The Philosophical Writings of Descartes*, transl. by J. Cottingham, R. Stoothoff, and D. Murdock, Cambridge University Press, Cambridge (1985), Vol. 1, p. 79. Complete translation (from which the quotations are taken): *The World: Le Monde*, transl. with introduction by M. S. Mahoney, Abaris Books, New York (1979).

The quotations at the beginning of the book (p. vi) are from Galileo's *Il Saggiatore* (1623; *Opere*, Vol. 6, p. 197; my translation), Kepler's *Astronomia Nova* (1609, *Gesammelte Werke*, Vol. 3, p. 47; my translation), Newton's *Principia*, p. 12, and Mach's *Mechanics*, p. 316.

REFERENCES

Introduction

1. *Mechanics.*
2. A. Einstein, *Ann. Phys.*, **55**, 241 (1918).
3. A. Einstein, *Ann. Phys.*, **49**, 769 (1916), translation in: *The Principle of Relativity* (collection of original papers by Einstein, Lorentz, Minkowski and others translated into English), Methuen (1923) (reissued by Dover, New York).
4. A. Einstein, *Sitzungsber. preuss. Akad., Wiss.*, **1**, 142 (1917). For a translation, see Ref. 3.
5. A. Einstein, 'Autobiographical notes', in: *Albert Einstein – Philosopher-Scientist*, ed. P. A. Schilpp, The Library of Living Philosophers, Inc., Evanston, Ill. (1949), reprinted as Harper Torchbook (1959), p. 29.
6. For a summary see: C. M. Will, *Theory and Experiment in Gravitational Physics*, Cambridge University Press, Cambridge (1981).
7. D. W. Sciama, *Mon. Not. R. Astron. Soc.*, **113**, 34 (1953); D. W. Sciama, P. C. Waylen, and R. C. Gilman, *Phys. Rev.*, **187**, 1762 (1969).
8. F. Hoyle and J. V. Narlikar, *Proc. R. Soc. Lond*, **A282**, 191 (1964); *Action at a Distance in Physics and Cosmology*, W. H. Freeman, New York (1966).
9. C. Brans and R. H. Dicke, *Phys. Rev.*, **124**, 925 (1961); reprinted in R. H. Dicke, *The Theoretical Significance of Experimental Relativity*, Blackie, London (1964).
10. H.-J. Treder, *Die Relativität der Trägheit*, Akademie-Verlag, Berlin (1972); see also: H.-J. Treder, *Gerlands Beitr. Geophys. (Germany)*, **82**, No. 1, 13 (1973); D.-E. Liebscher, *ibid.*, p. 3; D.-E. Liebscher and W. Yourgrau, *Ann. Phys. (Germany)*, **36**, No. 1, 20 (1979).
11. D. J. Raine, *Rep. Prog. Phys.*, **44**, 1151 (1981).
12. J. A. Wheeler, in: *Relativity, Groups, and Topology. Les Houches Summer School 1963*, eds. C. M. DeWitt and B. DeWitt, Gordon and Breach, New York (1964).
13. See especially the review articles by C. J. Isham and K. Kuchař in: *Quantum Gravity 2. A Second Oxford Symposium*, eds. C. J. Isham, R. Penrose, and D. W. Sciama, Clarendon Press, Oxford (1981).
14. This point is elaborated in: J. B. Barbour, 'Leibnizian time, Machian dynamics and quantum gravity,' in: *Quantum Concepts in Space and Time*, eds. R. Penrose and C. J. Isham, Clarendon Press, Oxford (1986).
15. See the book quoted in Ref. 14.

16. H. Bondi, *Cosmology,* Cambridge University Press, Cambridge (1960).
17. See Ref. 1.
18. This point is elaborated in: J. B. Barbour, 'Mach's Mach's Principles, Especially the Second,' in: *Gründlagenprobleme der modernen Physik* (Festschrift für Peter Mittelstaedt), eds. J. Nitsch, J. Pfarr, and E.-W. Stachow, Bibliographisches Institut, Mannheim (1981). See also Refs. 24 (1982) and 36.
19. A. Einstein, *Naturforsch. Gesellschaft, Zürich, Vierteljahrsschr.,* **58**, 284 (1913).
20. A. Einstein, *Naturwissenschaften,* 6-er Jahrgang, No. 48, 697 (1918) (passage on p. 699).
21. A. Einstein, *Ann. Phys.,* **17**, 891 (1905). For a translation, see Ref. 3.
22. H. Minkowski, *Phys. Zeitschrift,* **10**, 104 (1909). For a translation, see Ref. 3.
23. J. B. Barbour, *Nature,* **249**, 328 (1974). Serious misprints are corrected in *Nature,* **250**, 606 (1974).
24. J. B. Barbour and B. Bertotti, *Nuovo Cimento, B38,* 1 (1977); *Proc. R. Soc. Lond.,* **A382**, 295 (1982).
25. See Ref. 5.
26. E. Schrödinger, *What is Life? Mind and Matter,* Cambridge University Press, Cambridge (1980).
27. E. Wigner, *Symmetries and Reflections: Scientific Essays,* M. I. T. Press, Cambridge, Mass. (1967), pp. 153–99.
28. R. Arnowitt, S. Deser, and C. W. Misner, 'The dynamics of general relativity', in: L. Witten (ed.), *Gravitation: An Introduction to Current Research,* Wiley, New York (1962).
29. See Ref. 12 (and also the revision in Ref. 37 below).
30. A. Einstein, *The Meaning of Relativity,* Methuen, London (1922), p. 22.
31. T. S. Kuhn, *The Copernican Revolution,* Harvard University Press, Cambridge, Mass. (1957).
32. A. Koyré, *From the Closed World to the Infinite Universe,* Johns Hopkins University Press, Baltimore (1957).
33. M. Jammer, (a) *Concepts of Force: A Study in the Foundations of Dynamics,* Harvard University Press, Cambridge, Mass. (1957); (b) *Concepts of Mass in Classical and Modern Physics,* Harvard University Press, Cambridge, Mass. (1961); (c) *Concepts of Space: The History of Theories of Space in Physics,* 2nd edn., Harvard University Press, Cambridge, Mass. (1969).
34. *Westfall (1971).*
35. J. R. Tolkien, *The Lord of the Rings.*
36. This is discussed in the second reference of Ref. 24; see also: J. B. Barbour, *Br. J. Phil. Sci.,* **33**, 251 (1982).
37. This point is discussed by: J. A. Isenberg and J. A. Wheeler in: 'Inertia here is fixed by mass-energy there in every W model universe', in: *Relativity, Quanta, and Cosmology in the Development of the Scientific Thought of Albert Einstein,* eds. M. Pantaleo and F. deFinis, Johnson Reprint Corporation, New York, Vol. I, pp. 267–93; see also: J. A. Isenberg, *Phys. Rev D 24,* 251 (1981).

Chapter 1

1. *Principia,* pp. 1–28.
2. M. Jammer, *Concepts of Mass in Classical and Modern Physics,* Harvard University Press, Cambridge, Mass. 1961, pp. 87–9.

3. In the proof of Corollary V of the Axioms on pp. 20/1 of Ref. 1.
4. A Pais, *'Subtle is the Lord. . .'. The Science and Life of Albert Einstein,* Oxford University Press, Oxford (1982), p. vi.
5. *De Caelo,* pp. 223–5 (Book II, xiii, 294a).
6. M. R. Cohen and I. E. Drabkin, *A Source Book in Greek Science,* New York (1948), p. 211.
7. *De Caelo,* p. 171 (Book II, vi, 288a).
8. A. Maier, *Die Vorläufer Galilei's im 14. Jahrhundert,* Edizoni di Storia e Letteratura, Roma (1966), p. 137.
9. *Almagest,* p. 252 (HI, 416/7).
10. *Discori,* p. 179.
11. W. Gilbert, *De Magnete, Magneticisque Corporibus, et de Magno Magnete Tellure,* London (1600); transl. by P. F. Mottelay: *On the Loadstone and Magnetic Bodies, and on the Great Magnet the Earth,* Bernard Quaritch, London (1893).
12. *De Caelo,* p. 171 (Book II, vi, 288a).
13. *Plato IX Timaeus* (transl. by R. G. Bury), Loeb Classical Library, Heinemann, London (1981), p. 77.
14. *Almagest,* pp. 35–6 (HI, 5, 6).
15. J. L. E. Dreyer, *Tycho Brahe,* Adam and Charles Black, Edinburgh (1890), p. 14.
16. See Ref. 8, Chap. 1, Part 1, p. 23.
17. *Physics,* translator's footnote on pp. 274–5.
18. *Oxford English Dictionary,* entry for *phoronomy.*
19. M. Čapek, *The Concepts of Space and Time* (selections from the philosophical literature on space and time), D. Reidel Publishing Company, Dordrecht (1976).
20. *Clagett,* pp. 168–9.
21. Quoted from *Clagett,* p. 171.
22. d'Alembert, *Traité de Dynamique,* quoted from the article by A. Gabbey in *Descartes: Philosophy, Mathematics and Physics,* ed. S. Gaukroger, Harvester Readings in the History of Science and Philosophy, Vol. 1, Brighton (1980), p. 231.
23. *Mechanics,* pp. 287–8.
24. *De Caelo,* p. 283 (Book III, iii, 302a).
25. *Mechanics,* p. 281.
26. A. Einstein, *On the Method of Theoretical Physics.* The Herbert Spencer Lecture delivered at Oxford, 10 June 1933, Clarendon Press, Oxford (1933).
27. *The Taming of the Shrew,* Act. IV, Scene 5.

Chapter 2

1. D. D. Runes (ed.), *Dictionary of Philosophy,* Littlefield, Adams & Co., Totowa, New Jersey (1962).
2. J. Barnes, *The Presocratic Philosophers,* Routledge & Kegan Paul, London (1982).
3. G. S. Kirk, J. E. Raven, and M. Schofield, *The Presocratic Philosophers,* 2nd Edn, Cambridge University Press (1983).
4. For example: A. Koestler, *The Sleepwalkers. A History of Man's Changing Vision of the Universe,* Hutchinson, London (1959), reprinted by Penguin Books, London (1977), Part 1, Chap. 2.

5. C. Bailey, *The Greek Atomists and Epicurus,* The Clarendon Press, Oxford (1928), pp. 74–7; reproduced in the next reference.
6. M. Čapek, *The Concepts of Space and Time* (selections from the philosophical literature on space and time), D. Reidel Publishing Company, Dordrecht (1976).
7. *De Caelo*, p. 31 (Book I, iv, 271a).
8. *Clagett*, p. 422.
9. E. Schrödinger, *Mind and Matter,* Cambridge University Press (1958); reprinted as paperback with *What is Life*? (1980).
10. See Guthrie's introduction to Ref. 7.
11. *De Caelo*, p. 135–7 (Book II, i, 284b).
12. F. M. Cornford, in *Essays in Honour of Gilbert Murray,* Allen & Unwin, London (1936), pp. 215–35; reprinted in Ref. 6, pp. 3–16.
13. *Physics*, p. 333 (Book IV, vi, 213b).
14. F. A. Lange, *Geschichte des Materialismus*, Vol. 1, Baedeker, Iserlohn (1873), p. 12ff (my translation).
15. I. Newton, *Opticks*, 3rd edn, London (1730), Question 31; reprinted by Bell & Sons, London (1931), p. 400.
16. *Plato IX Timaeus* (transl. by R. G. Bury), Loeb Classical Library, Heinemann, London (1981), p. 131ff.
17. *Ibid.*, p. 135.
18. *Ibid.*, p. 155.
19. *Ibid.*, p. 156.
20. *Ibid.*, p. 217.
21. E. Wigner, *Symmetries and Reflections: Scientific Essays*, M.I.T. Press, Cambridge, Mass. (1967), p. 222.
22. *Dict. Sci. Biog.*, Vol. 4, p. 465. Article on Eudoxus.
23. Ref. 16, p. 137.
24. *De Caelo*, p. 269 (Book III, i, 300a).
25. *De Caelo*, p. 273 (Book III, ii, 300b).
26. *De Caelo*, p. 19 (Book I, iii, 269b).
27. *De Caelo*, pp. 11–12 (Book I, ii, 268b).
28. The following is taken from the *Oxford English Dictionary*.
29. *Almagest*, p. 36 (HI, 7).
30. *Physics*, p. 281 (Book IV, i, 208b).
31. *Physics*, p. 349 (Book IV, viii, 214b, 215a).
32. *Ibid.*, 215a.
33. *Physics*, pp. 279–81 (Book IV, i, 208b).
34. S. Sambursky, *The Physical World of the Greeks,* Routledge & Kegan Paul, London (1987), p. 111.
35. A. Einstein, 'Autobiographical notes', in: *Albert Einstein – Philosopher–Scientist*, ed. P. A. Schilpp, The Library of Living Philosophers, Inc., Evanston, Ill. (1949), reprinted as Harper Torchbook (1958), p. 29.
36. *De Caelo*, pp. 293–5 (Book III, iv, 303b).
37. *De Caelo*, pp. 357–9 (Book IV, iv, 312a).
38. *De Caelo*, pp. 187–9 (Book II, viii, 290a).
39. *Astr. Nova,* Chap. 33.
40. *De Caelo*, pp. 23–5 (Book I, iii, 270b).

41. *Dialogo*, pp. 58–9.
42. *De Caelo*, p. 133 (Book II, i, 284a).
43. *Physics*, p. 415 (Book IV, viii, 222b).
44. Letter from Hooke to Newton, 24 Nov. 1679; in: *The Correspondence of Isaac Newton*, Vol. 2, ed. H. W. Turnbull, Cambridge University Press, Cambridge (1960), pp. 297–8.
45. *Physics*, p. 277 (Book IV, i, 208a).
46. *Ibid.*, pp. 277–9.
47. *Physics*, p. 281 (Book IV, i, 208b–209a).
48. *Physics*, pp. 281–3 (Book IV, i, 209a).
49. *Ibid.*, p. 285.
50. Ref. 1, p. 243.
51. *Physics*, p. 307 (Book IV, iv, 211a).
52. *Physics*, p. 309 (Book IV, iv, 211b).
53. *Physics*, p. 313 (Book IV, iv, 212a).
54. *Physics*, p. 289 (Book IV, ii, 209b).
55. *Physics*, p. 309 (Book IV, iv, 211b).
56. *Physics*, p. 287 (Book IV, ii, 209a–209b).
57. *Physics*, pp. 313/5 (Book IV, iv, 212a).
58. Ref. 12.
59. *Physics*, pp. 321–5 (Book IV, v, 212a–212b).
60. *De Caelo*, p. 241ff (Book II, xiv).
61. Ref. 6, p. 39 (translation from: P. Duhem, *Le Système du Monde*, Vol. 1, pp. 317–19).
62. E. Grant, *Much Ado About Nothing. Theories of Space and Vacuum from the Middle Ages to the Scientific Revolution*, Cambridge University Press, Cambridge (1981).
63. P. E. Ariotti, in: *The Study of Time*, eds. J. T. Fraser and W. Lawrence, Proc. 2nd Conf. Int. Soc. Study of Time, Lake Yamanaka – Japan, Springer-Verlag, Berlin (1975), p. 69.
64. Ref. 16, p. 79.
65. *Physics*, p. 383 (Book IV, xi, 218b).
66. *Physics*, p. 377 (Book IV, x, 218b).
67. *Physics*, p. 383 (Book IV, xi, 218b).
68. *Physics*, p. 385 (Book IV, xi, 219a).
69. *Ibid.*, pp. 385–7.
70. *Physics*, p. 387 (Book IV, xi, 219a).
71. *Physics*, p. 391 (Book IV, xi, 219b).
72. *Physics*, p. 395 (Book IV, xi, 220a).
73. *Physics*, p. 421 (Book IV, xiv, 223a).
74. *Physics*, pp. 421–3 (Book IV, xiv, 223b).
75. *Physics*, p. 417 (Book IV, xiv, 222b–223a).
76. *Physics*, p. 399 (Book IV, xii, 220b).
77. *Physics*, p. 423 (Book IV, xiv, 223b).
78. *Ibid.*, p. 425.
79. Ref. 7, p. 159 (Book II, iv, 287a).
80. A. Maier, *Die Vorläufer Galilei's im 14. Jahrhundert*, Edizoni di Storia e Letteratura, Roma (1966), p. 10.

Chapter 3

1. For example: (a) J. L. E. Dreyer, *A History of Astronomy from Thales to Kepler*, Cambridge University Press (1906); reprinted by Dover, New York (1953); (b) T. L. Heath, *Aristarchus of Samos, The Ancient Copernicus, A History of Greek Astronomy to Aristarchus*, Oxford (1913); (c) A. Koestler, *The Sleepwalkers. A History of Man's Changing Vision of the Universe*, Hutchinson, London (1959), reprinted by Penguin Books, London (1977).
2. *HAMA*.
3. *Dict. Sci. Biog.*, articles on (a) Heraclides, Vol. 25, Suppl. 1, p. 202 (1978); (b) Apollonius, Vol. 1, p. 179 (1970); (c) Hipparchus, Vol. 25, Suppl. 1, p. 207 (1978); (d) Ptolemy, Vol. 11, p. 186 (1975).
4. *HAMA*, p. 271.
5. Ref. 1(a), pp. 141–2.
6. *HAMA*, pp. 675ff.
7. *Ibid.*, p. 644.
8. *Ibid.*, p. 262.
9. See Ref. 3(a).
10. *Almagest*, p. 555ff (HII, 450ff).
11. *HAMA*, p. 262.
12. *Ibid.*, Book II.
13. *Ibid.*, p. 4.
14. *Ibid.*, p. 5.
15. See Ref. 1(c), part 4.
16. *HAMA*, p. 22.
17. *Ibid.*, p. 3.
18. *Almagest*, p. 40 (HI, 15).
19. *De Caelo*, p. 255.
20. *Almagest*, p. 41ff.
21. *Almagest*, Book IV.
22. (a) *Dict. Sci. Biog.*, article on Aristarchus, Vol. I, p. 246 (1970); (b) *HAMA*, p. 634ff.
23. Ref. 3(c); *HAMA*, p. 322ff.
24. *Almagest*, p. 257 (HI, 425–7).
25. *HAMA*, p. 342.
26. *Almagest*, p. 48ff.
27. *HAMA*, p. 277ff.
28. *Ibid.*, p. 279.
29. *Almagest*, p. 153 (HI, 233).
30. Taken from C. W. Allen, *Astrophysical Quantities*, 3rd edn, Athlone Press, London (1976).
31. A. Koyré, *The Astronomical Revolution: Copernicus–Kepler–Borelli*, Methuen, London (1973).
32. F. Hoyle, *Nicolaus Copernicus. An Essay on his Life and Work*, Heinemann, London (1973).
33. *HAMA*, pp. 1095–103.
34. *Ibid.*, p. 4, p. 263.
35. *Oxford English Dictionary*, entry for *eccentricity*.

36. D. T. Whiteside, *J. Hist. Astronomy*, **5**, 1 (1974) (see also Ref. 33).
37. *Almagest*, p. 153 (HI, 233).
38. V. Peterson and O. Schmidt, *Centaurus*, **12**, 73 (1967).
39. *HAMA*, p. 14.
40. *Ibid.*, p. 1101.
41. T. S. Kuhn, *The Structure of Scientific Revolutions*, University of Chicago Press, Chicago (1962).
42. *HAMA*, p. 57.
43. *Almagest*, p. 555ff (HII, 450ff); see also Ref. 3(b) and *HAMA*, p. 263ff.
44. See Ref. 3(a) and *HAMA*, p. 694.
45. *HAMA*, p. 112.
46. *Almagest*, p. 420 (HII, 209).
47. *De Rev.*, pp. 17–18 (Book I, Chap. 9).
48. H. Spencer Jones, *General Astronomy*, Arnold & Co, London (1951), p. 118.
49. *Almagest*, p. 217, footnote 2.
50. *Almagest*, p. 217ff (HI, 351ff).
51. *Almagest*, Book V.
52. *HAMA*, p. 84ff; see also V. M. Petersen, *Centaurus*, **14**, 142 (1969).
53. *Almagest*, p. 216ff.
54. *HAMA*, p. 86.
55. *Ibid.*, p. 88.
56. *Almagest*, p. 174ff (HI, 269ff).
57. *Astr. Nova*, *passim*.
58. *Epitome*, §IV.5, p. 911.
59. *Almagest*, p. 206 (HI, 328).
60. *Almagest*, p. 422 (HII, 212).
61. *Almagest*, p. 480 (HII, 317).
62. *Almagest*, Book X, Sec. 3, p. 472ff (HII, 303).
63. N. Swerdlow and O. Neugebauer, *Mathematical Astronomy in Copernicus's De Revolutionibus* (Studies in the History of Mathematics and the Physical Sciences, Vol. 10), Springer, Berlin (1984), p. 80.
64. *Almagest*, pp. 600–1 (HII, 532–3).
65. *Principia*, p. 6.
66. *Mechanics*, p. 273.
67. *HAMA*, p. 61.
68. *Almagest*, p. 171 (HI, 262).
69. *Almagest*, p. 169 (HI, 258).
70. G. M. Clemence, *Astr. J.*, **53**, 169 (1948).
71. E. Mach. *Populär-Wissenschaftliche Vorlesungen* (5th edn), Barth, Leipzig (1923), p. 492.
72. *Almagest*, Book III, Sec. 9.
73. *Almagest*, p. 191 (HI, 302–3).
74. Ref. 1(b).
75. *HAMA*, p. 280.
76. Ref. 1(c), p. 72.
77. Ref. 1(c), p. 71.
78. *Almagest*, pp. 420–1 (HII, 209).
79. *Almagest*, pp. 421–2 (HII, 210–11).

80. *Almagest,* Toomer's Introduction, p. 1.
81. O. Pedersen, *A Survey of the Almagest* (Acta Historica Scientiarum Naturalium et Medicinalium, Vol. 30), Odense University Press (1974).
82. *Almagest,* Toomer's Preface, p. viii.
83. Ref. 22.
84. *Almagest,* pp. 43–5 (HI, 22–6).
85. *Richard II,* Act. III, Scene 2.

Chapter 4

1. S. L. Jaki, *Uneasy Genius: The Life and Work of Pierre Duhem,* Martinus Nijhoff, The Hague (1984).
2. As summarized by Anneliese Maier, *Die Vorläufer Galilei's im 14. Jahrhundert,* Edizoni di Storia e Letteratura, Roma (1966), p. 1.
3. P. Duhem, *Medieval Cosmology. Theories of Infinity, Place, Time, Void, and the Plurality of Worlds.* (Extracts from *Le Système du Monde,* edited and translated by R. Ariew), University of Chicago Press, Chicago (1985).
4. A. N. Whitehead, *Science and the Modern World: Lowell Lectures, 1925,* Macmillan, New York (1925), pp. 17–18.
5. See, for example, S. L. Jaki, *Science and Creation: From Eternal Cycles to an Oscillating Universe,* Scottish Academic Press, Edinburgh (1974).
6. Ref. 2, p. 1.
7. *Clagett.*
8. D. D. Runes (ed.), *Dictionary of Philosophy,* Littlefield, Adams and Co., New Jersey (1975), p. 17.
9. *Clagett,* Chap. 4.
10. *Discorsi,* Day 3.
11. *Clagett,* p. 248.
12. *Ibid.,* p. 331.
13. *Ibid.,* p. 340.
14. *Ibid.,* p. 340–1.
15. *Ibid.,* p. 346.
16. *Physics,* p. 351 (Book IV, viii, 215a).
17. *Ibid.,* p. 349.
18. *Aristotle V: Physics,* Books V–VIII (transl. by P. H. Wicksteed and F. M. Cornford), Loeb Classical Library, Heinemann, London (1980), pp. 417–19 (267a).
19. R. Sorabj (ed.), *Philoponus and the Rejection of Aristotelian Philosophy,* Duckworth (1987).
20. M. R. Cohen and I. E. Drabkin, *A Source Book in Greek Science,* Harvard University Press, Cambridge, Mass. (1948), pp. 221–3.
21. G. Galileo, *De Motu,* in: *Opere,* Vol. 1; translation in: *On Motion and On Mechanics* (transl. with introductions by I. E. Drabkin and S. Drake), University of Wisconsin Press, Madison (1960).
22. *Clagett,* p. 510ff.
23. *Ibid.,* p. 521.
24. *Ibid.,* p. 533.
25. *Ibid.,* p. 534–5.

26. *Ibid.*, p. 536.
27. *Ibid.*, p. 522.
28. *Ibid.*, p. 535.
29. Ref. 2, p. 137.
30. *Clagett*, p. 583.
31. *Ibid.*, p. 594–5.
32. *Ibid.*, p. 595.
33. *Ibid.*, p. 595.
34. *Ibid.*, p. 596.
35. *Ibid.*, p. 600.
36. *Ibid.*, p. 601.
37. *Ibid.*, p. 602–3.
38. *Ibid.*, p. 603.
39. *Ibid.*, p. 605.
40. *Ibid.*, p. 606.

Chapter 5

1. *De Rev.*, p. 8 (Book I, Introduction).
2. N. Copernicus, *Commentariolus*; transl. by E. Rosen in: *Three Copernican Treatises*, 3rd edn, Peter Smith, New York (1971); also translated with extensive commentary by N. M. Swerdlow, *Proc. Am. Phil. Soc.*, **117**, 423 (1973).
3. *De Rev.*, pp. 4–5 (Preface to Pope Paul III).
4. Bede, *A History of the English Church and People*, Penguin Books (1955), *passim*.
5. V. Roberts, *Isis*, **48**, 428 (1957).
6. E. S. Kennedy and V. Roberts, *Isis*, **50**, 227 (1959).
7. E. S. Kennedy, *Isis*, **57**, 365 (1966).
8. N. M. Swerdlow, *Proc. Am. Phil. Soc.*, **117**, 423 (1973).
9. N. M. Swerdlow and O. Neugebauer, *Mathematical Astronomy in Copernicus's De Revolutionibus* (Studies in the History of Mathematics and Physical Sciences, Vol. 10), Springer, Berlin (1984).
10. *De Rev.*, p. 5.
11. *Ibid.*, p. 4.
12. *Astr. Nova*, Introduction.
13. *De Rev.*, p. 5.
14. *Ibid.*, p. 22.
15. *Ibid.*, p. xvii.
16. *De Rev.*, p. 11 (Book I, Chap. 4).
17. *De Rev.*, pp. 11–12 (Book I, Chap. 5).
18. *De Rev.*, p. 16 (Book I, Chap. 8).
19. *HAMA*, p. 1097, footnote 3.
20. *De Rev.*, p. 22 (Book I, Chap. 10).
21. *Astr. Nova*, opposite dedication.
22. *De Rev.*, p. xvi.
23. *Astr. Nova*, Chap. 14.
24. *De Rev.*, pp. 147–69 (Book III, Chaps. 14–25).
25. Ref. 16.

26. *As recounted by Kepler, Astr. Nova*, Chaps. 7 and 19.
27. *De Rev.*, pp. 227–306 (Book V).
28. *De Rev.*, Book V, Chap. 22.
29. See the discussion in Ref. 9, pp. 483ff.
30. *Astr. Nova*, Chap. 14.
31. *De Rev.*, p. 5 (Preface to Pope Paul III).
32. *De Rev.*, p. 169 (Book III, Chap. 25).
33. O. Gingerich, *Proc. Amer. Phil. Soc.*, **117**, 513 (1973).
34. A. Koestler, *The Sleepwalkers. A History of Man's Changing Vision of the Universe*, Hutchinson, London (1959), reprinted by Penguin Books, London (1977), p. 194.
35. (a) *Oxford English Dictionary*, entry for *cosmos*; (b) *De Rev.*, p. 7 (Book I, Introduction, first paragraph).
36. J. Kepler, *Harmonice Mundi Libri V*, in: *Gesammelte Werke*, Vol. 6, ed. M. Caspar, C. H. Beck'sche Verlagsbuchhandlung, Munich (1940); translated into German by M. Caspar: *Weltharmonik*, Oldenbourg, Munich 1939.
37. *De Rev.*, p. 12 (Book I, Chap. 5).
38. *De Rev.*, p. 16 (Book I, Chap. 8).
39. *Ibid.*, pp. 15–16 (Book I, Chap. 8).
40. *De Rev.*, p. 21 (Book I, Chap. 10).
41. *Dict. Sci. Biog.*, Vol. 25, Suppl. 1, p. 207 (1978).
42. *Almagest*, p. 132 (HI, 192–3).
43. *Ibid.*
44. *HAMA*, p. 1084.
45. *Almagest*, p. 329ff (HII, 17ff).
46. *Almagest*, p. 321 (HII, 2–3).
47. *Almagest*, p. 325 (HII, 8).
48. E. Halley, *Philosophical Transactions*, No. 355, 736 (1718).
49. *De Rev.*, pp. 145–6 (Book III, Chap. 13).
50. *De Rev.*, p. 4 (Preface to Pope Paul III).
51. E. Rosen, *Arch. Intern. d'Histoire des Sciences*, **25**, 82 (1975); N. M. Swerdlow, *ibid.*, **26**, 108 (1976).
52. *De Rev.*, p. 5 (Preface to Pope Paul III).
53. *De Rev.*, p. 22 (Book I, Chap. 10).
54. *Astra. Nova*, Chap. 33.
55. T. S. Kuhn, *The Copernican Revolution*, Harvard University Press, Cambridge, Mass (1957); A. Koestler, Ref. 34.
56. *De Rev.*, p. 4 (Preface to Pope Paul III).
57. *Astr. Nova*, Chap. 1.
58. *Ibid.*
59. *De Rev.*, in the Preface to Pope Paul III.
60. Quoted by: P. Frank, *Einstein: His Life and Times*, New York (1947), p. 101.

Chapter 6

1. For biographical details, see *Dict. Sci. Biog.*, Vol. 2 (1970), p. 401ff; J. L. E. Dreyer, *Tycho Brahe*, Adam and Charles Black, Edinburgh (1890); A. Koestler,

The Sleepwalkers, A History of Man's Changing Vision of the Universe, Hutchinson, London (1959), republished by Penguin (1964).
2. A. Koyré, *From the Closed World to the Infinite Universe*, Johns Hopkins University Press, Baltimore (1957), p. 62ff and p. 90ff.
3. 'The status of astronomy', in: N. Jardine, *The Birth of History and Philosophy of Science. Kepler's A Defence of Tycho against Ursus with Essays on its Provenance and Significance*, Cambridge University Press, Cambridge (1984).
4. *Astr. Nova*, Chap. 18.
5. J. Kepler, *Mysterium Cosmographicum* (1596) (Preface to reader), in *Ges. Werke*, Vol. 1 (1938), p. 9ff.
6. *Astr. Nova*, Chap. 7.
7. *De Rev.*, p. 22 Book I, Chap. 10.
8. A. Koestler, Ref. 1.
9. Quoted from M. Caspar's introduction to his German translation of the *Mysterium Cosmographicum*, Benno Filser Verlag, Augsburg (1923), p. xxx.
10. E. Segrè, *From Falling Bodies to Radio Waves*, W. H. Freeman, New York (1984), p. 32.
11. J. Kepler, *Mysterium Cosmographicum Editio altera cum notis* (1621), *Ges. Werke*, Vol. 8 (1963).
12. I understand that a translation by W. Donahue is to be published by Cambridge University Press.
13. There is a translation into German by M. Caspar: *Neue Astronomie*, R. Oldenbourg, Munich–Berlin (1929).
14. See Ref. 9, p. xx.
15. *Astr. Nova*, Introduction.
16. M. Caspar, *Kepler* (transl. by C. Doris Hellman), Abelard–Schuman, London (1959) (first publ. in German in 1948).
17. *Tychonis Brahe Opera Omnia*, Vol. 8 (1925), pp. 52–5 (21 April 1598).
18. B. Stephenson, *Kepler's Physical Astronomy* (Studies in the History of Mathematics and the Physical Sciences, Vol. 13), Springer, Berlin (1987).
19. See Ref. 3, p. 20.
20. *Astr. Nova*, Chap. 21.
21. O. Gingerich, 'Kepler's treatment of redundant observations', in: *Internationales Kepler–Symposium*, eds. F. Krafft, K. Meyer, and B. Stickler, Gerstenberg, Hildesheim (1973), p. 307.
22. J. B. J. Delambre, *Histoire de l'Astronomie Moderne*, 2 vols., Paris (1821).
23. See Caspar's note on p. 404 of Ref. 13.
24. *HAMA*, p. 1096.
25. *Astr. Nova*, Chap. 19.
26. *Astr. Nova*, Chap. 22.
27. *Ibid.*
28. C. Wilson, 'Kepler's derivation of the elliptical path', *Isis*, **59**, 5 (1968).
29. E. Rosen (translator), *Kepler's Somnium: The Dream or Posthumous Work on Lunar Astronomy*, Madison (1967); J. Lear, *Kepler's Dream*, Berkeley (1965).
30. J. Kepler, *Epitome Astronomiae Copernicanae* (1618, 1620), in: *Ges. Werke*, Vol. 7 (1953), p. 316 (quoted from the English translation in: *Great Books of the Western World*, Vol. 16, ed. B. M. Hutchins, Encyclopaedia Britannica, Chicago (1952), pp. 915–6).

31. *Astr. Nova*, Chap. 26.
32. Coined in Ref. 5.
33. A. Koyré, *The Astronomical Revolution: Copernicus–Kepler–Borelli*, Methuen, London (1973), p. 165.
34. D. T. Whiteside, 'Keplerian planetary eggs, laid and unlaid, 1600–1605', *J. Hist. Astr.*, **5**, 1 (1974).
35. Caspar's introduction to Ref. 13.
36. E. J. Aiton, 'Infinitesimals and the area law', in the book cited in Ref. 21, p. 285; 'Kepler's second law of planetary motion', *Isis*, **60**, 75 (1969).
37. *Astr. Nova*, Chap. 60.
38. *Dict. Sci. Biog.*, Vol. 7, p. 289ff.
39. J. V. Field, *Kepler's Geometrical Cosmology*, Athlone Press (1988). See also A. Koyré, *From the Closed World to the Infinite Universe*, Johns Hopkins University Press, Baltimore (1957), Chap. 3; E. J. Aiton, *The Vortex Theory of Planetary Motions*, Macdonald, London (1972), Chap. 1.
40. M. Jammer, *Concepts of Mass in Classical and Modern Physics*, Harvard University Press, Cambridge, Mass. (1961), Chap. 5.
41. Ref. 5, Chap. 20.
42. In the General Scholium added at end of 2nd edn of the *Principia*.
43. *Astr. Nova*, Chap. 33.
44. *Ges. Werke*, Letters to David Fabricius, 11 Oct. 1605 (Vol. 15, p. 240) and 10 Nov. (1608) (Vol. 16, p. 194).
45. For example: U. Hoyer, 'Kepler's celestial mechanics', in: *Vistas in Astronomy*, Vol. 23 (1979), p. 69.
46. Ref. 30, Book IV, Part 2.
47. Ref. 44, near end of second letter.
48. J. Kepler, *Antwort auf Roeslini Diskurs* (1609), in: *Ges. Werke*, Vol. 4 (1941).
49. I. B. Cohen, 'Newton and Keplerian inertia: an echo of Newton's controversy with Leibniz', in: *Science, Medicine, and Society in the Renaissance. Essays to Honor Walter Pagel*, Vol. 2 (ed. A. J. Debus), Science History Publications, New York (1972), pp. 199–211.
50. In Definition 3 in the *Principia*.
51. Ref. 5, Chap. 21.
52. Ref. 11, notes to Chap. 21.
53. J. Kepler, *Harmonices Mundi Libri V* (1619), in: *Ges. Werke*, Vol. 6 (1940), Book V, Chap. 3.
54. Ref. 30, Book IV.
55. J. Treder, 'Die Dynamik der Kreisbewegungen der Himmelskörper und des freien Falls bei Aristoteles, Copernicus, Kepler and Descartes', *Studia Copernicana*, **14**, 105–150 (1975); 'Kepler und die Begründung der Dynamik', *Die Sterne*, **49**, 44–48 (1973).
56. *Astr. Nova*, Introduction.
57. J. Kepler, *De Stella Nova in Pede Serpentarii* (1606), in: *Ges. Werke*, Vol. 1 (1938) (English translation quoted from Ref. 2, p. 61).
58. Ref. 30, English translation, p. 855.
59. *Ibid.*, p. 888.
60. Ref. 2, p. 60.
61. *Ibid.*, p. 61.

62. Ref. 30, English translation, p. 855.
63. Ref. 2, p. 87.
64. *Astra. Nova*, Chap. 39.
65. Ref. 2, p. 86.
66. *Astr. Nova*, Chap. 7.
67. Ref. 2, p. 58.
68. Ref. 30, English translation, pp. 853–4.
69. G. J. Toomer, 'Hipparchus and Babylonian astronomy', in: *Memorial Volume for A. J. Sachs*, University Museum of Pennsylvania.
70. N. Swerdlow and O. Neugebauer, *Mathematical Astronomy in Copernicus's De Revolutionibus* (Studies in the History of Mathematics and the Physical Sciences, Vol. 10), Springer, Berlin (1984), p. 307.
71. J. L. Russell, 'Kepler's laws of planetary motion: 1609–1666', *Brit. J. Hist. Sci.*, **2**, 1–24 (1964).

Chapter 7

1. S. Drake, in: *Kepler: Four Hundred Years,* eds. A. Beer and P. Beer, *Vistas in Astronomy*, Vol. 18, Pergamon Press, Oxford (1975), p. 237.
2. *Dialogo*, p. 203.
3. S. Drake, *Galileo*, Oxford University Press, Oxford (1980), p. 22.
4. S. Drake, *Galileo at Work*, University of Chicago Press, Chicago, (1978), p. 41.
5. Ref. 4, p. 36ff.
6. G. Galileo, *Sidereus Nuncius* (1610), in: *Opere*, Vol. 3 (1892) (translated in Ref. 7).
7. S. Drake, *Discoveries and Opinions of Galileo* (translations of various works with an introduction), Doubleday Anchor Books, New York (1957), p. 57 (from translation of Ref. 6).
8. G. Galilei, *Delle Macchie Solari*, in: *Opere*, Vol. 6 (1895) (excerpts translated in Ref. 7, p. 87ff).
9. *Dialogo*.
10. *Discorsi*.
11. G. Galilei, *Il Saggiatore* (1623), in: *Opere*, Vol. 6, p. 197 (my translation).
12. *Discorsi*, p. 153.
13. E. Segrè, *From Falling Bodies to Radio Waves: Classical Physicists and Their Discoveries*, Freeman, New York (1985), p. 32.
14. S. Drake, Ref. 3, p. 12.
15. *Almagest*, p. 43 (HI, 21).
16. *De Rev.*, p. 18 (Book I, Chap. 9).
17. *Almagest*, p. 44 (HI, 23–4).
18. T. S. Kuhn, *The Copernican Revolution*, Harvard University Press, Cambridge, Mass. (1957).
19. *De Rev.*, p. 18 (Book I, Chap. 7).
20. *De Rev.*, p. 15 (Book I, Chap. 8).
21. *De Rev.*, p. 10 (Book I, Chap. 4).
22. *De Rev.*, p. 11 (Book I, Chap. 5).
23. *De Rev.*, p. 16 (Book I, Chap. 8).

24. *De Rev.*, p. 17 (Book I, Chap. 8).
25. *Dialogo*, p. 19.
26. *Ibid.*, p. 20.
27. *Ibid.*, p. 19.
28. *Ibid.*, p. 32.
29. *Ibid.*, p. 33.
30. *De Rev.*, p. 17 (Book I, Chap. 8).
31. *Dialogo*, p. 16.
32. *Ibid.*, p. 101.
33. G. Galilei, *De Motu*, in: *Opere*, Vol. 1 (1890); English translation in: *On Motion and On Mechanics* (transl. with introductions by I. E. Drabkin and S. Drake), University of Wisconsin Press, Madison (1960).
34. M. Jammer, *Concepts of Force: A Study in the Foundations of Dynamics*, Harvard University Press, Cambridge, Mass. (1957), pp. 60–1.
35. *Astr. Nova*, Chap. 34.
36. See Ref. 33, pp. 78–9 of English translation.
37. See Ref. 33, p. 65 of English translation.
38. *Ibid.*, pp. 65–6.
39. *Ibid.*, p. 66.
40. *Ibid.*, p. 73.
41. *Ibid.*, pp. 73–4.
42. Ref. 3, pp. 32–3.
43. *De Rev.*, p. 17 (Book I, Chap. 8).
44. Ref. 3, p. 33.
45. *Ibid.*
46. *Dialogo*, p. 51.
47. S. Drake, *Nuncius*, **3**, No. 1 (1988).
48. *Ibid.*, see also Ref. 4, p. 74ff (Chaps. 5 and 6).
49. *Discorsi*, p. 167.
50. *Ibid.*, p. 174.
51. *Clagett*, p. 541ff.
52. M. R. Cohen and I. E. Drabkin, *A Source Book in Greek Science*, Harvard University Press, Cambridge, Mass. (1948), p. 209.
53. *Clagett*, p. 555.
54. *Ibid.*, p. 556.
55. *Mechanics*, p. 332ff.
56. *Ibid.*, p. 171.
57. *Dialogo*, p. 164.
58. Ref. 4, p. 128.
59. *Discorsi*, p. 260.
60. Ref. 4, p. 129.
61. *Discorsi*, pp. 244–5.
62. *Ibid.*, pp. 248–9.
63. *Ibid.*, p. 250.
64. E. Wohlwill, 'Die Entdeckung des Beharrungsgesetzes', *Zeitschrift für Völkerpsychologie*, **14**, 365–410 (1884); **15**, 70–135, 337–87 (1884).
65. *Ibid.*

66. A. Koestler, The Sleepwalkers. *A History of Man's Changing Vision of the Universe*, Hutchison, London (1959), reprinted by Penguin Books, London (1977), pp. 454–5.
67. *Dialogo*, p. 455.
68. *Ibid.*, p. 52.
69. Ref. 7, pp. 113–4.
70. *Dialogo*, p. 146.
71. *Ibid.*, p. 148.
72. *Astr. Nova*, Introduction.
73. Quoted in the introduction to W. Gilbert, *On the Loadstone and Magnetic Bodies, and on the Great Magnet the Earth*, Bernard Quaritch, London (1893).
74. *Dialogo*, p. 142.
75. J. Herivel, *The Background to Newton's Principia*, Clarendon Press, Oxford (1965), p. 121.
76. *Dialogo*, p. 190ff.
77. See J. B. Barbour, *Contemporary Physics*, **26**, 397 (1985).
78. S. Drake, *Galileo Studies*, University of Michigan Press, Ann Arbor (1970), Chaps. 12 and 13.
79. *Dialogo*, p. 122.
80. *Ibid.*, p. 171.
81. *Ibid.*, p. 35.
82. *Ibid.*, p. 114.
83. *Ibid.*, p. 248.
84. *Ibid.*, p. 171.
85. *Ibid.*, p. 116.
86. *Ibid.*, p. 177–8.
87. *Ibid.*, p. 171.
88. *Ibid.*, p. 145.
89. *Dict. Sci. Biog.*, Vol. 5 (1972), p. 284ff.
90. *Dialogo*, p. 156.
91. See Ref. 64.
92. *Dialogo*, pp. 167–8.
93. *Ibid.*, p. 186.
94. Ref. 13, p. 31.
95. E. Whittaker, *From Euclid to Eddington, A Study of Conceptions of the External World*, Cambridge (1949), p. 58.
96. *Dialogo*, p. 186–7.
97. *Ibid.*, p. 19.
98. *Ibid.*, p. 20.
99. *Ibid.*, p. 327.
100. *De Rev.*, p. 17 (Book I, Chap. 8).
101. *Dialogo*, p. 356.
102. *Ibid.*, p. 130.
103. G. Galilei, *Discorso del Flusso e Reflusso del Mare* (1616), in: *Opere*, Vol. 5, p. 371.
104. S. Drake, 'The organizing theme of the Dialogue', *Atti dei Convegni Lincei*, **55**, 101–4 (1983).
105. *Dialogo*, p. 427 (see also Ref. 103, p. 382).

106. *Mechanics*, p. 264.
107. Ref. 64.
108. See *Encyclopaedia Britannica*, 15th edn, article on Plato.
109. *Discorsi*, pp. 153–4.
110. A. Koyré, 'A documentary history of the problem of fall from Kepler to Newton', *Trans. Amer. Phil. Soc.*, **45**, 329–95 (1955).
111. *Dialogo*, p. 166.
112. *Ibid.*
113. Drake's notes on the *Dialogo*, p. 477.

Chapter 8

1. *Encyclopaedia Britannica*, 15th edn, Vol. 5, article on Descartes, pp. 597–602.
2. Letter of Descartes to Mersenne, April 1634, in: *Oeuvres*, Vol. 4, p. 285ff.
3. Quoted from Santillana's intoduction to: G. Galilei, *Dialogue on the Great World Systems in the Salusbury Translation*, University of Chicago Press (1953), p. vii.
4. For details about Beeckman, see, for example, *Dict. Sci. Biog.*, Vol. 1 (1970), pp. 566ff.
5. *Principles*, §I.1, p. 3.
6. *Ibid.*, §I.43, p. 19.
7. See M. S. Mahoney's Introduction to his translation of *World*.
8. Quoted from: J. F. Scott, *The Scientific Work of René Descartes*, Taylor & Francis Ltd., London (1952), p. 161.
9. *Ibid.*
10. *World*, Chap. 7, pp. 75–6.
11. G. Galilei, *Il Saggiatore* (1623), in *Opere*, Vol. 16; quoted from S. Drake, *Discoveries and Opinions of Galileo*, Doubleday Anchor Books, New York (1957), pp. 276–7.
12. *World*, Chap. 5, pp. 39–41.
13. Descartes to Beaune, *Oeuvres*, Vol. 2, p. 542, quoted from Ref. 27 in Chap. 9.
14. See, for example, Mach's *Mechanics*, p. 141ff.
15. *Principles*, §111.30, p. 96.
16. *Principia*, p. 395–6.
17. *World*, Chap. 3.
18. *Ibid.*, p. 19.
19. *World*, Chap. 13, p. 171.
20. *World*, Chap. 11, p. 123.
21. *Principles*, §I.52, p. 23.
22. *Ibid.*, §I.53, pp. 23–4.
23. G. Berkeley, *Treatise Concerning the Principles of Human Knowledge* (1710).
24. *Astr. Nova*, Chap. 33.
25. *Principles*, §II.22, pp. 49–50.
26. A. Koyré, *From the Closed World to the Infinite Universe*, Johns Hopkins University Press, Baltimore (1957), p. 100.
27. *Ibid.*, p. 101ff.
28. R. Descartes, *La Géometrie* (first published in 1637 as an appendix to *Discours de la Méthode*), in: *Oeuvres*, Vol. 6; transl. by D. E. Smith and M. L. Latham as:

The Geometry of René Descartes, Open Court (1925), republished by Dover, New York (1954).
29. *Dict. Sci. Biog.*, Vol. 4, pp. 55ff.
30. *Mechanics*, Chap. V; see also E. Mach, *The Analysis of Sensations*, Dover Publications, New York (1959).
31. *Principles*, §I.48, p. 21.
32. B. Russell, *A History of Western Philosophy*, Simon and Schuster, New York (1945), p. 564.
33. *World*, Chap. 7, p. 71.
34. *Ibid.*, pp. 73–5.
35. *Ibid.*, p. 65.
36. *World*, Chap. 3, p. 15.
37. *World*, Chap. 7, p. 59.
38. S. L. Jaki, *Science and Creation: From Eternal Cycles to an Oscillating Universe*, Scottish Academic Press, Edinburgh (1974).
39. *World*, Chap. 7, p. 61.
40. *Ibid.*, p. 65.
41. *Principles*, §§II.46–52.
42. *World*, Chap. 7, p. 71.
43. *Principles*, §§II.37–52.
44. *World*, Chap. 6, pp. 53–5.
45. *Principles*, §IV.I, p. 181.
46. Quoted from E. Wohlwill, 'Die Entdeckung des Beharrungsgesetzes', *Zeitschrift für Völkerpsychologie*, **14**, 365–410 (1884); **15**, 70–135, 337–84 (1884).
47. A. Koyré, *Études Galiléennes*, Hermann, Paris (1966), p. 108.
48. Letter of Descartes to Mersenne, October 1638, *Oeuvres*, Vol. 2, p. 380.
49. W. Whewell, *History of the Inductive Sciences* (1857), p. 39.
50. *Mechanics*, p. 363.
51. *Oeuvres*, Vol. 1, pp. 270–2; quoted from *Principles*, p. 83.
52. Ref. 26, p. 143.
53. *World*, Chap. 6, p. 49.
54. *Ibid.*, p. 51.
55. *Ibid.*, p. 53.
56. *World*, Chap. 4, p. 25.
57. *World*, Chap. 7, p. 63.
58. *World*, p. xviii.
59. Ref. 47, p. 333.
60. *Principles*, §II.18, p. 48.
61. *Ibid.*, §II.13, p. 45.
62. *Ibid.*
63. *Ibid.*, §II.15, p. 46.
64. Ref. 26, Chap. 4, p. 104ff.
65. *Principles*, §II.25, p. 51.
66. *Ibid.*, §II.29, p. 53.
67. *Ibid.*, §II.30, pp. 53–4.
68. *Ibid.*, §III.38, p. 101.
69. *Ibid.*, §II.31, p. 54.

Chapter 9

1. C. Huygens, *Traité de la Lumière* (1690), in: *Oeuvres Complètes*, Vol. 20 (1937), p. 451ff.
2. C. Huygens, *Treatise on Light*, transl. by S. P. Thompson, Macmillan, London (1912).
3. E. Wohlwill, 'Die Entdeckung des Beharrungsgesetzes', *Zeitschrift für Völkerpsychologie*, **15**, 337 (1884).
4. *Dict. Sci. Biog.*, Vol. 5, p. 284ff.
5. See Ref. 4, Vol. 6, p. 597ff.
6. C. Huygens, *Horologium Oscillatorum* (1673) in: *Oeuvres Complètes*, Vol. 18 (1934), p. 87; German transl. in the series *Oswald's Klassiker der Exakten Wissenschaften*, No. 192, Leipzig (1913).
7. See the article on Richer in Ref. 4, Vol. 11, p. 423ff.
8. Asserted by Newton in Def. 7 in the *Principia*, p. 4; Hooke told Newton in a letter of 6 January 1679 that Halley had found his pendulum to go slower on the top of a mountain in St. Helena than at sea level (*Correspondence of Isaac Newton*, Vol. 2, ed. H. W. Turnbull, Cambridge University Press (1960), p. 310).
9. C. W. Misner, K. S. Thorne, and J. A. Wheeler, *Gravitation*, W. H. Freeman, San Francisco (1973), p. 762.
10. C. Huygens, *De Motu Corporum ex Percussione* (written 1656, publ. 1703), in *Oeuvres Complètes*, Vol. 16 (1929), p. 31; English transl. by R. J. Blackwell in *Isis*, **68**, 574–97 (1977); German transl. in the series *Oswald's Klassiker der Exakten Wissenschaften*, No. 138, Leipzig (1903).
11. C. Huygens, *De Vi Centrifuga* (written 1659, publ. 1703) in: *Oeuvres Complètes*, Vol. 16 (1929), p. 255; German transl. in the series *Oswald's Klassiker der Exakten Wissenschaften*, No. 138, Leipzig (1903).
12. C. Huygens, *Journal des Sçavans*, 18 March (1669), in: *Oeuvres Complètes*, Vol. 16 (1929), pp. 179–81.
13. C. Huygens, *Philosophical Transactions*, **4**, 927 (1669).
14. *Oeuvres Complètes*, Vol. 16 (1929), p. 27.
15. N. Armstrong as he stepped onto the moon.
16. *Principles*, §II.37–53, pp. 59–69.
17. *Ibid.*, §II.45.
18. A. Gabbey, *Studies in Hist. Phil. Sci.*, **3**, 373–85 (1973).
19. A. Gabbey, 'Force and inertia in the 17th century: Descartes and Newton', in: *Descartes: Philosophy, Mathematics and Physics*, ed. S. Gaukroger, (Harvester Readings in the History of Science and Philosophy, Vol. 1), Brighton (1980).
20. *Principles*, §II.53, p. 69.
21. Paper No. 42 in: *Gottfried Wilhelm Leibniz: Philosophical Papers and Letters*, ed. L. E. Loemker, Reidel, Dordrecht (1976) (see also Paper No. 24).
22. *Mechanics*, p. 285.
23. *Discorsi*, pp. 170–1.
24. *Mechanics*, p. 210ff.
25. See, for example, the articles by B. Carter and G. W. Gibbons in: *General Relativity. An Einstein Centenary Survey*, eds. S. W. Hawking and W. Israel, Cambridge University Press, Cambridge (1979).

26. G. W. Leibniz, *Acta Eruditorum,* March (1686); translated as Paper No. 34 in the book quoted in Ref. 21.
27. *Westfall* (1971), p. 148ff.
28. *Oeuvres Complètes,* Vol. 16, p. 92; translation quoted from Ref. 27, p. 149.
29. *Ibid.*, p. 93.
30. Quoted and translated from J. A. Schouten, *Jahresberichte der Deutschen Mathematiker – Vereinigung,* **29**, 136 (1920); also published in *Oeuvres Complètes,* Vol. 16 (1929), p. 233.
31. See Ref. 28, p. 96.
32. G. W. Leibniz, see the book quoted in Ref. 21, Paper No. 46.
33. *Oeuvres Complètes,* Vol. 16, p. 150; quoted from Ref. 27, p. 147.
34. H. More, *An Antidote to Atheism,* London (1652), quoted from p. 134 of Ref. 32 in Introduction.
35. G. Galilei, *De Motu* in: *Opere,* Vol. 1 (1890); English translation in: *On Motion and On Mechanics* (transl. with introductions by I. E. Drabkin and S. Drake), University of Wisconsin Press, Madison (1960).
36. L. Lange, *Die geschichtliche Entwicklung des Bewegungsbegriff und ihr vorausschichtliches Endergebnis,* Leipzig (1886).
37. *De Rev.*, Book I, chap. 8, p. 15.
38. *Ibid.*
39. *Dialogo*, p. 190ff.
40. *Dialogo*, p. 178.
41. *Principles,* §III.56–59, p. 112ff.
42. See Ref. 27, p. 172 and *Oeuvres Complètes,* Vol. 16, p. 304, 323–4.
43. Ref. 27.
44. *Oeuvres Complètes,* Vol. 19, p. 631; quoted from Ref. 27, p. 186.
45. *Oeuvres Complètes,* Vol. 10, p. 190; quoted from Ref. 27, p. 184.
46. *Oeuvres Complètes,* Vol. 10, p. 403; quoted from Ref. 27, p. 185.
47. Ref. 27, p. 188.

Chapter 10

1. *Discorsi*, pp. 153/4.
2. *Westfall (1980).*
3. J. Gascoigne, 'The universities and the scientific revolution: the case of Newton and Restoration Cambridge', *Hist. Sci.*, **23**, 391 (1985).
4. For more details on More see the article on him in *Dict. Sci. Biog.*, Vol. 9, pp. 509–10.
5. Ref. 2, p. 140.
6. I. Newton, Catalogue of Portsmouth Collection, Cambridge (1888), Sec. 1, Division xi, No. 41.
7. Ref. 2, p. 207.
8. Ref. 2, Chap. 14.
9. *Westfall (1971)*, opp. contents page.
10. See, for example, Ref. 2, p. 402ff.
11. (a) In Ref. 2 and in *Westfall (1971)*; (b) D. T. Whiteside, *Brit. J. Hist. Sci.*, **2**, 117 (1964); *Hist. Sci.*, **5**, 104 (1966); *J. Hist. Astr.*, **1**, 5 (1970).

12. Virtually all the relevant material can be found in: (a) J. W. Herivel, *The Background to Newton's 'Principia'*, Clarendon Press, Oxford (1965); (b) A. R. and M. B. Hall, *Unpublished Scientific Papers of Isaac Newton*, Cambridge University Press, Cambridge (1962); (c) Various letters in: *The Correspondence of Isaac Newton*, Vol. 2, ed. H. W. Turnbull, Cambridge University Press (1960). The letters are: Hooke to Newton, 24 Nov. 1679; Newton to Hooke, 28 Nov. 1679; Hooke to Newton, 6 Dec. 1679; Newton to Hooke, 13 Dec. 1679; Hooke to Newton, 6 Jan. 1679/80; ditto, 17 Jan. 1679/80; Newton to Crompton for Flamsteed, 28 Feb. 1680, p. 340; draft of Newton's letter [? to Crompton], ? April 1681, p. 358; Newton to Flamsteed, 16 April 1681; Newton to Flamsteed, 19 Sept. 1685, p. 419, and the Halley–Newton correspondence about Hooke's charge of plagiarism, commencing with Halley's letter of 22 May 1686 (pp. 431–47).
13. See, in particular Leibniz's *Specimen Dynamicum* (*Acta Eruditorum*, April 1695); a translation of an extended version is given as Paper No. 46 in Ref. 21 of Chap. 9.
14. Ref. 12(a), p. 133.
15. Ref. 12(a), p. 137.
16. Ref. 12(a), pp. 162–70.
17. *Ibid.*, p. 182.
18. *Ibid.*, p. 156 (§104).
19. *Ibid.*, p. 156 (§106).
20. *Ibid.*, p. 159 (§§119–21).
21. *Ibid.*, p. 142 (§9).
22. *Ibid.*, p. 143 (§10).
23. See Ref. 16.
24. J. L. Russell, *Brit. J. Hist. Sci.*, **9**, 25 (1976).
25. Ref. 12(a), p. 312.
26. *Principia*, p. 21.
27. *Ibid.*, p. 22ff.
28. *Ibid.*, p. 25.
29. *Discorsi*, pp. 264–5.
30. Ref. 12(a), pp. 146–7; see also Ref. 2, pp. 148–9.
31. G. A. Borelli, *Theoricae Mediceorum Planetarum a Causis Physicis Deductae*, Florence (1666).
32. A. Koyré, *The Astronomical Revolution: Copernicus–Kepler–Borelli*, Methuen, London (1973).
33. *Ibid.*, p. 480.
34. *Principia*, p. 47.
35. Ref. 12(a), p. 147.
36. *Ibid.*, pp. 129–30.
37. *Ibid.*, p. 129.
38. *Ibid.*, p. 195.
39. *Ibid.*, p. 130.
40. *Ibid.*, p. 121.
41. *Principia*, xvii.
42. Ref. 12(a), pp. 183–91.
43. *Dialogo*, p. 190ff.

44. Ref. 12(a), p. 196.
45. *Ibid.*, p. 196.
46. Huygens, *Oeuvres Complètes*, Vol. 16, p. 304; Ref. 27 in Chap. 9, p. 172.
47. C. W. Allen, *Astrophysical Quantities*, 3rd edn, Athlone Press, London (1976), p. 114.
48. Ref. 12(a), p. 11.
49. See the articles in *Dict. Sci. Biog.* on (a) Snel, Vol. 12, p. 500; (b) Norwood, Vol. 10, p. 151; (c) Picard, Vol. 10, p. 595.
50. Ref. 12(c), Hooke to Newton, 6 Jan. 1679/80, p. 309.
51. Ref. 12(a), Chap. 4, pp. 65–76.
52. W. Whiston, *Memoirs of the Life of Mr. William Whiston*, 2 vols., London (1749), Vol. 1, pp. 35–6.
53. See the Halley–Newton correspondence quoted in Ref. 12(c), esp. p. 436 and p. 445.
54. W. Stukeley, *Memoirs of Sir Isaac Newton's Life*, ed. A. H. White, London (1936), pp. 19–20.
55. H. Pemberton, *A View of Sir Isaac Newton's Philosophy*, Dublin (1728), Preface.
56. Ref. 6.
57. C. A. Wilson, *Arch. Hist. Exact Sci.*, **6**, 92 (1970).
58. J. W. Herivel, *Brit. J. Hist. Sci.*, **2**, 350 (1965).
59. Royal Society Register III, 14; see also T. Birch, *The History of the Royal Society of London*, 4 vols., London (1756–57), Vol. 2, p. 90.
60. J. L. Russell, *Brit. J. Hist. Sci.*, **2**, 1 (1964).
61. Robert Hooke, *Lectiones Cutleriane*, in R. T. Gunther, *Early Science in Oxford*, 14 vols., Oxford (1920–45), Vol. 8, pp. 27–8.
62. Ref. 2, p. 383.
63. *Dict. Sci. Biog.*, Vol. 6, p. 481ff.
64. Ref. 12(c), p. 297.
65. *Ibid.*, p. 301.
66. Ref. 2, p. 385.
67. Ref. 12(c), p. 305.
68. *Ibid.*, p. 307.
69. *Westfall (1971)*, p. 426ff.
70. *Principia*, p. 21.
71. Ref. 12(c); see also Ref. 12(a), Chap. 4.
72. Ref. 12(c), p. 301.
73. ULC. Add. 3968.9, 101^r, quoted from Ref. 11(b) (1964).
74. Ref. 11(b) (1964), p. 135.
75. Ref. 12(a), pp. 246–56.
76. *Dialogo*, p. 52.
77. Ref. 12(c), pp. 336–9.
78. *Ibid.*, pp. 340–7.
79. *Ibid.*, pp. 363–7.
80. Ref. 2, p. 394.
81. Ref. 12(c), pp. 358–62.
82. *Ibid.*, p. 361.
83. Ref. 2, pp. 395–6.
84. Ref. 12(c), p. 407.

85. *Ibid.*, p. 413.
86. *Ibid.*, p. 419.
87. *Ibid.*, pp. 421–2.
88. Ref. 2, p. 655ff.
89. See the article on Newton in *Dict. Sci. Biog.*, Vol. 10, p. 65.
90. *Discorsi*, p. 243.
91. *Principia*, xviii.
92. Ref. 2, p. 404.
93. Ref. 12(a), p. 277.
94. *Ibid.*, p. 279.
95. *Ibid.*, p. 284.
96. Ref. 12(c), p. 437.
97. Ref. 2, p. 404.
98. Ref. 12(a), pp. 294–326.
99. Ref. 2, p. 445.
100. *Ibid.*, p. 405.
101. *Ibid.*, p. 445.
102. Ref. 12(c), p. 431.
103. *Ibid.*, p. 309.
104. *Ibid.*, p. 313.
105. *Ibid.*, p. 438.
106. For an account of this work see the article on Picard in *Dict. Sci. Biog.*, Vol. 10, pp. 595–7.
107. Ref. 12(a), p. 302.
108. Ref. 12(c), p. 444.
109. *Principia*, p. 46.
110. Ref. 12(a), p. 277.
111. *Ibid.*, pp. 278–9.
112. *Ibid.*, p. 299.
113. H. More, *An Antidote to Atheism*, London (1652), quoted from p. 134 of A. Koyré, *From the Closed World to the Infinite Universe*, Johns Hopkins University Press, Baltimore (1957).
114. *Mechanics*, p. 305.
115. *Principia*, p. 14.
116. *Principia*, Book I, Lemma X, p. 34.
117. Ref. 12(a), p. 301.
118. *Ibid.*, pp. 304–20.
119. *Ibid.*, pp. 312–13.
120. *Principia*, p. 13.
121. *Ibid.*, p. 17.
122. *Ibid.*, p. 17.
123. *Ibid.*, p. 18.
124. *Ibid.*, p. 19.
125. *Ibid.*, p. 20.
126. *Ibid.*, p. 20.
127. *Ibid.*, xvii.
128. *Ibid.*, p. 397.
129. D. T. Whiteside: (a) 'Newtonian dynamics', *Hist. Sci.*, **5**, 104–17 (1966), esp.

pp. 108–10; (b) 'The mathematical principles underlying Newton's *Principia Mathematica*', *J. Hist. Astr.*, **1**, 116–38 (1970); (c) *The Mathematical Papers of Isaac Newton*, 8 vols., ed. D. T. Whiteside, Cambridge University Press, Cambridge (1967–81).

130. *Principia*, Book I, Prop. V, p. 47.
131. *Ibid.*, Prop. X, p. 53.
132. *Principia*, p. 423.
133. A. Gabbey, 'Newton and the libration of the rotating moon', in: *Newton and Halley Conference* (ed. N. Thrower), William Andrews Clark Memorial Library, Los Angeles (1986).
134. For a discussion of Newton's work on the moon see Cajori's note No. 23 on p. 649 of *Principia*.
135. *Dict. Sci. Biog.* (article on Newton by I. B. Cohen), Vol. 10 (1974), p. 78.
136. *Principia*, p. 203.
137. *Ibid.*, p. 164.
138. Ref. 12(a), pp. 301–2.
139. *Principia*, p. 378.
140. See Cajori's note No. 38 in *Principia*, pp. 661–2.
141. *Principia*, p. 393.
142. *Ibid.*, p. 394.
143. *Ibid.*, p. 397.
144. *Ibid.*, p. 408.
145. *Ibid.*, p. 411.
146. *Ibid.*, p. 411.
147. *Ibid.*, p. 412.
148. *Ibid.*, pp. 414–15.
149. *Ibid.*, p. 416.
150. T. S. Kuhn, *The Structure of Scientific Revolutions*, 2nd edn, University of Chicago Press, Chicago (1970).
151. *Principia*, p. xvii.
152. Quoted from: A. Koestler, *The Sleepwalkers. A History of Man's Changing Vision of the Universe*, Penguin Books (1977), p. 55.
153. *Principia*, p. 543ff.
154. I. B. Cohen, *Introduction to Newton's 'Principia'*, Cambridge University Press, Cambridge (1971).
155. A. Koyré, 'Newton and Descartes', in: *Newtonian Studies*, London (1965), pp. 53–114, esp. p. 65.

Chapter 11

1. M. Jammer, *Concepts of Space: The History of Theories of Space in Physics*, 2nd edn, Harvard University Press, Cambridge, Mass. (1969), p. 85ff.
2. A. Koyré, *From the Closed World to the Infinite Universe*, Johns Hopkins University Press, Baltimore (1957).
3. E. Grant, *Much Ado About Nothing. Theories of Space and Vacuum from the Middle Ages to the Scientific Revolution*, Cambridge University Press, Cambridge (1981).

4. Ref. 2, p. 159.
5. *Ibid.*, p. 149.
6. *Ibid.*, p. 144.
7. *Ibid.*, p. 144.
8. *Ibid.*, pp. 145–6.
9. *Ibid.*, p. 146.
10. *Mechanics*, p. 169.
11. E. Wohlwill, 'Die Entdeckung des Beharrungsgesetzes', *Zeitschrift für Völkerpsychologie*, **5**, 337 (1884).
12. J. W. Herivel, *The Background to Newton's Principia*, Clarendon Press, Oxford (1966), p. 35ff.
13. *Westfall (1980)*, p. 304.
14. A. Einstein, 'Prinzipielles zur allgemeinen Relativitätstheorie', *Ann. Phys.*, **55**, 241 (1918).
15. Ref. 12, p. 208.
16. See: *The Philosophical Works of Descartes*, transl. by E. S. Haldane and G. R. T. Ross, Cambridge University Press, Cambridge (1975), Vol. 1, p. 265.
17. See, for example, Ref. 13, p. 153.
18. A. R. Hall and M. B. Hall, *Unpublished Scientific Papers of Isaac Newton*, Cambridge University Press, Cambridge (1962).
19. See Cajori's note No. 52 in *Principia*, p. 669.
20. *Principia*, p. 546.
21. *Ibid.*, p. 547.
22. *Westfall (1971)*, Chap. 7, especially the end.
23. *Principia*, pp. 6–12.
24. *Ibid.*, p. 57.
25. I. Newton, Opticks, 3rd edn, London (1730), Query 28; reprinted by Bell & Sons, London (1931), p. 370.
26. *Ibid.*, Question 31, p. 401.
27. *Principia*, p. 20.
28. *Ibid.*, General Scholium (added in 2nd edition), p. 546.
29. *Almagest*, p. 169 (HI 259).
30. *The Correspondence of Isaac Newton*, Vol. 2, ed. H. W. Turnbull, Cambridge University Press (1960), p. 420.
31. *HAMA*, p. 67, Note 2.
32. Ref. 12, p. 309.
33. Ref. 13 (opp. title page) (Newton to Nathaniel Hawes, 25 May 1694).
34. See Ref. 2, p. 142ff.
35. *Principles*, §II.29.
36. Ref. 12, p. 310.
37. *Principia*, p. 424.
38. *Principia*, p. 429.
39. C. W. Allen, *Astrophysical Quantities*, 3rd edn, Athlone Press, London (1976), p. 112.
40. Cajori's note 41 in *Principia*, p. 664.
41. *Principia*, p. 401.
42. *Ibid.*, p. 419.
43. *Ibid.*, p. 574.

44. *Ibid.*, p. 543.
45. Ref. 25, Question 31, p. 402.
46. M. Keynes, 'Newton, the man', in: *Royal Society. Newton Tercentenary Celebrations 15–19 July, 1946*, Cambridge University Press, Cambridge (1947), p. 27.
47. H. Stein, 'Newtonian space–time', *Texas Quarterly*, **10**, 174 (1967).
48. J. Earman, 'Who's afraid of absolute space?', *Australasian J. Philosophy*, **48**, 287 (1970).

Chapter 12

1. C. Neumann, *Ueber die Principien der Galilei–Newton'schen Theorie*, Teubner, Leipzig (1870) (the quotations are my translation).
2. L. D. Landau and E. M. Lifshitz, *Mechanics*, 3rd edn, Pergamon Press, Oxford (1976), p. 98ff.
3. E. Mach, *History and Root of the Principle of the Conservation of Energy*, Open Court, Chicago (1911); German original: *Die Geschichte und die Wurzel des Satzes von der Erhaltung der Arbeit*, Prague (1872).
4. H. Streintz, *Die physikalischen Grundlagen der Mechanik*, Leipzig (1883).
5. L. Lange, 'Über das Beharrungsgesetz', *Berichte der math.–phys. Classe der Königl. Sächs. Gesellschaft der Wissenschaften* (1885).
6. L. Lange, *Die geschichtliche Entwicklung des Bewegungsbegriff und ihr vorausschichtliches Endergebnis*, Leipzig (1886).
7. *Principia*, p. 22.
8. *A Midsummer Night's Dream*, Act V, Sc. 1.
9. *The Tempest*, Act IV, Sc. 1.
10. B. Russell, *The Principles of Mathematics*, Allen and Unwin, London, pp. 489–93; quoted in Ref. 6 in Chap. 2, pp. 129–34.
11. G. M. Clemence, *Quart. J.R. Astr. Soc.*, **7**, 10 (1966).
12. C. A. Murray, *Contemp. Physics*, **20**, 211 (1979).
13. W. E. Carter and D. S. Robertson, *Sci. American*, **255**, No. 5, 44 (1986).
14. See for example: M. Jammer, *The Philosophy of Quantum Mechanics*, Wiley, New York (1974), §2.2.
15. *Oeuvres Complètes*, Vol. 16 (1929), pp. 189–251; see also Vol. 18 (1934), p. 657ff and Vol. 21 (1944), p. 503ff.
16. Quoted and translated from J. A. Schouten, *Jahresberichte der Deutschen Mathematiker – Vereinigung*, **29**, 136 (1920); also published in *Oeuvres Complètes*, Vol. 16 (1929), p. 233.
17. C. Huygens, *Horologium Oscillatorum* (1673) in: *Oeuvres Complètes*, Vol. 18 (1934), p. 87; German translation in the series *Oswald's Klassiker der Exakten Wissenschaften*, No. 192, Leipzig (1913).
18. *Oeuvres Complètes*, Vol. 16 (1944), p. 197.
19. A. Gabbey in: *Studies on Christiaan Huygens*, eds. H. J. M. Bos, M. J. S. Rudwick, H. A. M. Snelders, and R. P. W. Visser, Swets and Zeitlinger, Lisse (1980).
20. *Oeuvres Complètes*, Vol. 21 (1944), p. 504.
21. C. Huygens, *Discours de la Cause de la Pesanteur* (1692); *Oeuvres Complètes*, Vol. 21 (1944), p. 451ff.
22. G. W. Leibniz, *Mathematische Schriften* (ed. C. I. Gerhardt), 7 vols., Berlin and Halle (1849–55).

23. *Principia*, pp. xx–xxxiii.
24. M. Jammer, *Concepts of Mass in Classical and Modern Physics*, Harvard University Press, Cambridge, Mass. (1961).
25. I. B. Cohen, in: *Science, Medicine, and Society in the Renaissance*, 2 vols., (ed. A. Debus), New York (1972), Vol. 2, pp. 199–211.
26. A. R. and M. B. Hall, *Unpublished Scientific Papers of Isaac Newton*, Cambridge University Press, Cambridge (1962), p. 148.
27. A. Wallenquist, *The Penguin Dictionary of Astronomy*, Penguin (1966), pp. 106–7.
28. *Westfall (1971)*, p. 449.
29. J. W. Herivel, *The Background to Newton's 'Principia'*, Clarendon Press, Oxford (1965), p. 154.
30. Ref. 26, p. 150.
31. *Mechanics*, p. 237.
32. Ref. 24.
33. Ref. 29, p. 27.
34. Ref. 28, p. 348.
35. *Mechanics*, p. 271.
36. *Ibid.*, p. 306.
37. *Ibid.*, p. 167.
38. *Ibid.*, p. 269 and p. 271.
39. *Ibid.*, p. 296.
40. D. J. Raine, *Rep. Prog. Phys.*, **44**, 1151 (1981).
41. Ref. 26, p. 132.
42. Review article by K. Kuchař in: *Quantum Gravity 2. A Second Oxford Symposium*, eds. C. J. Isham, R. Penrose, and D. W. Sciama, Clarendon Press, Oxford (1981), p. 369.
43. *De Rev.*, Book I, Chap. 5.
44. M. Keynes, 'Newton, the man', in: *Royal Society. Newton Tercentenary Celebrations, 15–19 July, 1946*, Cambridge University Press, Cambridge (1947), p. 27.

INDEX

(Page numbers in italic signify Reference entries.